Lecture Notes
in Computational Science
and Engineering

62

Editors

Timothy J. Barth
Michael Griebel
David E. Keyes
Risto M. Nieminen
Dirk Roose
Tamar Schlick

Frank Graziani
Editor

Computational Methods in Transport: Verification and Validation

With 77 Figures and 33 Tables

Springer

Frank Graziani
Lawrence Livermore National Laboratory
7000 East Ave.
Livermore, CA 94550-9234
USA
graziani1@llnl.gov

ISBN 978-3-540-77361-0 e-ISBN 978-3-540-77362-7

Lecture Notes in Computational Science and Engineering ISSN 1439-7358

Library of Congress Control Number: 2008920256

Mathematics Subject Classification (2000): 85A25, 68U20, 65Z05, 65G99

Cover design: WMX Design GmbH, Heidelberg

Printed on acid-free paper

9 8 7 6 5 4 3 2 1

spinger.com

Preface

In a wide variety of applications, accurate simulation of particle transport is necessary whether those particles be photons, neutrinos, or charged particles. For inertial confinement fusion, where one is dealing with either direct drive through photon or ion beams or indirect drive via thermal photons in a hohlraum, the accurate transport of energy around and into tiny capsules requires high-order transport solutions for photons and electrons. In astrophysics, the life cycle of the stars, their formation, evolution, and death all require transport of photons and neutrinos. In planetary atmospheres, cloud variability and radiative transfer play a key role in understanding climate. These few examples are just a small subset of the applications where an accurate and fast determination of particle transport is required.

Computational Methods in Transport Workshop (CMTW) is devoted to providing a forum for interdisciplinary discussions among transport experts from a wide range of science, engineering, and mathematical disciplines. The goal is to advance the field of computational transport by exposing the methods used in a particular field to a wider audience, thereby opening channels of communication between practitioners in the field. The original concept for the workshop was born at the SCaLeS (scientific case for large scale simulation) meeting held in Washington, DC in June 2003. The discussions at SCaLeS were lively and informative, and it was clear that the opportunity to meet with experts outside of a particular subfield created new insights into the problems being discussed.

In 2004, the first CMTW was held at Granlibakken Conference Center in Lake Tahoe. A group of speakers and participants from academia, the national laboratories, and industry attended workshop. The workshop's goal of providing a forum, where researchers could discuss successes and failures of their methods across discipline boundaries in an invigorating and relaxing atmosphere, that would benefit the transport community at large, was not only met but also greatly exceeded.

As in 2004, the goal of CMTW in 2006 was to open channels of communication and cooperation so that (1) existing methods used in one field could be applied to other fields and (2) greater scientific resource could be brought to bear on the unsolved outstanding problems. In 2006, the focus was on a cross-cutting issue affecting all of the transport disciplines, verification and validation (V&V). To put it into as succinct a form as possible, are the equations being solved correctly, and are the correct equations being solved? A verification and validation program assures a scientist that a simulation code mirrors reality and is not just an expensive computer game. The workshop attempted to bring together experts from various scientific and engineering disciplines that use transport with experts in the verification and validation community.

The executive and scientific organizing committees worked extremely hard at putting together a conference designed to stimulate discussion and interaction among attendees. My deepest thanks go out to them for their long and hard work. In addition, the continued support of Lawrence Livermore National Laboratory and Mark Green and Stanley Osher of the Institute for Pure and Applied Mathematics (IPAM) has been the main reason why the Computational Methods in Transport Workshop exists.

Executive Committee:

Frank Graziani (Lawrence Livermore National Laboratory)
Mark Green (IPAM and UCLA)
David Keyes (Columbia University)
James McGraw (Lawrence Livermore National laboratory)
Stanley Osher (IPAM and UCLA)

Scientific Committee:

Marvin Adams (Texas A & M University)
John Castor (Lawrence Livermore National Laboratory)
Anthony Davis (Los Alamos National Laboratory)
Ivan Hubeny (University of Arizona)
Tom Manteuffel (University of Colorado at Boulder)
Tony Mezzacappa (Oak Ridge National Laboratory)

List of Contributors

John I. Castor
Lawrence Livermore National Lab
P.O. Box 808
Livermore, CA 94551, USA
castor1@llnl.gov

Anthony B. Davis
Los Alamos National Laboratory
Space & Remote Sensing Group
(ISR-2)
Los Alamos, NM 87545, USA
adavis@lanl.gov

R. Furfaro
Department of Aerospace and
Mechanical Engineering, University
of Arizona
Tucson, AZ, USA

Irene M. Gamba
The Department of Mathematics
and Institute of Computational
Engineering and Sciences, The
University of Texas at Austin
Austin, TX, USA
gamba@math.utexas.edu

B.D. Ganapol
Department of Aerospace and
Mechanical Engineering, University
of Arizona
Tucson, AZ, USA

N.A. Gentile
University of California
Lawrence Livermore National
Laboratory
P.O. Box 808
Livermore, CA 94550, USA
gentile1@llnl.gov

Frank Graziani
Lawrence Livermore National
Laboratory
Livermore, CA, 94550-9234, USA
graziani1@llnl.gov

Sri Harsha Tharkabhushanam
Institute of Computational
Engineering and Sciences, The
University of Texas at Austin
Austin, TX, USA
harsha@ices.utexas.edu

J.C. Helton
Department of Mathematics and
Statistics, Arizona State University
Tempe, AZ 85287-1804, USA
jchelto@sandia.gov

François M. Hemez
Los Alamos National Laboratory
X-1 Mail Stop B259
Los Alamos, NM 87545, USA
hemez@lanl.gov

Kayo Ide
Institute of Geophysics and Planetary Physics and Department of Atmospheric & Oceanic Sciences, University of California
Los Angeles, CA 90095, USA
kayo@atmos.ucla.edu

James R. Kamm
Los Alamos National Laboratory
Applied Science & Methods Development Group (X-1)
Los Alamos, NM 87545, USA
kammj@lanl.gov

David S. Miller
Lawrence Livermore National Laboratory
Livermore, CA 94550, USA
millerds@llnl.gov

B. Pinty
Global Environment Monitoring Unit, Institute of Environment and Sustainability, Joint Research Centre of the European Commission
Ispra, Italy

M. Saisana
European Commission-Joint Research Centre
Ispra, Italy

A. Saltelli
European Commission-Joint Research Centre
Ispra, Italy

Didier Sornette
Institute of Geophysics and Planetary Physics and Department of Earth & Space Sciences, University of California
Los Angeles, CA 90095, USA

Laboratoire de Physique de la Matière Condensée (CNRS UMR 6622) and Université de Nice-Sophia Antipolis
06108 Nice Cedex 2, France
and
D-MTEC, ETH Zürich
8032 Zürich, Switzerland
dsornette@ethz.ch

Charles Tong
Lawrence Livermore National Laboratory
Livermore, CA 94550-0808, USA

Jean-Luc Widlowski
Global Environment Monitoring Unit, Institute of Environment and Sustainability, Joint Research Centre of the European Commission
Ispra, Italy
Jean-Luc.Widlowski@jrc.it

Contents

Verification (Mostly) for High Energy Density Radiation Transport: Five Case Studies

J.I. Castor

Summary. In this review I will first describe some of the salient characteristics of high-energy-density physics (HEDP), and in particular the challenges of simulating HEDP radiation transport. The balance of the review is taken up with discussions of five case studies in HEDP radiation transport that illustrate some of the problems that are faced. The five case studies are: (1) the Marshak wave; (2) what I call the star-in-space problem; (3) the radiative shock wave; (4) the so-called crooked-pipe problem; and (5) a case that illustrates how severe the difficulty may be of choosing a sufficient set of discrete angles for 3-D radiation transport.

1 Introduction

In this review I will first describe some of the salient characteristics of high-energy-density physics (HEDP), and in particular the challenges of simulating HEDP radiation transport. The balance of the review is taken up with discussions of five case studies in HEDP radiation transport that illustrate some of the problems that are faced. The five case studies are: (1) the Marshak wave; (2) what I call the star-in-space problem; (3) the radiative shock wave; (4) the so-called crooked-pipe problem; and (5) a case that illustrates how severe the difficulty may be of choosing a sufficient set of discrete angles for 3-D radiation transport.

1.1 High-Energy-Density Physics Characteristics

HEDP is a cross-disciplinary subject, and overlaps high-energy astrophysics – active galaxies and quasars, X-ray binary stars – on one side and mechanical engineering on the other. While it may be true that "Close only counts in horseshoes, hand-grenades and astrophysics," the engineering applications make greater demands on the fidelity of the simulations.

One key characteristic of those problems is complexity. This may be the geometrical complexity of an object assembled from tens of pieces with complicated shapes, and their very inhomogeneous material properties such as density, pressure and opacity. An example is the design for an ignition target that is slated to be used in experiments on the National Ignition Facility, the large laser nearing completion at LLNL. This has pieces made of gold, copper, polystyrene, frozen and gaseous DT and other things, and the parts are not all axisymmetric.

In general the complex objects lead to 10^6-zone simulations to account for the spatial structure. The million zones may not be needed to give a useful rough picture of the problem, as in the astrophysical cases, but they are necessary when accuracy is a requirement. The computational demands are greatly intensified by including radiation transport in the modeling. The intensity field has three additional independent variables – two angles and the photon energy – compared to other hydrodynamic variables. If we think of tens of photon energies and tens to thousands of photon directions, the cost is seen to be 10^2–10^4 times that of the pure hydro calculation. Naturally, much of the work in rad-hydro simulations has gone into finding cost-cutting approximations. These have to be tested against benchmarks such as the five case studies presented here.

2 Case 1: The Marshak Wave

A problem that has considerable practical importance and which also reveals differences between the various approximations is the Marshak wave. Radiation at a fixed temperature is incident on a slab of initially cold material, as indicated in Fig. 1. With the passage of time the slab material is heated to a certain depth, which increases with time. The hot material may be much less opaque than the cold material, so the wave has a "bleaching" character,

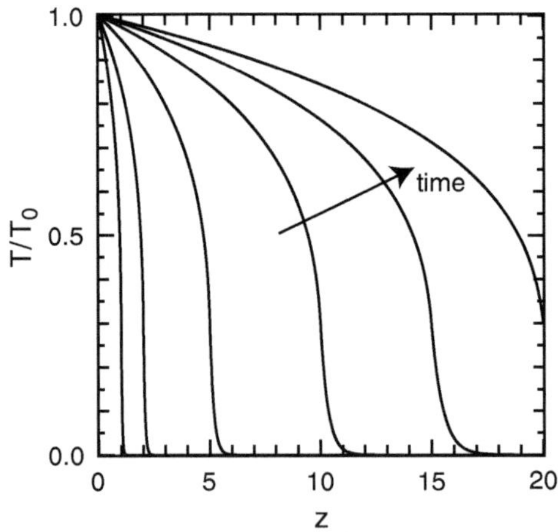

Fig. 1. The penetration of a Marshak wave into a slab of cold material. The slab is heated at the left, $z = 0$, side. The ordinate is material temperature in units of the driving temperature and the abscissa is the space coordinate, z

and the temperature is nearly uniform in the heated area. The goal of the simulations is to find the position of the wave front vs. time and the shape of the temperature profile.

2.1 1-D Diffusion and Related Methods

Gray radiation diffusion leads to the coupled equations for radiation energy density E and radiative flux F in the following generic form:

$$\frac{\partial E}{\partial t} + \frac{\partial F}{\partial z} = k(4\pi B - cE) \qquad \text{and} \qquad \frac{1}{3c}\frac{\partial F}{\partial t} + \frac{D}{3}\frac{\partial fcE}{\partial z} = -kF.$$

The different diffusion alternatives vary in their choice of the $\partial F/\partial t$ term and the factors D and f in the gradient term. The P_1 model includes $\partial F/\partial t$ without the 3, and omits both the D and f factors. Simple diffusion leaves out $\partial F/\partial t$ and the factor f. If the factor D is included it becomes flux-limited diffusion. The $P_{1/3}$ model of Olson et al. [OAH] omits the D and f factors but includes the factor 3 in the $\partial F/\partial t$ term. The variable-Eddington-factor (VEF) method omits the factor 3 in the $\partial F/\partial t$ term and the factor D but includes a suitable function for f. We will look to the results of Olson et al. [OAH] for the comparison of these methods on the gray Marshak problem.

There is a Range of Flux-Limiter Functions

For the flux-limited diffusion method the dimensionless limiter function D must be provided. This is given as a function of the dimensionless variable $R = |\nabla E|/(kE)$. A small value of R means that E varies only slightly in a distance of one mean free path; in this regime diffusion works well, and the limiter function D should be unity. When R is large the simple diffusion formula would imply a flux F that was much larger than cE, which is physically impossible. In this case D must be a small number, $\sim 3/R$, so that F will not violate the limit. Some of the possible functions are: (1) the "sum" limiter, $D = 3/(3 + R)$; (2) the "max" limiter, $D = 3/\max(3, R)$; (3) the Chapman–Enskog flux limiter of Levermore and Pomraning [LP81], which depends also on the single-scattering albedo ϖ; (4) Larsen's [L98] functions $D = [1 + (R/3)^n]^{-1/n}$ with an adjustable exponent $n > 1$. These are illustrated in Fig. 2. For the Levermore–Pomraning limiter the albedo ϖ was set to unity. The Larsen $n = 2$ limiter is seen to be close to Levermore–Pomraning.

2.2 Benchmark Results for the Su–Olson Linear Marshak Problem

Su and Olson [SO97] developed a Marshak test problem that is linear, but which has a full transport solution that is anaytic; this is achieved by making the material internal energy vary as T^4. The solutions using simple (not limited) diffusion, P_1 and a characteristic-ray VEF method are compared with

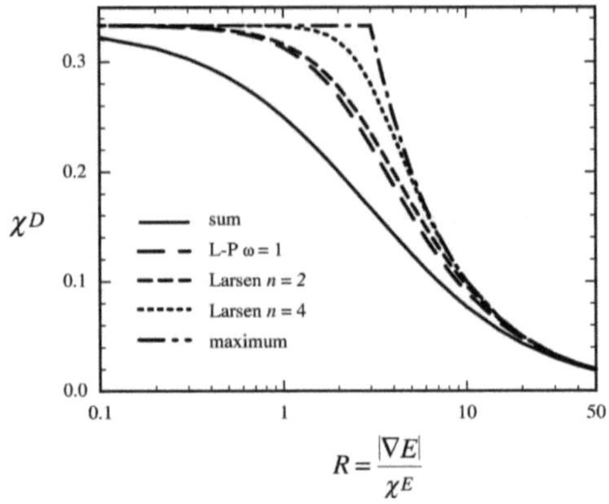

Fig. 2. Five different flux limiter functions compared by Olson et al. [OAH]. Abscissa: R defined in the text; ordinate: $D/3$ in terms of D in the text

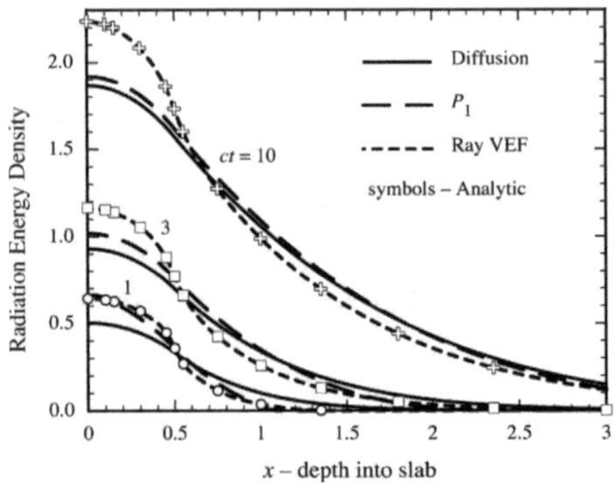

Fig. 3. Olson et al. [OAH] results for the energy density profiles in the linearized gray Marshak problem at selected times. The abscissa is optical depth, kz

the analytic answer in Fig. 3. While the VEF method is accurate in general, as expected, the diffusion wave races ahead at early time and slows down to the correct answer later. Both diffusion and P_1 are inaccurate at small optical depth. The flux-limited diffusion results are shown in Fig. 4. At middle time ($ct \approx$ mean free path) the flux limiter makes a difference and first "sum" is best and then later "max" and Larsen $n = 2$ are best.

Olson, Auer and Hall also show results for a Marshak wave with a realistic variation of opacity with temperature, $k \propto 1/T^3$. This adds the "bleaching" effect lacking in the linear model. The self-consistent VEF solution is presumably correct for this problem, and the other methods may be compared to it. Reference [OAH] illustrates the comparison with three flux-limited diffusion models, simple diffusion, P_1, and Minerbo's [M78] Eddington-factor formula. There is a noticeable spread in the models, with "sum" being too slow and Levermore–Pomraning (with $\varpi \neq 1$) too fast. The closest to the accurate answer are Larsen $n = 2$ and Levermore–Pomraning with $\varpi = 1$.

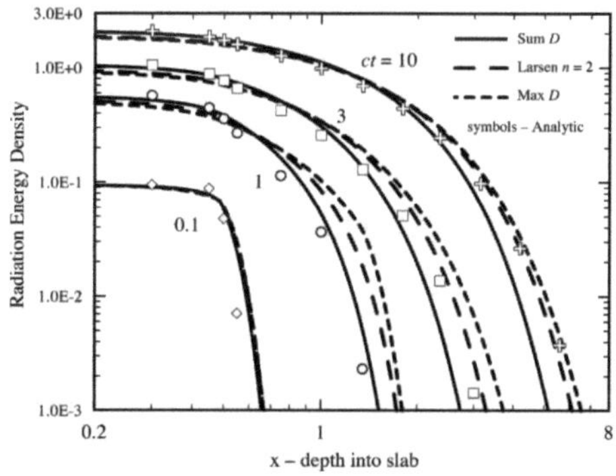

Fig. 4. Olson et al. [OAH] results as in Fig. 3 but with the sum, max and Larsen $n = 2$ flux limiters

2.3 Marshak Wave Summary

Some conclusions that may be drawn are these: No diffusion-like method is perfect. The "max" and "sum" flux limiters are the poorest. The ad hoc Eddington-factor methods are poor. The Levermore–Pomraning ($\varpi \neq 1$) method is poor, but the variation with $\varpi = 1$ is fairly good, on a par with Larsen's $n = 2$ method. P_1 and $P_{1/3}$ are good, but not better than Larsen $n = 2$ for the variable-opacity case. The self-consistent VEF method is accurate, as it should be. Implicit Monte Carlo does a good job on this problem, especially in the difference formulation; see Szöke, Brooks, McKinley and Daffin [SBMD].

3 Case 2: Star-in-Space Problem

This is a problem that arises in astrophysics, and it has variations in other disciplines. It is extremely simple: An opaque sphere radiates isotropically into a surrounding vacuum. The prototype is an extended stellar atmosphere; it also occurs in ICF hohlraum problems, where the source and receiver are reversed. Since it involves an infinite mean free path and a large dynamic range of radius, it is as far as it is possible to get from the diffusion regime. This problem elicits a variety of responses from the algorithms, almost all bad.

What goes wrong? Here are the spherical geometry, steady-state, moment equations in vacuum:

$$\frac{dF}{dr} + \frac{2}{r}F = 0 \qquad \text{and} \qquad \frac{dP}{dr} + \frac{3P - E}{r} = 0.$$

The first equation says $F \propto 1/r^2$, which is correct. The problem is the second equation, and in *any variation of diffusion* P is replaced by $E/3$, which implies E = constant, a very bad answer. Not only that, but *all* P_n

methods give the same result, E = constant, if the sphere radiates isotropically. Characteristic-ray methods will be exact in this case, if the rays properly sample the radiating disk.

3.1 What About the VEF Method(s)?

With an Eddington factor χ, the first moment equation becomes

$$\frac{dP}{dr} + \frac{3P - E}{r} = \frac{d\chi E}{dr} + \frac{(3\chi - 1)E}{r} = \frac{1}{q}\frac{dq\chi E}{dr} = 0 ,$$

and q is Auer's sphericality function defined by

$$\ln q = \int \left(3 - \frac{1}{\chi}\right)\frac{dr}{r} .$$

So the solution is $E \propto 1/(q\chi)$, and the rabbit in the hat is χ, the Eddington factor.

The exact value of the Eddington factor for this problem is found in this way: The half-angle subtended by the sphere of radius R at a radius $r > R$ is $\cos^{-1}\mu$, where $\mu = \sqrt{1 - R^2/r^2}$. The moments then become $cE = 2\pi I_0(1-\mu)$, $F = \pi I_0(1 - \mu^2)$ and $cP = 2\pi I_0(1 - \mu^3)/3$, so the Eddington factor is $\chi = (1 + \mu + \mu^2)/3$. With this χ, q turns out to be $q \propto 1/(1 - \mu^3)$, so $E \propto (1 - \mu)$, the right answer.

Streaming Solutions with Eddington Factor Formulae

There are several prescriptions for χ as a function of $f = F/cE$: Minerbo's, Levermore's, Kershaw's, and the exact relation based on $\chi = (1 + \mu + \mu^2)/3$ and $f = (1 + \mu)/2$. From $q\chi E$ = constant and r^2F = constant we infer that the quantity A defined by $A \equiv qR^2/r^2$ must obey

$$A = \frac{2\chi(.5)f}{\chi(f)} .$$

Taking $q = 1$ and $f = 1/2$ at $r = R$, we see that A obeys the differential equation

$$\frac{d\ln A}{d\ln r} = 1 - \frac{1}{\chi(f)} .$$

This is integrated by using the current A to find f from the relation above, evaluating $\chi(f)$, then evaluating the right-hand side to find dA/dr. This means that A decreases with r and therefore $f/\chi(f)$ must decrease with f, if f is to be physically reasonable.

In fact, none of the common Eddington factor prescriptions obey reasonableness, as it turns out. Figure 5 shows the A-f relation for the four Eddington-factor models mentioned above. The surface of the sphere, $r = R$,

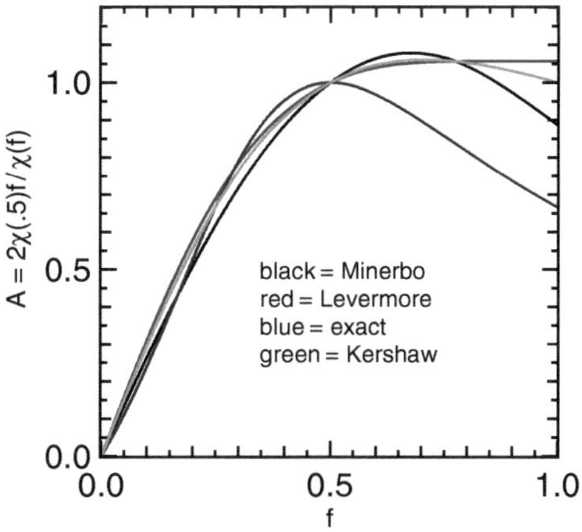

Fig. 5. The quantity A defined in the text is shown vs. the scaled flux, f, for four different Eddington factor prescriptions, as indicated. (See Plate 1 on page 301)

corresponds to $f = 1/2$, where $A = 1$; A must decrease as r increases, which leads to solutions with f decreasing from 1/2 for Minerbo's, Levermore's and Kershaw's functions; the regime $f \to 1$ for large r cannot be reached. The one that works is "exact," namely $\chi = (1-2f+4f^2)/3$; this function is unphysical in another sense, since it is not monotonic with f, and has a minimum value $= 1/4$ at $f = 1/4$.

3.2 The S_n Methods are Much Better

In S_n you solve these equations for the star-in-space problem:

$$\frac{\mu_i}{r^2}\frac{d(r^2 I_i)}{dr} + \frac{2}{rw_i}\left[\alpha_{i+1/2}I_{i+1/2} - \alpha_{i-1/2}I_{i-1/2}\right] = 0.$$

The I_i are zone-centered in μ and the $I_{i+1/2}$ are edge-centered; the αs are helper quantities derived from the quadrature set (see [LM93], Sects. 3 and 4); with diamond-difference in angle $I_i = (I_{i-1/2} + I_{i+1/2})/2$. Figure 6 compares the S_2, S_4, S_6 and S_8 results with the exact answer (red). We see that S_2 is off by a factor 2, a lot better than P_1, and S_8 is off by only 5%; the asymptotic behavior is $\propto 1/r^2$ for all n.

3.3 Star-in-Space Summary

Some methods give $E \sim$ constant instead of declining roughly as $1/r^2$: any kind of diffusion and P_n of any order. The characteristic-ray method (tangent rays) is exact. Ad hoc Eddington-factor closures behave unphysically for this problem. The S_n results are good: S_2 is up to a factor 2 off, but for larger n the error declines to a few percent.

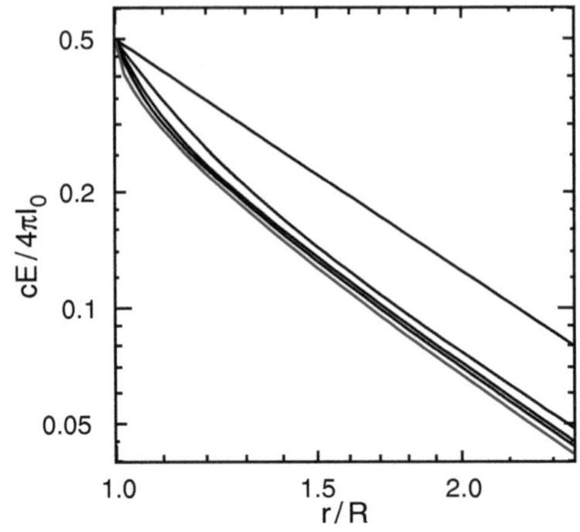

Fig. 6. Star-in-space solutions for the normalized energy density vs. radius are shown for S_n methods of different order in comparison with the exact result. The curves are for S_2 (*top*), S_4, S_6, S_8 and the exact solution (*bottom, in red*). (See Plate 2 on page 301)

4 Case 3: The Radiating Shock Problem

In a strong enough shock wave the radiation from shocked material preheats the material ahead of the shock, as illustrated in Fig. 7. This makes the peak temperature higher, but then there is a sharp cooling spike in which T comes down to the final value. The question is, what is the ratio f_s of the precursor temperature T_u to the final downstream temperature? This depends on a radiation-strength parameter, $Q \equiv 2\sigma_B u_s^5/\pi R^4 \rho_0$, where σ_B is the Stefan–Boltzmann constant, u_s is the shock speed relative to the pre-shock material, R is the gas constant and ρ_0 is the pre-shock density. There is disagreement on whether f_s gradually approaches 1 as $Q \to \infty$, or whether $f_s = 1$ above a critical Q, or some other behavior.

4.1 The Flow Model

The idealized model is ideal-gas ($\gamma = 4/3, p = \rho RT$), constant gray opacity, steady flow in slab geometry. We non-dimensionalize using ρ_0, u_s and R as units; radiation quantities are in units of $\rho_0 u_s^3/2$. The normalized specific volume is $\eta \equiv \rho_0/\rho$. The pre-shock temperature and pressure are supposed to be negligible. The steady flow equations give the non-dimensional temperature and flux

$$T = \eta(1-\eta) \qquad \text{and} \qquad F = (1-\eta)\left(\frac{\eta}{\eta_f} - 1\right).$$

Here η_f is the final downstream volume, $\eta_f \equiv (\gamma - 1)/(\gamma + 1)$.

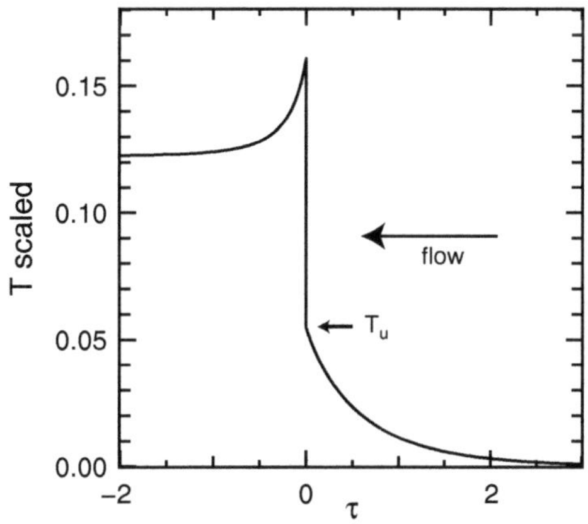

Fig. 7. The spatial profile of gas temperature is shown for a generic radiative shock. The shock moves from left to right in this picture, i.e., the flow is from right to left in the shock frame

4.2 The Non-Equilibrium Diffusion Approach

In the Eddington approximation the radiation moment equations are (with $B = Q[\eta(1-\eta)]^4$)

$$\frac{dF}{d\tau} = 4\pi(B - J) \qquad \text{and} \qquad \frac{4\pi}{3}\frac{dJ}{d\tau} = -F.$$

Substituting for F as a function of η and dividing the equations gives a single ODE for J vs. η. The integrations from far upsteam and downstream are matched at the shock, where η satisfies the jump conditions $\Delta F = 0$, $\Delta J = 0$. The results for various Qs are shown in Fig. 8. There is no shock that is exactly critical ($f_s = 1$), but f_s turns the corner sharply near $Q = 800$. This is consistent with Zel'dovich's and Raizer's statements ([ZR67], §§VII-14, 17, [Z57, R57]).

4.3 What About Exact Transport?

The VEF radiation moment equations are

$$\frac{dF}{d\tau} = 4\pi(B - J) \qquad \text{and} \qquad \frac{4\pi}{3}\frac{df_E J}{d\tau} = -F.$$

The Eddington factor f_E has to be found self-consistently by solving the transfer equation. For this I used two methods: (1) Feautrier [F64] solutions using 4-point Gaussian quadrature in each of five divisions of each hemisphere; (2) applying the exponential-integral kernels to the cubic spline interpolant of Planck function vs. τ. The left plot in Fig. 9 shows the non-trivial behavior of f_E in close-up, and the right plot compares the two methods, which agree

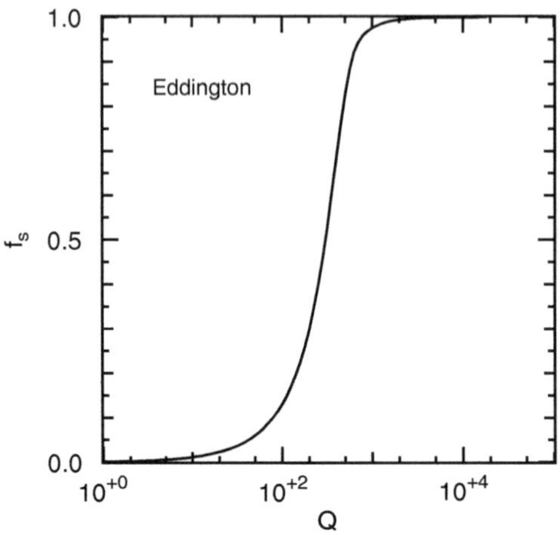

Fig. 8. The ratio f_s of the gas temperature just behind the shock to the final downstream temperature is shown vs. the dimensionless radiation strength Q defined in the text for the case that the Eddington approximation is used

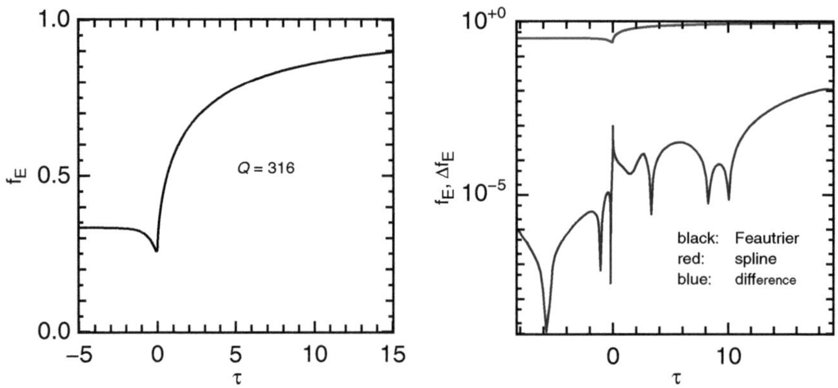

Fig. 9. The self-consistent Eddington factor is shown vs. optical depth τ measured from the shock front for a particular radiative shock, $Q = 316$. The flow goes from right to left in the figure. *Left*: values of f_E; *right*: difference between f_E computed from Feautrier and from spline-based quadrature. (See Plate 3 on page 302)

very well. After each RT solution the Eddington factor was re-computed and a relaxation was applied before the next solution of the steady flow equations. The combined equations give $K \equiv f_E J$ vs. η, and another integration yields τ. The convergence was found to be good for low and moderate Q, and less good for large Q.

The VEF results illustrated in Fig. 10 imply that there is a critical shock, and $f_s \to 1$ from above as $Q \to \infty$. The accurate VEF solutions show a modest 14% shift in Q for a given f_s, but also pass through $f_s = 1$ at $Q \approx 880$. The

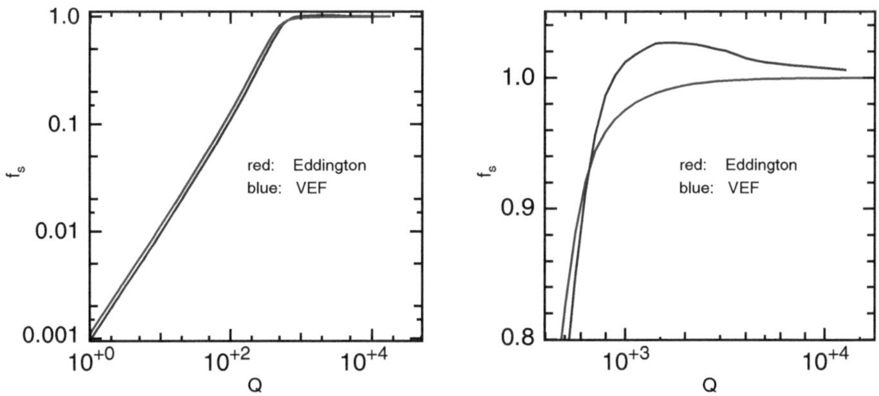

Fig. 10. As Fig. 8, but comparing Eddington approximation (*red*) with accurate transport (*blue*). (See Plate 4 on page 302)

14% shift is due to the Milne boundary-condition factor $2/\sqrt{3}$, as discussed in the footnote on p. 538 of [ZR67]. The VEF f_s results for $Q > 2000$ have a ± 0.001 uncertainty due to imperfect convergence.

The existence of shocks with $f_s > 1$ conflicts with the statements in [ZR67] and the proof in [Z57]. The proof in the latter reference, however, makes the assumption that the Eddington factor f_E has a constant value, which need not be 1/3. It is the non-constancy of f_E in the VEF calculations, where in fact f_E is substantially lower than 1/3 very near the shock, that allows J to be locally larger than $3K$, which in turn produces an elevation of T to a value slightly larger than T_f. Further work is clearly needed on this question.

Sincell, Gehmeyr and Mihalas [SGM] studied radiating shocks with the adaptive Titan code, which also uses the VEF method. The data for ambient density, molecular weight and gas law ($\gamma = 5/3$) were specified, and the velocity of the piston driving the shock took various values. Figure 11 shows the pre-shock temperature $T_u = f_s T_f$ vs. the final temperature T_f. The Sincell et al., results agree fairly well with the present VEF method. Both Sincell et al. and the present study show f_s crossing $f_s = 1$ and then tending asymptotically toward unity.

4.4 Radiating-Shock Summary

The difference between diffusion and transport is significant: the Eddington factor has strong spatial variation. There is a qualitative difference between Eddington and VEF results for the shock precursor temperature: For Eddington the precursor temperature rises monotonically as a fraction of the final temperature as shock strength increases, while for VEF the shock becomes critical (fraction 100%) at a certain strength. Some of the computational aspects of this problem are quite difficult, suggesting that it could be a good benchmark.

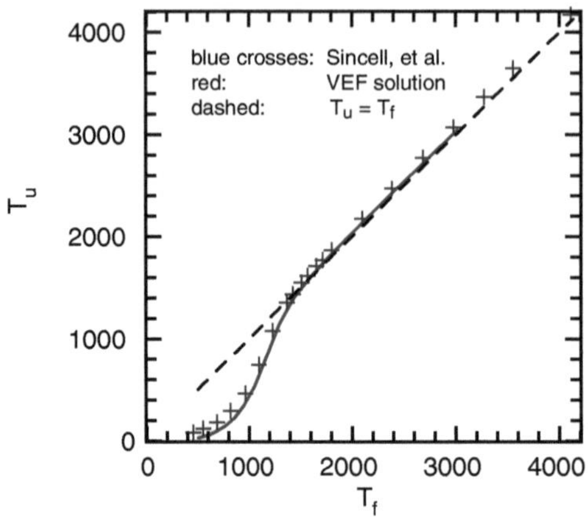

Fig. 11. Pre-shock temperature T_u is shown vs. final temperature T_f for the present VEF calculations (*red*) and the VEF results of [SGM] (*blue crosses*). The *dashed line* is $T_f = T_u$. (See Plate 5 on page 302)

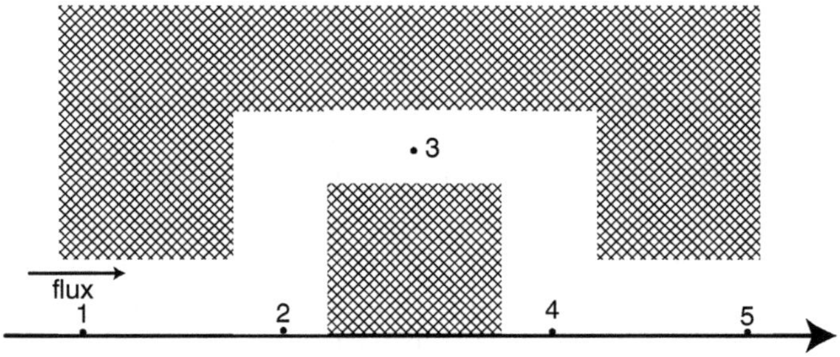

Fig. 12. Geometry of the crooked-pipe problem. The problem has axial symmetry. The shaded region is very opaque and the non-shaded parts are vacuum. Radiation enters at the left and diffusely reflects from the walls

5 Case 4: The Crooked Pipe

Isotropic radiation is incident from the left, beginning at $t = 0$, at the opening in a pipe that has a fat center section containing an obstructing plug; see Fig. 12. The pipe material is assumed to be very opaque, with a diffusely reflection surface (albedo = 1). The object is to find the time dependence of the radiation energy density at various places along the pipe, as marked in Fig. 12. It is known that diffusion and transport give quite different results for this problem. The precise definition of the problem is in [GL].

The accurate solutions of the crooked-pipe problem have been obtained with S_n and Monte Carlo methods. The comparison of S_n results for orders $n = 4$, $n = 8$ and $n = 16$ is shown in Fig. 13. The greatest sensitivity is at the first right-angle bend, where S_4 has a hard time getting the radiation arrival right.

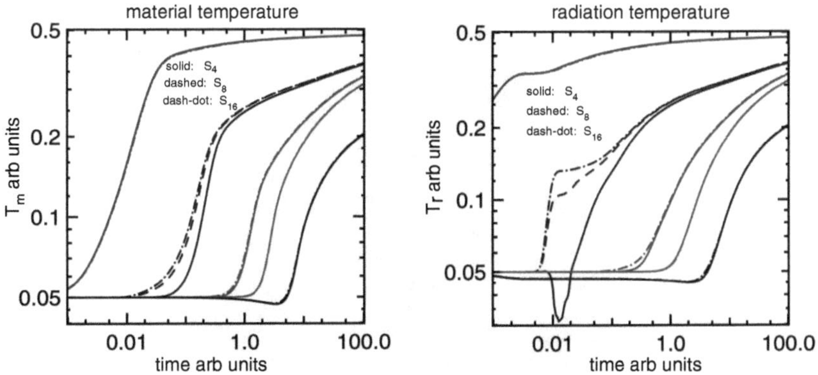

Fig. 13. Crooked-pipe results with S_4, S_8 and S_{16} are shown for the time histories of material temperature T_m (*left*) and radiation temperature T_r (*right*) at the five fiducial points in Fig. 12. Data from [PN06]. (See Plate 6 on page 303)

5.1 Crooked-Pipe Summary

This concludes my discussion of this useful transport test problem. Nick Gentile has published a description of his Implicit Monte Carlo Diffusion method, and a hybrid IMD/IMC method, including studies of the crooked-pipe problem, in [G01]. In Nick's following talk he will describe all this. Paul Nowak has studied this problem with 2-D S_n, and Nick will describe some of these results also. The short version is that IMD/IMC and S_n both perform well.

6 Case 5: Angle, Angles, Angles

We might say, "The three most important things about radiation transport in 2-D and 3-D are angles, angles and angles." Most of our S_n transport calculations are not converged in angles, that is, the set of discrete directions that is used, and doubling the angle set halves the error, which is none too small. Double angles means double running time, which is already (for S_8, 50 groups, 10^6 zones) $>\sim 10$ cpu-hours per cycle in 3-D. This provides a major incentive for keeping the geometry simple: 2-D, or 3-D with only large, well-defined structures.

How many angles are needed? In 1-D slab problems a few angles suffice; In 2-D problems with smooth source distributions or short mean free paths something like S_4 or S_8 might be OK; but it can get much worse. Figure 14 shows a realization of the angular distribution of the intensity at a point in an infinite 3-D medium with a broad-band random Gaussian emissivity with

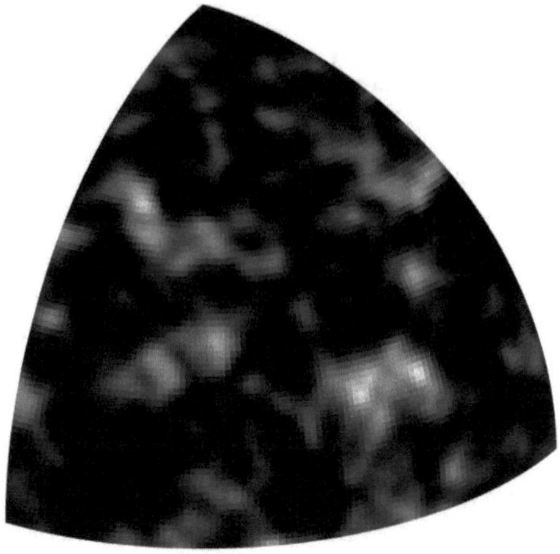

Fig. 14. Realization in one octant of the intensity field at a point in an infinite universe filled with uniform opacity and random emissivity, as described in the text. (See Plate 7 on page 303)

$|\mathbf{k}| \le 1.73$ and $\kappa\rho = 0.035$. The Fourier-space solution for the intensity I given a uniform opacity κ and a variable source function S is

$$\tilde{I}(\mathbf{n}) = \frac{\tilde{S}}{1 + i\mathbf{k}\cdot\mathbf{n}/(\kappa\rho)} \ .$$

A 3-dimensional array of wavevectors was defined, and the value of $\tilde{S}$ for each triple (k_x, k_y, k_z) was taken to be a sample of an independent normally-distributed complex variable with a variance derived from a $1/|\mathbf{k}|$ power spectrum. The sum of the individual $\tilde{I}(\mathbf{n})$ values leads to the picture. You can see that S_8, with only ten rays per octant, has no hope of describing this radiation field.

We can estimate the necessary angle resolution from $|\mathbf{k}|$ and $\kappa\rho$. We consider a single Fourier mode as described above, and construct the spherical harmonic expansion of the intensity. Figure 15 shows the fraction of the total power in harmonics above ℓ. A rule of thumb is that to see at least 90% of the power you need ℓs up to at least $k/(\kappa\rho) = k\lambda_p$, where λ_p is the radiation mean free path. The $\ell_{\max}$ equivalent to S_n order N is around $\ell_{\max} = N$. It varies because the quadrature set may not be exact for the maximal number of polynomials. According to this criterion, the example in Fig. 14 requires S_{50}. As we see, it is easy to construct a case for which $N > 60$ is needed for reasonable accuracy; this would have a disastrous cost, of course.

6.1 Angles Summary

The S_n order is a major cost factor in transport calculations; the time increases with $N(N+2)$. It is easy to construct a problem that would need $N >\approx 60$

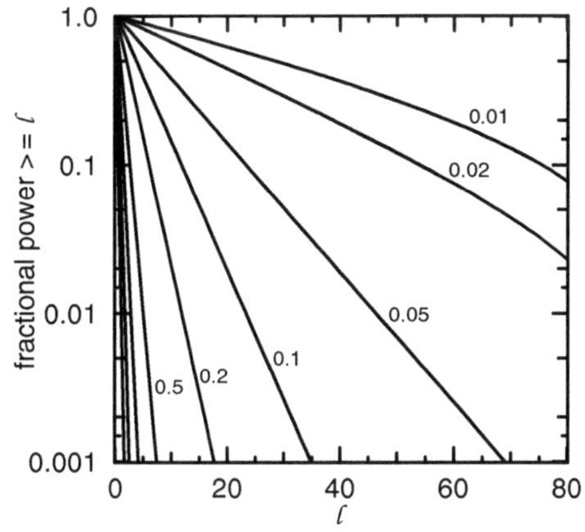

Fig. 15. The cumulative power distribution of the intensity field with respect to spherical harmonic order ℓ is shown vs. ℓ for the random emissivity problem with various values of $\kappa\rho/|\mathbf{k}|$, as indicated next to the curves

for an accurate solution; the solar chromosphere is of this sort So far, the best method of treating such problems is, "Don't do that."

7 Conclusion

The five case studies that have been discussed here illuminate different aspects or problem areas of radiation transport in HEDP. Case 1, the Marshak wave, describes a vitally important process in HEDP, since radiation ablation is an essential part of inertial fusion designs. In this case we are quite concerned to know what level of radiation transport is needed to compute the ablation correctly.

Case 2, the "Star-in-Space" problem, is a simplified representation of a difficult aspect of astrophysical and HEDP radiation transport: obtaining an accurate picture of the radiation from a distant, bright object. This case highlights the success and failure of otherwise respected algorithms.

The third case, the radiating shock, also occurs in both astrophysics and HEDP, and in this case the effects of exact transport are more subtle, but interestingly they produce one qualitative difference: whether the precursor heating can raise the material temperature above the final downstream temperature.

The crooked-pipe problem has been studied as a test of multi-dimensional transport in a complex geometry. It provides a benchmark for the accuracy of transport with different quadrature sets. The last case study is another way of testing convergence with respect to the angle quadrature set. The random emissivity problem can be made arbitrarily severe by letting the mean free path be larger and larger compared to the scale length of the spatial variations of emissivity. Radiation transport in random media does arise in applications such as solar physics, where so far, by necessity, rather low order quadratures have been used. This last case study may be regarded as a warning to transport code users of the need to examine the length scales of each problem before accepting a particular quadrature set.

Perhaps in the future these case studies may be part of a test suite used to perform verification and validation of radiation transport codes.

Acknowledgments

I am grateful to Prof. R. P. Drake for several discussions of the radiating shock problem, and to Drs. N. A. Gentile and P. F. Nowak for discussions of the crooked-pipe problem, and to Dr. P. F. Nowak for supplying his S_n results for this problem. I wish to acknowledge Elsevier Limited for permission to reproduce Figs. 2–4. This work was performed under the auspices of the U.S. Department of Energy by the University of California, Lawrence Livermore National Laboratory, under contract No. W-7405-Eng-48.

References

[F64] Feautrier, P.: Sur la resolution numerique de l'equations de transfert. C. R. Acad. Sci. Paris, Ser. B, **258**, 3189–3191 (1964)

[G01] Gentile, N.A.: Implicit Monte Carlo Diffusion – An Acceleration Method for Monte Carlo Time-Dependent Radiative Transfer Simulations. J. Comput. Phys., **172**, 543–571 (2001)

[GL] Graziani, F. and LeBlanc, J.: Lawrence Livermore National Laboratory report UCRL-MI-143393 (2000)

[L98] Larsen, E.W., private communication to Olson, Auer and Hall (1998)

[LP81] Levermore, C.D. and Pomraning, G.C.: A Flux-limited Diffusion Theory. Astrophys. J., **248**, 321–334 (1981)

[LM93] Lewis, E.E. and Miller, W.F., Jr.: Computational Methods of Neutron Transport. American Nuclear Society, La Grange Park, Illinois (1993)

[M78] Minerbo, G.N.: Maximum Entropy Eddington Factors. JQSRT, **20**, 541–545 (1978)

[PN06] Nowak, P.F.: private communication (2006)

[OAH] Olson, G.L., Auer, L.H. and Hall, M.L.: Diffusion, P_1, and Other Approximate Forms of Radiation Transport. JQSRT, **64**, 619–634 (2000)

[R57] Raizer, Yu.P.: On the Structure of the Front of Strong Shock Waves in Gases. Soviet Phys. JETP (English trans.) **5**, 1242–1248 (1957)

[SGM] Sincell, M.W., Gehmyer, M. and Mihalas, D.: The Quasi-Stationary Structure of Radiating Shock Waves. I. The One-Temperature Fluid. Shock Waves, **9**, 391–402 (1999)

[SBMD] Szöke, A., Brooks, E.D. III, McKinley, M.S. and Daffin, F.C.: Accurate and Efficient Radiation Transport in Optically Thick Media – by Means of the Symbolic Implicit Monte Carlo Method in the Difference Formulation. In: Graziani, F. (ed) Computational Methods in Transport. Springer, Berlin Heidelberg New York (2006)

[SO97] Su, B. and Olson, G.L.: An Analytical Benchmark for Non-Equilibrium Radiative Transfer in an Isotropically Scattering Medium. Ann. Nucl. Energy, **4**, 1035–1055 (1997)

[Z57] Zel'dovich, Ya.B.: Shock Waves of Large Amplitude in Air. Soviet Phys. JETP (English trans.) **5**, 919–927 (1957)

[ZR67] Zel'dovich, Ya.B. and Raizer, Yu.P.: Physics of Shock Waves and High-Temperature Hydrodynamic Phenomena. Academic Press, New York (1966–1967), reprinted Dover Publications, New York (2002)

A General Strategy for Physics-Based Model Validation Illustrated with Earthquake Phenomenology, Atmospheric Radiative Transfer, and Computational Fluid Dynamics

D. Sornette, A.B. Davis, J.R. Kamm, and K. Ide

Summary. Validation is often defined as the process of determining the degree to which a model is an accurate representation of the real world from the perspective of its intended uses. Validation is crucial as industries and governments depend increasingly on predictions by computer models to justify their decisions. In this article, we survey the model validation literature and propose to formulate validation as an iterative construction process that mimics the process occurring implicitly in the minds of scientists. We thus offer a formal representation of the progressive build-up of trust in the model, and thereby replace incapacitating claims on the impossibility of validating a given model by an adaptive process of constructive approximation. This approach is better adapted to the fuzzy, coarse-grained nature of validation. Our procedure factors in the degree of redundancy versus novelty of the experiments used for validation as well as the degree to which the model predicts the observations. We illustrate the new methodology first with the maturation of Quantum Mechanics as arguably the best established physics theory and then with several concrete examples drawn from some of our primary scientific interests: a cellular automaton model for earthquakes, an anomalous diffusion model for solar radiation transport in the cloudy atmosphere, and a computational fluid dynamics code for the Richtmyer–Meshkov instability*.

1 Introduction: Our Position with Respect to Previous Work on Validation and Related Concepts

1.1 Introductory Remarks and Outline

At the heart of the scientific endeavor, model building involves a slow and arduous selection process, which can be roughly represented as proceeding according to the following steps:

* This article is an augmented version of Ref. [1] by Sornette et al. that appeared in *Proceedings of the National Academy of Sciences* in 2007 (doi: 10.1073/pnas.0611677104), with an electronic supplement at URL http://www.pnas.org/cgi/content/full/0611677104/DC1. It is also available in preprint form at URL http://arxiv.org/abs/physics/0511219.

1. Start from observations and/or experiments.
2. Classify them according to regularities that they may exhibit: the presence of patterns, of some order, also sometimes referred to as structures or symmetries, is begging for "explanations" and is thus the nucleation point of modeling.
3. Use inductive reasoning, intuition, analogies, and so on, to build hypotheses from which a model[1] is constructed.
4. Test the model obtained in step 3 with available observations, and then extract predictions that are tested against new observations or by developing dedicated experiments.

The model is then rejected or refined by an iterative process, a loop going from step 1 to step 4. A given model is progressively validated by the accumulated confirmations of its predictions by repeated experimental and/or observational tests.

Building and using a model requires a language, i.e., a vocabulary and syntax, to express it. The language can be English or French for instance to obtain predicates specifying the properties of and/or relation with the subject(s). It can be mathematics, which is arguably the best language to formalize the relation between quantities, structures, space and change. It can be a computer language to implement a set of relations and instructions logically linked in a computer code to obtain quantitative outputs in the form of strings of numbers. In this later version, our primary interest here, validation must be distinguished from verification. Whereas *verification* deals with whether the simulation code correctly solves the model equations, *validation* carries an additional degree of trust in the value of the model vis-à-vis experiment and, therefore, may convince one to use its predictions to explore beyond known territories [2].

The validation of models is becoming a major issue as humans are increasingly faced with decisions involving complex tradeoffs in problems with large uncertainties, as for instance in attempts to control the growing anthropogenic burden on the planet within a risk-cost framework [3, 4] based on predictions of models. For policy decisions, national, regional, and local governments increasingly depend on computer models that are scrutinized by scientific agencies to attest to their legitimacy and reliability. Cognizance of this trend and its scientific implications is not lost on the engineering [5] and physics [6] communities.

Our purpose here is to clarify *from a physics-based perspective* what validation is and to propose a roadmap for the development of systematic approach to physics-based validation with broad applications. We will focus primarily on the needs of computational fluid dynamics and particle/radiation transport codes.

---

[1] By model, we understand an abstract conceptual construction based on axioms and logical relations developed to extract logical propositions and predictions.

In the remainder of this section, we first review different definitions and approaches found in the literature, positioning ourselves with respect to selected topics or practices pertaining to validation; we then show how the validation problem is related to the mathematical statistics of hypothesis testing and discuss some problems associated with emergent behaviors in complex systems. In Sect. 2, we list and describe qualitatively the elements required in our vision of model validation as an iterative process where one strives to build trust in the model going from one experiment to the next; however, one must also be prepared to uncover in the model a flaw, which may or may not be fatal. We offer in Sects. 3 and 4 our quantitative physics-based approach to model validation, where the relevance of the experiment to the validation process is represented explicitly. (An appendix explores the model validation problem more formally and in a broader context.) Section 5 demonstrates the general strategy for model validation using the historical development of quantum physics – a remarkably clear ideal case. Section 6 uses some research interests of the present authors to further illustrate the validation procedure using less-than-perfect models in geophysics, computational fluid dynamics (CFD), and radiative transfer. We summarize in Sect. 7.

1.2 Standardized Definitions

The following definitions are given by the American Institute of Aeronautics and Astronautics [7]:

- *Model*: A representation of a physical system or process intended to enhance our ability to predict, control and eventually to understand its behavior.
- *Calibration*: The process of adjusting numerical or physical modeling parameters in the computational model for the purpose of improving agreement with experimental data.
- *Verification*: The process of determining that a model implementation accurately represents the developer's conceptual description of the model and the solution of the model.
- *Validation*: The process of determining the degree to which a model is an accurate representation of the real world from the perspective of the intended uses of the model.

Figure 1, sometimes called a Sargent diagram [9, 10], shows where validation and several other of the above constructs and stages enter into a complete modeling project.

In the concise phasing of Roache [2], *"Verification consists in solving the equations right while validation is solving the right equations."* In the context of the validation of astrophysical simulation codes, Calder et al. [11] add: *"Verification and validation are fundamental steps in developing any new technology. For simulation technology, the goal of these testing steps is assessing the credibility of modeling and simulation."*

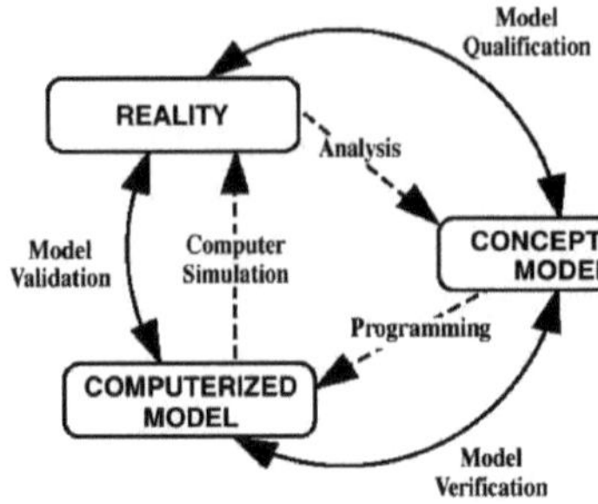

Fig. 1. The conventional position of validation in model construction according to Schlesinger [8] and Sargent [9,10]

Verifications of complex CFD codes are usually based on a suite of standard test problems in the field of fluid dynamics [11]. For compressible flow codes, these include Sod's test [12], the strong shock tube problem [13], the Sedov explosion problem [14], the interacting blast wave problem [15], a shock forced through a jump in mesh refinement, and so on.

Validations of complex CFD codes is usually done by comparison with experiments testing a variety of physical phenomena, including instabilities, turbulent mixing, shocks, etc. Validation requires that the numerical simulations recover the salient qualitative features of the experiments, such as the instabilities, their nonlinear development, the determination of the most unstable modes, and so on. See, for instance, Gnoffo et al. [16].

Considerable work on verification and validation of simulations has been done in the field of CFD, and in this literature the terms verification and validation have precise, technical meanings [2,7,9,10,17]. Verification is taken to mean demonstrating that a code or simulation accurately represents the conceptual model. Roache [18] stresses the importance of distinguishing between (1) verification of codes and (2) verification of calculations. The former is concerned with the correctness of the code. The later deals with the correctness of the physical equations used in the code. The programming and methods of solution can be correct (verification (1) successful) but they can solve erroneous equations (verification (2) failure). Validation of a simulation means demonstrating that the simulation appropriately describes Nature. The scope of validation is therefore much larger than that of verification and includes comparison of numerical results with experimental or observational data. In astrophysics, where it is difficult to obtain observations suitable for comparison to numerical simulations, this process can present unique challenges. Roache [op. cit.] goes on to offer the optimistic prognosis that *"the problems of Verification of Codes and Verification of Calculations are essentially solved for the case of structured grids, and for structured refinement of unstructured grids. It would appear that one higher level of algorithm/code development is required in order to claim a complete methodology for Verification of Codes and Calculations. I expect this to happen. Within 10 years, and likely much less, Verification of Codes and Calculations ought to be settled questions. I expect that Validation questions will always be with us."* We fully endorse this last sentence, as we will argue further on that validation is akin to the development of "trust" in theories of real phenomena, a never-ending quest.

1.3 Impossibility Statements

For these reasons, the possibility of validating numerical models of natural phenomena, often endorsed either implicitly or identified as reachable goals by natural scientists in their daily work, has been challenged; quoting from Oreskes et al. [19]: *"Verification and validation of numerical models of natural systems is impossible. This is because natural systems are never closed and because model results are always non-unique."* According to this view, the impossibility of "verifying" or "validating" models is not limited to computer models and codes but to all theories that rely necessarily on imperfectly measured data and auxiliary hypotheses. As Sterman [20] puts it: *"Any theory is underdetermined and thus unverifiable, whether it is embodied in a large-scale computer model or consists of the simplest equations."* Accordingly, many uncertainties undermine the predictive reliability of any model of a complex natural system in advance of its actual use.[2]

Such "impossibility" statements are reminiscent of other "impossibility theorems." Consider the mathematics of algorithmic complexity [25], which provides one approach to the study of complex systems. Following reasoning related to that underpinning Gödel's incompleteness theorem, most complex systems have been proved to be computationally irreducible, i.e., the only way to predict their evolution is to actually let them evolve in time. Accordingly, the future time evolution of most complex systems appears inherently unpredictable. Such sweeping statements turn out to have basically no practical value. This is because, in physics and other related sciences, one aims at predicting coarse-grained properties. Only by ignoring most of molecular detail, for example, did researchers ever develop the laws of thermodynamics, fluid dynamics and chemistry. Physics works and is not hampered by computational irreducibility because we only ask for approximate answers at some coarse-grained level [26]. By developing exact but *coarse-grained* procedures on computationally irreducible cellular automata, Israeli and Goldenfeld [27] have demonstrated that prediction may simply depend on finding the right level for describing the system. More generally, we argue that only coarse-grained scales are of interest in practice but their description requires "effective" laws which are in general based on finer scales. In other words, real understanding

[2] For further debate and commentary by Oreskes and her co-authors, see [21–23]; also noteworthy is the earlier paper by Konikov and Bredehoeft [24] for a statement about validation impossibility in the context of groundwater models.

must be rooted in the ability to predict coarser scales from finer scales, i.e., a real understanding solves the universal micro-macro challenge. Similarly, we propose that validation is possible, to some degree, as explained further on.

1.4 Validation and the Mathematical Statistics of Hypothesis Testing

Calder et al. [11] also write: *"We note that verification and validation are necessary but not sufficient tests for determining whether a code is working properly or a modeling effort is successful. These tests can only determine for certain that a code is not working properly."* This last statement is important because it points to a bridge between the problem of validation and some of the most central questions of mathematical statistics [28], namely, hypothesis testing and statistical significance tests. This connection has been made previously by several others authors [29–32]. In showing the usefulness of the concepts and framework of hypothesis testing, we depart from Oberkampf and Trucano [33] who mistakenly state that hypothesis testing is a true or false issue, only. Every test of significance begins with a "null" hypothesis H_0, which represents a theory that has been put forward, either because it is believed to be true or because it is to be used, but has not been proved.[3]

For example, in a clinical trial of a new drug, the null hypothesis might be: "the new drug is no better, on average, than the current drug." We would write H_0: "there is no difference between the two drugs on average." The alternative hypothesis H_1 is a statement of what a statistical hypothesis test is set up to establish. In the example of a clinical trial of a new drug, the alternative hypothesis might be that the new drug has a different effect, on average, to be compared to that of the current drug. We would write H_1: the two drugs have different effects, on average. The alternative hypothesis might also be that the new drug is better, on average, than the current drug. Once the test has been carried out, the final conclusion is always given in terms of the null hypothesis. We either "reject H_0 in favor of H_1" or "do not reject H_0." We never conclude "reject H_1," or even "accept H_1." If we conclude "do not reject H_0," this does not necessarily mean that the null hypothesis is true, it only suggests that there is not sufficient evidence against H_0 in favor of H_1; rejecting the null hypothesis then suggests that the alternative hypothesis may be true, or is at least better supported by the data. Thus, one can never prove that an hypothesis is true, only that it is wrong by comparing it with another hypothesis. One can also conclude that "hypothesis H_1 is not necessary and another, more parsimonious, one H_0 should be favored." The alternative hypothesis H_1 is not rejected, strictly speaking, but is found

[3] We refer the reader to V.J. Easton and J.H. McColl, *Statistics Glossary*, http://www.cas.lancs.ac.uk/glossary_v1.1/main.html, from which we have borrowed liberally for this brief summary.

unnecessary or redundant with respect to H_0. This is the situation when there are two (or several) alternative hypotheses H_0 and H_1, which can be composite, nested, or non-nested.[4]

Within this framework, the above-mentioned statement by Oreskes et al. [19] that verification and validation of numerical models of natural systems is impossible is hardly news: the theory of statistical hypothesis testing has taught mathematical and applied statisticians for decades that one can never prove an hypothesis or a model to be true. One can only develop an increasing trust in it by subjecting it to more and more tests that "do not reject it." We attempt to formalize below how such trust can be increased to lead to an asymptotic validation.

1.5 Code Comparison

The above definitions are useful in recasting the role of code comparison in verification and validation (Code Comparison Principle or CCP). Trucano et al. [35] are unequivocal on this practice: *"the use of code comparisons for validation is improper and dangerous."* We propose to interpret the meaning of CCP for code verification activities (which has been proposed in this literature) as parallel to the problem of hypothesis testing: Can one reject Code #1 in favor of Code #2? In this spirit, the CCP is nothing but a reformulation in the present context of the fundamental principle of hypothesis testing. Viewed in this way, it is clear why CCP is not sufficient for validation since validation requires comparison with experiments and several other steps described below. The analogy with hypothesis testing illuminates what CCP actually is: CCP allows the selection of one code among several codes (at least two) but does not help one to draw conclusions about the validity of a given code or model when considered as a unique entity independent of other codes or models.[5] Thus, the fundamental problem of validation is more closely associated with the other class of problems addressed by the theory of hypothesis testing, which consists in the so-called "tests of significance" where one considers only a single hypothesis H_0, and the alternative is "all the rest," i.e., all hypotheses that differ from H_0. In that case, the conclusion of a test can be the following: "this data sample does not contradict the hypothesis H_0," which is not the same as "the hypothesis H_0 is true." In other words, an hypothesis cannot be excluded because it is found sufficient at some confidence level for explaining the available data. This is not to say that the hypothesis is true. It is just that the available data is unable to reject said hypothesis.

[4] The technical difficulties of hypothesis testing depend on these nested structures of the competing hypotheses; see, for instance, Gourieroux and Monfort [34].

[5] We should stress that the Sandia Report [35] by Trucano et al. presents an even more negative view of code comparisons because it addresses the common practice in the simulation community that turns to code comparisons rather than bone fide verification or validation, without any independent referents.

Restating the same thing in a positive way, the result of a test of significance is that the hypothesis H_0 is "compatible with the available data."

It is implicit in the above discussion that, to compare codes quantitatively in a meaningful way, they must solve the same set of equations using different algorithms, and not just model the same physical system. Indeed, there is nothing wrong with "validating" a numerical implementation of a knowingly approximate approach to a given physical problem. For instance, a (duly verified) diffusion/P_1 transport code can be validated against a detailed Monte Carlo or S_n code. The more detailed model must in principle be validated against real-world data. In turn, it provides validation "data" to the coarser model. Naturally, the coarser (say, P_1 transport) model still needs to establish its relevance to the real world problem of interest, preferably by comparison with real observations, or at least be invoked only in regimes where it is known a priori to be sufficiently accurate based on comparison with a finer (say, Monte Carlo transport) model.

Two noteworthy initiatives in transport model comparison for non-nuclear applications are the Intercomparison of 3D Radation Codes (I3RC) [36] (i3rc.gsfc.nasa.gov) and the RAdiation Model Intercomparison (RAMI) [37, 38] (rami-benchmark.jrc.it). The former is focused on the challenge of 3D radiative transfer in the cloudy atmosphere while the later is about 3D radiative transfer inside plant canopies; both efforts are motivated by issues in remote sensing (especially from space) and radiative energy budget estimation (either in the framework of climate modeling or using observational diagnostics, which typically means more remote sensing).[6] Much has been learned by the modelers participating in these code comparison studies, and the models have been improved on average [39]. Although not connected so far to the engineering community that is at the forefront of V&V standardization and methodology, the I3RC and RAMI communities talk much about "testing,"

[6] In remote sensing science, transport theory (for photons) plays a central role and "validation" has a special meaning, namely, the estimation of uncertainty for remote sensing products based on "ground-truth," i.e., field measurements of the very same geophysical variables (e.g., surface temperature or reflectivity, vegetation productivity, soil moisture) that the satellite instrument is designed to quantify. These data are collected at the same location as the imagery, if possible, at the precision of a single pixel. This type of validation exercise will test both the "forward" radiation transport theory and its "inversion." Atmospheric remote sensing, particularly of clouds, poses a special challenge because, strictly-speaking, there is no counterpart of ground-truthing. One must therefore often make do with comparisons of ground-based and space-based remote-sensing (say, of the column-integrated aerosol burden) to quantify uncertainty in both operations. In situ measurements (temperature, humidity, cloud liquid water, etc.) from airborne platforms – balloon or aircraft – are always welcome but collocation is rarely close enough for point-to-point comparisons; statistical agreement is then all that is to be expected, and residuals provide the required uncertainty.

and sometimes "certification," and not so much about "verification" (which would be appropriate) or "validation" (which would not).

What about multi-physics codes such as those used routinely in astrophysics, nuclear engineering, or climate modeling? CCP, along with the stern warnings of Trucano et al. [35], applies here, too. Even assuming that all the model components are properly verified or even individually validated, the aggregated model is likely to be too complex to talk about clean verification through output comparison. Finding some level of agreement between two or more complex multi-physics models will naturally build confidence in the whole (community-wide) modeling enterprise. However, this is not to be interpreted as validation of any or all of the individual models.

There are many reasons for wanting to have not just one model on hand but a suite of more or less elaborate ones. A typical collection can range from the mathematically and physically exact but numerically intractable to the analytically solvable, possibly even on the proverbial back-of-an-envelope. We elaborate on and illustrate this kind of hierarchical modeling effort in Sect. 1.2 of the Appendix, offering it as an approach where model development is basically simultaneous with its validation.

1.6 Relations Between Validation, Calibration and Data Assimilation

As previously stated, validation can be characterized as the act of quantifying the credibility of a model to represent phenomena of interest. Virtually all such models contain numerical parameters, the precise values of which are not known a priori and, therefore, must be assigned. Calibration is the process of adjusting those parameters to optimize (in some sense) the agreement between the model results and a specific set of experimental data. Such data necessarily have uncertainties associated with them, e.g., due to natural variability in physical phenomena as well as to unavoidable imprecision of diagnostics. Likewise, there are intrinsic errors associated with the numerical methods used to evaluate many models, e.g., in the approximate solutions obtained from discretization schemes applied to partial differential equations. The approach of defensibly prescribing parameters for complex physical phenomena while incorporating the inescapable variability in these values is called "calibration under uncertainty," [40] a field that poses non-trivial challenges in its own right.

However calibration is approached, it must be undertaken using a set of data – ideally from specifically chosen calibration experiments/observations [41] – that differs from the physical configurations of ultimate interest (i.e., against which the model will be validated). In order to ensure that validation remains independent of calibration, it is imperative that these data sets be disjoint. In the case of large, complex, and costly experiments encountered in many real-world applications, it can be difficult to maintain a scientific "demilitarized zone" between calibration and validation. To not do so, however, risks undermining the scientific integrity of the associated modeling enterprise, the potential predictive power of which may rapidly wither as the validation study devolves into a thinly disguised exercise in calibration.

For complex systems, there are many choices to be made regarding experimental and numerical studies in both validation and calibration. The high-level approach of the Phenomena Identification and Ranking Table (PIRT) [42] can be used to heuristically characterize the nature of one's interest in complicated systems. This approach uses expert knowledge to identify the phenomenological components in a system of interest, to rank their (relative) perceived importance in the overall system, and to gauge the (relative) degree to which these component phenomena are perceived to be understood. This rough-and-ready approach can be used to target the choice of validation experiments for the greatest scientific payoff on fixed experimental and simulation budgets. To help guide calibration activities, one can apply the quantitative techniques of sensitivity analysis to rank the relative impact of input parameters on model outcome. Such considerations are particularly important for complex models containing many adjustable parameters, for which it may prove impossible to faithfully calibrate all input parameters.

Saltelli et al. [43, 44] have championed "sensitivity analysis" methods, which come in two basic flavors and many variations. One class of methods uses exact or numerical evaluation of partial derivatives of model output deemed important with respect to input parameters to seek regions of parameter space that might need closer examination from the standpoints of calibration and/or validation. If the model has time dependence, one can follow the evolution of how parameter choices influence the outcome. The alternate methodology uses adjoint dynamical equations to determine the relative importance of various parameters. The publications of Saltelli et al. provide numerous examples illustrating the value and practical impact of sensitivity analysis, as well as references to the wide scientific literature on this subject. The results of numerical studies guided by sensitivity analysis can be used both to focus experimental resources on high-impact experimental studies and to steer future model development efforts.

In dynamical modeling, initial conditions can be viewed as parameters and, as such, they need to be determined optimally from data. If the dynamical system in question is evolving continuously over time and data become available along the trajectory of the dynamical system, the problem of finding *a single* initial condition over the entire trajectory becomes increasingly and exceedingly difficult as the time window of the trajectory extends. In fact, it is practically impossible for the systems like the atmosphere or ocean whose dynamics is highly nonlinear, whose high-dimensional model is undoubtedly imperfect, and for which heterogeneous and sporadic data are subject to (poorly understood) errors.

Data assimilation is an approach that attends to this problem by breaking up the trajectory over (fixed-length) time windows and solving the initialization problem sequentially over one such window at a time as data become available. A novelty of data assimilation is that, rather than solving the initialization problem from scratch, it uses the model forecast as the first guess (the prior) of the initialization (optimization) problem. Once the optimization is completed, the optimal solution (the posterior) becomes the initial condition for the next model forecast.

This iterative Bayesian approach to data assimilation is most effective when the uncertainties in both the prior and the data are accurately quantified, as the system evolves over time and the data assimilation iterates one cycle after another. This is a non-trivial problem, because it requires the estimate of not only the model state but also the uncertainties associated with it, as well as the proper description of the uncertainties in data.

Numerical weather prediction (NWP) is one of the most familiar application areas of data assimilation – one with major societal impact. The considerable progress in skill of the NWP in recent decades has been due to improvements in all aspects of data assimilation [45], i.e., modeling of the atmosphere, quality and quantity of data, and data assimilation methods. At the time of writing, most operational NWP centers use the so-called the "three-dimensional variational method" (3D-Var) [46], which is an economical and accurate statistical interpolation scheme that does not include the effect of uncertainty in the forecast. Some centers have switched to the "four-dimensional variational method" (4D-Var) [47], which incorporates the evolution of uncertainty in a linear sense by use of the adjoint equations corresponding to the highly nonlinear model. These variational methods always call for the minimization of a cost function (cf. Appendix) that measures the difference between model results and observations throughout some relevant region of space and time. Currently active research areas in data assimilation include the effective and efficient quantification of the time-dependent uncertainties of both the prior and posterior in the analysis. To this end, the ensemble Kalman filter methods have recently received considerable attention motivated by future integration into operational environments [48–50]. As the importance of the uncertainties in data assimilation have become clear, many NWP centers perform ensemble prediction along with the single analysis obtained by the variational methods [51–53].

Clearly, considerable similarities exist between the data assimilation problem and the model validation problem. Can successful data assimilation be construed as validation of the model? In our opinion, that would be unjustified because the objectives are clearly different for these problems. As stated above, data assimilation admits the imperfection of the model. It explicitly makes use of the knowledge from the previous data assimilation cycle. As the initialization problem is solved iteratively over relatively short time windows, deviations of the model trajectory from the true evolution of the dynamical system in question tend to be small and data could be assimilated into the model without much discrepancy. Moreover, the operational centers perform careful quality-control of data to eliminate any isolated "outliers" with respect to the model trajectory. Thus, the data assimilation problem differs from the validation problem by design. Nevertheless, it is important to recognize that the resources offered by data assimilation can ensure that models perform well enough for their intended use.

1.7 Extension of the Meaning of Validation

A qualitatively new class of problems arise in fields such as the geosciences that deal with the construction of knowledge of a unique object, planet Earth, whose full scope and range of processes can be replicated or controlled neither in the laboratory nor in a supercomputer. This has led recently to championing the relevance of "systemic" (meaning "system approach") also called "complex system" approaches to the geosciences. In this framework, positive and negative feedbacks (and even more complicated nonlinear multiplicative noise processes) entangle many different mechanisms, whose impact on the overall organization can be neither assessed nor understood in isolation. How does one validate a model using the systemic approach? This very interesting and difficult question is at the core of the problem of validation.

How does one validate a model when it is making predictions on objects that are not fully replicated in the laboratory, either in the range of variables, of parameters, or of scales? For instance, this question is crucial

- In the scaling the physics of material and rock rupture tested in the laboratory to the scale of earthquakes;
- In the scaling the knowledge of hydrodynamical processes quantified in the laboratory to the length and time scales relevant to the atmospheric/oceanic weather and climate, not to mention astrophysical systems;
- In the science-based stewardship of the nuclear arsenal, where the challenge is to go from many component models tested at small scales in the laboratory to the full-scale explosion of an aging nuclear weapon.

The same issue arises in the evaluation of electronic circuits. In 2003, Allen R. Hefner, Founder and Chairman of the NIST/IEEE Working Group on Model Validation, writes in its description: *"The problem is that there is no systematic way to determine the range of applicability of the models provided within circuit simulator component libraries."* See full-page boxed text for the complete version of this interesting text, as well as [54]. This example of validation of electronic circuits is particularly interesting because it stresses the origin of the difficulties inherent in validation: the fact that the dynamics are nonlinear and complex with threshold effects and does not allow for a simple-minded analytic approach consisting in testing a circuit component by component. Extrapolating, this same difficulty is found in validating general circulation models of the Earth's climate or computer codes of nuclear explosions. The problem is thus fundamentally a "system" problem. The theory of systems, sometimes referred to as the theory of complex systems, is still in its infancy but has shown the existence of surprises. The biggest surprise may

NIST WORKING GROUP ON MODEL VALIDATION

Allen R. Hefner, Ph.D.
Semiconductor Electronics Division
National Institute of Standards and Technology
Gaithersburg, MD 20899, USA

The primary objective of the working group is to establish well-defined procedures for the comprehensive evaluation of circuit simulator models. The working group has been established because there is presently no systematic way to determine the range of applicability of the models provided within circuit simulator component libraries. In addition, the accuracy of component models varies from one circuit simulator to another due to the inclusion of different physical mechanisms in the generic models. The accuracy also varies due to the different methods used to determine the model parameters for specific device part numbers.

The complex behavior of electronic devices prohibits individual users of circuit simulators from evaluating the proper inclusion of model physics and from determining the validity of approximations made to simplify simulator implementation. In addition, software vendors are often reluctant to provide a complete description of their model equations and model parameters to users. Therefore, the goal of the NIST Working Group on Model Validation is to establish experimental test procedures that can be used to comprehensively evaluate circuit simulator models independently of the model equations, the simulator, or the model parameter extraction techniques.

Traditionally, most circuit simulator model evaluation has been performed by comparing simulations with measurements for the easily measured steady-state output characteristics and capacitance-voltage characteristics. However, this type of evaluation is of little use in determining the ability of the models to describe the more important dynamic characteristics. Presently, the only dynamic evaluation performed for circuit simulator models is for narrow ranges of conditions that tend to focus on those physical effects that are included in particular models and not on comprehensive evaluation of the model's ability to describe the dynamic behavior of the device for the full range of application conditions.

The primary tasks of the working group are to determine the complete range of dynamic conditions that must be described for each device type, and to then develop well-defined test procedures to evaluate the ability of the models to describe each type of dynamic condition. It is envisioned that each device type will require different test procedures, and that the circuit parameters for the test procedures will be determined for each part number based upon the device manufacturer's ratings and suggested application conditions. Furthermore, the model validation test procedures should evolve to account for new device variations, modeling requirements for new applications, and to prevent the development of models that are designed only to best fit the standard test procedures.

Typical model validation test procedures are expected to consist of test circuits that resemble application conditions, but that are simplified so that the test system is well characterizable and is able to isolate the important features of the device characteristics. Ultimately, standard characterization procedures will be established that specify the methods used to construct the test circuits to minimize the influence of parasitics, as well as the procedure for determining the circuit parameters based upon the device manufacturer's ratings. These standard test circuits could then be readily used to compare measured dynamic characteristics with those predicted by different circuit simulator component models. A database of the comparison results could also be maintained.

For the working group to be successful, expertise in the following areas will be essential: electronic component design and manufacturing, model development, software development, component characterization, and circuit and system design. The success of this working group could result in an improved understanding of the validity and limitations of existing circuit simulator component models. This could lead to the development of improved circuit simulator models and an increased confidence in the ability of the CAD tools to aid in the design of electronic systems.

be the phenomenon of "emergence" in which qualitatively new processes or phenomena appear in the collective behavior of the system, while they cannot be derived or guessed from the behavior of each element. The phenomenon of "emergence" is similar to the philosophical law on the "transfer of the quantity into the quality." How does one validate a model of such a system? Validation therefore requires an understanding of this emergence phenomenon.

From another angle, the problem is that of extrapolating a body of knowledge, which is firmly established only in some limited ranges of variables, parameters and scales, beyond this clear domain into a more fuzzy zone of unknowns. This problem has appeared and appears again and again in different guises in practically all scientific fields. A particularly notable domain of application is risk assessment; see, for instance, Kaplan and Garrick's classic paper on risks [55], and the instructive history of quantitative risk analysis in US regulatory practice [56], especially in the US nuclear power industry [57–60]. An acute question in risk assessment deals with the question of quantifying the potential for a catastrophic event (earthquake, tornado, hurricane, flood, huge solar mass ejection, large meteorite, industrial plant explosion, ecological disaster, financial crash, economic collapse, etc.) of amplitude never yet sampled from the knowledge of past history and present understanding.

To tackle this enduring question, each discipline has developed its own strategies, often being unaware of the approaches of others. Here, we attempt a formulation of the problem, and outline some general directions of attack, that hopefully will transcend the specificities of each discipline. Our goal is to formulate the validation problem in a way that may encourage productive crossings of disciplinary lines between different fields by recognizing the commonalities of the blocking points, and suggest useful guidelines.

2 Validation as a Constructive Iterative Process

In a generic exercise in model validation, one performs an experiment and, in parallel, runs the calculations with the available model. A comparison between the measurements of the experiment and the outputs of the model calculations is then performed. This comparison uses some metrics controlled by experimental feasibility, i.e., what can actually be measured. One then iterates by refining the model until (admittedly subjective) satisfactory agreement is obtained. Then, another set of experiments is performed, which is compared with the corresponding predictions of the model. If the agreement is still satisfactory without modifying the model, this is considered progress in the validation of the model. Iterating with experiments testing different features of the model corresponds to mimicking the process of construction of a theory in physics [61]. As the model is exposed to increasing scrutiny and testing, the testers develop a better understanding of the reliability (and limitations) of the model in predicting the outcome of new experimental and/or observational set-ups. This implies that *"validation activity should be organized like a project, with goals and requirements, a plan, resources, a schedule, and a documented record"* [6].

Extending previous work [29–32], we thus propose to formulate the validation problem of a given model as an iterative construction that embodies the often implicit process occurring in the minds of scientists:

1. One starts with an a priori trust quantified by the value V_{prior} in the potential value of the model. This quantity captures the accumulated evidence thus far. If the model is new or the validation process is just starting, take $V_{\text{prior}} = 1$. As we will soon see, the absolute value of V_{prior} is unimportant but its relative change is important.
2. An experiment is performed, the model is set-up to calculate what should be the outcome of the experiment, and the comparison between these predictions and the actual measurements is made either in model space or in observation space. The comparison requires a choice of metrics.
3. Ideally, the quality of the comparison between predictions and observations is formulated as a statistical test of significance in which an hypothesis (the model) is tested against the alternative, which is "all the rest." Then, the formulation of the comparison will be either "the model is rejected" (it is not compatible with the data) or "the model is compatible with the data." In order to implement this statistical test, one needs to attribute a likelihood $p(M|y_{\text{obs}})$ or, more generally, a metric-based "grade" that quantifies the quality of the comparison between the predictions of the model M and observations y_{obs}. This grade is compared with the reference likelihood q of "all the rest." Examples of implementations include the sign test and the tolerance interval methods.[7] In many cases, one does not have the luxury of a likelihood; one has then to resort to more empirical assessments of how well the model explains crucial observations. In the most complex cases, the outcome can be binary (accepted or rejected).
4. The posterior value of the model is obtained according to a formula of the type
$$V_{\text{posterior}}/V_{\text{prior}} = F\left[p(M|y_{\text{obs}}), q; c_{\text{novel}}\right] . \tag{1}$$
In this expression, $V_{\text{posterior}}$ is the posterior potential, or coefficient, of trust in the value of the model after the comparison between the prediction of the model and the new observations has been performed. By the action

[7] Pal and Makai [62] have used the mathematical statistics of hypothesis testing as a way to validate the correctness of code simulating the operation of a complex system with respect to a level of confidence for safety problems. The main conclusion is that the testing of the input variables separately may lead to incorrect safety related decisions with unforeseen consequences. They used two statistical methods: the sign test and the tolerance interval methods for testing more than one mutually dependent output variables. We propose to use these and similar tests delivering a probability level p which can then be compared with a predefined likelihood level q.

of $F[\cdots]$, $V_{\rm posterior}$ can be either larger or smaller than $V_{\rm prior}$: in the former case, the experimental test has increased our trust in the validity of the model; in the later case, the experimental test has signaled problems with the model. One could call $V_{\rm prior}$ and $V_{\rm posterior}$ the evolving "potential value of our trust" in the model or, loosely paraphrasing the theory of decision making in economics, the "utility" of the model [63].

The transformation from the potential value $V_{\rm prior}$ of the model before the experimental test to $V_{\rm posterior}$ after the test is embodied in the multiplier F, which can be either larger than 1 (towards validation) or smaller than 1 (towards invalidation). We postulate that F depends on the grade $p(M|y_{\rm obs})$, to be interpreted as proportional to the probability of the model M given the data $y_{\rm obs}$. It is natural to compare this probability with the reference likelihood q that one or more of all other conceivable models is compatible with the same data.

Our multiplier F depends also on a parameter $c_{\rm novel}$ that quantifies the importance of the test. In other words, $c_{\rm novel}$ is a measure of the impact of the experiment or of the observation, that is, how well the new observation explores novel "dimensions" of the parameter and variable spaces of both the process and the model that can reveal potential flaws. A fundamental challenge is that the determination of $c_{\rm novel}$ requires, in some sense, a pre-existing understanding of the physical processes so that the value of a new experiment can be fully appreciated. In concrete situations, one has only a limited understanding of the physical processes and the value of a new observation is only assessed after a long learning phase, after comparison with other observations and experiments, as well as after comparison with the model, making $c_{\rm novel}$ possibly self-referencing. Thus, we consider $c_{\rm novel}$ as a judgment-based weighting of experimental referents, in which judgment (for example, by a subject matter expert) is dominant in its determination. The fundamental problem is to quantify the relevance of a new experimental referent for validation to a given decision-making problem, given that the experimental domain of the test does not overlap with the application domain of the decision. Assignment of $c_{\rm novel}$ requires the judgment of subject matter experts, whose opinions will likely vary. This variability must be acknowledged (if not accounted for, however naively) in assigning $c_{\rm novel}$. Thus, providing an a priori value for $c_{\rm novel}$, as required in expression (1), remains a difficult and key step in the validation process. This difficulty is similar to specifying the utility function in decision making [63].

Repeating an experiment twice is a special degenerate case since it amounts ideally to increasing the size of the statistical sample. In such a situation, one should aggregate the two experiments 1 and 2 (yielding the relative likelihoods p_1/q and p_2/q respectively) graded with the same $c_{\rm novel}$ into an effective single test with the same $c_{\rm novel}$ and likelihood $(p_1/q)(p_2/q)$. This is the ideal situation, as there are cases where repeating an experiment may wildly increase the evidence of systemic uncertainty or demonstrate uncontrolled variability or other kinds of problems. When this occurs, this means that the assumption that there is no surprise, no novelty, in repeating the experiment is incorrect. Then, the two experiments should be treated so as to contribute two multipliers F's, because they reveal different kinds of uncertainty that can be generated by ensembles of experiments.

One experimental test corresponds to a entire loop $1-4$ transforming a given V_{prior} to a $V_{\text{posterior}}$ according to (1). This $V_{\text{posterior}}$ becomes the new V_{prior} for the next test, which will transform it into another $V_{\text{posterior}}$ and so on, according to the following iteration process:

$$V_{\text{prior}}^{(1)} \to V_{\text{posterior}}^{(1)} = V_{\text{prior}}^{(2)} \to V_{\text{posterior}}^{(2)} = V_{\text{prior}}^{(3)} \to \cdots \to V_{\text{posterior}}^{(n)}. \tag{2}$$

After n validation loops, we have a posterior trust in the model given by[8]

$$\frac{V_{\text{posterior}}^{(n)}}{V_{\text{prior}}^{(1)}} = F\left[p^{(1)}\left(M|y_{\text{obs}}^{(1)}\right), q^{(1)}; c_{\text{novel}}^{(1)}\right] \cdots F\left[p^{(n)}\left(M|y_{\text{obs}}^{(n)}\right), q^{(n)}; c_{\text{novel}}^{(n)}\right], \tag{3}$$

where the product is time-ordered since the sequence of values for $c_{\text{novel}}^{(j)}$ depend on preceding tests. Validation can be said to be asymptotically satisfied when the number of steps n and the final value $V_{\text{posterior}}^{(n)}$ are sufficiently high. How high is high enough is subjective and may depend on both the application and programmatic constraints. The concrete examples discussed below offer some insight on this issue. This construction makes clear that there is no absolute validation, only a process of corroborating or disproving steps competing in a global valuation of the model under scrutiny. The product (3) expresses the assumption that successive observations give independent multipliers. This assumption keeps the procedure simple because determining the dependence between different tests with respect to validation would be highly undetermined. We propose that it is more convenient to measure the dependence through the single parameter $c_{\text{novel}}^{(j)}$ quantifying the novelty of

[8] This sequence is reminiscent of a branching process: most of the time, after the first or second validation loop, the model will be rejected if $V_{\text{posterior}}^{(n)}$ becomes much smaller than $V_{\text{prior}}^{(1)}$. The occurrence of a long series of validation tests is specific to those rare models/codes that happen to survive. We conjecture that the nature of models and their tests make the probability of survival up to level n a power law decaying as a function of validation generation number n: $\Pr\left[V_{\text{posterior}}^{(n)} \geq V_{\text{prior}}^{(1)}\right] \sim 1/n^{\tau}$, for large n. The exponent $\tau = 3/2$ in mean-field branching processes [64]; being an ensemble average over random test outcomes, we expect this to be only an upper bound for actual validation processes. The four illustrative examples provided further on, augmented with a fifth one described in [1], yield $\tau \approx 0.85$ for $3 \leq n \leq 7$ with just one outlier. Although the sample of models is tiny, this illustrates our point.

the jth test with respect to those preceding it. In full generality, each new F multiplier should be a function of all previous tests.

The loop $1-4$ together with expression (1) are offered as an attempt to quantify the progression of the validation process. Eventually, when one has performed several approximately independent tests exploring different features of the model and of the validation process, $V_{\mathrm{posterior}}$ has grown to a level at which most experts will be satisfied and will believe in the validity of (i.e., be inclined to trust) the model. This formulation has the advantage of viewing the validation process as a convergence or divergence built on a succession of steps, mimicking the construction of a theory of reality.[9] Expression (3) embodies the progressive build-up of trust in a model or theory. This formulation provides a formal setting for discussing the difficulties that underlay the so-called impossibilities [19, 21] in validating a given model. Here, these difficulties are not only partitioned but quantified:

- In the definition of "new" non-redundant experiments (parameter c_{novel});
- In choosing the metrics and the corresponding statistical tests quantifying the comparison between the model and the measurements of this experiment (leading to the likelihood ratio p/q);
- In iterating the procedure so that the product of the gain/loss factors $F[\cdots]$ obtained after each test eventually leads to a clear-cut conclusion after several tests.

This formulation makes clear why and how one is never fully convinced that validation has been obtained: it is a matter of degree, of confidence level, of decision making, as in statistical testing. But this formulation helps in quantifying what new confidence (or distrust) is gained in a given model. It emphasizes that validation is an ongoing process, similar to the never-ending construction of a theory of reality.

The general formulation proposed here in terms of iterated validation loops is intimately linked with decision theory based on limited knowledge: the choice to "go ahead" and use the model is fundamentally a decision problem based on the accumulated confidence embodied in $V_{\mathrm{posterior}}$. The "go/no-go" decision must take into account conflicting requirements and compromise between different objectives. Decision theory was created by the statistician Abraham Wald in the late forties [65], but is based ultimately on game theory [63, 66]. Wald used the term *loss function*, which is the standard terminology used in mathematical statistics. In mathematical economics, the opposite of the loss (or cost) function gives the concept of the *utility function*, which quantifies (in a specific functional form) what is considered important and robust in the fit of the model to the data. We use $V_{\mathrm{posterior}}$ in an even more general sense than "utility," as a decision and information-based valuation

[9] It is conceivable that a new and radically different observation/experiment may arise and challenge the built-up trust in a model; such a scenario exemplifies how any notion of validation "convergence" is inherently local.

that supports risk-informed decision-making based on "satisficing"[10] (see the concrete examples discussed below).

It may be tempting to interpret the above formulation of the validation problem in terms of Bayes' theorem

$$p_{\text{posterior}}(M|\text{Data}) = \frac{p_{\text{prior}}(M) \times \Pr(\text{Data}|M)}{\Pr(\text{Data})} \tag{4}$$

where $\Pr(\text{Data}|M)$ is the likelihood of the data given the model M, and $\Pr(\text{Data})$ is the unconditional likelihood of the data. However, we can not make immediate sense of $\Pr(\text{Data})$. Only when a second model M' is introduced can we actually calculate

$$\Pr(\text{Data}) = p_{\text{prior}}(M)\ \Pr(\text{Data}|M) + p_{\text{prior}}(M')\ \Pr(\text{Data}|M'). \tag{5}$$

In other words, Bayes' formulation requires that we set a model/hypothesis in opposition to another or other ones, while we examine here the case of a single hypothesis in isolation.

We therefore stress that one should resist the urge to equate our V_{prior} and $V_{\text{posterior}}$ with p_{prior} and $p_{\text{posterior}}$ because the Vs are *not* probabilities. It is not possible to assign a probability to an experiment in an absolute way and thus Bayes' theorem is mute on the validation problem as we have chosen to formulate it. Rather, we propose that the problem of validation is fundamentally a problem of decision theory: at what stage is one willing to bet that the code will work for its intended use? At what stage are you ready to risk your reputation, your job, the lives of others, your own life on the fact that the model/code will predict correctly the crucial aspect of the real-life test? One must therefore incorporate ingredients of decision theory, not only fully objective probabilities. Coming from a Bayesian perspective, p_{prior} and $p_{\text{posterior}}$ could then be called the potential *value* or *trust* in the model/code or, as we prefer, to move closer to the application of decision theory in economics, the *utility* of the model/code [63].

To summarize the discussion so far, expression (1) may be reminiscent of a Bayesian analysis, however, it does not manipulate probabilities. (Instead, they appear as *independent* variables, viz., $p(M|y_{\text{obs}})$ and q.) In the Bayesian methodology of validation [69, 70], only comparison between models can be performed due to the need to remove the unknown probability of the data in Bayes' formula. In contrast, our approach provides a value for each single model independently of the others. In addition, it emphasizes the importance of quantifying the novelty of each test and takes a more general view on how to use the information provided from the goodness-of-fit. The valuation (1) of a model uses probabilities as partial inputs, not as the qualifying criteria for

---

[10] In economics, *satisficing* is a behavior that attempts to achieve at least some minimum level of a particular variable, but that does not strive to achieve its maximum possible value. The verb "to satisfice" was coined by Herbert A. Simon in his theory of bounded rationality [67, 68].

model validation. This does not mean, however, that there are not uncertainties in these quantities or in the terms F, q or $c_{\rm novel}$ and that aleatory and epistemic uncertainties[11] are ignored, as discussed below.

3 Desirable Properties of the Multiplier of the Validation Step

The multiplier $F\left[p(M|y_{\rm obs}), q; c_{\rm novel}\right]$ should have the following properties:

1. If the statistical test(s) performed on the given observations is (are) passed at the reference level q, then the posterior potential value is larger than the prior potential value: $F > 1$ (resp. $F \leq 1$) for $p > q$ (resp. $p \leq q$), which can be written succinctly as $\ln F / \ln(p/q) > 0$.
2. The larger the statistical significance of the passed test, the larger the posterior value. Hence
$$\frac{\partial F}{\partial p} > 0,$$
for a given q. There could be a saturation of the growth of F for large p/q, which can be either that $F < \infty$ as $p/q \to \infty$ or of the form of a concavity requirement
$$\frac{\partial^2 F}{\partial p^2} < 0$$
for large p/q: obtaining a quality of fit beyond a certain level should not be attempted.
3. The larger the statistical level at which the test(s) performed on the given observations is (are) passed, the larger the impact of a "novel" experiment on the multiplier enhancing the prior into the posterior potential value of the model: $\partial F / \partial c_{\rm novel} > 0$ (resp. ≤ 0), for $p > q$ (resp. $p \leq q$).

A very simple multiplier that obeys this these properties (not including the saturation of the growth of F) is given by

$$F\left[p(M|y_{\rm obs}), q; c_{\rm novel}\right] = \left(\frac{p}{q}\right)^{c_{\rm novel}}, \tag{6}$$

and is illustrated in the upper panel of Fig. 2 as a function of p/q and $c_{\rm novel}$. This form provides an intuitive interpretation of the meaning of the experiment impact parameter $c_{\rm novel}$. A non-committal evaluation of the novelty of a test would be $c_{\rm novel} = 1$, thus $F = p/q$ and the chain (3) reduces to a product of normalized likelihoods, as in standard statistical tests. A value $c_{\rm novel} > 1$

[11] For an in-depth discussion on aleatory versus systemic (a.k.a. epistemic) uncertainties, see for example *Review of Recommendations for Probabilistic Seismic Hazard Analysis: Guidance on Uncertainty and Use of Experts* [71], available at http://www.nap.edu/catalog/5487.html.

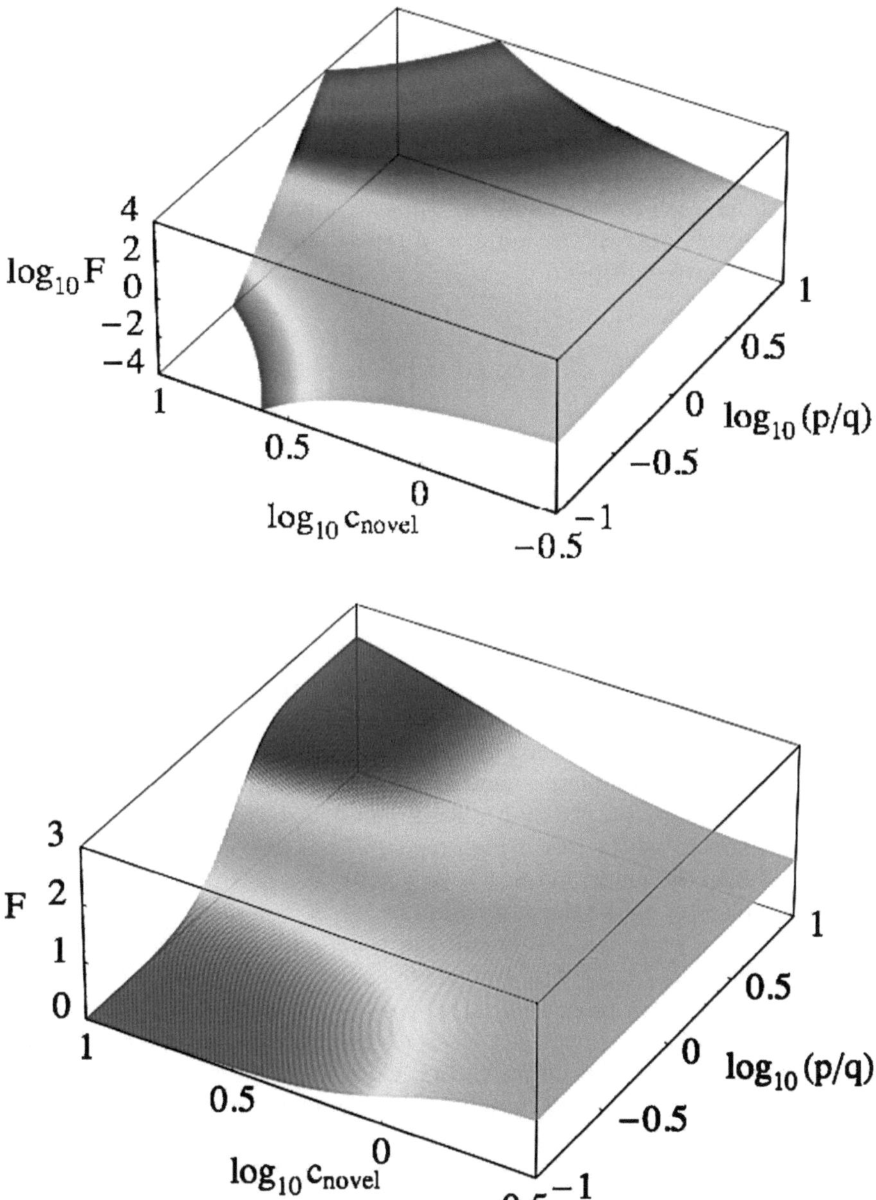

Fig. 2. The multipliers defined by (6) and (7) are plotted as functions of p/q and c_{novel} in the upper and lower panels, respectively. Note the vertical log scale used for the multiplier (6) in the top panel. (See Plate 8 on page 304)

(resp. <1) for a given experiment describes a nonlinearly rapid (resp. slow) updating of our trust V as a function of the grade p/q of the model with respect to the observations. In particular, a large value of $c_{\rm novel}$ corresponds to the case of "critical" tests.[12] Note that the parameterization of $c_{\rm novel}$ in (6) should account for the decreased novelty noted above occurring when the same experiment is repeated two or more times. The value of $c_{\rm novel}$ should be reduced for each repetition of the same test; moreover, the value of $c_{\rm novel}$ should approach unity as the number of repetitions increases.

An alternative multiplier,

$$F\left[p(M|y_{\rm obs}), q; c_{\rm novel}\right] = \left[\frac{\tanh\left(\frac{p}{q} + \frac{1}{c_{\rm novel}}\right)}{\tanh\left(1 + \frac{1}{c_{\rm novel}}\right)}\right]^4 , \qquad (7)$$

is plotted in the lower panel of Fig. 2 as a function of p/q and $c_{\rm novel}$. It emphasizes that F saturates as a function of p/q and $c_{\rm novel}$ as either one or both of them grow large. A completely new experiment corresponds to $c_{\rm novel} \to \infty$ so that $1/c_{\rm novel} = 0$ and thus F tends to $[\tanh(p/q)/\tanh(1)]^4$, i.e., $V_{\rm posterior}/V_{\rm prior}$ is only determined by the quality of the "fit" of the data by the model quantified by p/q. A finite $c_{\rm novel}$ thus implies that one already takes a restrained view on the usefulness of the experiment since one limits the amplitude of the gain $= V_{\rm posterior}/V_{\rm prior}$, whatever the quality of the fit of the data by the model. The exponent 4 in (7) has been chosen so that the maximum confidence gain F is equal to $\tanh(1)^{-4} \approx 3$ in the best possible situation of a completely new experiment ($c_{\rm novel} = \infty$) and perfect fit ($p/q \to \infty$). In contrast, the multiplier F can be arbitrarily small as $p/q \to 0$ even if the novelty of the test is high ($c_{\rm novel} \to \infty$). For a finite novelty $c_{\rm novel}$, a test that fails the model miserably ($p/q \approx 0$) does not necessarily reject the model completely: unlike the expression in (6), F remains greater than zero. Indeed, if the novelty $c_{\rm novel}$ is small, the worst-case multiplier (attained for $p/q = 0$) is $\left[\tanh\left(1/c_{\rm novel}\right) / \tanh\left(1 + 1/c_{\rm novel}\right)\right]^4 \approx 1 - 6.9\,{\rm e}^{-2/c_{\rm novel}}$, which is only slightly less than unity if $c_{\rm novel} \ll 1$. In short, this formulation does not heavily weight unimportant tests, as seems intuitively appropriate.

In the framework of decision theory, expression (1) with one of the specific expressions in (6) or (7) provides a parametric form for the utility or decision "function" of the decision maker. It is clear that many other forms of the utility function can be used, however, with the constraint of keeping the salient features of expression (1) with (6) or (7), in terms of the impact of a new test given past tests, and the quality of the comparison between the

[12] A momentous example is the Michelson–Morley experiment for the Theory of Special Relativity. For the Theory of General Relativity, it was the observation during the famous 1919 solar eclipse of the bending of light rays from distant stars by the Sun's mass and the elegant explanation of the anomalous precession of the perihelion of Mercury's orbit.

model predictions and the data. This indetermination is helpful since it mirrors the inherent variability of the validation landscape. For instance, what comprises adequate validation for phenomena at one (e.g., macro-)scale may prove inadequate for related phenomena at another (e.g., micro-)scale.

Finally, we remark that the proposed form for the multiplier (7) contains an important asymmetry between gains and losses: the failure to a single test with strong novelty and significance[13] cannot be compensated by the success of all the other tests combined. In other words, a single test is enough to reject a model. This encapsulates the common lore that reputation gain is a slow process requiring constancy and tenacity, while its loss can occur suddenly with one single failure and is difficult to re-establish. We believe that the same applies to the build-up of trust in and, thus, validation of a model.

4 Practical Guidelines for Determining p/q and c_{novel}

These two crucial elements of a validation step are conditioned by four basic problems, over which one can exert at least partial control. In particular, they address the two sources of uncertainty: "reducible" or epistemic (i.e., due to lack of knowledge) and "irreducible" or aleatory (i.e., due to variability inherent in the phenomenon under consideration). In a nutshell, as becomes clearer below, the comparison between p and q is more concerned with the aleatory uncertainty while c_{novel} deals in part with the epistemic uncertainty. In the following, as in the two examples (6) and (7), we consider that p and q enter only in the form of their ratio p/q. This should not generally be the case but, given the many uncertainties, this restriction simplifies the analysis by removing one degree of freedom.

1. *How to model?* This addresses model construction and involves the structure of the elementary contributions, their hierarchical organization, and requires dealing with uncertainties and fuzziness. This concerns the epistemic uncertainty.
2. *What to measure?* This relates to the nature of c_{novel}: ideally, one should target adaptively the observations to "sensitive" parts of the system and the model (as, e.g., Palmer et al. [72] did for atmospheric dynamics). Targeting observations could be directed by the desire to access the most "relevant" information as well as to get information that is the most reliable, i.e., which is contaminated by the smallest errors. This is also the stance of Oberkampf and Trucano [33]: *"A validation experiment is conducted for the primary purpose of determining the validity, or predictive accuracy, of a computational modeling and simulation capability. In other*

[13] See, e.g., the impact of localized seismicity on faults in the case of the Olami–Feder–Christensen model discussed below, or that of the "leverage" effect in quantitative finance for the Multifractal Random Walk model described and evaluated in [1].

words, a validation experiment is designed, executed, and analyzed for the purpose of quantitatively determining the ability of a mathematical model and its embodiment in a computer code to simulate a well-characterized physical process." In practice, we view c_{novel} as an estimate of the importance of the new observation and the degree of "surprise" it brings to the validation step. Being the cornerstone of our formal approach to validation, we eventually want to see its determination grounded in sensitivity and/or PIRT analysis (cf. Sect. 1.6). The epistemic uncertainty alluded to above is partially addressed in the choice of the empirical data and its rating with c_{novel} (see the examples of application discussed below).

3. *How to measure?* For given measurements or experiments, the problem is to find the "optimal" metric or cost function (involved in the quality-of-fit measure p) for the intended use of the model. The notion of optimality needs to be defined. It could capture a compromise between fitting best the important features of the data (what is "important" may be decided on the basis of previous studies and understanding or other processes, or programmatic concerns), and minimizing the extraction of spurious information from noise. This requires one to have a precise idea of the statistical properties of the noise. If such knowledge is not available, the cost function should be chosen accordingly. The choice of the cost function involves the choice of how to look at the data. For instance, one may want to expand the measurements at multiple scales using wavelet decompositions and compare the prediction and observations scale by scale, or in terms of multifractal spectra of the physical fields estimated from these wavelet decompositions [73] or from other methods. The general idea here is that, given complex observation fields, it is appropriate to unfold the data on a variety of "metrics," which can then be used in the comparison between observations and model predictions. The question is then: How well is the model able to reproduce the salient multi-scale and multifractal properties derived from the observations? The physics of turbulent fields and of complex systems have offered many such new tools with which to unfold complex fields according to different statistics. Each of these statistics offers a metric to compare observations with model predictions and is associated with a cost function focusing on a particular feature of the process. Since these metrics are derived from the understanding that turbulent fields can be analyzed using these metrics that reveal strong constraints in their organization, these metrics can justifiably be called "physics-based." In practice, p, and eventually p/q, has to be inferred as an estimate of the degree of matching between the model output and the observation. This can be done following the concept of fuzzy logic in which one replaces the yes/no pass test by a more gradual quantification of matching [74,75]. We thus concur with Oberkampf and Barone [76], while our general methodology goes beyond. Note that this discussion relates primarily to the aleatory uncertainty.

4. *How to interpret the results?* This question relates to defining the test and the reference probability level q that any other model (than the one under scrutiny) can explain the data. The interpretation of the results should aim at detecting the "dimensions" that are missing, misrepresented or erroneous in the model (systemic/epistemic uncertainty). What tests can be used to betray the existence of hidden degrees of freedom and/or dimensions? This is the hardest problem. It can sometimes possess an elegant solution when a given model is embedded in a more general one. Then, the limitation of the more restricted model becomes clear from the vantage of the more general model.

We refer to the Appendix for further thoughts on these four basic steps in model construction and validation in a broader context than our present formulation.

We now illustrate our algorithmic approach to model validation using the historical development of quantum mechanics and three examples based on the authors' research activities. In these crude but revealing examples, we will use the form (7) and consider three finite values: $c_{\text{novel}} = 1$ (marginally useful new test), $c_{\text{novel}} = 10$ (substantially new test), and $c_{\text{novel}} = 100$ (important new test). When a likelihood test is not available, we propose to use three possible marks: $p/q = 0.1$ (poor fit), $p/q = 1$ (marginally good fit), and $p/q = 10$ (good fit). Extreme values (c_{novel} or p/q are 0 or ∞) have already been discussed. Due to limited experience with this approach, we propose these ad hoc values in the following examples of its application.

5 Illustration with the Development of Quantum Mechanics

Quantum mechanics (QM) offer a vivid incarnation of how a model can turn progressively into a theory held "true" by almost all physicists. Since its birth, QM has been tested again and again because it presents a view of "reality" that is shockingly different from the classical view experienced at the macroscopic scale. QM prescriptions and predictions often go against (classically-trained) intuition. Nevertheless, we can state that, by a long and thorough process of confirmed predictions of QM in experiments, fueled by the imaginative set-up of paradoxes, QM has been validated as a correct description of nature. It is fair to say that the overwhelming majority of physicists have developed a strong trust in the validity of QM. That is, if someone comes up with a new test based on a new paradox, most physicists would bet that QM will come up with the right answer with a very high probability. It is thus by the on-going testing and the compatibility of the prediction of QM with the observations that QM has been validated. As a consequence, one can use it with strong confidence to make predictions in novel directions. This is ideally the situation one would like to attain for the problem of validation of all models, those discussed in the following section in particular. We now give a very partial list of selected tests that established the trust of physicists in QM.

1. Pauli's exclusion principle states that no two identical fermions (particles with non-integer values of spin) may occupy the same quantum state simultaneously [77]. It is one of the most important principles in quantum physics, primarily because the three types of particle from which ordinary matter is made, electrons, protons, and neutrons, are all subject to it. With $c_{\rm novel} = 100$ and perfect agreement in numerous experiments ($p/q = \infty$), this leads to $F^{(1)} = 2.9$.
2. The EPR paradox [78] was a thought experiment designed to prove that quantum mechanics was hopelessly flawed: according to QM, a measurement performed on one part of a quantum system can have an instantaneous effect on the result of a measurement performed on another part, regardless of the distance separating the two parts. Bell's theorem [79] showed that quantum mechanics predicted stronger statistical correlations between entangled particles than the so-called local realistic theory with hidden variables. The importance of this prediction requires $c_{\rm novel} = 100$ at a minimum. The QM prediction turned out to be correct, winning over the hidden-variables theories [80, 81] ($p/q = \infty$), leading again to $F^{(2)} = 2.9$.
3. The Aharonov–Bohm effect predicts that a magnetic field can influence an electron that, strictly speaking, is located completely beyond the field's range, again an impossibility according to non-quantum theories ($c_{\rm novel} = 100$). The Aharonov-Bohm oscillations were observed in ordinary (i.e., not superconducting) metallic rings, showing that electrons can maintain quantum mechanical phase coherence in ordinary materials [82, 83]. This yields $p/q = \infty$ and thus $F^{(3)} = 2.9$ yet again.
4. The Josephson effect provides a macroscopic incarnation of quantum effects in which two superconductors are predicted to preserve their long-range order across an insulating barrier, for instance, leading to rapid alternating currents when a steady voltage is applied across the superconductors. The novelty of this effect again warrants $c_{\rm novel} = 100$ and the numerous verifications and applications (for instance in SQUIDs, Superconducting QUantum Interference Devices) argues for $p/q = \infty$ and thus $F^{(4)} = 2.9$, as usual.
5. The prediction of possible collapse of a gas of atoms at low temperature into a single quantum state is known as Bose-Einstein condensation, again so much against classical intuition ($c_{\rm novel} = 100$). Atoms are indeed bosons (particles with integer values of spin), which are *not* subjected to the Pauli exclusion principle evoked in the above test #1 of QM. The first such Bose-Einstein condensate was produced using a gas of rubidium atoms cooled to $1.7 \cdot 10^{-7}$ K [84] ($p/q = \infty$), leading once more to $F^{(4)} = 2.9$.

6. There have been several attempts to develop a paradox-free nonlinear QM theory, in the hope of eliminating Schrödinger's cat paradox, among other embarrassments. The nonlinear QM predictions diverge from those of orthodox quantum physics, albeit subtly. For instance, if a neutron impinges on two slits, an interference pattern appears, which should, however, disappear if the measurement is made far enough away ($c_{\mathrm{novel}} = 100$). Experimental tests of the neutron prediction rejected the nonlinear version in favor of the standard QM [85] ($p/q = \infty$), leading to $F^{(6)} = 2.9$.
7. In addition, measurements at the National Institute of Standards and Technology (NIST) in Boulder, CO, on frequency standards have been shown to set limits of order 10^{-21} on the fraction of the energy of the rf transition in ^{9}Be ions that could be due to nonlinear corrections to quantum mechanics [86]. We assign $c_{\mathrm{novel}} = 10$, with $p/q = 10$, to this result, leading to $F^{(7)} = 2.4$. Although less than $F^{(1-6)}$, this is still an impressive score.

Combining the multipliers according to (3) leads to $V^{(8)}_{\mathrm{posterior}}/V^{(1)}_{\mathrm{prior}} \simeq 1400$, which is of course only a lower limit given the many other validation tests not mentioned here.

Tests of QM are ongoing [87]. But given the presumably huge amount of trust physicists have in QM which we tried to quantify, why do physicists still feel the need to put QM to the "validation" test? This raises the question whether we can ever establish a sense of sufficiency for validation. Our position is that this reflects a quixotic quest for absolute truth – and also a taste for surprises – that most scientists can relate to. Perhaps, by continuing to test QM, a new insight or an anomaly will be uncovered which may help progress in the understanding of reality.

6 Three Examples Drawn from the Authors' Research Interests

6.1 The Olami–Feder–Christensen (OFC) Sand-Pile Model of Earthquakes

This is perhaps the simplest sand-pile model of self-organized criticality, which exhibits a phenomenology resembling real seismicity [88]. Figure 3 shows a "stress" map generated by the OFC model immediately after a large avalanche (main shock) at two magnifications, to illustrate the rich organization of almost synchronized regions [89]. To validate the OFC model, we examine the properties and prediction of the model that can be compared with real seismicity, together with our assessment of their c_{novel} and quality-of-fit. We are careful to state these properties in an ordered way, as specified in the above sequences (2)–(3).

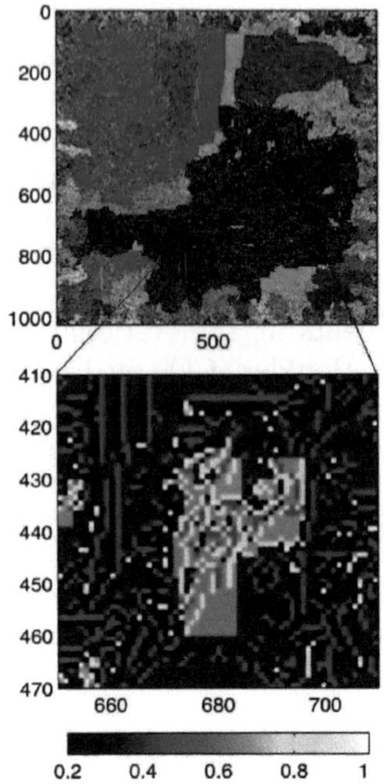

Fig. 3. Map of the "stress" field generated by the OFC model immediately after a large avalanche (main shock) at two magnifications. The *upper panel* shows the whole grid of size 1,024 and the *lower plot* represents a subset of the grid delineated by the square in the upper plot. Adapted from [90]. (See Plate 9 on page 305)

1. The statistical physics community recognized the discovery of the OFC model as an important step in the development of a theory of earthquakes: without a conservation law (which was thought before to be an essential condition), it nevertheless exhibits a power law distribution of avalanche sizes resembling the Gutenberg-Richter law [88]. On the other hand, many other models with different mechanisms can explain observed power law distributions [91]. We thus attribute only $c_{\text{novel}} = 10$ to this evidence. Because the power law distribution obtained by the model is of excellent quality for a certain parameter value ($\alpha \approx 0.2$), we formally take $p/q = \infty$ (perfect fit). Expression (7) then gives $F^{(1)} = 2.4$.
2. Prediction of the OFC model concerning foreshocks and aftershocks, and their exponents for the inverse and direct Omori laws. These predictions are twofold [90]: (i) the finding of foreshocks and aftershocks with similar qualitative properties, and (ii) their inverse and direct Omori rates. The first aspect, deserves a large $c_{\text{novel}} = 100$ as the observation of foreshocks and aftershocks came as a rather big surprise in such sand-pile models [92]. The clustering in time and space of the foreshocks and aftershocks are qualitatively similar to real seismicity [90], which warrants $p/q = 10$, and thus $F^{(2a)} = 2.9$. The second aspect is less important than the first one

($c_{\text{novel}} = 1$). Since the exponents are only qualitatively reproduced (but with no formal likelihood test available), we therefore take $p/q = 0.1$. This leads to $F^{(2b)} = 0.47$.
3. Scaling of the number of aftershocks with the main shock size (productivity law) [90]: $c_{\text{novel}} = 10$ as this observation is rather new but not completely independent of the Omori law. The fit is good so we grant a grade $p/q = 10$ leading to $F^{(3)} = 2.4$.
4. Power law increase of the number of foreshocks with the main shock size [90]: this is not observed in real seismicity, probably because this property is absent or perhaps due to a lack of quality data. This test is therefore not very selective ($c_{\text{novel}} = 1$) and the large uncertainties suggest a grade $p/q = 1$ (to reflect the different viewpoints on the absence of effect in real data) leading to $F^{(4)} = 1$ (neutral test).
5. Most aftershocks are found to nucleate at "asperities" located on the main shock rupture plane or on the boundary of the avalanche, in agreement with observations [90]: $c_{\text{novel}} = 10$ and $p/q = 10$ leading to $F^{(5)} = 2.4$.
6. Earthquakes cluster on spatially localized geometrical structures known as faults. This property is arguably central to the physics of seismicity ($c_{\text{novel}} = 100$), but absolutely not reproduced by the OFC model ($p/q = 0.1$). This leads to $F^{(6)} = 4 \cdot 10^{-4}$.

Combining the multipliers according to (3) up to test #5 leads to $V^{(6)}_{\text{posterior}}/V^{(1)}_{\text{prior}} = 18.8$, suggesting that the OFC model is validated as a useful model of the statistical properties of seismic catalogs, at least with respect to the properties which have been examined in these first five tests. Adding the crucial last test strongly fails the model since $V^{(7)}_{\text{posterior}}/V^{(1)}_{\text{prior}} = 7.5 \cdot 10^{-3}$. The model can not be used as a realistic predictor of seismicity. The results of our quantitative validation process indicate that it can nevertheless be useful to illustrate certain statistical properties and to help formulate new questions and hypotheses.

6.2 An Anomalous Diffusion Model for Solar Radiation in Cloudy Atmospheres

To improve our modeling skill for climate dynamics, it is essential to reduce the significant uncertainty associated with clouds. In particular, estimation of the radiation budget in the presence of clouds needs improvement since current operational models for the most part ignore all variability below the scale of the climate model's grid ($\sim$100 km). A considerable effort has therefore been expended to derive more realistic mean-field radiative transfer models [93], mostly by considering only the one-point variability of clouds, that is, irrespective of their actual structure as captured by 2-point (or higher) correlation statistics. However, it has been widely recognized that the Earth's cloudiness is fractal over a wide range of scales [94]. This is the motivation for modeling

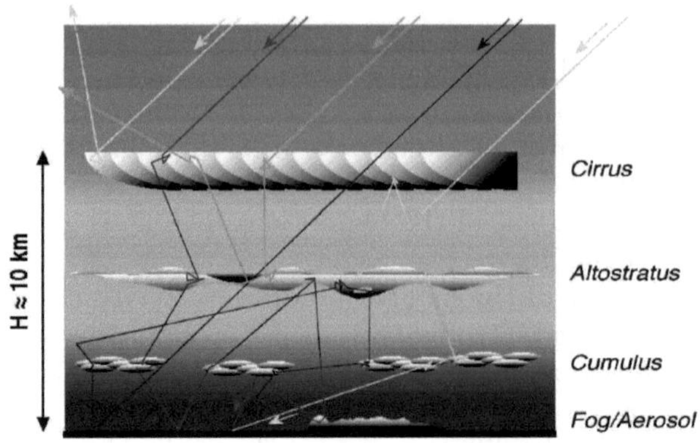

Fig. 4. The anomalous diffusion model of solar photon transport at non-absorbing wavelengths in the cloudy atmosphere. In this model, solar beams follow convoluted Lévy walks, which are characterized by frequent small steps (inside clouds) and occasional large jumps (between clouds or between clouds and the surface). The partition between small and large jumps is controlled by the Lévy index α (the PDF of the jump sizes ℓ has a tail decaying as a power law $\sim 1/\ell^{1+\alpha}$). Reproduced from [95]. (See Plate 10 on page 305)

the paths of solar photons at non-absorbing wavelengths in the cloudy atmosphere as Lévy walks [91], which are characterized by frequent small steps (inside clouds) and occasional large jumps (typically between clouds) as represented schematically in Fig. 4. These (on-average downward) paths start at the top of the highest clouds and end in escape to space or in absorption at the surface, respectively, cooling and warming the climate system. In contrast with most other mean-field models for solar radiative transfer, this diffusion model with anomalous scaling can be subjected to a battery of observational tests.

1. The original goal of this phenomenological model, which accounts for the clustering of cloud water droplets into broken and/or multi-layered cloudiness, was to predict the increase in steady-state flux transmitted to the surface compared to what would filter through a fixed amount of condensed water in a single unbroken cloud layer [96]. This property is common to all mean-field photon transport models that do anything at all about unresolved variability [93]. Thus, we assign only $c_{\text{novel}} = 1$ to this test and, given that all models in this class are successful, we have to take $p/q = 1$, hence $F^{(1)} = 1$. The outcome of this first test is neutral.
2. The first real test for this model occurred in the late 1990s, when it became possible to accurately estimate the mean total path cumulated by solar radiation that reaches the surface. This breakthrough was enabled by access to spectroscopy at medium (high) resolution of oxygen bands (lines) [97,98]. There was already remote sensing technology to infer simultaneously cloud optical depth, which is column-integrated water

in g (or cm^3) per cm^2 multiplied by the average cross-section for scattering or absorption in cm^2 per g (or cm^3). The observed trends between mean path and optical depth were explained only by the new model in spite of relatively large instrumental error bars. So we assign $c_{\text{novel}} = 100$ to this highly discriminating test and $p/q = 10$ (even though other models were generally not in a position to compete), hence $F^{(2)} = 2.9$.

3. Another test was proposed using time-dependent photon transport with a source near the surface (cloud-to-ground lightning) and a detector in space (aboard the US DOE FORTÉ satellite) [99]. The quantity of interest is the observed delay of the light pulse (due to multiple scattering in the cloud system) with respect to the radio-frequency pulse (which travels in a straight line). There was no simultaneous estimate of cloud optical depth, so assumptions had to be made (informed by the fact that storm clouds are at once thick and dense). Because of this lack of an independent measurement, we assign only $c_{\text{novel}} = 10$ to the observation and $p/q = 1$ to the model performance. Indeed, this test is arguably only about the finite horizontal extent of the rain clouds resulting from deep convection: one can exclude only most simplistic cloud models based on uniform plane-parallel slabs. So, again we obtain $F^{(3)} = 1$ for an interesting but presently neutral test that needs refinement.
4. Min et al. [100] developed an oxygen-line spectrometer with sufficient resolution to estimate not just the *mean* path but also its *root-mean-square* (RMS) value. They found the prediction by Davis and Marshak [101] for normal diffusion to be an extreme (envelop) case for the empirical scatter plot of mean vs. RMS path, and this is indicative that the anomalous diffusion model will cover the bulk of the data. Because of some overlap with item #2, we assign $c_{\text{novel}} = 10$ to the test and $p/q = 10$ for the model performance since the anomalous diffusion model had not yet made a prediction for the RMS path (although we note that other models have yet to make one for the mean path). We therefore receive $F^{(4)} = 2.4$.
5. Using similar data but a different normalization than Min et al., more amenable to model testing, Scholl et al. [102] observed that the RMS-to-mean ratio for solar photon path is essentially constant whether the cloud structure (according to mm-wave radar profiles) is complex or not (respectively, diffusion is normal or anomalous). This is a remarkable empirical finding to which we assign $c_{\text{novel}} = 100$. The new mean- and RMS-path data was explained by Scholl et al. by creating an ad hoc hybrid between normal diffusion theory (which indeed has a prediction for the RMS path [101]) and its anomalous counterpart (which still has none). This modification of the basic model can be viewed as significant, meaning that we are in principle back to validation step 1 with the new model. However, this exercise uncovered something quite telling about the original anomalous diffusion model, namely, that its simple asymptotic (large optical depth) form used in all the above tests is not generally valid: for typical cloud covers, the pre-asymptotic terms computed explicitly for the

normal diffusion case prove to be important irrespective of whether the diffusion is normal or not. Consequently, in its original form (resulting in a simple scaling law for the mean path with respect to cloud thickness and optical depth), the anomalous diffusion model fails to reproduce the new data even for the mean path. (Consequently, previous fits yielded only "effective" anomaly parameters and were misleading if taken too literally.) So we assign $p/q = 0.1$ at best for the original model, hence $F^{(5)} = 4 \cdot 10^{-4}$.

Thus, $V^{(6)}_{\text{posterior}}/V^{(1)}_{\text{prior}} = 3 \cdot 10^{-3}$, a fatal blow for the anomalous diffusion in its simple asymptotic form, even though $V^{(5)}_{\text{posterior}}/V^{(1)}_{\text{prior}} = 7.0$ which would have been interpreted as close to a convincing validation.

This is not the end of the story, of course. The original model has already spawned Scholl et al.'s empirical hybrid and a formalism based on integral (in fact, pseudo-differential) operators has been proposed [103] that extends the anomalous *diffusion* model to pre-asymptotic regimes. More recently, a model for anomalous *transport* (i.e., where angular details matter) has been proposed that fits all of the new oxygen spectroscopy results [95].

In summary, the first and simplest incarnation of the anomalous diffusion model for solar photon transport ran its course and demonstrated the power of oxygen-line spectroscopy as a test for the performance of radiative transfer models required in climate modeling for large-scale average responses to solar illumination. Eventually, new and interesting tests will become feasible when we obtain dedicated oxygen-line spectroscopy from space with NASA's Orbiting Carbon Observatory (OCO) mission planned for launch in 2008. Indeed, we already know that the asymptotic scaling for reflected photon paths [104] is different from their transmitted counterparts [101] in standard diffusion theory for both mean and RMS.

6.3 A Computational Fluid Dynamics Model for Shock-Induced Mixing

So far, our examples of models for complex phenomena have hailed from quantum and statistical physics. In the latter case, they are stochastic models composed of: (1) simple code (hence rather trivial verification procedures) to generate realizations, and (2) analytical expressions for the ensemble-average properties (that are used in the above validation exercises). We now turn to gas dynamics codes which have a broad range of applications, from astrophysical and geophysical flow simulation to the design and performance analysis of engineering systems. Specifically, we discuss the validation of the "Cuervo" code developed at Los Alamos National Laboratory [105, 106] for use as a simulation tool in the complex physics of compressible mixing. This software generates solutions of the Euler equations for flows of inviscid, non-heat-conducting, compressible gas. Cuervo has been verified against a suite of test problems including, e.g., those discussed by Liska and Wendroff [107]. As clearly stated by Oberkampf and Trucano [33] however, such verification

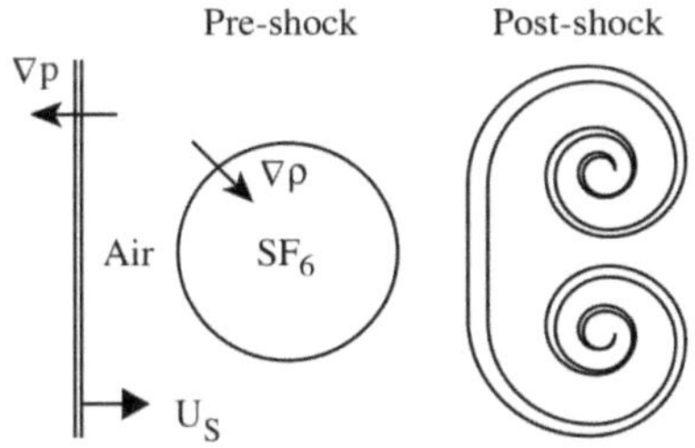

Fig. 5. The interactions between weakly shocked (Mach number ≈1.2) light gas (air) and a column of dense gas (SF_6). The Richtmyer–Meshkov instability occurs from the mismatch between the pressure gradient (at the shock front) and the density gradient (between the light and dense gases), which acts as a source of baroclinic vorticity. The column of dense gas "rolls up" into a double-spiral form under the action of the evolving vorticity

differs from and does not guarantee validation against experimental data. A standard validation scenario involves the Richtmyer–Meshkov (RM) instability [108, 109], which arises when a density gradient in a fluid is subjected to an impulsive acceleration, e.g., due to passage of a shock wave (see Fig. 5). Evolution of the RM instability is nonlinear and hydrodynamically complex and hence defines an excellent problem-space to assess CFD code performance for more general mixing scenarios.

In the series of shock-tube experiments described in [110], RM dynamics are realized by preparing one or more cylinders with approximately identical axisymmetric Gaussian concentration profiles of dense sulfur hexaflouride (SF_6) in air. This (or these) vertical "gas cylinder(s)" is (are) subjected to a weak shock – Mach number ≈1.2 – propagating horizontally, i.e., perpendicular to the axis of the gas cylinders. The ensuing dynamics are largely governed by the mismatch of the density gradient between the gases (with the density of SF_6 approximately five times that of air) and the pressure gradient through the shock wave; this mismatch acts as the source for baroclinic vorticity generation. Moreover, the flow evolution is strongly two-dimensional up to the final times considered. Visualization of the density field is obtained using a planar laser-induced fluorescence (PLIF) technique, which provides high-resolution quantitative concentration measurements in a plane that cross-cuts the cylinders. The velocity field is diagnosed using particle image velocimetry (PIV), based on correlation measurements of small-scale particles that are seeded in the initial flow field. Careful post-processing of images from 130 μs to 1,000 μs after shock passage yields planar concentration and velocity with error bars.

1. This RM flow is dominated at early times by a vortex pair. Later, secondary instabilities rapidly transition the flow to a mixed state. We rate $c_{\text{novel}} = 10$ for the observations of these two instabilities. The Cuervo code correctly captures these two instabilities, best observed and modeled with a single cylinder. At this qualitative level, we rate $p/q = 10$ (good fit), which leads to $F^{(1)} = 2.4$.

2. Older data for two-cylinder experiments acquired with a fog-based technique (rather than PLIF) showed two separated spirals associated with the primary instability, but the Cuervo code predicted the existence of a material bridge between those structures. This previously unobserved connection was subsequently diagnosed experimentally with the improved observational technique, i.e., the simulation code was truly predictive of this phenomenon. Using $c_{\rm novel} = 10$ and $p/q = 10$ yields $F^{(2)} = 2.4$.
3. The evolution of the total power as a function of time offers another useful metric. The numerical simulation quantitatively accounts for the exponential growth of the power with time, within the experimental error bars. Using $c_{\rm novel} = 10$ and $p/q = 10$ yields $F^{(3)} = 2.4$.
4. The concentration power spectrum as a function of wavenumber for different times provides another way (in the Fourier domain) to present the information of the hierarchy of structures already visualized in physical space ($c_{\rm novel} = 1$). The Cuervo code correctly accounts for the low wavenumber part of the spectrum but underestimates the high wavenumber part (beyond the deterministic-stochastic transition wavenumber) by a factor 2 to 5. We capture this by setting $p/q = 0.1$, which yields $F^{(4)} = 0.47$.

Combining the multipliers according to (3) leads to $V^{(5)}_{\rm posterior}/V^{(1)}_{\rm prior} = 6.5$, a significant gain, but still not sufficient to compellingly validate the Cuervo code for inviscid shock-induced hydrodynamic instability simulations, at least in 2D. Clearly, validation against this single set of experiments is inadequate to address all intended uses of a CFD code such as Cuervo.[14]

6.4 Discussion

The above three examples illustrate the utility of representing the validation process as a succession of steps, each of them characterized by the two parameters $c_{\rm novel}$ and p/q. The determination of $c_{\rm novel}$ requires expert judgment and that of p/q a careful statistical analysis, which is beyond the scope of the present report (see [76] for a detailed case study). The parameter q is ideally imposed as a confidence level, say 95% or 99% as in standard statistical tests. In practice, it may depend on the experimental test and requires a case-by-case examination.

The uncertainties of $c_{\rm novel}$ and of p/q need to be assessed. Indeed, different statistical estimations or metrics may yield different p/q's and different experts will likely rate differently the novelty $c_{\rm novel}$ of a new test. As a result, the trust gain $V^{(n+1)}_{\rm posterior}/V^{(1)}_{\rm prior}$ after n tests necessarily has a range of possible values that grows geometrically with n. In certain cases, a drastic difference can be obtained by a change of $c_{\rm novel}$. For instance, if instead of

[14] Intricate experiments with three gas cylinders have since been performed [111] and others are currently under way to further challenge compressible flow codes.

attributing $c_{\text{novel}} = 100$ to the sixth OFC test, we put $c_{\text{novel}} = 10$ (resp. 1) while keeping $p/q = 0.1$, $F^{(6)}$ is changed from $4 \cdot 10^{-4}$ to $4 \cdot 10^{-3}$ (resp. 0.47). The trust gain then becomes $V^{(7)}_{\text{posterior}}/V^{(1)}_{\text{prior}} = 0.07$ (resp. $\simeq 9$). For the sixth OFC test, $c_{\text{novel}} = 1$ is arguably unrealistic, given the importance of faults in seismology. The two possible choices $c_{\text{novel}} = 100$ and $c_{\text{novel}} = 10$ then give similar conclusions on the invalidation of the OFC model. In our examples, $V^{(n+1)}_{\text{posterior}}/V^{(1)}_{\text{prior}}$ provides a qualitatively robust measure of the gain in trust after n steps; this robustness has been built-in by imposing a coarse-grained quality to p/q and c_{novel}.

7 Summary

The validation of numerical simulations continues to become more important as computational power grows, as the complexity of modeled systems increases, and as increasingly important decisions are influenced by computational models. We have proposed an iterative, constructive approach to validation using *quantitative measures* and *expert knowledge* to assess the relative state of validation of a model instantiated in a computer code. In this approach, the increase/decrease in validation is mediated through a function that incorporates the results of the model vis-à-vis the experiment together with a measure of the impact of that experiment on the validation process. While this function is not uniquely specified, it is not arbitrary: certain asymptotic trends, consistent with heuristically plausible behavior, must be observed. In four fundamentally different examples, we have illustrated how this approach might apply to a validation process for physics or engineering models. We believe that the multiplicative decomposition of trust gains or losses (given in 3), using a suitable functional prescription (such as 7), provides a reasoned and principled description of the key elements – and fundamental limitations – of validation. It should be equally applicable to biological and social sciences, especially since it is built upon the decision-making processes of the latter. We believe that our procedure transforms the paralyzing criticisms in Popper's style that "we cannot validate, we can only invalidate" [19] into a practical constructive algorithm. This strategy addresses specifically both problems of distinguishing between competing models and transforming the vicious circle of lack of suitable data into a virtuous spiral path: each cycle is marked by a quantified increment of the evolving trust we put in a model based on the novelty and relevance of new data and the quality of fits.

We have also surveyed and commented extensively on the V&V literature. We hope this digest will help the reader as much as its collation helped us deepen our understanding of the challenge of model validation, including a new perspective on some of our own work. We close with these far-reaching thoughts by Patrick J. Roache [112]:

> *In an age of spreading pseudoscience and anti-rationalism, it behooves those of us who believe in the good of science and engineering to be above reproach whenever possible. Public confidence is further eroded with every error we make. Although many of society's problems can be solved with a simple change of values, major issues such as radioactive waste disposal and environmental modeling require technological solutions that necessarily involve computational physics. As Robert Laughlin [113] noted in this magazine,* "there is a serious danger of this power [of simulations] being misused, either by accident or through deliberate deception." *Our intellectual and moral traditions will be served well by conscientious attention to verification of codes, verification of calculations, and validation, including the attention given to building new codes or modifying existing codes with specific features that enable these activities.*

Acknowledgments

This work was supported by the LDRD 20030037DR project "Physics-Based Analysis of Dynamic Experimentation and Simulation" and the US DOE Atmospheric Radiation Measurement (ARM) program. We acknowledge stimulating interactions and discussions with the other members of the project, including Bob Benjamin, Mike Cannon, Karen Fisher, Andy Fraser, Sanjay Kumar, Vladilen Pisarenko, Kathy Prestridge, Bill Rider, Chris Tomkins, Kevin Vixie, Peter Vorobieff, and Cindy Zoldi. We were particularly impressed by Timothy G. Trucano as our reviewer for [1], who provided an extremely insightful and helpful report, with many thoughtful comments and constructive suggestions. As authors, we count ourselves very fortunate to have had such a strong audience to scrutinize and improve our contribution.

Appendix: A More Formal Look at the Role of Validation in the Modeling Enterprise

We deal with models that possess two aspects: a conceptual part based on the physical laws of nature (such as the Navier–Stokes conservation equations for fluid dynamics) and a computational part (like in CFD). Mathematically, a model along with observations are defined formally, as described in Sect. 1.1 below:

- The model $\mathcal{M}$ maps the set $\{A\}$ of parameters and of initial and boundary conditions to a forecast of state variables in a formal vector X_f.
- An observation projection $\mathcal{G}$ maps the true dynamics or physics in X_t to raw measurements y_o.

Such definitions may seem abstract and of little use but they are important foundations to build a comprehensive roadmap for physically-based model validation.

In the following section, we refine the above definitions and introduce a few more operators and quantities. In Sect. 1.2, we revisit the key steps in a validation loop with this notation in hand. Finally, we discuss some fundamental limitations on model validation in Sect. 1.3 using some of our own research in time-series analysis for illustration.

1.1 Definitions

Let us denote $X_{\mathrm{t}}(\boldsymbol{r},t)$ the true physical field. Observations $y_{\mathrm{o}}(\boldsymbol{r},t)$ are obtained via a possibly nonlinear operator $\mathcal{G}$ acting on $X_{\mathrm{t}}(\boldsymbol{r},t)$:

$$y_{\mathrm{o}}(\boldsymbol{r},t) = \mathcal{G}\{X_{\mathrm{t}}(\boldsymbol{r}',t')\}. \tag{1.1}$$

The observations at position $\boldsymbol{r}$ and time t may be a combination of past values obtained over some finite region, hence our use of $(\boldsymbol{r}',t')$ which are different from $(\boldsymbol{r},t)$. The operator $\mathcal{G}$ may thus be non-local and (causally) time-dependent. In addition, any measurement has noise and uncertainties. Therefore, $\mathcal{G}$ is a stochastic operator. The simplest specification beyond ignoring noise is to consider an additive noise.

A model $\mathcal{M}$ provides a forecast $X_{\mathrm{f}}(\boldsymbol{r},t)$ either in the actual future or in terms of what will lead (via another operator) to the value of the measurements beyond a certain fiducial point in time. This is expressed by

$$X_{\mathrm{f}}(\boldsymbol{r},t) = \mathcal{M}\left(\{A\}\right). \tag{1.2}$$

$\mathcal{M}$ is the model operator, which contains for instance the equation of states, the formulation in terms of ODEs, PDEs, discrete maps and so on, which are supposed to embody the known physics of the underlying processes. $\{A\}$ contains the parameters of the model as well as the boundary and initial conditions. The model operator $\mathcal{M}$ has a non-random part. It can also contain an additive or multiplicative noise component to represent the forecast errors as well as possible intrinsic stochastic components of the dynamics. The forecast errors may stem from computational errors, numerical instabilities and uncertainties, the existence of multiple branches in the solution and so on. The simplest specification is again to consider an additive noise.

The output $\mathcal{M}\left(\{A\}\right)$ of the model is translated into physical quantities that can be compared with the observation via another operator $\mathcal{H}$, which models mathematically and in code the observation process. In general, one would like to compare $y_{\mathrm{o}}(\boldsymbol{r},t)$ given by (1.1) with $\mathcal{H}\left[X_{\mathrm{f}}(\boldsymbol{r},t)\right]$, that is, $\mathcal{G}\{X_{\mathrm{t}}(\boldsymbol{r}',t')\}$ with $\mathcal{H}\left[\mathcal{M}\left(\{A\}\right)\right]$. The intended use of the model is key to "objective model validation," because it turns "subjectiveness" of the model validation into an "object" using hypothesis testing and decision theory. To implement this idea, it is natural to introduce a cost function (see below) for the intended use of the model:

$$\mathcal{C}\left(\mathcal{G}\{X_{\mathrm{t}}(\boldsymbol{r}',t')\};\mathcal{H}\left[\mathcal{M}\left(\{A\}\right)\right]\right),$$

which is a measure of how well the model accounts for the observations. In this expression, the cost function is evaluated in the "physical space" of observations/measurements. An alternative is to evaluate the cost function in the "model space," i.e.,

$$\mathcal{C}\left(\mathcal{G}^{-1}\{\mathcal{H}\{X_{\mathrm{t}}(\boldsymbol{r}',t')\}\};\mathcal{M}\left(\{A\}\right)\right),$$

where $\mathcal{G}^{-1}$ is the formal inverse operator to $\mathcal{G}$ which maps observations y_{o} onto the model space X_{f}. In data assimilation, explicit form of $\mathcal{G}^{-1}$ does not exist in general due to rank deficiency. However, such alternative representation within the linear theory corresponds to the duality between Kalman filtering and the 3D-Var [114].

We propose to define the validation problem as a decision problem in which one uses the loss function to infer/decide how much confidence one feels in the reliability of the model to function in the range in which it is supposed to apply. The interesting and challenging situation occurs when this range extends beyond the region of parameter space in which all reasonably stringent controls have been performed. Validation requires the build-up of trust in the model or code so that it is believed to be resilient and to work in complex real situations combining the simple regimes that have been tested. The cost function is just an alternative way of constructing the statistical test that provides the probability level p defined in the main text.

1.2 Four Recurring Types of Problems in Physically-Based Model Validation

Our overarching goal is to advocate approaches to validation that are grounded in physics. The term "physics-based" embodies two strategies:

(a) Use physical reasoning to improve modeling, target experiments and loss functions, and detect missed "dimensions"; and
(b) Use concepts from statistical physics to formulate (in the spirit of Brown and Sethna [115]) a validation process of complex models with complex data in the form of an N-body problem.

Following this roadmap, we find ourselves asking the same four questions again and again:

1. *How to model?* (the question of model construction)
2. *What to measure?* (the question of estimating c_{novel} in the main text)

3. *How to measure it?* (the question of choosing and estimating the cost function or "metric")
4. *How to interpret the results?* (the question of estimating p in the main text)

We view these four defining questions as the crucial steps within the validation loop described in Sects. 2–4 of the main text.

Problem 1: Targeting Model Development (How to Model?)

Our discussion so far may give the impression that the modeling step is "homogeneous." It may actually be advantageous to develop a hierarchical modeling framework. In this respect, Oden et al. [116] proposed to use hierarchical modeling as a mathematical structure that can be useful in directing validation studies. In this construction, a class of models of events of interest is defined in which one identifies a "fine" model that possesses a level of sophistication high enough to adequately capture the event of interest with good accuracy. This model may be intractable, even computationally. Hierarchical modeling consists in identifying a family of coarse models that are solvable. Using the fine model output as a datum, the error in the solution of ever coarser models can be estimated and controlled, with the goal of obtaining a model best suited for the simulation goal at hand. The essential components of this program are the following [116]:

1. Experimental data are collected to fully characterize the fine model.
2. Quantities $\mathcal{G}(X)$ of interest are specified as the essential physical entity to be predicted in the simulation (for instance in the form of the probability of the predicted values of the quantity).
3. The coarsest model is used to extract a preliminary estimate of $\mathcal{G}(X)$ and modeling and approximation errors are computed.
4. If the estimated error exceeds the prescribed tolerance, the model is enhanced and the calculation is repeated until a model yielding results within the preset bounds is obtained.
5. The truncation error of the perturbation expansion is estimated: if the total error exceeds a preset tolerance, the data set and the fine model definition must be updated; if not, the predicted $\mathcal{G}(X)$ and the probability that it will take on values in a given interval are produced as output.

A concrete implementation of this program has been performed by Israeli and Goldenfeld [27]. Using elementary cellular automata as an example, Israeli and Goldenfeld show how to coarse-grain cellular automata in all categories of Wolfram's exhaustive classification [117]. The main discovery is that computationally irreducible physical processes can be predictable and even computationally reducible at a coarse-grained level of description. The resulting coarse-grained cellular automaton constructed with the coarse-graining procedure emulate the large-scale behavior of the original systems without accounting for small-scale details. These results remind us that it is advantageous to develop a view of complex physical processes at different scales, as the predictability may depend on the scale of observation.

A related approach has been discussed recently by Brown and Sethna [115], who consider models defined in terms of a set of nonlinear ODEs applied to systems that have large numbers of poorly known parameters, simplified dynamics, and uncertain connectivity. They call models possessing these three features, "sloppy models." Sloppy models characterize many other high-dimensional multi-parameter nonlinear models. Brown and Sethna propose to use the maximum likelihood method to frame the problem of parameter estimation and model validation in the form of statistical ensemble method. In our language, the problem boils down to a study of the cost function $\mathcal{C}$ and its stiff and soft directions determined from the eigenvalue problem of the Hessian of $\mathcal{C}$ (with respect to the parameters of the model). In practice, Brown and Sethna propose to estimate the Hessian of $\mathcal{C}$ in terms of the so-called "Levenberg–Marquardt" Hessian (thus called because of its use of that popular minimization algorithm); that quantity is defined simply as a sum of pairwise products of first-order derivatives of the residuals with respect to the model parameters. Stiff modes correspond to large eigenvalues. Similar to a decomposition in principal components, retaining the stiff modes allows one to get a more robust signature of the coarse-grained properties of the dynamics. This constitutes a concrete implementation of our Problem 4 below on "targeting model errors." This procedure also addresses the problem of defining the operator $\mathcal{H}$ that selects the output of the model for comparison to the experimental data.

There is an interesting avenue for research here: rather than performing the principal component decomposition in one step, it may be advantageous to perform a series of sub-system analysis, or cluster analysis, retaining the stiff modes of each sub-system and then aggregating them at the next level of the hierarchy.

Problem 2: Targeting the Observations (What to Measure?)

The objective here is to find $\mathcal{G}$ (and the associated $\mathcal{H}$) that reveals the most about model critical behavior. The problem has been addressed specifically in these terms by Palmer et al. [72] to target adaptive observations to "sensitive" parts of the atmosphere. Targeting observations could be directed by the desire to get access to the most relevant information that is also the most reliable (e.g., contaminated by the smallest errors). It may be worth mentioning that targeting the observations depends not only on $\mathcal{G}$, but also $\mathcal{M}$, $\{A\}$, as well as $\mathcal{C}$ (along with its own parameters discussed below). The targeting of the observations is the problem of maximizing the coefficient c_{novel} introduced in the main text so that the new experiment/observation explores novel dimensions of the parameter and variable spaces of both the process and the model that can best reveal potential flaws that could compromise the important applications. In general, one targets observations by developing experiments that are thought to provide, in some sense, the most relevant tests of the physics.

Oberkampf and Trucano (2002) [33] suggest that traditional experiments could generally be grouped into three categories:

1. Experiments that are conducted primarily for the purpose of improving the fundamental understanding of some physical process;
2. Experiments conducted primarily for constructing or improving mathematical models of fairly well-understood flows;
3. Experiments that determine or improve the reliability, performance, or safety of components, subsystems, or complete systems.

These authors argue that validation experiments constitute a fourth type of experiment: *"A validation experiment is conducted for the primary purpose of determining the validity, or predictive accuracy, of a computational modeling and simulation capability. In other words, a validation experiment is designed, executed, and analyzed for the purpose of quantitatively determining the ability of a mathematical model and its embodiment in a computer code to simulate a well-characterized physical process."* This leads them to propose the following guidelines:

- *Guideline #1*: A validation experiment should be jointly designed by experimentalists, model developers, code developers, and code users working closely together throughout the program, from inception to documentation, with complete candor about the strengths and weaknesses of each approach.
- *Guideline #2*: A validation experiment should be designed to capture the essential physics of interest, including all relevant physical modeling data and initial and boundary conditions required by the code.
- *Guideline #3*: A validation experiment should strive to emphasize the inherent synergism between computational and experimental approaches.
- *Guideline #4*: Although the experimental design should be developed cooperatively, independence must be maintained in obtaining both the computational and experimental results.
- *Guideline #5*: A hierarchy of experimental measurements of increasing computational difficulty and specificity should be made, for example, from globally integrated quantities to local measurements.
- *Guideline #6*: The experimental design should be constructed to analyze and estimate the components of random (precision) and bias (systematic) experimental errors.

Problem 3: Targeting the Cost Function (How to Estimate the Penalty on Imperfect Models and Measurements Using Their Discrepancies?)

For given measurements or experiments, that is, for given $\mathcal{G}$, the problem is to find the optimal cost function $\mathcal{C}$ for the intended use of the model. The notion of optimality needs to be defined. It could capture a compromise between the following requirements:

- Fit best the important features of the data (what is "important" may be decided on the basis of previous studies and understanding or other processes, or programmatic concerns), and
- Minimize the extraction of spurious information from noise, which requires one to have a precise idea of the statistical properties of the noise (if such knowledge is not available, the cost function should take this into account).

The choice of the cost function involves the choice of how to look at the data. For instance, one may want to expand the measurements at multiple scales using wavelet decompositions and compare the prediction and observations scale by scale, or in terms of multifractal spectra of the physical fields estimated from these wavelet decomposition or from other methods. The general idea here is that, given complex observation fields, it is appropriate to "project" the data onto a variety of "metrics" designed to detect and characterize phenomena of particular interest. For instance, wavelet-based scaling properties can be used in the comparison between observations and model predictions; the question is then: How well is the model/code able to reproduce the salient multi-scale properties derived from the observations? The physics of turbulent fields and of complex systems have offered many such new tools to unfold complex fields according to different statistics. Each of these statistics provides a basis for a metric to compare observations with model predictions. Such statistics thus lead to a cost function focusing on a particular feature of the process. These metrics are derived from the understanding that turbulent fields can be analyzed using them, revealing strong constraints in their organization (spatial structure and temporal evolution). These metrics can therefore be described as "physics-based."

Furthermore, the choice of the cost function should take into account that the diagnostics of the experiments may lead to spurious results [11]. For example, in laser-driven shock experiments, because the laser-induced fluorescence method illuminates the mixing zone with a planar sheet of light, this diagnostic can lead to aliasing of long-wavelength structures into short-wavelength features in the images, thus affecting the interpretation of observed small-scale structures in the mixing zone. Also, because of the dynamic limits on diagnostic resolution, the formation of small-scale structure cannot be completely determined.

As emphasized by Noam Chomsky in his own field of work [118], the danger with the Popperian strategy [119] is that one might prematurely reject a theory based on "falsification" using data that are themselves poorly understood. For instance, lack of quality control for the experiments can result in premature rejection of the model. On these issues, Stein [120] discusses means for controlling and for understanding sample selection and variability, which can compromise conclusions drawn from validation tests.

The problem of the choice of the cost function $\mathcal{C}$ seems, however, to be of less importance than Problem 2 above and Problem 4 below. In fact, almost all classical results on the limit properties of efficiency of statistical inference are valid (and proved) for a whole general family of cost functions $\mathcal{C}(\cdot;\cdot)$ satisfying the following conditions (see, e.g., Ibragimov and Hasminskii [121]):

(a) $\mathcal{C}(x,y) = c(|x-y|)$;
(b) $c(z)$ is a positive monotonically increasing function (including, e.g., power-law functions $|z|^q$, with $q > 0$);
(c) $c(z)$ should not increase too fast (its mean with respect to the Gaussian distribution must remain finite).

Thus, statistical limit theorems are proved for the whole class of different power-law cost functions (including the classic choice $q = 2$).

As an example, it may be appropriate to consider the cost function in the following form. Let us assume we are interested in some functional

$$Z(R,T|\mathcal{G}\{X_{\mathrm{t}}(\boldsymbol{r},t)\}, \boldsymbol{r} \in D(R), t \le T)$$

depending on the past true physical field $X_{\mathrm{t}}(\boldsymbol{r},t)$ in some region $D(R)$. In this case, the cost function can be chosen as

$$\mathcal{C}\left(Z\left[R,T|\mathcal{G}\{X_{\mathrm{t}}(\boldsymbol{r},t)\}, \boldsymbol{r}{\in}D(R), t \le T\right]; Z\left[R,T|\mathcal{H}\{X_{\mathrm{f}}(\boldsymbol{r},t)\}, \boldsymbol{r}{\in}D(R), t \le T\right]\right) \tag{1.3}$$

where $\mathcal{C}(\cdot;\cdot)$ is some function satisfying above conditions (a)–(c). The formulation (1.3) for $\mathcal{C}(\cdot;\cdot)$ should not only be a function of $\mathcal{G}$ and $\mathcal{M}$, but also of those parameters that correspond to our best guess for the uncertainties, errors and noise. Indeed, in most cases, we can never know real uncertainties, errors and noise in $\mathcal{G}$ and $\mathcal{M}$ (or even $\mathcal{H}$). Hence, we must parameterize them based on our best guess. In data assimilation (described in the main text in relation to model calibration and validation), the accuracy of such parameterization is known to influence the results significantly.

Generalizations to (1.3) allowing for different fields in the two sets of variables in $\mathcal{C}$ are needed for some problems, such as in validation of meteorological models. For instance, consider a model state vector X (dimension is on the order of 10^6) which is computed on a fixed spatial grid. In general, the locations of the observations are not on the computational grid (for example, consider measurements with weather balloons released from the surface). Thus, the observation Y is a function of X, but is not an attempt to estimate X itself. Hence, if the cost function is quadratic, it has the form $(Y - H(X))^{\mathrm{T}} O^{-1} (Y - H(X))$ where H acts on the interpolation function to pick up the model variable at the grid points close to the observed location, and O is related to the error covariance. Let us imagine a validation case using satellite infrared images for Y and atmospheric radiative state for X. Observations are quasi-uniform in space at a given time; at each time, available observations and their quality (represented by O) may change, however. In this case, the cost function must take into account the mapping between X and Y so that we have $\mathcal{C}(X,Y) = C(|H(X)-Y|)$ rather than $\mathcal{C}(X,Y) = C(|X-Y|)$; therefore $(Y - H(X))^{\mathrm{T}} O^{-1} (Y - H(X))$ when $\mathcal{C}$ is quadratic. In addition, for heterogeneous observations (satellite images, weather balloon measurements, airplane sampling, and so on), cost functions should take into account all these data into account such as

$$\mathcal{C}(x,y) = \mathcal{C}_{\text{satellite}}(x,y) + \mathcal{C}_{\text{balloon}}(x,y) + \mathcal{C}_{\text{airplane}}(x,y) + \cdots$$

and each $\mathcal{C}$ may have a complex idiosyncratic observation function H. See Courtier et al. [122] for a discussion on cost functions for atmospheric models and observation systems.

Problem 4: Targeting Model Errors (How to Interpret the Results?)

The problem here is to find the "dimensions" of the model that are missing, misrepresented or erroneous. The question of how to interpret the results thus leads to the discussion of the missing or misrepresented elements in the model. What tests can be used to betray the existence of hidden degrees of freedom and/or dimensions?

This is the hardest problem of all. It can sometimes find an elegant solution when a given model is embedded in a more general model. Then, the limitation of the "small" model becomes clear from the vantage of the more general model. Well-known examples include the following.

- Newtonian mechanics as part of special relativity, when $v \ll c$ where v (resp. c) is the velocity of the body (resp. of light).
- Classical mechanics as part of quantum mechanics when $h/mc \ll L$ (where h is Planck's constant, m and L are the mass and size of the body and h/mc is the associated Compton wavelength).
- Eulerian hydrodynamics as part of Navier–Stokes hydrodynamics with its rich phenomenology of turbulent motion (when the Reynolds number goes to infinity, equivalently, viscosity goes to zero).
- Classical thermodynamics as part of statistical physics of $N \gg 1$ particles or elements, where phase transitions and thermodynamic phases emerge in the limit $N \to \infty$.

The challenge of targeting model errors is to develop diagnostics of missing dimensions even in absence of a more encompassing model. This could be done by adding random new dimensions to the model and studying its robustness.

In what sense can one detect that a model is missing some essential ingredient, some crucial mechanisms, or that the number of variables or dimensions is inadequate? To use a metaphor, this question is similar to asking ants living and walking on a plane to gain awareness that there is a third dimension.[15]

1.3 Fundamental Limits on Model Validation

Before, while and after engaging in model validation, it is wise to reflect frequently and systematically on what is not known. Two examples using the formalization introduced in Sect. 1.1 are:

Ignorance on the Model $\mathcal{M}(\{A\})$

As quoted in the main text, Roache [2] states, in a nutshell, that validation is about solving the right equations for the problem of immediate concern. How do we know the right equations?

Consider, for instance, point vortex models, and let us perform "twin experiments," i.e., (1) first generate the "simulated observations" by a "true" point vortex system that are unknown to the make-believe observer and modeler; (2) use the procedure of Sect. 1.1 and construct a "validated" point vortex system. The problem is that, even before we start model validation, we are already using one of the most critical pieces of information, which is that the system is based on point vortices. Similar criticism for the use of "simulated observations" has been raised in data assimilation studies using OSSEs (Observing-System Simulation Experiments). This criticism is crucial for model validation.

For this unavoidable issue of model errors, we suggest that one needs a hierarchy of investigations:

1. Look at the statistical or global properties of the time series and/or fields generated by the models as well as from the data, such as distributions, correlation functions, n-point statistics, fractal and multifractal properties of the attractors and emergent structures, in order to characterize how much of the data our model fits. Part of this approach is the use of maximum likelihood theory to determine the most probable value of the parameters of the model, conditioned on the realization of the time series.
2. We can bring to bear on the problem the modern methods of computational intelligence (or machine learning), including pattern classification

[15] This question (raised already by the German philosopher Kant) actually has an answer that has been studied and solved by Ehrenfest in 1917 [123] (see also Whitrow's 1956 article [124]). This answer is based on the analysis of several fundamental physical laws in $\mathbb{R}^n$ spaces and comparing their predictions as a function of n. The value $n = 3$ turns out to be very special! Thus, ants studying gravitation or electro-magnetic fields will see that there is more to space than their plane.

and recognition methods ranging from the already classical ones (e.g., neural networks, K-means) to the most recent advances (e.g., support vector machines, "random forests").

3. Lastly, a qualification of the model is obtained by testing and quantifying how well it predicts the "future" beyond the interval used for calibration/initialization.

Levels of Ignorance on the Observation $\mathcal{G}$

- *First level*: The characteristics of the noise are known, such as its distribution, covariance, and maybe higher-order statistics.
- *Second level*: It may happen that the statistical properties of the noise are poorly known or constrained.
- *Third level*: A worse situation is when some noise components are not known to exist and are thus simply not considered in the treatment. For instance, imagine that one forgets in climate modeling about the impact of biological variability in time and space in the distribution of CO_2 sequestration sites.
- *Fourth level*: Finally, there is the representation error in $\mathcal{G}$ itself, i.e., how $\mathcal{G}$ is modeled mathematically in $\mathcal{H}$.

Consequences of the Sensitivity to Initial Conditions and Nonlinearity in the Model

Even an accurate forecast is limited by the inherent predictability of the system. In the same way, validation may be hindered by limited access to testing. The predictability of a system refers to the fundamental limits of prediction for a system. For instance, if a system is pure noise, there is no possibility of forecasting it better than chance. Similarly, there may be limits in the possibilities of testing the performance of a model because of limits in measurements, limits in access to key parameters for instance. With such limitations, it may be impossible to fully validate a model.

A well-known source that limits predictability is the property of sensitivity to initial conditions, which is one of the ingredients leading to chaotic behavior. Validation has to be made immune to this sensitivity upon initial conditions, by using a variety of methods, including the properties of attractors, their invariant measures, the properties of Lyapunov exponents, and so on. Pisarenko and Sornette [125] have shown that the sensitivity to initial conditions leads to a limit of testability in simple toy models of chaotic dynamical systems, such as the logistic map. They addressed the possibility of applying standard statistical methods (the least square method, the maximum likelihood estimation method, the method of statistical moments for estimation of parameters) to deterministically chaotic low-dimensional dynamic system containing an additive dynamical noise. First, the nature of the system is found to require that any statistical method of estimation combines the estimation of the structural parameter with the estimation of the initial value. This is potentially an important lesson for such a class of systems. In addition, in such systems, one needs a trade-off between the need of using a large number of data points in the statistical estimation method to decrease the bias (i.e., to guarantee the consistency of the estimation) and the unstable nature of dynamical trajectories with exponentially fast loss of memory of the initial condition. In this simple example, the limit of testability is reflected in the absence of theorems on the consistency and efficiency of maximum likelihood estimation (MLE) methods [125]. We can use MLE with sometimes good practical results in controlled situations for which past experience has been accumulated but there is no guarantee that the MLE will not go astray in some cases.

This work has also shown that the Bayesian approach to parameter estimation of chaotic deterministic systems is incorrect and probably suboptimal. The Bayesian approach usually assumes non-informative priors for the structural parameters of the model, for the initial value and for the standard deviation of the noise. This approach turns out to be incorrect, because it amounts to assuming a stochastic model, thus referring to quite another problem, since the correct model is fundamentally deterministic (only with the addition of some noise).

This negative conclusion on the use of the Bayesian approach should be contrasted with the Bayesian approach of Hanson and Hemez [126] to model the plastic-flow characteristics of a high-strength steel by combining data from basic material tests. The use of a Bayesian approach to this later problem seems warranted because the priors reflect the intrinsic heterogeneity of the samples and the large dispersion of the experiments. In this particular problem concerning material properties, the use of Bayesian priors is warranted by the fact that the structural parameters of the model can be viewed as drawn from a population. It is very important to stress this point: Bayesian approaches to structural parameter determination are justified only in problems with random distributions of the parameters. For the previous problem of deterministic nonlinear dynamics, it turns out to be fundamentally incorrect. We therefore view proper partition of the problem at hand between deterministic and random components as an essential part of validation.

Extrapolating Beyond the Range of Available Data

In the previous discussion, the limit of testability is solely due to the phenomenon of sensitive dependence upon initial conditions, as the model is assumed to be known (the logistic map in the above example). In general, we do not have such luxury. Let us illustrate the much more difficult problem by two examples stressing the possibility for the existence of "indistinguishable states." Consider a map f_1 that generates a time series. Assuming that f_2 is unknown a priori, let us construct/constrain the map f_2 whose initial condition and parameters can be tuned in such a way that trajectories of f_2 can follow data of f_1 for a while, but eventually the two maps diverge. Suppose that the time series of f_1 is too short to explore the range expressing the divergence between the two maps. How can we (in-)validate f_2 as a incorrect model of f_1?

This problem arises in the characterization of the tail of distributions of stochastic variables. For instance, Malevergne et al. [127] have shown that, based on available data, the best tests and efforts can not distinguish between a power law tail and a stretched exponential distribution for financial returns. The two classes of models are indistinguishable, given the amount of data. This fundamental limitation has unfortunately severe consequences, because choosing one or the other models involves different predictions for the frequency of very large losses that lie beyond the range sampled by historical data (the $f_1 - f_2$ problem). The practical consequences are significant, in terms of the billions of dollars banks should put (or not) aside to cover large market swings that are outside the data set available from the known past history.

This example illustrates a crucial aspect of model validation, namely that it requires the issuance of predictions outside the domain of parameters and/or of variables that has been tested "in-sample" to establish the (calibrated or "tuned") model itself.

References

1. D. Sornette, A.B. Davis, K. Ide, K.R. Vixie, V. Pisarenko, and J.R. Kamm. Algorithm for model validation: Theory and applications. *Proc. Nat. Acad. Sci.*, 104:6562–6567, 2007.
2. P.J. Roache. *Verification and Validation in Computational Science and Engineering.* Hermosa Publishers, Albuquerque, 1998.
3. R. Costanza, R. d'Arge, R. deGroot, S. Farber, M. Grasso, B. Hannon, K. Limburg, S. Naeem, R.V. O'Neill, J. Paruelo, R.G. Raskin, P. Sutton, and M. vandenBelt. The value of the world's ecosystem services and natural capital. *Nature*, 387:253–260, 1997.
4. S. Pimm. The value of everything. *Nature*, 387:231–232, 1997.
5. I. Babuska and J.T. Oden. Verification and validation in computational engineering and science: Basic concepts. *Comput. Methods Appl. Mech. Eng.*, 193:4057–4066, 2004.
6. D.E. Post and L. G. Votta. Computational science demands a new paradigm. *Phys. Today*, 58:35–41, 2005.
7. AIAA. Guide for the verification and validation of computational fluid dynamics simulations. Technical Report AIAA G-077-1998, American Institute of Aeronautics and Astronautics, 1998.
8. S. Schlesinger. Terminology for model credibility. *Simulation*, 32(3):103–104, 1979.
9. R.G. Sargent. Verification and validation of simulation models. In D.J. Medeiros, E.F. Watson, J.S. Carson, and M.S. Manivannan (eds.), *Proceedings of the 1998 Winter Simulation Conference*, pp. 121–130. 1998.

10. R.G. Sargent. Some approaches and paradigms for verifying and validating simulation models. In B.A. Peters, J.S. Smith, D.J. Medeiros, and M.W. Rohrer (eds.), *Proceedings of the 2001 Winter Simulation Conference*, pp. 106–114. 2001.
11. A.C. Calder, B. Fryxell, T. Plewa, R. Rosner, L.J. Dursi, V.G. Weirs, T. Dupont, H.F. Robey, J.O. Kane, B.A. Remington, R.P. Drake, G. Dimonte, M. Zingale, F.X. Timmes, K. Olson, P. Ricker, P. MacNeice, and H.M. Tufo. On validating an astrophysical simulation code. *Astrophys. J. Suppl. Ser.*, 143:201–229, 2002.
12. G. Sod. A survey of several finite difference methods for systems of nonlinear hyperbolic conservation laws. *J. Comput. Phys.*, 27:1–31, 1978.
13. W.J. Rider. An adaptive Riemann solver using a two-shock approximation. *Comput. Fluids*, 28:741–777, 1999.
14. L.I. Sedov. *Similarity and Dimensional Methods in Mechanics.* Academic Press, New York, NY, 1959.
15. P. Woodward and P. Colella. The numerical simulation of two-dimensional fluid flow with strong shocks. *J. Comput. Phys.*, 54:115–173, 1984.
16. P.A. Gnoffo, R.D. Braun, K.J. Weilmuenster, R.A. Mitcheltree, W.C. Engelund, and R.W. Powell. Prediction and validation of Mars Pathfinder hypersonic aerodynamic data base. In *Proceedings of 7th AIAA/ASME Joint Thermophysics and Heat Transfer Conference, June 15–18, 1998, Albuquerque, NM*, 1998.
17. P.J. Roache. *Fundamentals of Computational Fluid Dynamics.* Hermosa Publishers, Albuquerque, 1998.
18. P.J. Roache. Recent contributions to verification and validation methodology. In *Proceedings of the Fifth World Congress on Computational Mechanics, Vienna, Austria*, 2002.
19. N. Oreskes, K. Shrader-Frechette, and K. Belitz. Verification, validation and confirmation of numerical models in the Earth sciences. *Science*, 263:641–646, 1994.
20. J.D. Sterman. The meaning of models. *Science*, 264:329–330, 1994.
21. E.J. Rykiel, Jr. The meaning of models. *Science*, 264:330–331, 1994.
22. N. Oreskes, K. Belitz, and K. Shrader-Frechette. The meaning of models – Response. *Science*, 264:331, 1994.
23. N. Oreskes. Evaluation (not validation) of quantitative models. *Environmental Health Perspective Supplements*, 106(Suppl. 6):1453–1460, 1998.
24. L.F. Konikow and J.D. Bredehoeft. Groundwater models cannot be validated. *Adv. Water Res.*, 15:75–83, 1992.
25. G.J. Chaitin. *Algorithmic Information Theory.* Cambridge University Press, New York, NY, 1987.
26. M. Buchanan. Revealing order in the chaos. *New Scientist*, 2488, 2005.
27. N. Israeli and N. Goldenfeld. On computational irreducibility and the predictability of complex physical systems. *Phys. Rev. Lett.*, 92:74105–74108, 2004.
28. A.A. Borovkov. *Mathematical Statistics.* Taylor & Francis, Amsterdam, The Netherlands, 1998.
29. H.W. Coleman and F. Stern. Uncertainties and CFD code validation. *J. Fluids Eng.*, 119:795–803, 1997.

30. R.G. Hills and T.G. Trucano. Statistical validation of engineering and scientific models: Background. Technical Report SAND99-1256, Sandia National Laboratory, 1999.
31. R.G. Hills and T.G. Trucano. Statistical validation of engineering and scientific models with application to CTH. Technical Report SAND2001-0312, Sandia National Laboratory, 2000.
32. R.G. Easterling. Measuring the predictive capability of computational models: Principles and methods, issues and illustrations. Technical Report SAND2001-0243, Sandia National Laboratory, 2001.
33. W.L. Oberkampf and T.G. Trucano. Verification and validation in computational fluid dynamics. *Progr. Aerospace Sci.*, 38:209–272, 2002.
34. C. Gourieroux and A. Monfort. Testing non-nested hypotheses. In R.F. Engle and D. McFadden (eds.), *Handbook of Econometrics*, Vol. IV, Chap. 44, pp. 2583–2637. Elsevier Science B.V., Amsterdam, The Netherlands, 1994.
35. T.G. Trucano, M. Pilch, and W.L. Oberkampf. On the role of code comparisons in verification and validation. Technical Report SAND2003-2752, Sandia National Laboratory, 2003.
36. R.F. Cahalan, L. Oreopoulos, A. Marshak, K.F. Evans, A.B. Davis, R. Pincus, K.H. Yetzer, B. Mayer, R. Davies, T.P. Ackerman, H.W. Barker, E.E. Clothiaux, R.G. Ellingson, M.J. Garay, E. Kassianov, S. Kinne, A. Macke, W. O'Hirok, P.T. Partain, S.M. Prigarin, A.N. Rublev, G.L. Stephens, F. Szczap, E.E. Takara, T. Varnai, G.Y. Wen, and T.B. Zhuravleva. The I3RC: Bringing together the most advanced radiative transfer tools for cloudy atmospheres. *Bull. Am. Meteor. Soc.*, 86(9):1275–1293, 2005.
37. B. Pinty, N. Gobron, J.L. Widlowski, S.A.W. Gerstl, M.M. Verstraete, M. Antunes, C. Bacour, F. Gascon, J.P. Gastellu, N. Goel, S. Jacquemoud, P. North, W.H. Qin, and R. Thompson. RAdiation Transfer Model Intercomparison (RAMI) exercise. *J. Geophys. Res.*, 106(D11):11937–11956, 2001.
38. B. Pinty, J.L. Widlowski, M. Taberner, N. Gobron, M.M. Verstraete, M. Disney, F. Gascon, J.P. Gastellu, L. Jiang, A. Kuusk, P. Lewis, X. Li, W. Ni-Meister, T. Nilson, P. North, W. Qin, L. Su, S. Tang, R. Thompson, W. Verhoef, H. Wang, J. Wang, G. Yan, and H. Zang. RAdiation Transfer Model Intercomparison (RAMI) exercise: Results from the second phase. *J. Geophys. Res.*, 109(D6):D06210, doi:10.1029/2003JD004252, 2004.
39. J.L. Widlowski, M. Taberner, B. Pinty, V. Bruniquel-Pinel, M. Disney, R. Fernandes, J.P. Gastellu-Etchegorry, N. Gobron, A. Kuusk, T. Lavergne, S. Leblanc, P.E. Lewis, E. Martin, M. Mottus, P.R.J. North, W. Qin, M. Robustelli, N. Rochdi, R. Ruiloba, C. Soler, R. Thompson, W. Verhoef, M.M. Verstraete, and D. Xie. Third RAdiation Transfer Model Intercomparison (RAMI) exercise: Documenting progress in canopy reflectance models. *J. Geophys. Res.*, 112(D9):D09111, doi:10.1029/2006JD007821, 2007.
40. T.G. Trucano, L.P. Swiler, T. Igusa, W.L. Oberkampf, and M. Pilch. Calibration, validation, and sensitivity analysis: What's what. *Reliab. Eng. Syst. Saf.*, 91:1331–1357, 2006.
41. W.L. Oberkampf. What are validation experiments? *Exp. Tech.*, 25:35–40, 2002.
42. T.G. Trucano, M. Pilch, and W.L. Oberkampf. General concepts for experimental validation of ASCI code applications. Technical Report SAND2002-0341, Sandia National Laboratory, 2002.

43. A. Saltelli, K. Chan, and E.M. Scott (eds.). *Sensitivity Analysis.* Wiley, Chichester, 2000.
44. A. Saltelli, S. Tarantola, F. Campolongo, and M. Ratto. *Sensitivity Analysis in Practice.* Wiley, Chichester, UK, 2004.
45. E. Kalnay. *Atmospheric Modeling, Data Assimilation, and Predictability.* Cambridge University Press, New York, NY, 2003.
46. D.F. Parrish and J.F. Derber. The National Meteorological Center's spectral statistical-interpolation analysis system. *Mon. Weather Rev.*, 120:1747–1763, 1992.
47. F. Rabier, H. Jarvinen, E. Klinker, J.-F. Mahfouf, and A. Simmons. The ECMWF implementation of four-dimensional variational assimilation (4D-Var): I. Experimental results with simplified physics. *Q. J. Roy. Meteorol. Soc.*, 126:1143–1170, 2000.
48. P.L. Houtekamer and H.L. Mitchell. Ensemble Kalman filtering. *Q. J. Roy. Meteorol. Soc.*, 131:3269–3289, 2005.
49. I. Szunyogh, E.J. Kostelich, G. Gyarmati, E. Kalnay, B.R. Hunt, E. Ott, E. Satterfield, and J.A. Yorke. A local ensemble transform Kalman filter data assimilation system for the ncep global model. *Tellus A: Dynamic Meteorology and Oceanography*, 59:doi: 10.1111/j.1600–0870.2007.00274, 2007. In press.
50. J.S. Whitaker, T.M. Hamill, X. Wei, Y. Song, and Z. Toth. Ensemble data assimilation with the NCEP global forecast system. *Mon. Weather Rev.*, 136(2):463–482, 2007.
51. Z. Toth and E. Kalnay. Ensemble forecasting at NMC: The generation of perturbations. *Bull. Am. Meteorol. Soc.*, 74:2317–2330, 1993.
52. P.L. Houtekamer and J. Derome. Methods for ensemble prediction. *Mon. Weather Rev.*, 123:2181–2196, 1995.
53. F. Molteni, R. Buizza, T.N. Palmer, and T. Petroliagis. The ECMWF ensemble prediction system: Methodology and validation. *Q. J. Roy. Meteorol. Soc.*, 122:73–119, 1996.
54. D.W. Berning and A.R. Hefner, Jr. IGBT model validation. *IEEE Ind. Appl. Mag.*, 4(6):23–34, 1998.
55. S. Kaplan and B.J. Garrick. On the quantitative definition of risk. *Risk Anal.*, 1:11–27, 1981.
56. R.P. Rechard. Historical background on performance assessment for the Waste Isolation Pilot Plant. *Reliab. Eng. Syst. Saf.*, 69:5–46, 2000.
57. R.L. Keeney and D. von Winterfeldt. Eliciting probabilities from experts in complex technical problems. *IEEE Trans Eng. Manage.*, 38:191–201, 1991.
58. J.C. Helton. Treatment of uncertainty in performance assessments for complex systems. *Risk Anal.*, 14:483–511, 1994.
59. J.C. Helton. Uncertainty and sensitivity analysis in performance assessment for the Waste Isolation Pilot Plant. *Comp. Phys. Commun.*, 117:156–180, 1999.
60. J.C. Helton, D.R. Anderson, G. Basabilvazo, H.-N. Jow, and M. G. Marietta. Conceptual structure of the 1996 performance assessment for the Waste Isolation Pilot Plant. *Reliab. Eng. Syst. Saf.*, 69:151–165, 2000.
61. F. Dyson. *Infinite in All Directions.* Penguin, London, UK, 1988.
62. L. Pal and M. Makai. Remarks on statistical aspects of safety analysis of complex systems. 2003.
63. J. von Neumann and O. Morgenstein. *Theory of Games and Economic Behavior.* Princeton University Press, Princeton, NJ, 1944.

64. T.E. Harris. *The Theory of Branching Processes.* Dover, New York, NY, 1988.
65. A. Wald. *Statistical Decision Functions.* Wiley, New York, NY, 1950.
66. S. Kotz and N. Johnson. *Breakthroughs in Statistics (Foundations and Theory)*, Vol. 1. Springer, Berlin Heidelberg New York, Germany, 1993.
67. H.A. Simon. A behavioral model of rational choice. *Q. J. Econ.*, 69(1):99–118, 1955.
68. H.A. Simon. Rational decision making in business organizations. *Am. Econ. Rev.*, 69(4):493–513, 1979.
69. R. Zhang and S. Mahadevan. Bayesian methodology for reliability model acceptance. *Reliab. Eng. Syst. Saf.*, 80:95–103, 2003.
70. S. Mahadevan and R. Rebba. Validation of reliability computational models using Bayesian networks. *Reliab. Eng. Syst. Saf.*, 87:223–232, 2005.
71. Board on Earth Sciences and Resources, Panel on Seismic Hazard Evaluation, Committee on Seismology, Commission on Geosciences, Environment, and Resources, National Research Council. *Review of Recommendations for Probabilistic Seismic Hazard Analysis: Guidance on Uncertainty and Use of Experts.* National Academy Press, Washington, DC, 1997.
72. T.N. Palmer, R. Gelaro, J. Barkmeijer, and R. Buizza. Singular vectors, metrics and adaptive observations. *J. Atmos. Sci.*, 55:633–653, 1998.
73. J.-F. Muzy, E. Bacry, and A. Arnéodo. The multifractal formalism revisited with wavelets. *Int. J. Bifurcat. Chaos*, 4:245–302, 1994.
74. R.E. Bellman and L.A. Zadeh. Decision-making in a fuzzy environment. *Manage. Sci.*, 17:141–164, 1970.
75. L.A. Zadeh. Knowledge representation in fuzzy logic. In R.R. Yager and L.A. Zadeh (eds.), *An Introduction to Fuzzy Logic Applications in Intelligent Systems*, pp. 2–25. Kluwer Academic, Norwell, 1992.
76. W.L. Oberkampf and M.F. Barone. Measures of agreement between computation and experiment: Validation metrics. Technical Report SAND2005-4302, Sandia National Laboratory, 2005.
77. M. Massimi. *Pauli's Exclusion Principle: The Origin and Validation of a Scientific Principle.* University of Cambridge Press, Cambridge, UK, 2005.
78. A. Einstein, B. Podolsky, and N. Rosen. Can quantum-mechanical description of physical reality be considered complete? *Phys. Rev.*, 47:777–780, 1935.
79. J.S. Bell. On the Einstein–Podolsky–Rosen paradox. *Physics*, 1:195–290, 1964.
80. A. Aspect, C. Imbert, and G. Roger. Absolute measurement of an atomic cascade rate using a two-photon coincidence technique. Application to the $4p^2\ {}^1S_0 \rightarrow 4s4p\ {}^1P_1 \rightarrow 4s^2\ {}^1S_0$ cascade of calcium exited by a two photon absorption. *Opt. Commun.*, 34:46–52, 1980.
81. N.J. Rarity and J.P. Tapster. Experimental violation of Bell's inequality based on phase and momentum. *Phys. Rev. Lett.*, 64:2495–2498, 1990.
82. R. Webb, S. Washburn, C. Umbach, and R. Laibowitz. Observation of h/e Aharonov–Bohm oscillations in normal-metal rings. *Phys. Rev. Lett.*, 54:2696–2969, 1985.
83. B. Schwarzschild. Currents in normal-metal rings exhibit Aharonov–Bohm effect. *Phys. Today*, 39:17–20, 1986.
84. M.H. Anderson, J.R. Ensher, M.R. Matthews, C.E. Wieman, and E.A. Cornell. Observation of Bose–Einstein condensation in a dilute atomic vapor. *Science*, 296:198–201, 1995.
85. R. Gähler, A.G. Klein, and A. Zeilinger. Neutron optical test of nonlinear wave mechanics. *Phys. Rev. A*, 23:1611–1617, 1981.

86. S. Weinberg. Precision tests of quantum mechanics. *Phys. Rev. Lett.*, 62:485–488, 1989.
87. A.J. Leggett. Testing the limits of quantum mechanics: Motivation, state of play, prospects. *J. Phys. Condens. Matter*, 14:R415–R451, 2002.
88. Z. Olami, H.J.S. Feder, and K. Christensen. Self-organized criticality in a continuous, nonconservative cellular automaton modeling earthquakes. *Phys. Rev. Lett.*, 68:1244–1247, 1992.
89. B. Drossel. Complex scaling behavior of nonconserved self-organized critical systems. *Phys. Rev. Lett.*, 89:238701–238704, 2002.
90. A. Helmstetter, S. Hergarten, and D. Sornette. Properties of foreshocks and aftershocks of the non-conservative self-organized critical Olami-Feder-Christensen model. *Phys. Rev. E*, 70:046120, 2004.
91. D. Sornette. *Critical Phenomena in Natural Sciences*. Springer, Berlin Heidelberg New York, Germany, 2004.
92. S. Hergarten and H.J. Neugebauer. Foreshocks and aftershocks in the Olami-Feder-Christensen model. *Phys. Rev. Lett.*, 88:2385011–2385014, 2002.
93. H.W. Barker and A.B. Davis. Approximation methods in atmospheric 3D radiative transfer, Part 2: Unresolved variability and climate applications. In A. Marshak and A.B. Davis (eds.), *3D Radiative Transfer in Cloudy Atmospheres*, pp. 343–383. Springer, Berlin Heidelberg New York, Germany, 2005.
94. S. Lovejoy. Area-perimeter relation for rain and cloud areas. *Science*, 216: 185–187, 1982.
95. A.B. Davis. Effective propagation kernels in structured media with broad spatial correlations, illustration with large-scale transport of solar photons through cloudy atmospheres. In F. Graziani (ed.), *Computational Transport Theory – Granlibakken 2004*, pp. 84–140. Springer, Berlin Heidelberg New York, 2006.
96. A.B. Davis and A. Marshak. Lévy kinetics in slab geometry: Scaling of transmission probability. In M.M. Novak and T.G. Dewey (eds.), *Fractal Frontiers*, pp. 63–72. World Scientific, Singapore, 1997.
97. K. Pfeilsticker. First geometrical pathlengths probability density function derivation of the skylight from spectroscopically highly resolving oxygen A-band observations, 2. Derivation of the Lévy-index for the skylight transmitted by mid-latitude clouds. *J. Geophys. Res.*, 104:4101–4116, 1999.
98. Q.-L. Min, L.C. Harrison, and E.E. Clothiaux. Joint statistics of photon pathlength and cloud optical depth: Case studies. *J. Geophys. Res.*, 106:7375–7385, 2001.
99. A.B. Davis, D.M. Suszcynski, and A. Marshak. Shortwave transport in the cloudy atmosphere by anomalous/Lévy photon diffusion: New diagnostics using FORTÉ lightning data. In *Proceedings of 10th Atmospheric Radiation Measurement (ARM) Science Team Meeting, San Antonio, Texas, March 13–17, 2000*, 2000.
100. Q.-L. Min, L.C. Harrison, P. Kiedron, J. Berndt, and E. Joseph. A high-resolution oxygen A-band and water vapor band spectrometer. *J. Geophys. Res.*, 109:D02202–D02210, 2004.
101. A.B. Davis and A. Marshak. Space-time characteristics of light transmitted through dense clouds: A Green function analysis. *J. Atmos. Sci.*, 59:2714–2728, 2002.

102. T. Scholl, K. Pfeilsticker, A.B. Davis, H.K. Baltink, S. Crewell, U. Löhnert, C. Simmer, J. Meywerk, and M. Quante. Path length distributions for solar photons under cloudy skies: Comparison of measured first and second moments with predictions from classical and anomalous diffusion theories. *J. Geophys. Res.*, 111:D12211–D12226, 2006.
103. S.V. Buldyrev, M. Gitterman, S. Havlin, A.Ya. Kazakov, M.G.E. da Luz, E.P. Raposo, H.E. Stanley, and G.M. Viswanathan. Properties of Lévy flights on an interval with absorbing boundaries. *Physica A*, 302:148–161, 2001.
104. A.B. Davis, R.F. Cahalan, J.D. Spinehirne, M.J. McGill, and S.P. Love. Off-beam lidar: An emerging technique in cloud remote sensing based on radiative Green-function theory in the diffusion domain. *Phys. Chem. Earth (B)*, 24: 177–185 (Erratum 757–765), 1999.
105. W.J. Rider, J.A. Greenough, and J.R. Kamm. Combining high-order accuracy with non-oscillatory methods through monotonicity preservation. *Int J. Num. Methods Fluids*, 47:1253–1259, 2005.
106. W.J. Rider, J.A. Greenough, and J.R. Kamm. Accurate monotonicity- and extrema-preserving methods through adaptive nonlinear hybridizations. *J. Comput. Phys.*, 225:1827–1848, 2007.
107. R. Liska and B. Wendroff. Comparison of several difference schemes on 1D and 2D test problems for the Euler equations. *SIAM J. Sci. Comput.*, 25:995–1017, 2003.
108. R.D. Richtmyer. Taylor instability in shock acceleration of compressible fluids. *Commun. Pure Appl. Math.*, 13:297–319, 1960.
109. E.E. Meshkov. Instability of the interface of two gases accelerated by a shock wave. *Izv. Akad. Nauk SSSR, Mekh. Zhidk. Gaza*, 5:151–158, 1969.
110. R.F. Benjamin. An experimenter's perspective on validating codes and models with experiments having shock-accelerated fluid interfaces. *Comput. Sci. Eng.*, 6:40–49, 2004.
111. S. Kumar, G. Orlicz, C. Tomkins, C. Goodenough, K. Prestridge, P. Vorobieff, and R.F. Benjamin. Stretching of material lines in shock-accelerated gaseous flows. *Phys. Fluids*, 17:82107–82117, 2005.
112. P.J. Roache. Building PDE codes to be verifiable and validatable. *Comput. Sci. Eng.*, 6:30–38, 2004.
113. R.B. Laughlin. The physical basis of computability. *Comput. Sci. Eng.*, 4:27–30, 2002.
114. K. Ide, P. Courtier, M. Ghil, and A.C. Lorenc. Unified notation for data assimilation: Operational, sequential and variational. *J. Meteorol. Soc. Jpn.*, 75(1B):181–189, 1997.
115. K.S. Brown and J.P. Sethna. Statistical mechanical approaches to models with many poorly known parameters. *Phys. Rev. E*, 68:21904–21912, 2003.
116. J.T. Oden, J.C. Browne, I. Babuska, K.M. Liechti, and L.F. Demkowicz. A computational infrastructure for reliable computer simulations. In P.M.A. Sloot, D. Abramson, A.V. Bogdanov, J. Dongarra, A.Y. Zomaya, and Yu.E. Gorbachev (eds.), *Computational Science – ICCS 2003, International Conference on Computational Science, Melbourne, Australia and St. Petersburg, Russia, June 2–4, 2003. Proceedings, Part IV*, Vol. 2660 of *Lecture Notes in Computer Science*, pp. 385–390, Springer, Berlin Heidelberg New York, 2003.
117. S. Wolfram. *Cellular Automata and Complexity.* Westview Press, Boulder, 2002.

118. N. Chomsky. *Lectures on Government and Binding.* Foris, Dordrecht, The Netherlands, 1981.
119. K.R. Popper. *The Logic of Scientific Discovery.* Basic Books, New York, 1959.
120. R.M. Stein. *Benchmarking Default Prediction Models: Pitfalls and Remedies in Model Validation (Technical Report 020305).* Moody's KMV, New York, NY, 2002.
121. I.A. Ibragimov and R.Z. Hasminskii. *Statistical Estimation: Asymptotic Theory.* Springer, Berlin Heidelberg New York, 1981.
122. P. Courtier, E. Andersson, W. Heckley, J. Pailleux, D. Vasiljevic, M. Hamrud, A. Hollingsworth, F. Rabier, and M. Fisher. The ECMWF implementation of three-dimensional variational assimilation (3D-Var), Part 1: Formulation. *Q. J. Roy. Meteor. Soc.*, 124:1783–1807, 1998.
123. H.A. Ehrenfest. In what way does it become manifest in the fundamental laws of physics that space has three dimensions? In M.J. Klein (ed.), *Collected Scientific Papers.* Interscience, New York, 1959.
124. G.J. Whitrow. Why physical space has three-dimensions. *Br. J. Philos. Sci.*, 6:13–31, 1956.
125. V.F. Pisarenko and D. Sornette. On statistical methods of parameter estimation for deterministically chaotic time-series. *Phys. Rev. E*, 69:36122–36133, 2004.
126. K.M. Hanson and F.M. Hemez. Uncertainty quantification of simulation codes based on experimental data. In *Proceedings of 41th AIAA Aerospace Sciences Meeting and Exhibit*, pp. 1–10. American Institute of Aeronautics and Astronautics, Reston, 2003.
127. Y. Malevergne, V.F. Pisarenko, and D. Sornette. Empirical distributions of log-returns: Between the stretched exponential and the power law? *Quant. Finance*, 5(4):379–401, 2005.

Spectral Solvers to Non-Conservative Transport for Non-Linear Interactive Systems of Boltzmann Type

I.M. Gamba and S.H. Tharkabhushanam

Summary. In several technological areas there is increasingly a need for analysis of problems involving a vast range of physical scales. For example, compressible Navier Stokes equations are completely adequate for understanding gas fluid flows where the length scales are large compared to the local mean free path λ of fluid molecules. If the flow scales are not large compared to λ, either because λ itself is large if the flow is in low density state (rarefied) or because boundary conditions produce inhomogeneities resolved at small flow scales, a kinetic theory model from statistical formulations and their simulations becomes necessary.

1 Introduction

In several technological areas there is increasingly a need for analysis of problems involving a vast range of physical scales. For example, compressible Navier Stokes equations are completely adequate for understanding gas fluid flows where the length scales are large compared to the local mean free path λ of fluid molecules. If the flow scales are not large compared to λ, either because λ itself is large if the flow is in low density state (rarefied) or because boundary conditions produce inhomogeneities resolved at small flow scales, a kinetic theory model from statistical formulations and their simulations becomes necessary.

As a consequence, in recent years non-equilibrium time dependent kinetic systems have received significant attention in such diverse areas as electron transport in solids, biological transport, granular and energy dissipative flows. These non-conservative systems exhibit a common feature: their steady or self similar states are given by statistical *stationary non-equilibrium states* (SNS), meaning they are far deviated from Gaussian probability distributions. Traditionally, macroscopic fluid approximations from particle systems, such as Navier–Stokes, Euler and drift diffusion equations, have been derived as hydrodynamics limits under the assumption of Gaussian equilibrium states, and then are no longer adequate near non-equilibrium statistical regimes.

These kinetic or statistical levels of modeling, involving a vast range of scales, wherein the smallest scales require an atomistic view while the largest are well modeled by continuum equations, occur in many areas such as rarefied gas dynamics, rapid granular flows, astrophysical models of galactic interactions, reacting gas mixtures, hot electron transport in submicron semiconductors and nano-devices, etc. Such problems are typically computationally intensive. Possible approaches to their solution are either based on implementations of uniform (but costly) atomistic/particle approach, usually via a Monte Carlo statistical simulation that permits the inclusion of complex physics; or simulations of continuum equations, including the modifications with higher order terms (e.g., as the Burnett equations in gas dynamics) even though their validity for large gradient areas may remain questionable.

A current approach under development calls for a systematic approach of hybridization, where an atomistic model is used in one area and a continuum approach in another. The issues of the hybridization then relate on the localization and resolution of interfaces between regions and how the exchange of information between regions is performed. It should be noted that even 'pure' atomistic approaches may be hybrid to the extent of using fixed or specified boundary conditions that have continuum properties to inject/remove particles.

For example, in the context of charged transport, kinetic models predict transport behavior of charged particle systems in semi-conducting materials or ionic charges in biological channels, where the electrostatic potentials are regulated by boundary driven forces or background space inhomogeneities. Similarly, strongly active plasma regions due to electric and magnetic fields require the modeling of kinetic transport coupled to the full Maxwell system that models the Lorentz accelerating force.

While all these processes of dynamically adaptive multi scale interface transitions are part of the fundamental component for a successful hybridization, we advocate in these work and report on recent developments regarding deterministic schemes and corresponding initial value and boundary value solvers to models that involve deterministic kinetic statistical transport equations such as the Boltzmann equation for rarefied gas dynamics, or Vlasov Maxwell systems for plasmas, or Boltzmann Poisson systems for transport modeling in solids, among many problems.

Our goal is to discuss alternative methods to stochastic based solvers such as Monte Carlo used on deterministic based models, even though they are of non-local nature.

We describe in this manuscript recent developments for particular problems as illustrative examples of the issues involved, due mainly to the nature of interactions. It consists on implantations of a Spectral-Lagrangian based method recently developed to simulate the transient behavior of the space homogeneous, non necessarily conservative, Boltzmann equation for binary interactions.

Another example related to deterministic simulations is of one and two space dimensional (three dimensional in velocity space) short channel based nano scale semiconductor devices modeled by a Boltzmann–Poisson system subject to external applied voltage bias and space inhomogeneous background impurities modeling the non-equilibrium statistical states associated to the short based channel size. A deterministic method using WENO based methods was proposed and tested in [21, 22], and more recently, an alternative method based on Discontinuous Galerkin schemes was developed [25,26] where unstructured grids were employed.

2 The Non-Linear Boltzmann Equation for Binary Particle Interactions

In a microscopic description of a rarefied gas, all particles are assumed to be traveling in a straight line with a fixed velocity until they enter into a collision. In such dilute flows, binary collisions are often assumed to be the main mechanism of the particle interactions. The statistical effect of such collisions can be modeled by collision terms of the Boltzmann or Enskog transport equation type, where the kinetic dynamics of the gas are subject to the Molecular Chaos assumption. The nature of these interactions could be either elastic, inelastic or coalescing. They could either be isotropic or anisotropic, depending on their intramolecular potentials reflected in collision rates which are modeled as functions of powers of relative velocities and scattering angle, at the time of the interaction. When the rate of collisions is independent of the relative speed, the interaction is referred to as of Maxwell type. When it corresponds to relative speed to a positive power less than unity, they are referred to as Variable Hard Potentials (VHP) and when the rate of collisions is proportional to the relative speed, it is referred to as hard spheres.

Then, the Boltzmann Transport Equation (an integro-differential transport equation) describes the evolution of a single point probability distribution function $f(x, v, t)$ which is defined as the probability of finding a particle at position x with velocity (kinetic) v at time t. The mathematical and computational difficulties associated to the Boltzmann equation are due to the non local-non linear nature of the collision operator, which is usually modeled as a multi linear integral form in d-dimensional velocity space and unit sphere S^{d-1}.

It is important to keep in mind that this non-linear model of statistical transport entices the *Molecular Chaos assumption* which basically states that the particle velocities are uncorrelated before their interactions by pairs, and that probability of three or more particles interacting at the same time is negligible (rarefied anszat).

From the computational point of view, of the well-known and well-studied methods developed in order to solve this equation is an stochastic based method called "Direct Simulation Monte-Carlo" (DSMC) developed initially by Bird [2] and Nanbu [48] and more recently by [53, 54]. This method is usually employed as an alternative to hydrodynamic solvers to model the evolution of moments or hydrodynamic quantities. Is is based on the assumption of stochasticity of interactions as such of *Molecular Chaos*. In particular, this method have been shown to converge to the solution of the classical non-linear Boltzmann equation in the case of mono atomic rarefied gases [56]. One of the main drawbacks of such methods is the inherent statistical fluctuations in the numerical results, which becomes very expensive or unreliable in presence of non-stationary flows or non equilibrium statistical states, where more information is desired about the evolving probability distribution. Currently, there is extensive work from Rjasanow and Wagner [54] and references therein, to determine accurately the high-velocity tail behavior of the distribution functions from DSMC data.

In contrast, a deterministic method computes approximations of the probability distribution function using the Boltzmann equation, as well as approximations to the observables like density, momentum, energy, etc., can be calculated from this probability distribution.

There are currently two deterministic approaches to the computations of non-linear Boltzmann, one is the the well known discrete velocity models and the second a spectral based method, both implemented for simulations of elastic interactions i.e., energy conservative evolution.

Discrete velocity models were developed by Broadwell [19] and mathematically studied by Illner, Cabannes, Kawashima among many authors [20,41,42]. More recently these models have been studied for many other applications on kinetic elastic theory in [7, 24, 39, 44, 58]. These models have not adapted to inelastic collisional problems up to this point according to our best knowledge.

Spectral based models, which are the ones of our choice in this work, have been developed by Pareschi et al. [34], and later by Bobylev and Rjasanow [16] and Pareschi and Russo [52]. These methods are supported by the ground breaking work of Bobylev [4] using the Fourier Transformed Boltzmann Equation to analyze its solutions in the case of Maxwell type of interactions.

After the introduction of the inelastic Boltzmann equation for Maxwell type interactions and the use of the Fourier transform for its analysis by Bobylev, Carrillo and this authors in [6], the spectral based approach is becoming the most suitable tool to deal with deterministic computations of kinetic models associated with Boltzmann non-linear binary collisional integral, both for elastic or inelastic interactions.

Our current approach involves spectral-Lagrangian based methods recently developed by this author in collaboration with Sri Harsha Tharkabhushanam [38]. These type of spectral methods for the non-linear Boltzmann are originally inspired by the work of Bobylev and Rjasanow [18] who developed a method using the Fast Fourier Transform (FFT) for Maxwell type of interactions and then for Hard-Sphere interactions [17] using generalized Radon and X-ray transforms via FFT. Simultaneously, L. Pareschi and B. Perthame [51] developed a similar scheme using FFT for Maxwell type of interactions.

Later, I. Ibragimov and S. Rjasanow [40] developed a numerical method to solve the space homogenous Boltzmann Equation on a uniform grid for a Variable Hard Potential interactions with elastic collisions which has been a great inspiration to our work in [38], We mention that, most recently, Filbet and Russo [29,30] implemented a method to solve the space inhomogenous Boltzmann equation using the previously developed spectral methods in [51,52]. The afore mentioned work in developing deterministic solvers for non-linear BTE have been restricted to elastic, conservative interactions.

Our current approach in [38], based on a modified version of the work in [16] and [40], works for elastic or inelastic collisions and energy dissipative non-linear Boltzmann type models for variable hard potentials. We do not use periodic representations for the distribution function. The only restriction of the current method is that it requires that the distribution function at any time step be Fourier transformable. The required conservation properties of the distribution function are enforced through a Lagrange multiplier constrained optimization problem with the desired conservation quantities set as the constraints. Such corrections to the distribution function to make it conservative are very small but crucial for the evolution of the probability distribution function according to the Boltzmann equation.

This Lagrange optimization problem gives the freedom of not conserving the energy, independent of the collision mechanism, as long momentum is conserved. Such a technique plays a major role as it gives the option of computing energy dissipative solutions by just eliminating one constraint in the corresponding optimization problem. The current method can be easily implemented in any dimension. A novel aspect of the presented approach here lays on a new method that uses the Fourier Transform as a tool to simplify the computation of the collision operator that works, both for elastic and inelastic collisions. It is based on an integral representation of the Fourier Transform of the collision kernel as used in [16]. If N is the number of discretizations in one direction of the velocity domain in d-dimensions, the total number of operations required to solve for the collision integral are of the order of $nN^{2d}\log(N) + O(N^{2d})$. And this number of operations remains the same for elastic/inelastic, isotropic/anisotropic VHP type of interactions. However, when the differential cross section is independent of the scattering angle, the integral representation kernel is further reduced by an exact closed integrated form that is used to save in computational number of operations to $O(N^d \log(N))$. This reduction is possible when computing hard spheres in d+2 dimensions or Maxwell type models in 2-dimensions. Nevertheless, the method can be employed without much changes for the other case. In particular the method becomes $O(P^{d-1}N^d \log(N))$, where P, the number of each angular discretizations is expected to be much smaller than N used for energy discretizations.

Our numerical study in [38] was performed for several examples of well establish behavior associated to solutions of energy dissipative space homogenous collisional models under heating sources that secure existence of stationary states with positive and finite energy. We shall consider heating sources corresponding to randomly heated inelastic particles in a heat bath, with and without friction; elastic or inelastic collisional forms with anti-divergence terms due to dynamically (self-similar) energy scaled solutions [14,36] and a particularly interesting example of inelastic collisions added to a slow down linear process that can be derived as a weakly coupled heavy/light binary mixture. On this particular case, when Maxwell type interactions are considered, it has been shown that [11,13], on one hand dynamically energy scaled solutions exist, they have a close, explicit formula in Fourier space for a particular choice of parameters and their corresponding anti Fourier transform in probability space exhibit a singularity at the origin and power law high energy tails, while remaining integrable and with finite energy. On the other hand they are stable within a large class of initial states. We used this particular example to benchmark our computations by spectral methods by comparing the dynamically scaled computed solutions to the explicit one self similar one.

In the next subsection we recall some properties of the Boltzmann Transport equation and describe various models associated with the elastic or inelastic Boltzmann equation for which our scheme has been tested and discuss some aspects of the actual numerical method and its discretization, and present some computational output of our simulations. Readers should see [38] for a more complete presentation.

2.1 The Space Homogeneous Non Conservative Boltzmann Equation: Preliminaries

The initial value problem associated to space homogenous Boltzmann Transport Equation modeling the statistical (kinetic) evolution of a single point probability distribution function $f(v,t)$ for Variable Hard Potential (VHP) interactions is given by

$$\begin{aligned}\frac{\partial}{\partial t}f(v,t) &= Q(f,f)(v,t)\\ &= \int_{w\in\mathbb{R}^d,\sigma\in S^{d-1}} \left[J_\beta f('v,t)f('w,t) - f(v,t)f(w,t)\right] B(|u|,\mu)\, d\sigma dw\\ f(v,0) &= f_0(v)\,, \end{aligned} \tag{1}$$

where the initial probability distribution $f_0(v)$ is assumed integrable. It may or may not have finite initial energy $\mathrm{e}_0 = \int_{\mathbb{R}^d} f_0(v)|v|^2 dv$ and the velocity interaction law, written in center of mass and relative velocity coordinates is

$$u = v - w: \text{ the relative velocity}$$

$$v' = v + \frac{\beta}{2}(|u|\sigma - u), \qquad w' = w - \frac{\beta}{2}(|u|\sigma - u)\ ,$$

$$\mu = \cos(\theta) = \frac{u \cdot \sigma}{|u|} : \text{ the cosine of the scattering angle,} \tag{2}$$

$$B(|u|, \mu) = |u|^{\lambda}\, b(\cos(\theta)) \qquad \text{with } 0 \leq \lambda \leq 1,$$

$$\omega_{d-2} \int_0^{\pi} b(\cos\theta) \sin^{d-2}\theta < K : \textit{ Grad cut-off assumption}$$

$$\beta = \frac{1+e}{2} : \text{ the energy dissipation parameter,}$$

where the parameter $e \in [0,1]$ is the restitution coefficient corresponding from sticky to elastic interactions.

We denote by $'v$ and $'w$ the pre-collision velocities corresponding to v and w. In the case of micro-reversible (elastic) collisions one can replace $'v$ and $'w$ with v' and w' respectively in the integral part of (1). We assume the differential cross section function $b(\frac{u \cdot \sigma}{|u|})$ integrable with respect to the post-collisional specular reflection direction σ in the $d-1$ dimensional sphere, referred as the *Grad cut-off assumption*, and that $b(\cos\theta)$ is renormalized such that

$$\begin{aligned} \int_{S^{d-1}} b\left(\frac{u \cdot \sigma}{|u|}\right) d\sigma &= \omega_{d-2} \int_0^{\pi} b(\cos\theta) \sin^{d-2}\theta \, d\theta \\ &= \omega_{d-2} \int_{-1}^{1} b(\mu)(1-\mu^2)^{(d-3)/2} d\mu = 1 \,, \end{aligned} \tag{3}$$

where the constant ω_{d-2} is the measure of the $d-2$ dimensional sphere and the corresponding scattering angle is θ is defined by $\cos\theta = \frac{\sigma \cdot u}{|u|}$.

The parameter λ regulates the collision frequency as a function of the relative speed $|u|$. It accounts for inter particle potentials defining the collisional kernel, and they are referred as Variable Hard Potentials (VHP) whenever $0 < \lambda < 1$, Maxwell Molecules type interactions (MM) for $\lambda = 0$ and Hard Spheres (HS) for $\lambda = 1$. In particular case of three dimensions they take one of the following forms, depending on the collision frequency parameter:

Maxwell type of interactions:

$$B(|u|, \mu) = \frac{1}{4\pi} b(\theta) \,, \tag{4}$$

i.e., the collision kernel is independent of the relative velocity of the colliding particles and we have assumed the Grad's cut-off assumption.

Hard-Sphere interactions:

$$B(|u|, \mu) = \frac{a^2}{4} |u| \,, \tag{5}$$

where the constant a is the diameter of the spherical particles.

Variable Hard Potential (VHP) interactions:

$$B(|u|,\mu) = C_\lambda(\sigma)|u|^\lambda\,, \tag{6}$$

with $\lambda \in [0,1]$ and $C_{\lambda,}(\sigma)$ is the differential cross section, depending on the scattering angle, λ. If $C_{\lambda,}(\sigma)$ is independent of the scattering angle we call the interactions isotropic. Otherwise we referred to them as anisotropic Variable Hard Potential interactions.

For classical case of elastic collisions, it has been established that the Cauchy problem for the space homogeneous Boltzmann equation has a unique solution in the class of integrable functions with finite energy (i.e. $C^1(L^1_2(\mathbb{R}^d))$), it is regular if initially so, and $f(.,t)$ converges in $L^1_2(\mathbb{R}^d)$ to the Maxwellian distribution $M_{\rho,V,\mathrm{e}}(v)$ associated to the $d+2$-moments of the initial state $f(v,0) = f_0(v) \in L^1_2(\mathbb{R}^d)$. In addition, if the initial state has Maxwellian decay, this property will remain with a Maxwellian decay globally bounded in time ([35]), as well as all derivatives is initial so (see [1]).

Depending on their nature, collisions either conserve density, momentum and energy (elastic) or density and momentum (inelastic) or density (elastic – linear Boltzmann operator), depending on the number of collision invariants the operator $Q(f,f)(t,v)$ has. In the case of the classical Boltzmann equation for rarefied (elastic) mono-atomic gases, the collision invariants are exactly $d+2$, that is, according to the Boltzmann theorem, the number of polynomials in velocity space v that generate $\phi(v) = A + \mathbf{B}\cdot\mathbf{v} + \mathbf{C}|\mathbf{v}|^2$, with $C \le 0$. In particular, one obtains the following *conserved quantities*

$$\text{density}\quad \rho(t) = \int_{v\in\mathbb{R}^d} f(v,t)dv \tag{7}$$

$$\text{momentum}\quad j(t) = \int_{v\in\mathbb{R}^d} vf(v,t)dv \tag{8}$$

$$\text{energy}\quad \mathrm{e}(t) = \frac{1}{2\rho(t)}\int_{v\in\mathbb{R}^d} |v|^2 f(v,t)dv \tag{9}$$

Of significant interest from the statistical view point are the evolution of moments or observables, at all orders. They are defined by the dynamics of the corresponding time evolution equation for the velocity averages, given by

$$\frac{\partial}{\partial t}M_j(t) = \int_{v\in\mathbb{R}^d} f(v,t)v^{\oslash j}dv = \int_{v\in\mathbb{R}^d} Q(f,f)(v,t)v^{\oslash j}dv, \tag{10}$$

Thus, according to (9), for the classical elastic Boltzmann equation, the first $d+2$ moments are conserved, meaning, $M_j(t) = M_{0,j} = \int_{v\in\mathbb{R}^d} f_0(v)v^{\oslash j}dv$ for $j = 0,1$; and $\mathrm{e}(t) = \mathrm{tr}(M_2)(t) = \mathrm{e}_0 = \int_{v\in\mathbb{R}^d} f_0(v)|v|^2dv$. Higher order moments or observables of interest are

$$\text{Momentum Flow}\quad M_2(t) = \int_{\mathbb{R}^d} vv^T f(v,t)dv$$

$$\text{Energy Flow}\quad r(t) = \frac{1}{2\rho(t)}\int_{\mathbb{R}^d} v|v|^2 f(v,t)dv$$

$$\begin{aligned} \text{Bulk Velocity} \quad V(t) &= \frac{m(t)}{\rho(t)} \\ \text{Internal Energy} \quad \mathrm{e}(t) &= \frac{1}{2\rho}(tr(M_2) - \rho|V|^2) \\ \text{Temperature} \quad T(t) &= \frac{2\mathrm{e}(t)}{\mathbf{k}d} \end{aligned} \tag{11}$$

with **k**-Boltzmann constant.
We finally point out that, in the case of Maxwell molecules ($\lambda = 0$), it is possible to write recursion formulas for higher order moments of all orders ([5] for the elastic case, and [6] in the inelastic case) which, in the particular case of isotropic solutions depending only on $|v|^2/2$, take the form

$$\begin{aligned} m_n(t) = \int_{\mathbb{R}^d} |v|^{2n} f(v,t)dv = e^{-\lambda_n t} m_n(0) \\ + \sum_{k=1}^{n-1} \frac{1}{2(n+1)} \binom{2n+2}{2k+1} B_\beta(k, n-k) \int_0^t m_k(\tau)\, m_{n-k}(\tau)\, e^{-\lambda_n(t-\tau)}\, d\tau\,; \end{aligned} \tag{12}$$

with

$$\lambda_n = 1 - \frac{1}{n+1}\left[\beta^{2n} + \sum_{k=0}^{n} (1-\beta)^{2k}\right],$$

$$B_\beta(k, n-k) \;=\; \beta^{2k} \int_0^1 s^k (1 - \beta(2-\beta)s)^{n-k} ds\,,$$

for $n \geq 1$, $0 \leq \beta \leq 1$, where $\lambda_0 = 0$, $m_0(t) = 1$, and $m_n(0) = \int_{\mathbb{R}^d} |v|^{2n} f_0(v)dv$.

2.2 Boltzmann Collisional Models with Heating Sources

A collisional model associated to the Boltzmann transport equation (1)–(3), can be modified in order to accomodate for an energy or 'heat source' like term $\mathcal{G}(f(t,v))$, where $\mathcal{G}$ is a differential or integral operator. In these cases, it is possible to obtain stationary states with finite energy as for the case of inelastic interactions. In such general framework, the corresponding initial value problem model is

$$\begin{aligned} \frac{\partial}{\partial t} f(v,t) &= \zeta(t)\, Q(f,f)(v,t) + \mathcal{G}(f(t,v))\,, \\ f(v,0) &= f_0(v)\,, \end{aligned} \tag{13}$$

where the collision operator $Q(f,f)(v,t)$ is as in (1) and $\mathcal{G}(f(t,v))$ models a 'heating source' due to different phenomena. The term $\zeta(t)$ may represent a mean field approximation that allows from proper time rescaling. See [6] and [14] for several examples for these type of models and additional references.

Following the work initiated in [14] and [13] on Non-Equilibrium Stationary States (NESS), our computational approach we shall present several computational simulations of non-conservative models for either elastic or inelastic collisions associated to (13) of the Boltzmann equation with 'heating' sources. In all the cases we have addressed one can see that stationary states with finite energy are admissible, but they may not be Maxwellian distributions.

Of this type of model we show computational output for three cases, first one is the pure diffusion thermal bath due to a randomly heated background [36, 49, 57], in which case

$$\mathcal{G}_1(f) = \mu \Delta f, \tag{14}$$

where $\mu > 0$ is a constant. The second example relates to self-similar solutions of (13) for $\mathcal{G}(f) = 0$ [27, 45], but dynamically rescaled by

$$f(v,t) = \frac{1}{v_0^d(t)} \tilde{f}(\tilde{v}(v,t), \tilde{t}(t)), \quad \tilde{v} = \frac{v}{v_0(t)}, \tag{15}$$

where

$$v_0(t) = (a + \eta t)^{-1}, \quad \tilde{t}(t) = \frac{1}{\eta} \ln\left(1 + \frac{\eta}{a} t\right), \quad a,\, \eta > 0. \tag{16}$$

Then, the equation for $\tilde{f}(\tilde{v}, \tilde{t})$ coincides (after omitting the tildes) with (13), for

$$\mathcal{G}_2(f) = -\eta \, \mathrm{div}(vf), \qquad \eta > 0\,. \tag{17}$$

Of particular interest for dynamical time-thermal speed rescaling is the case of collisional kernels corresponding to Maxwell type of interactions. Since the second moment of the collisional integral is a linear function of the energy, the energy evolves exponentially with a rate proportional to the energy production rate, that is

$$\frac{d}{dt}\mathrm{e}(t) \;=\; \lambda_0 \, \mathrm{e}(t), \qquad \text{or equivalently } \mathrm{e}(t) = \mathrm{e}(0)\, e^{\lambda_0 t}\,, \tag{18}$$

with λ_0 the energy production rate. Therefore the corresponding rescaled variables and (15) and (13), (17) for the study of long time behavior of rescaled solutions are

$$f(v,t) = \mathrm{e}^{-\frac{d}{2}}(t)\, \tilde{f}\left(\frac{v}{\mathrm{e}^{\frac{1}{2}}(t)}\right) \;=\; \left(\mathrm{e}(0)e^{\lambda_0 t}\right)^{-\frac{d}{2}} \tilde{f}\left(v \left(\mathrm{e}(0)e^{\lambda_0 t}\right)^{-\frac{1}{2}}\right), \tag{19}$$

and $\tilde{f}$ satisfies the self-similar (13)

$$\mathcal{G}_{2'}(f) = -\lambda_0 x f_x, \qquad \text{where } x = v\mathrm{e}^{-\frac{1}{2}}(t) \text{ is the self-similar variable}. \tag{20}$$

We note that it has been shown that these dynamically self-similar states are stable under very specific scalings for a large class of initial states [11].

The last source type we consider is given by a model, related to weakly coupled mixture modeling slowdown (cooling) process [13] given by an elastic model in the presence of a thermostat given by Maxwell type interactions of particles of mass m having the Maxwellian distribution

$$M_{\mathcal{T}}(v) = \frac{m}{(2\pi\mathcal{T})^{d/2}} e^{\frac{-m|v|^2}{2\mathcal{T}}},$$

with a constant reference background or thermostat temperature $\mathcal{T}$ (i.e., the average of $\int M_{\mathcal{T}}\, dv = 1$ and $\int |v|^2 M_{\mathcal{T}}\, dv = \mathcal{T}$). Define

$$Q_L(f)(v,t) \doteq \int_{w\in\mathbb{R}^d, \sigma\in S^{d-1}} B_L(|u|,\mu) f('v,t) M_{\mathcal{T}}('w) - f(v,t) M_{\mathcal{T}}(w)]\; d\sigma dw \tag{21}$$

Then the corresponding evolution equation for $f(v,t)$ is given by

$$\begin{aligned} \frac{\partial}{\partial t} f(v,t) &= Q(f,f)(v,t) + \Theta Q_L(f)(v,t) \\ f(v,0) &= f_0(v) \end{aligned} \tag{22}$$

where $Q(f,f)$, defined as in (1), is the classical collision integral for elastic interactions (i.e., $\beta = 1$), so it conserves density, momentum and energy. The second integral term in (22) is a linear collision integral which conserves just the density (but not momentum or energy) since

$$\begin{aligned} u &= v - w \qquad \text{the relative velocity} \\ v' &= v + \frac{m}{m+1}(|u|\sigma - u), \qquad w' = w - \frac{1}{m+1}(|u|\sigma - u) \end{aligned} \tag{23}$$

The coupling constant Θ depends on the initial density, the coupling constants and on m. The collision kernel B_L of the linear part may not be the same as the one for the non-linear part of the collision integral, however we assume that the *Grad cut-off assumption* (3) is satisfied and that, in order to secure mass preservation, the corresponding differential cross section functions b_N and b_L, the non-linear and linear collision kernels respectively, satisfy the renormalized condition

$$\int_{S^{d-1}} b_N\left(\frac{u\cdot\sigma}{|u|}\right) + \Theta b_L\left(\frac{u\cdot\sigma}{|u|}\right) d\sigma \;=\; 1 + \Theta\,. \tag{24}$$

This last model describes the evolution of binary interactions of two sets of particles, heavy and light, in a weakly coupled limit, where the heavy particles have reached equilibrium. The heavy particles set constitutes the background or thermostat for the second set of particles. It is the light particle distribution that is modeled by (22). Indeed, $Q(f,f)$ corresponds to all the collisions which the light particles have with each other and the second linear integral term corresponds to collisions between light and heavy particles at equilibrium given by a classical distribution $M_{\mathcal{T}}(v)$. In this binary 3-dimensional, mixture scenario, collisions are assumed to be isotropic, elastic and the interactions kernels of Maxwell type.

In the particular case of equal mass (i.e., $m = 1$), the model is of particular interest for the development of numerical schemes and simulations benchmarks. Even though the local interactions are reversible (elastic), it does not conserve the total energy. In such case, there exits a special set of explicit, in spectral space, self-similar solutions which are attractors for a large class of initial states. When considering, in the case of Maxwell type of interactions in three dimensions i.e., $B(|u|, \mu) = b(\mu)$, a cooling background process corresponding to a time temperature transformation, i.e., $\mathcal{T} = \mathcal{T}(t)$ such that $\mathcal{T}(t) \to 0$ as $t \to 0$, the models have self similar asymptotics [11,13] for a large class of initial states. Such long time asymptotics corresponding to dynamically scaled solutions of (22), in the form of (20), yields interesting behavior in $f(v,t)$ for large time, converging to states with power like decay tails in v. In particular, such solution $f(v,t)$ of (22) will lose moments as time grows, even if the initial state has all moments bounded. (see [11, 13] for the analytical proofs and [38] for simulations).

2.3 Collision Integral Representation

One of the main points of spectral numerical method for the computation of the non-linear Boltzmann equation lays in the representation of the collision integral in Fourier space by means of the weak form. Since for a suitably regular test function $\psi(v)$, the weak form of the collision integral takes the form (suppressing the time dependence in f)

$$\int_{v\in\mathbb{R}^d} Q(f,f)\psi(v)dv = \int_{(w,v,\sigma)\in\mathbb{R}^d\times\mathbb{R}^d\times S^{d-1}} f(v)f(w)[\psi(v') - \psi(v)]B(|u|,\mu)d\sigma dw dv,$$

with the velocity relations as given in (3). In particular, taking $\psi(v) = e^{-ik.v}$, where k is the Fourier variable, then the Fourier Transform of the collision integral (see [38])

$$\widehat{Q}(k) = \int_{(u,v)\in\mathbb{R}^d\times\mathbb{R}^d} f(v)f(v-u)e^{-ik.v}G_{\lambda,\beta}dudv\,, \tag{25}$$

with

$$G_{\lambda,\beta}(u,k) = \int_{\sigma\in S^{d-1}} C_\lambda(\sigma)|u|^\lambda[e^{-i\frac{\beta}{2}k.(|u|\sigma-u))} - 1]d\sigma\,, \tag{26}$$

or, equivalently,

$$\widehat{Q}(k) = \int_{u\in\mathbb{R}^d} G_{\lambda,\beta}(u,k)[f(v)f(v-u)]\widehat{\ }du\,. \tag{27}$$

with

$$G_{\lambda,\beta}(u,k) = |u|^{\lambda}\left[e^{i\frac{\beta}{2}k.u}\int_{\sigma\in S^{d-1}} C_{\lambda}(\sigma)e^{-i\frac{\beta}{2}|u|k.\sigma}d\sigma - \omega_{d-2}\right]. \quad (28)$$

which works for both isotropic and anisotropic interactions. For the former type, a simplification ensues due to the fact the $C_{\lambda}(\sigma)$ is independent of $\sigma \in S^{d-1}$:

$$G_{\lambda,\beta}(u,k) = C_{\lambda}\omega_{d-2}\,|u|^{\lambda}\left[e^{i\frac{\beta}{2}k.u}\operatorname{sinc}\left(\frac{\beta|u||k|}{2}\right) - 1\right]. \quad (29)$$

Thus, it is seen that the dependence on σ i.e., the integration over the unit sphere S^{d-1} is completely done independently by the closed form expression given by (29) in the case of isotropic collisions. In the case of anisotropic collisions, the dependence of C_{λ} on σ is again isolated into a separate integral over the unit sphere S^{d-1} as given in (28). The above expression can be transformed for elastic collisions $\beta = 1$ into a form suggested in [40].
To obtain the collision integral back in velocity space, one takes the inverse Fourier Transform of $\widehat{Q}_{\lambda,\beta}[f](k)$ i.e.,

$$Q_{\lambda,\beta}(f,f)(v) = \mathcal{F}^{-1}\left[\int_{u\in\mathbb{R}^d} G_{\lambda,\beta}(u,k)[f(v)f(v-u)]\widehat{\ }du\right], \quad (30)$$

where $G_{\lambda,\beta}(u,k)$ is defined in (28), $\beta \in [0,1], \lambda \in [0,1]$. For all of the numerical simulations in [38] it was assumed that the collisions are of Variable Hard Potential type.

3 Numerical Method and Discretization

It can be seen from (30) that computation of $Q_{\lambda,\beta}(f,f)(v)$ involves computing Fourier Transforms of $f(v)f(v-u)$ for each u with respect to v, multiplying this result by the kernel $G_{\lambda,\beta}(u,k)$ for each u and k; and finally one Inverse Fourier Transform with respect to k.

Because of the extensive use of Fourier Transforms, a Fast Fourier Transform has been employed. Using such FFT solvers the total number of operations in computing the collision intergal reduces to the order of $3N^{2d}\log(N) + O(N^{2d})$. As observed, the proposed scheme works for both elastic and inelastic collisions. In [29, 30] and [47], a spectral method based on a Carleman-like representation of the collision integral was proposed. A decoupling assumption on the collision kernel is made for simplification which results in a constant (unit) kernel for the cases of Maxwell interactions (2-D in velocity) and Hard-Sphere interactions (3-D in velocity). Under this assumption, the kernel can be written as a finite sum by identifying a convolution structure in them. If P is the number of angular discretizations in both the angles, the total number of operations was found to have reduced from $O(N^{2d})$ (Classical Spectral) to $O(P^{d-1}N^d \log(N))$ (Fast Spectral), where d-dimensionality in velocity, P is usually small number in comparison to N. For more general cases of collision kernel (non-constant), their methods can be extended easily but the operation complexity might not be $O(P^{d-1}N^d \log(N))$. As a note, the method proposed in the current work can also be extended to lower dimensions in velocity space.

The discretization in velocity space is assumed such that the two interacting velocities and the corresponding relative velocity satisfy v, w, and $w \in [-L, L)^d$ while $k \in [-L_k, L_k)^d$, where the length L is chosen such that $u = v - w \in [-L, L)^d$ through an assumption that $supp(f) \in [-L, L)^d$.

Since the computations correspond to space homogeneous problems, the computed distribution does not lose mass if L is chosen initially large given that the initial momentum is conserved (there is no convection in space homogeneous problems), and is renormalized to zero mean velocity. A uniform grid is assumed in the velocity space and in the fourier space with h_v and h_k be the grid element sizes respectively. h_v and h_k are chosen such that $h_v h_k = \frac{2\pi}{N}$, where N = number of discretizations of v and k in each direction. Thus the space discretization is

$$h_v = \frac{2L}{N} \qquad h_k = \frac{2\pi}{N} h_v^{-1} = \frac{\pi}{L} = \frac{2L_k}{N} . \tag{31}$$

and

$$\begin{aligned} v^{(j_1,j_2,\ldots,j_d)} &\in \{(\zeta_1, \zeta_2, \ldots, \zeta_d) : \zeta_l = -L + j_l h_v; l = 1, 2, \ldots, d\} \\ w^{(j_1,j_2,\ldots,j_d)} &\in \{(\zeta_1, \zeta_2, \ldots, \zeta_d) : \zeta_l = -L + j_l h_v; l = 1, 2, \ldots, d\} \\ u^{(j_1,j_2,\ldots,j_d)} &\in \{(\zeta_1, \zeta_2, \ldots, \zeta_d) : \zeta_l = -L + j_l h_v; l = 1, 2, \ldots, d\} \\ k^{(j_1,j_2,\ldots,j_d)} &\in \{(\zeta_1, \zeta_2, \ldots, \zeta_d) : \zeta_l = -L_k + j_l h_k; l = 1, 2, \ldots, d\} \end{aligned} \tag{32}$$

for $j_i = 0, 1, 2, \ldots, N-1; i = 1, 2, \ldots d$.

For time discretization we have used either second-order Runge–Kutta scheme or a Euler forward step method. For actual numerical computations, the time step was $dt = 0.1 \times$ *the time between consecutive collisions* (which depends on the collision frequency). During the standard process of non-dimensionalization of the Boltzmann equation time between collisions comes up and has been fixed during the discretization process. With time discretizations taken as $t^n = ndt$, the discrete version of the Runge–Kutta scheme is given by

$$\begin{aligned} f^0(v^j) &= f_0(v^j) \\ \tilde{f}(v^j) &= f^{t^n}(v^j) + \frac{dt}{2} Q_{\lambda,\beta}\left[f^{t^n}(v^j), f^{t^n}(v^j)\right] \\ f^{t^{n+1}}(v^j) &= f^{t^n}(v^j) + dt Q_{\lambda,\beta}\left[\tilde{f}(v^j), \tilde{f}(v^j)\right] . \end{aligned} \tag{33}$$

The corresponding Forward Euler scheme with smaller time step is given by

$$\tilde{f}(v^j) = f^{t^n}(v^j) + dt Q\left(f^{t^n}, f^{t^n}\right). \tag{34}$$

3.1 Conservation by Lagrange Multipliers

Due to the spectral and time discretizations and the use of Fast Fourier Transform, the accuracy of the proposed method relies heavily on the size of the grid and the number of points taken in each velocity/Fourier space directions. Because of these issues, we observe that the computed $Q_{\lambda,\beta}[f,f](v)$ does not conserve the quantities it is supposed to, i.e. ρ, m, e for elastic collisions, or ρ for Linear Boltzmann Integral, and ρ, m for inelastic collisions. Even though the difference between the computed (discretized) collision integral and the continuous one is not great, it is nevertheless essential that this issue be resolved. To remedy this, a simple constrained lagrange multiplier method is employed where the constraints are the required conservation properties. Let $M = N^d$, the total number of discretizations of the velocity space. Assume that the classical Boltzmann collision operator is being computed. So ρ, m and e are conserved. Let ω_j be the integration weights where $j = 1, 2, \ldots, M$. Let

$$\tilde{f} = \left(\tilde{f}_1\ \tilde{f}_2\ ..\ \tilde{f}_M\right)^T$$

be the distribution vector at the computed time step and

$$f = \left(f_1\ f_2\ ..\ f_M\right)^T$$

be the corrected distribution vector with the required moments conserved. Let

$$C = d+2\left\{\overbrace{\begin{pmatrix} - & \omega_j & - \\ - & v_j\omega_j & - \\ - & |v_j|^2\omega_j & - \end{pmatrix}}^{M}\right.$$

and

$$a = \overbrace{\left(\rho\ m_1\ m_2\ m_3\ e\right)}^{5}{}^T$$

be the vector of conserved quantities. Using the above vectors, the conservation can be written as a constrained optimization problem:

$$(*)\begin{cases} \|\tilde{f} - f\|_2^2 \to min \\ Cf = a;\ \ C \in \mathbb{R}^{d+2\times M}, f \in \mathbb{R}^M, a \in \mathbb{R}^{d+2} \end{cases}.$$

To solve $(*)$, one can employ the Lagrange multiplier method as follows. Let $\lambda \in \mathbb{R}^{d+2}$ be the Lagrange multiplier vector. Then, the scalar objective function to be optimized is given by

$$L(f,\lambda) = \sum_{j=1}^{M} \left|\tilde{f}_j - f_j\right|^2 + \lambda^T(Cf - a)\,. \tag{35}$$

Equation (35) can actually be solved explicitly for the corrected distribution value and the resulting equation of correction be implemented numerically in the code. Taking the derivative of $L(f,\lambda)$ with respect to $f_j, j = 1, \ldots, M$ and $\lambda_i, i = 1, \ldots, d+2$ i.e., gradients of L,

$$\begin{aligned} \frac{\partial L}{\partial f_j} &= 0; j = 1, \ldots, M \\ &\Rightarrow \\ f &= \tilde{f} + \frac{1}{2} C^T \lambda . \end{aligned} \tag{36}$$

And

$$\begin{aligned} \frac{\partial L}{\partial \lambda_1} &= 0; 1 = 1, \ldots, d+2 \\ &\Rightarrow \\ Cf &= a \end{aligned} \tag{37}$$

i.e., retrieves the constraints.
Solving for λ,

$$CC^T \lambda = 2(a - C\tilde{f}) . \tag{38}$$

with CC^T symmetric since C is the integration matrix and so it is positive definite. In particular CC^T has an inverse and the value of λ can be computed by

$$\lambda = 2(CC^T)^{-1}(a - C\tilde{f}) . \tag{39}$$

Substituting λ into (36) yields,

$$f = \tilde{f} + C^T (CC^T)^{-1}(a - C\tilde{f}) . \tag{40}$$

Using equation for the Forward Euler scheme (34), the complete scheme is given by $\left(f^{t^n}(v^j) = f_j^n\right) \forall j$:

$$\begin{aligned} \tilde{f}_j &= f_j^n + dtQ(f_j^n, f_j^n) \\ f_j^{n+1} &= \tilde{f}_j + C^T (CC^T)^{-1}(a - C\tilde{f}) . \end{aligned} \tag{41}$$

Then, it follows

$$f_j^{n+1} = f_j^n + dt \left[\mathbb{I} - C^T \left(CC^T\right)^{-1} C\right] Q(f_j^n, f_j^n) , \tag{42}$$

with $\mathbb{I}$-$N \times N$ identity matrix. Letting $\Lambda_N(C) = \mathbb{I} - C^T (CC^T)^{-1} C$, one obtains

$$f_j^{n+1} = f_j^n + dt\Lambda_N(C) Q(f_j^n, f_j^n) . \tag{43}$$

where we expect the required observables are conserved and the solution approaches a stationary state, since $\lim_{n\to\infty} \|\Lambda_N(C)\, Q(f_j^n, f_j^n)\|_\infty = 0$.

Identity (43) summarizes the whole conservation process. As described previously, setting the conservation properties as constraints to a Lagrange multiplier optimization problem ensures that the required observables are conserved. Also, the optimization method can be extended to have the distribution function satisfy more (higher order) moments from (12). In this case, $a(t)$ will include entries of $m_n(t)$ from (12).

We point out that for the linear Boltzmann collision operator used in the mixture problem conserves density and not momentum(unless one computes isotropic solutions) and energy. For this problem, the constraint would just be the density equation. For inelastic collisions, density and momentum are conserved and for this case the constraint would be the energy and momentum equations. And for the elastic Boltzmann operator, all three quantities (density, momentum and energy) are conserved and thus they become the constraints for the optimization problem.

4 Numerical Results

In [38] we present several numerical simulations of the models described above. Here we have selected a few and we refer the reader to that publication for a complete presentation.

We show next several computed examples of 3-D in velocity and time for initial value problems associated with non-conservative models where some analysis is available, as are exact moment formulas for Maxwell type of interactions as well as qualitative analysis for solutions of Variable Hard Potentials models. We plot our numerical results versus the exact available solutions in several cases. Because all the computed problems converge to an isotropic long time state, we choose to plot the distribution function in only one direction, which is chosen to be the one with the initial anisotropies. All examples considered in this manuscript are assumed to have isotropic, Variable Hard Potential collision kernels, i.e., differential cross section independent from scattering angle. We simulate the homogeneous problem associated to the following problems for different choices of the parameters β and λ, and the Jacobian J_β and 'heating' force term $\mathcal{G}(f)$.

4.1 Conservative Model for Micro-Reversible Interactions

Here we shall show results for elastic collisions of Maxwell type and also we test the classical model of hard spheres elastic collisions in three dimensions.

First we start with elastic collisions of Maxwell type, i.e. when the collision frequency is independent of the local velocity vector. In particular one considers the initial value problem (1, 3), with $B(|u|, \mu) = \frac{1}{4\pi}|u|^\lambda$. In (1, 3), the value of the parameters are $\beta = 1, J_\beta = 1$ and $\lambda = 0$ with the pre-collision velocities defined from (3). In this case, for a general initial state with finite mass, mean and kinetic energy, there is no exact formula for the evolving distribution function. However there are exact formulas for all the statistical moments (observables). Thus, the numerical method is compared with the known analytical moments for different discretizations in the velocity space.

The initial states we take are convex combinations of two shifted Maxwellian distributions, renormalized to unit mass, which enables the first five moment equations corresponding to the collision invariants to be computed from those of the initial state. All higher order moments are computed using the classical moments recursion formulas for Maxwell type of interactions (11). In particular, it is possible to obtain the exact evolution of moments as functions of time. Thus, in our example

$$\rho(t) = \rho_0 = 1 \quad \text{and} \quad V(t) = V_0 = [0,1,0]^T,$$

and the complete evolution of the second moment tensor (12) is given by

$$M(t) = \begin{pmatrix} 5 & -2 & 0 \\ -2 & 3 & 0 \\ 0 & 0 & 1 \end{pmatrix} e^{-t/2} + \frac{1}{3}\begin{pmatrix} 8 & 0 & 0 \\ 0 & 11 & 0 \\ 0 & 0 & 8 \end{pmatrix} (1 - e^{-t/2}),$$

the energy flow (11) given by

$$r(t) = \frac{1}{2}\begin{pmatrix} -4 \\ 13 \\ 0 \end{pmatrix} e^{-t/3} + \frac{1}{6}\begin{pmatrix} 0 \\ 43 \\ 0 \end{pmatrix} (1 - e^{-t/3}) - \frac{1}{6}\begin{pmatrix} 12 \\ 4 \\ 0 \end{pmatrix} (e^{-t/2} - e^{-t/3}),$$

and the kinetic temperature is conserved, so $T(t) = T_0 = \frac{8}{3}$.

These moments along with their numerical approximations for different discretizations in velocity space are plotted in Fig. 1.

For the case of hard-sphere Elastic collisions set $\beta = 1$, $J_\beta = 1$ and $\lambda = 1$ in (1, 3), with the post-collision velocities given by (1). Unlike Maxwell type of interactions, there is no explicit expression for the moment equations and neither is there any explicit solution expression. The expected behavior of the moments is somewhat similar to that of the Maxwell type of interactions case except that in this case, the moments evolve to the equilibrium a bit faster than in the former case i.e., Fig. 2. This is a known effect due to the fact that the collision frequency depends on the relative velocity indicating that the higher the relative velocity the more collisions, and so the approach to equilibrium is faster. We also plot the time evolution of distribution functions, for both Maxwell type and hard-sphere interactions, starting from the convex combination of Maxwellians as described above in Fig. 3.

We finally benchmark the elastic interactions with the Bobylev–Krook–Wu (BKW) explicit solution formula for solution to the initial value problem (1) for elastic, Maxwell type of interactions ($\beta = 1$, $\lambda = 0$), which was derived in [3] and independently in [43] for initial states that have at least $2+\delta$-moments bounded. It is not of self-similar type, but it can be shown to converge to a Maxwellian distribution. This solution takes the form

$$f(v,t) = \frac{e^{-|v|^2/(2K\eta^2)}}{2(2\pi K\eta^2)^{3/2}} \left(\frac{5K-3}{K} + \frac{1-K}{K^2}\frac{|v|^2}{\eta^2} \right), \tag{44}$$

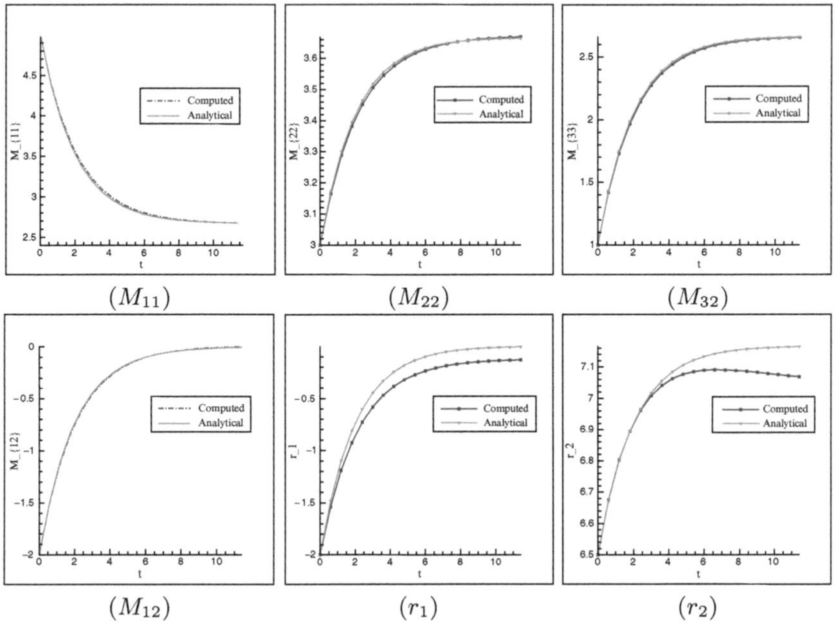

Fig. 1. Maxwell type of elastic interactions (number of modes N = 40): *Top row*: momentum flow M_{11}, M_{22}, M_{33}. *Bottom row*: momentum flow M_{12} and energy flow r_1, r_2. (See Plate 11 on page 306)

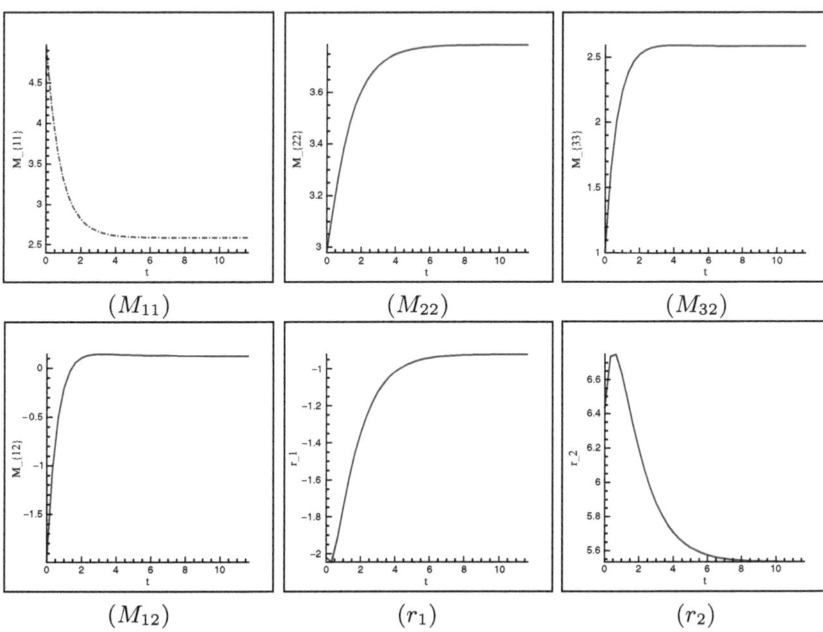

Fig. 2. Hard spheres elastic interactions in 3-dimensions (number of modes N = 40): *Top row*: momentum flow M_{11}, M_{22}, M_{33}. *Bottom row*: momentum flow M_{12} and energy flow r_1, r_2. (See Plate 12 on page 306)

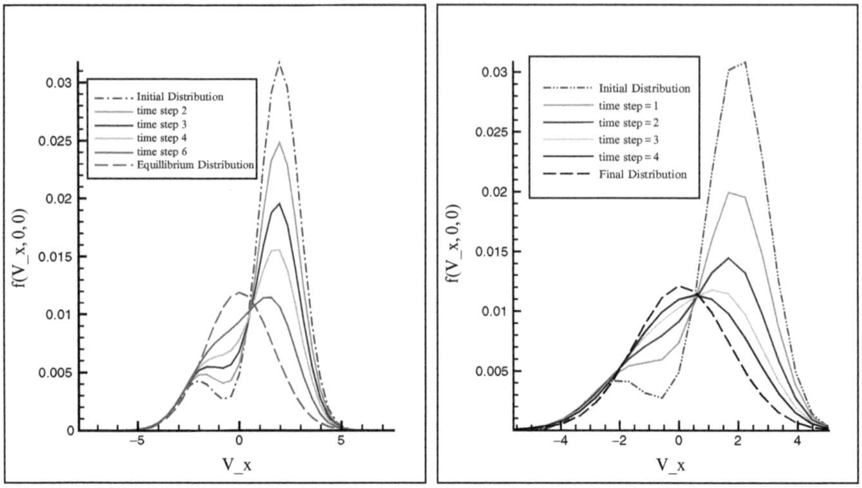

Fig. 3. Evolution of the distribution function for elastic interactions. *Left*: Maxwell type interactions, N = 40. *Right*: Hard-sphere interactions, N = 32. (See Plate 13 on page 307)

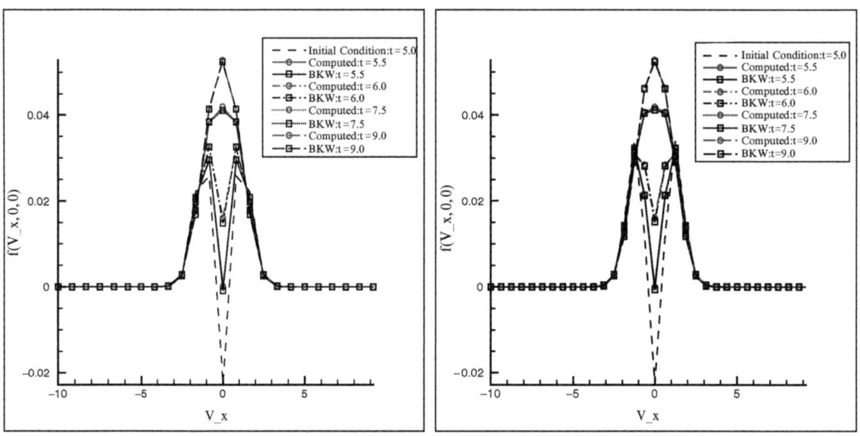

($N = 24$) ($N = 32$)

Fig. 4. BKW, $\rho, E(t)$ conserved. (See Plate 14 on page 307)

where $K = 1 - e^{-t/6}$ and η = initial distribution temperature. It is interesting that it is negative for small values of t. Since, in order to obtained a physically meaning probability distribution, f must be non-negative, the BKW solution is physically relevant for any $K \geqslant \frac{3}{5}$ or $t \geqslant t_0 \equiv 6ln(\frac{5}{2}) \sim 5.498$. In order to test the accuracy of our solver, set the initial distribution function to be the BKW solution, the numerical approximation to the BKW solution and the exact solution are plotted for different values of N at various time steps in Fig. 4.

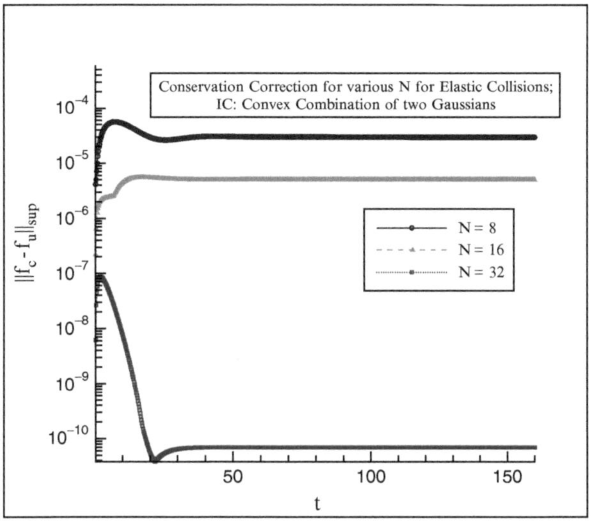

Fig. 5. Maxwell type elastic collisions: conservation correction for elastic collisions. (See Plate 15 on page 308)

We close this subsection addressing the issue of accuracy of conservation. Thus, in order to check the conservation accuracy of the method, we let f_u-unconserved distribution given as input to the conservation routine and f_c-conserved distribution resulting from the conservation routine. With a convex combination of two Gaussians as input, the numerical method is allowed to run and $\|f_c - f_u\|_\infty$ is plotted for all times for different values of N in Fig. 5. For t approaching the final time, the largest value of N gives the smallest conservation correction as expected. A comprehensive study of spectral accuracy and convergence of our Spectral-Lagrangian approach is underway by the two authors and will be available in a forthcoming publication.

4.2 Non-Conservative Model for Slow Down Processes or Micro-Irreversible Interactions

This is the setting where the Spectral-Lagrangian approach makes its most innovative contribution. No other existing deterministic method can compute the distribution function in the case of inelastic collisions. Our approach can handle the simulation of this phenomena without major modifications with respect to the elastic interactions case, and with the exactly same number of operations. The approximations also works for all sorts of Variable Hard Potential interactions with 'heating' forcing terms. We first look at the special case of Maxwell ($\lambda = 0$) type of inelastic ($\beta \neq 1$) collisions in a space homogenous Boltzmann Eqs. (1), (3). In this particular scenario it was shown in [6] that solutions converge to the a Dirac delta distribution localized at the initial mean velocity. In addition an explicit equation for the kinetic temperature or kinetic energy still can be derived. Using the weak form of the Boltzmann equation, one can obtain the ODE governing the evolution of the kinetic energy

$$K'(t) = \beta(1-\beta)\left(\frac{|V|^2}{2} - K(t)\right) \tag{45}$$

where V is the conserved (constant) bulk velocity of the distribution function. In particular the kinetic energy as computed in (refmoments3)

$$K(t) = K(0)e^{-\beta(1-\beta)t} + \frac{|V|^2}{2}\left(1 - e^{-\beta(1-\beta)t}\right), \tag{46}$$

where $K(0) =$ kinetic energy at time $t = 0$. We use this analytical form of the second moment to compare with its numerical approximation for accuracy. In Fig. 6 we plot the kinetic energy and the general evolution of the distribution in an inelastic collision environment. In the conservation routine (constrained Lagrange multiplier method), energy is not used as a constraint and just density and momentum equations are used for constraints. Figure 6 shows the numerical accuracy of the method even though the energy (plotted quantity) is not being conserved as part of the constrained optimization method.

Next we simulate a model corresponding to inelastic interaction in a randomly excited heat bath with constant temperature η. The corresponding Boltzmann transport equation is given by (13), (14). Also, in the case of Maxwell type of inelastic interactions, the kinetic temperature of the solutions has a closed form that is derived using methods similar to the ones used in the previous section. Indeed, the evolution equation for kinetic temperature as a function of time is, according to (12),

$$\frac{dT}{dt} = 2\eta - \zeta\pi C_0(1-e^2)T\,. \tag{47}$$

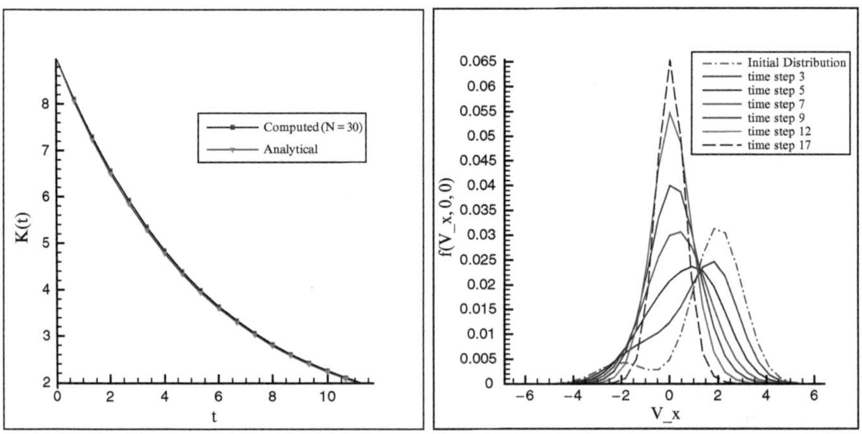

Fig. 6. Inelastic: kinetic energy (*left*) and $f(v,t)$ (*right*). (See Plate 16 on page 308)

whose solution is

$$T(t) = T_0 e^{-\zeta\pi C_0(1-e^2)t} + T_\infty^{MM}[1 - e^{-\zeta\pi C_0(1-e^2)t}], \tag{48}$$

where

$$T_0 = \frac{1}{3}\int_{v\in\mathbb{R}^3} |v|^2 f(v)dv \qquad \text{and} \qquad T_\infty^{MM} = \frac{2\eta}{\zeta\pi C_0(1-e^2)}.$$

As it can be seen from the expression for T, the presence of the diffusion term pushes the temperature to an equilibrium value of $T_\infty^{MM} > 0$ if $\eta > 0$. Also note that if the interactions were elastic and the diffusion coefficient positive, then $T_\infty^{MM} = +\infty$, so there would be no equilibrium states with finite kinetic temperature. These properties was also shown, in [36], to hold even in the case of inelastic collisions where $T_\infty^{HS} > 0$ is shown to exist, even though there not an explicit ODE for the evolution of the corresponding $T^{HS}(t)$. Similar time asymptotic behavior is expected in the case of hard-sphere interactions. However, the time evolution of the kinetic temperature is a non-local integral which does not satisfy a close ODE form.

The proposed numerical method for the calculation of the collision integral is tested for these two cases. We compared with the analytical expression (48) for different initial data, the corresponding computed kinetic temperatures for Maxwell type interactions in Fig. 7. The asymptotic behavior is observed in the case of hard-sphere interactions in Fig. 7. The conservation properties for this case of inelastic collisions with a diffusion term are set exactly like in the previous subsection (inelastic collisions without the diffusion term).

We close by mentioning another important application of our method to Maxwell type of Elastic interactions for a *slow down process.* This model was analytically addressed in [13] and [11], and the setting for the numerical simulations and computational output has been fully developed by the authors in [38], so we refer to the reader to these publications for a comprehensive reading. We just mention here a few comments regarding the presented numerical output.

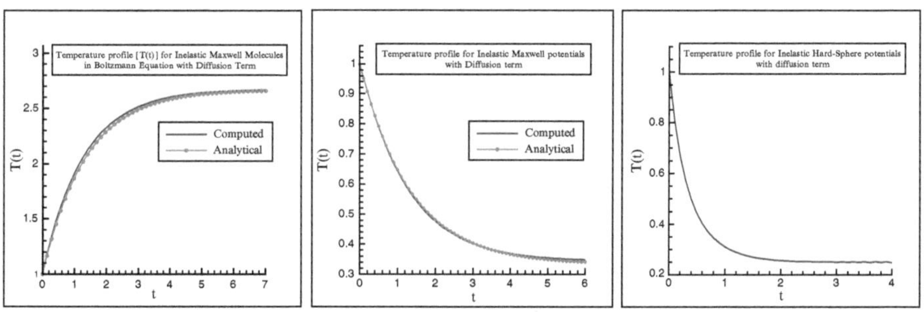

$(T_\infty^{MM} > T_0)$ $(T_\infty^{MM} < T_0)$ $(T_\infty^{HS} < T_0)$

Fig. 7. Energy decay for inelastic collisions with a diffusion term, N = 16. *Left and center*: Maxwell type interactions, *right*: hard-sphere interactions. (See Plate 17 on page 309)

If considering (22) for isotropic interactions with $\beta = 1, J_\beta = 1$ and $B(|u|, \mu) = \frac{1}{4\pi}$, the second term is the linear collision integral which conserves just the density and not the momentum and energy. Whereas the first (classical) collision integral from (22) conserves density, momentum and energy. The distribution $M(v)$ in (22) refers to the Maxwellian defined by $M_{\mathcal{T}}(v) = e^{\frac{-|v|^2}{(2\mathcal{T})}} \frac{1}{(2\pi\mathcal{T})^{3/2}}$, where $\mathcal{T}$ is the fixed background kinetic temperature. As the model indicates, any initial distribution function would converge to the background distribution $M_{\mathcal{T}}$ and this behavior is well captured by the numerical method. Indeed, in Fig. 8, a convex combination of two Maxwellians is given as the initial condition for this model simulation. In addition, it was shown in the above cited references that there is a dynamically scaled (self-similar) solution

$$f_{\mathcal{T}}^{ss}(v,t) = \frac{\sqrt(2)}{\pi^{5/2}} \int_0^\infty \frac{1}{(1+s^2)^2} \frac{e^{-|v|^2/2\bar{\mathcal{T}}}}{\bar{\mathcal{T}}^{\frac{3}{2}}} ds \tag{49}$$

where

$$\bar{\mathcal{T}} = \mathcal{T} + as^2 e^{\frac{-2t}{3}},$$

with finite energy for the particular choice of parameters $\mu = \frac{2}{3}, \theta = \frac{4}{3}$ modeling the energy dissipation rate and coupling constant respectively. The numerical distribution has been compared with the analytical solution $f_{\mathcal{T}}^{ss}$, given in (4.2) as $t \to \infty$ for a unit background temperature. For a positive background temperature $\mathcal{T}$, the solution to the slow down process problem converges to a Maxwellian $M_{\mathcal{T}}$. From Fig. 8, it can be seen that the numerical method is quite accurate and the computed distribution is in very good agreement with the analytical self-similar distribution $f_{\mathcal{T}}^{ss}$ from (4.2). Similar agreement

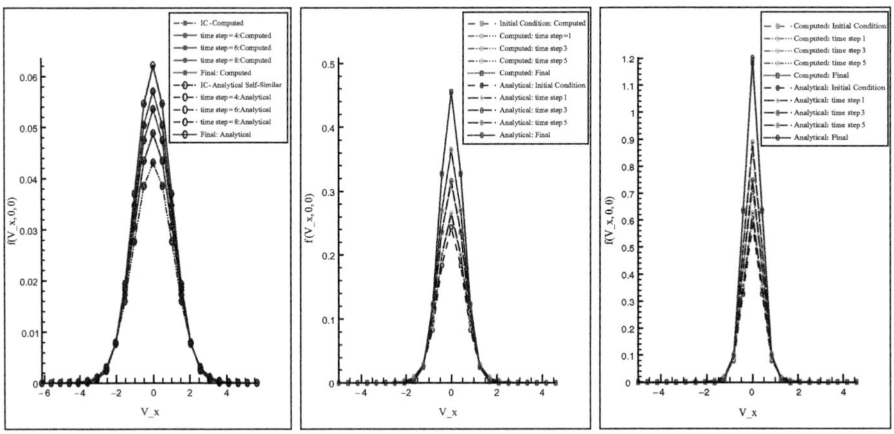

$\mathcal{T} = 1$ $\mathcal{T} = 0.25$ $\mathcal{T} = 0.125$

Fig. 8. Maxwell type collisions, slow down process with $\theta = 4/3, \mu = 2/3, N = 24$ evolving computed $f(v,t)$ vs. $f_{\mathcal{T}}^{ss}$ at different values of $\mathcal{T}$. (See Plate 18 on page 309)

has been observed for different constant values of $\mathcal{T}$ approaching 0 (Fig. 8). In addition interesting asymptotics occur when the background temperature $\mathcal{T} \to 0$ at the same rate of the decay of the kinetic energy. The corresponding self-similar solutions have power-like high energy tails and infinitely many particles at zero energies which occur only when $\mathcal{T} = 0$ as shown in [11]. For the numerical implementation, there needs to be a mechanism designed in order to achieve it. Since, letting $\mathcal{T} = 0$ in the numerical method created an instability, a new methodology was devised to counter for this effect: Instead of letting $\mathcal{T} = 0$, one can let $\mathcal{T} = \zeta e^{-\alpha t}$ thus ensuring that the thermostat temperature $\to 0$. So, taking

$$\bar{T} = \zeta e^{-\alpha t} + as^2 e^{\frac{-2t}{3}} . \qquad (50)$$

In our simulations, we take $\zeta = 0.25$ and the value of α need to be chosen exactly as $\alpha = \mu(1) = 2/3$, the energy dissipation rate, in order to recover the self-similar asymptotics corresponding to $\mathcal{T} = 0$. In Fig. 9 we compare the evolution of the computed pdf $f(v,t)$ for initial states that approach the self-similar solutions, both for positive $\mathcal{T}$ and for $\mathcal{T} = 0.25e^{-2t/3}$, where 2/3 is the energy dissipation rate for the slow down model under consideration.

Finally we mention some comments related to the ability to better capture the power-like behavior using the numerical method, as the background temperature $\mathcal{T} \to 0$ as the energy decay rate ($\mathcal{T} = e^{-2t/3} = e^{-\mu t}$). In fact (see [11,13]), the parameter μ is related the spectral properties of the Fourier transformed equation and it has been shown that, for this particular case of a slow down process model for $\mathcal{T} = 0$, the corresponding self-similar solutions can not have finite moments $m_q = \int f_{ss}|v|^{2q}dv$ for any $q \geq 1.5$. In particular

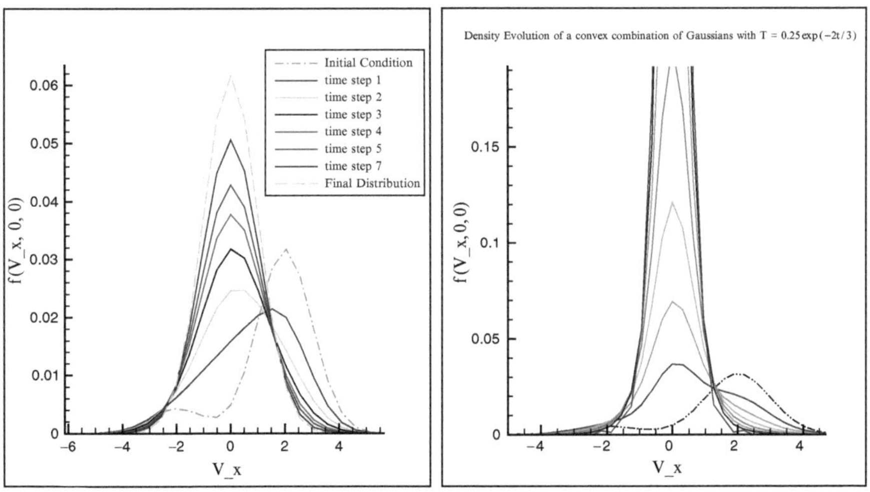

Fig. 9. Maxwell type collisions slow down process. Evolving of the pdf $f(v,t)$. *Left:* $\mathcal{T} = 1$, $N = 24$, *right* $\mathcal{T} = 0.25e^{-2t/3}$, $N = 32$. (See Plate 19 on page 310)

one obtains, for the dynamically scaled solution $f(v,t)$ the asymptotics

$$e^{-2/3tq} \int_{v\in\mathbb{R}^3} f(v)|v|^{2q} dv \to m_q \,.$$

In particular, one know that the quantities $e^{-2/3tq} \int_{v\in\mathbb{R}^3} f(v)|v|^{2q} dv$ will become unbounded for large time t. In order to illustrate this effect, we looked at different values of $q = 1, 1.3, 1.45, 1.5, 1.55, 1.7, 2.0$, m_q, which has been plotted for different values of $N = 10, 14, 16, 18, 22, 26$. Even though, according to the theory $m_q : q = 1.5, 1.55, 1.7, 2$ is supposed to blow-up, numerically it is seen that as the value of N increases, the blow-up point in $m_q : q = 1.5, 1.55, 1.7, 2$ shifts to the right (as expected) i.e., the blow-up occurs for larger final time. Because of the fact that an FFTW package has been used and the velocity domain is truncated which results in the distribution function taking small negative values for large velocities, there are numerical errors associated with the FFT method. In particular we believe this is what causes the $m_q : q = 1.5, 1.55, 1.7, 2$ to peak and then relax back to the original value as given in Fig. 10. Numerically, as the velocity domain is truncated and only a finite number of Fourier modes (N) are used to approximate the function $f(t,v)$, then the solution has oscillations which make the distribution function negative at large times. As m_q requires to take larger order moments of the computed self-similar asymptotics, the negative parts of oscillations result in the large negative moment values for the above mentioned values of N. In these cases, the negative oscillation values of $f(t,v)$ coincide with large velocity values used in getting m_q. As can be seen from Fig. 10, as time progresses (and as the background temperature $\mathcal{T}$ decreases to 0), the moments

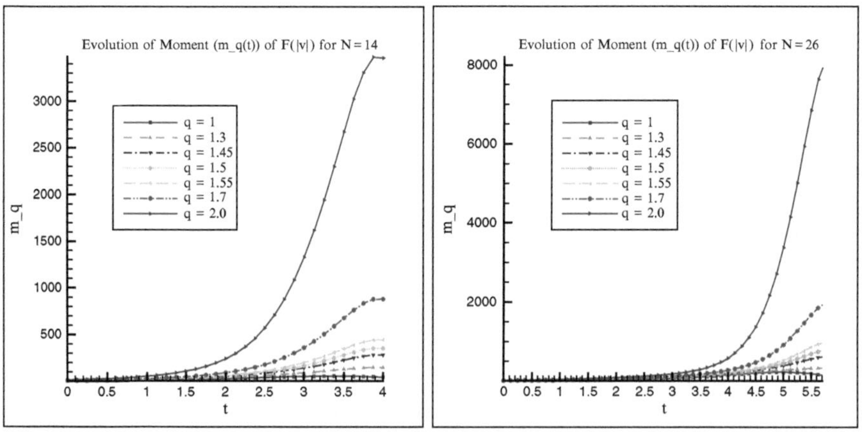

$N = 14$ $N = 26$

Fig. 10. Rescaled moments evolutions $m_q(t)$ for $T = e^{-2t/3}$. (See Plate 20 on page 310)

$m_q, q \geq 1.5$ start to blow up as predicted. The value $q = 1.5$ is the threshold value, as any moment $m_{q>1.5}(t) \rightarrow \infty$ as $t \rightarrow \infty$ and such a behavior is clearly exhibited by the computed distribution function. Figure 10 indicates the spectral accuracy of the numerical method since the analytically expected behavior (blow up) in $m_q(t)$ is replicated by the computed distribution with increasing values of N, the number of Fourier modes.

5 Conclusions and Future Work

The Spectral-Lagrangian numerical method works for elastic and inelastic variable hard potential interactions. This is first of its kind as no additional modification is required to compute for elastic and inelastic collisions. Comparison with known analytical special solutions for a suitable choice of parameters (moment equations for elastic BTE, BKW self-similar solution, attracting Bobylev–Cercignani–Gamba [11] self-similar solutions for elastic collisions in a slow down process), allows for an accurate benchmarking on the proposed scheme and subsequent simulations, since the computed vs the explicit ones are found to be very close. The numerical method employs a Fast Fourier Transform for faster evaluation of the collision integral. Even though the method has been implemented for a uniform grid in velocity space, it can even be implemented for a non-uniform velocity grid. The only challenge in this case is computing the Fast Fourier Transform on such a non-uniform grid. The integration over the unit sphere is avoided completely and only a simple integration over a regular velocity grid is needed. Even though a trapezoidal rule has been used as an integration rule, better integration rules like a gaussian quadrature can be used as well to get better accuracy. For time discretization, a simple second-order Runge Kutta scheme is used. The proposed method has a big advantage over other non-deterministic methods as the exact distribution function can actually be computed instead of just the averages.

The convergence and error results of the Fourier Transform method are being looked at. Implementation of the space inhomogeneous case is underway for simulation of elastic gases in 2-D in space such as the Couette flow problem posed in cylindrical symmetry in physical space.

Acknowledgments

Both authors are partially supported under the NSF grant DMS-0507038. Support from the Institute of Computational Engineering and Sciences and the University of Texas Austin is also gratefully acknowledged.

References

1. R. J. Alonso, I. M. Gamba, Propagation of l^1 and l^∞ maxwellian weighted bounds for derivatives of solutions to the homogeneous elastic Boltzmann equation, submitted for publication. Journal de Mathematiques Pures et Appliquees, to appear (2008).
2. G. A. Bird, Molecular Gas Dynamics, Clarendon Press, Oxford, 1994.
3. A. V. Bobylev, Exact solutions of the Boltzmann equation, (Russian) Dokl. Akad. Nauk SSSR 225 (1975) 1296–1299.
4. A. V. Bobylev, Exact solutions of the nonlinear Boltzmann equation and the theory of relaxation of a maxwellian gas, Translated from Teoreticheskaya i Mathematicheskaya Fizika 60 (1984) 280–310.
5. A. V. Bobylev, The theory of the nonlinear spatially uniform Boltzmann equation for maxwell molecules, Soviet Sci. Rev. Sect. C Math. Phys. Rev. 7 (1988) 111–233.
6. A. V. Bobylev, J. A. Carrillo, I. M. Gamba, On some properties of kinetic and hydrodynamic equations for inelastic interactions, J. Stat. Phys. 98 (2000) 743–773.
7. A. V. Bobylev, C. Cercignani, Discrete velocity models without nonphysical invariants, J. Stat. Phys. 97 (1999) 677–686.
8. A. V. Bobylev, C. Cercignani, Exact eternal solutions of the Boltzmann equation, J. Stat. Phys. 106 (2002) 1019–1038.
9. A. V. Bobylev, C. Cercignani, Moment equations for a granular material in a thermal bath, J. Stat. Phys. 106 (2002) 547–567(21).
10. A. V. Bobylev, C. Cercignani, Self-similar asymptotics for the Boltzmann equation with inleastic and elastic interactions, J. Stat. Phys. 110 (2003) 333–375.
11. A. V. Bobylev, C. Cercignani, I. M. Gamba, On the self-similar asymptotics for generalized non-linear kinetic maxwell models, Commun. Math. Phys. (to appear) URL `http://arxiv.org/abs/math-ph/0608035`
12. A. V. Bobylev, C. Cercignani, G. Toscani, Proof of an asymptotic property of self-similar solutions of the Boltzmann equation for granular materials, J. Stat. Phys. 111 (2003) 403–417.
13. A. V. Bobylev, I. M. Gamba, Boltzmann equations for mixtures of maxwell gases: Exact solutions and power like tails, J. Stat. Phys. 124 (2006) 497–516.
14. A. V. Bobylev, I. M. Gamba, V. Panferov, Moment inequalities and high-energy tails for Boltzmann equations with inelastic interactions, J. Stat. Phys. 116 (2004) 1651–1682.
15. A. V. Bobylev, M. Groppi, G. Spiga, Approximate solutions to the problem of stationary shear flow of smooth granular materials, Eur. J. Mech. B Fluids 21 (2002) 91–103.
16. A. V. Bobylev, S. Rjasanow, Difference scheme for the Boltzmann equation based on the fast fourier transform, Eur. J. Mech. B Fluids 16:22 (1997) 293–306.
17. A. V. Bobylev, S. Rjasanow, Fast deterministic method of solving the Boltzmann equation for hard spheres, Eur. J. Mech. B Fluids 18:55 (1999) 869–887.
18. A. V. Bobylev, S. Rjasanow, Numerical solution of the Boltzmann equation using fully conservative difference scheme based on the fast fourier transform, Trans. Theory Stat. Phys. 29 (2000) 289–310.
19. J. E. Broadwell, Study of rarefied shear flow by the discrete velocity method, J. Fluid Mech. 19 (1964) 401–414.

20. H. Cabannes, Global solution of the initial value problem for the discrete Boltzmann equation, Arch. Mech. (Arch. Mech. Stos.) 30 (1978) 359–366.
21. J. A. Carrillo, I. M. Gamba, A. Majorana, C.-W. Shu, A weno-solver transients of Boltzmann-poisson system for semiconductor devices. performance and monte carlo comparisons, J. Comput. Phys. 184 (2003) 498-525.
22. J. A. Carrillo, I. M. Gamba, A. Majorana, C.-W. Shu, 2d non-stationary Boltzmann-poisson systems: Dsmc versus a weno-Boltzmann scheme, J. Comput. Phys. 214 (2006) 55–80.
23. C. Cercignani, Shear flow of a granular material, J. Stat. Phys. 102 (2001) 1407–1415.
24. C. Cercignani, H. Cornille, Shock waves for a discrete velocity gas mixture, J. Stat. Phys. 99 (2000) 115–140.
25. Y. Cheng, I. M. Gamba, A. Majorana, C.-W. Shu, Discontinuous galerkin solver for Boltzmann-poisson transients, J. Comput. Elect. (to appear)
26. Y. Cheng, I. M. Gamba, A. Majorana, C.-W. Shu, Discontinuous galerkin solver for the semiconductor Boltzmann equation, Simul. Semicond. Proc. Devices 12 (2007) 257–260.
27. M. H. Ernst, R. Brito, Driven inelastic maxwell models with high energy tails, Phys. Rev. E 65(4) (2002) 040301.
28. M. H. Ernst, R. Brito, Scaling solutions of inelastic Boltzmann equations with over-populated high energy tails, J. Stat. Phys. 109 (2002) 407–432.
29. F. Filbet, C. Mouhot, L. Pareschi, Solving the Boltzmann equation in nlogn, SIAM J. Sci. Comput. 28 (2006) 1029–1053.
30. F. Filbet, G. Russo, High order numerical methods for the space non homogeneous Boltzmann equation, J. Comput. Phys. 186 (2003) 457–480. URL `citeseer.ist.psu.edu/filbet03high.html`
31. F. Filbet, G. Russo, High order numerical methods for the space non homogeneous Boltzmann equation, J. Comput. Phys. 186 (2003) 457–480.
32. N. Fournier, S. Mischler, A Boltzmann equation for elastic, inelastic and coalescing collisions, J. Math. Pures Appl. 84 (2005) 1173–1234.
33. M. Frigo, S. G. Johnson, Fast fourier transform of the west. URL `www.fftw.org`
34. E. Gabetta, L. Pareschi, G. Toscani, Relaxation schemes for nonlinear kinetic equations, SIAM J. Num. Anal. 34 (1997) 2168–2194.
35. I. M. Gamba, V. Panferov, C. Villani, Upper maxwellian bounds for the spatially homogeneous Boltzmann equation, submitted for publication. URL `http://arxiv.org/abs/math/0701081`
36. I. M. Gamba, V. Panferov, C. Villani, On the Boltzmann equation for diffusively excited granular media, Commun. Math. Phys. 246 (2004) 503–541(39).
37. I. M. Gamba, S. Rjasanow, W. Wagner, Direct simulation of the uniformly heated granular Boltzmann equation, Math. Comput. Model. 42 (2005) 683–700.
38. I. M. Gamba, S. H. Tharkabhushanam, Spectral-lagrangian methods for collisional models of non-equilibrium statistical states, Journal of Computational Physics, to appear.
39. M. Herty, L. Pareschi, M. Seaid, Discrete-velocity models and relaxation schemes for traffic flows, SIAM J. Sci. Comput. 28 (2006) 1582–1596.
40. I. Ibragimov, S. Rjasanow, Numerical solution of the Boltzmann equation on the uniform grid, Computing 69 (2002) 163–186.
41. R. Illner, On the derivation of the time-dependent equations of motion for an ideal gas with discrete velocity distribution, J. de Mecanique 17 (1978) 781–796.

42. S. Kawashima, Global solution of the initial value problem for a discrete velocity model of the Boltzmann equation, Proc. Jpn. Acad. Ser. A Math. Sci. 57 (1981) 19–24.
43. K. Max, W. T. Tsun, Formation of maxwellian tails, Phys. Rev. Lett. 36 (1976) 1107–1109.
44. L. Mieussens, Discrete-velocity models and numerical schemes for the Boltzmann-bgk equation in plane and axisymmetric geometries, J. Comput. Phys. 162 (2000) 429–466.
45. J. M. Montanero, A. Santos, Computer simulation of uniformly heated granular fluids, Gran. Matt. 2 (2000) 53–64.
46. S. J. Moon, M. D. Shattuck, J. Swift, Velocity distributions and correlations in homogeneously heated granular media, Phys. Rev. E 64 (2001) 031303.
47. C. Mouhot, L. Pareschi, Fast algorithms for computing the Boltzmann collision operator, Math. Comput. 75 (2006) 1833–1852.
48. K. Nanbu, Direct simulation scheme derived from the Boltzmann equation i.monocomponent gases, J. Phys. Soc. Jpn. 52 (1983) 2042–2049.
49. T. V. Noije, M. Ernst, Velocity distributions in homogeneously cooling and heated granular fluids, Gran. Matt. 1:57(1998).
50. V. Panferov, S. Rjasanow, Deterministic approximation of the inelastic Boltzmann equation, Unpublished manuscript.
51. L. Pareschi, B. Perthame, A fourier spectral method for homogenous Boltzmann equations, Trans. Theory Stat. Phys. 25 (2002) 369–382.
52. L. Pareschi, G. Russo, Numerical solution of the Boltzmann equation. i. spectrally accurate approximation of the collision operator, SIAM J. Num. Anal. (online) 37 (2000) 1217–1245.
53. S. Rjasanow, W. Wagner, A stochastic weighted particle method for the Boltzmann equation, J. Comput. Phys. (1996) 243–253.
54. S. Rjasanow, W. Wagner, Stochastic Numerics for the Boltzmann Equation, Springer, Berlin Heidelberg New York, 2005.
55. C. Villani, Handbook of Fluid dynamics, chap. A Review of Mathematical Topics in Collisional Kinetic Theory, Elsevier, Amsterdam, 2003, pp. 71–306. URL `http://www.umpa.ens-lyon.fr/ cvillani`
56. W. Wagner, A convergence proof for bird's direct simulation monte carlo method for the Boltzmann equation, J. Stat. Phys. (1992) 1011–1044.
57. D. R. M. Williams, F. MacKintosh, Driven granular media in one dimension: Correlations and equation of state, Phys. Rev. E 54 (1996) 9–12.
58. Y. Zheng, H. Struchtrup, A linearization of mieussens's discrete velocity model for kinetic equations, Eur. J. Mech. B Fluids 26 (2007) 182–192.

The Art of Analytical Benchmarking

B.D. Ganapol and R. Furfaro

Summary. The most primitive numerical method of solving the radiative transfer equation forms the basis of an algorithm to generate high quality numerical solutions. We apply convergence acceleration to the discrete ordinates balance equation derived from a diamond difference approximation. Wynn-epsilon and Romberg accelerations then rearrange low order solutions to produce a high order solution. Published semi-analytical benchmarks from the C.E. Siewert portfolio establish the credibility of the new converged discrete ordinates method (CDOM).

1 Introduction

Given the complex nature of solving the radiative transfer equation for engineering applications, we seldom consider analytical methods as our first choice for establishing highly accurate solutions. Couple this with current powerful and expansive computational resources, the "dumbing down" of engineering curricula, and the increasing number of activities competing for our time and attention, analytical thought will soon become outdated. While this may be somewhat of an over exaggeration, unfortunately, there is some truth to this statement. Thus, primarily because of the intellectual overhead associated with developing analytical solutions, we desire an alternative approach to establishing highly accurate solutions to the equations of radiative transfer. The primary intent of this presentation is to alert the transport community to the notion that increased computational resources and the demands on our time also enable the creation of innovative numerical algorithms. In particular, we consider a case study in radiative transfer to demonstrate the art of convergence acceleration as a method to generate benchmarks for the radiative transfer equation.

A suitable development of numerical methods to solve the transport equation requires the method's developer to deliver a reliable product. In particular, the method performs as intended, is dependable, robust and failsafe. Here, failsafe refers to performance indicators of proper as well as improper operation. More importantly, today's user community holds the code developer and implementation team accountable for product performance. Code developers should therefore follow a documented path of certification to ensure the highest level of confidence. The certification process should reveal code limitations and establish suitable code performance through review and audit of the software design, algorithm verification and model validation.

This presentation is concerned with code verification as applied to the radiative transfer equation. Here, verification refers to confirmation that a numerical algorithm is operating properly at an acceptable level of accuracy. Obviously, full verification is impossible, since, if possible, there would be no need for the algorithm in the first place. Hence, the practical objective of verification is to provide an acceptable level of accuracy and code limitations with regard to an application. Accepted modalities of code verification for radiative transfer applications include, monitoring particle conservation, comparison to physical and numerical intuition, similar model and method comparisons, creation of manufactured solutions and benchmarks or standards. We will be concerned with establishing purely numerical benchmarks rather than traditional semi-analytical benchmarks. The difference being a semi-analytical benchmark begins with an analytical solution representation; while, a numerical benchmark begins with a discretized formulation. Specifically, we demonstrate how benchmark quality is indeed possible from the fully discretized discrete ordinates balance equation in space and angle for a variety of radiative transfer problems. In addition, the proposed method, in principle, will enable transport solutions to any desired accuracy, thus eliminating the uncertainty associated with the conventional discrete ordinates method (DOM).

We exploit a little known solution concept in transport theory. The concept is to extrapolate low order solutions to give a higher order solution. We use the Wynn-epsilon extrapolation for the angular discretization; while, Romberg and Wynn-epsilon extrapolations apply to the spatial discretization. In essence, the discrete ordinates solutions are "mined" to extract higher order solutions. In addition, we apply the Wynn-epsilon acceleration and course mesh rebalance to the inner iteration to accelerate convergence for relatively thick media. Effectively, with extrapolation, we redefine the solution to the radiative transfer equation in discrete ordinates form as the acceleration of sequence solutions to their limit.

As presented in the next section, plane geometry, with one angle (inclination) transport, forms the theoretical setting considered. We follow this by the numerical diamond difference discrete ordinates formulation to obtain the discrete ordinates balance equation. The final section presents evaluations of a series of previously established radiative transfer benchmarks and demonstrates how the converged discrete ordinates method (CDOM) can indeed provide high quality benchmark results.

2 The Theoretical Setting

We begin with the 1D radiative transfer equation in a heterogeneous medium with anisotropic scattering and a prescribed temperature distribution

$$\left[\mu\frac{\partial}{\partial x}+\sigma\left(x\right)\right]I\left(x,\mu\right) = \sigma_s\left(x\right)\int_{-1}^{1}d\mu' f\left(\mu',\mu\right)I\left(x,\mu'\right)+\sigma\left(x\right)\left[1-\omega\left(x\right)\right]\frac{\sigma_{sb}T\left(x\right)^4}{\pi}, \tag{1a}$$

where the single scatter albedo is

$$\omega\left(x\right)\equiv\frac{\sigma_s\left(x\right)}{\sigma\left(x\right)},$$

and σ_{sb} is the Stefan–Boltzmann constant. The total width of the participating medium is a, and $[a_i, a_{i+1}]$ denote the material boundaries for R material regions. The outer surface conditions at $x=0$ and a are for $\mu>0$

$$I\left(0,\mu\right)=g_1\left(\mu\right)+\varepsilon_1\left(\frac{\sigma_{sb}T_1^4}{\pi}\right)+\rho_s I\left(0,-\mu\right)+2\rho_d\int_0^1 d\mu'\mu' I\left(0,-\mu'\right) \tag{1b}$$

$$I\left(a,-\mu\right)= g_2\left(\mu\right)+\varepsilon_2\left(\frac{\sigma_{sb}T_2^4}{\pi}\right)+\rho_s I\left(a,\mu\right)+2\rho_d\int_0^1 d\mu'\mu' I\left(a,\mu'\right), \tag{1c}$$

where ρ_s, ρ_d are the specular and diffuse reflectances, and $\varepsilon_1, \varepsilon_2$ are the surface emmisivities. In the usual way, by defining the optical depth

$$\tau\equiv\int_0^x dx'\sigma\left(x'\right),$$

Equation (1a) becomes

$$\left[\mu\frac{\partial}{\partial\tau}+1\right]I\left(\tau,\mu\right)=\omega\left(\tau\right)\int_{-1}^{1}d\mu' f\left(\mu',\mu\right)I\left(\tau,\mu'\right)+S\left(\tau\right), \tag{2a}$$

where

$$I\left(\tau,\mu\right)\equiv I\left(x\left(\tau\right),\mu\right)$$
$$S\left(\tau\right)\equiv\left[1-\omega\left(\tau\right)\right]\frac{\sigma_{sb}T\left(\tau\right)^4}{\pi};$$

and is subject to the following surface boundary conditions for $\mu > 0$:

$$I(0,\mu) = g_1(\mu) + \varepsilon_1\left(\frac{\sigma_{sb}T_1^4}{\pi}\right) + \rho_s I(0,-\mu) + 2\rho_d \int_0^1 d\mu'\mu' I(0,-\mu') \quad (2b)$$

$$I(\Delta,-\mu) = g_2(\mu) + \varepsilon_2\left(\frac{\sigma_{sb}T_2^4}{\pi}\right) + \rho_s I(\Delta,\mu) + 2\rho_d \int_0^1 d\mu'\mu' I(\Delta,\mu'). \quad (2c)$$

The total optical depth is

$$\Delta \equiv \int_0^a dx'\sigma(x').$$

In general, we assume the source at $x = 0$ to be composed of singular and regular contributions

$$g_1(\mu) \equiv q_{s0}\delta(\mu - \mu_0) + s_1(\mu).$$

To treat the singular component, we partition the flux into corresponding uncollided and collided contributions

$$I(\tau,\mu) = I_{s0}(\tau,\mu) + I_c(\tau,\mu),$$

where the uncollided component satisfies

$$\left[\mu\frac{\partial}{\partial\tau} + 1\right] I_{s0}(\tau,\mu) = 0, \quad (3a)$$

with conditions for $\mu > 0$

$$I_{s0}(0,\mu) = q_{s0}\delta(\mu - \mu_0) + \rho_s I_{s0}(0,-\mu) \quad (3b)$$

$$I_{s0}(\Delta,-\mu) = \rho_s I_{s0}(\Delta,\mu). \quad (3c)$$

In addition, the "collided" component satisfies

$$\left[\mu\frac{\partial}{\partial\tau} + 1\right] I_c(\tau,\mu) = \left\{\begin{array}{l} \omega(\tau)\int_{-1}^{1} d\mu' f(\mu',\mu) I_c(\tau,\mu') \\ +\omega(\tau)\int_{-1}^{1} d\mu' f(\mu',\mu) I_{s0}(\tau,\mu') \end{array}\right\} + S(\tau) \quad (4a)$$

and for $\mu > 0$

$$I_c(0,\mu) = s_1(\mu) + \varepsilon_1\left(\frac{\sigma_{sb}T_1^4}{\pi}\right) + \rho_s I_c(0,-\mu) + 2\rho_d \left\{\begin{array}{l} \int_0^1 d\mu'\mu' I_{s0}(0,-\mu') \\ + \int_0^1 d\mu'\mu' I_c(0,-\mu') \end{array}\right\} \quad (4b)$$

$$I_c(\Delta,-\mu)=\varepsilon_2\left(\frac{\sigma_{sb}T_2^4}{\pi}\right)+\rho_s I_c(\Delta,\mu)+2\rho_d\left\{\begin{array}{l}\int_0^1 d\mu'\mu' I_{s0}(\Delta,\mu')\\+\int_0^1 d\mu'\mu' I_c(\Delta,\mu')\end{array}\right\}. \tag{4c}$$

We solve (3) to give for $\mu>0$

$$\begin{aligned}I_{s0}(\tau,\mu)&=q_{s0}\frac{e^{-\tau/\mu_0}}{1-\rho_s^2e^{-2\Delta/\mu_0}}\delta(\mu-\mu_0)\\I_{s0}(\tau,-\mu)&=q_{s0}\frac{\rho_s e^{-(2\Delta-\tau)/\mu_0}}{1-\rho_s^2e^{-2\Delta/\mu_0}}\delta(\mu-\mu_0).\end{aligned} \tag{5}$$

With these expressions introduced into the scattering source for the collided component, (4a) becomes

$$\left[\mu\frac{\partial}{\partial\tau}+1\right]I_c(\tau,\mu)=\omega(\tau)\int_{-1}^{1}d\mu' f(\mu',\mu)I_c(\tau,\mu')+\omega(\tau)S_{s0}(\tau,\mu)+S(\tau) \tag{6a}$$

with

$$S_{s0}(\tau,\mu)\equiv\frac{q_{s0}}{1-\rho_s^2e^{-2\Delta/\mu_0}}\left[f(\mu_0,\mu)e^{-\tau/\mu_0}+\rho_s f(-\mu_0,\mu)e^{-(2\Delta-\tau)/\mu_0}\right],$$

and the associated boundary conditions are ($\mu>0$)

$$\begin{aligned}I_c(0,\mu)&=s(\mu)+\varepsilon_1\left(\frac{\sigma_{sb}T_1^4}{\pi}\right)+2\rho_d S_{s1}+\rho_s I_c(0,-\mu)\\&\quad+2\rho_d\int_0^1 d\mu'\mu' I_c(0,-\mu')\end{aligned} \tag{6b}$$

$$\begin{aligned}I_c(\Delta,-\mu)&=g_2(\mu)+\varepsilon_2\left(\frac{\sigma_{sb}T_2^4}{\pi}\right)+2\rho_d S_{s2}+\rho_s I_c(\Delta,\mu)\\&\quad+2\rho_d\int_0^1 d\mu'\mu' I_c(\Delta,\mu'),\end{aligned} \tag{6c}$$

where

$$\begin{aligned}S_{s1}&\equiv q_{s0}\frac{\mu_0}{1-\rho_s^2e^{-2\Delta/\mu_0}}\\S_{s2}&\equiv q_{s0}\frac{\mu_0\rho_s e^{-2\Delta/\mu_0}}{1-\rho_s^2e^{-2\Delta/\mu_0}}.\end{aligned}$$

Equations (6) are now the equations to which we apply the CDOM.

3 The Numerical Setting

3.1 Error Terms

Spatial Discretization Error

We begin with the discretization of the optical depth by partitioning the medium into n intervals $[\tau_j, \tau_{j+1}]$ as shown Fig. 1. The discretization is such that each material region contains an integral number of intervals. In the jth interval, the center is $\tau_{j+1/2}$. The interval width is

$$h_j \equiv \tau_{j+1} - \tau_j = \sigma_j h_x \tag{7}$$

where $h_x \equiv x_{j+1} - x_j$ is the actual spatial interval width. Through a change of variable, we can normalize each region width to 1. Thus, for N spatial intervals in each region, we have

$$h_x = \frac{1}{N}. \tag{8}$$

To arrive at the discrete ordinate balance equations, we integrate (6a) over interval $[h_j, h_{j+1}]$ to give the (exact) balance equation

$$\mu\left[I_c\left(\tau_{j+1},\mu\right) - I_c\left(\tau_j,\mu\right)\right] + \int\limits_{h_j} d\tau I_c\left(\tau,\mu\right) = \int\limits_{h_j}^{d\tau} Q_j\left(\tau,\mu\right) + \omega_j \int\limits_{h_j} d\tau S_0\left(\tau,\mu\right) \tag{9a}$$

where

$$Q_j\left(\tau,\mu\right) = \omega_j Q_{cs}\left(\tau,\mu\right) + S\left(\tau\right) \tag{9b}$$

$$Q_{cs}\left(\tau,\mu\right) = \int\limits_{-1}^{1} d\mu' f\left(\mu',\mu\right) I_c\left(\tau,\mu'\right). \tag{9c}$$

We perform the integral over $S_0\left(\tau,\mu\right)$ analytically

$$\int\limits_{h_j} d\tau S_0\left(\tau,\mu\right) \equiv S_{0,j}\left(\mu\right) =$$

$$= q_{s0}\frac{\mu_0}{1-\rho_s^2 e^{-2\Delta/\mu_0}}\left\{\begin{array}{l} f\left(\mu_0,\mu\right)\left[e^{-\tau_j/\mu_0} - e^{-\tau_{j+1}/\mu_0}\right] \\ +\rho_s f\left(-\mu_0,\mu\right)\left[e^{-(2\Delta-\tau_{j+1})/\mu_0} - e^{-(2\Delta-\tau_j)/\mu_0}\right] \end{array}\right\}. \tag{9d}$$

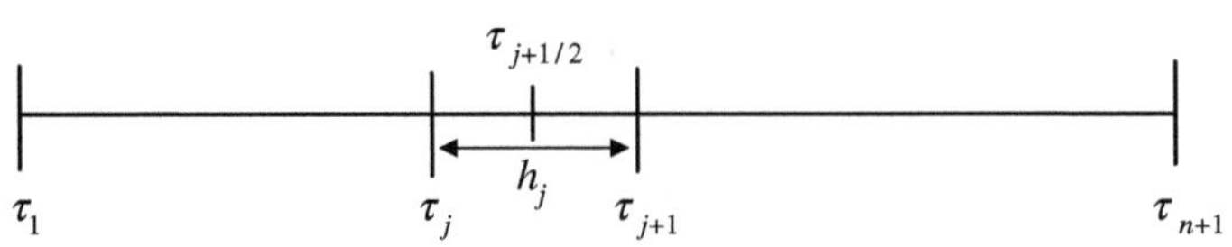

Fig. 1. Discretization of the optical depth

Closure of the exact balance equation requires that we relate the integral of the intensity over the interval somehow to the interval edge intensities $I_{c,j}, I_{c,j+1}$. The desired relationship, called the diamond difference, results from the following series of manipulations. First, we assume a Taylor series in interval j

$$I_c(\tau,\mu) = \sum_{n=0}^{\infty} \frac{I_c^{(n)}\left(\tau_{j+1/2},\mu\right)}{n!}\left(\tau - \tau_{j+1/2}\right)^n \tag{10}$$

which is relative to the midpoint of the jth interval. Upon integration, there results

$$\int_{h_j} d\tau I_c(\tau,\mu) = h_j I_c\left(\tau_{j+1/2},\mu\right) + \sum_{k=1}^{\infty} \frac{I_c^{(2k)}\left(\tau_{j+1/2},\mu\right)}{(2k+1)!}\frac{h_j^{2k+1}}{2^{2k}}. \tag{11}$$

Next, we require an interpretation of the first term in terms of the edge intensities. By specifying the Taylor series representation at the cell edges

$$I_c(\tau_{j+1},\mu) = \sum_{n=0}^{\infty} \frac{I_c^{(n)}\left(\tau_{j+1/2},\mu\right)}{n!}\frac{h_j^n}{2^n} \tag{12a}$$

$$I_c(\tau_j,\mu) = \sum_{n=0}^{\infty} (-1)^n \frac{I_c^{(n)}\left(\tau_{j+1/2},\mu\right)}{n!}\frac{h_j^n}{2^n}, \tag{12b}$$

and adding, we have

$$I_c(\tau_{j+1},\mu) + I_c(\tau_j,\mu) = 2I_c\left(\tau_{j+1/2},\mu\right) + \sum_{k=1}^{\infty} \frac{I_c^{(2k)}\left(\tau_{j+1/2},\mu\right)}{(2k)!}\frac{h_j^{2k}}{2^{2k}}. \tag{13}$$

Upon solving for $I_c\left(\tau_{j+1/2},\mu\right)$ from the last equation and substituting into (11), we have the following (exact) relation between the average and edge fluxes

$$\int_{h_j} d\tau I_c(\tau,\mu) = \frac{h_j}{2}\left[I_c(\tau_{j+1},\mu) + I_c(\tau_j,\mu)\right] - \sum_{k=1}^{\infty}\left[\frac{k}{2^{2k-1}(2k+1)!}\right] I_c^{(2k)}\left(\tau_{j+1/2},\mu\right) h_j^{2k+1} \tag{14}$$

complete with error term. When the summation is ignored, the diamond difference relation follows. Similarly,

$$\int_{h_j} d\tau Q_j(\tau,\mu) = \frac{\omega_j h_j}{2}\left[Q_{cs}(\tau_{j+1},\mu) + Q_{cs}(\tau_j,\mu)\right] + \frac{h_j}{2}\left[S(\tau_{j+1}) + S(\tau_j)\right] - \sum_{k=1}^{\infty} \frac{k}{2^{2k-1}(2k+1)!}\left[\omega_j Q_{cs}^{(2k)}\left(\tau_{j+1/2},\mu\right) + S^{(2k)}\left(\tau_{j+1/2}\right)\right] h_j^{2k+1} \tag{15}$$

where

$$Q_{cs}(\tau,\mu)=\sum_{n=0}^{\infty}\frac{Q_{cs}^{(n)}\left(\tau_{j+1/2},\mu\right)}{n!}\left(\tau-\tau_{j+1/2}\right)^{n}$$

and

$$S(\tau)=\sum_{n=0}^{\infty}\frac{S^{(n)}\left(\tau_{j+1/2}\right)}{n!}\left(\tau-\tau_{j+1/2}\right)^{n}.$$

Upon substitution of (14) and (15) into (9a), the exact balance equation becomes

$$\begin{aligned}\left[\mu+\frac{h_j}{2}\right]I_c\left(\tau_{j+1},\mu\right)-\left[\mu-\frac{h_j}{2}\right]I_c\left(\tau_j,\mu\right)=&\frac{\omega_j h_j}{2}\left[Q_{cs}\left(\tau_{j+1},\mu\right)+Q_{cs}\left(\tau_j,\mu\right)\right]\\&+\frac{h_j}{2}\left[S\left(\tau_{j+1}\right)+S\left(\tau_j\right)\right]+\omega_j S_{0,j}(\mu)+E_{R,j}(\mu)\end{aligned}\tag{16a}$$

which remains exact with the error term

$$E_{R,j}\left(\tau_{j+1/2},\mu\right)\equiv\sum_{k=1}^{\infty}\left[\frac{k}{2^{2k-1}(2k+1)!}\right]\left[\begin{array}{l}I_c^{(2k)}\left(\tau_{j+1/2},\mu\right)-S^{(2k)}\left(\tau_{j+1/2}\right)\\-\omega_j Q_{cs}^{(2k)}\left(\tau_{j+1/2},\mu\right)\end{array}\right]h_j^{2k+1}.\tag{16b}$$

When the derivatives in this term are recast in terms of x rather than τ, the error can be expressed in terms of powers of h_x

$$E_{R,j}\left(\tau_{j+1/2},\mu\right)=\sigma_j\sum_{k=1}^{\infty}a_{j,k}(\mu)h_x^{2k+1}\tag{17}$$

where

$$a_{j,k}(\mu)\equiv\frac{k}{2^{2k-1}(2k+1)!}\left[I_c^{(2k)}\left(\tau_{j+1/2},\mu\right)-S^{(2k)}\left(\tau_{j+1/2}\right)-\omega_j Q_{cs}^{(2k)}\left(\tau_{j+1/2},\mu\right)\right].$$

$E_{R,j}$ represents the local spatial discretization error. This form is more convenient for theoretical error manipulations (not done here, however).

Angular Discretization Error

The local error associated with the angular discretization results from the quadrature approximation of the scattering source

$$Q_{cs}(\tau,\mu)=\int_0^1 d\mu' f(\mu',\mu)I_c(\tau,\mu')+\int_{-1}^0 d\mu' f(\mu',\mu)I_c(\tau,\mu'),$$

and evaluation through shifted Gauss/Legendre quadrature. Thus [1],

$$Q_{cs}(\tau,\mu) = \bar{Q}_{cs}(\tau,\mu) + E_{Q,N}(\tau,\mu) \tag{18a}$$

with

$$\bar{Q}_{cs}(\tau,\mu) \equiv \sum_{m'=1}^{2N} \varpi_{m'} f(\mu_{m'},\mu) I_c(\tau,\mu_{m'}) \tag{18b}$$

$$E_{Q,N}(\tau,\mu) \equiv \frac{2^{2N+1}(N!)^4}{(2N+1)[(2N)!]^3} \left\{ \begin{array}{l} [f(\mu',\mu) I_c(\tau,\mu')]^{(2N)}_{\mu'=\xi^+} \\ + [f(\mu',\mu) I_c(\tau,\mu')]^{(2N)}_{\mu'=\xi^-} \end{array} \right\} \tag{18c}$$

and

$$0 \le \xi^+ \le 1 \quad -1 \le \xi^- \le 0.$$

ϖ_m, μ_m are the weights and quadrature points of the shifted Gauss/Legendre quadrature.
When we include the quadrature approximation in (16a), there results

$$\left[\mu + \frac{h_j}{2}\right] I_c(\tau_{j+1},\mu) - \left[\mu - \frac{h_j}{2}\right] I_c(\tau_j,\mu) = \frac{\omega_j h_j}{2}\left[\bar{Q}_{cs}(\tau_{j+1},\mu) + \bar{Q}_{cs}(\tau_j,\mu)\right]$$
$$+ \frac{h_j}{2}[S(\tau_{j+1}) + S(\tau_j)] + \omega_j S_{0,j}(\mu) + E_{Tot,j}(\mu) \tag{19a}$$

where

$$E_{Tot,j}(\mu) \equiv E_{R,j}(\tau_{j+1/2},\mu) + \frac{\omega_j h_j}{2} E^{Q}_{j,N}(\mu) \tag{19b}$$

and

$$E^{Q}_{j,N}(\mu) \equiv E_{Q,N}(\tau_{j+1},\mu) + E_{Q,N}(\tau_j,\mu). \tag{19c}$$

The last expression represents the total local error from both spatial and angular discretization.

3.2 The Discrete Ordinate Balance Equation

We obtain the discrete ordinates balance equation from (19a), evaluated at a particular inclination μ_m, by ignoring the error term $E_{Tot,j}$

$$\begin{aligned} &\left[\mu_m + \frac{h_j}{2}\right] I_{c,m,j+1} - \left[\mu_m - \frac{h_j}{2}\right] I_{c,m,j} \\ &= \frac{\omega_j h_j}{2}\left[\bar{Q}_{cs,m,j+1} + \bar{Q}_{c,m,j}\right] + \frac{h_j}{2}[S_{j+1} + S_j] + \omega_j S_{0,m,j}, \end{aligned} \tag{20}$$

with

$$I_{c,m,j} \equiv I_c(\tau_j,\mu_m).$$

The most straightforward solution to this equation is via (inner) iteration, where we assume a known scattering source from the last iteration

$$\begin{aligned}&\left[\mu_m + \frac{h_j}{2}\right] I^k_{c,m,j+1} - \left[\mu_m - \frac{h_j}{2}\right] I^k_{c,m,j}\\&= \frac{\omega_j h_j}{2}\left[\bar{Q}^{k-1}_{cs,m,j+1} + \bar{Q}^{k-1}_{cs,m,j}\right] + \frac{h_j}{2}\left[S_{j+1} + S_j\right] + \omega_j S_{0,m,j}.\end{aligned} \tag{21}$$

Thus, in directions $0 \le \mu$

$$I^k_{c,m,j+1} = \gamma^+_j(\mu_m) I^k_{c,m,j} + \frac{W^{k-1}_j(\mu_m)}{\left[\mu_m + \frac{h_j}{2}\right]} \tag{22a}$$

with

$$\gamma^+_j(\mu_m) \equiv \frac{\left[\mu_m - \frac{h_j}{2}\right]}{\left[\mu_m + \frac{h_j}{2}\right]}$$

$$W^k_j(\mu_m) \equiv \frac{\omega_j h_j}{2}\left[\bar{Q}^k_{cs,j+1}(\mu_m) + \bar{Q}^k_{cs,j}(\mu_m)\right] + \frac{h_j}{2}\left[S_{j+1} + S_j\right] + \omega_j S_{0,m,j}$$

and in directions $\mu < 0$

$$I^k_{c,m,j} = \gamma^-_j(\mu_m) I^k_{c,m,j+1} - \frac{W^{k-1}_j(\mu_m)}{\left[\mu_m - \frac{h_j}{2}\right]} \tag{22b}$$

$$\gamma^-_j(\mu_m) \equiv \frac{\left[\mu_m + \frac{h_j}{2}\right]}{\left[\mu_m - \frac{h_j}{2}\right]}.$$

The DOM approximation is therefore

$$I_c(\tau_j, \mu_m) \cong I^{conv}_{c,m,j} \tag{23}$$

where I^{conv} is the converged angular intensity of the inner iteration sweeps.

Special Case $\mu = 0$

Often the case $\mu = 0$ is of interest and is most easily treated by setting $\mu = 0$ in (6a) to give

$$I_c(\tau, 0) = \omega(\tau)\int_{-1}^{1} d\mu' f(\mu', 0) I_c(\tau, \mu') + \omega(\tau) S_{s0}(\tau, 0) + S(\tau)$$

at any position $\tau > 0$. At τ_j, we have

$$I_{c,j}(0) = \omega_j \sum_{m'=1}^{2N} f(\mu_{m'}, 0) I_{c,m',j}^{conv} + \omega_j S_{s0,j}(0) + S_j. \tag{24}$$

after applying the quadrature approximation. Equation (24) avoids numerical instabilities.

Angular Intensity Interpolation

We can also view (21) as an interpolation formula to determine the intensity at any μ. To do this, one notes that (21) is a two-point recurrence with the RHS known from the last iteration. We can show that for any $\mu_m = \mu$ in $0 \leq \mu$ and $1 \leq j \leq n+1$

$$I_{c,j}(\mu) = \beta_j^+(\mu) I_{c,1}(\mu) + \sum_{l=1}^{j-1} \left[\frac{\beta_j^+(\mu) \beta_{l+1}^{+^{-1}}(\mu)}{a_l^+(\mu)} \right] W_l^{conv}(\mu) \tag{25a}$$

with

$$\beta_j^+(\mu) \equiv \prod_{l=1}^{j-1} \gamma_l^+(\mu)$$

$$a_j^{\pm} \equiv \mu \pm \frac{h_j}{2}$$

$$W_j^{conv}(\mu) \equiv \frac{\omega_j h_j}{2} \left[\bar{Q}_{cs,j+1}^{conv}(\mu) + \bar{Q}_{cs,j}^{conv}(\mu) \right] + \frac{h_j}{2} \left[S_{j+1} + S_j \right] + \omega_j S_{0,j}(\mu);$$

and similarly, for $\mu < 0$

$$I_{c,j}(\mu) = \beta_j^-(\mu) I_{c,n}(\mu) + \sum_{l=j}^{n-1} \left[\frac{\beta_j^-(\mu) \beta_{l+1}^{-^{-1}}(\mu)}{a_l^-(\mu)} \right] W_l^{conv}(\mu) \tag{25b}$$

with

$$\beta_j^-(\mu) \equiv \prod_{l=j}^{n-1} \gamma_l^-(\mu).$$

Thus, at any spatial position j, (25) give the intensity at any μ once the incoming intensity at either surface ($j = 1, n$) is known.

Iteration for the General Boundary Conditions

In this section, we consider how to determine the incoming and outgoing intensities at the free surfaces for general boundary conditions

$$\begin{aligned} I_{c,1}(\mu) &= s_1(\mu) + \rho_s I_{c,1}(-\mu) + 2\rho_d R_{cf} \\ I_{c,n}(-\mu) &= g_2(\mu) + \rho_s I_{c,n}(\mu) + 2\rho_d T_{cn}, \end{aligned} \tag{26a}$$

where the reflectance and transmittance are

$$R_{cf} \equiv \int_0^1 d\mu' \mu' I_{c,1}(-\mu) \tag{26b}$$

$$T_{cn} \equiv \int_0^1 d\mu' \mu' I_{c,n}(\mu). \tag{26c}$$

To find $I_{c,1}(\pm\mu)$ and $I_{c,n}(\pm\mu)$, which are the boundary angular intensities, (25) are evaluated at the surfaces ($j = n$ and 1, respectively)

$$\begin{aligned} &I_{c,1}^k(\mu) - \rho_{1s}\beta_1^-(\mu) I_{c,n}^k(-\mu) \\ &= \left\{ \tilde{g}_1(\mu) - \rho_s \sum_{l=1}^{n-1} \left[\frac{\beta_j^-(\mu)\beta_{l+1}^{-^{-1}}(\mu)}{a_l^-(\mu)} \right] W_l^{k-1}(\mu) + 2\rho_d R_{cf}^{k-1} \right\} \\ &\quad - \rho_{1s}\beta_n^-(\mu) I_{c,1}^k(\mu) + I_{c,n}^k(-\mu) \\ &= \left\{ \tilde{g}_2(\mu) + \rho_s \sum_{l=1}^{n-1} \left[\frac{\beta_j^+(\mu)\beta_{l+1}^{+^{-1}}(\mu)}{a_l^+(\mu)} \right] W_l^{k-1}(\mu) + 2\rho_d T_{cn}^{k-1} \right\}, \end{aligned} \tag{27}$$

with

$$\begin{aligned} \tilde{g}_1(\mu) &\equiv s_1(\mu) + \varepsilon_1 \left(\frac{\sigma_{sb} T_1^4}{\pi} \right) + 2\rho_d S_{s1} \\ \tilde{g}_2(\mu) &\equiv g_2(\mu) + \varepsilon_2 \left(\frac{\sigma_{sb} T_2^4}{\pi} \right) + 2\rho_d S_{s2}. \end{aligned}$$

We then solve this 2 by 2 set of equations for the desired quantities $I_{c,1}^k$ and $I_{c,n}^k$.

4 "Mining" the Discrete Ordinates Solution

The "solution mining" concept will now be introduced to improve upon the accuracy of the conventional DOM of the previous section. We can appreciate mining through the acceleration of a sequence of solutions to their limit.

4.1 Proposed Accelerations

Acceleration of the Inner Iterations

At the heart of the DOM is the inner transport sweep iteration. Indeed, its simplicity is unique to the DOM method and makes it an efficient transport solver. As indicated above, the sweep iterations proceed until convergence. It is well known that for optically thick, highly scattering media, convergence of the inner iterations can be extremely slow. For this reason, transport method's developers have devised various acceleration schemes to achieve improved convergence. Since the inner iterations form a sequence of angular intensities or fluxes, which presumably have limits, the concept of sequence acceleration toward those limits using the Wynn-epsilon (Wε) algorithm [2] becomes possible.

The non-linear Wϵ accelerator is an acceleration of a sequence towards its limit through the following non-linear recurrence:

$$\begin{aligned} &\varepsilon_{-1}^{(k)} = 0, \; \varepsilon_0^{(k)} = q_k \\ &\varepsilon_{s+1}^{(k)} = \varepsilon_{s-1}^{(k+1)} + \left[\varepsilon_s^{(k+1)} - \varepsilon_s^{(k)}\right]^{-1} \end{aligned} \tag{28}$$

where q_k is the original sequence element, e.g., the intensity for inner iteration k. We can express (28) as the tableau shown in Fig. 2, where the last term in each column forms a $[2s/2s]$ Pade' approximate [2]. The even-column elements of the lower diagonal indicated is interrogated for convergence. A dramatic reduction in the required number of inner iterations is possible with the Wε convergence acceleration [3].

Acceleration of the Spatial Discretization

Recall (16a) where we have ascribed the trapezoidal rule to the balance equation with a local error of order h_x^3 and an error tail composed of only odd powers of h. One can show that the angular intensity in a homogeneous medium will therefore inherit an even error power series with leading order h_x^2. The appearance of the even order error series suggests the application of Romberg acceleration [4] to successively eliminate higher order error terms. In the Romberg acceleration, we sequentially halve the intervals with a DOM

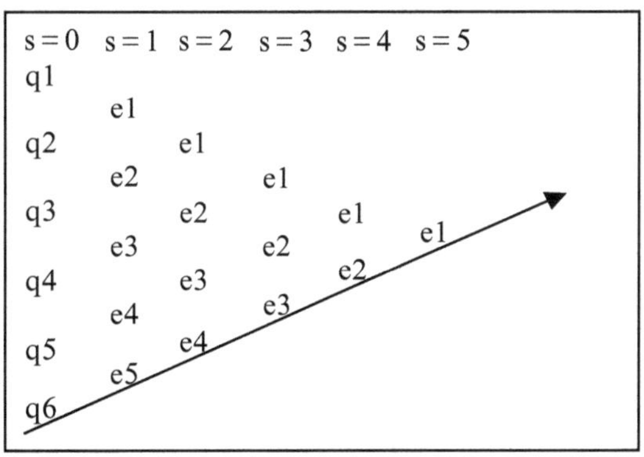

Fig. 2. Wynn-epsilon tableau

evaluation at each reduction. The following algorithm gives the Romberg acceleration toward convergence:

$$D_k^{l+1} = D_k^l + \left[\frac{D_k^l - D_{k-1}^l}{4^{k+1} - 1}\right] \tag{29}$$

where D_k^0 is any desired quantity such as intensity or flux always at the same τ for all uniform mesh intervals $h_x/2^{l-1}, l = 1, 2, \ldots$. We interrogate the sequence $\left[D_0^0, D_1^1, \ldots D_l^l\right]$ for convergence over the spatial discretization index l. A maximum of $l = 11$ is allowed. The initial mesh spacing h_x is the slab thickness divided by the number of intervals N_e defining the total number of spatial edits. Thus, the total number of spatial intervals is

$$n = 2^{l-1} N_e. \tag{30}$$

Two additional accelerations are possible. One applies the Wε acceleration directly to the DOM sequence itself. Also, we can apply the Wε acceleration to the Romberg sequence. The iteration strategy is to perform all three accelerations simultaneously and take the first that converges.

Acceleration in the Angular Quadrature Order

We obtain a sequence of solutions in the quadrature order by imposing iteration on the quadrature order $2N$. Thus, included in the CDOM algorithm is an "outer" iteration to convergence. Again, advantage can possibly be gained by filtering the sequence found through the Wε algorithm, where we create a sequence by successively increasing $2N$ by 2.

Course Mesh Rebalance

Requiring an overall particle balance at each iteration produces a more efficient convergence – a well-known fact. To affect what is called course mesh rebalance (CMR), the inner iteration is split into two steps. The first "half step" is

$$\begin{aligned} &\left[\mu_m + \frac{h_j}{2}\right] I_{c,m,j+1}^{l+1/2} - \left[\mu_m - \frac{h_j}{2}\right] I_{c,m,j}^{l+1/2} \\ &= \frac{\omega_j h_j}{2} \left[\bar{Q}_{cs,m,j+1}^{l-1} + \bar{Q}_{cs,m,j}^{l-1}\right] + \frac{h_j}{2} \left[S_{j+1} + S_j\right] + \omega_j S_{0,m,j} \end{aligned} \tag{31a}$$

leading to the definition of the CMR factor f

$$I_{c,m,j}^{l} = f I_{m,j}^{l+1/2}. \tag{31b}$$

We find f by requiring an overall balance for a uniform mesh h

$$\int_{-1}^{1} d\mu\mu I_c(a,\mu) - \int_{-1}^{1} d\mu\mu I_c(0,\mu) + \int_{0}^{a} dx\left[\sigma(x) - \sigma_s(x)\right] I_c(x)$$
$$= \int_{0}^{a} dx\sigma_s(x) S_{s0}(x) + \int_{0}^{a} dx S(x),$$

which also is

$$T_f + R_f + \sum_{j=1}^{n} \left[\sigma_j - \sigma_{sj}\right]\left[I_{c,j+1}(x) + I_{c,j}(x)\right]$$
$$= \sum_{m=n/2+1}^{2N} \varpi_m \mu_m I_{c,m,1} + q_0\mu_0 \sum_{j=1}^{n} \omega_j \left[e^{-\tau_j/\mu_0} - e^{-\tau_{j+1}/\mu_0}\right] + \frac{h}{2}\sum_{j=1}^{n}\left[S_{j+1} + S_j\right] \tag{32}$$

where R_f, T_f are the reflectance and transmittance. Finally, substituting (31b) gives

$$f\left\{T_f^{l+1/2} + R_f^{l+1/2} + \sum_{j=1}^{n}\left[\sigma_j - \sigma_{sj}\right]\left[I_{c,j+1}^{l+1/2} + I_{c,j}^{l+1/2}\right]\right\}$$
$$= \sum_{m=N/2+1}^{2N} \varpi_m \mu_m I_{c,m,1} + q_0\mu_0 \sum_{j=1}^{n} \omega_j \left[e^{-\tau_j/\mu_0} - e^{-\tau_{j+1}/\mu_0}\right] + \frac{h}{2}\sum_{j=1}^{n}\left[S_{j+1} + S_j\right] \tag{33}$$

CMR can have a dramatic affect on the convergence of the inner iteration.

4.2 Benchmark Results

In this section, we compare the results of CDOM to previously published benchmarks to demonstrate that we can generate equally valid benchmarks for a variety of radiative transfer scenarios. The comparison benchmarks, six in all, come from the C.E. Siewert portfolio and cover a range of participating media. In the cases treated, we provide additional accuracy and indicate when there is disagreement with the published result. We report only CDOM results. All computational times quoted are for a 2.4 MHz Dell PC.

Benchmark 1: No Reflection

The first benchmark considers both a homogeneous and a six-slab heterogeneous water medium.

Table 1. Scattering coefficients for Benchmark 1

l	$(2l+1)\, f_l$
0	1.00000
1	2.00916
2	1.56339
3	0.67407
4	0.22215
5	0.04725
6	0.00671
7	0.00068
8	0.00005

(a) Homogeneous Water Slab

The assumed scattering law [5], given in Table 1, derives from Mie theory for spherical particles with size parameter $\alpha = 2$ and index of refraction $n = 1.33$ for a water medium. We assume the single slab medium to be of either 1 or 10 optical thickness and with albedos ranging from 0.9 to 0.9999.

We also assume isotropic incidence on the left free surface. The reflectance and transmittance, defined here as

$$R_f = \frac{1}{J_0} \int_0^1 d\mu \mu I\,(0, -\mu)$$

$$T_n = \frac{1}{J_0} \int_0^1 d\mu \mu I\,(\Delta, \mu)\,,$$

with

$$J_0 = \int_0^1 d\mu \mu I\,(0, \mu),$$

given by the CDOM are displayed in Table 2. The CDOM results required under a second to compute. We include one additional digit with respect to the *Fn* benchmark values [5], which we round to 5-digits. Several entries (shaded) are discrepant in the last digit in comparison to the *Fn* results when rounded. The difference in the first entry is within the quoted accuracy of the *Fn* results; while, the other two discrepancies are, presumably, a result of rounding a more accurate value to 5-digits.

(b) Heterogeneous Water Slab

For this case, we assume the six slabs with the albedos and optical thicknesses given in Table 3. Table 4 gives the reflectances and transmittances. For this

Table 2. Reflectances and transmittances for Benchmark 1a

ω	Δ	R_f	T_n
0.9	1	1.71913E−01	6.54267E−01
0.9	10	2.90702E−01	3.29360E−02
0.99	1	2.26618E−01	7.53677E−01
0.99	10	6.22062E−01	2.10784E−01
0.999	1	2.33104E−01	7.64899E−01
0.999	10	7.06945E−01	2.73441E−01
0.9999	1	2.33765E−01	7.66035E−01
0.9999	10	7.16914E−01	2.81090E−01

Table 3. Data for Benchmark 1b

r	ω_r	τ_r
1	0.65	1.0
2	0.70	2.0
3	0.75	3.0
4	0.80	4.0
5	0.85	5.0
6	0.90	6.0

Table 4. Reflectances and transmittances for Benchmark 1b

β	L	$R_{f(CDOM)}$	$R_{f(ANISN)}$	$T_{n(CDOM)}$	$T_{n(ANISN)}$
0	0	2.27952E−01	2.280E−01	6.17135E−07	6.099E−07
1	0	2.14837E−01	2.148E−01	7.18092E−07	7.098E−07
2	0	2.07856E−01	2.079E−01	7.90572E−07	7.815E−07
0	8	1.00045E−01	1.001E−01	7.41883E−05	7.391E−05
1	8	8.05816E−02	8.059E−02	8.54303E−05	8.512E−05
2	8	7.05278E−02	7.051E−02	9.30733E−05	9.274E−05

case, we assume an incident source of the form μ^β. In addition to anisotropic scattering ($L = 8$), we consider isotropic scattering ($L = 0$). On rounding, only one transmittance does not comply with the *Fn* results in the last place. Also displayed are the original discrete ordinate values obtained by the *ANISN* neutron transport code.

While accurate for the reflectance, the original *ANISN* DOM results are quite inaccurate compared to the CDOM for the transmittance indicating the advantage of the CDOM. Again, all CDOM values are accurate to within one digit in the last place in comparison to the *Fn* results. The computation of this table was less than 8 s.

Benchmark 2: Single Slab with Reflection/Reflectance and Transmittance

Following the analysis found in [6], the solution to (6a) with volume (ε) and surface ($\varepsilon_1 = \varepsilon_2$) emission is

$$I_c(\tau,\mu) = \varepsilon\left(\frac{\sigma_{sb}T^4}{\pi}\right)\phi(\tau,\mu) + \varepsilon_1\left(\frac{\sigma_{sb}T_1^4}{\pi}\right)\theta(\tau,\mu) + \varepsilon_2\left(\frac{\sigma_{sb}T_2^4}{\pi}\right)\theta(\tau,-\mu) \tag{34}$$

where $\theta(\tau,\mu)$ and $\phi(\tau,\mu)$ satisfy (6a) for the assumed volume and surface emissions respectively with boundary conditions

$$\begin{aligned}\theta(0,\mu) &= 1 + \rho_s\theta(0,-\mu) + 2\rho_d\int_0^1 d\mu'\mu'\theta(0,-\mu') \\ \theta(\Delta,-\mu) &= \rho_s\theta(\Delta,\mu) + 2\rho_d\int_0^1 d\mu'\mu'\theta(\Delta,\mu')\end{aligned} \tag{35a}$$

and

$$\phi(0,\mu) = \rho_s\phi(0,-\mu) + 2\rho_d\int_0^1 d\mu'\mu'\phi(0,-\mu'). \tag{35b}$$

(a) Volume Emission ($\varepsilon = 1, \varepsilon_1 = \varepsilon_2 = 0$)

Table 5 gives the scattering phase function assumed for this case. The reflectance parameters ρ_s and ρ_d vary from 0.0 to 0.25 and 0.0 to 0.5 respectively.

The reflectance and transmittances, in the form

Table 5. Scattering coefficients for Benchmark 2

l	$(2l+1)\,f_l$
0	1.00000
1	2.35789
2	2.76628
3	2.20142
4	1.24514
5	0.51215
6	0.16096
7	0.03778
8	0.03778
9	0.00081
10	0.00000

$$R_f = \int_0^1 d\,\mu\mu\Psi\left(0,-\mu\right)$$

$$T_n = \int_0^1 d\,\mu\mu\Psi\left(\Delta,\mu\right),$$

where Ψ is either θ or ϕ, are given in Table 6 for a medium of optical thickness 1. All rounded values to 4-digits at $\tau = 0$ agree with the *Fn* benchmark. In addition, the benchmark for ρ_s and ρ_d both 0.25 at $\tau = 0.5$ agree with those in [7] except for the last value (shaded) which again is, presumably, a result of rounding a more accurate value. The computational time for Table 6 was less than 1.5 s.

(b) Surface Emission $(\varepsilon = 0, \varepsilon_1 = \varepsilon_2 = 1)$

For this case, the volume emission is absent and the surfaces emit. The reflectances and transmittances, given in Table 7, are in complete agreement with the rounded values of [7]. The computational time for Table 7 was less than 1 s.

Table 6. Reflectances and transmittances for Benchmark 2a

ω	ρ_s	ρ_d	$\tau = 0$	$\tau = 0.5$
0.2	0.0	0.5	4.16418E−01	3.52327E−01
	0.5	0.0	4.13970E−01	3.49030E−01
	0.25	0.25	4.15263E−01	3.50763E−01
0.8	0.0	0.5	2.34151E−01	1.89952E−01
	0.5	0.0	2.32265E−01	1.87470E−01
	0.25	0.25	2.33300E−01	1.88825E−01
0.95	0.0	0.5	8.49423E−02	6.81158E−02
	0.5	0.0	8.46200E−02	6.74886E−02
	0.25	0.25	8.47998E−02	6.78355E−02

Table 7. Reflectances and transmittances for Benchmark 2b

ω	ρ_s	ρ_d	$\tau = 0$	$\tau = 1.0$
0.2	0.0	0.5	2.45792E−02	1.42584E−01
	0.5	0.0	2.86340E−02	1.43425E−01
	0.25	0.25	2.65643E−02	1.42910E−01
0.8	0.0	0.5	1.63859E−01	3.67839E−01
	0.5	0.0	1.69562E−01	3.65908E−01
	0.25	0.25	1.66648E−01	3.66752E−01
0.95	0.0	0.5	2.98222E−01	5.31893E−01
	0.5	0.0	3.03038E−01	5.27722E−01
	0.25	0.25	3.00632E−01	5.29769E−01

Benchmark 3: (Benchmark 2)-Interior Partial Currents

(a) Volume Emission ($\varepsilon = 1, \varepsilon_1 = \varepsilon_2 = 0$)

This benchmark is identical to Benchmark 2 but with emphasis now on the partial currents. Tables 8 and 9 display the interior partial currents for volume emission with ρ_s and ρ_d both 0.25 and for a medium of optical depth 1 with $\omega = 0.2$, 0.8 and 0.95 at $\tau = 0.0(0.1)0.5$. To rounding error, all values agree with the *Fn* results of [7].

(b) Surface Emission ($\varepsilon = 0, \varepsilon_1 = \varepsilon_2 = 1$)

Similarly, Tables 10 and 11 give the interior partial currents for emitting surfaces. Only three partial currents differ from those of [7] and all within the estimated error.

All tables in this section required less than 25 s of total computational time.

Benchmark 4: Angular Intensities/Conventional

This benchmark tests the capability of the CDOM to determine angular intensities accurately for a conventional case. In general, it has been found that the DOM does not give angular intensities accurately. This is a result of the interference between the angular and spatial discretizations. For extreme forward

Table 8. Partial currents for Benchmark 3a

ω	0.2		0.8	
τ	q^-	q^+	q^-	q^+
0.00E+00	4.15263E−01	2.07632E−01	2.33300E−01	1.16650E−01
0.10E+00	4.05947E−01	2.48495E−01	2.26579E−01	1.34341E−01
0.20E+00	3.94991E−01	2.81080E−01	2.18586E−01	1.49938E−01
0.30E+00	3.82312E−01	3.08181E−01	2.09600E−01	1.64068E−01
0.40E+00	3.67678E−01	3.31128E−01	1.99682E−01	1.76982E−01
0.50E+00	3.50763E−01	3.50763E−01	1.88824E−01	1.88824E−01

Table 9. Partial currents for Benchmark 3a

ω	0.95	
τ	q^-	q^+
0.00E+00	8.47998E−02	4.23999E−02
0.10E+00	8.21971E−02	4.83726E−02
0.20E+00	7.91059E−02	5.37844E−02
0.30E+00	7.56592E−02	5.87987E−02
0.40E+00	7.18975E−02	6.34732E−02
0.50E+00	6.78354E−02	6.78354E−02

Table 10. Partial currents for Benchmark 3b

ω	0.2		0.8	
τ	q^-	q^+	q^-	q^+
0.00E+00	2.65643E−02	5.13282E−01	1.66648E−01	5.83324E−01
0.10E+00	2.78210E−02	4.40240E−01	1.64464E−01	5.51972E−01
0.20E+00	2.99057E−02	3.82282E−01	1.63931E−01	5.24469E−01
0.30E+00	3.25910E−02	3.34204E−01	1.64358E−01	4.99474E−01
0.40E+00	3.58441E−02	2.93555E−01	1.65514E−01	4.76448E−01
0.50E+00	3.96899E−02	2.58785E−01	1.67281E−01	4.55070E−01
0.60E+00	4.41886E−02	2.28800E−01	1.69587E−01	4.35122E−01
0.70E+00	4.94334E−02	2.02786E−01	1.72390E−01	4.16441E−01
0.80E+00	5.55593E−02	1.80113E−01	1.75654E−01	3.98897E−01
0.90E+00	6.27701E−02	1.60285E−01	1.79346E−01	3.82377E−01
1.00E+01	7.14548E−02	1.42910E−01	1.83376E−01	3.66752E−01

Table 11. Partial currents for Benchmark 3b

ω	0.95	
τ	q^-	q^+
0.00E+00	3.00632E−01	6.50316E−01
0.10E+00	2.94994E−01	6.35054E−01
0.20E+00	2.90502E−01	6.21311E−01
0.30E+00	2.86602E−01	6.08439E−01
0.40E+00	2.83098E−01	5.96198E−01
0.50E+00	2.79876E−01	5.84453E−01
0.60E+00	2.76855E−01	5.73107E−01
0.70E+00	2.73964E−01	5.62079E−01
0.80E+00	2.71120E−01	5.51286E−01
0.90E+00	2.68201E−01	5.40612E−01
1.00E+01	2.64884E−01	5.29769E−01

peaked scattering and thick media, discretization leads to negative angular intensities. In common codes such as *ANISN*, a negative flux fixup is used, where, if the source is not negative but the flux is, the intensity is set to zero with total conservation maintained. Hence, in general, we do not anticipate highly accurate results from a DOM for the angular intensity for these extreme cases. Here, however we consider a conventional case with isotropic scattering in a homogeneous medium to demonstrate the ability of the CDOM to, nevertheless, generate benchmark accurate angular intensities.

(a) Single Slab: Isotropic Scattering

The first case considered is for basic isotropic scattering in a homogeneous slab of optical thickness 1 with a normal beam source and an albedo of 0.9.

Table 12. Exiting angular intensities for Benchmark 4a

μ	$I(0,-\mu)$	$I(\Delta,\mu)$
0.0	3.593711E−01	1.861376E−01
0.1	3.726693E−01	2.240330E−01
0.2	3.633064E−01	2.483276E−01
0.3	3.442406E−01	2.575817E−01
0.4	3.212669E−01	2.552022E−01
0.5	2.981513E−01	2.466790E−01
0.6	2.765449E−01	2.355361E−01
0.7	2.569735E−01	2.236334E−01
0.8	2.394718E−01	2.118832E−01
0.9	2.238850E−01	2.007043E−01
1.0	2.100008E−01	1.902647E−01

Table 13. Exiting angular intensities for Benchmark 4b

μ	$I(0,-\mu)$	$I(\Delta,\mu)$
0.0	3.59379E−01	1.51520E−01
0.1	3.40126E−01	2.03073E−01
0.2	2.97523E−01	2.44242E−01
0.3	2.51677E−01	2.70302E−01
0.4	2.08230E−01	2.81066E−01
0.5	1.69677E−01	2.80822E−01
0.6	1.36504E−01	2.73029E−01
0.7	1.08350E−01	2.59822E−01
0.8	8.45877E−02	2.42453E−01
0.9	6.45644E−02	2.21667E−01
1.0	4.76807E−02	1.97932E−01

The exiting angular radiances, shown in Table 12, are in complete agreement with those of the *Fn* calculation of [8] when rounded to 5 decimal places. The computational time for this table was less than 5 s.

(b) Single Water Slab

The zeroth component of the azimuthally varying intensity is the intensity for the corresponding azimuthally symmetric case. Hence, we can compare the CDOM with the zeroth component of the azimuthal intensity [9] for a water slab of optical thickness of 1 and the phase function of Table 1. The albedo is 0.95 and we assume a beam source at 0.5 inclination. Table 13 displays the exiting angular flux. We achieve agreement with [9] to 4-places except for one value, which again is a consequence of rounding a more accurate number.

Thus, the CDOM can indeed produce benchmark accuracy even for the angular intensity – at least for non-extreme cases. The time to compute Table 13 was under 7 s.

Benchmark 5: Angular Intensities/Extreme Cases

Here, we consider extreme cases of extensive scattering kernels and large optical depth. In particular, we attempt two highly anisotropic cases–for the Haze L and Cloud C1 phase functions. These are phase functions with 82 and 299 scattering coefficients respectively [10] and indeed represent extreme scattering because of the shear number of scattering coefficients. All scattering coefficients can be found in [10].

(a) Haze L and Cloud C1: Intensities

The first case is for a cloud of optical thickness 1 with the Haze L scattering phase function and albedos of 0.9 and 1.0. We assume a normal beam source. Table 14 presents the integrated intensities. All values are in agreement with the benchmark. Table 14 required less than 3 s to construct.

The Cloud C1 case combines extreme scattering and an optically thick medium. Table 15 gives the integrated intensities. Several values do not round properly but are still within the accuracy of the results reported in [10]. We note that no special numerical procedures are used to obtain the results quoted.

For the Haze L and Cloud C1 phase functions, the maximum quadrature order was less than 60. The C1 intensities required less than 14 minutes to calculate indicating the extreme nature of this case.

Table 14. Integrated intensities for Haze L

τ/ω	0.9	1
0.000000E+00	3.017927E+00	2.968370E+00
5.000000E−02	3.000612E+00	2.968370E+00
1.000000E−01	2.983141E+00	2.968370E+00
2.000000E−01	2.947929E+00	2.968370E+00
5.000000E−01	2.841777E+00	2.968370E+00
7.500000E−01	2.754687E+00	2.968370E+00
1.000000E+00	2.671270E+00	2.968370E+00

Table 15. Integrated intensities for Cloud C1

τ/ω	0.9	1
0.000000E+00	2.766288E+00	4.798519E−01
3.200000E+00	1.605240E+00	4.798519E−01
6.400000E+00	8.298688E−01	4.798519E−01
1.280000E+01	1.965584E−01	4.798520E−01
3.200000E+01	2.245788E−03	4.798520E−01
4.800000E+01	5.354412E−05	4.798520E−01
6.400000E+01	1.473343E−06	4.798520E−01

Table 16. Angular intensities for Haze L

$\|\mu\|$	$\omega = 0.9$		$\omega = 1.0$	
	$\tau = 0$	$\tau = 1$	$\tau = 0$	$\tau = 1$
0.0	5.1748**9**E−02	7.93127E−02	6.9877**8**E−02	1.14808E−01
0.1	6.5529**8**E−02	1.08189E−01	9.2933**0**E−02	1.56976E−01
0.2	6.6956**1**E−02	1.24212E−01	9.7081**3**E−02	1.76818E−01
0.3	6.1541**9**E−02	1.35712E−01	8.9968**0**E−02	1.88301E−01
0.4	5.35113E−02	1.48268E−01	7.8148**0**E−02	2.00019E−01
0.5	4.56379E−02	1.67516E−01	6.60942E−02	2.19633E−01
0.6	3.91308E−02	2.00701E−01	5.5821**3**E−02	2.55983E−01
0.7	3.4384**0**E−02	2.61672E−01	4.8005**3**E−02	3.25125E−01
0.8	3.14480E−02	3.86921E−01	4.2731**6**E−02	4.68658E−01
0.9	3.0180**4**E−02	7.17745E−01	3.9782**0**E−02	8.46084E−01
1.0	2.7971**9**E−02	2.97766E+00	3.6145**4**E−02	3.38091E+00

(b) Haze L: Angular Intensity

Table 16 gives the exiting angular intensity for a medium with the Haze L phase function and albedos of 0.9 and 1.0. Bold indicates the incorrect digits. Here, we see that the CDOM is not able to produce the desired benchmark quality for the left boundary exiting angular intensity in the 14 minutes that it took to generate Table 16. However, we obtain the right boundary intensity to within the accuracy of [10]. Apparently, the extreme scattering nature of the problem is responsible for generating a significant number of negative angular intensities that degrade the calculation. There are several possible ways of improving the present calculation, which we consider in a future effort.

When we apply the CDOM to find the exiting angular intensities for the Cloud C1 case, it fails at the left boundary, but does reproduce the right boundary exiting angular intensity to at lease 4-digits except for μ near zero. Here, the large optical thickness is responsible for the failure. Again, this issue will be considered in a future effort.

Benchmark 6: Inhomogeneous Atmosphere

The final benchmark is for an inhomogeneous atmosphere [11], where we solve the transport equation in the following form:

$$\left[\mu\frac{\partial}{\partial\tau}+1\right]I\left(\tau,\mu\right)=\omega_s\left(\tau\right)\int_{-1}^{1}d\mu' f\left(\mu',\mu\right)I\left(\tau,\mu'\right) \tag{36a}$$

with boundary conditions for $\mu < 0$

$$\begin{aligned} I\left(0,\mu\right) &= g\left(\mu\right) \\ I\left(a,\mu\right) &= 0. \end{aligned} \tag{36b}$$

$\omega(\tau)$ represents a continuous variation of the scattering opacity. The example we consider here is for an exponential atmosphere

$$\omega(\tau) = \omega_{s0} e^{-\tau/s}$$

and

$$f(\mu', \mu) = \frac{1}{2}.$$

The DOM equations are simply as before with

$$\sigma_{sj} = \omega_{sj} \equiv \omega_{s0} e^{-\tau_j/s}.$$

Each interval $[x_j, x_{j+1}]$ now has a different scattering opacity.

Tables 17 and 18 provide the exiting angular intensity to 6-places for either an isotropically entering intensity at the left boundary or a beam inclined at $\mu_0 = 0.9$. The value of s varies from 1 to infinity and the optical depth assumed is 5. When rounded, all values conform to the results obtained previously by the Fn method [11].

Tables 19 and 20 display the reflectance and transmittance from a medium of various optical thicknesses, albedos and s-values. When rounded to 5-digits perfect agreement is achieved with the benchmark of [11] for both isotropic and beam incidence.

Tables 21 shows the ratio of the exiting to the incoming partial currents for the inhomogeneous atmosphere with a Haze L scattering phase function and $s = 1$ and for an optical thickness of 1. This example is included just to provide a new calculation not in the Siewert portfolio.

5 Conclusions

We have successfully generated benchmark solutions for the radiative transfer equation from the most fundamental entirely discretized discrete ordinates numerical scheme. The new scheme couples discretization with convergence acceleration to give highly accurate solutions. We demonstrate the scheme's accuracy on a variety of benchmarks for integrated and angular quantities alike considered by C.E. Siewert. The algorithm performs best for integrated quantities however. The importance of this presentation is in its simplicity in that solutions for a straightforward numerical scheme are simply filtered to distill accuracy. Mathematically sophisticated solutions like those of Siewert are therefore not required and the method can be applied to transport problems which do not have analytical representations.

While some of my colleagues have criticized this approach as being too crude and inelegant, they fail to appreciate that convergence acceleration applies to any 1D linear transport equation enabling benchmarks for any consistent numerical method. It is anticipated that convergence acceleration will become an integral part of numerical solutions of the transport equation in the future.

Table 17. Isotopic Incidence ($\omega_0 = 1, a = 5$) for Benchmark 6 $I(0, -\mu)$

$\mu \setminus s$	1	10	100	1000	∞
0.00E+00	6.667348E−01	7.934811E−01	8.808091E−01	9.064392E−01	9.100787E−01
1.00E−01	5.311198E−01	7.339771E−01	8.502824E−01	8.831861E−01	8.878361E−01
2.00E−01	4.432796E−01	6.863201E−01	8.252313E−01	8.640984E−01	8.695811E−01
3.00E−01	3.803062E−01	6.441806E−01	8.018095E−01	8.460571E−01	8.522993E−01
4.00E−01	3.329644E−01	6.064737E−01	7.794641E−01	8.285668E−01	8.355032E−01
5.00E−01	2.960902E−01	5.725684E−01	7.579949E−01	8.114313E−01	8.189957E−01
6.00E−01	2.665635E−01	5.419715E−01	7.373042E−01	7.945521E−01	8.026761E−01
7.00E−01	2.423895E−01	5.142668E−01	7.173350E−01	7.778786E−01	7.864927E−01
8.00E−01	2.222347E−01	4.890964E−01	6.980513E−01	7.613936E−01	7.704294E−01
9.00E−01	2.051742E−01	4.661525E−01	6.794306E−01	7.451046E−01	7.544962E−01
1.00E+00	1.905464E−01	4.451700E−01	6.614585E−01	7.290351E−01	7.387212E−01
			$I(a, \mu)$		
0.00E+00	8.030175E−04	8.570625E−03	5.829085E−02	8.565376E−02	8.992131E−02
1.00E−01	6.925156E−06	6.370178E−03	6.902380E−02	1.062863E−01	1.121640E−01
2.00E−01	9.642324E−06	7.618264E−03	8.056730E−02	1.236342E−01	1.304190E−01
3.00E−01	1.623389E−05	9.148214E−03	9.191767E−02	1.401229E−01	1.477008E−01
4.00E−01	4.385804E−05	1.111853E−02	1.034181E−01	1.562191E−01	1.644969E−01
5.00E−01	1.693724E−04	1.372493E−02	1.152300E−01	1.721134E−01	1.810044E−01
6.00E−01	5.734651E−04	1.718322E−02	1.274434E−01	1.879034E−01	1.973240E−01
7.00E−01	1.512818E−03	2.168015E−02	1.400980E−01	2.036372E−01	2.135073E−01
8.00E−01	3.243693E−03	2.733098E−02	1.531877E−01	2.193265E−01	2.295707E−01
9.00E−01	5.960356E−03	3.416621E−02	1.666679E−01	2.349553E−01	2.455039E−01
1.00E+00	9.771232E−03	4.214234E−02	1.804676E−01	2.504892E−01	2.612789E−01

Table 18. Beam incidence ($\mu_0 = 0.5, \omega_0 = 1, a = 5$) for Benchmark 6

$\mu \setminus s$	1	10	100	1000	∞
			$I(0, -\mu)$		
0.00E+00	7.217807E−01	9.431546E−01	1.114466E+00	1.166508E+00	1.173936E+00
1.00E−01	6.565125E−01	9.934316E−01	1.218162E+00	1.284565E+00	1.294009E+00
2.00E−01	5.763730E−01	9.847209E−01	1.253070E+00	1.331460E+00	1.342588E+00
3.00E−01	5.095507E−01	9.598856E−01	1.265540E+00	1.354896E+00	1.367578E+00
4.00E−01	4.552278E−01	9.284923E−01	1.265513E+00	1.364864E+00	1.378979E+00
5.00E−01	4.107847E−01	8.947202E−01	1.257645E+00	1.365987E+00	1.381406E+00
6.00E−01	3.739616E−01	8.606428E−01	1.244580E+00	1.360879E+00	1.377465E+00
7.00E−01	3.430438E−01	8.273272E−01	1.227975E+00	1.351191E+00	1.368804E+00
8.00E−01	3.167594E−01	7.953147E−01	1.208942E+00	1.338062E+00	1.356562E+00
9.00E−01	2.941621E−01	7.648580E−01	1.188266E+00	1.322330E+00	1.341581E+00
1.00E+00	2.745398E−01	7.360488E−01	1.166522E+00	1.304642E+00	1.324517E+00
			$I(a, \mu)$		
0.00E+00	2.474590E−03	1.952941E−02	1.247844E−01	1.822209E−01	1.911679E−01
1.00E−01	2.122586E−05	1.449753E−02	1.477248E−01	2.260764E−01	2.384157E−01
2.00E−01	2.921378E−05	1.730683E−02	1.723724E−01	2.629144E−01	2.771564E−01
3.00E−01	4.708215E−05	2.072314E−02	1.965586E−01	2.978755E−01	3.137789E−01
4.00E−01	1.021753E−04	2.505392E−02	2.209670E−01	3.319022E−01	3.492692E−01
5.00E−01	2.685494E−04	3.057846E−02	2.457912E−01	3.652499E−01	3.838965E−01
6.00E−01	6.503474E−04	3.742222E−02	2.708890E−01	3.977945E−01	4.175441E−01
7.00E−01	1.327597E−03	4.548436E−02	2.958607E−01	4.291575E−01	4.498409E−01
8.00E−01	2.324916E−03	5.448664E−02	3.201914E−01	4.588660E−01	4.803241E−01
9.00E−01	3.616122E−03	6.407325E−02	3.433892E−01	4.864918E−01	5.085779E−01
1.00E+00	5.144399E−03	7.389372E−02	3.650657E−01	5.117266E−01	5.343084E−01

Table 19. Isotropic incidence for Benchmark 6

$\omega = 0.7$

a	0.1		1.0	
s	Rf	Tn	Rf	Tn
1	5.302838E−02	8.845570E−01	1.520711E−01	2.936584E−01
10	5.548774E−02	8.872306E−01	2.115999E−01	3.582626E−01
100	5.574310E−02	8.875086E−01	2.209660E−01	3.698121E−01
1000	5.576873E−02	8.875366E−01	2.219593E−01	3.710559E−01
∞	5.577158E−02	8.875397E−01	2.220704E−01	3.711952E−01
a	5.0		10.0	
1	1.552404E−01	2.335285E−03	1.552404E−01	9.184045E−06
10	2.354144E−01	6.472446E−03	2.354182E−01	3.254501E−05
100	2.540170E−01	1.132676E−02	2.540444E−01	1.395811E−04
1000	2.562637E−01	1.227427E−02	2.562999E−01	1.869446E−04
∞	2.565193E−01	1.238924E−02	2.565566E−01	1.937485E−04

$\omega = 0.9$

a	0.1		1.0	
s	Rf	Tn	Rf	Tn
1	7.062334E−02	9.018287E−01	2.189199E−01	3.282836E−01
10	7.403291E−02	9.055263E−01	3.302108E−01	4.473806E−01
100	7.438771E−02	9.059117E−01	3.502872E−01	4.717553E−01
1000	7.442334E−02	9.059504E−01	3.524676E−01	4.744440E−01
∞	7.442730E−02	9.059547E−01	3.527120E−01	4.747459E−01
a	5.0		10.0	
1	2.243149E−01	2.611686E−03	2.243149E−01	1.017730E−05
10	3.954198E−01	1.366072E−02	3.954451E−01	8.001507E−05
100	4.637066E−01	4.314191E−02	4.645435E−01	1.647856E−03
1000	4.749787E−01	5.221898E−02	4.765320E−01	3.495611E−03
∞	4.763385E−01	5.342145E−02	4.780157E−01	3.855583E−03

$\omega = 1.0$

a	0.1		1.0	
s	Rf	Tn	Rf	Tn
1	7.990311E−02	9.109426E−01	2.588910E−01	3.494328E−01
10	8.384110E−02	9.152081E−01	4.125061E−01	5.124277E−01
100	8.425135E−02	9.156532E−01	4.428626E−01	5.488583E−01
1000	8.429255E−02	9.156979E−01	4.462173E−01	5.529462E−01
∞	8.429713E−02	9.157029E−01	4.465941E−01	5.534059E−01
a	5.0		10.0	
1	2.658917E−01	2.782464E−03	2.658917E−01	1.079057E−05
10	5.311817E−01	2.287391E−02	5.312681E−01	1.489699E−04
100	7.259719E−01	1.369394E−01	7.419349E−01	1.575189E−02
1000	7.840733E−01	1.981468E−01	8.544887E−01	8.688997E−02
∞	7.923427E−01	2.076573E−01	8.832549E−01	1.167451E−01

Table 20. Beam ($\mu_0 = 0.9$) incidence

$\omega = 0.7$

a	0.1		1.0	
s	Rf	Tn	Rf	Tn
1	3.309380E−02	9.275856E−01	1.150052E−01	3.944460E−01
10	3.468275E−02	9.293223E−01	1.684174E−01	4.575478E−01
100	3.484782E−02	9.295030E−01	1.770217E−01	4.689107E−01
1000	3.486439E−02	9.295212E−01	1.779368E−01	4.701352E−01
∞	3.486623E−02	9.295232E−01	1.780392E−01	4.702724E−01
a	5.0		10.0	
1	1.188532E−01	4.446596E−03	1.188532E−01	1.699277E−05
10	1.969791E−01	1.018496E−02	1.969850E−01	5.156079E−05
100	2.167452E−01	1.694133E−02	2.167860E−01	2.105005E−04
1000	2.191854E−01	1.825306E−02	2.192390E−01	2.801689E−04
∞	2.194638E−01	1.841210E−02	2.195191E−01	2.901588E−04

$\omega = 0.9$

a	0.1		1.0	
s	Rf	Tn	Rf	Tn
1	4.408075E−02	9.384623E−01	1.662831E−01	4.241791E−01
10	4.628182E−02	9.408615E−01	2.657189E−01	5.379318E−01
100	4.651098E−02	9.411117E−01	2.840748E−01	5.613243E−01
1000	4.653399E−02	9.411368E−01	2.860740E−01	5.639056E−01
∞	4.653655E−02	9.411396E−01	2.862981E−01	5.641954E−01
a	5.0		10.0	
1	1.726166E−01	4.709362E−03	1.726166E−01	1.792315E−05
10	3.402434E−01	1.873953E−02	3.402779E−01	1.106826E−04
100	4.142466E−01	5.543707E−02	4.153208E−01	2.122732E−03
1000	4.268633E−01	6.664227E−02	4.288439E−01	4.468772E−03
∞	4.283928E−01	6.812488E−02	4.305298E−01	4.924801E−03

$\omega = 1.0$

a	0.1		1.0	
s	Rf	Tn	Rf	Tn
1	4.987650E−02	9.442008E−01	1.971015E−01	4.421926E−01
10	5.241766E−02	9.469669E−01	3.340119E−01	5.960840E−01
100	5.268253E−02	9.472557E−01	3.616773E−01	6.306027E−01
1000	5.270912E−02	9.472847E−01	3.647429E−01	6.344769E−01
∞	5.271208E−02	9.472879E−01	3.650873E−01	6.349127E−01
a	5.0		10.0	
1	2.051742E−01	4.868730E−03	2.051743E−01	1.848815E−05
10	4.661525E−01	2.915451E−02	4.662623E−01	1.911489E−04
100	6.794306E−01	1.630986E−01	6.984363E−01	1.877013E−02
1000	7.451046E−01	2.344321E−01	8.284016E−01	1.028156E−01
∞	7.544962E−01	2.455038E−01	8.619630E−01	1.380370E−01

Table 21. Haze L phase function for incident beam source ($\mu_0 = 0.9, a = 1$)

$\lvert\mu\rvert$	$\omega = 0.9$		$\omega = 1.0$	
	$\tau = 0$	$\tau = 1$	$\tau = 0$	$\tau = 1$
0.00E+00	1.304722E−01	7.512956E−04	7.217807E−01	2.474590E−03
1.00E−01	1.346405E−01	8.297773E−06	6.565125E−01	2.122586E−05
2.00E−01	1.130603E−01	1.538700E−05	5.763730E−01	2.921378E−05
3.00E−01	9.128601E−02	3.529741E−05	5.095507E−01	4.708215E−05
4.00E−01	7.362053E−02	1.090506E−04	4.552278E−01	1.021753E−04
5.00E−01	6.034836E−02	3.781164E−04	4.107847E−01	2.685494E−04
6.00E−01	5.078105E−02	1.201698E−03	3.739616E−01	6.503474E−04
7.00E−01	4.408727E−02	3.422209E−03	3.430438E−01	1.327597E−03
8.00E−01	3.919713E−02	9.091735E−03	3.167594E−01	2.324916E−03
9.00E−01	3.584911E−02	2.101427E−02	2.941621E−01	3.616122E−03
1.00E+00	3.549864E−02	2.573960E−02	2.745398E−01	5.144399E−03

References

1. M. Abramowitz and I. Stegun, *Handbook of Mathematical Functions*, Dover, NY, 1965.
2. G.A. Baker and P. Graves-Morris, *Pade' Approximants*, Cambridge University Press, New York, 1996.
3. B. Ganapol, Mining the multigroup-discrete ordinates algorithm for high quality solutions, *Mathematics and Computation, Supercomputing, Reactor Physics and Nuclear and Biological Applications*, Avignon, France, September 12–15, 2005, American Nuclear Society, LaGrange Park, IL, 2005.
4. W.H. Press, B.P. Flannery, S.A. Teukolsky and W.T. Vetterling, Romberg Integration, *Numerical Recipes in FORTRAN: The Art of Scientific.*
5. C. Devaux, P. Grandjean, Y. Ishiguro and C.E. Siewert, On multi-Region Problems in Radiative Transfer, *Astrophys. Space Sci.* **62** (1979) 223–233.
6. C.E. Siewert, J.R. Maiorino and M.N. Ozisik, The use of the FN method for radiative transfer problems with reflective boundary conditions, *JQSRT* **25** (1990) 565–573.
7. M. Benassi, R.M. Cotta and C.E. Siewert, The PN method for radiative transfer problems with reflective boundary conditions, *JQSRT*, **30** (1983) 547–553.
8. R.C.Y. Chin, G.W. Hedstrom and C.E. Siewert, On the use of the FN method with splines for radiative transfer problems, *JQSRT*, **39** (1988) 206–217.
9. C. Devaux and C.E. Siewert, The FN method for radiative transfer problems without azimuthal symmetry, *ZAMP* **31** (1980) 592–604.
10. R.D.M. Garcia and C.E. Siewert, Benchmark results in radiative transfer, *TTSP* **14** (1983) 437–483.
11. R.D.M. Garcia and C.E. Siewert, Radiative transfer in finite inhomogeneous plane-parallel atmospheres, *JQSRT* **27** (1982) 141–148.

Implicit Monte Carlo Radiation Transport Simulations of Four Test Problems

N.A. Gentile

Summary. Radiation transport codes, like almost all codes, are difficult to develop and debug. It is helpful to have small, easy to run test problems with known answers to use in development and debugging. It is also prudent to re-run test problems periodically during development to ensure that previous code capabilities have not been lost. We describe four radiation transport test problems with analytic or approximate analytic answers. These test problems are suitable for use in debugging and testing radiation transport codes. We also give results of simulations of these test problems performed with an Implicit Monte Carlo photonics code.

1 The Units Used for the Simulations Described in this Work

We describe test problems used in the development and debugging of an implicit Monte Carlo (IMC) radiation transport package used in the Kull [GKR98] and Alegra [BM04] inertial confinement fusion codes. (Details of the IMC algorithm can be found in [FC71].) The test problems were run using cgs units, with temperature in keV.

In these units:

- The speed of light $c = 2.9979 \times 10^{10}\,\mathrm{cm\,s^{-1}}$
- Boltzmann's constant $k = 1.6022 \times 10^{-8}\,\mathrm{erg\,keV^{-1}}$
- Planck's constant $h = 6.6262 \times 10^{-27}\,\mathrm{erg\,s}$
- The radiation constant (the a in aT^4) $a = 1.3720 \times 10^{14}\,\mathrm{erg\,cm^{-3}\,keV^{-4}}$
- The Stefan–Boltzmann constant $\sigma_{SB} = ac/4 = 1.0283 \times 10^{24}\,\mathrm{erg\,cm^{-2}\,s^{-1}\,keV^{-4}}$

2 A Gray Infinite Medium Problem with a Matter Energy Source Allowing an Analytic Answer

Here we describe a test problem with no spatial dependence. A source of energy which heats the matter, constant in space and time, acts on a medium with a constant opacity and an equation of state (EOS) specified so as to linearize the coupled radiation and matter energy equations. The linearized equations have an analytic solution for the matter and radiation energy density. An IMC simulation is compared to the analytic answer.

The time-dependent transport equation for gray photons in the absence of scattering in an infinite medium is [P73]

$$\frac{1}{c}\frac{\partial I}{\partial t} = -\sigma I + \frac{c\sigma a T_m^4}{4\pi}, \tag{1}$$

where σ is the absorption cross section in inverse length units and T_m is the matter temperature. The transport equation is coupled to the material energy balance equation [P73]

$$\frac{\partial e_m}{\partial t} = \sigma \int I \mathrm{d}\Omega - c\sigma a T^4 + \frac{\rho}{4\pi}\frac{\mathrm{d}\epsilon}{\mathrm{d}t}. \tag{2}$$

Here, e_m is the matter energy density in units of energy per volume, ϵ is the specific energy in units of energy per mass, ρ is the mass density, and c_v is the heat capacity in units of energy per mass per temperature. The matter energy source is specified as the time rate of change of specific energy, in units of energy per mass per time.

In this section, we will use the following EOS relating e_m and T_m:

$$e_m = \frac{\alpha}{4} T^4, \tag{3}$$

where α is a constant. This EOS has $c_v = \alpha T_m^3/\rho$.

The radiation energy density is defined as

$$e_r = \frac{1}{c}\int I \mathrm{d}\Omega. \tag{4}$$

If we integrate (1) and (2) over Ω, and employ the EOS described in (3) and the definition of energy density (4), we get the following equations for the radiation energy density and matter energy density:

$$\frac{\partial e_r}{\partial t} = -c\sigma e_r + c\sigma \frac{4a}{\alpha} e_m \tag{5}$$

$$\frac{\partial e_m}{\partial t} = c\sigma e_r - c\sigma \frac{4a}{\alpha} e_m + \rho \frac{\mathrm{d}\epsilon}{\mathrm{d}t}. \tag{6}$$

With constant ρ and σ, these coupled ordinary differential equations are linear. The solution can be found in, for example, [BD77]. It is expressed most conveniently in terms of the following quantities:

$$
\begin{aligned}
\tau &\equiv c\sigma t \\
\beta &\equiv \frac{4a}{\alpha} \\
\lambda &\equiv -(1+\beta) \\
e_{total}(t) &\equiv e_r(t) + e_m(t) \\
&= e_r(0) + e_m(0) + \rho\frac{\mathrm{d}\epsilon}{\mathrm{d}t} \\
K_0 &\equiv -\left[\frac{\beta\rho}{c\sigma\lambda^2}\frac{\mathrm{d}\epsilon}{\mathrm{d}t} + \frac{\beta e_{total}(0)}{\lambda}\right] \\
K_\lambda &\equiv e_r(0) - K_0.
\end{aligned} \tag{7}
$$

In terms of these quantities, we get:

$$
e_r(\tau) = K_0 - \frac{\beta}{c\sigma\lambda}\rho\frac{\mathrm{d}\epsilon}{\mathrm{d}t}\tau + K_\lambda e^{\lambda\tau} \tag{8}
$$

$$
e_m(\tau) = e_{total}(\tau) - e_r(\tau) = e_{total}(0) + \rho\frac{\mathrm{d}\epsilon}{\mathrm{d}t}\frac{\tau}{c\sigma}. \tag{9}
$$

The IMC simulation of this test problem used the following parameters:

- $\alpha = 4.0a = 5.4881 \times 10^{14}\,\mathrm{erg\,cm^{-3}\,keV^{-4}}$
- $\sigma = 1.0\ \mathrm{cm}^{-1}$
- $\mathrm{d}\epsilon/\mathrm{d}t = 1.0 \times 10^{24}\,\mathrm{erg\,g^{-1}\,s^{-1}}$
- $T_m(0) = 0.1$ keV.

The mesh was a unit cube with sides of 1 cm and reflecting boundary conditions on all six faces. The simulation used $\Delta t = 1.0 \times 10^{-12}\,\mathrm{s}$, which is the equivalent of $\Delta\tau = 2.9979 \times 10^{-2}$. The simulation was run for 1,000 time steps with a constant Δt, to a time of $1.0 \times 10^{-9}\,\mathrm{s}$ ($\tau = 29.979$).

The results of the IMC simulation compared to the analytic answer are shown in Fig. 1. The IMC simulation used 10^4 particles and took approximately 1 min on a Pentium 4 processor. The radiation and matter temperature are derived from the analytic expressions for the matter energy density given by (8) and (9): $T_m = \sqrt[4]{4e_m/\alpha}$ and $T_r = \sqrt[4]{e_r/\mathrm{a}}$. The IMC results for both T_r and T_m match the analytic results very well.

3 A Cube with a Face Source Allowing an Approximate Analytic Answer

A problem similar to the preceding one can be obtained by replacing the matter energy source with a constant temperature source on one face. (The other five faces remain reflecting boundaries.) This problem is no longer an infinite medium problem. However, if we ignore the spatial variation in matter and

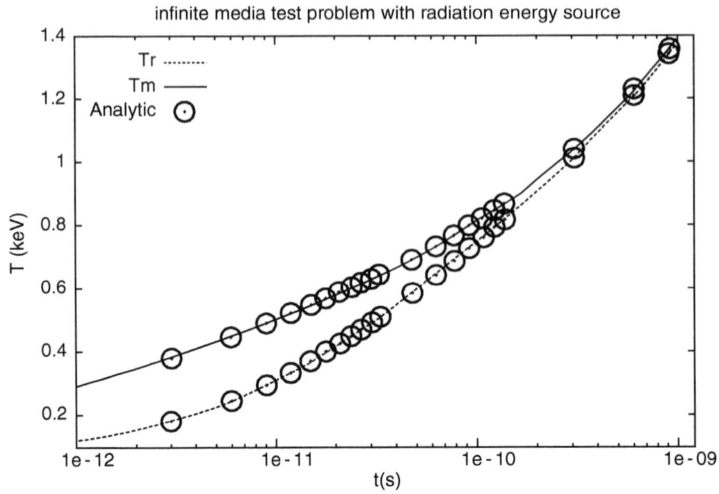

Fig. 1. The IMC results for the linearized infinite medium test problem and the analytic answer for the matter and radiation temperature, derived from the energy density given by (8) and (9). Temperature is plotted vs. time. The temperature is in units of keV and the time is in units of seconds

radiation energy density in the cube, it can be approximated as an infinite medium problem. This assumption is justified if the light travel time across the cube is short compared to the time scale on which the matter and radiation energy densities are changing. In that case, T_m and T_r can reach an approximately constant value across the cube.

The radiation energy input per unit area from a temperature face source with temperature T_s is $\sigma_{SB} T_s^4$. If we assume that the absorption opacity σ is large enough that most source photons are absorbed before being reflected back out of the cube, then we can approximate the energy leaving the problem through the source face per unit area as $\sigma_{SB} T_m^4$.

With these assumptions, (5) and (6) become

$$\frac{\partial e_r}{\partial t} = -c\sigma e_r + c\sigma \frac{4a}{\alpha} e_m + \frac{1}{4}\frac{ac}{L}[T_s^4 - e_m/(\alpha/4)] \tag{10}$$

$$\frac{\partial e_m}{\partial t} = c\sigma e_r - c\sigma \frac{4a}{\alpha} e_m . \tag{11}$$

Here L is the length of the cube.

These coupled linear ODEs can be solved by the techniques found in, for example, [BD77]. In order to cast the solution into a relatively easily useable form, we need to define several quantities. We start by noting that the asymptotic solution as $t \to \infty$ on physical grounds must satisfy $T_m = T_r = T_s$. Then we make the following definitions:

$$
\begin{aligned}
\hat{e_r}(t) &\equiv \frac{e_r(t)}{e_r(t\to\infty)} = \frac{e_r(t)}{aT_s^4} \\
\hat{e_m}(t) &\equiv \frac{e_m(t)}{e_m(t\to\infty)} = \frac{e_m(t)}{\alpha/4T_s^4} \\
\tau &\equiv c\sigma t \\
\beta &\equiv \frac{4a}{\alpha} \\
\lambda_\pm &\equiv \frac{1}{2}\left[-(1+\beta) \pm \sqrt{(1+\beta)^2 - \frac{\beta}{\sigma L}}.\right]
\end{aligned}
\tag{12}
$$

(Note that λ is real if $\sigma L \gg \beta$. This is consistent with the assumption that σ was large enough that most of the entering source photons are absorbed.)

Using these definitions, the solutions are

$$
\begin{aligned}
\hat{e_r}(\tau) &= 1 + C_{r+}e^{\lambda_+\tau} + C_{r-}e^{\lambda_-\tau} \\
\hat{e_m}(\tau) &= 1 + C_{m+}e^{\lambda_+\tau} + C_{m-}e^{\lambda_-\tau}
\end{aligned}
\tag{13}
$$

with

$$
\begin{aligned}
C_{m\pm} &\equiv \frac{C_{r\pm}}{1+\lambda_\pm/\beta} \\
C_{r+} &\equiv \hat{e_r}(0) - 1 - C_{r-} \\
C_{r-} &\equiv \frac{\hat{e_m}(0) - 1 - \frac{\hat{e_r}(0)-1}{1+\lambda_+/\beta}}{\frac{1}{1+\lambda_-/\beta} - \frac{1}{1+\lambda_+/\beta}}.
\end{aligned}
\tag{14}
$$

Results of the IMC simulation are plotted in Fig. 2. The mesh used was a unit cube with a temperature face source on $z = 0$ and reflecting boundaries on the other faces. The simulation used $\Delta t = 1.0 \times 10^{-12}$ s, which is the equivalent of $\Delta\tau = 2.9979 \times 10^{-2}$. The simulation was run for 1,000 time steps with a constant Δt, to a time of 1.0×10^{-9} s ($\tau = 29.979$). The simulation used 10^4 particles and took approximately 1 min on a Pentium 4 processor. The matter temperature calculated by IMC agrees reasonably well with (13) at all times. The radiation temperature from the IMC simulation is lower than that derived from (13) at early times, but begins to agree reasonably well after about $t = 2.0 \times 10^{-11}$ s.

4 Graziani's Spherical Multigroup Prompt Spectrum Test Problem

Graziani [G07] has developed an analytic solution for a time-dependent multigroup radiation transport test problem in spherical geometry. The test consists of a sphere, held at a fixed temperature T_s, embedded in an infinite medium,

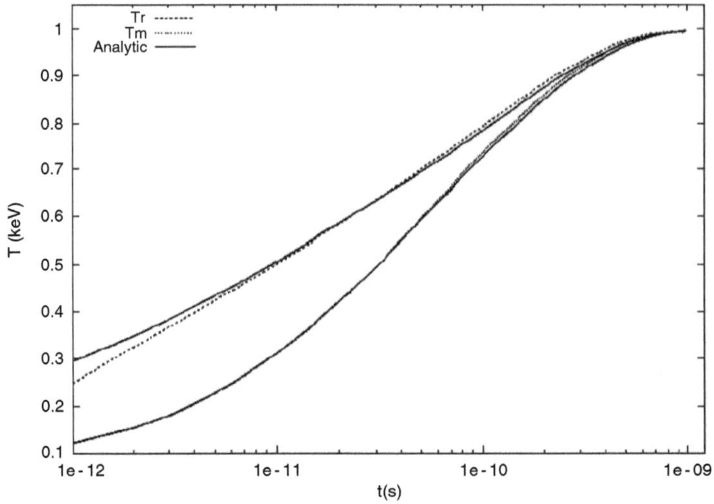

Fig. 2. The IMC results for the linearized face source test problem and the approximate analytic answer for the matter and radiation temperature, derived from the energy density given by (13) and (14). Temperature is plotted vs. time. The temperature is in units of keV and the time is in units of seconds

the temperature of which is held fixed at T_c. The embedded sphere begins radiating into the surrounding medium at $t = 0$. The medium has a multigroup opacity, which is constant in time because the temperature is held fixed. As radiation from the embedded sphere passes through the surrounding medium, photons in groups with different opacities are absorbed in varying amounts. The radiation energy density at a given point in the medium some distance from the sphere at time $t > 0$ is the sum of two contributions: a Planckian at T_c from the local material, and an attenuated Planckian at T_s from the part of the sphere at a distance less than ct from the given point. This is illustrated in Fig. 3.

Graziani has constructed an analytic expression for the radiation energy density in each group at a given point in the medium as a function of time. He refers to this as the prompt spectrum because it is the correct expression for the radiation energy density at times that are short compared to the time scale for the medium to change temperature. Graziani has derived analytic expressions for both diffusion and transport. In this paper, we will discuss only the transport solution.

Here we give the analytic expression for the prompt spectrum at a fiducial point r outside the embedded sphere of radius R at time t [G07]. The multigroup absorption opacity is denoted by σ_ν to indicate that it is a function of the frequency group.

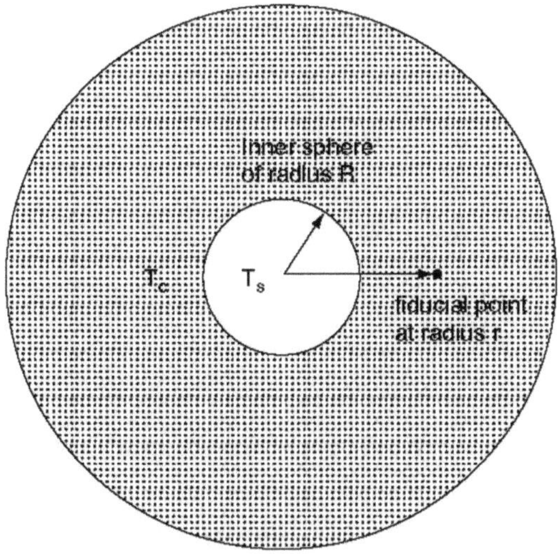

Fig. 3. The geometry of the Graziani spherical prompt spectrum problem. An inner sphere of radius R is embedded in an infinite medium. The temperature of the embedded sphere is held fixed at T_s. The temperature of the medium outside the sphere is held at T_c. Photons from the embedded sphere begin to radiate into the medium at t = 0. The radiation energy density at the fiducial point at radius r is the sum of a Planckian at T_c and a time-dependent contribution from the fraction of photons from the source (a Planckian at T_s) that make it out to radius r without being absorbed

$$e_r(r,t,\nu) = B_\nu(T_c) + [B_\nu(T_s) - B_\nu(T_c)]F_\nu(r,t) \tag{15}$$

In this expression, e_r is the radiation energy density, and $B_\nu(T)$ is the Planck function,

$$B_\nu(T) = \frac{2h\nu^3/c^2}{exp(h\nu/kT) - 1}. \tag{16}$$

$F_\nu(r,t)$ is given by

$$F_\nu(r,t) = 0 \tag{17}$$

for $ct < r - R$,

$$F_\nu(r,t) = \frac{R}{4r}\left\{\left[1 + \frac{r}{R} - \frac{1}{R\sigma_\nu}\right]e^{-\sigma_\nu(r-R)} + \left[\frac{1}{R\sigma_\nu} - \frac{R}{ct}\left(\frac{r^2}{R^2} - 1\right)\right]e^{-\sigma_\nu ct} - R\sigma_\nu\left(\frac{r^2}{R^2} - 1\right)[E_1(\sigma_\nu(r-R)) - E_1(\sigma_\nu ct)]\right\} \tag{18}$$

for $r - R < ct < \sqrt{r^2 - R^2}$, and

$$F_\nu(r,t) = \frac{R}{4r}\left\{ \left[1 + \frac{r}{R} - \frac{1}{R\sigma_\nu}\right] e^{-\sigma_\nu(r-R)} + \left[\frac{1}{R\sigma_\nu} - \sqrt{\frac{r^2}{R^2} - 1}\right] e^{-\sigma_\nu\sqrt{r^2-R^2}} - R\sigma_\nu\left(\frac{r^2}{R^2} - 1\right) \times \left[E_1(\sigma_\nu(r-R)) - E_1(\sigma_\nu\sqrt{r^2-R^2})\right] \right\} \tag{19}$$

for $\sqrt{r^2 - R^2} < ct < \infty$.

In these expressions for F, E_1 is the exponential integral:

$$E_1(\mathrm{x}) = \int_1^\infty \frac{e^{xt}}{t}\mathrm{d}t = \int_x^\infty \frac{e^{-u}}{u}\mathrm{d}u. \tag{20}$$

See [CTh68] for a rational function approximation, and [PTVF02] for a series approximation.

Details of the IMC simulation follow. The embedded sphere radius R was taken to be 0.02 cm. The solution was calculated for a fiducial point at radius $r = 0.04$ cm. $T_s = 0.3$ keV and $T_c = 0.03$ keV. The density of the material was $0.0916\,\mathrm{g\,cm^{-3}}$ and was held fixed throughout the simulation. The heat capacity was set to $10^{50}\,\mathrm{erg\,g^{-1}\,keV^{-1}}$. This caused the material temperature to remain constant. Since the density and temperature of the material do not change, the opacity was constant also, and only one value of the opacity for each group was needed for the simulation. (The values of the opacity given in column 5 in Table 1 can be employed in simulations in order to compare to the analytic answer given in column 4.) The IMC simulation used a tabular opacity calculated for plastic in which some hydrogen atoms were replaced by bromine. This material was frequently used in ICF simulations. Fifty groups were used. The group boundaries were spaced logarithmically between 3.0×10^{-3} and 30.0 keV. The opacity of the material at $T = 0.03$ keV and a density of $\rho = 0.0916\,\mathrm{g\,cm^{-3}}$ is shown in Fig. 4.

The problem was run to a time of 1.4×10^{-12} s using a time step of $\Delta t = 10^{-13}$ s. (Because the radiation energy density in the IMC algorithm is centered at the half time steps, the answer generated by the IMC simulation corresponds to a time of 1.35×10^{-12} s.)

The analytic solution of this problem is spherical. However, it is not necessary to run a whole sphere, or run the problem in spherical coordinates; it is also not necessary to simulate the inner sphere. It can be replaced by a face temperature source with a temperature of T_s. The outer boundary of the problem does not have to extend very far past the fiducial point. The outer boundary should have a face temperature source with a temperature of T_c imposed on it to avoid edge effects.

Table 1. Results of the IMC simulation of the Graziani spherical prompt spectrum problem

Group	ν	e_r	Analytic e_r	σ_ν
0	3.28943e−03	3.60690e+06	6.56128e+06	9.16000e+04
1	3.95477e−03	6.06118e+06	9.37744e+06	9.06781e+04
2	4.75468e−03	1.12608e+07	1.33706e+07	6.08939e+04
3	5.71638e−03	9.30579e+06	1.90091e+07	4.08607e+04
4	6.87260e−03	3.25408e+07	2.69304e+07	2.72149e+04
5	8.26269e−03	3.50625e+07	3.79893e+07	1.86425e+04
6	9.93393e−03	6.14524e+07	5.33093e+07	1.24389e+04
7	1.19432e−02	7.57451e+07	7.43291e+07	8.19288e+03
8	1.43589e−02	1.15741e+08	1.02827e+08	5.79710e+03
9	1.72632e−02	2.20469e+08	1.40887e+08	5.14390e+03
10	2.07549e−02	1.95919e+08	1.90765e+08	5.20350e+03
11	2.49529e−02	2.11776e+08	2.54559e+08	8.69569e+03
12	3.00000e−02	3.48662e+08	3.33604e+08	6.67314e+03
13	3.60679e−02	3.96097e+08	4.27461e+08	4.15912e+03
14	4.33632e−02	4.44384e+08	5.32466e+08	2.62038e+03
15	5.21340e−02	6.73583e+08	6.40012e+08	1.64328e+03
16	6.26789e−02	6.76189e+08	7.35205e+08	1.01613e+03
17	7.53566e−02	7.34447e+08	7.97303e+08	6.19069e+02
18	9.05986e−02	6.17024e+08	8.04394e+08	3.75748e+02
19	1.08923e−01	7.15130e+08	7.43877e+08	2.97349e+02
20	1.30955e−01	5.83979e+08	6.06784e+08	8.21172e+02
21	1.57442e−01	4.41725e+08	4.32675e+08	4.01655e+03
22	1.89287e−01	2.62729e+08	2.59951e+08	4.54828e+03
23	2.27573e−01	1.46260e+08	1.26955e+08	3.50487e+03
24	2.73603e−01	3.47654e+07	4.83345e+07	3.02359e+03
25	3.28943e−01	1.30942e+07	1.36588e+07	4.34203e+03
26	3.95477e−01	3.79204e+06	2.70428e+06	2.98594e+03
27	4.75468e−01	8.89654e+05	3.50392e+05	1.55364e+03
28	5.71638e−01	0.00000e+00	2.74523e−12	9.42213e+02
29	6.87260e−01	0.00000e+00	1.31590e+05	5.76390e+02
30	8.26269e−01	1.40974e+07	1.70905e+07	3.52954e+02
31	9.93393e−01	4.24349e+08	3.95663e+08	2.09882e+02
32	1.19432e+00	2.30338e+09	2.30389e+09	1.26546e+02
33	1.43589e+00	5.60175e+09	5.46696e+09	7.80087e+01
34	1.72632e+00	2.15155e+09	2.16292e+09	9.97421e+01
35	2.07549e+00	3.88801e+08	3.78948e+08	1.48848e+02
36	2.49529e+00	7.93594e+08	7.67169e+08	8.22907e+01
37	3.00000e+00	5.99171e+08	5.58890e+08	4.86915e+01
38	3.60679e+00	2.18511e+08	2.12224e+08	2.91258e+01
39	4.33632e+00	5.48765e+07	4.60720e+07	1.68133e+01
40	5.21340e+00	5.30714e+06	5.53396e+06	9.92194e+00
41	6.26789e+00	1.19953e+06	3.63829e+05	5.18722e+00
42	7.53566e+00	0.00000e+00	1.18769e+04	2.24699e+00

Table 1. Continued

Group	ν	e_r	Analytic e_r	σ_ν
43	9.05986e+00	0.00000e+00	1.68621e+02	1.29604e+00
44	1.08923e+01	0.00000e+00	9.26939e−01	7.46975e−01
45	1.30955e+01	0.00000e+00	1.62634e−03	8.43058e−01
46	1.57442e+01	0.00000e+00	7.11158e−07	2.43746e+00
47	1.89287e+01	0.00000e+00	6.49680e−11	1.50509e+00
48	2.27573e+01	0.00000e+00	8.32408e−16	9.01762e−01
49	2.73603e+01	0.00000e+00	1.00856e−21	5.38182e−01

e_r is the result of an IMC simulation, and analytic e_r results from (15) and (19), with $r = 0.04\,\text{cm}$, $t = 1.35 \times 10^{-12}\,\text{s}$, and σ_ν as given in column 5. The value of ν for each group is the rms value of the group bounds, which were logarithmically distributed between $3.0 \times 10^{-3}\,\text{eV}$ and $30.0\,\text{keV}$. The energy density is in units of $\text{erg}\,\text{cm}^{-3}\,\text{keV}^{-1}$ and the opacity is in units of $1\,\text{cm}^{-1}$

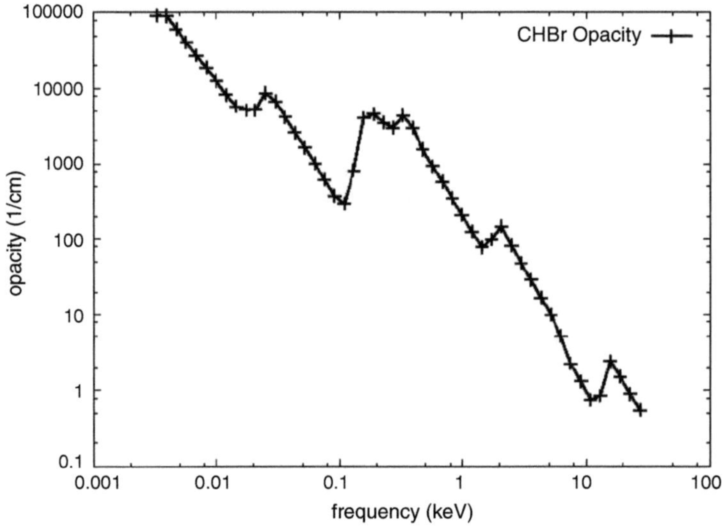

Fig. 4. The opacity of brominated plastic used in the Graziani prompt spectrum simulation as a function of frequency. The opacity is in units of $1\,\text{cm}^{-1}$ and the frequency is in units of keV

The IMC simulation was run in Cartesian coordinates on a 3D grid of 11 hexagons. These were stacked in the z direction. The first zone had the inner face at z = 0.02 cm, the radius of the inner sphere. The center of the tenth zone is the fiducial point location at $z = 0.04\,\text{cm}$. This requires that $9.5\Delta z = 0.02\,\text{cm}$; thus $\Delta z = 2.10526316 \times 10^{-3}\,\text{cm}$. The x and y coordinates of the nodes of the zone were on lines radiating from $z = 0$ at a 10 degree angle. Reflecting boundary conditions are imposed on the faces that are not in the x-y plane. This mesh is pictured in Fig. 5.

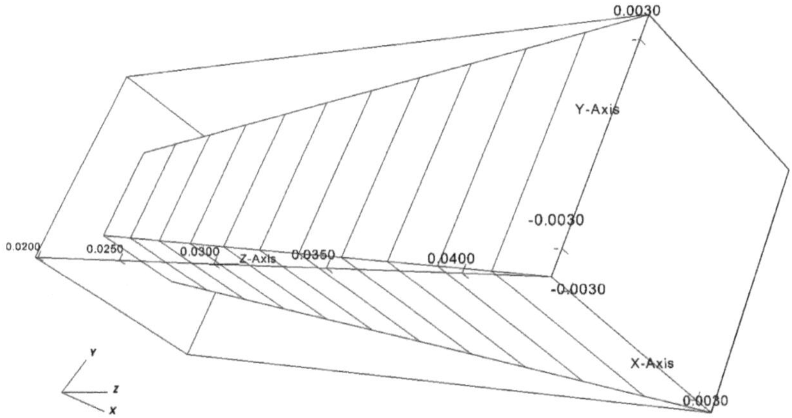

Fig. 5. The mesh used in the IMC simulation of the spherical Graziani prompt spectrum problem. The mesh has eleven zones in the z direction. The innermost face is at $z = 0.02\,\mathrm{cm}$ and the outermost face is at $z = 0.043157895\,\mathrm{cm}$, with $\Delta z = 2.10526316 \times 10^{-3}\,\mathrm{cm} = 0.02/9.5\,\mathrm{cm}$. The center of the tenth zone is the fiducial point location at $z = 0.04\,\mathrm{cm}$

The results of the simulation, the analytic answer, and the opacity are given for each group in Table 1.

The results of the IMC simulation, and the analytic answer, are shown in Fig. 6. (This is a plot of column 3 and column 4 vs. column 2 of Table 1.) This plot shows the radiation energy density in $\mathrm{erg\,cm^{-3}\,keV^{-1}}$ as a function of frequency in keV. The energy density has two maxima. The first, centered around 0.1 keV, is a Planckian formed from emission of the local material at $T_c = 0.03\,\mathrm{keV}$. The opacity of the material is very high at frequencies less than 0.5 keV, so almost no photons from the source get to the fiducial point in this range. The second maxima, centered around 1.0 keV, is formed by photons from the source that have passed through the outer material. The Planckian source, at $T_s = .3\,\mathrm{keV}$, radiates at a peak frequency of approximately 1 keV. Since the opacity of the material is low for frequencies equal to or greater than 1 keV, compared to the distance from the inner sphere to the fiducial point, higher frequency photons from the source can get to the fiducial point and contribute to the energy density in some of the higher frequency groups.

The IMC simulation used 8×10^5 particles and took about 2 min on four Pentium processors. The simulation answers are in good agreement with the analytic results. Some statistical noise is evident in groups with less energy in them. This is expected, since regions with lower energy density are represented by fewer particles.

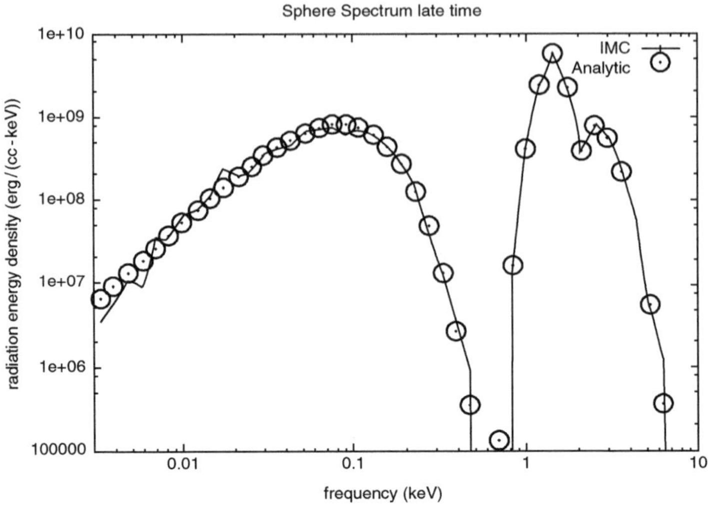

Fig. 6. The IMC results for the Graziani prompt spectrum problem and the analytic answer, (15) and (19). Radiation energy density is plotted vs. frequency. Values of energy density at frequencies below 0.5 keV are from the thermal emission at T_c by the local matter. Values of energy density at frequencies above 0.5 keV are from higher frequency photons from the source, which have not been fully absorbed. These photons have frequencies at which the brominated plastic opacity has a low opacity and hence a large mean free path (see Fig. 4). The energy density is in units of erg cm^{-3} keV^{-1} and the frequency is in units of keV

5 A Slab Version of Graziani's Prompt Spectrum Test Problem

The analytic answer to a slab version of Graziani's test problem can be derived from (18) and (19). Define $d \equiv r - R$, the distance of the fiducial point from the embedded sphere. Then let both r and R grow without bound, while keeping d fixed. In this limit, the embedded sphere becomes a plane source. The solutions become functions of d:

$$e_r(r,t,\nu) = B_\nu(T_c) + B_\nu(T_s) - B_\nu(T_c)F_\nu(d,t) \tag{21}$$

with $F_\nu(d,t)$ given by

$$F_\nu(r,t) = 0, ct < d \tag{22}$$

$$F_\nu(d,t) = \frac{1}{2}\left\{e^{-\sigma_\nu d} - \frac{d}{ct}e^{-\sigma_\nu ct} - \sigma_\nu d\left[E_1(\sigma_\nu d) - E_1(\sigma_\nu ct)\right]\right\} \tag{23}$$

for $d < ct < \infty$.

Since there is always some part of the source plane that is not in causal contact with the fiducial point, there is no steady-state solution for the slab case.

The slab version of Graziani's test problem was simulated using the same material as the spherical version. The same opacities and group structure were used. A face temperature source with a temperature of T_s is imposed at $z = 0$. The outer boundary of the problem had a face temperature source with a temperature of T_c imposed on it to avoid edge effects.

The mesh used had 11 zones spanning $[0.0, z = 2.3157895 \times 10^{-2}\,\text{cm}]$. The center of the tenth zone is the fiducial point location at $z = 0.02\,\text{cm}$, i.e., $d = 0.02\,\text{cm}$. The zones had extent 1 cm in the ignorable x and y directions.

As in the spherical case, the simulation was run to a time of $1.4 \times 10^{-12}\,\text{s}$ using a time step of $\Delta t = 10^{-13}\,\text{s}$, and the answer generated by the IMC simulation corresponds to a time of $1.35 \times 10^{-12}\,\text{s}$.

The results of the simulation, the analytic answer, and the opacity for the slab case are given for each group in Table 2.

The results of the IMC simulation of the slab problem and the analytic answer, are shown in Fig. 7. (This is a plot of column 3 and column 4 vs. column 2 of Table 2.) This plot shows the radiation energy density in $\text{erg}\,\text{cm}^{-3}\,\text{keV}^{-1}$ as a function of frequency in keV. As in the sphere case, the energy density has two maxima. The height of the maxima centered near 1 keV is higher in the sphere case, because the plane source emits many more photons toward the fiducial point than a spherical source at the same distance.

Because the source is stronger in the slab case, the IMC simulation used more particles and took slightly longer. The simulation used 2×10^6 particles and took about 3 min on four Pentium processors. As in the sphere case, the simulation answers are in good agreement with the analytic results, and some statistical noise is evident in groups with lower energy density.

Acknowledgment

This work was performed under the auspices of the U.S. Department of Energy by University of California Lawrence Livermore National Laboratory under contract No. W-7405-ENG-48.

Table 2. Results of the IMC simulation of the Graziani slab prompt spectrum problem

group	ν	e_r	analytic e_r	σ
0	3.28943e−03	0.00000e+00	6.56128e+06	9.16000e+04
1	3.95477e−03	0.00000e+00	9.37744e+06	9.06781e+04
2	4.75468e−03	6.08837e+06	1.33706e+07	6.08939e+04
3	5.71638e−03	2.21914e+07	1.90091e+07	4.08607e+04
4	6.87260e−03	0.00000e+00	2.69304e+07	2.72149e+04
5	8.26269e−03	3.39862e+07	3.79893e+07	1.86425e+04
6	9.93393e−03	9.77224e+07	5.33093e+07	1.24389e+04
7	1.19432e−02	5.35097e+07	7.43291e+07	8.19288e+03
8	1.43589e−02	1.28744e+08	1.02827e+08	5.79710e+03
9	1.72632e−02	9.27517e+07	1.40887e+08	5.14390e+03
10	2.07549e−02	1.57907e+08	1.90765e+08	5.20350e+03
11	2.49529e−02	2.83336e+08	2.54559e+08	8.69569e+03
12	3.00000e−02	3.82734e+08	3.33604e+08	6.67314e+03
13	3.60679e−02	5.08110e+08	4.27461e+08	4.15912e+03
14	4.33632e−02	4.12233e+08	5.32466e+08	2.62038e+03
15	5.21340e−02	4.44628e+08	6.40012e+08	1.64328e+03
16	6.26789e−02	9.17212e+08	7.35205e+08	1.01613e+03
17	7.53566e−02	8.23536e+08	7.97306e+08	6.19069e+02
18	9.05986e−02	6.79309e+08	8.05117e+08	3.75748e+02
19	1.08923e−01	9.75612e+08	7.49915e+08	2.97349e+02
20	1.30955e−01	8.02509e+08	6.06784e+08	8.21172e+02
21	1.57442e−01	3.78322e+08	4.32675e+08	4.01655e+03
22	1.89287e−01	2.77227e+08	2.59951e+08	4.54828e+03
23	2.27573e−01	1.14425e+08	1.26955e+08	3.50487e+03
24	2.73603e−01	5.79367e+07	4.83345e+07	3.02359e+03
25	3.28943e−01	1.37617e+07	1.36588e+07	4.34203e+03
26	3.95477e−01	2.32763e+06	2.70428e+06	2.98594e+03
27	4.75468e−01	0.00000e+00	3.50392e+05	1.55364e+03
28	5.71638e−01	0.00000e+00	2.75095e+04	9.42213e+02
29	6.87260e−01	0.00000e+00	2.85002e+05	5.76390e+02
30	8.26269e−01	3.22636e+07	3.91809e+07	3.52954e+02
31	9.93393e−01	9.34736e+08	9.85459e+08	2.09882e+02
32	1.19432e+00	6.30907e+09	6.33419e+09	1.26546e+02
33	1.43589e+00	1.64366e+10	1.64600e+10	7.80087e+01
34	1.72632e+00	6.11433e+09	6.22834e+09	9.97421e+01
35	2.07549e+00	9.91868e+08	1.00872e+09	1.48848e+02
36	2.49529e+00	2.36965e+09	2.28836e+09	8.22907e+01
37	3.00000e+00	1.77814e+09	1.80809e+09	4.86915e+01
38	3.60679e+00	7.47867e+08	7.26508e+08	2.91258e+01
39	4.33632e+00	1.37272e+08	1.64067e+08	1.68133e+01
40	5.21340e+00	1.77538e+07	2.01749e+07	9.92194e+00
41	6.26789e+00	7.71867e+05	1.34876e+06	5.18722e+00
42	7.53566e+00	0.00000e+00	4.44994e+04	2.24699e+00

Table 2. Continued

group	ν	e_r	analytic e_r	σ
43	9.05986e+00	0.00000e+00	6.33976e+02	1.29604e+00
44	1.08923e+01	0.00000e+00	3.49212e+00	7.46975e−01
45	1.30955e+01	0.00000e+00	6.12484e−03	8.43058e−01
46	1.57442e+01	0.00000e+00	2.66266e−06	2.43746e+00
47	1.89287e+01	0.00000e+00	2.44078e−10	1.50509e+00
48	2.27573e+01	0.00000e+00	3.13420e−15	9.01762e−01
49	2.73603e+01	0.00000e+00	3.80253e−21	5.38182e−01

e_r is the result of an IMC simulation, and analytic e_r results from (21) and (23), with $d = 0.02\,\text{cm}$, $t = 1.35 \times 10^{-12}\,\text{s}$, and σ_ν as given in column 5. The value of ν for each group is the rms value of the group bounds, which were logarithmically distributed between 3.0×10^{-3} and $30.0\,\text{keV}$. The energy density is in units of $\text{erg}\,\text{cm}^{-3}\,\text{keV}^{-1}$ and the opacity is in units of $1\,\text{cm}^{-1}$

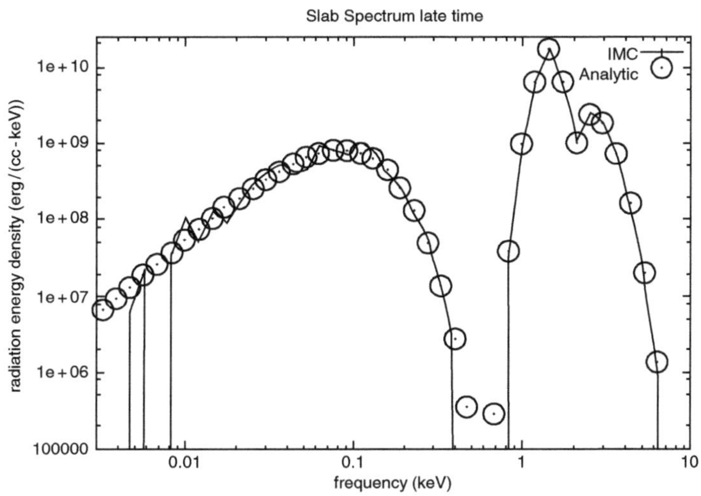

Fig. 7. The results for the slab version of the Graziani prompt spectrum problem and the analytic answer, (21) and (23). Radiation energy density is plotted vs. frequency. Values of energy density at frequencies below 0.5 keV are from the thermal emission at T_c by the local matter. Values of energy density at frequencies above 0.5 keV are from higher frequency photons from the source, which have not been fully absorbed. These photons have frequencies at which the brominated plastic opacity has a small opacity and hence a large mean free path (see Fig. 4). The energy density is in units of $\text{erg}\,\text{cm}^{-3}\,\text{keV}^{-1}$ and the frequency is in units of keV

References

[P73] G. C. Pomraning, *Equations of Radiation Hydrodynamics*, in International Series of Monographs in Natural Philosophy, Ed. by D. ter Harr (Pergamon, New York, 1973), Vol. 54

[BD77] W. E. Boyce and R. C. DiPrima, *Elementary Differential Equations and Boundary Value Problems*, 3rd edn., (Wiley, New York, 1977)

[FC71] J. A. Fleck, Jr. and J. D. Cummings, An implicit Monte Carlo scheme for calculating time and frequency dependent nonlinear radiation transport. J. Comput. Phys., **31**, 313–342 (1971)

[GKR98] N. A. Gentile, N. Keen, and J. Rathkopf, The KULL IMC Package, Tech. Rep. UCRL-JC-132743, Lawrence Livermore National Laboratory, Livermore, CA, 1998

[BM04] T. A. Brunner and T. A. Mehlhorn, A User's Guide to Radiation Transport in ALEGRA-HEDPP, Tech. Rep. SAND-2004-5799, Sandia National Laboratories, Albuquerque, NM, 2004

[G07] F. Graziani, The Prompt Spectrum of a Radiating Sphere, Tech. Rep. UCRL-TR-232619-DRAFT, Lawrence Livermore National Laboratory, Livermore, CA, 2007

[CTh68] W. Cody and H. Thatcher, Rational chebyshev approximations for the exponential integral E1(x), Math. Comput., **22**, 641–649 (1968)

[PTVF02] W. Press, S. A. Tuekolsky, W. T. Vetterling, B. P. Flannery, *Numerical Recipies in C++* (Cambridge University Press, United Kingdom, 2002)

The Prompt Spectrum of a Radiating Sphere: Benchmark Solutions for Diffusion and Transport

F. Graziani

Summary. An analytic 1D spherical time dependent solution for the spectrum of a sphere radiating into a surrounding semi-infinite material of arbitrary opacity but uniform material temperature is presented. Two solutions are presented. One is for the frequency dependent or multi-group diffusion equation and the other is for the transport equation. The solution is exact for non-participating media. It is also valid for participating media in the prompt or early phase where the gradient of the matter temperature in the surrounding medium is near zero. A companion paper presents a comparison between the analytic and numerical solutions for several realistic opacities. The solution should prove useful for benchmarking multi-group diffusion and transport codes.

1 Introduction

Determining the quantitative accuracy of a numerical algorithm in a physics or engineering code is an important step in assessing the believability of any result which the code may yield. Analytic solutions represent an absolute measure or benchmark to which numerical solutions may be compared. For example, problems such as the shock tube, Noh problem, and Sedov blast wave [1] have provided generations of computational physicists and engineers with the standard test suite which all hydrodynamic algorithms must pass.

Similarly, the need for analytic solutions of both the diffusion and Boltzmann descriptions of particle transport are great. For example, in modern multi-physics computer codes, the particle transport model usually dominates overall CPU time. In addition, typically, particle transport algorithms must be accurate for a wide variety of mesh distortions which also cover a wide range of optical depths. Analytic solutions to the Boltzmann and diffusion descriptions of particle transport have historically been difficult to find. Even in the simplest case of the 1D grey diffusion approximation, code developers of algorithms have been limited to either linear diffusion or the Marshak wave [2] as benchmark problems. Su and Olson [3], made a significant advance by solving the 1D slab grey non-equilibrium (two temperature) extension of the Marshak problem analytically for constant opacity and a specific heat going as T^3. Taken together, the linear diffusion, Marshak [2] and Su and Olson [3] solutions have proven useful in assessing the spatial and temporal accuracy of grey diffusion algorithms [4,5].

In general the types of problems in radiative transfer we are ultimately interested in verifying are functions of space, time, angle, and energy group. The multigroup transport and diffusion equations, in the absence of scattering, for photons traveling in a participating medium in local thermodynamic equilibrium are given by

$$\frac{1}{c}\frac{\partial \psi_\nu(x,\Omega,t)}{\partial t} + \Omega \bullet \nabla \psi_\nu(x,\Omega,t) = \sigma_\nu B_\nu(T) - \sigma_\nu \psi_\nu(x,\Omega,t) \tag{1}$$

Where the Planck function is defined by

$$B_\nu(T) = \frac{2h\nu^3}{c^2}\left(e^{\frac{h\nu}{kT}} - 1\right)^{-1}$$

In 1D spherical geometry, which will be the focus of this paper, azimuthal symmetry is assumed and we define

$$I_\nu(x,\mu,t) = 2\pi\psi_\nu(x,\mu,t)$$
$$\times \frac{1}{c}\frac{\partial I_\nu(x,\Omega,t)}{\partial t} + \mu\frac{\partial I_\nu(r,\mu,t)}{\partial r} + \frac{(1-\mu^2)}{r}\frac{\partial I_\nu(r,\mu,t)}{\partial \mu} = \frac{\sigma_\nu b_\nu(T)}{2} - \sigma_\nu I_\nu(x,\Omega,t) \tag{2}$$

Where the modified Planck function is defined by

$$b_\nu(T) = 4\pi B_\nu(T)$$

The spectrum is defined by

$$\varepsilon_\nu(r,t) = \int_{-1}^{+1} d\mu I_\nu(r,\mu,t) \tag{3}$$

and the flux

$$F_\nu(r,t) = \int_{-1}^{+1} d\mu\, \mu I_\nu(r,\mu,t) \tag{4}$$

Using (3),

$$\frac{1}{c}\frac{\partial \varepsilon_\nu(x,t)}{\partial t} = \nabla\cdot\left(\frac{1}{3\sigma_\nu}\nabla\varepsilon_\nu(x,t)\right) + \sigma_\nu b_\nu(T) - \sigma_\nu\varepsilon_\nu(x,t) \tag{5}$$

Equation (5) defines the multi-group diffusion equation in the absence of flux-limiting.

Again, in 1D we have

$$\nabla \cdot \left(\frac{1}{3\sigma_\nu} \nabla \varepsilon_\nu(x,t) \right) \equiv \frac{\partial}{\partial r} \left[\frac{1}{3\sigma_\nu} \frac{\partial \varepsilon_\nu(r,t)}{\partial r} \right] + \frac{2}{r} \left[\frac{1}{3\sigma_\nu} \frac{\partial \varepsilon_\nu(r,t)}{\partial r} \right] \tag{6}$$

The material temperature in (1) and (5) is given by T. The absorptivity is given by σ_ν which is in general a function of the material density and temperature.

If the medium the photons travel through is participating, meaning the material itself both absorbs and emits radiation, then there is a material energy balance equation associated with both (1) and (5) which also must be solved. As we will not need the material energy balance equation in our analysis, it is not quoted in this paper. However, interested readers may look at the review paper by Graziani [6] or the work of Bowers and Wilson [7].

As mentioned above, the existence of analytic solutions to either the Boltzmann or diffusion descriptions of radiative transfer have historically been made by invoking the grey approximation. Except for the picket fence multi-group diffusion solution for the two temperature Marshak wave given by Su and Olson [8] and the extension by Bolstad and Shestakov [9], there are no benchmark solutions for the multigroup equations the author is aware of. In particular, as far as the author is aware, there are no analytic solutions of the time dependent radiation spectrum itself. Of course, multigroup transport or diffusion codes can always be benchmarked against the above grey diffusion results by making the opacities independent of frequency. However, this assumption fails to address the central feature of multigroup diffusion: strong frequency dependence of the opacities yielding time evolving non-Planckian radiation spectra. The need for a benchmark problem whose solution is a space and time dependent spectrum is clearly needed.

2 The Prompt Spectrum: Problem Definition

It is clear that a benchmark problem which yields an answer for the time dependent spectrum would be useful. However, the multigroup equations are even a more difficult challenge than solving the coupled photon-matter equations in the grey limit. This is because the matter equation is an integro-differential equation with all parts of the spectrum contributing simultaneously to the matter temperature. Given this high degree of non-linearity the hope for an analytic solution to the multigroup diffusion equations seems hopeless. However, there are situations in which this non-linear coupling between matter and radiation can be simplified and the resulting equations linearized. If we focus the attention on the material opacity and ignore scattering, then the general trend of the absorption opacity is that it decreases strongly with frequency. Consider a hot source of radiation in contact with a cold material. Imagine an observer being in the cold material at some distance r from the source with a piece of diagnostic equipment that measures photon number and photon frequency. If the spectrum of the hot source consists of some distribution of frequencies (e.g. Planckian), then the observer will register the high frequency photons arriving first since it is they that have the largest mean free path. An interesting feature of these photons is that their large mean free path also implies that their coupling to the matter is very weak. In fact, it is so weak that the transport of these high energy photons takes place in a medium that is non-participating. That is, during this prompt phase, matter and radiation decouple and the material temperature is effectively frozen at its initial value. These high frequency photons make up the prompt or early part of the spectrum as measured by the observer. It is the low frequency portion of the source spectrum that couples strongly to the material temperature and hence causes the material temperature to eventually rise. Figure 1a shows the material and radiation temperatures as a function of time as measured by the observer in CH foam who is 1 mm from a Planckian source. Figure 1b shows the spectrum as a function of frequency at the same location. An equivalent argument concerning the prompt spectrum and its decoupling from the material response can be made if we consider a material with a very large heat capacity.

In light of the above discussion, consider a sphere of radius R, and temperature T_S which is surrounded by a uniform semi-infinite medium characterized by a heat capacity C_V an absorption opacity σ_ν and an initial temperature $T_m < T_S$. We wish to ask the following question; what is the time dependent prompt spectrum $\varepsilon_\nu(r,t)$ at a radius $r > R$ in the absence of any hydrodynamic motion?

3 The Prompt Spectrum: Mathematical Derivation

3.1 Diffusion

In a 1D spherical geometry, the multi-group diffusion equation is given by

$$\frac{\partial \varepsilon_\nu(r,t)}{\partial t} = \frac{\partial}{\partial r}\left[\frac{1}{3\sigma_\nu}\frac{\partial \varepsilon_\nu(r,t)}{\partial r}\right] + \frac{2}{r}\left[\frac{1}{3\sigma_\nu}\frac{\partial \varepsilon_\nu(r,t)}{\partial r}\right] + \sigma_\nu\big(b_\nu(T) - \varepsilon_\nu(r,t)\big) \quad (7)$$

With the specified initial and boundary conditions

$$\begin{aligned} \varepsilon_\nu(r,t=0) &= b_\nu(T_m) \\ \varepsilon_\nu(r=R,t) &= b_\nu(T_S) \\ \varepsilon_\nu(r\to\infty,t) &= 0 \end{aligned} \quad (8)$$

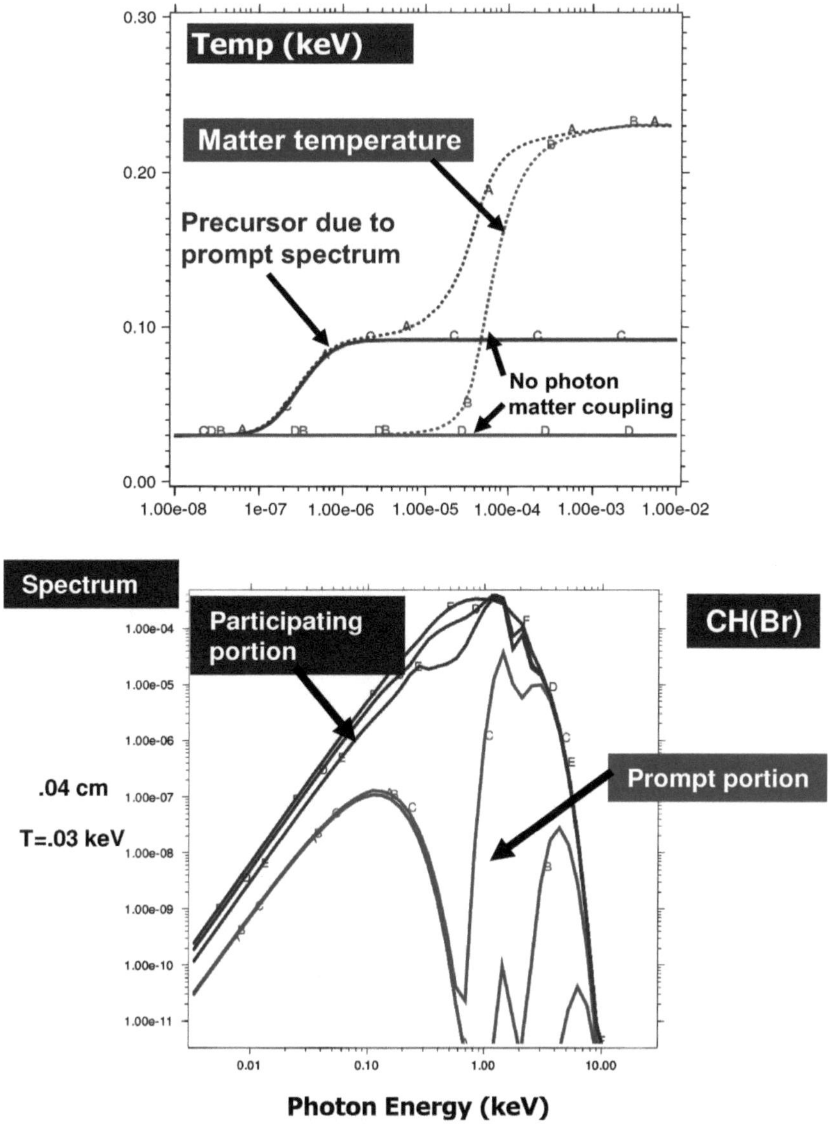

Fig. 1. Figure (1a) [top], CH material and radiation temperatures as a function of time, Figure (1b) [bottom] time-snapshots of the radiation spectrum in the CH, showing the prompt portion. (See Plate 21 on page 311)

It is assumed that the material is initially emitting with a Planckian distribution at a temperature T_m and that the sphere is radiating with a Planckian distribution at a temperature T_S. The choice of Planck functions for the initial and boundary conditions is merely for convenience. The solution described below can be carried out with general forms for (8). The only requirement is that the solution vanishes at infinity. For the prompt spectrum, the material temperature is fixed at the initial matter temperature and this implies that the opacity is a function only of the initial matter temperature. Hence, we can write

$$\frac{\partial \varepsilon_\nu(r,t)}{\partial t} = \frac{1}{3\sigma_\nu}\left[\frac{\partial^2 \varepsilon_\nu(r,t)}{\partial r^2} + \frac{2}{r}\frac{\partial \varepsilon_\nu(r,t)}{\partial r}\right] + \sigma_\nu\left(b_\nu(T_m) - \varepsilon_\nu(r,t)\right) \quad (9)$$

At this point, we have reduced the multi-group problem to a linear partial differential equation (PDE) with known sink and source terms. Our strategy is to reduce (9) to an ordinary differential equation by performing a Laplace transform in time. Applying the Laplace transform defined by

$$U_\nu(r,p) = \int_0^\infty dt e^{-pt} \varepsilon_\nu(r,t) \quad (10)$$

to (9) we obtain

$$\frac{c}{3\sigma_\nu}\left[\frac{\partial^2 U_\nu(r,p)}{\partial r^2} + \frac{2}{r}\frac{\partial U_\nu(r,p)}{\partial r}\right] - (p + c\sigma_\nu)\, U_\nu(r,p) + \left(1 + \frac{c\sigma_\nu}{p}\right) b_\nu(T_m) = 0 \quad (11)$$

This must obey the modified boundary condition,

$$U_\nu(r = R, p) = \frac{b_\nu(T_S)}{p} \quad (12)$$

We seek a homogeneous and particular solution of (11) such that

$$U_\nu(r,p) = U_\nu^H(r,p) + U_\nu^P(r,p) \quad (13)$$

Clearly, the particular solution is simply given by

$$U_\nu^P(r,p) = \frac{b_\nu(T_m)}{p} \quad (14)$$

Whereas the homogeneous solution obeys

$$\frac{c}{3\sigma_\nu}\left[\frac{\partial^2 U_\nu^H(r,p)}{\partial r^2} + \frac{2}{r}\frac{\partial U_\nu^H(r,p)}{\partial r}\right] - (p + c\sigma_\nu)\, U_\nu^H(r,p) = 0 \quad (15)$$

Specifying the ansatz $U_\nu^H(r,p) \equiv \frac{\Phi_\nu(r,p)}{r}$. Substituting this expression into (15), yields

$$\frac{c}{3\sigma_\nu}\left[\frac{\partial^2 \Phi_\nu(r,p)}{\partial r^2}\right] - (p + c\sigma_\nu)\, \Phi_\nu(r,p) = 0 \quad (16)$$

Using the fact that $U_\nu^H(r,p) \equiv \frac{\Phi_\nu(r,p)}{r}$ and requiring that $\Phi_\nu(r,p)$ be finite as $r \to \infty$, we obtain

$$U_\nu^H(r,p) = \frac{A}{r}\exp\left(-r\sqrt{\frac{3\sigma_\nu}{c}}\,(p + c\sigma_\nu)\right) \tag{17}$$

Combining the homogeneous and particular solutions

$$U_\nu^H(r,p) = \frac{b_\nu(T_m)}{p} + \frac{A}{r}\exp\left(-r\sqrt{\frac{3\sigma_\nu}{c}\,(p + c\sigma_\nu)}\right) \tag{18}$$

Using the modified boundary condition to determine A gives us the final solution in Laplace transform space

$$\begin{aligned} U_\nu^H(r,p) &= \frac{b_\nu(T_m)}{p} \\ &+ \left(\frac{R}{r}\right)\frac{(b_\nu(T_S) - b_\nu(T_m))}{p}\exp\left(-(r-R)\sqrt{\frac{3\sigma_\nu}{c}\,(p + c\sigma_\nu)}\right) \end{aligned} \tag{19}$$

The space and time solution to the prompt spectrum (as predicted from multi-group diffusion) is given by the inverse Laplace transform of (19).

$$\varepsilon_\nu(r,t) = b_\nu(T_m) + \left(\frac{R}{r}\right)(b_\nu(T_S) - b_\nu(T_m))\,F_\nu(r,t) \tag{20}$$

Where the function $F_\nu(r,t)$ is defined by

$$\begin{aligned} F_\nu(r,t) &= \frac{1}{2}\left\{\exp\left(-\sqrt{3}\sigma_\nu\,(r-R)\right)\mathrm{Erfc}\left[\frac{1}{2}\sqrt{\frac{3\sigma_\nu}{ct}}\,(r-R) - \sqrt{\sigma_\nu ct}\right]\right\} \\ &+ \frac{1}{2}\left\{\exp\left(-\sqrt{3}\sigma_\nu\,(r-R)\right)\mathrm{Erfc}\left[\frac{1}{2}\sqrt{\frac{3\sigma_\nu}{ct}}\,(r-R) + \sqrt{\sigma_\nu ct}\right]\right\} \end{aligned} \tag{21}$$

$\mathrm{Erfc}[x] = \frac{2}{\sqrt{\pi}}\int_x^\infty dy\,\exp(-y^2)$ is the complementary error function.

Equations (20) and (21) define the prompt spectrum of a sphere diffusing into a material. It is valid for all radii and time and opacities provided the heating of the material is negligible. Equivalently, for code verification purposes, the solution is valid when the material possesses infinite heat capacity. The interpolation function $F_\nu(r,t)$, has the following properties

$$\begin{aligned} F_\nu(r = R, t) &= 1 \\ F_\nu(r, t = 0) &= 0 \\ F_\nu(r, t \to \infty) &= \exp\left(-\sqrt{3}\sigma_\nu\,(r-R)\right) \end{aligned} \tag{22}$$

3.2 Transport

The Specific Intensity

The transport description of the prompt radiation spectrum at a position r at a time t is given by the following equation with absorption and emission terms. Namely,

$$\frac{1}{c}\frac{\partial I_\nu(r,\mu,t)}{\partial t} + \mu\frac{\partial I_\nu(r,\mu,t)}{\partial r} + \frac{(1-\mu^2)}{r}\frac{\partial I_\nu(r,\mu,t)}{\partial \mu} + \sigma_\nu I_\nu(r,\mu,t) = \frac{\sigma_\nu}{2} b_\nu(T_m) \tag{23}$$

$$\begin{aligned} I_\nu(r,\mu,t=0) &= \frac{b_\nu(T_m)}{2} \quad \forall\mu \\ I_\nu(r=R,\mu,t) &= b_\nu(T_S) \quad \sqrt{1-\frac{R^2}{r^2}} \le \mu \le 1 \\ I_\nu(r=\Lambda,\mu,t) &= \frac{1}{2}b_\nu(T_m) \quad \sqrt{1-\frac{R^2}{r^2}} \le \mu \le 0 \\ I_\nu(r=\Lambda,\mu,t) &= \frac{1}{2}b_\nu(T_m) \qquad -1 \le \mu \le 0 \end{aligned} \tag{24}$$

The initial and boundary conditions described in (24), describe a sphere of radius R embedded in a larger sphere of radius Λ. The smaller sphere radiates in the forward plane with a Planckian at a source temperature. The first set of limits on the angle is the result of the solid angle subtended by a sphere of radius R as seen by an observer positioned at a radius r. The second and third set of limits is a consequence of applying an inwardly directed source of Planckian radiation at the surface of the outer sphere. This boundary condition effectively makes the medium semi-infinite.

As in the diffusion example, the modified transport equation, (23), has the time dependent material temperature replaced by the initial material temperature in both the emission term (Planck function) and in the opacities. Therefore, we arrive at a linearization of the non-linear problem which is much more amenable to solution.

The strategy for solving (23) follows lines similar to the diffusion example. That is, we first perform a Laplace transform in time to reduce the number of dependent variables in the PDE. By applying the Laplace transform, (23) is reduced to a first order linear PDE in two dependent variables, where the method of characteristics can be easily applied.

Applying the Laplace transform in time to (23) yields,

$$\begin{aligned} &\mu\frac{\partial \Phi_\nu(r,\mu,p)}{\partial r} + \frac{(1-\mu^2)}{r}\frac{\partial \Phi_\nu(r,\mu,p)}{\partial \mu} \\ &+ \left(\sigma_\nu + \frac{p}{c}\right)\Phi_\nu(r,\mu,p) = \left(\frac{\sigma_\nu}{2p} + \frac{1}{2c}\right) b_\nu(T_m) \end{aligned} \tag{25}$$

This obeys the modified boundary condition,

$$\Phi_\mathrm{v}(\mathrm{r}=\mathrm{R},\mu,\mathrm{t}) = \frac{\mathrm{b_v(T_S)}}{2p}\ \sqrt{1-\frac{\mathrm{R}^2}{\mathrm{r}^2}} \le \mu \le 1$$

$$\Phi_\mathrm{v}(\mathrm{r}=\Lambda,\mu,\mathrm{p}) = \frac{\mathrm{b_v(T_m)}}{2\mathrm{p}}\ 0 \le \mu \le \sqrt{1-\frac{\mathrm{R}^2}{\mathrm{r}^2}} \tag{26}$$

$$\Phi_\mathrm{v}(\mathrm{r}=\Lambda,\mu,\mathrm{p}) = \frac{\mathrm{b_v(T_m)}}{2\mathrm{p}} - 1 \le \mu \le 0$$

Where

$$\Phi_\mathrm{v}(\mathrm{r},\mu,\mathrm{p}) = \int_0^\infty \mathrm{dt}e^{-\mathrm{pt}}\mathrm{I_v}(\mathrm{r},\mu,\mathrm{t})$$

Transforming to a new set of coordinates (characteristics) defined by

$$\begin{aligned}\xi &= \mathrm{r}\\ \eta &= \mathrm{r}\sqrt{1-\mu^2}\end{aligned} \tag{27}$$

Equation (25) becomes the relatively simple set of equations

$$\frac{\partial\Phi_\nu(\xi,\eta,\mathrm{p})}{\partial\xi} + \frac{\left(\sigma_\mathrm{v}+\frac{\mathrm{p}}{\mathrm{c}}\right)}{\sqrt{1-\left(\frac{\eta}{\xi}\right)^2}}\Phi_\nu(\xi,\eta,\mathrm{p}) = \frac{\frac{1}{2}\left(\sigma_\mathrm{v}+\frac{\mathrm{p}}{\mathrm{c}}\right)}{\sqrt{1-\left(\frac{\eta}{\xi}\right)^2}}\mathrm{b}_\nu(\mathrm{T_m}) \tag{28}$$

Equation (28) can be easily solved to yield

$$\Phi_\nu\left(\xi,\eta,\mathrm{p}\right) = \frac{\mathrm{b_v(T_m)}}{2\mathrm{p}} + \frac{\left[\mathrm{b_v(T_S)} - \mathrm{b_v(T_m)}\right]}{2\mathrm{p}}\exp\left(-\mathrm{A}\left(\xi\right)\right), \sqrt{1-\frac{\mathrm{R}^2}{\mathrm{r}^2}} \le \mu \le 1$$

$$\Phi_\nu\left(\xi,\eta,\mathrm{p}\right) = \frac{\mathrm{b_v(T_m)}}{2\mathrm{p}}, -1 \le \mu \le \sqrt{1-\frac{\mathrm{R}^2}{\mathrm{r}^2}} \tag{29}$$

$$\mathrm{A}\left(\xi\right) = \left[\sigma_\mathrm{v}+\frac{\mathrm{p}}{\mathrm{c}}\right]\mathrm{g}\left(\xi,\eta\right) \quad \text{and where}$$

$$\mathrm{g}\left(\xi,\eta\right) = \sqrt{\xi^2-\eta^2} - \sqrt{\mathrm{R}^2-\eta^2}$$

The inverse Laplace transform of (29) is easily performed by noting that

$$\begin{aligned}&\frac{\mathrm{b_v(T)}}{\mathrm{p}} \Rightarrow \mathrm{b_v(T)}\\ &\frac{\exp\left(-\frac{\mathrm{p}}{\mathrm{c}}\mathrm{g}\left(\xi,\eta\right)\right)}{\mathrm{p}} \Rightarrow \Theta\left(\mathrm{t}-\frac{\mathrm{g}\left(\xi,\eta\right)}{\mathrm{c}}\right)\end{aligned} \tag{30}$$

Transforming back to the original r, μ coordinates, the solution to the transport equation becomes,

$$I_\nu(r,\mu,t) = \frac{b_\nu(T_m)}{2} + \frac{[b_\nu(T_S) - b_\nu(T_m)]}{2}\exp\left(-\sigma_\nu g(r,\mu)\right)\Theta\left(t - \frac{g(r,\mu)}{c}\right),$$

$$\sqrt{1-\frac{R^2}{r^2}} \le \mu \le 1$$

$$I_\nu(r,\mu,t) = \frac{b_\nu(T_m)}{2}, \quad -1 \le \mu \le \sqrt{1-\frac{R^2}{r^2}} \tag{31}$$

The function $g(r,\mu)$ can be written

$$g(r,\mu) = \mu r - \sqrt{R^2 - r^2 + \mu^2 r^2} \quad \text{where} \quad g(r=R,\mu) = 0 \tag{32}$$

The function $g(r,\mu)$ represents the path length traveled by a photon which has been emitted at the surface of the sphere in a direction μ. The solution for the radiative specific intensity equation (31) therefore has a simple interpretation namely; the solution is a linear combination of the initial emission function and the attenuation of the radiating spherical source along the optical path between sphere surface and the spatial point (r,μ). The Θ functions enforce causality. This is exactly the type of result to be expected from a characteristic interpretation of the transport equation.

Unfortunately we are not done. From the specific intensity solution equation (31) we still must construct the spectrum $\varepsilon_\nu(r,t)$. In fact, the construction of the spectrum is actually a more challenging analytic problem to solve than the solution of the transport equation itself. The derivation of the specific intensity solution has been straightforward because in the absence of non-linear interactions with the material, the derivation described above is based on a simple characteristics interpretation of the first order differential representation of the transport equation which is quite amenable to an analytic solution.

The Spectrum

The spectrum is defined by

$$\varepsilon_\nu(r,t) = \int_{-1}^{+1} d\mu I_\nu(r,\mu,t) = \int_{\sqrt{1-\frac{R^2}{r^2}}}^{+1} d\mu I_\nu(r,\mu,t) + \int_{-1}^{\sqrt{1-\frac{R^2}{r^2}}1} d\mu I_\nu(r,\mu,t) \tag{33}$$

To simplify the notation, we define

$$I_\nu^\pm(r,\mu,t) = \frac{b_\nu(T_m)}{2} + J_\nu^\pm(r,\mu,t) \tag{34}$$

where

$$J_\nu^+(r,\mu,t) = \frac{[b_\nu(T_S) - b_\nu(T_m)]}{2} \times \exp(-\sigma_\nu g(r,\mu))\,\Theta\left(t - \frac{g(r,\mu)}{c}\right), \quad \sqrt{1 - \frac{R^2}{r^2}} \le \mu \le 1 \tag{35}$$

$$J_\nu^-(r,\mu,t) = 0, \quad -1 \le \mu \le \sqrt{1 - \frac{R^2}{r^2}}$$

Using (33)–(35), we obtain

$$\varepsilon_\nu(r,t) = B_\nu(T_m) + \int_{\sqrt{1-\frac{R^2}{r^2}}}^{+1} d\mu J_\nu^+(r,\mu,t) \tag{36}$$

We show in detail how the "+" component integral is evaluated. The derivation is quite involved and readers seeking the final solution should skip to the end of this sub-section.

Focusing our attention on the "+" component, we have

$$\begin{aligned} E_\nu^+(r,t) &\equiv \int_{\sqrt{1-\frac{R^2}{r^2}}}^{+1} d\mu J_\nu^+(r,\mu,t) \\ &= \frac{[b_\nu(T_S) - b_\nu(T_m)]}{2} \int_{\sqrt{1-\frac{R^2}{r^2}}}^{+1} d\mu \exp(-\sigma_\nu g(r,\mu))\,\Theta\left(t - \frac{g(r,\mu)}{c}\right) \end{aligned} \tag{37}$$

We now transform the integration variable into an integration over the photon path length. Namely,

$$\begin{aligned} \tau &= g(r,\mu) \\ d\mu &= \frac{1}{2r}\left[1 - \frac{(r^2 - R^2)}{\tau^2}\right] d\tau \end{aligned} \tag{38}$$

Equation (37) becomes,

$$E_\nu^+(r,t) = \frac{[b_\nu(T_S) - b_\nu(T_m)]}{4r} \int_{r-R}^{\sqrt{r^2-R^2}} d\tau \left[\frac{(r^2 - R^2)}{\tau^2} - 1\right] \exp(-\sigma_\nu \tau)\,\Theta\left(t - \frac{\tau}{c}\right) \tag{39}$$

We now make use of the integral definition of the Heaviside step-function $\Theta(x)$ given by

$$\Theta(x) = \frac{1}{2\pi i} \int_{\Gamma - i\infty}^{\Gamma + i\infty} ds \frac{e^{sx}}{s} \tag{40}$$

Γ is a contour defined to lie to the left of any pole in the complex plane. We substitute (40) into (39) and exchange the order of the integrations to obtain

$$\begin{aligned} E_\nu^{+}(r,t) = & \frac{[b_\nu(T_S) - b_\nu(T_m)]}{4r} \\ & \times \frac{1}{2\pi i} \int_{\Gamma - i\infty}^{\Gamma + i\infty} ds \frac{e^{st}}{s} \int_{r-R}^{\sqrt{r^2-R^2}} d\tau \left[\frac{(r^2 - R^2)}{\tau^2} - 1 \right] \exp\left(- \left(\sigma_\nu + \frac{s}{c} \right) \tau \right) \end{aligned} \tag{41}$$

We now perform the integration over the τ variable. It proves useful to decompose the integration over τ into a difference between semi-infinite integrals.

$$\begin{aligned} \int_{r-R}^{\sqrt{r^2-R^2}} d\tau F(\tau) = \int_{r-R}^{\infty} d\tau F(\tau) - \int_{\sqrt{r^2-R^2}}^{\infty} d\tau F(\tau) E_\nu^{+}(r,t) = & \frac{[b_\nu(T_S) - b_\nu(T_m)]}{4r} \\ \times \frac{(r^2 - R^2)}{2\pi i} \int_{\Gamma - i\infty}^{\Gamma + i\infty} ds \frac{e^{st}}{s} \Bigg\{ \int_{r-R}^{\infty} \frac{d\tau}{\tau^2} \exp\left(- \left(\sigma_\nu + \frac{s}{c} \right) \tau \right) & \\ - \int_{\sqrt{r^2-R^2}}^{\infty} \frac{d\tau}{\tau^2} \exp\left(- \left(\sigma_\nu + \frac{s}{c} \right) \tau \right) \Bigg\} - & \frac{[b_\nu(T_S) - b_\nu(T_m)]}{4r} \\ \frac{1}{2\pi i} \int_{\Gamma - i\infty}^{\Gamma + i\infty} ds \frac{e^{st}}{s} \frac{\left[\exp\left(- \left(\sigma_\nu + \frac{s}{c} \right) (r - R) \right) - \exp\left(- \left(\sigma_\nu + \frac{s}{c} \right) \sqrt{r^2 - R^2} \right) \right]}{\left(\sigma_\nu + \frac{s}{c} \right)} & \end{aligned} \tag{42}$$

The integrals over τ can now be evaluated by making use of the definition of the Exponential Integral function. That is,

$$\begin{aligned} \int_{a}^{\infty} \frac{d\tau}{\tau^2} \exp(-\alpha\tau) &= \frac{\exp(-\alpha a)}{a} + \alpha \mathrm{Ei}[-\alpha a] \\ \mathrm{Ei}[x] &= \int_{-x}^{\infty} \frac{dy}{y} \exp(-y), \; x < 0 \end{aligned} \tag{43}$$

The former integral in (43) is easily derived using integration by parts applied to the definition of the Exponential Integral. Using (43), and defining $\Omega(s) = \sigma_v + \frac{s}{c}$ we immediately arrive at,

$$E_v^+(r,t) = \frac{[b_v(T_S) - b_v(T_m)]}{4r} \times \frac{(r^2 - R^2)}{2\pi i} \int_{\Gamma - i\infty}^{\Gamma + i\infty} ds \frac{e^{st}}{s} \left\{ \begin{array}{l} \left[\frac{\exp(-(r-R)\Omega(s))}{(r-R)} + \Omega(s) \, Ei\left[-(r-R)\Omega(s)\right] \right] \\ - \left[\frac{\exp\left(-\sqrt{r^2-R^2}\Omega(s)\right)}{\sqrt{r^2-R^2}} + \Omega(s) \, Ei\left[-\sqrt{r^2-R^2}\Omega(s)\right] \right] \end{array} \right\}$$
$$- \frac{[b_v(T_S)-b_v(T_m)]}{4r} \frac{1}{2\pi i} \int_{\Gamma - i\infty}^{\Gamma + i\infty} ds \frac{e^{st}}{s} \frac{\left[\exp(-(r-R)\Omega(s)) - \exp\left(-\sqrt{r^2-R^2}\Omega(s)\right)\right]}{\Omega(s)} \tag{44}$$

A more compact notation for (3.2) can be used when we make use of the definition of the inverse Laplace transform.

$$L^{-1}[f(s)] = \frac{1}{2\pi i} \int_{\Gamma - i\infty}^{\Gamma + i\infty} ds e^{st} f(s) \tag{45}$$

Therefore, (3.2) can be written as a sum of Laplace transforms as follows

$$E_v^+(r,t) = \frac{[b_v(T_S) - b_v(T_m)]}{4r} (r^2 - R^2)$$
$$\left\{ \begin{array}{l} \left(L^{-1}\left[\frac{\exp(-(r-R)\Omega(s))}{s(r-R)}\right] + L^{-1}\left[\frac{\Omega(s) Ei[-(r-R)\Omega(s)]}{s}\right] \right) \\ - \left(L^{-1}\left[\frac{\exp\left(-\sqrt{r^2-R^2}\Omega(s)\right)}{s\sqrt{r^2-R^2}}\right] + L^{-1}\left[\frac{\Omega(s) Ei\left[-\sqrt{r^2-R^2}\Omega(s)\right]}{s}\right] \right) \end{array} \right\}$$
$$- \frac{[b_v(T_S)-b_v(T_m)]}{4r} \left\{ L^{-1}\left[\frac{[\exp(-(r-R)\Omega(s))]}{s\Omega(s)}\right] - L^{-1}\left[\frac{\left[\exp\left(-\sqrt{r^2-R^2}\Omega(s)\right)\right]}{s\Omega(s)}\right] \right\} \tag{46}$$

The first and third inverse transforms in (46) can be easily evaluated. Define the function

$$F[t;\alpha,\beta] = \frac{1}{2\pi i} \int_{\Gamma - i\infty}^{\Gamma + i\infty} ds \frac{e^{st}}{s+\beta} \exp(-\alpha s) = \Theta(t-\alpha) \exp(-\beta(t-\alpha)) \tag{47}$$

$$L^{-1}\left[\frac{\exp(-(r-R)\Omega(s))}{s(r-R)}\right] = \frac{\exp(-(r-R)\sigma_v)}{(r-R)} F\left[t; \frac{(r-R)}{c}, 0\right]$$
$$L^{-1}\left[\frac{\exp\left(-\sqrt{r^2-R^2}\Omega(s)\right)}{s\sqrt{r^2-R^2}}\right] = \frac{\exp\left(-\sqrt{r^2-R^2}\sigma_v\right)}{\sqrt{r^2-R^2}} F\left[t; \frac{\sqrt{r^2-R^2}}{c}, 0\right] \tag{48}$$

The last two inverse Laplace transforms in (46) can also be easily evaluated. Decomposing the integrands contained in the inverse Laplace transforms of the last two terms of (46) into partial fractions, allows us to write for the (r − R) term,

$$L^{-1}\left[\frac{\exp\left(-\left(r-R\right)\Omega\left(s\right)\right)}{s\Omega\left(s\right)}\right]$$
$$=\frac{\exp\left(-\left(r-R\right)\sigma_{v}\right)}{\sigma_{v}}\left\{L^{-1}\left[\frac{\exp\left(-\frac{(r-R)s}{c}\right)}{s}\right]-L^{-1}\left[\frac{\exp\left(-\frac{(r-R)s}{c}\right)}{s+c\sigma_{v}}\right]\right\} \tag{49}$$

A similar expression of course is valid for the $\sqrt{r^2-R^2}$ term. Each of the inverse Laplace transforms contained in (49) is known from (47). Therefore, we can write

$$L^{-1}\left[\frac{\exp\left(-\left(r-R\right)\Omega\left(s\right)\right)}{s\Omega\left(s\right)}\right]$$
$$=\frac{\exp\left(-\left(r-R\right)\sigma_{v}\right)}{\sigma_{v}}\left\{F\left[t;\frac{(r-R)}{c},0\right]-F\left[t;\frac{(r-R)}{c},c\sigma_{v}\right]\right\}$$
$$L^{-1}\left[\frac{\exp\left(-\sqrt{r^2-R^2}\Omega\left(s\right)\right)}{s\Omega\left(s\right)}\right] \tag{50}$$
$$=\frac{\exp\left(-\sqrt{r^2-R^2}\sigma_{v}\right)}{\sigma_{v}}\left\{F\left[t;\frac{\sqrt{r^2-R^2}}{c},0\right]-F\left[t;\frac{\sqrt{r^2-R^2}}{c},c\sigma_{v}\right]\right\}$$

The final inverse Laplace transforms that need to be evaluated involve terms that look like $L^{-1}\left[\frac{\Omega(s)Ei\left[-(r-R)\Omega(s)\right]}{s}\right]$ and $L^{-1}\left[\frac{\Omega(s)Ei\left[-\sqrt{r^2-R^2}\Omega(s)\right]}{s}\right]$. Therefore, calculating these last unknown contributions can be reduced to determining,

$$L^{-1}\left[Ei\left[-\alpha-\beta s\right]\right] \text{ and } L^{-1}\left[\frac{Ei\left[-\alpha-\beta s\right]}{s}\right] \tag{51}$$

Using the definition of the $Ei\left(x\right)$ function and the definition of the inverse Laplace transform we have,

$$L^{-1}\left[Ei\left[-\alpha-\beta s\right]\right]=\frac{1}{2\pi\pi}\int_{\Gamma-i\infty}^{\Gamma+i\infty}ds e^{st}\int_{\alpha+\beta s}^{\infty}\frac{dy}{y}\exp\left(-y\right) \tag{52}$$

By translating the y variable in (52), defining $z=y-(\alpha+\beta s)$ and changing the order of integration, we obtain for (52),

$$L^{-1}\left[\mathrm{Ei}\left[-\alpha-\beta s\right]\right]=\frac{\exp\left(-\alpha\right)}{\beta}\int_{0}^{\infty}dz\exp\left(-z\right)\frac{1}{2\pi\pi}\int_{\Gamma-i\infty}^{\Gamma+i\infty}ds\; e^{st}\frac{\exp\left(-\beta s\right)}{s+\frac{\alpha}{\beta}+\frac{z}{\beta}} \tag{53}$$

Using the definition of the F function in (47), we obtain,

$$\begin{aligned} L^{-1}\left[\mathrm{Ei}\left[-\alpha-\beta s\right]\right] &= \frac{\exp\left(-\alpha\right)}{\beta}\int_{0}^{\infty}dz\exp\left(-z\right)F\left[t;\beta,\frac{\alpha}{\beta}+\frac{z}{\beta}\right] \\ &= \frac{\exp\left(-\frac{\alpha t}{\beta}\right)\Theta\left(t-\beta\right)}{t} \end{aligned} \tag{54}$$

The identical procedure is followed for determining the second of the two expressions in (51). The only complication is the extra factor of s appearing in the denominator. This is easily handled by decomposing the integrand into partial fractions. The final results is,

$$L^{-1}\left[\frac{\mathrm{Ei}\left[-\alpha-\beta s\right]}{s}\right]=\Theta\left(t-\beta\right)\left(\mathrm{Ei}\left(-\alpha\right)-\mathrm{Ei}\left(-\frac{\alpha t}{\beta}\right)\right) \tag{55}$$

Combining (48)–(50), (54), and (55) yields,

$$\begin{aligned} E_{v}^{+}(r,t) &= \frac{\left[b_{v}(T_{S})-b_{v}(T_{m})\right]}{4r} \\ &\left\{\begin{aligned} &(r^{2}-R^{2})\left(\frac{e^{(-(r-R)\sigma_{v})}}{(r-R)}F\left[t;\frac{(r-R)}{c},0\right]-L^{-1}\left[\frac{\Omega(s)\mathrm{Ei}[-(r-R)\Omega(s)]}{s}\right]\right) \\ &-(r^{2}-R^{2})\left(\frac{e^{\left(-\sqrt{r^{2}-R^{2}}\sigma_{v}\right)}}{\sqrt{r^{2}-R^{2}}}F\left[t;\frac{\sqrt{r^{2}-R^{2}}}{c},0\right]-L^{-1}\left[\frac{\Omega(s)\mathrm{Ei}\left[-\sqrt{r^{2}-R^{2}}\Omega(s)\right]}{s}\right]\right) \\ &-\left(\frac{e^{(-(r-R)\sigma_{v})}}{\sigma_{v}}\left\{F\left[t;\frac{(r-R)}{c},0\right]-F\left[t;\frac{(r-R)}{c},c\sigma_{v}\right]\right\}\right) \\ &+\left(\frac{e^{\left(-\sqrt{r^{2}-R^{2}}\sigma_{v}\right)}}{\sigma_{v}}\left\{F\left[t;\frac{\sqrt{r^{2}-R^{2}}}{c},0\right]-F\left[t;\frac{\sqrt{r^{2}-R^{2}}}{c},c\sigma_{v}\right]\right\}\right) \end{aligned}\right\} \end{aligned} \tag{56}$$

Where,

$$\begin{aligned} &L^{-1}\left[\frac{\Omega\left(s\right)\mathrm{Ei}\left[\left(r-R\right)\Omega\left(s\right)\right]}{s}\right] \\ &\quad=\Theta\left(t-\frac{(r-R)}{c}\right)\left[\frac{e^{(-c\sigma_{v}t)}}{ct}+\sigma_{v}\left(\mathrm{Ei}\left(\left(r-R\right)\sigma_{v}\right)-\mathrm{Ei}\left(c\sigma_{v}t\right)\right)\right] \\ &L^{-1}\left[\frac{\Omega\left(s\right)\mathrm{Ei}\left[\sqrt{r^{2}-R^{2}}\Omega\left(s\right)\right]}{s}\right] \\ &\quad=\Theta\left(t-\frac{\sqrt{r^{2}-R^{2}}}{c}\right)\left[\frac{e^{(-c\sigma_{v}t)}}{ct}+\sigma_{v}\left(\mathrm{Ei}\left(\sqrt{r^{2}-R^{2}}\sigma_{v}\right)-\mathrm{Ei}\left(c\sigma_{v}t\right)\right)\right] \end{aligned} \tag{57}$$

This completes the "+" mode contribution to the spectrum.
The spectrum itself can be divided into three regions

(I) ct < (r − R)
(II) (r − R) $< ct < \sqrt{r^2 - R^2}$
(III) $\sqrt{r^2 - R^2} < ct < \infty$

In order to see how the solution behaves in each of the three time domains we focus on the behavior of $F[t;\alpha,\beta]$ and the inverse Laplace transforms of the Exponential Integral function in each domain.

$$F_\nu(r,t) = 0 \tag{58}$$

Therefore the solution for the prompt spectrum is simply $\varepsilon_\nu(r,t) = B_\nu(T_m)$, as would be expected.

$$F_\nu(r,t) = \frac{R}{4r}\left\{\left[1 + \frac{r}{R} - \frac{1}{R\sigma_\nu}\right]e^{-\sigma_\nu(r-R)} + \left[\frac{1}{R\sigma_\nu} - \frac{R}{ct}\left(\frac{r^2}{R^2} - 1\right)\right]e^{-\sigma_\nu ct} - R\sigma_\nu\left(\frac{r^2}{R^2} - 1\right)\left[E_i(\sigma_\nu(r-R)) - E_i(\sigma_\nu ct)\right]\right\} \tag{59}$$

$$F_\nu(r,t) = \frac{R}{4r}\left\{\left[1 + \frac{r}{R} - \frac{1}{R\sigma_\nu}\right]e^{-\sigma_\nu(r-R)} + \left[\frac{1}{R\sigma_\nu} - \sqrt{\frac{r^2}{R^2} - 1}\right]e^{-\sigma_\nu\sqrt{r^2-R^2}} - R\sigma_\nu\left(\frac{r^2}{R^2} - 1\right) \times \left[E_i(\sigma_\nu(r-R)) - E_i(\sigma_\nu\sqrt{r^2 - R^2})\right]\right\} \tag{60}$$

Acknowledgments

This work was performed under the auspices of the U.S. Department of Energy by University of California, Lawrence Livermore National Laboratory under contract No. W-7405-Eng-48.

References

1. Noh, J. Comp. Phys. 1987; **72**: 78.
2. Marshak, R.E. Phys. Fluids (1958); **1**: 24.
3. Su, B. and Olson, G.L. JQSRT (1999); **62**: 279.
4. See papers in Graziani, F. "Computational Methods in Transport: Granlibakken 2004 (Lecture Notes in Computational Science and Engineering)" (2005), Springer-Verlag.

5. See Miller, D.S. in this volume.
6. See Graziani, F. in Mezzacappa, A. and Fuller, G. "Open Issues in Core Collapse Supernova Theory (Proceedings from the Institute for Nuclear Theory)" (2006), World Scientific.
7. Bowers, R. and Wilson, J. "Numerical Modeling in Applied Physics" (1991), Jones and Bartlett.
8. Su, B. and Olson, G. Ann Nucl Energy (1997); **24**: 1035.
9. Shestakov, A.I. and Bolstad, J.H. JQSRT (2005); **91**: 133.

Some Verification Problems with Possible Transport Applications

D.S. Miller

Introduction

Verification is an essential part of the process whereby a code developer builds a sense of correctness and reliability in the answers his code produces. Unfortunately, many codes lack the machinery to make use of the method of manufactured solutions (*Verification of Computer Codes in Computational Science and Engineering*, Chapman and Hall/CRC, Boca Raton, FL, 2003) and so are reliant for verification on problems which possess analytic solutions. The authors present here three problems they derived for their own verification purposes. We hope they may be useful to others in their search for computational correctness.

1 Open Radiation Boundary Test

This problem is meant to test the correctness of implementation for open radiation diffusion boundary conditions. It is a pure radiation diffusion flow problem with no hydrodynamics or material coupling. The solution is exact and simple to compute. Although it does require an iterative root solve, the iteration is easy to perform and generally quick to converge. Consider a 1-D slab geometry. The slabs are infinite in extent in the y and z directions. On the left is a closed radiation boundary at $x = 0$. On the right is an open boundary at $x = L$. Outside the open boundary there is an incident radiative flux at temperature T_{out}. The radiation temperature inside the problem is initially at T_{in}. The Rosseland opacity k_{ross} and the density ρ are constants. The speed of light is c. The Planck opacity is zero. There is no flux limiter. Then the 1-D radiation diffusion equation is

$$\frac{\partial \phi}{\partial t} - D\frac{\partial^2 \phi}{\partial x^2} = 0 \tag{1}$$

where $D = \dfrac{c}{3\rho\kappa_{ross}}$ is the constant diffusion coefficient. Here $\phi = T^4$. We seek a general time and space dependent solution for $\phi(x,t)$. If we define $\phi_{in} = T_{in}^4$ and $\phi_{out} = T_{out}^4$ then the boundary and initial conditions are

$$\begin{aligned} \phi(x,t=0) &= \phi_{in} \\ \tfrac{\partial\phi(x=0,t)}{\partial x} &= 0 \\ \phi(x=L,t) + \tfrac{2D}{c}\tfrac{\partial\phi(x=L,t)}{\partial x} &= \phi_{out} \end{aligned} \quad . \tag{2}$$

We begin by taking the Laplace transform of (1) and (2) getting

$$s\phi(x,s) - \phi_{in} - D\frac{\partial^2\phi(x,s)}{\partial x^2} = 0 \tag{3}$$

$$\frac{\partial\phi(0,s)}{\partial x} = 0 \tag{4}$$

and

$$\phi(L,s) + \frac{2D}{c}\frac{\partial\phi(L,s)}{\partial x} = \frac{\phi_{out}}{s} \tag{5}$$

where s is now the Laplace transformed time variable. The general solution to (3) is

$$\phi(x,s) = A\exp(\beta x) + B\exp(-\beta x) + \frac{\phi_{in}}{s} \tag{6}$$

where A and B are as yet unknown functions of s and β is defined as

$$\beta = \sqrt{\frac{s}{D}}. \tag{7}$$

If we apply the boundary condition at x =0 we get $A = B$ so that (6) becomes

$$\phi(x,s) = A[\exp(\beta x) + \exp(-\beta x)] + \frac{\phi_{in}}{s}. \tag{8}$$

Applying the boundary condition at $x = L$ we have that

$$A = \frac{-(\phi_{in} - \phi_{out})}{2s[\cosh(\beta L) + (\frac{2D}{c})\beta\sinh(\beta L)]} \tag{9}$$

so that

$$\phi(x,s) = \frac{\phi_{in}}{s} - \frac{(\phi_{in} - \phi_{out})}{s}\frac{\cosh(\beta x)}{[\cosh(\beta L) + (2D/c)\beta\sinh(\beta L)]}. \tag{10}$$

To obtain our final solution we must compute the Laplace inversion of (10). We accomplish this by directly computing the Bromwich Inversion integral

$$\phi(x,t)=\frac{1}{2\pi i}\int_{\gamma-i\infty}^{\gamma+i\infty}\left\{\frac{\phi_{in}}{s}-\frac{(\phi_{in}-\phi_{out})\cosh(x\sqrt{s/D})}{s\left[\cosh(L\sqrt{s/D})+(2/c)\sqrt{sD}\sinh(L\sqrt{s/D})\right]}\right\}e^{st}ds. \tag{11}$$

This at first appears daunting but may be done exactly using the residue theorem. A branch cut is taken along the positive real axis and the integrals along the branch cut cancel. There is a simple pole at s = 0 and an infinite number of simple poles at $s = -R_n$ where the R_n are the roots of

$$\cos\left(L\sqrt{\frac{R_n}{D}}\right)-\frac{2\sqrt{DR_n}}{c}\sin\left(L\sqrt{\frac{R_n}{D}}\right)=0. \tag{12}$$

This expression can be rewritten as a fixed point iteration to obtain the values of R_n as

$$R_n^{new}=\frac{D}{L^2}\left[n\pi+\arctan\left(\frac{c}{2\sqrt{DR_n^{old}}}\right)\right]^2. \tag{13}$$

With an initial guess of 1.0, (13) will rapidly converge to the nth root of (12). The final solution we seek is just the sum over the residues resulting in

$$\phi(x,t)=\phi_{out}+(\phi_{in}-\phi_{out})\sum_{n=1}^{\infty}\frac{2\exp(-R_n t)}{\left[1+\frac{Lc}{2D}+\frac{2LR_n}{c}\right]}\frac{\cos\left(x\sqrt{\frac{R_n}{D}}\right)}{\cos\left(L\sqrt{\frac{R_n}{D}}\right)}. \tag{14}$$

If the value of $T_{out} > T_{in}$, then a heat wave will propagate from right to left at a rate given by (14) and reach a steady state $T = T_{out}$. Likewise, if $T_{in} > T_{out}$ then a cooling wave will propagate from right to left and reach steady state $T = T_{out}$.

This test problem bears similarities to the excellent radiation diffusion test problem from Su and Olson [SO96]. But while the solution presented here is not as intricate as Su-Olson and includes less physics, it is much easier to evaluate. If one were interested in testing the boundary conditions in isolation from the other physics this problem would be a better choice.

2 Spherical Heat Flow Test

What follows is the derivation of an analytic solution for a pure heat conduction problem which should be useful for verification purposes. Consider a sphere of radius R at a constant temperature T_0. We seek a solution to the homogeneous heat diffusion equation in spherical coordinates (exterior to the hot sphere)

$$\rho C\frac{\partial T}{\partial t}-\frac{1}{r^2}\frac{\partial}{\partial r}\left(\kappa r^2\frac{\partial T}{\partial r}\right)=0 \tag{15}$$

subject to the initial and boundary conditions

$$\begin{aligned} T(t=0, r>R) &= 0 \\ T(t, r \leq R) &= T_0 \,. \\ T(t, r=\infty) &= 0 \end{aligned} \tag{16}$$

In (15), C is the specific heat, ρ is the density, and κ is the conduction coefficient. Specify temperature dependent forms for the specific heat and conduction coefficients as

$$\begin{aligned} \kappa &= \kappa_0 T^n \\ C &= C_0 T^n \end{aligned} \tag{17}$$

where κ_0 and C_0 are constants and n is some exponent not necessarily an integer. If we substitute (17) into (15) and define

$$\Phi = T^{n+1} \tag{18}$$

we have

$$\rho C_0 \frac{\partial \Phi}{\partial t} - \frac{1}{r^2}\frac{\partial}{\partial r}\left(\kappa_0 r^2 \frac{\partial \Phi}{\partial r}\right) = 0. \tag{19}$$

Take the Laplace transform of (19) to get

$$\frac{\partial}{\partial r}\left(r^2 \frac{\partial \Phi(s,r)}{\partial r}\right) - r^2 s \frac{\rho C_0}{\kappa_0} \Phi(s,r) = 0. \tag{20}$$

Now the transformed boundary conditions are

$$\begin{aligned} \Phi(s, r=R) &= \frac{\Phi_0}{s} = \frac{T_0^{n+1}}{s} \,. \\ \Phi(s, r=\infty) &= 0 \end{aligned} \tag{21}$$

This ODE has the solution (for r > R)

$$\Phi(s,r) = \frac{\Phi_0 R}{sr} \exp\left\{-(r-R)\sqrt{\frac{s\rho C_0}{\kappa_0}}\right\}. \tag{22}$$

Perform the Laplace inversion to get

$$T(t,r) = \left\{ \begin{matrix} T_0 \\ T_0 \left[\frac{R}{r} Erfc\left\{\frac{(r-R)}{2}\sqrt{\frac{\rho C_0}{\kappa_0 t}}\right\}\right]^{\left(\frac{1}{n+1}\right)} \end{matrix} \right\} \text{ for } \begin{matrix} r \leq R \\ r > R \end{matrix} \tag{23}$$

For $n = 0$ this is an easy computational problem. But at large n it will strain the diffusion codes ability to accurately resolve the gradients in the material properties. This is mainly an issue because the diffusion coefficient is not a constant but will vary in space and time. Approximations or averaging schemes used in the construction of the diffusion matrix will be sorely tested for large n and typically will require a finer mesh for accurate answers. This problem was specifically constructed to test the heat conduction in a spherical geometry and in isolation from other physics in response to difficulties running another verification problem – the Coggeshall problem number 8 [CO91].

3 Coupled Multi-Temperature Diffusion Test

This is a modification of the problem in Sect. 2 meant to exercise more of the diffusion code as well as the code coupling the radiation and matter temperatures. It has radiation diffusion as well at electron and ion conduction. It is a 1-D slab geometry infinite in extent in the y and z directions. On the left at $x = x_0$ the radiation and material temperatures are fixed at T_0. On the right boundary the problem extends to $x = infinity$.Hydrodynamics is turned off and there is no flux limiter on the radiation diffusion or the matter conduction. The density ρ and the radiation diffusion coefficient D are constants. The radiation and material temperature update equations are

$$\sigma \frac{\partial T_{rad}^4}{\partial t} - \sigma D \frac{\partial^2 T_{rad}^4}{\partial x^2} = c\rho\sigma\kappa_P (T_{mat}^4 - T_{rad}^4) \tag{24}$$

$$\rho C_{ve} \frac{\partial T_{mat}}{\partial t} - \frac{\partial}{\partial x}\left(K_e \frac{\partial T_{mat}}{\partial x}\right) = c\rho\sigma\kappa_P (T_{rad}^4 - T_{mat}^4) + k_{ie}(T_{ion} - T_{mat}) \tag{25}$$

$$\rho C_{\nu i} \frac{\partial T_{ion}}{\partial t} - \frac{\partial}{\partial x}\left(K_i \frac{\partial T_{iont}}{\partial x}\right) = k_{ie}(T_{mat} - T_{ion}) \tag{26}$$

where c is the speed of light, k_{ie} is the electron-ion coupling rate, σ the radiation constant, K_e and K_i are conduction coefficients, C_{ve} and C_{vi} are specific heats, κ_p is the Planck opacity. The boundary and initial conditions are

$$\begin{aligned} T_{mat}(x = x_0, t) &= T_{rad}(x = x_0, t) = T_{ion}(x = x_0, t) = T_0 \\ T_{mat}(x = \infty, t) &= T_{rad}(x = \infty, t) = T_{ion}(x = \infty, t) = 0. \\ T_{mat}(x > x_0, t = 0) &= T_{rad}(x > x_0, t = 0) = T_{ion}(x > x_0, t = 0) = 0 \end{aligned} \tag{27}$$

We now impose that

$$\begin{aligned} K_e &= K_{e0} T_{mat}^3 \\ K_i &= K_{i0} T_{ion}^3 \end{aligned} \tag{28}$$

and

$$\begin{aligned} C_{\nu e} &= C_e T_{mat}^3 \\ C_{\nu i} &= C_i T_{ion}^3 \end{aligned} \tag{29}$$

where $K_{e0}, K_{i0}, C_e, C_i,$ are constants. We also impose $\kappa_P = k_{ie} = \infty$. This guarantees $T_{mat} = T_{rad} = T_{ion}$. We can now combine (28) and (29) with (24)–(26) to get

$$(4\sigma + \rho(C_e + C_i)) \frac{\partial \phi}{\partial t} - (4\sigma D + K_{e0} + K_{i0}) \frac{\partial^2 \phi}{\partial x^2} = 0 \tag{30}$$

with

$$\phi = T^4. \tag{31}$$

The diffusion of energy can be split between the ion conduction, electron conduction and radiation diffusion to whatever extent we wish by simply modifying the coefficients. The radiation and material temperatures are locked together and we can solve (30) to get the time and space dependence of ϕ. Take the Laplace transform of (27) and (30) to get

$$\phi(x,s)(4\sigma + \rho(C_e + C_i))s - (4\sigma D + K_{i0} + K_{e0})\frac{\partial^2\phi(x,s)}{\partial x^2} = 0 \tag{32}$$

and

$$\begin{aligned} \phi(x = x_0, t) &= \frac{\phi_0}{s} = \frac{T_0^4}{s}. \\ \phi(x = \infty, t) &= 0 \end{aligned} \tag{33}$$

This has the solution

$$\phi(x,s) = \frac{\phi_0}{s}\exp\left\{-\frac{1}{2}(x - x_0)\sqrt{s}\sqrt{\frac{4\sigma + \rho(C_e + C_i)}{4\sigma D + K_{e0} + K_{i0}}}\right\}. \tag{34}$$

Now we do the Laplace inversion of (34) to get

$$T(x,t) = \left\{\begin{matrix} T_0 \\ T_0 Erfc\left\{\frac{(x - x_0)}{2\sqrt{t}}\sqrt{\frac{4\sigma + \rho(C_e + C_i)}{4\sigma D + K_{e0} + K_{i0}}}\right\}^{1/4} \end{matrix}\right\} \text{ for } \begin{matrix} x \le x_0 \\ x > x_0 \end{matrix}. \tag{35}$$

One should be careful not to dismiss this problem as trivial just because the radiation and matter have the same temperature. This is simply part of the solution. In order to get the answer correct one must compute the ion conduction, electron conduction, radiation diffusion and the matter-to-radiation coupling correctly. An error in any one of these places will give an incorrect result.

4 Conclusion

While the body of verification problems in the literature is growing, it is still far smaller than the authors would like [KS03]. And while the problems presented here are not terribly complicated, nor are they trivial. Also they are exact, easy to compute, and easy to implement. And from simple to sophisticated there is a role to play for all verification problems. Should a code fail on the more complicated verification solutions the simpler ones may provide a way of isolating what is causing the problem. And with every exact solution a computer code computes our confidence in the unverifiable solutions grows. The verification solutions presented above have proven useful to the authors and we hope others will find them so as well.

Acknowledgment

This work was performed under the auspices of the Lawrence Livermore National Security, LLC, (LLNS) under Contract No. DE-AC52-07NA27344.

References

[KS03] Knupp, P., Salari, K.: Verification of Computer Codes in Computational Science and Engineering, Chapman and Hall/CRC, Boca Raton, FL, (2003)

[SO96] Su, B., Olson, G.L.: Benchmark Results for the Non-Equilibrium Marshak Diffusion Problem, J. Quant. Specrosc. Radiat. Transfer **56**, 3, 337–351 (1996)

[CO91] Coggeshall, S.V.: Analytic Solutions of Hydrodynamic Equations, Phys. Fluids A, **3**, 757–769 (1991)

Canopy Reflectance Model Benchmarking: RAMI and the ROMC

J.-L. Widlowski, B. Pinty, and The RAMI participants

Summary. Canopy radiative transfer (RT) models simulate radiative transfer processes in the solar domain at or near the Earth's terrestrial surface, i.e., within plant canopies and over bare soil surfaces. Such models are capable of simulating the transmitted, reflected and absorbed radiation fluxes, as well as, the angular distribution of the reflected light above vegetated surfaces in the optical domain of the solar spectrum. This directional aspect is important since the brightness with which a given canopy target appears in Earth Observation data (pixel) depends not only on the structural and spectral properties of the target area but also on the viewing and illumination geometry at the time of measurement. The quantitative interpretation of remote sensing data thus hinges more and more on the exploitation of RT models: In operational contexts they are used to pre-compute Look-Up-Tables of quantities required by the inversion algorithms, whereas, in off-line applications they may be directly inverted against the measurements to yield sets of state variables that describe the properties of the observed target. In both cases, the accuracy and reliability of the solutions to the inverse problems are determined by the performance of the RT model as well as the remote sensing instrument. Reducing the uncertainty of RT models will thus increase the quality of the information derived from remote sensing data (which is of interest to scientists and policy makers), and augment the reliability of RT model studies in forward mode (which is of interest to space agencies when testing new sensor concepts or mission strategies). This understanding has led to a series of model intercomparison projects (MIP) aiming either to document the spread of currently available simulation models, or, else to assess and benchmark the quality of their simulation results, e.g. (Bull. Am. Meteorol. Soc. 73:1962–1970, 1998; J. Geophys. Res. 106:11937–11956, 2001; J. Geophys. Res. 109:D06210, 2004; J. Geophys. Res. 110, 2005; Bull. Am. Meteorol. Soc. 86:1275–1293, 2005; J. Geophys. Res., 2007). Among these MIPs the RAdiation transfer Model Intercomparison (RAMI) activity focuses on the proper representation of the radiative processes occurring in vegetated environments in the optical domain of the solar spectrum.

Launched for the first time in 1999 the triennial RAMI community exercise encourages the systematic evaluation of canopy reflectance models on a voluntary basis. The first phase of RAMI focused on documenting the spread among radiative transfer simulations over a small set of primarily 1-D canopies (J Geophys Res

106:11937–11956, 2001). The positive response of the various RAMI-1 participants and the subsequent improvements made to a series of RT models promoted the launching of the second phase of RAMI (RAMI-2) in 2002. Here the number of test cases was expanded to focus further on the performance of RT models dealing with structurally complex 3-D plant environments. The main outcomes of RAMI-2 included (1) an increase in the number of participating models, (2) a better agreement between the model simulations in the case of the structurally simple scenes inherited from RAMI-1, and (3) the need to reduce the sometimes substantial differences between some of the 3-D RT models over complex heterogeneous scenes (J. Geophys. Res. 109:D06210, 2004). The latter issue was noted as one of the challenges that future intercomparison activities would have to face if some sort of reliable "surrogate truth" was to be derived for other RT models to be compared against. The third phase of RAMI (RAMI-3) was launched in 2005 and investigated the self-consistency, the relative and – to a limited extend – also the absolute performance of canopy RT models. RAMI-3 showed significant progress in the mutual agreement between RT models when compared to RAMI-2. In particular for 3-D Monte Carlo (MC) models the dispersion between simulated bidirectional reflectance factor (BRF) quantities was less than 1%, which supported the usage of these models in the generation of a "surrogate truth" data set covering all of the RAMI test cases (J. Geophys. Res., 2007). The availability of such a reference data set, in turn, lead to the development of the RAMI on-line model checker (ROMC), an open-access web-based interface allowing model developers and users to evaluate canopy RT models independently via the internet.

1 The Organisation of RAMI

Figure 1 is a schematic diagram of the temporal evolution of activities during a RAMI phase. In essence there are three separate activities involved:

1. *RAMI preparation*: Several months before the official launch of a new RAMI phase the RAMI coordinators – currently the Joint Research Centre of the European Commission – design a set of experiments and measurements that should be of interest to the RT modelling community as a whole. Test cases may be tailored around long standing issues (e.g., topography), or more advanced topics (e.g., horizontal radiation fluxes, local sampling). This proposal is then submitted to the RAMI advisory body (RAB), a small group of well-known scientists in the field of radiative transfer modelling and/or model intercomparison activities. Once a consensus on the number and types of test cases has been achieved, they are presented in great detail on the RAMI website (`http://rami-benchmark.jrc.ec.europa.eu/`. The RAMI phase is then officially launched by inviting scientists from around the world with a known interest in canopy RT modelling to visit the RAMI website.
2. *Active submission*: Over a period of several months RT model simulation results will be accepted by the RAMI coordinators. During this time interested participants will visit the RAMI website, implement the proposed

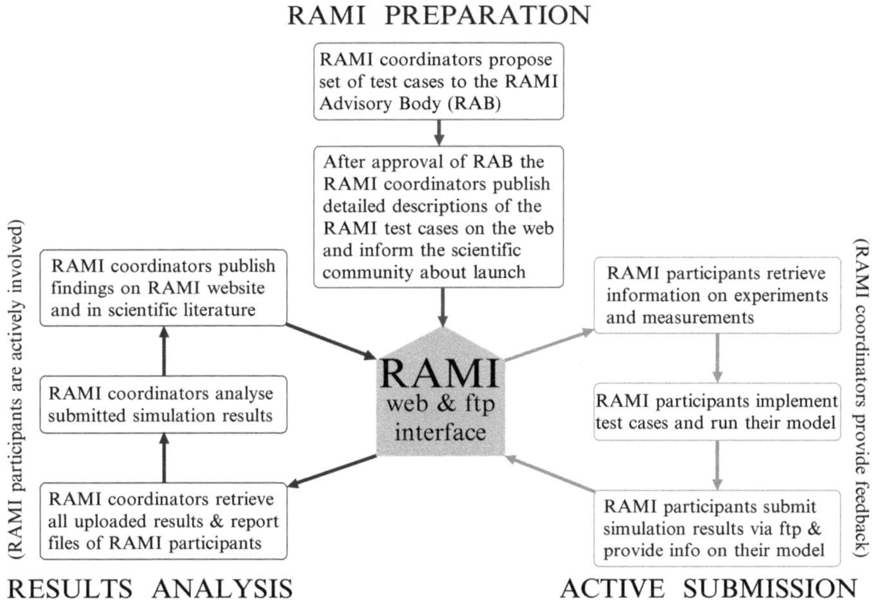

Fig. 1. The three main parts of a RAMI phase: RAMI phase preparation (*red*), RAMI active submission (*green*), and RAMI results analysis (*blue*). (See Plate 22 on page 312)

test cases with their own model(s), and simulate the prescribed RAMI "measurements". The resulting data sets have to be formatted in accordance with the RAMI formatting conventions and file-naming rules that are described in great detail on the RAMI website. Once this is done, the simulation results are then sent via ftp protocol to the RAMI coordinators (again following the detailed instructions available on the website). In case of difficulties or obvious errors/deviations of the submitted results RAMI participants receive substantial feedbacks. Multiple submissions are allowed but it is only the latest version of a file that is included in the subsequent analysis part of a RAMI phase.

3. *Results analysis*: As soon as the deadline for the active submission period has passed, the RAMI coordinators collect all the submitted model simulations and perform a detailed analysis of these results. In case of inconsistencies or noticeable deviations a dialogue is opened with the concerned participants to shed light on the causes of their model's behaviour. The results of the RAMI phase are eventually presented during an international scientific meeting and subsequently posted (in all detail) on the RAMI website. At the same time, a summary of the findings of a given RAMI phase is published in the refereed scientific literature, with all RAMI participants being actively involved in the drafting of the manuscript [PGW01, PWT04, WTP07].

One of the traits of RAMI is to increase the number of test cases by including a few new experiments (and measurements) from one phase to another. This strategy serves a triple purpose, namely, (a) to present RT models that underwent further development after participating in a previous RAMI phase with the opportunity to rerun the same experiments again, (b) to allow the evaluation of RT models under an increasingly comprehensive set of structural, spectral, viewing and illumination conditions, and (c) to tailor new sets of RAMI experiments and measurements around scientific questions emerging in the context of RT modelling and the quantitative interpretation of remotely sensed data. Indeed, such an approach guarantees that every phase will contain at least some test cases for which the simulation results cannot be known a prior:

- During RAMI-1 a large ensemble of structurally homogeneous leaf canopies, as well as a small set of structurally heterogeneous "floating spheres" test cases were introduced. In both cases the foliage was represented either as finite-sized (discrete) or infinitesimally small (turbid medium) scatterers with spectral properties relating to those typically observed in the red and near-infrared (NIR) spectral domains. In the case of the homogeneous leaf canopies a so called "purist corner" was also proposed where the spectral leaf and soil properties were such as to test model performance in the limit of conservative scattering conditions, i.e., the soil brightness ($\alpha = 1$) and the single-scattering albedo ($r_L + t_L = 1$) are unity, and the leaf reflectance (r_L) is equal to the leaf transmittance (t_L).
- For RAMI-2 this set of 660 test cases was expanded to 715 by including (1) a "Gaussian hill" experiment featuring a coniferous forest stand above a snow covered background and (2) a "real zoom-in" experiment, composed of a series of precisely located spherical and cylindrical envelopes filled with discrete, oriented scatterers, where simulations had to be carried out for a series of successively finer spatial resolutions (from 270 to 30 m target areas) centered in the middle of the overall $270 \times 270\,\mathrm{m}^2$ scene.
- Within RAMI-3 the simulation scenarios increased by a further 37% with respect to RAMI-2. Three new experiments were proposed: (1) a conservative scattering scenario for the heterogeneous "floating spheres" test cases originally introduced during RAMI-1, (2) a "coniferous forest" scene analogous to the "Gaussian hill" canopy introduced during RAMI-2 but without the topography, and (3) a "birch stand" populated with trees of variable sizes and spectral properties – intended primarily to enhance the degree of structural realism among the RAMI test cases.

Figure 2 provides a graphical overview of the various test cases so far included in RAMI. Scene depictions in colour refer to spectral leaf, wood and soil conditions that are reminiscent of those in the red and near-infrared (NIR) spectral domains. Scenes in grey relate to conservative scattering conditions (purist corner).

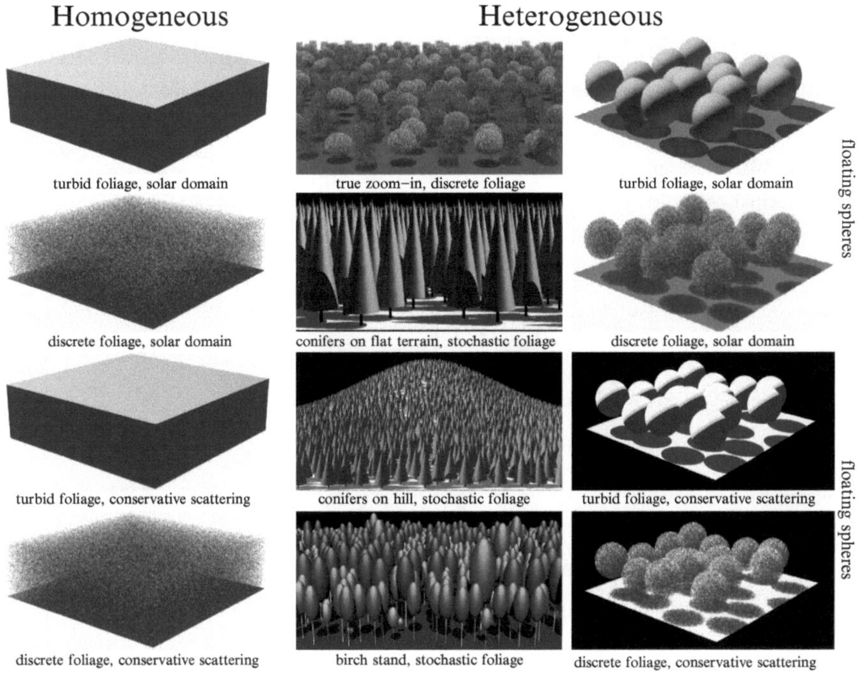

Fig. 2. Overview of the structurally homogeneous (*left panels*) and heterogeneous (*middle and right panels*) test cases featuring in the third phase of RAMI. Canopy foliage may be represented through infinitesimally small scatterers (turbid medium), or, through finite sized scatterers having deterministic (discrete) or statistic (stochastic) locations. The spectral properties of both foliage and background material may reflect either values typical for the red and near-infrared spectral region (*colour panels*) or for purely conservative scattering conditions (*grey panels*). (See Plate 23 on page 313)

A standard set of 11 "measurements" is prescribed for every test case in RAMI. These measurements include the total spectral BRF, in both the principal and the cross plane, together with the corresponding contributions due to the single-uncollided radiation scattered once by the soil only, the single-collided radiation by the leaves or trees only, and the radiation multiply collided by the leaves/trees/soil system. Three flux quantities are also routinely asked for, namely, the spectral albedo of the canopy (i.e., the directional hemispherical reflectance), the total transmission down to the underlying background, and, the total absorption of radiation in the vegetation layer. In addition to these standard measurements, RAMI-3 introduced two new measurement types, e.g., horizontal fluxes and local canopy transmission transects, that applied, however, only to selected test cases. In general the RAMI measurements, whether directional or hemispherical, have to be carried out with respect to a reference plane located at the top-of-canopy (TOC) height level.

In general, the majority of BRF simulations tends to cluster around some sort of common estimate. This can be seen in Fig. 3 which shows BRF simulations for structurally homogeneous (1-D) canopies that feature point-like

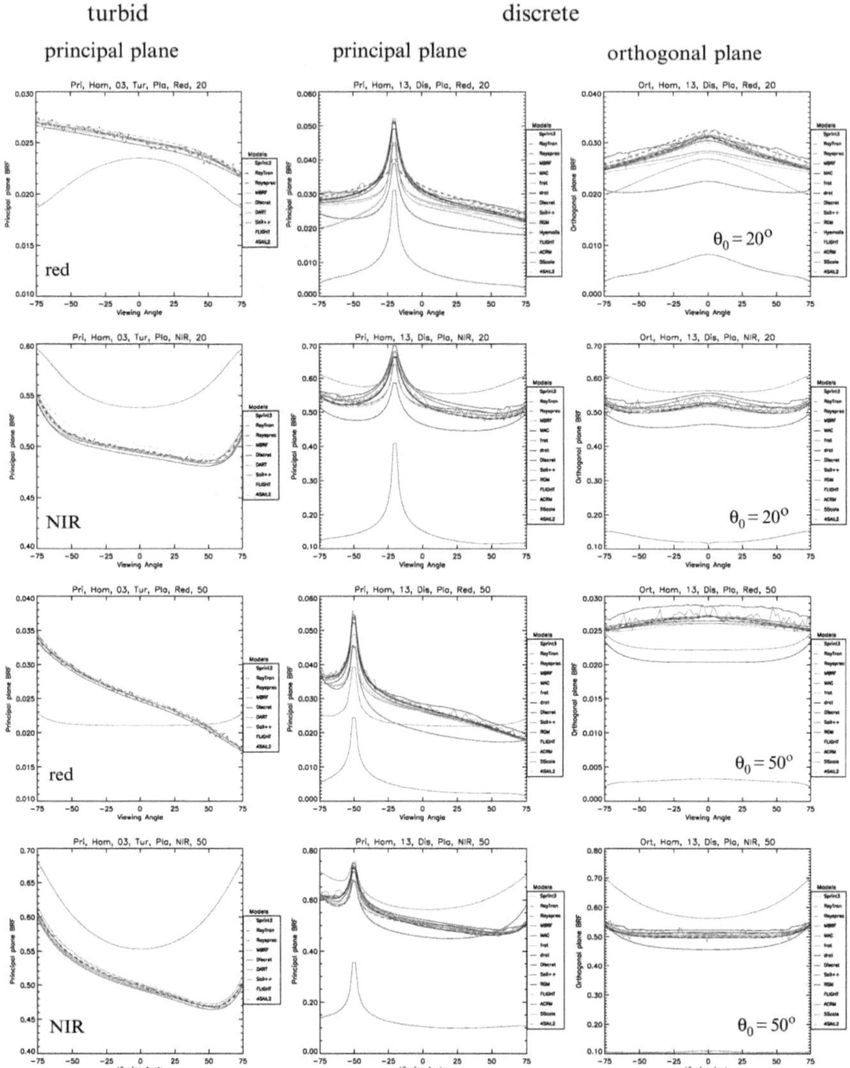

Fig. 3. BRF simulations in the red (*first and third row*) and near-infrared (*second and fourth row*) spectral band for structurally homogeneous canopies composed of infinitesimally small (*left column*) and finite-sized (*middle and right column*) scatterers. Simulations pertain to the principal plane (*left and middle column*) and to the orthogonal plane (*right column*) for solar zenith angles (θ_0) of 20° (*top two rows*) and 50°(*bottom two rows*). Note the close agreement between the 3-D Monte Carlo models: `FLIGHT`, `DART`, `drat Rayspread`, `raytran`, and `Sprint3`, respectively. (See Plate 24 on page 314)

(turbid) or finite-sized (discrete) foliage elements randomly distributed within the volume of a horizontally infinite vegetation layer bounded by some TOC level, as well as a lower flat background surface. More specifically, BRF simulations in the principal (first and second column) and orthogonal (third column) observation planes for solar zenith angles (θ_0) of 20° (top two rows) and 50° (bottom two rows) are shown for spectral conditions resembling those in the red (first and third row) and NIR (second and fourth row) spectral domains. Notice the prominent rise of BRF values when the viewing direction coincides with the illumination direction in the case of canopies with finite-sized scatterers (middle column). This is known as the hot spot effect and arises due to the absence of shadows in the field of view of the observer under those viewing conditions [Ger88, Ver88, Kuu91]. Similar patterns are shown in Fig. 4 which documents BRF simulations for structurally heterogeneous "floating spheres" environments where the discrete foliage elements are randomly distributed within a series of spherical volumes that are themselves freely floating above an underlying flat background surface. More specifically, BRF simulations along the principal plane in the red (first column), NIR (second column) and for conservative scattering conditions (third column) are shown for SZA = 20° (top two rows) and 50° (bottom two rows). In both Figs. 3 and 4 one should notice the close agreement between the `FLIGHT` [Nor96], `DART` [GDP96, GMG04], `drat` [Lew99, SLD01], `Rayspread` [WLP06], `raytran` [GV98], and `Sprint3` [TG98] Monte Carlo models for both the structurally homogeneous and heterogeneous test cases.

During RAMI-3 a total of 464,816 (2,112) individual BRF (flux) simulations were received by the RAMI coordinators. In order to pursue the analysis of these data beyond a mere visual comparison a protocol is needed that permits the quantitative evaluation of RT model simulations despite the fact that the true solution is in general not known. [OSB94], and many others since, maintain that – under these latter conditions – the complete validation/verification of a computer simulation model is quite impossible, and that any such endeavour should focus instead on showing the opposite, that is, the onset of flaws in a model's behaviour. In the context of RAMI such an undertaking is, however, complicated by the fact that canopy RT models differ both in the way they implement/approximate the radiation transfer equations and in how they represent the actual canopy structure. Several studies e.g., [GN97, WPG01, PWG02, LG02, CRH03, RSK03] have shown that canopy structure affects the radiative properties of 3-D vegetation targets and most notably their reflectance field. Thus as soon as the internal structural representation of a RT model deviates from the canopy architecture presented on the RAMI website it becomes no longer possible to unambiguously determine the causes behind the deviations of that model's simulations when compared to the output results of other RT models. As a consequence RAMI does not include RT models featuring a 1-D representations of plant canopies when evaluating the performance of models simulating the RT of heterogeneous (3-D) vegetation canopies. Similarly, 3-D models that are only

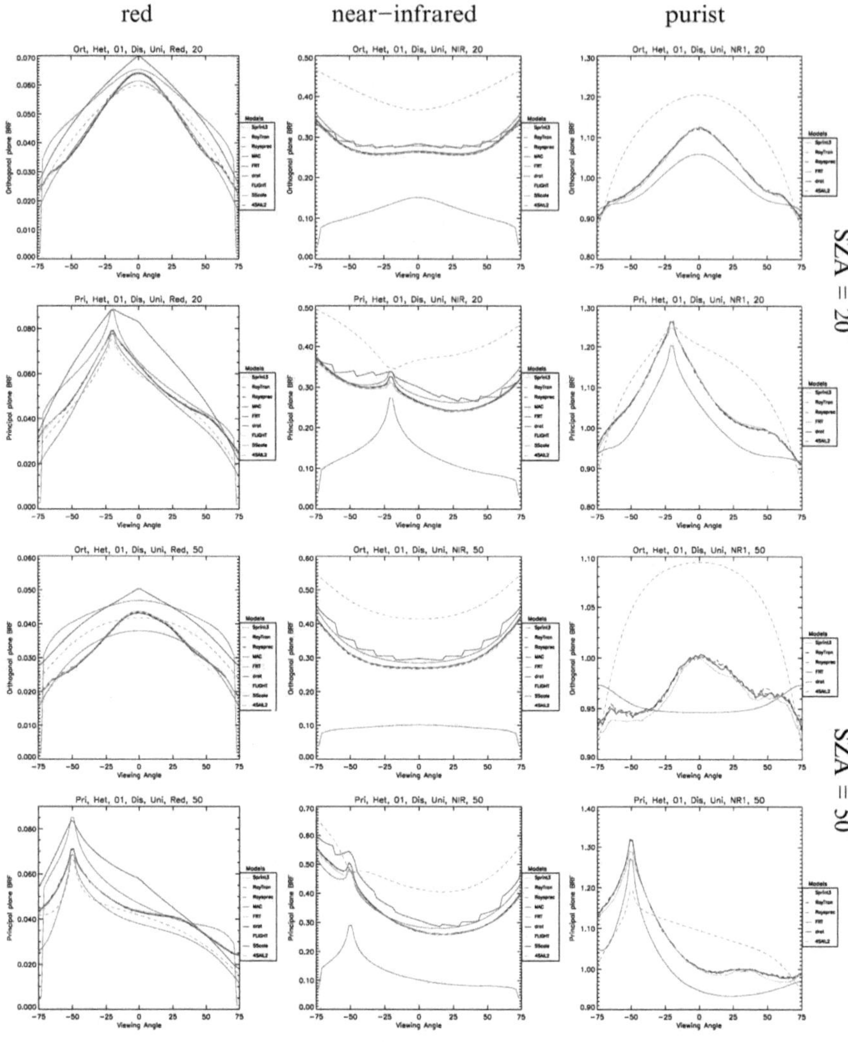

Fig. 4. BRF simulations in the orthogonal (*first and third row*) and principal (*second and fourth row*) plane over discrete "floating spheres" test cases with optical properties typical for the red (*left column*) and near-infrared (*middle column*) spectral regimes, as well as, for conservative scattering conditions (*right column*). The top (*bottom*) two rows relate to a solar zenith angle (SZA) of 20° (50°). Note the close agreement among the 3-D Monte Carlo models: `FLIGHT`, `drat`, `Rayspread`, `raytran`, and `Sprint3`, respectively. (See Plate 25 on page 315)

capable of representing heterogeneous canopy architectures (i.e., geometric optics (GO) models) are not included in the evaluation of model performance over structurally homogeneous (1-D) canopies. Furthermore, in an attempt to account for the presence of minor structural differences amid the RT models participating in RAMI thresholding criteria were adopted that relate to the actual usage of these models. More specifically, the absolute calibration accuracy of the latest generation of space borne sensors (~3%) was used as a yardstick for RT model performance in cases where a "reference" solution was available.

RAMI-3 introduced a three-step procedure to identify incongruous RT models: (1) by assessing the absence of inconsistencies in the internal RT formulation of a model, (2) by verifying the accurate and reliable performance of a model in the limited number of cases where analytical solutions are available, and (3) by comparing the output of a model against a "surrogate truth" that was established from "credible" candidates within the ensemble of available RT simulations. Obviously the latter can only be meaningful if sufficient consensus exists among the simulation results of the "credible" RT models, in particular those that are known to minimise the number of simplifications/approximations in their radiative transfer formulation, and, feature accurate representations of the prescribed canopy architecture. The objective of this three-step procedure lies in identifying RT models that deviate from the norm rather than boosting the credibility of those models that do not differ. In fact, conformity with the anticipated outcome in each one of the above steps is not proof of a model's physical correctness. Hence any claims regarding the credibility of a model's performance should be avoided, or – if they have to be made – should always be limited to the set of prescribed conditions under which the models were actually tested.

2 The RAMI Protocol

This section will present the above three-step invalidation procedure using examples derived from the third phase of RAMI.

2.1 Model Self-Consistency

It is difficult to offer meaningful interpretations as to why the output of a given RT model may be different from simulation results of other models without verification of the model's internal consistency. Energy conservation, for example, is one of the key principles to ensure, and this both with respect to hemispherically integrated (fluxes) and directional (BRFs) quantities.

Energy Conservation

The solar radiation entering a plant canopy is partitioned into an absorbed A, a reflected R and a transmitted T fraction such that all incident photons are accounted for. Energy conservation at the scale of radiatively independent volumes [PGW04, WPL06] thus requires that $A + R + (1 - \alpha)T = 1$, where α

is the soil brightness. The capacity of a given model (m) to conserve energy can be described using:

$$\Delta_{\mathrm{F}}(m) = \frac{1}{N_{\mathrm{F}}(m)} \sum_{\lambda=1}^{N_\lambda^m} \sum_{\zeta=1}^{N_\zeta^m} \sum_{i=1}^{N_{\Omega_i}^m} [A_m(\lambda,\zeta,i) + R_m(\lambda,\zeta,i) + [1-\alpha(\lambda)]\, T_m(\lambda,\zeta,i)\,] \; - \; 1$$

where $N_{\mathrm{F}}(m) = N_\lambda^m + N_\zeta^m + N_{\Omega_i}^m$ is the total number of spectral λ, structural ζ, and illumination Ω_i conditions for which flux simulations were performed by model m. During RAMI-3 [WTP07] showed that RT models, which provided all these radiative flux simulations, were compliant with energy conservation to within 3%. One should note, however, that the above check will never identify deficiencies in models applying the principle of energy conservation to close their radiation budget, for example, by deriving canopy absorption on the basis of simulations of the reflectance and transmission properties of the canopy, e.g., `1/2-discret` [GPV97], `Sail++` [Ver02] and `2-Stream` [PLD06].

BRF Consistency

[PGW01] noted that internal inconsistencies in one or more submodules of a given RT model may compensate each other and lead to apparently correct overall BRF estimates. The same authors recommend the evaluation of BRF components as well as the total BRFs generated by a model. Since the RAMI format specifications ask for all radiative quantities to be provided with a precision of six decimal places, the implicit error associated with a given measurement is thus of the order of 10^{-6}. The average absolute difference $\Delta\rho$ between the total BRF (ρ_{tot}) and the sum of the BRF contributions due to the single uncollided (ρ_{uc}), the single-collided (ρ_{co}), and the multiple-collided (ρ_{mlt}) radiation components should thus be of a similar magnitude when defined as follows:

$$\Delta\rho(m) = \frac{1}{N_\rho(m)} \sum_{\lambda=1}^{N_\lambda^m} \sum_{\zeta=1}^{N_\zeta^m} \sum_{v=1}^{N_{\Omega_v}^m} \sum_{i=1}^{N_{\Omega_i}^m} \left| \rho_{tot}^m(\lambda,\zeta,v,i) \times - \left[\rho_{uc}^m(\lambda,\zeta,v,i) + \rho_{co}^m(\lambda,\zeta,v,i) + \rho_{mlt}^m(\lambda,\zeta,v,i)\right]\right|$$

where $N_\rho(m) = N_\lambda^m + N_\zeta^m + N_{\Omega_v}^m + N_{\Omega_i}^m$ is the total number of BRFs that were generated by model m for different spectral λ, structural ζ, viewing Ω_v, and illumination Ω_i conditions. During RAMI-3 the average absolute deviation $\Delta\rho$ was, in general, less than 10^{-5}. Three models (`5Scale` [LC01], `frat` (not published yet), and `FLIGHT`), however, featured larger deviations with $\Delta\rho$(`5Scale`) reaching 0.0027 for the homogeneous discrete purist corner. These deviations, although relatively small in terms of the magnitude of the total BRF, and, often related to the configuration of the model in its day to day usage, are nevertheless significant in the context of a model intercomparison exercise like RAMI since – by their statistical nature – they seem to indicate that, bar operator errors, some of the RT models do not conserve energy when partitioning the total BRF into its various subcomponents.

Spectral Ratio of the Single-Uncollided BRF

Model self-consistency can also be evaluated across different wavelengths. The ratio $\rho_{uc}(\lambda_1)/\rho_{uc}(\lambda_2)$ of the single-uncollided BRF components in the red and NIR spectral regimes, for example, relates to the differing amounts of radiation that have been scattered once by the underlying background (and never interacted with the canopy foliage) at these two wavelengths (λ_1 and λ_2). In the case of Lambertian soils, this spectral ratio must be a directionally invariant constant equal to the ratio of the soil albedos at the wavelengths of interest, i.e., $\alpha(\lambda_1)/\alpha(\lambda_2)$. Ensemble-averaging over a variety of structure ζ and illumination Ω_i conditions ($N_S = N_\zeta^m + N_{\Omega_i}^m$) then provides an indication of the average deviation from spectral consistency for any model m:

$$\Delta_S(m, \Omega_v) = \frac{\alpha(\lambda_1)}{\alpha(\lambda_2)} - \left[\frac{1}{N_S(m)} \sum_{\zeta=1}^{N_\zeta^m} \sum_{i=1}^{N_{\Omega_i}^m} \frac{\rho_{uc}^m(\lambda_1, \zeta, \Omega_v, i)}{\rho_{uc}^m(\lambda_2, \zeta, \Omega_v, i)} \right]$$

RAMI-3 showed that all models – with the exception of the `Sprint3` model in the "floating spheres" case – provided a spectral ratio for the single-uncollided BRF component that remained relatively constant and very close to the correct value up to view zenith angles of about 65–70°.

2.2 Absolute Model Performance

Exact analytical solutions to the radiative transfer equation do not exist for the vast majority of conceivable vegetation canopies. In some cases, however, the structural and spectral properties of vegetated surfaces may be such that it becomes possible to predict at least some of their radiative properties analytically. Within the available set of RAMI test cases there are at least two different types of absolute model evaluations that can be performed: The first one relates to single-collided BRF components of structurally homogeneous turbid medium canopies with uniform leaf normal distributions (LND), and the second to the reflected and absorbed energy fluxes in the various conservative scattering (purist corner) scenarios.

Homogeneous Turbid Uniform Canopy

Structurally homogeneous leaf canopies with azimuthally invariant uniform LNDs are characterised by a constant probability of foliage interception irrespective of the direction of propagation in that medium [Ros81, Ver87]. In addition, turbid media – with their infinitesimally small scatterers – satisfy the far field approximation and thus never yield a hot spot, i.e., a localised increase in the BRF around the retro-reflection direction of the incident illumination, [Ger88, Ver88, Kuu91]. The single-uncollided BRF component of such a canopy can be written as:

$$\rho_{uc}(\Omega_i, \Omega_v) = \alpha \exp\left[\frac{-\mathrm{LAI}\,(\mu_i + \mu_v)}{2\,\mu_i\,\mu_v}\right]$$

where α is the albedo of the Lambertian soil, $\mu = \cos\theta$ is the cosine of the illumination (i) or view (v) zenith angle $0 \le \theta \le \pi/2$, and LAI is the leaf area index of the canopy. Similarly the single-collided BRF component of such a canopy can be written as:

$$\rho_{co}(\Omega_i, \Omega_v) = \frac{2\,\Gamma(\Omega_i \to \Omega_v)\left[1 - \exp\frac{-\mathrm{LAI}\,(\mu_i+\mu_v)}{2\,\mu_i\,\mu_v}\right]}{\mu_i + \mu_v}$$

where the canopy scattering phase function is given by [SM88]:

$$\Gamma(\Omega_i \to \Omega_v) = \frac{r_L + t_L}{3\pi}\left(\sin\beta - \beta\cos\beta\right) + \frac{t_L}{3}\cos\beta$$

and β is the phase angle between the illumination and viewing direction:

$$\cos\beta = \cos\theta_i\cos\theta_v + \sin\theta_i\sin\theta_v\cos|\phi_i - \phi_v|$$

and r_L (t_L) is the reflectance (transmittance) of the foliage elements. Figure 5 shows the mean absolute error between RT model simulations and the above analytical formulations for the single-uncollided (left panel) and single-collided (right panel) BRF components of a turbid medium canopy with uniform LND and Lambertian scattering laws. The averaging was performed over BRF simulations in the principal and orthogonal planes, as well as, for illumination zenith angles of 20° and 50°. With the exception of `MBRF` [QX97] all RT models lie within 0.0025 of the truth in the single-collided case (right panel). In the single-uncollided case (left panel) the agreement between the participating RT models and the analytical solution is ten times better still, i.e., all models lie within 2.5×10^{-4} of the analytical solutions. This is impressive since the magnitude of ρ_{co} (ρ_{uc}) along the orthogonal plane was typically around 0.017 (0.003) in the red and 0.16 (0.005) in the NIR.

Purist Corner Fluxes

The RAMI purist corner allows to assess the performance of RT models against a known absolute reference, q^{truth} since – under conservative scattering conditions – all of the energy that enters a canopy system has to leave it again, i.e., $q^{truth} \equiv R^{truth} = 1$ and by the same token the canopy absorption is zero, i.e., $q^{truth} \equiv A^{truth} = 0$. The mean absolute deviation ϵ_q (from an ensemble of N different flux measurements) thus can be defined as:

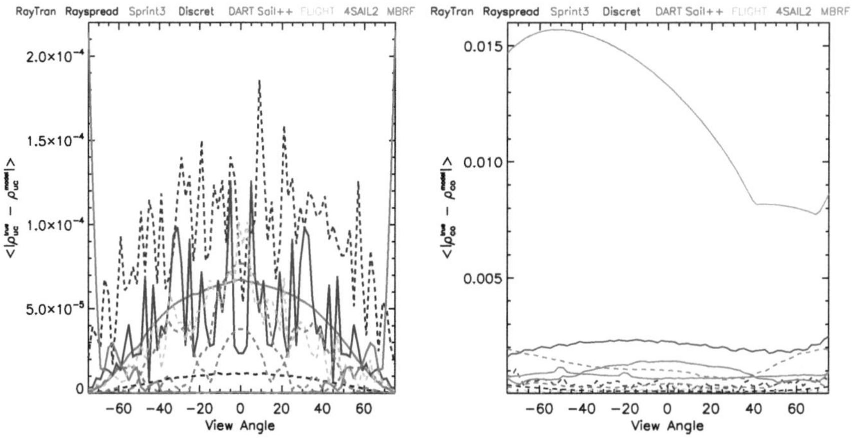

Fig. 5. The mean absolute error between model simulations and the analytical formulation of the single-collided, ρ_{co} (*right panel*) and the single-uncollided, ρ_{uc} (*left panel*) BRF components of a homogeneous turbid medium canopy with uniform LND and Lambertian scattering laws. For any view zenith angle the averaging was performed over the principal and orthogonal plane, as well as, for illumination zenith angles of 20° and 50°. (See Plate 26 on page 316)

$$\epsilon_q = \frac{1}{N}\sum_i^N \left|q_i^{truth} - q_i^{model}\right|$$

[WTP07] found that during RAMI-3 most participating models submitted the correct solutions. They recalled, however, that the lack of deviations from the "truth" is not a proof of the physical correctness of a model since, for example, hard-encoded programming statements may be contained inside the computer code that do account for the eventuality of situations for which the exact solution is known. In this way, the actual model would not be executed – to compute canopy reflectance and absorption here – but sidestepped to generate the anticipated results.

2.3 Relative Model Performance

Without access to absolute reference standards the evaluation of RT models has to rely on relative model intercomparison. [PGW01] argued that under those conditions model benchmarking on the basis of statistical moments, derived from the entirety of participating models, may be biased in the presence of outliers. Instead they proposed a relative evaluation scheme where the RT simulations of individual models are compared against those from all other participating models over as large as possible a set of conditions. In this way, RT models that are consistently different from others can be identified [PWT04]. Three different types of relative intercomparison metrics will be proposed here: model-to-model deviations, model-to-ensemble deviations, and deviations from model-derived surrogate truths.

Model-to-Model Deviations

The differences in the BRF simulations between two models (c and m), when averaged over a variety of spectral (λ), structural (ζ), viewing (Ω_v) and illumination (Ω_i) conditions, can be defined as:

$$\delta_{m\leftrightarrow c} = \frac{200}{N} \sum_{\lambda=1}^{N_\lambda} \sum_{\zeta=1}^{N_\zeta} \sum_{v=1}^{N_{\Omega_v}} \sum_{i=1}^{N_{\Omega_i}} \left| \frac{\rho_m(\lambda,\zeta,v,i)-\rho_c(\lambda,\zeta,v,i)}{\rho_m(\lambda,\zeta,v,i)+\rho_c(\lambda,\zeta,v,i)} \right|$$

where $N = N_\lambda + N_\zeta + N_{\Omega_v} + N_{\Omega_i}$ is the total number of BRF simulations that have been performed by both models c and m, and $\delta_{m\leftrightarrow c}$ is expressed in percent.

Figure 6 depicts a series of two-dimensional grids containing information on the various model-to-model BRF differences (blue–red colour scheme in the lower right half of each panel), as well as, the percentage of the total number of BRFs over which the $\delta_{m\leftrightarrow c}$ values were derived (black–green colour scheme in upper left half of each panel). More specifically, $\delta_{m\leftrightarrow c}$ is shown for those models having submitted the total (top left), multiple-collided (top right), single-uncollided (bottom left), and single-collided (bottom right) BRF data for structurally homogeneous canopies with finite-sized foliage representations in the solar domain. The blue colour scale increments in steps of 2%, the green colour scale in steps of 10%, and the red also in steps of 10% with the bright red colour indicating values larger than 50%. The maximum number of BRF simulations included in the computation of $\delta_{m\leftrightarrow c}$ was 1,216. To illustrate the reading of the various panels in Fig. 6 let us consider, for example, the total BRFs of the `Hyemalis` [SS00, HBS03] and `Sprint3` models (top left panel): Their model-to-model difference value, which lies between 10 and 20% (light red colour), has been obtained from less than 10% of the total number of BRF simulations (dark green colour) and thus may not be too representative. On the other hand, the $\delta_{m\leftrightarrow c}$ of the `ACRM` [Kuu01] and `Sprint3` models (same top left panel) lies somewhere between 6 and 8% and has been established using 100% of the possible BRFs. In general, the majority of models in the discrete and turbid homogeneous cases agree rather well with each other ($\delta_{m\leftrightarrow c} < 10\%$). This behaviour is also present for the various BRF components with the exception of the single-uncollided BRF component (ρ_{uc}) in the discrete homogeneous case where the various implementations/approximations of the hot spot phenomenon have increased the differences amid the simulated BRFs. In the multiple scattered BRF case the `FLIGHT` model – which did not update its simulations during RAMI-3 – shows slightly diverging values with respect to most other RT models. These are caused by a Lambertian assumption governing the angular distribution of higher orders of scattered radiation in simulation results originally submitted during RAMI-1.

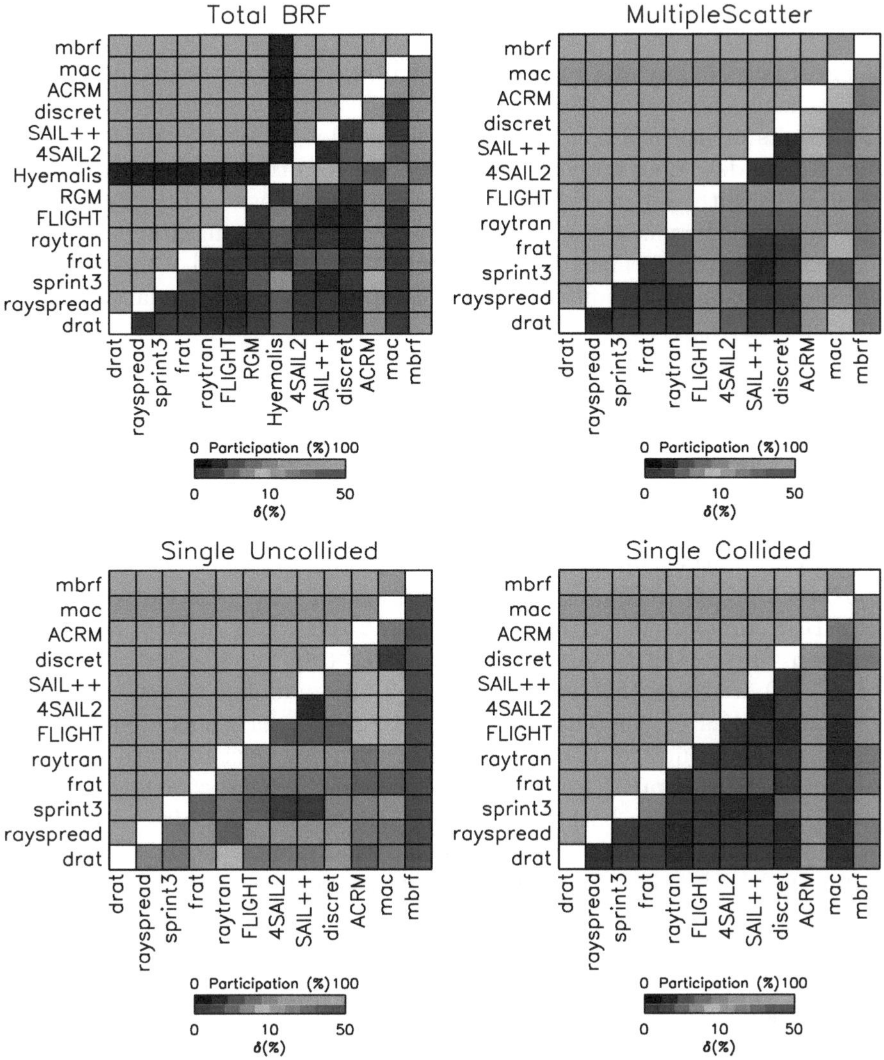

Fig. 6. Model-to-model differences $\delta_{m\leftrightarrow c}$ of the total (*top left*), multiple-collided (*top right*), single-uncollided (*bottom left*), and single-collided (*bottom right*) BRF simulations for structurally homogeneous canopies with finite-sized foliage representations in the solar domain. The lower right half of every panel indicates $\delta_{m\leftrightarrow c}$ in (%) (*blue–red colour scheme*), whereas the *top left* half indicates the percentage of available test cases that pairs of models performed together (*black–green colour scheme*). The green colour scale increments in steps of 10%, the blue in steps of 2% (up to $\delta_{m\leftrightarrow c} = 10\%$), and the red in steps of 10% (with a *bright red colour* indicating $\delta_{m\leftrightarrow c} > 50\%$). (See Plate 27 on page 317)

Model-to-Ensemble Deviations

In the absence of any absolute reference truth, the output from individual RT models may also be compared to ensemble averages computed from simulation results of other RT models, as proposed by [PGW01, PWT04]. In this way, RT models that are very different from all other models can be identified and – although not wrong in any absolute sense – they may then be excluded from further iterations of the ensemble averaging process, if this is deemed appropriate. Thus, for any spectral (λ), structural (ζ), viewing (v), and illumination (i) condition one can compute the mean deviation from all other models, δ_m (%):

$$\delta_m(\lambda, \zeta, v, i) = \frac{200}{N_c} \sum_{c=1;c\neq m}^{N_c} \left| \frac{\rho_m(\lambda, \zeta, v, i) - \rho_c(\lambda, \zeta, v, i)}{\rho_m(\lambda, \zeta, v, i) + \rho_c(\lambda, \zeta, v, i)} \right|$$

where N_c is the number of models with which the output of model m is to be compared. One way to analyse such δ_m statistics is to bin them over a variety of conditions in order to yield a histogram of model-to-ensemble deviations. Figure 7, for example, shows a variety of δ_m histograms generated from total BRF simulations of the `DART`, `drat`, `FLIGHT`, `frt` [KN00], `Rayspread`, `raytran`, and `Sprint3` models in the case of the "birch stand" scene (i.e., a 100×100 m^2 scene that is composed of a large number of non-overlapping tree-like entities of different sizes and spectral properties randomly located across a planar surface representing the underlying background). These histograms were generated from BRF simulations along the orthogonal plane and principal plane in both the red and NIR spectral domains and for solar zenith angles of 20° and 50°. One will notice that model dispersion is typically around 1–2% with the maximum deviation reaching 6% in the principal plane

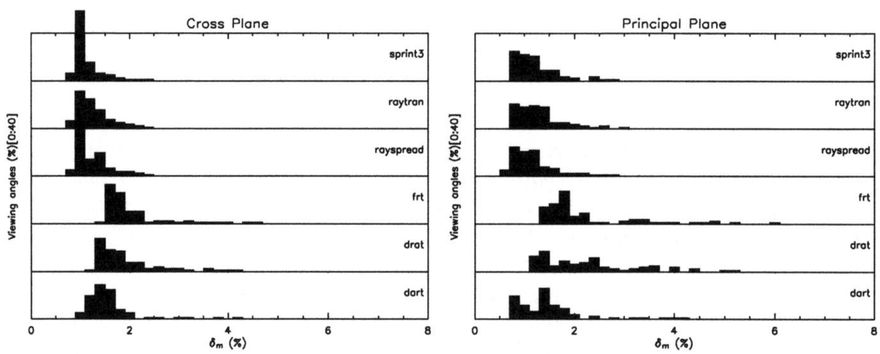

Fig. 7. Histograms of model-to-ensemble deviations δ_m derived from model simulations along the orthogonal (*left panel*) and principal (*right panel*) planes over the "birch stand" scene in the red and NIR spectral band and for solar zenith angles (SZA) of 20° and 50°

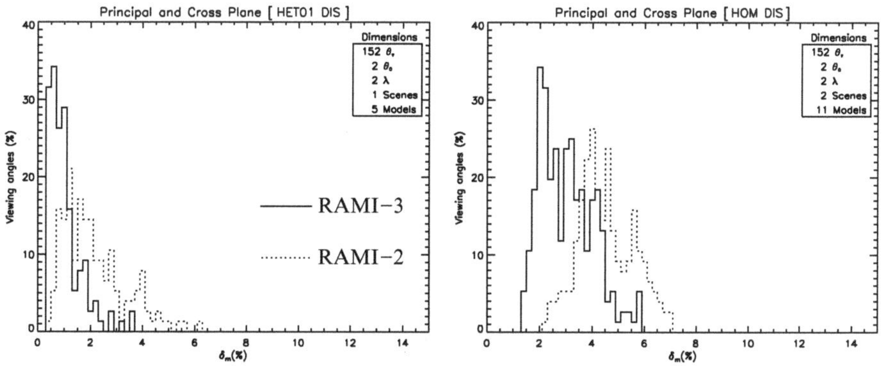

Fig. 8. The envelope of a series of histograms of model-to-ensemble differences, δ_m (%) (similar to those depicted in Fig. 7) for non-outlier RT models participating in the discrete "floating spheres" (*left panel*) and discrete homogeneous (*right panel*) test cases of RAMI-3 (*solid line*) and RAMI-2 (*dotted line*). BRF simulations in the principal and orthogonal planes using illumination zenith angles of 20° and 50° in both the red and NIR spectral domain contributed to the generation of these histograms

(**frt**). Figure 8 shows the outer envelope of such δ_m histograms generated from BRF simulations pertaining to structurally homogeneous (left panel) and "floating spheres" (right panel) canopies composed of finite-sized leaves. One will notice that the agreement between the RT models in RAMI-3 (solid line) is better than the corresponding agreement of models during RAMI-2 three years before (dotted line).

Alternatively one may define an overall indicator of model-to-ensemble differences, $\bar{\delta}_m$ (%) by averaging the above $\delta_m(\lambda, \zeta, v, i)$ over appropriate sets ($\bar{N}$) of spectral λ, structural ζ, viewing v and illumination i conditions:

$$\bar{\delta}_m = \frac{1}{\bar{N}} \sum_{\lambda=1}^{N_\lambda} \sum_{\zeta=1}^{N_\zeta} \sum_{v=1}^{N_{\Omega_v}} \sum_{i=1}^{N_{\Omega_i}} \delta_m(\lambda, \zeta, v, i)$$

Figure 9 documents $\bar{\delta}_m$ (%) for the six 3-D Monte Carlo models: **DART**, **drat**, **FLIGHT**, **Rayspread**, **raytran** and **Sprint3**. Statistics are provided for total BRF simulations over structurally heterogeneous (top) and homogeneous (bottom) discrete canopy representations. The various model names are listed on the top of each table (one per column). The experiment identifier is provided to the left, whereas the spectral regime is indicated to the right of each table column. Light (dark) grey fields indicate incomplete (no) data submission. The green–yellow–red colour scheme represents the overall model-to-ensemble difference, $\bar{\delta}_m$ (%) quantifying the dispersion that exists between a given model m and the mean of all other models that have performed the complete set of prescribed total BRF simulations for a given experiment/spectral

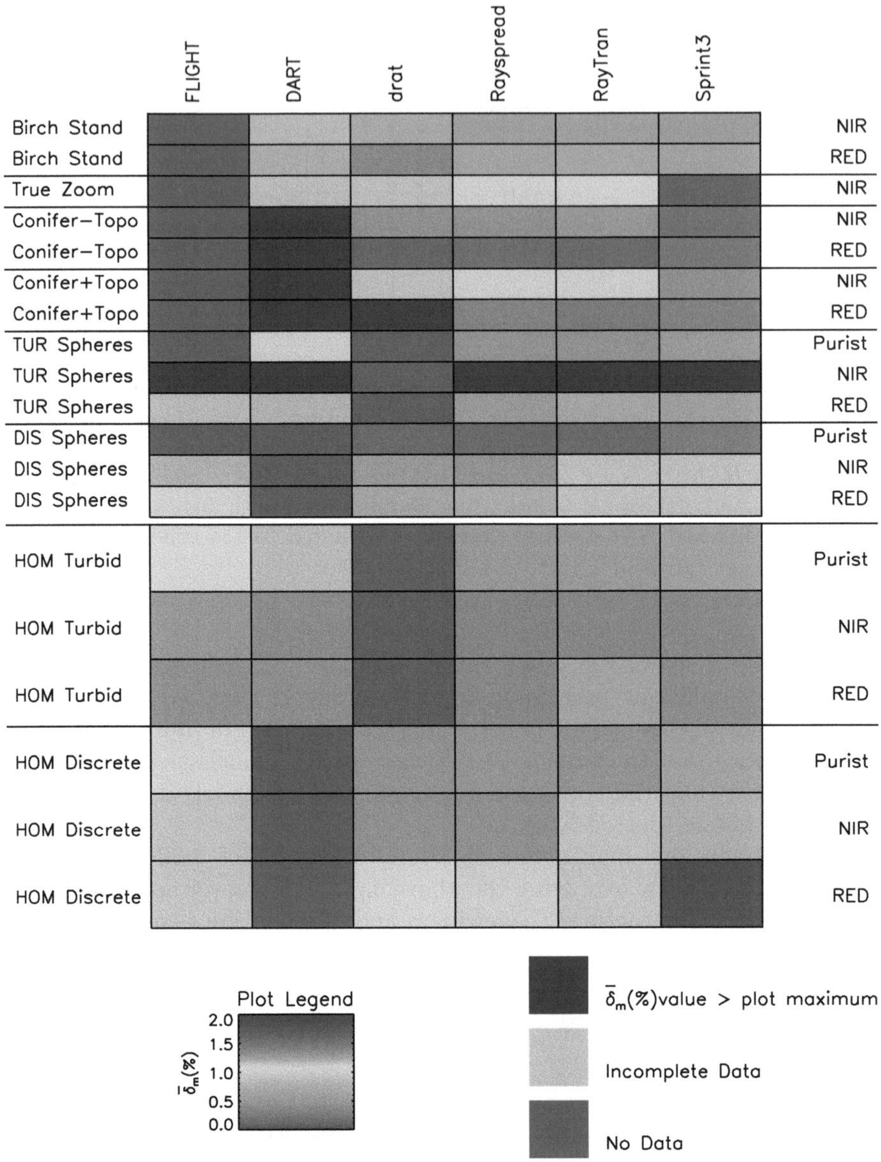

Fig. 9. Performance and participation of the "credible" 3-D Monte Carlo models during RAMI-3 for structurally heterogeneous (*top table*) and homogeneous (*bottom table*) discrete canopy representation. Model names are listed on the top of the table (one per column). The experiment identifier is provided to the left, the spectral regime to the right of each table column. *Light* (*dark*) *grey fields* indicate incomplete (no) data submission. The *green–yellow–red colour scheme* represent the integrated model-to-ensemble difference, $\bar{\delta}$ (%) obtained with respect to the mean of all models that have performed the complete set of prescribed total BRF simulations for any given experiment/spectral regime combination. (See Plate 28 on page 318)

regime combination. One should note that the predominant hue in the $\bar{\delta}_m$ colours of any given row in Fig. 9 depends both on the degree and manner in which the models are dispersed around the main cluster of simulation results. As such, the dark red colour ($\bar{\delta}_m > 2\%$) of the DART model in the coniferous forest scenes with and without topography is due to the fact that this model is somewhat different from all others here (thus raising the $\bar{\delta}_m$ values of all models as well). Similarly, the dark red colours, that occur for all of the participating RT models in the turbid medium "floating spheres" case, arise because here the simulations of the DART model are sufficiently far away from the other five models such as to increase the distance of any given model from the common mean by more than 2% of the BRF of that mean. In all of these cases, the agreement of the five model drat, FLIGHT, Rayspread, raytran and Sprint3 is substantially better if DART is removed. Finally, one should also note the large amount of (light and dark) grey patches in Fig. 9 which indicates that a significant number of experiments were not completed or not submitted by certain models.

Table 1 shows the values of the overall model dispersion indicator $\bar{\delta}_m$ (%) obtained from the total BRF simulations of these six 3-D Monte Carlo models in both RAMI-2 and RAMI-3. With the exception of the total BRF simulations of DART all other $\bar{\delta}_m$ values improved between RAMI-2 and RAMI-3, meaning that a smaller dispersion exists between the BRF values of the latest version of these models. The average dispersion between the total BRF simulations of the six 3-D Monte Carlo models was found to have almost halved from RAMI-2 (1.37 %) to RAMI-3 (0.72 %) in the discrete case, and in the turbid medium case it improved by a factor of ~7 from RAMI-2 (6.36 %) to RAMI-3 (0.91 %).

Model-to-Surrogate-Truth Deviations

Monte Carlo RT models allow for explicit 3-D representations of complex canopy architectures by describing these environments with (sometimes

Table 1. Model-to-ensemble dispersion statistics, $\bar{\delta}_m$ (%) for six 3-D Monte Carlo models participating in RAMI-2 and/or RAMI-3

Model	BRF	Discrete scenes		Turbid scenes	
name	type	RAMI-2	RAMI-3	RAMI-2	RAMI-3
DART	ρ_{tot}	–	–	1.42	1.46
drat	ρ_{tot}	1.92	0.55	–	–
FLIGHT	ρ_{tot}	1.26	0.97	9.63	1.06
Rayspread	ρ_{tot}	–	0.55	–	0.64
raytran	ρ_{tot}	1.31	0.60	1.06	0.69
Sprint3	ρ_{tot}	1.29	1.01	9.66	0.69

In each case, $\bar{\delta}_m$ was computed from all available structural, spectral, illumination and viewing conditions

Boolean combinations of) sufficiently small geometric building blocks of known radiative properties. Solving the radiative transfer equation for such 3-D environments can be achieved through a stochastic sampling of the surface-leaving radiation field [DLN00]. Since this is a time consuming undertaking – in particular for complex 3-D scenes – the current generation of 3-D MC models differ primarily in the amount of deterministic detail that is used when constructing a scene, and, in the approach and extent to which ray trajectories are sampled within the 3-D media. Both Fig. 6 and Table 1 indicate that the 3-D Monte Carlo models, `DART`, `drat`, `FLIGHT`, `Rayspread`, `raytran` and `Sprint3` are generally in very close agreement with each other. In particular the numbers in Table 1 support their usage in attempts to provide a "surrogate truth" estimate against which further RT model comparisons may then be carried out. One way to obtain such a "surrogate truth" estimate is by simply averaging the radiative quantities (X) obtained from a set of $N_{3D}^{credible}$ credible 3-D MC models for any given spectral (λ), structural (ζ), viewing (v), and illumination (i) conditions:

$$X_{ref}(\lambda, \zeta, v, i) = \frac{1}{N_{3D}^{credible}} \sum_{n=1}^{N_{3D}^{credible}} X_{3D}(\lambda, \zeta, v, i; n)$$

where the precise number and names of the 3-D MC models that feature within $N_{3D}^{credible}$ is selected from among the following models: `DART`, `drat`, `FLIGHT`, `Rayspread`, `raytran` and `Sprint3`. The selection procedure for "credible" 3-D MC models is applied to every RAMI experiment and measurement type separately and is described in detail in [WTP07]. A synoptic table featuring the names of the various "credible" 3-D Monte Carlo models that contribute toward the computation of X_{ref} is available (for each one of the RAMI-3 experiments and measurement types individually) from: `http://romc.jrc.ec.europa.eu/WWW/PAGES/ROMC_Home/RAMIREF.html`.

During RAMI-3, [WTP07] used χ^2 statistics (described in detail in Sect. 3) to document model deviations with respect to the ROMC "surrogate truth". In this way it was possible to document model "equivalency" to within 3% with the reference solution, and also to show that even for models, which do not belong to the set of "credible" 3-D MC models, the continuing participation in RAMI can lead to measurable improvements in their performance.

3 The RAMI On-Line Model Checker

With the availability of a "surrogate truth" data set, it becomes possible to develop a web-based interface allowing for the on-line evaluation of RT models even outside the frame of a RAMI phase. The ROMC was developed at the Joint Research Centre of the European Commission to allow model owners, developers and also customers of such models to evaluate the performance of a given RT model. This is done by comparing the simulation results of a model against the corresponding "surrogate truth" data derived from the "credible" 3-D MC models identified during RAMI-3. Access to the ROMC can be obtained either via the RAMI website or directly using the URL `http://romc.jrc.ec.europa.eu/`. After providing a username and valid email address the ROMC users can provide up to three different model names. Each one of these registered models can then be evaluated (1) in *debug mode*, which allows to *repeatedly* compare the output of a RT model to that of one or more experiments and/or measurements from RAMI, i.e., models are checked under conditions where the simulation results are readily available on the RAMI website, and, (2) in *validate mode*, which enables the *once-only* testing of the RT model against a continuously changing set of test cases that are similar but not quite equivalent to those from RAMI, i.e., models are checked under conditions where the solutions are not known a prior and the experiments cannot be repeated on purpose.

- In *debug mode* users may choose to execute one particular experiment and/or measurement from the set of RAMI-3 test cases *ad infinitum*, or, at least until they are satisfied with the performance of their model. Detailed descriptions of the structural, spectral, illumination and measurement conditions are available. Once the model simulation results are generated, they can be uploaded via the web-interface, and – provided they adhere to the ROMC filenaming and formatting conventions – this process will result in a series of graphical files being made available for all test cases. In debug mode users may not only download these ROMC results but also an ASCII file containing the actual "surrogate truth" data.
- In *validate mode* users may choose between structurally homogeneous and/or heterogeneous "floating spheres" canopies to verify the performance of their model. These test cases are similar but not quite identical to those featuring in RAMI-3. The actual set of test cases will be drawn at random from a large list of possible ones, such that it is unlikely to obtain the same test case twice, i.e., in all likelihood one will not "know" the solution a prior. Again, the "surrogate truth" was derived from simulations generated by models belonging to the same set of 3-D MC models as was the case for debug mode. In validate mode the reference data will not be available for downloading. The procedure for data submission, on the other hand, is identical to that of debug mode, and – provided that all ROMC formatting and filenaming requirements were applied – will also lead to a results page featuring a variety of intercomparison graphics.

Users may download their ROMC results either as jpeg formatted images from the ROMC website, or else, opt for receiving them via email in postscript form. Both the debug and validate mode ROMC results graphs feature a reference number and many ROMC graphs are available in colour, as well as, in black and white. In general, ROMC graphs have different appearances depending on the context in which they were generated. In debug mode, for example, the ROMC reference number – which is always located in the top left corner of a graph – has a smaller font size than in validate mode. Furthermore, all debug mode graphs feature a two-line footer informing readers that: "*ROMC DEBUG mode compares RT model simulations against already published RAMI results. To obtain unambiguous proof of an RT model's performance use the ROMC VALIDATE mode*". ROMC users that have submitted model simulation results for different models, or, multiple versions of the same model may explore the ROMC's "interactive" graph plotting capabilities. Figure 10 provides a series of examples of ROMC graphs generated in debug mode (top row), validate mode (middle row), and using the "interactive" ROMC capabilities when analysing simulation results generated in debug mode (bottom row). In these graphs the "surrogate truth" data set is referred to as "ROMCREF" where applicable. The list of currently available ROMC graph types includes:

- *Data plots*, featuring both the model and the reference BRFs either along the principal or the orthogonal plane. Three grey-coloured envelopes corresponding to 1, 2.5 and 5% of the BRF of the "surrogate truth" solution surround the reference solution (for an example of such a data plot featuring BRFs in the principal plane see the right panel in the top row of Fig. 10).
- *1 to 1 plots*, where the model data (X_{usr}) is plotted against the reference data (X_{ref}). These graphs also feature the root-mean-square (RMS) error:

$$RMS = \sqrt{\frac{1}{N}\sum_{n=1}^{N}\left[X_{ref}(n) - X_{usr}(n)\right]^2}$$

as well as the signal-to-noise ratio (SNR):

$$SNR = \frac{\sqrt{\frac{1}{N}\sum_{n=1}^{N}\left[X_{ref}(n)\right]^2}}{RMS}$$

and the number of datapoints N. In the case of BRF quantities these graph types are available both for individual test case and for all submitted test cases combined. In the case of flux quantities (A, R and T) the 1 to 1 graphs always include simulation results from multiple test cases. The middle panel in the top row of Fig. 10 provides an example of a 1 to 1 graph with data points from a large ensemble of ROMC test cases, whereas the bottom right panel of Fig. 10 displays the same graph type for BRF simulations generated by three RT models (belonging to the same user) when applied to a single test case.
- *Histograms*, showing the distribution of the differences between model and reference BRFs. These graph types are generated for individual test cases as well as for all submitted test cases together, and include information on the RMS error and the number of contributing experiments, or, data points. For examples, see the right panel in the middle row of Fig. 10 (where

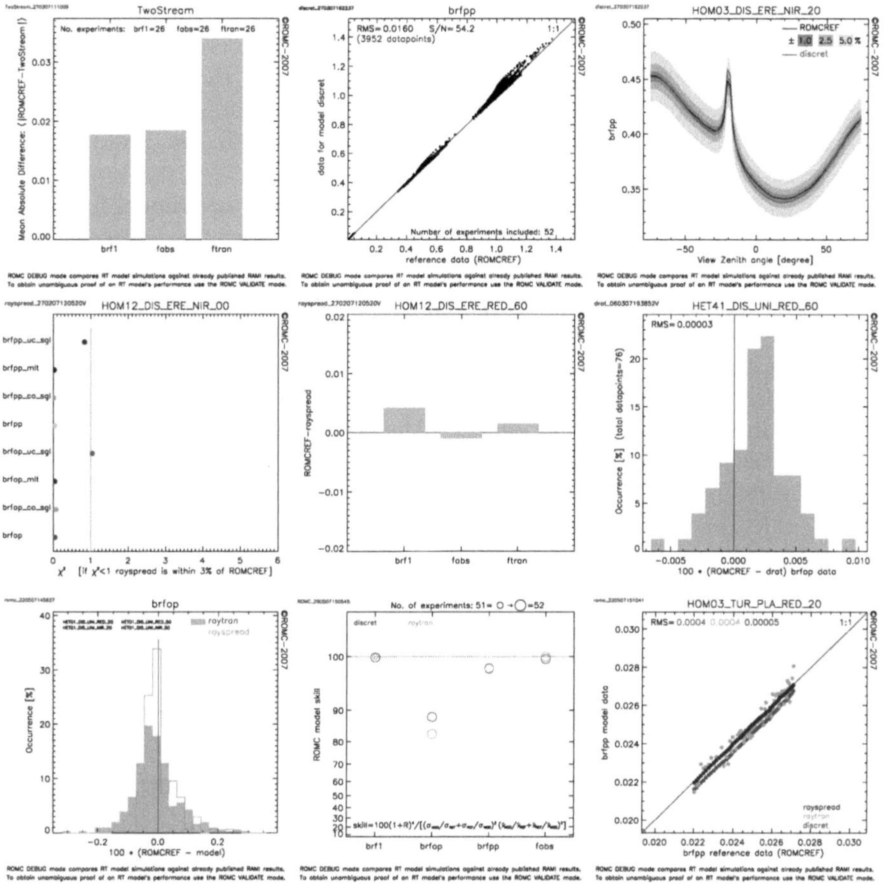

Fig. 10. Examples of currently generated ROMC graph types generated in "debug mode" (*top row*), "validate mode" (*middle row*), and using the "interactive" ROMC capabilities when analysing simulation results generated in "debug mode" (*bottom row*). For more information see text. (See Plate 29 on page 319)

a histogram of BRF differences for a single test case is shown), and, the left panel in the bottom row of Fig. 10 (where the histograms of two RT models are shown that did perform multiple test cases).

- χ^2 *graphs*, quantifying the deviations of a model from the "surrogate truth". This is done using the following metric [PWT04]:

$$\chi^2 = \frac{1}{N-1} \sum_{\lambda=1}^{N_\lambda} \sum_{\zeta=1}^{N_\zeta} \sum_{v=1}^{N_{\theta_v}} \sum_{i=1}^{N_{\theta_i}} \frac{\left[X_{usr}(\lambda,\zeta,v,i) - X_{ref}(\lambda,\zeta,v,i)\right]^2}{\sigma^2(\lambda,\zeta,v,i)}$$

where $N = N_\lambda + N_\zeta + N_{\theta_v} + N_{\theta_i}$ is the total number of simulations covering a variety of spectral λ, structural ζ, viewing v and illumination i conditions (if applicable), and, $\sigma(\lambda, \zeta, v, i) = f \cdot X_{ref}(\lambda, \zeta, v, i)$ corresponds to a fraction f of the "surrogate truth" obtained from the "credible" 3-D Monte Carlo models. By default $f = 0.03$ which reflects the absolute calibration accuracy of current space borne instruments like MISR [BCA02] and MERIS [KST02], among others (see the left panel in the middle row of Fig. 10 for an example of a χ^2 graph).

- *Bar charts*, depicting the deviation of the reference and model fluxes. For individual test cases absolute differences are documented, whereas, when multiple test cases are dealt with together then the average absolute deviation is reported Examples of these graph types can be seen in the central (individual test case) and top left (multiple test cases combined) panels of Fig. 10, respectively.
- *Skill plots*, displaying the ability of the model to match the pattern (evaluated in terms of correlation, R), the variability (assessed in terms of the ratio of standard deviations, $\sigma_{usr}/\sigma_{ref}$), and, the mean amplitude of the reference data (evaluated in terms of the ratio of the two means, $\overline{X}_{usr}/\overline{X}_{ref}$). Skill is defined as $100 \times S$, where S is the following metric:

$$S = \frac{(1+R)^4}{\left(\frac{\overline{X}_{usr}}{\overline{X}_{ref}} + \frac{\overline{X}_{ref}}{\overline{X}_{usr}}\right)^2 \left(\frac{\sigma_{usr}}{\sigma_{ref}} + \frac{\sigma_{ref}}{\sigma_{usr}}\right)^2}$$

and the correlation between the two data sets having N datapoints is:

$$R = \frac{\frac{1}{N}\sum_{n=1}^{N}\left(X_{usr}(n) - \overline{X}_{usr}\right)\left(X_{ref}(n) - \overline{X}_{ref}\right)}{\sigma_{usr}\ \sigma_{ref}}$$

The ROMC skill scores are defined from zero (least skillful) to 100 (most skillful) assuming that a perfect match is possible. Particular emphasis is given to the correct matching of the "shape" or pattern of the reference data, e.g., the skill of a model is zero if its BRFs are perfectly anticorrelated with the ROMC "surrogate truth" ($R = -1$). The middle panel in the bottom row of Fig. 10 shows the skills of multiple models (colours) when performing various measurement types over structurally heterogeneous canopies.
- *Taylor diagrams*, providing a concise statistical summary of how well patterns match each other in terms of their correlation (R), their root-mean-square difference (ΔRMS), and the ratio of their variances ($\sigma^2_{usr}/\sigma^2_{ref}$) [Tay06]. This is achieved by separating the overall RMS error (see above) into a contribution arising from the mean bias ($\overline{RMS} = \overline{X}_{usr} - \overline{X}_{ref}$), and a contribution due to the root-mean-square differences between the two data sets (ΔRMS):

$$RMS^2 = \overline{RMS}^2 + \Delta RMS^2$$

such that the root-mean-square difference can be written as:

$$\Delta RMS^2 = \sigma_{usr}^2 + \sigma_{ref}^2 - 2\,\sigma_{usr}\,\sigma_{ref}\,R$$

where the relation of ΔRMS to the correlation coefficient (R), and the standard deviations of both the user (σ_{usr}) and reference (σ_{ref}) data sets is similar to that expressed by the law of cosines for the relationship between two adjacent sides of a triangle together with the angle between them, and the third (opposite) side of the triangle [Tay06]. This finding then allows to construct a two-dimensional diagram allowing three different kinds of information to be retrieved from any single point on it.

Figure 11 provides an example of a ROMC generated Taylor diagram showing how the `raytran` model matches the "surrogate truth"

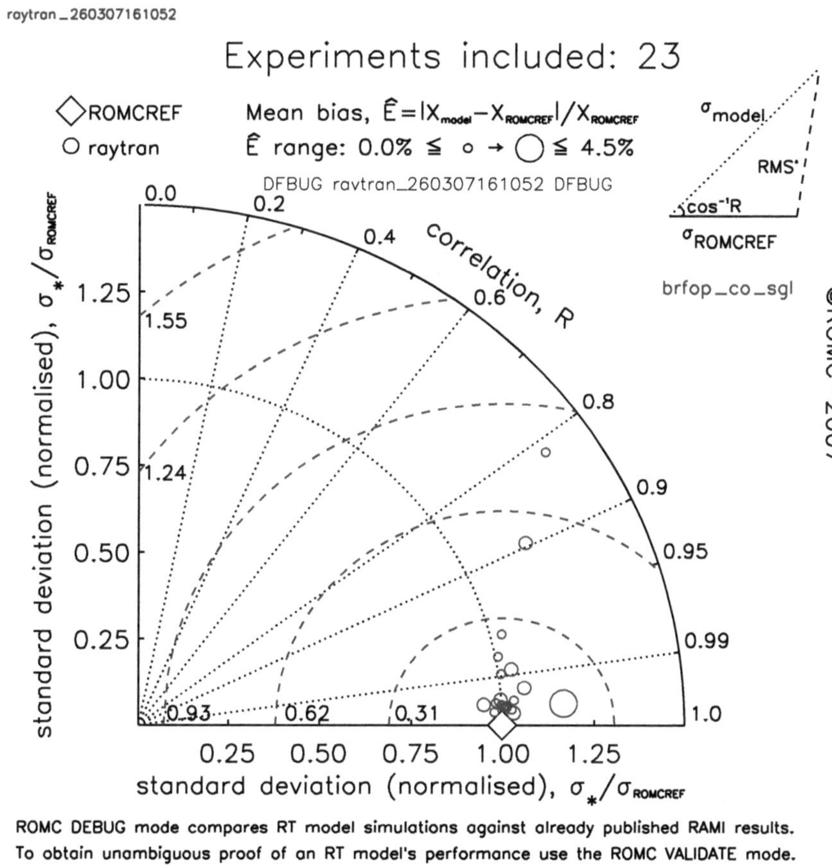

Fig. 11. Example of a ROMC generated Taylor diagram in "debug mode" depicting the performance of the `raytran` model in terms of its ability to match the pattern (correlation R) and variability (normalised standard deviation) of the single-collided BRF anisotropy of the reference solution. The normalised RMS difference error (*dashed circular arcs*) and the mean bias (*symbol size*) are also indicated. (See Plate 30 on page 320)

for simulations of the single-collided BRF along the orthogonal plane (`brfop_co_sgl`) over 23 different heterogeneous canopy targets in "debug mode". Rather than using the actual variances of the data sets, Fig. 11 normalises the above equation for ΔRMS by the variance of the reference data set. As such the reference solution will always be located at unit distance along the abscissa (big diamond symbol). The closer the 23 simulations of `raytran` (red circles) are to the diamond symbol the better their matching with the "surrogate truth" data set. More specifically, the radial distances from the origin to the red circular points are proportional to the normalised-standard-deviation (that is the ratio of the standard deviation of the users data to that of the reference data). `raytran` simulations that correspond to (red circular) points located outside the dotted circular arc at unit normalised-standard-deviation feature more variability than the reference data set of the ROMC. Conversely, if the model's (red circular) points are located inside the circular arc at unit normalised-standard-deviation then the model generated data set exhibits less variability than the "surrogate truth" data set. The angle between the abscissa and the line connecting the origin to the (red circular) points in Fig. 11 is a measure of the correlation coefficient, R between a model's RT simulations and the reference data set of the ROMC (the labels of the correlation coefficient are given along the outer circular rim of the Taylor diagram). As expected the ROMC reference data set (big diamond) matches itself to perfection ($R = 1$). Last but not least, the dashed circular lines that originate from the (big diamond) symbol of the reference data set correspond to the normalised RMS difference between the two data sets.
Despite the advantages of Taylor diagrams one should bear in mind, however, they do not allow to differentiate between two data sets that differ only by a constant offset. To address this issue the ROMC generates Taylor diagrams that vary the size of the plotting symbol if $\overline{RMS}$ exceeds 3% of the reference solution (and no negative correlations occurred). As such one can observe in Fig. 11 that at least on one occasion the `raytran` generated single-collided BRF data exhibited a mean bias of 4.5% with respect to the "surrogate truth" solution. Overall, however, most of the simulations of the `raytran` model fall very close to the location of the "surrogate truth" (big diamond) data set in Fig. 11, and this both in term of correlation and normalised RMS difference.

Users of ROMC are encouraged to utilize only ROMC results that were obtained in validate mode for publications. Those obtained in debug mode, obviously, do not qualify as proof regarding the performance of a RT model since all simulation results may readily be viewed on the RAMI website.

4 Conclusion

Since its first phase in 1999, RAMI has served as a vehicle to document the performance of the latest generation of RT models by charting both their capabilities and weaknesses under a variety of spectral and structural conditions. It is now estimated that about 60–65% of all currently available canopy reflectance models have voluntarily participated in the RAMI initiative. This would not have been possible without the continuing support and active involvement of scientists from the canopy RT modelling community that (1) acknowledged the necessity for quality assurance of RT models if these are to be applied to the interpretation of remotely sensed data, (2) voluntarily contributed to the establishment of benchmarking scenarios against which future developments of RT models may be evaluated, and (3) agreed to publish their RT model simulations in the refereed scientific literature prior to actually knowing the results of these intercomparison tests. During RAMI-3 it has for the first time been possible to demonstrate a general convergence of the ensemble of submitted RT simulations (with respect to RAMI-2), and to document the unprecedented level of agreement that now exists between the participating 3-D Monte Carlo models. These positive developments do not only further the confidence that may be placed in the quality of canopy reflectance models, but they also pave the way for addressing new and challenging issues, most notably, in the context of supporting field validation efforts of remotely sensed products. The latter is of prime importance given the abundance of global surface products from the current fleet of instruments, like MISR, MODIS, MERIS, etc.

Acknowledgments

The coordination of the RAMI activity since 1999 would not have been possible without the support of the European Commission, and more specifically, the Global Environment Monitoring unit of the Institute for Environment and Sustainability in the DG Joint Research Centre. The continuing involvement of the RT modelling community in the RAMI activity is appreciated, in particular the active contributions of Drs. M. Antunes, C. Bacour, V. Bruniquel-Pinel, M. Disney, R. Fernandes, F. Gascon, J-P. Gastellu-Etchegorry, S. A. W. Gerstl, N. Gobron, N. Goel, S. Jacquemoud, L. Jiang, A. Kuusk, T. Lavergne, S. Leblanc, P. E. Lewis, X. Li, E. Martin, M. Mõttus, W. Ni-Meister, T. Nilson, P. R. J. North, W. Qin, M. Robustelli, N. Rochdi, R. Ruiloba, C. Soler, L. Su, M. Taberner, S. Tang, R. Thompson, W. Verhoef, H. Wang, J. Wang, M. M. Verstraete, D. Xie, G. Yan, and H. Zang. Furthermore, the stimulating exchanges with the various scientists of the RAMI Advisory Body (RAB), as well as those involved with the I3RC are also gratefully acknowledged. Last but not least a special thank you goes also to Michaela Saisana.

References

[BCA02] Bruegge, C. J., N. L. Chrien, R. R. Ando, D. J. Diner, W. A. Abdou, M. C. Helmlinger, S. H. Pilorz, and K. J. Thome, Early validation of the multi-angle imaging spectroradiometer (MISR) radiometric scale, *IEEE Transactions on Geoscience and Remote Sensing*, *40*, 1477–1492, 2002.

[CRH03] Chopping, M. J., A. Rango, K. M. Havstad, F. R. Schiebe, J. C. Ritchie, T. J. Schmugge, A. N. French, L. Su, L. McKee, and R. Davis, Canopy attributes of desert grasslands and transition communities derived from multi-angular airborne imagery, *Remote Sensing Environment*, *85*, 339–354, 2003.

[DLN00] Disney, M. I., P. Lewis, and P. R. J. North, Monte Carlo raytracing in optical canopy reflectance modelling, *Remote Sensing Reviews*, *18*, 163–196, 2000.

[GDP96] Gastellu-Etchegorry, J.-P., V. Demarez, V. Pinel, and F. Zagolski, Modeling radiative transfer in heterogeneous 3-d vegetation canopies, *Remote Sensing of Environment*, *58*, 131–156, 1996.

[GMG04] Gastellu-Etchegorry, J.-P., E. Martin, and F. Gascon, Dart: A 3D model for simulating satellite images and studying surface radiation budget, *International Journal of Remote Sensing*, *25*, 73–96, 2004.

[GN97] Gerard, F. F. and P. R. J. North, Analyzing the effect of structural variability and canopy gaps on forest BRDF using a geometric-optical model, *Remote Sensing of Environment*, *62*, 46–62, 1997.

[Ger88] Gerstl, S. A. W., Angular reflectance signature of the canopy hotspot in the optical regime, in *4th Intl. Coll. on Spectral Signatures of Objects in Remote Sensing, Aussois, France*, p. 129, ESA report SP-287, 1988.

[GPV97] Gobron, N., B. Pinty, M. M. Verstraete, and Y. Govaerts, A semi-discrete model for the scattering of light by vegetation, *Journal of Geophysical Research*, *102*, 9431–9446, 1997.

[GV98] Govaerts, Y. and M. M. Verstraete, Raytran: A Monte Carlo ray tracing model to compute light scattering in three-dimensional heterogeneous media, *IEEE Transactions on Geoscience and Remote Sensing*, *36*, 493–505, 1998.

[HBS03] Helbert, J., B. Berthelot, and C. Soler, Hyemalis: Un simulateur d'images de paysages tridimensionnels complexes, in *Revue Française de Photogrammétrie et de Télédétection*, 173 / 174, pp. 27–35, Société Française de Photogrammétrie et de Télédétection, 2003.

[KST02] Kneubühler, M., M. Schaepman, K. Thome, F. Baret, and A. Müller, Calibration and validation of Envisat MERIS. Part 1: Vicarious calibration at Rail Road valley Playa (NV), in *Proceedings of MERIS level 2 validation Workshop, ESRIN, Frascati, Italy, December 9–13*, 2002.

[Kuu91] Kuusk, A., The hot spot effect in plant canopy reflectance, in *Photon-Vegetation Interactions*, edited by R. Myneni and J. Ross, pp. 139–159, Springer, Berlin Heidelberg New York, 1991.

[Kuu01] Kuusk, A., A two-layer canopy reflectance model, *Journal of Quantitative Spectroscopy and Radiative Transfer*, *71*, 1–9, 2001.

[KN00] Kuusk, A. and T. Nilson, A directional multispectral forest reflectance model, *Remote Sensing of Environment*, *72*, 244–252, 2000.

[LC01] Leblanc, S. G. and J. M. Chen, A windows graphic user interface (GUI) for the five-scale model for fast BRDF simulations, *Remote Sensing Reviews, 19*, 293–305, 2001.

[Lew99] Lewis, P., Three-dimensional plant modelling for remote sensing simulation studies using the botanical plant modelling system, *Agronomie - Agriculture and Environment, 19*, 185–210, 1999.

[LG02] Lovell, J. L. and R. D. Graetz, Analysis of POLDER-ADEOS data for the Australian continent: The relationship between BRDF and vegetation structure, *International Journal of Remote Sensing, 23*, 2767–2796, 2002.

[Nor96] North, P. R. J., Three-dimensional forest light interaction model using a Monte Carlo method, *IEEE Transactions on Geoscience and Remote Sensing, 34*, 946–956, 1996.

[OSB94] Oreskes, N., K. Shrader-Frechette, and K. Belitz, Verification, validation, and confirmation of numerical models in the earth sciences, *Science, 263*, 641–646, 1994.

[PGW01] Pinty, B., N. Gobron, J.-L. Widlowski, S. A. W. Gerstl, M. M. Verstraete, M. Antunes, C. Bacour, F. Gascon, J.-P. Gastellu, N. Goel, S. Jacquemoud, P. North, W. Qin, and R. Thompson, The radiation transfer model intercomparison (RAMI) exercise, *Journal of Geophysical Research, 106*, 11937–11956, 2001.

[PWG02] Pinty, B., J.-L. Widlowski, N. Gobron, and M. M. Verstraete, Uniqueness of Multi-angular Information – Part 1: A surface heterogeneity indicator from MISR, *IEEE Transactions on Geoscience and Remote Sensing, 40*, 1560–1573, 2002

[PWT04] Pinty, B., J.-L. Widlowski, M. Taberner, N. Gobron, M. M. Verstraete, M. Disney, F. Gascon, J.-P. Gastellu, L. Jiang, A. Kuusk, P. Lewis, X. Li, W. Ni-Meister, T. Nilson, P. North, W. Qin, L. Su, R. Tang, R. Thompson, W. Verhoef, H. Wang, J. Wang, G. Yan, and H. Zang, The radiation transfer model intercomparison (RAMI) exercise: Results from the second phase, *Journal of Geophysical Research, 109*, D06210, doi:10.1029/2004JD004252, 2004b.

[PGW04] Pinty, B., N. Gobron, J.-L. Widlowski, T. Lavergne, and M. M. Verstraete, Synergy between 1-D and 3-D radiation transfer models to retrieve vegetation canopy properties from remote sensing data, *Journal of Geophysical Research, 109*, D21205, doi:10.1029/2004JD005214, 2004a.

[PLD06] Pinty, B., T. Lavergne, R. E. Dickinson, J.-L. Widlowski, N. Gobron, and M. M. Verstraete, Simplifying the interaction of land surfaces with radiation for relating remote sensing products to climate models, *Journal of Geophysical Research, 111*(D02116), D02,116, doi:10.1029/2005JD005,952, 2006.

[QX97] Qin, W. and Y. Xiang, An analytical model for bidirectional reflectance factor of multicomponent vegetation canopies, *Science in China (Series C), 40*, 305–315, 1997.

[RSK03] Rautiainen, M., P. Stenberg, N. T, and A. Kuusk, The effect of crown shape on the reflectance of coniferous stands, *Remote Sensing of Environment, 89*, 41–52, 2003.

[Ros81] Ross, J., *The Radiation Regime and Architecture of Plant Stands*, Dr. W. Junk, Boston, 1981.

[SLD01] Saich, P., P. Lewis, M. Disney, and G. Thackrah, Comparison of Hymap/E-SAR data with models for optical reflectance and microwave scattering from vegetation canopies, in *Proceedings of Third International Workshop on Retrieval of Bio- and Geo-Physical Parameters from SAR data for Land Applications*, p. 427, 2001.

[SM88] Shultis, J. K. and R. B. Myneni, Radiative transfer in vegetation canopies with anisotropic scattering, *Journal of Quantitative Spectroscopy and Radiation Transfer*, *39*, 115–129, 1988.

[SS00] Soler, C. and F. Sillion, Hierarchical instantiation for radiosity, in *Rendering Techniques '00*, edited by B. Peroche and H. Rushmeier, pp. 173–184, Springer, Wien, Berlin Heidelberg New York, 2000.

[Tay06] Taylor, K. E., Summarizing multiple aspects of model performance in a single diagram, *Journal of Geophysical Research*, *106*, 7183–7192, 2006.

[TG98] Thompson, R. L. and N. S. Goel, Two models for rapidly calculating bidirectional reflectance: Photon spread (ps) model and statistical photon spread (sps) model, *Remote Sensing Reviews*, *16*, 157–207, 1998.

[Ver02] Verhoef, W., Improved modelling of multiple scattering in leaf canopies: The model SAIL++, in *Proceedings of the First Symposium on Recent Advances in Quantitative Remote Sensing, Torrent, Spain, September 2002*, edited by A. Sobrino, pp. 11–20, 2002.

[Ver87] Verstraete, M. M., Radiation transfer in plant canopies: Transmission of direct solar radiation and the role of leaf orientation, *Journal of Geophysical Research*, *92*, 10, 985–10, 995, 1987.

[Ver88] Verstraete, M. M., Radiation transfer in plant canopies: Scattering of solar radiation and canopy reflectance, *Journal of Geophysical Research*, *93*, 9483–9494, 1988.

[WPG01] Widlowski, J.-L., B. Pinty, N. Gobron, M. M. Verstraete, and A. B. Davis, Characterization of surface heterogeneity detected at the MISR/TERRA subpixel scale, *Geophysical Research Letters*, *28*, 4639–4642, 2001.

[WLP06] Widlowski, J.-L., T. Lavergne, B. Pinty, M. M. Verstraete, and N. Gobron, Rayspread: A virtual laboratory for rapid BRF simulations over 3-D plant canopies, in *Computational Methods in Transport*, edited by G. Frank, pp. 211–231, ISBN–10 3–540–28, 122–3, Lecture Notes in Computational Science and Engineering Series, 48, Springer, Berlin Heidelberg New York, 2006.

[WPL06] Widlowski, J.-L., B. Pinty, T. Lavergne, M. M. Verstraete, and N. Gobron, Horizontal radiation transport in 3-D forest canopies at multiple spatial resolutions: Simulated impact on canopy absorption, *Remote Sensing of Environment*, *103*, 379–397, doi:10.1016/j.rse.2006.03.014, 2006.

[WTP07] Widlowski, J.-L., M. Taberner, B. Pinty, V. Bruniquel-Pinel, M. Disney, R. Fernandes, J.-P. Gastellu-Etchegorry, N. Gobron, A. Kuusk, T. Lavergne, S. Leblanc, P. Lewis, E. Martin, M. Mõttus, P. J. R. North, W. Qin, M. Robustelli, N. Rochdi, R. Ruiloba, C. Soler, R. Thompson, W. Verhoef, M. M. Verstraete, and D. Xie, The third radiation transfer model intercomparison (RAMI) exercise: Documenting progress in canopy reflectance modelling, *Journal of Geophysical Research*, 2007.

Uncertainty and Sensitivity Analysis for Models of Complex Systems

J.C. Helton

Summary. Sampling-based methods for uncertainty and sensitivity analysis are reviewed. The following topics are considered: (1) Definition of probability distributions to characterize epistemic uncertainty in analysis inputs, (2) Generation of samples from uncertain analysis inputs, (3) Propagation of sampled inputs through an analysis, (4) Presentation of uncertainty analysis results, and (5) Determination of sensitivity analysis results.

1 Introduction

Uncertainty analysis and sensitivity analysis are essential parts of analyses for complex systems [1–12] Specifically, uncertainty analysis refers to the determination of the uncertainty in analysis results that derives from uncertainty in analysis inputs, and sensitivity analysis refers to the determination of the contributions of individual uncertain analysis inputs to the uncertainty in analysis results. The uncertainty under consideration here is often referred to as epistemic uncertainty; alternative designations for this form of uncertainty include state of knowledge, subjective, reducible, and type B [13–21]. Epistemic uncertainty derives from a lack of knowledge about the appropriate value to use for a quantity that is assumed to have a fixed value in the context of a particular analysis. In the conceptual and computational organization of an analysis, epistemic uncertainty is generally considered to be distinct from aleatory uncertainty, which arises from an inherent randomness in the behavior of the system under study [13–22]. Alternative designations for aleatory uncertainty include variability, stochastic, irreducible, and type A.

A number of approaches to uncertainty and sensitivity analysis have been developed, including differential analysis [23–28], response surface methodology [29–35], Monte Carlo analysis [36–48], and variance decomposition procedures [49–53]. Overviews of these approaches are available in several reviews [54–62].

The focus of this presentation is on Monte Carlo (i.e., sampling-based) approaches to uncertainty and sensitivity analysis. Sampling-based approaches to uncertainty and sensitivity analysis are both effective and widely used [41,43,45,47,48,56,57]. Analyses of this type involve the generation and exploration of a mapping from uncertain analysis inputs to uncertain analysis results. The underlying idea is that analysis results $\mathbf{y}(\mathbf{x}) = [y_1(\mathbf{x}), y_2(\mathbf{x}), \ldots, y_{nY}(\mathbf{x})]$ are functions of uncertain analysis inputs $\mathbf{x} = [x_1, x_2, \ldots, x_{nX}]$. In turn, uncertainty in $\mathbf{x}$ results in a corresponding uncertainty in $\mathbf{y}(\mathbf{x})$. This leads to two questions: (1) What is the uncertainty in $\mathbf{y}(\mathbf{x})$ given the uncertainty in $\mathbf{x}$?, and (2) How important are the individual elements of $\mathbf{x}$ with respect to the uncertainty in $\mathbf{y}(\mathbf{x})$? The goal of uncertainty analysis is to answer the first question, and the goal of sensitivity analysis is to answer the second question. In practice, the implementation of an uncertainty analysis and the implementation of a sensitivity analysis are closely connected on both a conceptual and a computational level.

The following sections summarize the five basic components that underlie the implementation of a sampling-based uncertainty and sensitivity analysis: (1) Definition of distributions $D_1, D_2, \ldots, D_{nX}$ that characterize the epistemic uncertainty in the components $x_1, x_2, \ldots, x_{nX}$ of $\mathbf{x}$ (Sect. 2), (2) Generation of a sample $\mathbf{x}_1, \mathbf{x}_2, \ldots, \mathbf{x}_{nS}$ from the $\mathbf{x}$'s in consistency with the distributions $D_1, D_2, \ldots, D_{nX}$ (Sect. 3), (3) Propagation of the sample through the analysis to produce a mapping $[\mathbf{x}_i, \mathbf{y}(\mathbf{x}_i)], i = 1, 2, \ldots, nS$, from analysis inputs to analysis results (Sect. 4), (4) Presentation of uncertainty analysis results (i.e., approximations to the distributions of the elements of $\mathbf{y}$ constructed from the corresponding elements of $\mathbf{y}(\mathbf{x}_i), i = 1, 2, \ldots, nS$) (Sect. 5), and (5) Determination of sensitivity analysis results (i.e., exploration of the mapping $[\mathbf{x}_i, \mathbf{y}(\mathbf{x}_i)], i = 1, 2, \ldots, nS$) (Sect. 6). Space limitations in this presentation preclude the presentation of detailed examples of the indicated analysis components; however, extensive examples can be found in the published descriptions of an uncertainty and sensitivity analysis carried out for the Waste Isolation Pilot Plant (e.g., Refs. [63–65]). The presentation then ends with a concluding summary (Sect. 7).

Only probabilistic characterizations of uncertainty are considered in this presentation. Alternative uncertainty representations (e.g., evidence theory, possibility theory, fuzzy set theory, interval analysis) are an active area of research [66–73] but are outside the intended scope of this presentation.

This presentation is a lightly edited version of a prior workshop presentation [74] and is intended to introduce the reader to sampling-based procedures for uncertainty and sensitivity analysis. More extensive information on these procedures is available in five technical reports [75–79] and a number of additional presentations derived from these reports [44–48,80,81].

2 Characterization of Uncertainty

Definition of the distributions $D_1, D_2, \ldots, D_{nX}$ that characterize the epistemic uncertainty in the components $x_1, x_2, \ldots, x_{nX}$ of $\mathbf{x}$ is the most important part of a sampling-based uncertainty and sensitivity analysis as these distributions determine both the uncertainty in $\mathbf{y}$ and the sensitivity of the elements of $\mathbf{y}$ to the elements of $\mathbf{x}$. The distributions $D_1, D_2, \ldots, D_{nX}$ are typically defined through an expert review process [82–88], and their development can constitute a major analysis cost. A possible analysis strategy is to perform an initial exploratory analysis with rather crude definitions for $D_1, D_2, \ldots, D_{nX}$ and use sensitivity analysis to identify the most important analysis inputs; then, resources can be concentrated on characterizing the uncertainty in these inputs and a second presentation or decision-aiding analysis can be carried out with these improved uncertainty characterizations.

The scope of an expert review process can vary widely depending on the purpose of the analysis, the size of the analysis, and the resources available to carry out the analysis. At one extreme is a relatively small study in which a single analyst both develops the uncertainty characterizations (e.g., on the basis of personal knowledge or a cursory literature review). At the other extreme, is a large analysis on which important societal decisions will be based and for which uncertainty characterizations are carried out for a large number of variables by teams of outside experts who support the analysts actually performing the analysis.

Given the breadth of analysis possibilities, it is beyond the scope of this presentation to provide an exhaustive review of how the distributions $D_1, D_2, \ldots, D_{nX}$ might be developed. However, as general guidance, it is best to avoid trying to obtain these distributions by specifying the defining parameters (e.g., mean and standard deviation) for a particular distribution type. Rather, distributions can be defined by specifying selected quantiles (e.g., 0.0, 0.1, 0.25, ..., 0.9, 1.0) of the corresponding cumulative distribution function (CDF), which should keep the individual supplying the information in closer contact with the original sources of information or insight than is the case when a particular named distribution is specified (Fig. 1a). Distributions from multiple experts can be aggregated by averaging (Fig. 1b).

3 Generation of Sample

Several sampling strategies are available, including random sampling, importance sampling, and Latin hypercube sampling [36, 47]. Latin hypercube sampling is very popular for use with computationally demanding models because its efficient stratification properties allow for the extraction of a large amount of uncertainty and sensitivity information with a relatively small sample size [89–91].

Latin hypercube sampling operates in the following manner to generate a sample of size nS from the distributions $D_1, D_2, \ldots, D_{nX}$ associated with the elements of $\mathbf{x} = [x_1, x_2, \ldots, x_{nX}]$. The range of each x_j is exhaustively divided into nS disjoint intervals of equal probability and one value x_{ij} is randomly selected from each interval. The nS values for x_1 are randomly paired without replacement with the nS values for x_2 to produce nS pairs.

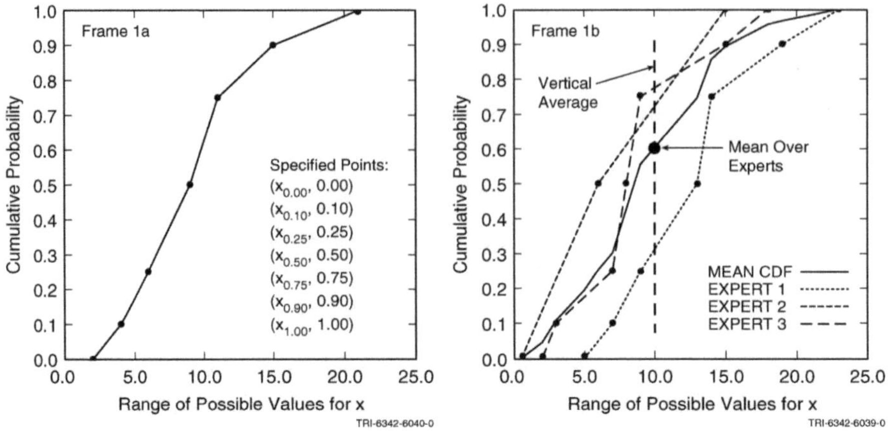

Fig. 1. Characterization of epistemic uncertainty: (**a**) Construction of CDF from specified quantile values (Fig. 4.1, Ref. [76]), and (**b**) Construction of mean CDF by vertical averaging of CDFs defined by individual experts with equal weight (i.e., $1/nE = 1/3$, where $nE = 3$ is the number of experts) given to each expert (Fig. 4.2, Ref. [76])

These pairs are then randomly combined without replacement with the nS values for x_3 to produce nS triples. This process is continued until a set of nS nX-tuples $\mathbf{x}_i = [x_{i1}, x_{i2}, \ldots, x_{i,nX}], i = 1, 2, \ldots, nS$, is obtained, with this set constituting the Latin hypercube sample (LHS) (Fig. 2).

Latin hypercube sampling is a good choice for a sampling procedure when computationally demanding models are being studied. The popularity of Latin hypercube sampling recently led to the original article being designated a *Technometrics* classic in experimental design [92]. When the model is not computationally demanding, many model evaluations can be performed and random sampling works as well as Latin hypercube sampling.

Control of correlations is an important aspect of sample generation. Specifically, correlated variables should have correlations close to their specified values, and uncorrelated variables should have correlations close to zero. In general, the imposition of complex correlation structures is not easy. However, Iman and Conover have developed a broadly applicable procedure to impose rank correlations on sampled values that (1) is distribution free (i.e., does not depend on the assumed marginal distributions for the sampled variables), (2) can impose complex correlation structures involving multiple variables, (3) works with both random and Latin hypercube sampling, and (4) preserves the intervals used in Latin hypercube sampling [93, 94]. Details on the implementation of the procedure are available in the original reference [93]; illustrative results are provided in Fig. 3.

Unlike simple random sampling, the size of an LHS cannot be increased by simply adding one sample element at a time. However, recently developed

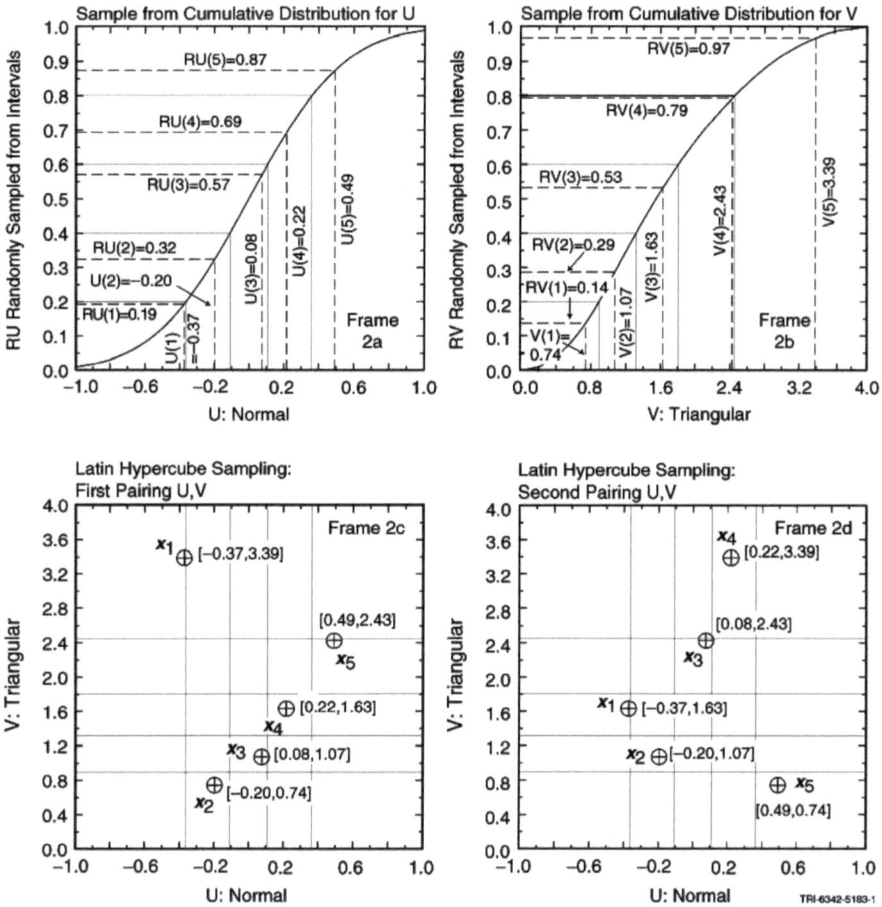

Fig. 2. Example of Latin hypercube sampling to generate a sample of size $nS = 5$ from $\mathbf{x} = [U, V]$ with U normal on $[-1, 1]$ (mean $= 0.0; 0.01$quantile $= -1; 0.99$quantile $= 1$) and V triangular on [0, 4] (mode $= 1$): (**a, b**) Upper frames illustrate sampling of values for U and V, and (**c, d**) Lower frames illustrate two different pairings of the sampled values of U and V in the construction of a Latin hypercube sample (Fig. 5.3, Ref. [76])

techniques provide a means to retain the elements of an initial LHS in an expanded LHS [95–97]. This can be important in a computationally demanding analysis in which it is desired both to increase the size of an LHS and also to retain already performed calculations in the analysis. Further, the stability of results obtained with Latin hypercube sampling for a given sample size can be assessed with a replicated sampling technique developed by R.L. Iman [98,99].

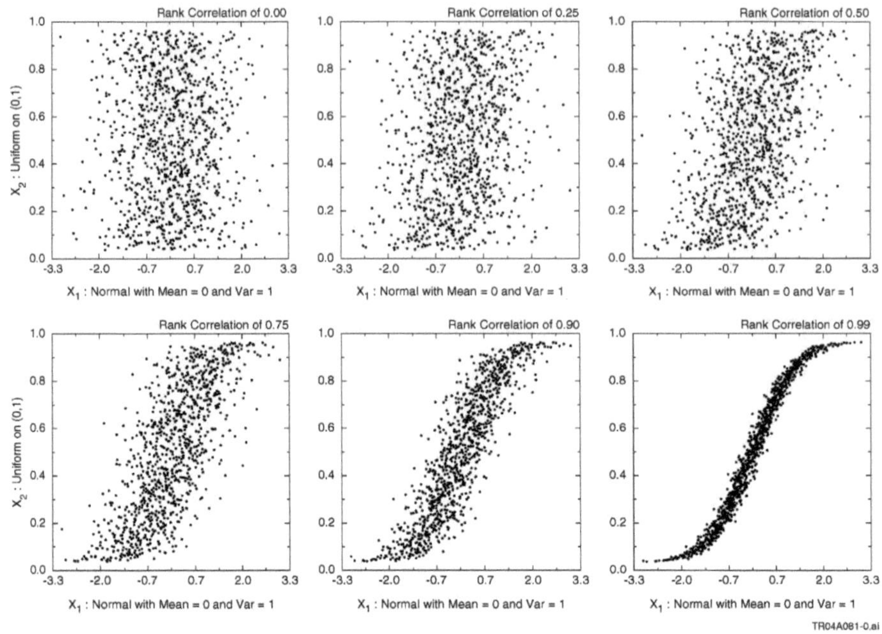

Fig. 3. Examples of rank correlations of 0.00, 0.25, 0.50, 0.75, 0.90 and 0.99 imposed with the Iman/Conover restricted pairing technique for an LHS of size $nS = 1{,}000$ (Fig. 5.1, Ref. [77])

4 Propagation of Sample Through the Analysis

Propagation of the sample through the analysis to produce the mapping $[\mathbf{x}_i, \mathbf{y}(\mathbf{x}_i)], i = 1, 2, \ldots, nS$, from analysis inputs to analysis results is often the most computationally demanding part of a sampling-based uncertainty and sensitivity analysis. The details of this propagation are analysis specific and can range from very simple for analyses that involve a single model to very complicated for large analyses that involve complex systems of linked models [65, 100].

When a single model is under consideration, this part of the analysis can involve little more than putting a DO loop around the model that (1) supplies the sampled input to the model, (2) runs the model, and (3) stores model results for later analysis. When more complex analyses with multiple models are involved, considerable sophistication may be required in this part of the analysis. Implementation of such analyses can involve (1) development of simplified models to approximate more complex models, (2) clustering of results at model interfaces, (3) reuse of model results through interpolation or linearity properties, and (4) complex procedures for the storage and retrieval of analysis results.

5 Presentation of Uncertainty Analysis Results

Presentation of uncertainty analysis results is generally straight forward and involves little more than displaying the results associated with the already calculated mapping $[\mathbf{x}_i, \mathbf{y}(\mathbf{x}_i)], i = 1, 2, \ldots, nS$. Presentation possibilities include means and standard deviations, density functions, cumulative distribution function (CDFs), complementary cumulative distribution functions (CCDFs), and box plots [47, 56]. Presentation formats such as CDFs (Fig. 4a), CCDFs (Fig. 4a) and box plots (Fig. 4b) are usually preferable to means and standard deviations because of the large amount of uncertainty information that is lost in the calculation of means and standard deviations. Owing to their flattened shape, box plots are particularly useful when it is desired to the display and compare the uncertainty in a number of related variables.

The representational challenge is more complex when the analysis outcome of interest is a function rather than a scalar. For example, a system property that is a function of time is a common analysis outcome. As another example, a CCDF that summarizes the effects of aleatory uncertainty is a standard analysis outcome in risk assessments. An effective display format for such analysis outcomes is to use two plot frames, with first frame displaying the analysis results for the individual sample elements and the second frame displaying summary results for the outcomes in the first frame (e.g., quantiles and means) (Fig. 5).

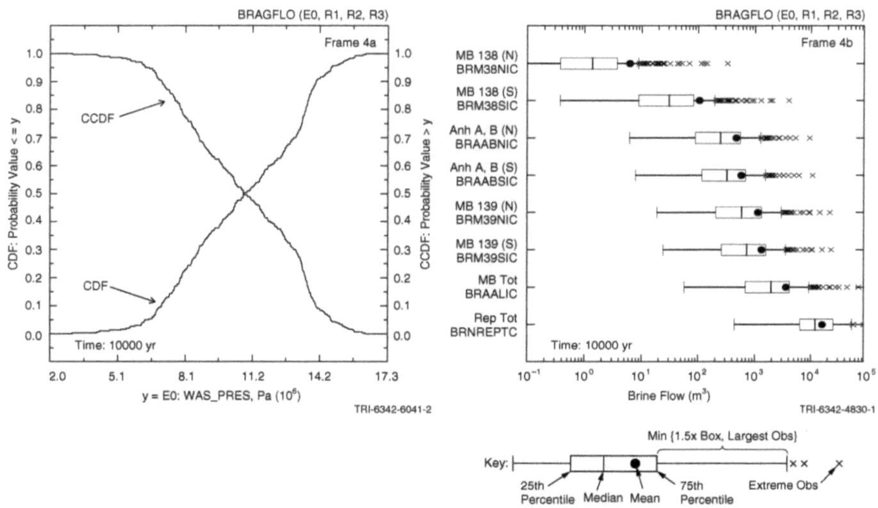

Fig. 4. Representation of uncertainty in scalar-valued analysis results: (**a**) CDFs and CCDFs (Fig. 7.2, Ref. [76]), and (**b**) box plots (Fig. 7.4, Ref. [76])

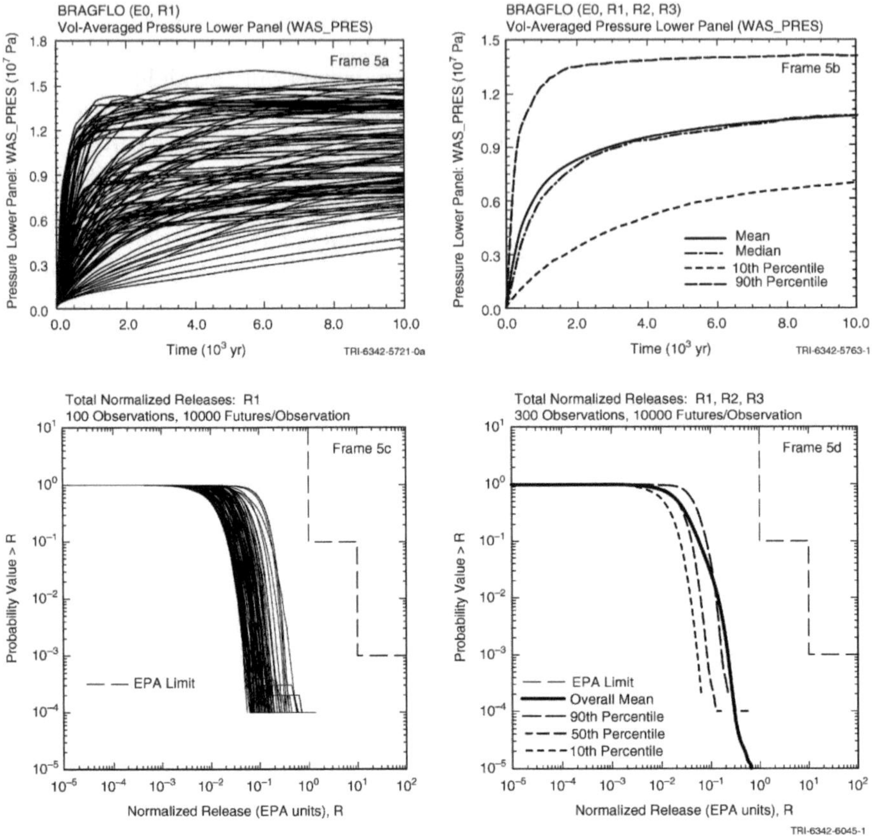

Fig. 5. Representation of uncertainty in analysis results that are functions: (**a, b**) Pressure as a function of time (Figs. 7.5, 7.9, Ref. [76]), and (**c, d**) Effects of aleatory uncertainty summarized as a CCDF (Fig. 10.5, Ref. [76])

6 Determination of Sensitivity Analysis Results

Determination of sensitivity analysis results is usually more demanding than the presentation of uncertainty analysis results due to the need to actually explore the mapping $[\mathbf{x}_i, \mathbf{y}(\mathbf{x}_i)], i = 1, 2, \ldots, nS$, to assess the effects of individual components of $\mathbf{x}$ on the components of $\mathbf{y}$. A number of approaches to sensitivity analysis that can be used in conjunction with a sampling-based uncertainty analysis are listed and briefly summarized below. In this summary, (1) x_j is an element of $\mathbf{x} = [x_1, x_2, \ldots, x_{nX}]$, (2) y_k is an element of $\mathbf{y}(\mathbf{x}) = [y_1(\mathbf{x}), y_2(\mathbf{x}), \ldots, y_{nY}(\mathbf{x})]$, (3) $\mathbf{x}_i = [x_{i1}, x_{i2}, \ldots, x_{i,nX}], i = 1, 2, \ldots, nS$, is a random or Latin hypercube sample from the possible values for $\mathbf{x}$ generated in consistency with the joint distribution assigned to the x_j, (4) $\mathbf{y}_i = \mathbf{y}(\mathbf{x}_i)$ for $i = 1, 2, \ldots, nS$, and (5) x_{ij} and y_{ik} are elements of $\mathbf{x}_i$ and $\mathbf{y}_i$, respectively.

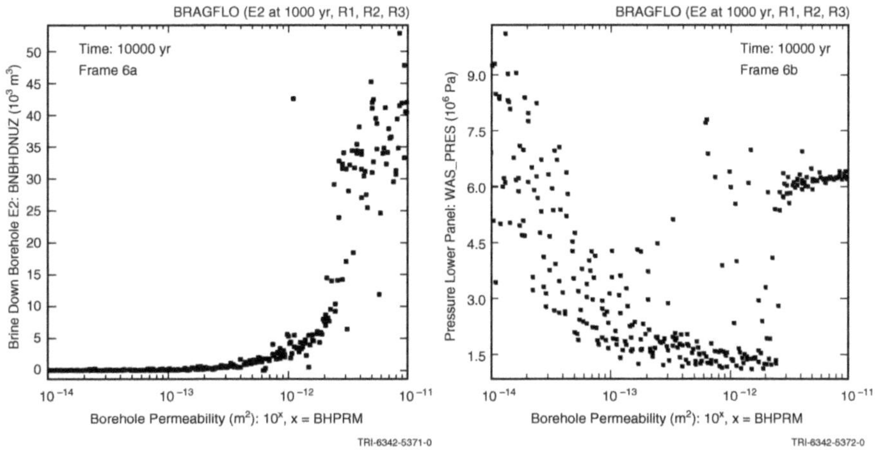

Fig. 6. Examples of scatterplots obtained in a sampling-based uncertainty/ sensitivity analysis (Figs. 8.1, 8.2, Ref. [76])

Scatterplots. Scatterplots are plots of the points $[x_{ij}, y_{ik}]$ for $i = 1, 2, \ldots, nS$ and can reveal nonlinear or other unexpected relationships (Fig. 6). In many analyses, scatterplots provide all the information that is needed to understand the sensitivity of analysis results to the uncertainty in analysis inputs. Further, scatterplots constitute a natural starting point in a complex analysis that can help in the development of a sensitivity analysis strategy using one or more additional techniques. Additional information: Sect. 6.6.1, Ref. [45]; Sect. 6.1, Ref. [48].

Cobweb Plots. Cobweb plots are plots of the points $[\mathbf{x}_i, y_{ik}] = [x_{i1}, x_{i2}, \ldots, x_{i,nX}, y_{ik}]$ for $i = 1, 2, \ldots, nS$ and provide a two-dimensional representation for $[\mathbf{x}_i, y_{ik}]$, which is a $nX + 1$ dimensional quantity. Specifically, values for the y_{ik} and also for the elements $\mathbf{x}_{ij}$ of $\mathbf{x}_i$ appear on the ordinate of a cobweb plot and the variables themselves are designated by fixed locations on the abscissa. Then, the values $y_{ik}, i = 1, 2, \ldots, nS$, for y_k and the values $x_{ij}, i = 1, 2, \ldots, nS$, for each x_j are plotted above the locations for y_k and x_j on the abscissa and each $nX + 1$ dimensional point $[\mathbf{x}_i, y_{ik}]$ is represented by a line connecting the values for the individual components of $[\mathbf{x}_i, y_{ik}]$. Cobweb plots provide more information in a single plot frame than a scatterplot but are harder to read. Additional information: Sect. 11.7, Ref. [101].

Correlation. A correlation coefficient (CC) provides a measure of the strength of the linear relationship between x_j and y_k. The CC between x_j and y_k is equal to the standardized regression coefficient (SRC) in a linear regression relating y_k to x_j and is also equal in absolute value to the square root of the R^2 value associated with the indicated regression. When calculated with raw (i.e., untransformed) data, the CC is often referred to as the Pearson CC. Additional information: Sect. 6.6.4, Ref. [45]; Sect. 6.2, Ref. [48].

Regression Analysis. Regression analysis provides an algebraic representation of the relationships between y_k and one or more x_j's. Regression analysis is usually performed in a stepwise fashion, with initial inclusion of the most important x_j, then the two most important x_j's, and so on until no more x_j's that significantly affect y_k can be identified. Variable importance is indicated by order of selection in the stepwise process, changes in R^2 values as additional variables are added to the regression model, and SRCs for the x_j's in the final regression model (Table 1). A display of regression results in the form shown in Table 1 is very unwieldy when results at a sequence of times are under consideration. In this situation, a more compact display of regression results is provided by plotting time-dependent SRCs (Fig. 7a). Additional information: Sects. 6.6.2, 6.6.3, 6.6.5, Ref. [45]; Sect. 6.3, Ref. [48].

Table 1. Example of stepwise regression analysis to identify uncertain variables affecting the uncertainty in pressure at 10,000 yr in Fig. 5a (Table 8.6, Ref. [76])

Step[a]	Variable[b]	SRC[c]	R^{2d}
1	*WMICDFLG*	0.718	0.508
2	*HALPOR*	0.466	0.732
3	*WGRCOR*	0.246	0.792
4	*ANHPRM*	0.129	0.809
5	*SHRGSSAT*	0.070	0.814
6	*SALPRES*	0.063	0.818

[a] Steps in stepwise regression analysis
[b] Variables listed in the order of selection in regression analysis
[c] SRCs for variables in final regression model
[d] Cumulative R^2 value with entry of each variable into regression model

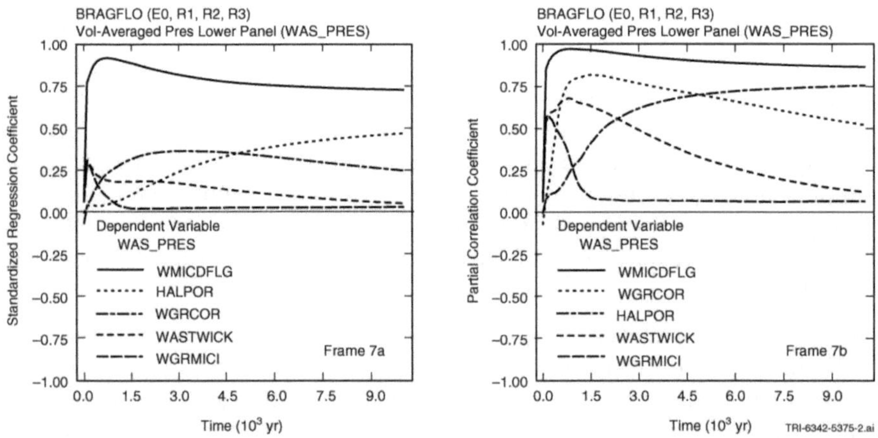

Fig. 7. Time-dependent sensitivity analysis results for uncertain pressure curves in Fig. 5a: (**a**) SRCs as a function of time, and (**b**) PCCs as a function of time (Fig. 8.3, Ref. [76])

Partial Correlation. A partial correlation coefficient (PCC) provides a measure of the strength of the linear relationship between y_k and x_j after the linear effects of all other elements of $\mathbf{x}$ have been removed. Similarly to SRCs, PCCs can be determined as a function of time for time-dependent analysis results (Fig. 7b). Additional information: Sect. 6.6.4, Ref. [45]; Sect. 6.4, Ref. [48].

Rank Transformations. A rank transformation replaces values for y_k and x_j with their corresponding ranks. Specifically, the smallest value for a variable is assigned a rank of 1; next largest value is assigned a rank of 2; tied values are assigned their average rank; and so on up to the largest value, which is assigned a rank of nS. Use of the rank transformation converts a nonlinear but monotonic relationship between y_k and x_j to a linear relationship and produces rank (i.e., Spearman) correlations, rank regressions, standardized rank regression coefficients (SRRCs) and partial rank correlation coefficients (PRCCs). In the presence of nonlinear but monotonic relationships between the x_j and y_k, the use of the rank transform can substantially improve the resolution of sensitivity analysis results (Table 2). Additional information: Sect. 6.6.6, Ref. [45]; Sect. 6.6, Ref. [48]; Ref. [102].

Tests for Patterns Based on Gridding. Analyses on raw and rank-transformed data can fail when the underlying relationships between the x_j and y_k are nonlinear and nonmonotonic (Fig. 8). The scatterplot in Fig. 6b is for the pressure at 10,000 yr in Fig. 8a versus the uncertain variable *BHPRM*. The analyses with PRCCs summarized in Fig. 8b fail at later times because the pattern appearing in Fig. 6b is too complex to be captured with a regression analysis based on raw or rank-transformed data. An alternative analysis strategy for situations of this type is to place grids on the scatterplot for y_k and x_j and then perform various statistical tests to determine if the distribution of points

Table 2. Comparison of stepwise regression analyses with raw and rank-transformed data for variable *BRAALIC* in Fig. 4b (Table 8.8, Ref. [76])

Step[a]	Raw data			Rank-transformed data		
	Variable[b]	SRC[c]	R^{2d}	Variable[b]	SRRC[e]	R^{2d}
1	*ANHPRM*	0.562	0.320	*WMICDFLG*	−0.656	0.425
2	*WMICDFLG*	−0.309	0.423	*ANHPRM*	0.593	0.766
3	*WGRCOR*	−0.164	0.449	*HALPOR*	−0.155	0.802
4	*WASTWICK*	−0.145	0.471	*WGRCOR*	−0.152	0.824
5	*ANHBCEXP*	−0.120	0.486	*HALPRM*	0.143	0.845
6	*HALPOR*	−0.101	0.496	*SALPRES*	0.120	0.860
7				*WASTWICK*	−0.010	0.869

[a] Steps in stepwise regression analysis
[b] Variables listed in order of selection in regression analysis
[c] SRCs for variables in final regression model
[d] Cumulative R^2 value with entry of each variable into regression model
[e] SRRCs for variables in final regression model

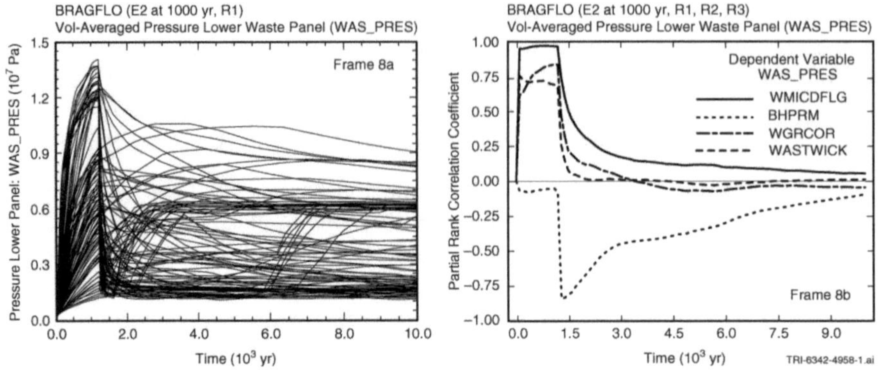

Fig. 8. Illustration of failure of a sensitivity analysis based on rank-transformed data: (**a**) pressures as a function of time and (**b**) PRCCs as a function of time (Fig. 8.7, Ref. [76])

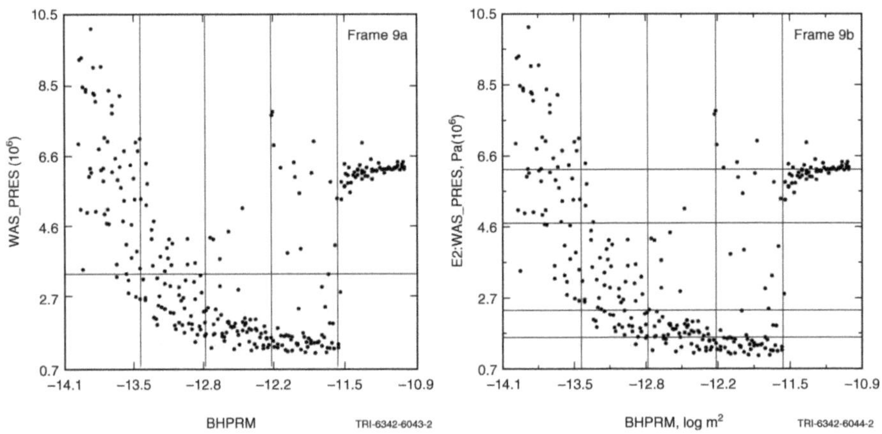

Fig. 9. Grids used to test for nonrandom patterns: (**a**) Partitioning of range of x_j for tests based on common means and common distributions and ranges of x_j and y_k for test based on common medians (Fig. 8.8, Ref. [76]), and (**b**) Partitioning of ranges of x_j and y_k for tests of no influence (Fig. 8.9, Ref. [76])

across the grid cells appears to be nonrandom. Appearance of a nonrandom pattern indicates that x_j has an effect on y_k. Possibilities include (1) tests for common means and common distributions for values of y_k based on partitioning the range of x_j (Fig. 9a) and (2) tests for common medians and no influence based on partitioning the ranges of x_j and y_k (Figs. 9a, b). Additional information: Ref. [44]; Sects. 6.6.8 and 6.6.9, Ref. [45]; Sects. 6.6 and 6.7, Ref. [48].

Nonparametric Regression. Nonparametric regression seeks more general models than those obtained by least squares regression and can succeed in situations such as the one illustrated in Fig. 8 where regression and correlation analysis based on raw and rank-transformed data fail. Nonparametric regression attempts to find models that are local in the approximation to the relationship between y_k and multiple x_j's, and, as a result, are better at capturing complex nonlinear relationships than models obtained with traditional regression or rank regression. Nonparametric regression models can be constructed in a stepwise manner with incremental changes in R^2 values with the addition of successive variables to the model providing an indication of variable importance. Additional information: Sect. 6.8, Ref. [48]; Refs. [79–81, 103–105].

Tests for Patterns Based on Distance Measures. Tests based on distance measures consider relationships within the scatterplot for y_k and x_j such as the distribution of distances between nearest neighbors and provide a way to identify nonrandom relationships between y_k and x_j. A positive feature of these tests is the avoidance of the problem of defining an appropriate grid as is the case with grid-based methods. Additional information: Sect. 6.11, Ref. [48]; Refs. [106–109].

Trees. Tree-based sensitivity analyses search for relationships between y_k and multiple x_j's by successively subdividing the sample elements $\mathbf{x}_i$ on the basis of observed effects of individual x_j's on y_k. Additional information: Refs. [110, 111].

Two-Dimensional Kolmogorov–Smirnov Test. The two-dimensional Kolmogorov–Smirnov test provides a way to test for nonrandom patterns in the scatterplot for y_k and x_j that does not require the imposition of a grid. Additional information: Sect. 6.10, Ref. [48]; Refs. [112–114].

Squared Differences of Ranks. The squared difference of ranks procedure seeks to identify the presence of nonlinear relationship between y_k and x_j and is based on squared differences of consecutive ranks of y_k when the values of y_k have been ordered by the corresponding values of x_j. Additional information: Sect. 6.9, Ref. [48]; Ref. [115].

Top-Down Concordance with Replicated Samples. This procedure uses the top-down coefficient of concordance and replicated (i.e., independently generated) samples. Sensitivity analysis with some appropriate technique is used to rank variable importance for each sample. The top-down coefficient is then used to identify important variables by seeking variables with similar rankings across all replicates. Additional information: Sect. 6.12, Ref. [48]; Refs. [89, 116].

Variance Decomposition. The variance decomposition procedure proposed by Sobol' and others is formally defined by high-dimensional integrals involving the x_j and $y_k(\mathbf{x})$. This procedure provides a decomposition of the variance $V(y_k)$ of y_k in terms of the contributions V_j of individual x_j's to $V(y_k)$ and also the contributions of various interactions between the x_j to $V(y_k)$. In practice, the indicated decomposition is obtained with sampling-based methods. Two samples from $\mathbf{x}$ of size nS are required to estimate all V_j; $nX+2$ samples of size nS are required to estimate all V_j and also the contributions of each of the x_j's and its interactions with other elements of $\mathbf{x}$ to $V(y_k)$. This procedure is very appealing but can be computationally demanding as more samples and probably larger samples are required than with other sampling-based approaches to sensitivity analysis. Software for sampling-based variance decomposition is available as part of the SIMLAB package [117]. Additional information: Sect. 6.13, Ref. [48]; Refs. [49–53, 117].

7 Summary

Sampling-based uncertainty and sensitivity analysis is widely used, and as a result, is a fairly mature area of study. However, there still remain a number of important challenges and areas for additional study. For example, there is a need for sensitivity analysis procedures that are more effective at revealing nonlinear relations than those currently in use. Among the approaches to sensitivity analysis listed in the preceding section, nonparametric regression [80, 81, 103–105], the two-dimensional Kolmogorov–Smirnov test [112–114], tests for nonmonotonic relations [115], tests for nonrandom patterns [44, 106–109], and complete variance decomposition [49–53, 117] have not been as widely used as some of the other approaches and merit additional investigation and use.

As another example, sampling-based procedures for uncertainty and sensitivity analysis usually use probability as the model, or representation, for uncertainty. However, when limited information is available with which to characterize uncertainty, probabilistic characterizations can give the appearance of more knowledge than is really present. Alternative representations for uncertainty such as evidence theory and possibility theory merit consideration for their potential to represent uncertainty in situations where limited information is available [66–73].

Finally, a significant challenge is the education of potential users of uncertainty and sensitivity analysis about (1) the importance of such analyses and their role in both large and small analyses, (2) the need for appropriate separation of aleatory and epistemic uncertainty in the conceptual and computational implementation of analyses of complex systems [13, 15, 17–19], (3) the need for a clear conceptual view of what an analysis is intended to represent and a computational design that is consistent with that view [118], (4) the role that uncertainty and sensitivity analysis plays in model and analysis verification, and (5) the importance of avoiding deliberately conservative assumptions if meaningful uncertainty and sensitivity analysis results are to be obtained [119–123].

As indicated in the Introduction, more detail on the topics introduced in this presentation are available in five technical reports [75–79] and a number of additional presentations derived from these reports [44–48, 80, 81]. Conceptual ideas on the organization of large analyses involving aleatory and epistemic uncertainty are elaborated on in several additional presentations [13, 118, 124–126].

Acknowledgments

Work performed for Sandia National Laboratories (SNL), which is a multiprogram laboratory operated by Sandia Corporation, a Lockheed Martin Company, for the United States Department of Energy's National Security Administration under contract DE-AC04-94AL-85000. This work is a product of the Epistemic Uncertainty Project directed by W. L. Oberkampf as part of the Accelerated Strategic Computing Initiative (ASCI). Review at SNL provided by A. A. Giunta and C. J. Sallaberry. Editorial support provided by F. Puffer, K. Best and J. Ripple of Tech Reps, a division of Ktech Corporation.

References

1. Christie, M.A., J. Glimm, J.W. Grove, D.M. Higdon, D.H. Sharp, and M.M. Wood-Schultz. 2005. Error Analysis and Simulations of Complex Phenomena, *Los Alamos Science* **29**, 6–25.
2. Sharp, D.H. and M.M. Wood-Schultz. 2003. QMU and Nuclear Weapons Certification: What's Under the Hood?, *Los Alamos Science* **28**, 47–53.
3. Wagner, R.L. 2003. Science, Uncertainty and Risk: The Problem of Complex Phenomena, *APS News.* **12**(1), 8.
4. Oberkampf, W.L., S.M. DeLand, B.M. Rutherford, K.V. Diegert, and K.F. Alvin. 2002. Error and Uncertainty in Modeling and Simulation, *Reliability Engineering and System Safety* **75**(3), 333–357.
5. Risk Assessment Forum. 1997. *Guiding Principles for Monte Carlo Analysis*, EPA/630/R-97/001. Washington DC: U.S. Environmental Protection Agency.
6. NCRP (National Council on Radiation Protection and Measurements). 1996. *A Guide for Uncertainty Analysis in Dose and Risk Assessments Related to Environmental Contamination*, NCRP Commentary No. 14. Bethesda, MD: National Council on Radiation Protection and Measurements.
7. Draper, D. 1995. Assessment and Propagation of Model Uncertainty, *Journal of the Royal Statistical Society Series B* **57**, 45–97.
8. NRC (National Research Council). 1994. *Science and Judgment in Risk Assessment*, Washington, DC: National Academy Press.
9. NRC (National Research Council). 1993. *Issues in Risk Assessment.* Washington, DC: National Academy Press.
10. U.S. EPA (U.S. Environmental Protection Agency). 1993. *An SAB Report: Multi-Media Risk Assessment for Radon, Review of Uncertainty Analysis of Risks Associated with Exposure to Radon*, EPA-SAB-RAC-93-014. Washington, DC: U.S. Environmental Protection Agency.
11. IAEA (International Atomic Energy Agency). 1989. *Evaluating the Reliability of Predictions Made Using Environmental Transfer Models*, Safety Series No. 100. Vienna: International Atomic Energy Agency.
12. Beck, M.B. 1987. Water-Quality Modeling: A Review of the Analysis of Uncertainty, *Water Resources Research* **23**(8), 1393–1442.
13. Helton, J.C. 1997. Uncertainty and Sensitivity Analysis in the Presence of Stochastic and Subjective Uncertainty, *Journal of Statistical Computation and Simulation* **57**(1–4), 3–76.

14. Helton, J.C. and D.E. Burmaster. 1996. Guest Editorial: Treatment of Aleatory and Epistemic Uncertainty in Performance Assessments for Complex Systems, *Reliability Engineering and System Safety* **54**(2–3), 91–94.
15. Paté-Cornell, M.E. 1996. Uncertainties in Risk Analysis: Six Levels of Treatment, *Reliability Engineering and System Safety* **54**(2–3), 95–111.
16. Winkler, R.L. 1996. Uncertainty in Probabilistic Risk Assessment, *Reliability Engineering and System Safety* **54**(2–3), 127–132.
17. Hoffman, F.O. and J.S. Hammonds. 1994. Propagation of Uncertainty in Risk Assessments: The Need to Distinguish Between Uncertainty Due to Lack of Knowledge and Uncertainty Due to Variability, *Risk Analysis* **14**(5), 707–712.
18. Helton, J.C. 1994. Treatment of Uncertainty in Performance Assessments for Complex Systems, *Risk Analysis* **14**(4), 483–511.
19. Apostolakis, G. 1990. The Concept of Probability in Safety Assessments of Technological Systems, *Science* **250**(4986), 1359–1364.
20. Haan, C.T. 1989. Parametric Uncertainty in Hydrologic Modeling, *Transactions of the ASAE* **32**(1), 137–146.
21. Parry, G.W. and P.W. Winter. 1981. Characterization and Evaluation of Uncertainty in Probabilistic Risk Analysis, *Nuclear Safety* 22(1), 28–42.
22. Kaplan, S. and B.J. Garrick. 1981. On the Quantitative Definition of Risk, *Risk Analysis* **1**(1), 11–27.
23. Cacuci, D.G. 2003. *Sensitivity and Uncertainty Analysis*, Vol. 1: *Theory.* Boca Raton, FL: Chapman and Hall/CRC Press.
24. Turányi, T. 1990. Sensitivity Analysis of Complex Kinetic Systems. Tools and Applications, *Journal of Mathematical Chemistry* **5**(3),203–248.
25. Rabitz, H., M. Kramer, and D. Dacol. 1983. Sensitivity Analysis in Chemical Kinetics, *Annual Review of Physical Chemistry.* Vol. 34. Eds. B.S. Rabinovitch, J.M. Schurr, and H.L. Strauss. Palo Alto, CA: Annual Reviews Inc, pp. 419–461.
26. Lewins, J. and M. Becker, eds. 1982. *Sensitivity and Uncertainty Analysis of Reactor Performance Parameters.* Vol. 14. New York, NY: Plenum Press.
27. Frank, P.M. 1978. *Introduction to System Sensitivity Theory.* New York, NY: Academic Press.
28. Tomovic, R. and M. Vukobratovic. 1972. *General Sensitivity Theory.* New York, NY: Elsevier.
29. Myers, R.H. 1999. Response Surface Methodology – Current Status and Future Directions, *Journal of Quality Technology* **31**(1), 30–44.
30. Andres, T.H. 1997. Sampling Methods and Sensitivity Analysis for Large Parameter Sets, *Journal of Statistical Computation and Simulation* **57**(1)–4, 77–110.
31. Kleijnen, J.P.C. 1997. Sensitivity Analysis and Related Analyses: A Review of Some Statistical Techniques, *Journal of Statistical Computation and Simulation* 57(1–4), 111–142.
32. Kleijnen, J.P.C. 1992. Sensitivity Analysis of Simulation Experiments: Regression Analysis and Statistical Design, *Mathematics and Computers in Simulation* **34**(3–4), 297–315.
33. Morton, R.H. 1983. Response Surface Methodology, *Mathematical Scientist* **8**, 31–52.
34. Mead, R. and D.J. Pike. 1975. A Review of Response Surface Methodology from a Biometric Viewpoint, *Biometrics* **31**, 803–851.

35. Myers, R.H. 1971. *Response Surface Methodology.* Boston, MA: Allyn and Bacon.
36. McKay, M.D., R.J. Beckman, and W.J. Conover. 1979. A Comparison of Three Methods for Selecting Values of Input Variables in the Analysis of Output from a Computer Code, *Technometrics* 21(2), 239–245.
37. Iman, R.L. and W.J. Conover. 1980. Small Sample Sensitivity Analysis Techniques for Computer Models, with an Application to Risk Assessment, *Communications in Statistics: Theory and Methods* **A9**(17), 1749–1842.
38. Iman, R.L., J.C. Helton, and J.E. Campbell. 1981. An Approach to Sensitivity Analysis of Computer Models, Part 1. Introduction, Input Variable Selection and Preliminary Variable Assessment, *Journal of Quality Technology* **13**(3), 174–183.
39. Iman, R.L., J.C. Helton, and J.E. Campbell. 1981. An Approach to Sensitivity Analysis of Computer Models, Part 2. Ranking of Input Variables, Response Surface Validation, Distribution Effect and Technique Synopsis, *Journal of Quality Technology* **13**(4), 232–240.
40. Saltelli, A. and J. Marivoet. 1990. Non-Parametric Statistics in Sensitivity Analysis for Model Output. A Comparison of Selected Techniques, *Reliability Engineering and System Safety* **28**(2), 229–253.
41. Iman, R.L. 1992. Uncertainty and Sensitivity Analysis for Computer Modeling Applications, *Reliability Technology – 1992, The Winter Annual Meeting of the American Society of Mechanical Engineers, Anaheim, California, November 8–13, 1992.* Eds. T.A. Cruse. Vol. 28, pp. 153–168. New York, NY: American Society of Mechanical Engineers, Aerospace Division.
42. Saltelli, A., T.H. Andres, and T. Homma. 1993. Sensitivity Analysis of Model Output. An Investigation of New Techniques, *Computational Statistics and Data Analysis* **15**(2), 445–460.
43. Blower, S.M. and H. Dowlatabadi. 1994. Sensitivity and Uncertainty Analysis of Complex Models of Disease Transmission: an HIV Model, as an Example, *International Statistical Review* **62**(2), 229–243.
44. Kleijnen, J.P.C. and J.C. Helton. 1999. Statistical Analyses of Scatterplots to Identify Important Factors in Large-Scale Simulations, 1: Review and Comparison of Techniques, *Reliability Engineering and System Safety* **65**(2), 147–185.
45. Helton, J.C. and F.J. Davis. 2000. Sampling-Based Methods, *Sensitivity Analysis.* Eds. A. Saltelli, K. Chan, and E.M. Scott, pp. 101–153. New York, NY: Wiley.
46. Helton, J.C. and F.J. Davis. 2002. Illustration of Sampling-Based Methods for Uncertainty and Sensitivity Analysis, *Risk Analysis* **22**(3), 591–622.
47. Helton, J.C. and F.J. Davis. 2003. Latin Hypercube Sampling and the Propagation of Uncertainty in Analyses of Complex Systems, *Reliability Engineering and System Safety* **81**(1), 23–69.
48. Helton, J.C., J.D. Johnson, C.J. Sallaberry, and C.B. Storlie. 2006. Survey of Sampling-Based Methods for Uncertainty and Sensitivity Analysis, *Reliability Engineering and System Safety* **91**(10–11), 1175–1209.
49. Li, G., C. Rosenthal, and H. Rabitz. 2001. High-Dimensional Model Representations, *The Journal of Physical Chemistry* **105**(33), 7765–7777.
50. Rabitz, H. and O.F. Alis. 1999. General Foundations of High-Dimensional Model Representations, *Journal of Mathematical Chemistry* **25**(2–3), 197–233.

51. Saltelli, A., S. Tarantola, and K.P.-S. Chan. 1999. A Quantitative Model-Independent Method for Global Sensitivity Analysis of Model Output, *Technometrics* **41**(1), 39–56.
52. Sobol', I.M. 1993. Sensitivity Estimates for Nonlinear Mathematical Models, *Mathematical Modeling & Computational Experiment* **1**(4), 407–414.
53. Cukier, R.I., H.B. Levine, and K.E. Shuler. 1978. Nonlinear Sensitivity Analysis of Multiparameter Model Systems, *Journal of Computational Physics* **26**(1), 1–42.
54. Iman, R.L. and J.C. Helton. 1988. An Investigation of Uncertainty and Sensitivity Analysis Techniques for Computer Models, *Risk Analysis* **8**(1), 71–90.
55. Ronen, Y. 1988. *Uncertainty Analysis.*, Boca Raton, FL: CRC Press, Inc.
56. Helton, J.C. 1993. Uncertainty and Sensitivity Analysis Techniques for Use in Performance Assessment for Radioactive Waste Disposal, *Reliability Engineering and System Safety* **42**(2–3), 327–367.
57. Hamby, D.M. 1994. A Review of Techniques for Parameter Sensitivity Analysis of Environmental Models, *Environmental Monitoring and Assessment* **32**(2), 135–154.
58. Saltelli, A., K. Chan, and E.M. Scott (eds). 2000. *Sensitivity Analysis.* New York, NY: Wiley.
59. Frey, H.C. and S.R. Patil. 2002. Identification and Review of Sensitivity Analysis Methods, *Risk Analysis* **22**(3), 553–578.
60. Ionescu-Bujor, M. and D.G. Cacuci. 2004. A Comparative Review of Sensitivity and Uncertainty Analysis of Large-Scale Systems–I: Deterministic Methods, *Nuclear Science and Engineering* **147**(3), 189–2003.
61. Cacuci, D.G. and M. Ionescu-Bujor. 2004. A Comparative Review of Sensitivity and Uncertainty Analysis of Large-Scale Systems–II: Statistical Methods, *Nuclear Science and Engineering* **147**(3), 204–217.
62. Saltelli, A., M. Ratto, S. Tarantola, and F. Campolongo. 2005. Sensitivity Analysis for Chemical Models, *Chemical Reviews* **105**(7), 2811–2828.
63. Helton, J.C., D.R. Anderson, H.-N. Jow, M.G. Marietta, and G. Basabilvazo. 1999. Performance Assessment in Support of the 1996 Compliance Certification Application for the Waste Isolation Pilot Plant, *Risk Analysis* **19**(5), 959–986.
64. Helton, J.C. 1999. Uncertainty and Sensitivity Analysis in Performance Assessment for the Waste Isolation Pilot Plant, *Computer Physics Communications* **117**(1–2), 156–180.
65. Helton, J.C. and M.G. Marietta. 2000. Special Issue: The 1996 Performance Assessment for the Waste Isolation Pilot Plant, *Reliability Engineering and System Safety* **69**(1–3), 1–451.
66. Baudrit, C. and D. Dubois. 2006. Practical Representations of Incomplete Probabilistic Knowledge, *Computational Statistics & Data Analysis* **51**(1), 86–108.
67. Dubois, D. and H. Prade. 2006. Possibility Theory and Its Applications: A Retrospective and Prospective View, *CISM Courses and Lectures.* **482**, 89–109.
68. Klir, G.J. 2006. *Uncertainty and Information: Foundations of Generalized Information Theory.* New York, NY: Wiley-Interscience.
69. Ross, T.J. 2004. *Fuzzy Logic with Engineering Applications.* 2nd edn. New York, NY: Wiley.

70. Helton, J.C., J.D. Johnson, and W.L. Oberkampf. 2004. An Exploration of Alternative Approaches to the Representation of Uncertainty in Model Predictions, *Reliability Engineering and System Safety* **85**(1–3), 39–71.
71. Ross, T.J., J.M. Booker, and W.J. Parkinson (eds.). 2002. *Fuzzy Logic and Probability Applications: Bridging the Gap*. Philadelphia, PA: Society for Industrial and Applied Mathematics.
72. Klir, G.J. and M.J. Wierman. 1999. *Uncertainty-Based Information*, New York, NY: Physica-Verlag.
73. Bardossy, G. and J. Fodor. 2004. *Evaluation of Uncertainties and Risks in Geology*, New York, NY: Springer.
74. Helton, J.C. 2004. Uncertainty and Sensitivity Analysis for Models of Complex Systems, *Verification & Validation of Complex Models for Design and Performance Evaluation of High-Consequence Engineering Systems*. Eds. J.T. Fong and R. deWit. Gaithersburg, MD: U.S. National Institute of Standards and Technology. pp. 237–263.
75. Kleijnen, J.P.C. and J.C. Helton. 1999. *Statistical Analyses of Scatterplots to Identify Important Factors in Large-Scale Simulations*, SAND98-2202. Albuquerque, NM: Sandia National Laboratories.
76. Helton, J.C. and F.J. Davis. 2000. *Sampling-Based Methods for Uncertainty and Sensitivity Analysis*, SAND99-2240. Albuquerque, NM: Sandia National Laboratories.
77. Helton, J.C. and F.J. Davis. 2002. *Latin Hypercube Sampling and the Propagation of Uncertainty in Analyses of Complex Systems*, SAND2001-0417. Albuquerque, NM: Sandia National Laboratories.
78. Helton, J.C., J.D. Johnson, C.J. Sallaberry, and C.B. Storlie. 2006. *Survey of Sampling-Based Methods for Uncertainty and Sensitivity Analysis*, SAND2006-2901. Albuquerque, NM: Sandia National Laboratories.
79. Storlie, C.B. and J.C. Helton. 2006. *Multiple Predictor Smoothing Methods for Sensitivity Analysis*, SAND2006-4693. Albuquerque, NM: Sandia National Laboratories.
80. Storlie, C.B. and J.C. Helton. 2008. Multiple Predictor Smoothing Methods for Sensitivity Analysis: Description of Techniques, *Reliability Engineering and System Safety* 93(1), 28–54.
81. Storlie, C.B. and J.C. Helton. 2008. Multiple Predictor Smoothing Methods for Sensitivity Analysis: Example Results, *Reliability Engineering and System Safety* 93(1), 55–77.
82. Hora, S.C. and R.L. Iman. 1989. Expert Opinion in Risk Analysis: The NUREG-1150 Methodology, *Nuclear Science and Engineering* **102**(4), 323–331.
83. Thorne, M.C. and M.M.R. Williams. 1992. A Review of Expert Judgement Techniques with Reference to Nuclear Safety, *Progress in Nuclear Safety* **27**(2–3), 83–254.
84. Budnitz, R.J., G. Apostolakis, D.M. Boore, L.S. Cluff, K.J. Coppersmith, C.A. Cornell, and P.A. Morris. 1998. Use of Technical Expert Panels: Applications to Probabilistic Seismic Hazard Analysis, *Risk Analysis* **18**(4), 463–469.
85. McKay, M. and M. Meyer. 2000. Critique of and Limitations on the Use of Expert Judgements in Accident Consequence Uncertainty Analysis, *Radiation Protection Dosimetry* **90**(3), 325–330.
86. Ayyub, B.M. 2001. *Elicitation of Expert Opinions for Uncertainty and Risks*, Boca Raton, FL: CRC Press.

87. Cooke, R.M. and L.H.J. Goossens. 2004. Expert Judgement Elicitation for Risk Assessment of Critical Infrastructures, *Journal of Risk Research* **7**(6), 643–656.
88. Garthwaite, P.H., J.B. Kadane, and A. O'Hagan. 2005. Statistical Methods for Eliciting Probability Distributions, *Journal of the American Statistical Association* **100**(470), 680–700.
89. Helton, J.C., F.J. Davis, and J.D. Johnson. 2005. A Comparison of Uncertainty and Sensitivity Analysis Results Obtained with Random and Latin Hypercube Sampling, *Reliability Engineering and System Safety* **89**(3), 305–330.
90. Helton, J.C., J.D. Johnson, M.D. McKay, A.W. Shiver, and J.L. Sprung. 1995. Robustness of an Uncertainty and Sensitivity Analysis of Early Exposure Results with the MACCS Reactor Accident Consequence Model, *Reliability Engineering and System Safety* **48**(2), 129–148.
91. Iman, R.L. and J.C. Helton. 1991. The Repeatability of Uncertainty and Sensitivity Analyses for Complex Probabilistic Risk Assessments, *Risk Analysis* **11**(4), 591–606.
92. Morris, M.D. 2000. Three Technometrics Experimental Design Classics, *Technometrics* **42**(1), 26–27.
93. Iman, R.L. and W.J. Conover. 1982. A Distribution-Free Approach to Inducing Rank Correlation Among Input Variables, *Communications in Statistics: Simulation and Computation* **B11**(3), 311–334.
94. Iman, R.L. and J.M. Davenport. 1982. Rank Correlation Plots for Use with Correlated Input Variables, *Communications in Statistics: Simulation and Computation* **B11**(3), 335–360.
95. Tong, C. 2006. Refinement Strategies for Stratified Sampling Methods, *Reliability Engineering and System Safety* **91**(10–11), 1257–1265.
96. Sallaberry, C.J., J.C. Helton, and S.C. Hora. 2006. *Extension of Latin Hypercube Samples with Correlated Variables*, SAND2006-6135. Albuquerque, NM: Sandia National Laboratories.
97. Sallaberry, C.J., J.C. Helton, and S.C. Hora. 2008. Extension of Latin Hypercube Samples with Correlated Variables, *Reliability Engineering and System Safety* 93(7), 1047–1059.
98. Iman, R.L. 1981. Statistical Methods for Including Uncertainties Associated with the Geologic Isolation of Radioactive Waste which Allow for a Comparison with Licensing Criteria, *Proceedings of the Symposium on Uncertainties Associated with the Regulation of the Geologic Disposal of High-Level Radioactive Waste*. NUREG/CP-0022; CONF-810372. Eds. D.C. Kocher. Gatlinburg, TN: Washington, DC: US Nuclear Regulatory Commission, Directorate of Technical Information and Document Control. pp. 145–157.
99. Helton, J.C., M.-A. Martell, and M.S. Tierney. 2000. Characterization of Subjective Uncertainty in the 1996 Performance Assessment for the Waste Isolation Pilot Plant, *Reliability Engineering and System Safety* **69**(1–3), 191–204.
100. Breeding, R.J., J.C. Helton, E.D. Gorham, and F.T. Harper. 1992. Summary Description of the Methods Used in the Probabilistic Risk Assessments for NUREG-1150, *Nuclear Engineering and Design* **135**(1), 1–27.
101. Cooke, R.M. and J.M. van Noortwijk. 2000. Graphical Methods, *Sensitivity Analysis*. Eds. A. Saltelli, K. Chan, and E.M. Scott. New York, NY: Wiley. pp. 245–264.
102. Iman, R.L. and W.J. Conover. 1979. The Use of the Rank Transform in Regression, *Technometrics* **21**(4), 499–509.

103. Hastie, T.J. and R.J. Tibshirani. 1990. *Generalized Additive Models.* London: Chapman & Hall.
104. Simonoff, J.S. 1996. *Smoothing Methods in Statistics.* Berlin, Heidelberg, New York: Springer.
105. Bowman, A.W. and A. Azzalini. 1997. *Applied Smoothing Techniques for Data Analysis.* Oxford: Clarendon.
106. Ripley, B.D. 1979. Tests of Randomness for Spatial Point Patterns, *Journal of the Royal Statistical Society* **41**(3), 368–374.
107. Diggle, P.J. and T.F. Cox. 1983. Some Distance-Based Tests of Independence for Sparsely-Sampled Multivariate Spatial Point Patterns, *International Statistical Review* **51**(1), 11–23.
108. Zeng, G. and R.C. Dubes. 1985. A Comparison of Tests for Randomness, *Pattern Recognition* **18**(2), 191–198.
109. Assunçao, R. 1994. Testing Spatial Randomness by Means of Angles, *Biometrics* **50**, 531–537.
110. Mishra, S., N.E. Deeds, and B.S. RamaRao. 2003. Application of Classification Trees in the Sensitivity Analysis of Probabilistic Model Results, *Reliability Engineering and System Safety* **79**(2), 123–129.
111. Breiman, L., J.H. Friedman, R.A. Olshen, and C.J. Stone. 1984. *Classification and Regression Trees.* Belmont, CA: Wadsworth Intl.
112. Peacock, J.A. 1983. Two-Dimensional Goodness-Of-Fit Testing in Astronomy, *Monthly Notices of the Royal Astronomical Society* **202**(2), 615–627.
113. Fasano, G. and A. Franceschini. 1987. A Multidimensional Version of the Kolmogorov–Smirnov Test, *Monthly Notices of the Royal Astronomical Society* **225**(1), 155–170.
114. Garvey, J.E., E.A. Marschall, and R.A. Wright. 1998. From Star Charts to Stoneflies: Detecting Relationships in Continuous Bivariate Data, *Ecology* **79**(2), 442–447.
115. Hora, S.C. and J.C. Helton. 2003. A Distribution-Free Test for the Relationship Between Model Input and Output when Using Latin Hypercube Sampling, *Reliability Engineering and System Safety* **79**(3), 333–339.
116. Iman, R.L. and W.J. Conover. 1987. A Measure of Top-Down Correlation, *Technometrics* **29**(3), 351–357.
117. Saltelli, A., S. Tarantola, F. Campolongo, and M. Ratto. 2004. *Sensitivity Analysis in Practice.* New York, NY: Wiley.
118. Helton, J.C. 2003. Mathematical and Numerical Approaches in Performance Assessment for Radioactive Waste Disposal: Dealing with Uncertainty, *Modelling Radioactivity in the Environment.* Ed. E.M. Scott. New York, NY: Elsevier Science. pp. 353–390.
119. Diaz, N.J. 2003. Realism and Conservatism, Remarks by Chairman Diaz at the 2003 Nuclear Safety Research Conference, October 20, 2003, *NRC News.* No. S-03–023. Washington, D.C.: U.S. Nuclear Regulatory Commission.
120. Paté-Cornell, E. 2002. Risk and Uncertainty Analysis in Government Safety Decisions, *Risk Analysis* **22**(3), 633–646.
121. Caruso, M.A., M.C. Cheok, M.A. Cunningham, G.M. Holahan, T.L. King, G.W. Parry, A.M. Ramey-Smith, M.P. Rubin, and A.C. Thadani. 1999. An Approach for Using Risk Assessment in Risk-Informed Decisions on Plant-Specific Changes to the Licensing Basis, *Reliability Engineering and System Safety* **63**, 231–242.

122. Sielken, R.L., Jr., R.S. Bretzlaff, and D.E. Stevenson. 1995. Challenges to Default Assumptions Stimulate Comprehensive Realism as a New Tier in Quantitative Cancer Risk Assessment, *Regulatory Toxicology and Pharmacology* **21**, 270–280.
123. Nichols, A.L. and R.J. Zeckhauser. 1988. The Perils of Prudence: How Conservative Risk Assessments Distort Regulation, *Regulatory Toxicology and Pharmacology* **8**, 61–75.
124. Helton, J.C. and C.J. Sallaberry. 2007. *Illustration of Sampling-Based Approaches to the Calculation of Expected Dose in Performance Assessments for the Proposed High Level Radioactive Waste Repository at Yucca Mountain, Nevada*, SAND2007-1353. Albuquerque, NM: Sandia National Laboratories.
125. Helton, J.C., D.R. Anderson, G. Basabilvazo, H.-N. Jow, and M.G. Marietta. 2000. Conceptual Structure of the 1996 Performance Assessment for the Waste Isolation Pilot Plant, *Reliability Engineering and System Safety* **69**(1–3), 151–165.
126. Helton, J.C. and R.J. Breeding. 1993. Calculation of Reactor Accident Safety Goals, *Reliability Engineering and System Safety* **39**(2), 129–158.

A Brief Overview of the State-of-the-Practice and Current Challenges of Solution Verification

F.M. Hemez and J.R. Kamm

Summary. The main goal of solution verification is to assess the convergence of numerical predictions as a function of discretization variables such as element size Δx or time step Δt. The challenge is to verify that the approximate solutions of the discretized conservation laws or equations-of-motion converge to the solution of the continuous equations. In the case of code verification where the continuous solution of a test problem is known, common practice is to obtain several discrete solutions from successively refined meshes or grids; calculate norms of the solution error; and verify the rate with which discrete solutions converge to the continuous solution. With solution verification, where the continuous solution is unknown, common practice is to obtain several discrete solutions from successively refined meshes or grids; extrapolate an estimate of the continuous solution; verify the rate-of-convergence; and estimate numerical uncertainty bounds. The formalism proposed to verify the convergence of discrete solutions derives from postulating how truncation error behaves in the asymptotic regime of convergence. This publication summarizes the state-of-the-practice but also challenges some of the commonly accepted views of verification. Examples are given from the disciplines of computational hydro-dynamics that involve the calculation of smooth or discontinuous solutions of non-linear, hyperbolic equations such as the 1D Burgers equation; and engineering mechanics that involve the calculation of smooth solutions of linear or non-linear, elliptic equations. A non-exhaustive list of topics that warrant further research includes: extending the state-of-the-practice to non-scalar quantities (curves, multiple-dimensional fields); studying the coupling between space and time discretizations; defining a reference mesh for the estimation of solution error; and developing technology to verify adaptive mesh refinement calculations in computational engineering and physics. *(Approved for unlimited, public release, LA-UR-06-8078, Unclassified.)*

1 Introduction

In computational engineering and physics, partial differential equations that govern the evolution of state variables, such as energy, pressure, velocity, or displacement, are discretized for implementation and resolution on finite-digit computer arithmetic. The challenge of code and solution verification is to assess the extent to which approximate solutions of the discretized equations converge to the exact solution of the continuous equations. In addition to assessing how "closely" discrete solutions match the continuous solution, it is often of great importance to verify that the observed rate-of-convergence matches the rate that should theoretically be provided by the numerical method or solver. Simply speaking, verification is the first "V" of the Verification and Validation (V&V) of predictions obtained from numerical simulations [1].

Solution verification can be defined as a *scientifically rigorous and quantitative process for assessing the mathematical consistency between continuum and discrete variants of partial differential equations used to represent a reality of interest* [2]. Verification involves comparing numerical solutions obtained from calculations performed on successively refined meshes or grids to a reference. The main difficulty is that the exact solution of the continuous equations is not always known and available to define this reference. Continuous solutions can be derived analytically in a few special cases only that feature a simple geometry, smooth dynamics, linearized operators, or combinations of the above. Well-known examples include the single degree-of-freedom damped harmonic oscillator, linear vibration of simple geometries, Poiseuille flow of 2D incompressible fluids, and the 1D Riemann problem for ideal gases. These are problems that have been extensively studied and for which exact or highly-accurate solutions of the continuous equations can be obtained. The overall number of verification problems that offer closed-form solutions for practical problems is quite small.

In general one talks of *code verification* when the solution of the continuous equations can be derived in closed form, which provides an "exact" reference to which the discrete solutions are compared. By "closed form," we mean either an analytical formula or a simplified representation that can be accurately evaluated (e.g., reduction from a partial differential equation to an ordinary differential equation). If the system of equations and its initial and boundary conditions are complicated enough, then a continuous solution cannot be derived analytically and one talks of *solution verification.* In this latter case no exact reference is available to calculate the solution error due to discretization, which makes verification very challenging.

Even though the case of code verification is briefly mentioned, the discussion presented in this publication focuses on solution verification. The dominant paradigm for solution verification postulates that the numerical method provides discrete solutions that converge to the (unknown) continuous solution with a specific rate-of-convergence. For example, a finite element analysis featuring quadratic elements should provide discrete solutions that converge with a rate $p = 2$. Common practice is to obtain several discrete solutions from successively refined meshes or grids; estimate an extrapolation of the continuous solution; verify the rate-of-convergence; and estimate numerical uncertainty bounds. Techniques such as Richardson extrapolation and the Grid Convergence Index are available to extrapolate the numerical approximations and quantify solution error, respectively. Studying the convergence of solutions is generally restricted to scalar predictions to simplify the equations involved, as shown below, and provide closed-form derivations. This would mean verifying the convergence, for example, of peak values obtained at given locations in the computational domain instead of verifying convergence of the entire fields predicted by the simulations.

Successes of code and solution verification include the development of a formalism to study the convergence of discrete solutions for a wide range of applications, from linear, non-dissipative, elliptic equations to non-linear, shocked, hyperbolic equations. It means that the convergence, for example, of solutions for the resonant frequencies of a modal analysis can be studied with the same tools as the convergence of solutions for a blast wave propagating in ideal gases. The failures of code and calculation verification are embodied by the many restrictions imposed on the way an analysis is typically carried out. They include restricting studies to scalar quantities; further restricting them to spatial-only convergence; and not always accounting for how discrete solutions are constructed by the numerical methods, and the specific properties that result.[1]

The publication offers an introduction to code and solution verification. It is restricted to well-established techniques and does not discuss recent developments that attempt to push the boundaries of the current state-of-the-practice. (For a partial account of recent developments see, for example, [3–9].) This work is contributed to the proceedings of the 2006 Granlibakken Workshop on Computational Methods in Transport in an attempt to raise awareness in the transport community about convergence-related issues that ubiquitously present themselves in numerical simulations.

2 The Asymptotic Regime of Convergence

Verifying the convergence of discrete solution hinges on the concept of *asymptotic regime of convergence*. The choice of element or cell size, denoted by Δx, in a computational model can be thought of as being a one-dimensional discretization domain. Different choices of Δx values within the discretization domain induce different behaviors of the overall numerical error. By definition

[1] It is common practice, for example, to study the convergence of discrete solutions in the sense of a generic norm, such as L^1 or L^2, or restrict the analysis to scalar responses defined point-wise. Doing so may or may not be appropriate depending on characteristics of the numerical method and how it constructs the discrete solutions. A finite element solution, for example, never satisfies the equations-of-motion in a point-wise sense. It is defined in the sense of the energy norm consistent with the weak formulation of the equations. Assessing convergence of point-wise, scalar responses is inconsistent with the way finite element solutions are constructed by the approximation method, which may or may not lead to erroneous conclusions.

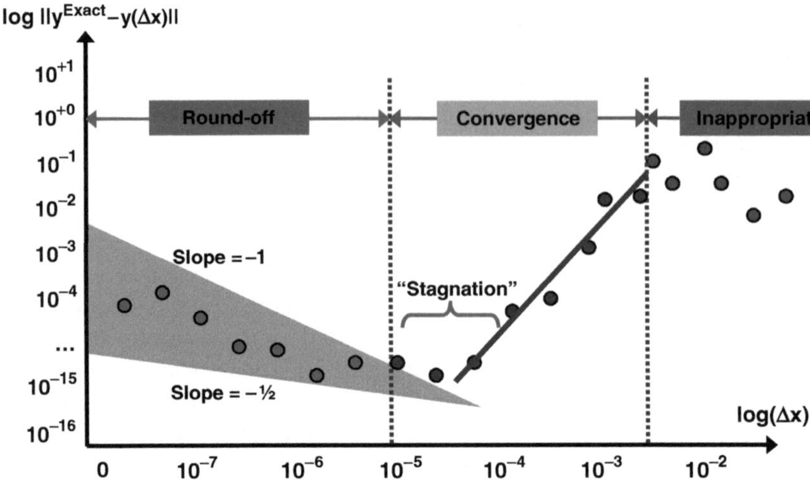

Fig. 1. Illustration of the asymptotic regime of convergence. (See Plate 31 on page 321)

the asymptotic regime is the region of the discretization domain where truncation error dominates the overall production of numerical error.

Figure 1 gives a simplified illustration of the main three regimes of discretization, color-coded such that *red* is a region where the discretization is inappropriate; *green* denotes the asymptotic regime of convergence; and *grey* is a region where round-off errors accumulate.

Going from right (larger values of Δx) to left (smaller values of Δx), the first domain shown is color red, where the choice of element or cell size is not even appropriate to solve the discrete equations. This is, for example, the case when elements are too coarse to resolve important geometrical features of a contact between components; a constraint of numerical stability is violated; or the cell size of a computational fluid dynamics grid is too coarse to capture a characteristic scale over which mixing between two materials occurs. Although it could be argued that discrete solutions originating from this regime of discretization are useful to indicate overall trends, their numerical quality should be questioned with the highest degree of suspicion.

The second region shown in color green is where truncation error dominates the numerical error of discrete solutions: it is the asymptotic regime of convergence. Because truncation error dominates, the numerical error can be reduced (that is, solution accuracy can be improved) by performing a calculation with smaller element or cell sizes. We have just encapsulated the basic principle of conducting a mesh or grid refinement study.

Because truncation dominates within the asymptotic regime of convergence, the behavior of numerical error can be modeled mathematically using an equation such as:

$$\varepsilon_{Y}\left(\Delta x\right) = \left\| y^{Exact} - y\left(\Delta x\right)\right\| = \beta_1\left(\Delta x^{p}\right) + O\left(\Delta x^{p}\right) \quad (1)$$

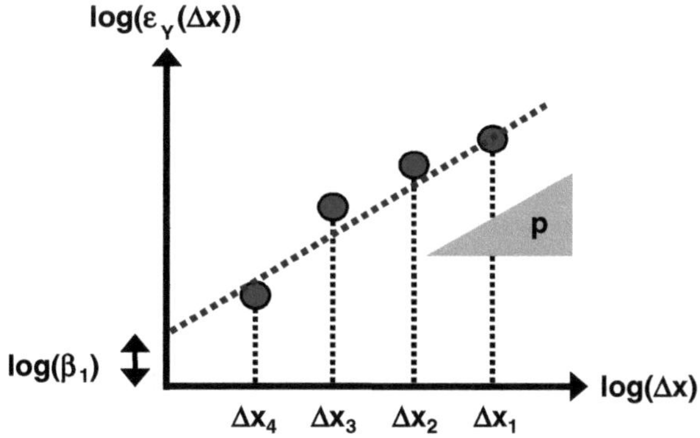

Fig. 2. Illustration of solution error $\varepsilon_Y(\Delta x) = \|y^{Exact} - y(\Delta x)\|$ vs. mesh size Δx. (See Plate 32 on page 321)

where ε_Y denotes the difference, estimated in the sense of a user-defined norm $\| \bullet \|$, between the exact solution y^{Exact} of the continuous partial differential equations and the discrete solution $y(\Delta x)$ obtained with an element or cell size Δx. In (1), the pre-factor β_1 represents a regression coefficient. The exponent p characterizes the rate with which the solution error ε_Y is reduced when the cell size decreases, $\Delta x \to 0$. Once again, because the asymptotic regime of convergence is the region of discretization where truncation dominates, the value of exponent p should match the rate-of-convergence of the numerical method implemented. For example, a finite element calculation that uses linear shell elements should exhibit a rate equal to $p = 1$. Likewise a second-order accurate Gudonov finite difference scheme in fluid dynamics should exhibit a rate equal to $p = 2$.

Figure 2 suggests the meaning of the two parameters $(\beta_1; p)$ of (1). If the solution error ε_Y is represented graphically versus cell size Δx on a log–log scale, then the exponent p is the slope of a straight line that best-fits the data and $\log(\beta_1)$ is the value of the intercept at $\Delta x = 0$.

The last region shown in color grey in Fig. 1 is a limiting case for asymptotic convergence due to the fact that, with finite arithmetic implemented in a computer, round-off effects eventually start to accumulate as $\Delta x \to 0$. Round-off could then accumulate to the point where it supplants truncation error as the dominant mechanism that produces numerical error.

Understanding the asymptotic regime of convergence is important for two reasons. First, the derivations given in this paper, which provide the foundation of solution verification, are valid only within the asymptotic regime. Second, verifying that the discretization Δx used in the calculation leads to a discrete solution in the asymptotic regime provides a strategy to reduce the solution error. If the solution error is too large for the intended application and needs to be reduced, then performing a calculation with a smaller element or cell size reduces it. This is, however, true only within the asymptotic regime of convergence.

3 State-of-the-Practice to Verify the Convergence of Solutions

This section provides a summary of the main equations written to assess the convergence of discrete solutions (or lack thereof). The discussion introduces the current state-of-the-practice in computational engineering and physics without paying tribute to some of the more advanced techniques or recent developments. The reader is referred to [4] and [8–10], to list only a few, that discuss issues that extend beyond the limited scope of this section.

The starting point to assess the convergence of discrete solutions is to postulate an equation, usually referred to as the *solution error Ansatz model* that describes how the numerical error behaves in the asymptotic regime of convergence. We have explained why (1) is an appropriate choice (see Sect. 2).

The distinction needs to be emphasized between *code verification*, where the exact solution y^{Exact} of the continuous equations is known analytically (or provided by an approximate yet highly-accurate solution procedure), and *solution verification* where y^{Exact} is unknown. In the former case the left-hand side of (1), that is, $\varepsilon_Y(\Delta x) = \|y^{Exact} - y(\Delta x)\|$, can be computed given a discrete solution $y(\Delta x)$ obtained from a computational grid with element or cell size Δx. The exact and discrete solutions can be any scalar quantity, curve, or multi-dimensional field and the only unknowns are the two parameters $(\beta_1; p)$ of (1). Two discrete solutions, one obtained from a "coarse" grid of cell size Δx_C and another one obtained from a "fine" grid of cell size Δx_F (where $\Delta x_C = R\Delta x_F$ and $R > 1$), suffice to estimate the pair of unknowns $(\beta_1; p)$. The two equations are:

$$\begin{cases} \varepsilon_Y(\Delta x_C) = \left\|y^{Exact} - y(\Delta x_C)\right\| \approx \beta_1 (\Delta x_C^p) \\ \varepsilon_Y(\Delta x_F) = \left\|y^{Exact} - y(\Delta x_F)\right\| \approx \beta_1 (\Delta x_F^p) \end{cases} \tag{2}$$

and the solution for parameters $(\beta_1; p)$ is given by:

$$p = \frac{\log\left(\frac{\varepsilon_Y(\Delta x_C)}{\varepsilon_Y(\Delta x_F)}\right)}{\log(R)}, \quad \beta_1 = \frac{\varepsilon_Y(\Delta x_C)}{(\Delta x_C^p)} = \frac{\varepsilon_Y(\Delta x_F)}{(\Delta x_F^p)} \tag{3}$$

where R is the refinement ratio, that is, $R = \Delta x_C/\Delta x_F > 1$. Running code verification test problems is useful for two purposes. First, it assesses the performance of the code on specific test problems for which an exact solution y^{Exact} is known, which helps code developers and analysts to ensure that no programming error is present. Second, it can serve as a benchmarking exercise to assess the rate-of-convergence of the numerical method or solver implemented in the code. The main drawback, however, is that the exact solution of the continuous equations must be known. It seriously restricts the application of code verification to a few test problems that likely do not cover the full range of mechanics or physics implemented in the code.

We now proceed to discuss the general case of solution verification, that is, the assessment of convergence of discrete solutions as the cell size is refined,

$\Delta x \to 0$. The difficulty is that the exact solution y^{Exact} of the continuous equations is unknown. It means that the solution procedure described by (3) cannot be implemented since $\varepsilon_Y(\Delta x)$ cannot be computed.

To circumvent this difficulty, two assumptions are made. The first assumption is to replace the exact solution y^{Exact} by a "reference" solution denoted by $y^{Reference}$ and selected by the analyst. A typical practice is to run a calculation with a very fine grid, obtain a highly refined solution of the discretized partial differential equations, and assume that it is close enough to the exact-but-unknown solution y^{Exact}. The solution error is then defined as:

$$\varepsilon_Y(\Delta x) \approx \left\| y^{Reference} - y(\Delta x) \right\| \tag{4}$$

and the same solution procedure applies, that is, (3) using definition (4).

Situations may occur where this practice is not satisfactory because one does not know a priori what may or may not constitute a good-quality reference, one that would be "close enough" to the exact-but-unknown solution y^{Exact}. In these situations it is customary to augment the first assumption with a second one whereby (4) is specialized to scalar quantities. The standard error Ansatz equation takes the functional form of one of the following equations:

$$\varepsilon_Y(\Delta x) = y^{Reference} - y(\Delta x) \approx \beta_1 (\Delta x^p) \tag{5a}$$

$$\varepsilon_Y(\Delta x) = \left| y^{Reference} - y(\Delta x) \right| \approx \beta_1 (\Delta x^p) \tag{5b}$$

where the difference between the two versions is that convergence of discrete solutions to the reference solution $y^{Reference}$ must be monotonic in the case of (5a). Equation (5b) is somewhat more general because monotonic convergence is not a prerequisite. Changing the definition of the solution error $\varepsilon_Y(\Delta x)$ from a norm between multi-dimensional fields, such as shown in (1) or (4), to a difference between scalars, such as shown in (5), is essential to obtain an analytical solution for the triplet of unknowns ($y^{Reference}$; β_1; p).

The case of non-monotonic convergence is dealt with in [8, 11]. An illustration is given in Fig. 3. Four levels of refinement are applied to the finite element mesh of a plate and the fifth resonant frequency ω_5 is extracted from the mass and stiffness matrices. Figure 3 shows ω_5 as a function of element size Δx. Even though this calculation is linear, well-behaved, and ω_5 represents an integral quantity (ratio of internal energy to kinetic energy for the fifth mode shape), it can be observed that convergence of ω_5 is not monotonic. This is a situation where a solution error Ansatz model based on (5a) would not be appropriate.[2]

[2] Note that, to analyze the data of Fig. 3, proposing a solution error Ansatz model based on (5b) may not be appropriate either. It depends on whether a reference solution $y^{Reference}$ can be found such that the error $|y^{Reference} - y(\Delta x)|$ converges monotonically as a function of discretization size Δx. (See Refs. [8] and [11] for details.)

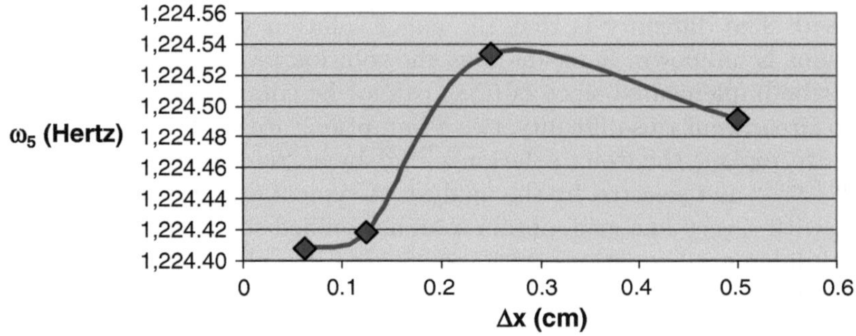

Fig. 3. Convergence of a resonant frequency versus element size. (See Plate 33 on page 322)

For simplicity, and because it corresponds to practices most often encountered in computational engineering and physics, the discussion is now limited to the case of monotonic convergence of scalar-valued solutions, that is, (5a). With the essential assumptions of restricting the analysis to scalar solutions, replacing the exact solution y^{Exact} by a reference $y^{Reference}$, and considering the case of monotonic-only convergence, the solution error Ansatz model is re-written without loss of generality as:

$$y(\Delta x) = y^{Reference} + \beta_1 (\Delta x^p) + O(\Delta x^p) \tag{6}$$

Because (6) features three unknowns ($y^{Reference}$; β_1; p), a minimum of three equations are needed. These are provided by discrete solutions obtained from a coarse grid (with element or cell size Δx_C), medium grid (with Δx_M), and fine grid (with Δx_F). Neglecting the influence of higher-order terms leads to the system of equations:

$$\begin{cases} y(\Delta x_C) \approx y^{Reference} + \beta_1 (\Delta x_C^p) \\ y(\Delta x_M) \approx y^{Reference} + \beta_1 (\Delta x_M^p) \\ y(\Delta x_F) \approx y^{Reference} + \beta_1 (\Delta x_F^p) \end{cases} \tag{7}$$

The system of (7) is solved for the triplet ($y^{Reference}$; β_1; p) by combining the equations to eliminate two unknowns, $y^{Reference}$ and β_1. The value of the rate-of-convergence is obtained by solving the following non-linear equation:

$$p(\log(R_{MF})) + \log(1 - R_{CM}^p) - \log(1 - R_{MF}^p) = \log\left(\frac{y(\Delta x_M) - y(\Delta x_C)}{y(\Delta x_F) - y(\Delta x_M)}\right) \tag{8}$$

where R_{CM} is the refinement ratio from coarse-to-medium grids and R_{MF} is the refinement ratio from medium-to-fine grids, that is, $R_{CM} = \Delta x_C/\Delta x_M$ and $R_{MF} = \Delta x_M/\Delta x_F$ ($R_{CM} > 1$ and $R_{MF} > 1$). There is no closed-form solution to the non-linear equation (8) when the refinement ratios R_{CM} and R_{MF} are different, and the value of p that satisfies the equation must be obtained through numerical techniques. With a constant refinement ratio, (8) yields the well-known solution:

$$p = \frac{\log\left(\frac{y(\Delta x_M) - y(\Delta x_C)}{y(\Delta x_F) - y(\Delta x_M)}\right)}{\log(R)} \tag{9}$$

where R denotes the constant refinement ratio, that is, $R = R_{CM} = R_{MF}$. The next step is to extrapolate the three discrete solutions $y(\Delta x_C)$, $y(\Delta x_M)$, and $y(\Delta x_F)$ to a reference $y^{Reference}$ that serves the purpose of substitute for the exact-but-unknown solution y^{Exact} of the continuous partial differential equations. This is commonly referred to as Richardson extrapolation. It can be verified that any combination of two grids provides the same value for $y^{Reference}$:

$$\begin{aligned} y^{Reference} &= y(\Delta x_F) + \frac{y(\Delta x_F) - y(\Delta x_M)}{R^p - 1} = y(\Delta x_M) + \frac{y(\Delta x_M) - y(\Delta x_C)}{R^p - 1} \\ &= y(\Delta x_C) + R^p \frac{y(\Delta x_M) - y(\Delta x_C)}{R^p - 1} \\ &= y(\Delta x_C) + R^{2p} \frac{y(\Delta x_F) - y(\Delta x_M)}{R^p - 1} \end{aligned} \tag{10}$$

where the rate-of-convergence is the value calculated from (9). Finally, the regression coefficient β_1 can be back-calculated from any one of (7):

$$\beta_1 = \frac{y(\Delta x_C) - y(\Delta x_M)}{\Delta x_M^p (R^p - 1)} = \frac{y(\Delta x_M) - y(\Delta x_F)}{\Delta x_F^p (R^p - 1)} = \frac{y(\Delta x_C) - y(\Delta x_F)}{\Delta x_F^p (R^{2p} - 1)} \tag{11}$$

An illustration is given by verifying the convergence of discrete solutions for the one-dimensional Burgers equation initialized with an arc-tangent ($\tan^{-1}$) initial condition at time $t = 0$:

$$\begin{aligned} \frac{\partial}{\partial t}(y(x;t)) + \frac{\partial}{\partial x}\left(\frac{1}{2}y^2(x;t)\right) = 0, -\frac{L}{2} \le x \le \frac{L}{2}, 0 \le t \le 1, \\ y(x;t=0) = y_o(x) \end{aligned} \tag{12}$$

where $y_o(x) = (L/\pi)\tan^{-1}(\omega x)$. Coefficients $L = 3\,\text{cm}$ and $\omega = 5\,\text{cm}^{-1}$ are used for the numerical application. Equation (12) is solved in conservation form using the Lax–Wendroff integration scheme [12]. The non-linear and hyperbolic nature of the Burgers equation evolves the initially smooth jump $y_o(x)$ into a sharp discontinuity, therefore, reducing convergence to first-order [13].

Figure 4 illustrates the convergence of the discontinuity jump, denoted as $[y]_S$ and defined as the value of the solution to the left of the discontinuity minus the value to the right, as a function of cell size Δx. Four calculations are performed with grid sizes listed in Table 1 and a constant refinement ratio equal to $R = 4$. Also shown are values of the rate-of-convergence obtained from

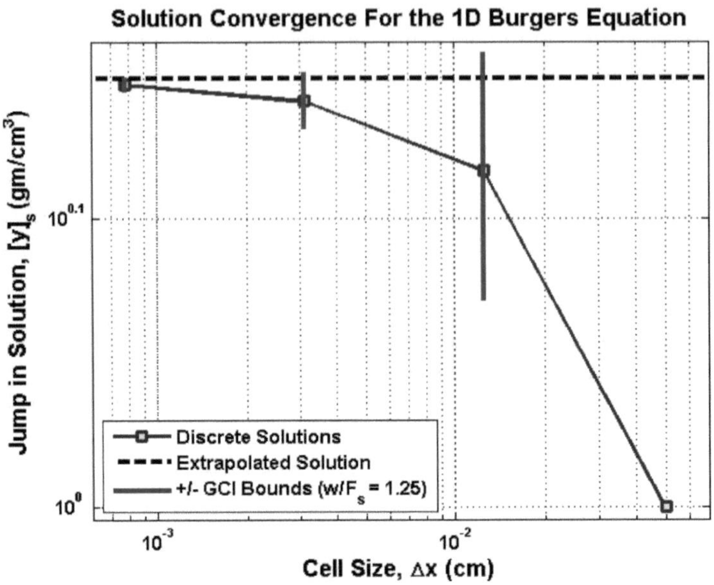

Fig. 4. Convergence of a discontinuity $[y]_S$ as a function of cell size Δx. (See Plate 34 on page 322)

Table 1. Results of a grid refinement study for the 1D Burgers equation

Grid resolution	Cell size, Δx(cm)	Density jump, $(y)_S$(gm cm^{-3})	Rates-of-convergence, p (XC; C; M)	(C; M; F)
X-coarse (XC)	5.00	0.9995	1.021	N/A
Coarse (C)	1.25	1.3083		1.008
Medium (M)	0.3125	1.3832		
Fine (F)	0.078125	1.4018	N/A	

Richardson extrapolation of the solution based on grids (C; M; F) = 1.4080 gm.cm^{-3}.

(9) and Richardson extrapolation obtained from (10). It is clear from Fig. 4 that convergence of the discrete solutions is monotonic. The observed rates-of-convergence match, for all practical purpose, the theoretical value of p = 1 as one would expect to get for this shocked (discontinuous) prediction $[y]_S$.

In summary (9–11) provide expressions for the triplet of unknowns ($y^{Reference}$; β_1; p) in the case of a constant mesh or grid refinement ratio. Clearly this analysis technique is appropriate only to the extent that the assumptions upon which it relies are satisfied. There are four essential assumptions summarized to conclude this section. (1) The cell sizes used (such as Δx_C, Δx_M, and Δx_F) provide discrete solutions located in the asymptotic regime of convergence. (2) The analysis is restricted to scalar quantities. (3) Because it is unknown, the exact solution y^{Exact} of the continuous equations is replaced by a reference $y^{Reference}$ to be determined during the analysis. (4) Convergence of the discrete solutions $y(\Delta x)$ to the reference $y^{Reference}$ is monotonic.

4 The Grid Convergence Index

The Grid Convergence Index (GCI) was proposed by Patrick Roache in 1994 to report the results of grid convergence studies in computational fluid dynamics [14]. Its main advantage is to provide a simple, standardized metric independent of the order of the numerical method and refinement strategy. In other words the GCI obtained from analyzing a first-order numerical scheme with grid "halving" (that is, $p = 1$ and $R = 2$) can be compared to the GCI obtained from analyzing a second-order numerical method with non-constant mesh refinement.

In simple terms the GCI is a scaled version of the relative difference between two discrete solutions $y(\Delta x_M)$ and $y(\Delta x_F)$ obtained from running the same calculation with two element or grid sizes Δx_M and Δx_F (where, as noted before, $\Delta x_M = R\Delta x_F$). Scaling by a factor $F_S/(R^p - 1)$ is applied to account for the specific values of the rate-of-convergence and grid refinement ratio:

$$GCI = \frac{F_S}{R^p - 1} \left| \frac{y(\Delta x_F) - y(\Delta x_M)}{y(\Delta x_F)} \right| \tag{13}$$

where F_S is the so-called "safety factor" whose value depends on the application, characteristics of the code used, and smoothness of the solution. Values $F_S = 1.25$ and $F_S = 3$ have been proposed for various applications in computational fluid dynamics and solid mechanics.[3] Small values of the GCI, such as 1% or less, indicate that the discrete solutions are converging to the continuous solution y^{Exact} and the Richardson extrapolation $y^{Reference}$ accurately estimates y^{Exact}.

Figure 4 illustrates the bounds of solution error obtained for the Burgers equation (12). The GCI values calculated with $F_S = 1.25$ are listed in Table 2. The vertical, solid lines shown in Fig. 4 in color red are solution error bounds defined somewhat arbitrarily as +/− one GCI value. (Note that, since the GCI is defined as a one-sided estimate of solution uncertainty, the uncertainty is represented twice in Fig. 4.) The small GCI values listed in Table 2 provide evidence of asymptotic convergence.

Another attractive property of the GCI, besides providing a standardized metric of solution error, is that the ratio between two successive GCI values

[3] A justification of scaling in (13) is briefly presented. Note that for second-order accuracy and grid halving (that is, $p = 2$ and $R = 2$) the denominator is equal to $(R^p - 1) = 3$. A safety factor $F_S = 3$ simplifies definition (13) to the relative difference between the discrete solutions $y(\Delta x_M)$ and $y(\Delta x_F)$. The utility of prefactor $F_S/(R^p - 1)$ is therefore to scale the relative solution error when $(p; R) \neq (2; 2)$ so that the resulting GCI value can be directly compared to one that would be obtained with $(p; R) = (2; 2)$.

Table 2. Grid convergence indices for the 1D Burgers equation (with $F_S = 1.25$)

Grid resolution	Cell Size, Δx(cm)	Density jump, $(y)_S$(gm cm^{-3})	GCI Values, GCI (XC; C)	(C; M)	(M;F)
X-coarse (XC)	5.00	0.9995	9.83%	N/A	N/A
Coarse (C)	1.25	1.3083		2.26%	N/A
Medium (M)	0.3125	1.3832	N/A		0.55%
Fine (F)	0.078125	1.4018	N/A	N/A	

should be equal to the value of R^p when the discrete solutions are located in the asymptotic regime of convergence, that is:

$$\frac{GCI_{C,M}}{GCI_{M,F}} \approx R^p \tag{14}$$

where $GCI_{C,M}$ denotes the value calculated from solutions $y(\Delta x_C)$ and $y(\Delta x_M)$, while $GCI_{M,F}$ is the value calculated from solutions $y(\Delta x_M)$ and $y(\Delta x_F)$. This diagnostic of asymptotic convergence does not depend on the safety factor F_S. In fact one only needs triplets of simulation results ($y(\Delta x_C)$; $y(\Delta x_M)$; $y(\Delta x_F)$), such as those given in Tables 1 and 2, to calculate the diagnostic:

$$\frac{GCI_{C,M}}{GCI_{M,F}} = \left|\frac{y(\Delta x_M) - y(\Delta x_C)}{y(\Delta x_F) - y(\Delta x_M)}\right| \cdot \left|\frac{y(\Delta x_F)}{y(\Delta x_M)}\right| \tag{15}$$

The numerical application using the three finest grids for the Burgers equation (12) gives a ratio of GCI values equal to 4.08, which is in excellent agreement with the theoretical value of R^p for $p = 1.008$ (see Table 1) and $R = 4$, further confirming the hypothesis of asymptotic convergence.

The selection of an appropriate safety factor F_S is a topic that generates heated debate. Some practitioners advocate keeping it constant while others prefer to turn it into a variable to account for better or worse-than-expected convergence, such as proposed by Fred Stern [15]:

$$F_S = \frac{R^{p^*} - 1}{R^p - 1} \tag{16}$$

where p^* and p are the theoretical and observed rates-of-convergence, respectively.

It is often claimed that defining solution error bounds as $|y^{Exact} - y(\Delta x_F)| \leq$ GCI where the safety factor is equal to $F_S = 3$ defines a *"confidence interval of grid convergence uncertainty at the 95% confidence level."* We emphasize that there is no rigorous justification for such practice and, to the best of our understanding, ***the claim that the GCI defines a probability or confidence interval at any confidence level is simply wrong***. Nevertheless, GCI values can be used to define conservative bounds of the solution error, such as those shown graphically in Fig. 4.

5 Application of Solution Verification to a Finite Element Calculation

This section gives an example of solution verification for a linear Hertz contact problem. The problem is solved with an axi-symmetric finite element model analyzed using the general-purpose software package HKS/ABAQUS version 6.4.3. Figure 5 illustrates the contact between a rigid hemisphere (shown as a solid, blue line) and deformable medium.

Dimensions of the deformable, elastic medium are one meter-by-one meter. Material properties are $E = 10^{+3}\,N\,m^{-2}$ (modulus of elasticity) and $\nu = 0.3$ (Poisson's ratio). The hemisphere is moved down under displacement control by 1 cm beyond the initial contact with the deformable medium shown as point C in Fig. 5. The contact condition is the HKS/ABAQUS default hard contact model with small sliding allowed and no friction. For the purpose of this illustration, it is assumed that one is interested predicting stress values σ_{YY} in the vertical direction on the top surface of the deformable medium, that is, at the location of finite elements that could potentially come into contact with the rigid hemisphere. Stress values σ_{YY} are given in units of $N\,m^{-2}$ (Pa) at the final simulation time of $T = 1$ s.

Four meshes are analyzed with 20, 40, 80, and 160 finite elements in each spatial direction to discretize the deformable medium. Figure 6 illustrates the stress fields σ_{YY} predicted by two of the four meshes analyzed, the extra-coarse and fine meshes. The four results are referred to as the extra-coarse, coarse, medium, and fine solutions. Their respective element sizes are $\Delta x = 5$, 2.5, 1.25, and 0.625 cm, hence, providing a refinement ratio of $R = 2$. The elements

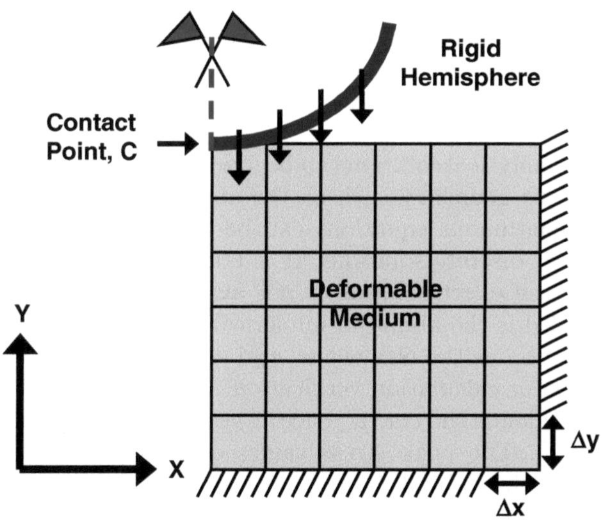

Fig. 5. Illustration of the Hertz contact problem. (See Plate 35 on page 323)

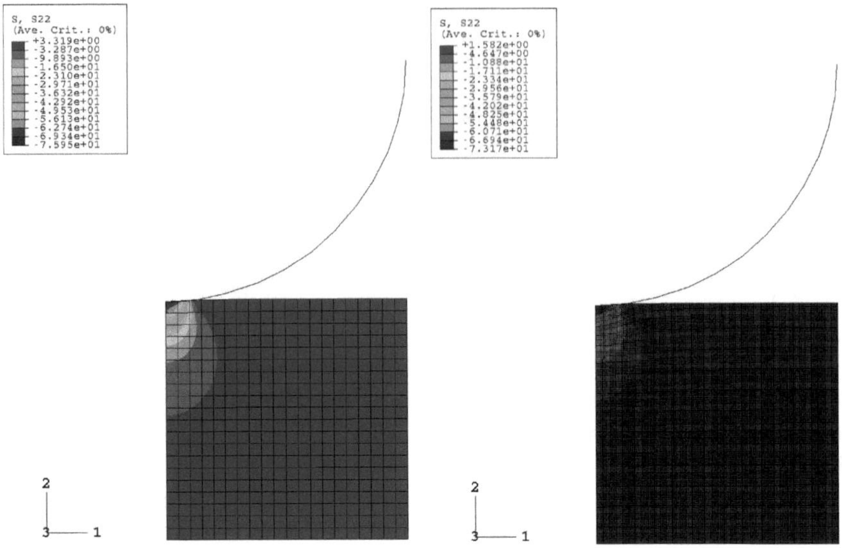

(**a**) Extra-Coarse mesh with 20 elements (**b**) Fine mesh with 160 elements

Fig. 6. Stress fields σ_{YY} for the Hertz contact problem. (See Plate 36 on page 323)

are quadratic quadrilaterals and the aspect ratio is kept uniform, constant, and equal to one for all meshes. Simply put, $\Delta x = \Delta y$ as suggested by Figs. 5 and 6.

Figure 7 shows the stress values (σ_{YY}) vs. location (x) of top-row finite elements. Each curve is obtained from one of the four meshes. Such "view-graph norm" comparison of curves indicates visually that the prediction seems to converge as the discretization is refined. An analysis of solution convergence using quantitative metrics, such as those defined in the previous sections, offers the important advantage over the view-graph norm of assessing convergence rigorously.

Note that an analytical solution can be found for the Hertz contact problem applied to a simple geometry such as the one used here. That is, an exact solution of the continuous equations can be derived and used for assessing the convergence of discrete solutions. It is not what is done here; instead, it is assumed that the exact solution is not available to us. Consequently, all that can be studied is the ability of numerical approximations to converge to an unknown solution as the element or grid size is refined, $\Delta x \to 0$, which is known as solution or calculation verification.

Tables 3–5 catalogue the convergence of peak stress prediction (in absolute value) at point C. The peak stress values extracted from the four curves of Fig. 7 are listed in Table 3. These are the only data needed to compute (9–11). Table 4 lists the rates-of-convergence calculated from (9) and Table 5 lists values of the GCI calculated from (13) with a safety factor equal to $F_S = 1.25$.

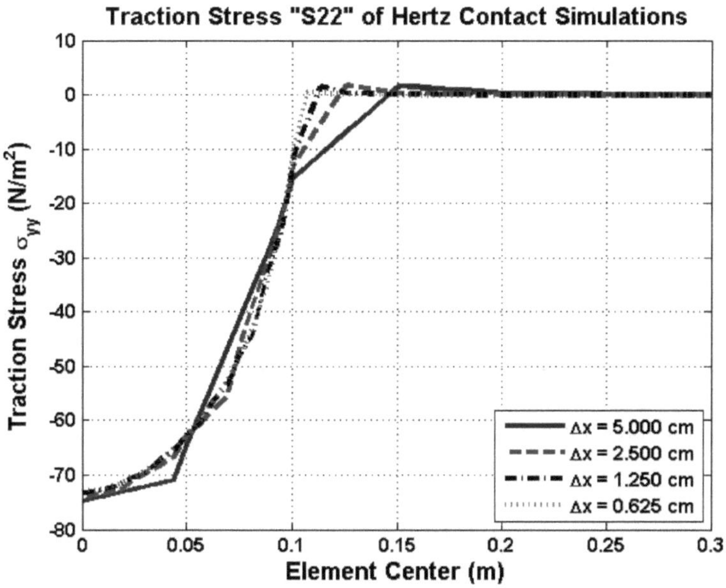

Fig. 7. Stress values (σ_{YY}) vs. location (x) of top-row finite elements. (See Plate 37 on page 324)

Table 3. Stress values $|\sigma_{yy}|$ predicted at point C

Quantity	Extra-coarse	Coarse	Medium	Fine
Size, Δx	5.000 cm	2.500 cm	1.250 cm	0.625 cm
Stress $\lvert\sigma_{yy}\rvert$ at Point C	74.82 N m^{-2}	74.00 N m^{-2}	73.26 N m^{-2}	73.09 N m^{-2}

Refinement ratio, R = 2.

Table 4. Rates-of-convergence observed using triplets of discrete solutions

Triplets Used	Extra-coarse	Coarse	Medium	Fine
Triplet (XC; C; M)		p = 0.15		(Not used.)
Triplet (C; M; F)	(Not used.)		p = 2.09	

XC = Extra-coarse; C = Coarse; M = Medium; F = Fine. Theoretical value, $p^* = 2$.

Table 5. Grid convergence indices observed using pairs of discrete solutions

Pairs used	Extra-coarse	Coarse	Medium	Fine
Pair (XC; C)	GCI = 0.458%		Not used	Not used
Pair (C; M)	Not used	GCI = 0.416%		Not used
Pair (M; F)	Not used	Not used	GCI = 0.099%	

XC = Extra-coarse; C = Coarse; M = Medium; F = Fine. Obtained with $F_S = 1.25$.

It is observed from Table 4 that the rate-of-convergence obtained with the triplet of solutions (XC; C; M) does not match the theoretical value of $p^* = 2$.[4] This result makes sense because, with the extra-coarse discretization ($\Delta x = 5\,cm$), only two finite elements are in contact with the rigid hemisphere at the final simulation time of $T = 1\,s$, as shown in Figure 6a. This is apparently not enough to provide an accurate solution. Table 4 also indicates that the rate-of-convergence becomes quadratic when the triplet of solutions (C; M; F) is used instead of the previous one (XC; C; M). It suggests that this set of results is indeed in the presence of asymptotic convergence. Alternatively it can be stated that for this problem the asymptotic regime of convergence, where truncation error dominates the overall numerical quality of discrete solutions, is reached for finite elements whose sizes are no larger than $\Delta x = 2.5\,cm$, or areas $\Delta x \Delta y \leq 6.25\,cm^2$.

Values of the GCI listed in Table 5 are less than 1%, which is an indication of small numerical error. Of particular interest is the ratio between two successive GCI values such as obtained, for example, with the pairs of solutions (XC; C) and (C; M). From Table 5 the ratio of GCI values with the pairs of solutions (XC; C) and (C; M) is equal to $0.458\%/0.416\% = 1.1$, which does not indicate asymptotic convergence. The next ratio is $0.416\%/0.099\% = 4.2$, which nearly equals $R^p = 2^{2.09} = 4.3$, confirming the hypothesis of asymptotic convergence.

6 Discussion of the Challenges of Code and Solution Verification

Clearly the success of code and solution verification is embodied by the formalism outlined in Sects. 3 and 4. It leads to a wide range of applicability, from linear, non-dissipative, elliptic equations to non-linear, shocked, hyperbolic equations. Failures, on the other hand, result from the many restrictions imposed on the way an analysis is typically carried out. A non-exhaustive list of eight challenges encountered during the verification of general-purpose codes is briefly discussed in this section. These issues are listed below in no particular order.

Challenge 1. Obviously the first challenge of solution verification is to search for evidence of the asymptotic regime of convergence. The values of discretization variables, such as grid size Δx or time step Δt, that lead to discrete solutions located in the asymptotic regime are not known a priori. This is a "catch-22" situation because the analysis technique outlined in Sect. 3 is valid only when solutions are located in the asymptotic regime but one cannot know if it is the case before the analysis has been completed. There may

[4] A rate-of-convergence equal to 2 is expected because quadratic shape functions are used to define the finite elements of the four meshes used in this illustration, and the solution predicted is "smooth."

be no fundamental answer to this dilemma but addressing some of the issues discussed below may help define a more rigorous formalism to search for the asymptotic regime of convergence.

Challenge 2. Derivations of Sect. 3 are currently restricted to the analysis of scalar quantities. What prevents the analysis of curves or multi-dimensional fields is that norms of the solution error $\varepsilon_Y(\Delta x) = ||y^{Reference} – y(\Delta x)||$ must be evaluated. This is possible only if the reference solution is known. One could, if it is known, use the exact solution of the continuous partial differential equations, that is, $y^{Reference} = y^{Exact}$; or obtain an approximate but highly-accurate reference, for example, by running another code. Examples of techniques available to formulate an exact or highly-accurate reference include the method of manufactured solutions [16] and the method of nearby solutions [17]. In cases where the reference solution is unknown and one is not willing to assume it, the only option is to restrict the analysis to scalar predictions. Doing so reduces the norm to a difference such as $|y^{Reference} – y(\Delta x)|$ or $(y^{Reference} – y(\Delta x))$ and a closed-form solution can be obtained for the triplet of unknowns $(y^{Reference}; \beta_1; p)$ of the Ansatz (5). An outstanding challenge is to develop methods to verify the convergence of discrete solutions when these are curves or multi-dimensional fields. Exploiting functional data analysis, such as suggested in [18], is a promising direction of research and there may be others worth exploring.

Challenge 3. The third challenge is the definition of the reference mesh or grid used to calculate norms of the solution error. Should solution errors be calculated on the coarsest grid, that is, on the basis of the largest cell or element size $\Delta x = \Delta x_{Max}$, or should solution errors be calculated on the finest grid, that is, on the basis of the smallest cell or element size $\Delta x = \Delta x_{Min}$? The two options are indeed to rely on either a coarse or fine grid. In the first case the reference solution $y^{Reference}$ (or y^{Exact}, if known) is defined on a coarse grid, with element or cell size Δx_C, and discrete solutions $y(\Delta x)$ obtained from meshes or grids where $\Delta x \leq \Delta x_C$ are averaged to calculate errors $(y^{Reference} – y(\Delta x))$ defined on the coarse grid. In the second case the exact or reference solution is defined on a fine grid, with cell size Δx_F, and discrete solutions, obtained from grids where $\Delta x_F \leq \Delta x$, are interpolated to calculate errors $(y^{Reference} – y(\Delta x))$ defined on the fine grid. Details of how to construct these averaging or extrapolation schemes are generally not published in the literature or documented in analysis codes [18]. Although not discussed, it is postulated that these details can have a very significant (and sometimes detrimental) effect on one's ability to verify solution convergence. In addition to being a practical question of how to implement (9–11) in a computer code, this challenge also raises a fundamental issue.

When solving the conservation laws or equations-of-motion relies on a weak formulation of the equations, should the convergence of discrete solutions be studied using an error norm that is different from the weak norm? A weak solution, such as one obtained with the finite element method, is defined in the sense of a specific norm of the Sobolev $W^{m,n}$ space. It can therefore be argued that verifying the convergence of a solution using anything other that its weak norm introduces an inconsistency with properties of the discrete solution.[5] If this assertion is correct, then the common practice of studying convergence with scalar (point-wise) quantities provides an incorrect picture of asymptotic convergence. It is unclear whether this issue is relevant to the study of the regime of asymptotic convergence but, at the very least, the question should be debated in the computational engineering and physics communities. (For further discussion of this issue, and a potential step forward using functional data analysis, see [18].)

Challenge 4. The fourth challenge is to understand the role played by discretization variables other than space Δx in defining the regime of asymptotic convergence. Also questioned is the effect that hardwired or user-defined parameters of the numerical methods may have on the convergence of discrete solutions. Observation of the solution error as $\Delta x \to 0$ sometimes indicates that the error "stagnates" or converges in an oscillatory manner as opposed to monotonically, which suggests that effects other than spatial discretization alone can influence the definition of the asymptotic regime. The challenge is to extend space-only Ansatz equations of solution error to multiple discretizations that include, for example, both cell size Δx and time step Δt, or other parameters that present themselves in the equations of other physics, such as particle or radiation transport.

Space-only error Ansatz formulations such as (1), for example, do not account for a potential coupling between spatial and temporal discretization or, in the case of solution self-similarity, a potential dependence of solution error on the cell size-to-time step ratio $(\Delta x/\Delta t)$. The fourth challenge is to extend conventional error Ansatz equations to multiple dimensions:

$$\varepsilon_Y(\Delta x) = \left\| y^{Exact} - y(\Delta x;\Delta t) \right\| = \beta_1 (\Delta x)^p + \beta_2 (\Delta t)^q + \beta_3 (\Delta x)^r (\Delta t)^s + O(\Delta x^p;\Delta t^q) \tag{17}$$

where the symbols p, q, and (r;s) denote rates of convergence in space, time, and space–time coupling, respectively, and $(\beta_1;\beta_2;\beta_3)$ are three (unknown) regression coefficients. Studies of the convergence of discrete solutions in

[5] Three examples of potential inconsistencies are given in the context of a finite element calculation in engineering mechanics. The first inconsistency may be to study the convergence of scalar quantities $y_k(\Delta x)$ when these are defined locally, where the subscript $(\bullet)_k$ identifies a single element or node of the mesh. Because, by construction, finite element solutions $y(\Delta x)$ satisfy the equations-of-motion on average and in the sense of the weak norm, it may not make much sense studying the convergence of an element-wise or point-wise quantity $y_k(\Delta x)$. Likewise the reference solution $y^{Reference}$ given by Richardson extrapolation in (10) is defined with a norm that is different from the norm of a weak formulation. The third inconsistency may be to study the convergence of discrete solutions using L^p norms of the error $\left\| y^{Reference} - y(\Delta x) \right\|$ because finite element solutions $y(\Delta x)$ are constructed in $W^{m,n}$ spaces, not L^p spaces.

space–time (see, for example, [4, 9]), remain exploratory and restricted to a few code verification test problems in hydro-dynamics.

Hardwired or user-defined parameters of codes for computational engineering and physics can also have a significant effect on the convergence of discrete solutions. Examples of parameters are the various options that a numerical solver may include, tolerances to assess convergence or singularity, type and amount of artificial viscosity, bulk viscosity, and parameters that mitigate the occurrence of hourglass modes. While the scientific community is still debating whether such parameters should be kept fixed or varied for solution verification, there is evidence that their optimal settings ("optimal" in the sense of producing a high-quality solution) can be mesh-dependent. It suggests that the proper approach to formulate code or solution verification in the context of parameter uncertainty may be to collect evidence of bounded convergence in addition to the verification of the rate-of-convergence.

Challenge 5. The fifth challenge is presented by the use of automated meshing software that, while simplifying the difficult task of re-meshing a computational domain, can also approximate the geometry of the domain in a somewhat uncontrolled manner. Such approximations can be detrimental to the assessment of solution convergence because they may change the geometry of the computational domain from one level of mesh or grid refinement to the next. This, in turn, changes the definition of the boundary condition, which changes the (known or unknown) exact y^{Exact} or reference $y^{\mathrm{Reference}}$ solutions. Analysts should be careful not to change the geometry of the domain and definition of its boundary and initial conditions when coarsening or refining a computational mesh or grid.

Challenge 6. Solution error Ansatz models, such as (5), are usually written with constant element or cell sizes Δx. Defining characteristic values of the discretization variables, such as cell size Δx or time step Δt, when these are adjusted during the numerical simulation is not addressed. Cell sizes change, for example, during an Adaptive Mesh Refinement (AMR) calculation in hydrodynamics or computational fluid dynamics. Logically equivalent to this is the finite element h-refinement, which likewise makes it difficult to define a characteristic element size. Refinement strategies attempt to concentrate a higher density of elements or cells where they are needed the most to improve the numerical quality of the solution, that is, in regions of discontinuity or high gradient. They can lead to very significant computational savings relative to refining the mesh or grid uniformly across the computational domain. Time stepping strategies are also routinely implemented to adapt the choice of time step and match an accuracy or stability criterion. Examples are Courant-limited runs and implicit time integration solvers. Efforts are needed to define characteristic values of the discretization variables in cases of space or time-adaptive calculations, and derive appropriate solution error Ansatz models.

Challenge 7. The role currently played in code and solution verification by techniques such as the Method of Manufactured Solutions (MMS),

Modified Equation Analysis (MEA), and various approaches to a posteriori error estimation is, at best, marginal. The MMS is useful to define an arbitrarily-complex solution that can serve as reference $y^{Reference}$ for studying the convergence of discrete solutions in the context of either code or solution verification [16, 19]. MEA is an analytical technique to study the behavior of truncation error due to spatial or temporal discretization for finite difference schemes or finite volume methods [20]. The field of a posteriori error estimation develops diagnostics of numerical quality defined from one or several discrete solutions and calculated element-by-element or cell-by-cell [21]. Analysts who may be concerned about the numerical quality of discrete solutions have yet to take full advantage of these powerful analysis techniques. How to make a connection, however, with the verification of general-purpose codes remains a serious challenge because, if nothing else, of implementation difficulties.

Challenge 8. One final question is to understand the extent to which Ansatz models of solution error can be used to estimate confidence bounds of the unknown solution of the continuous partial differential equations. This is, of course, the Holy Grail of solution verification. The current state-of-the-practice in computational engineering and physics relies on the Grid Convergence Index (GCI) to define bounds of solution uncertainty such as, for example, $\|y^{Exact} - y(\Delta x)\| \leq GCI$, with a constant or varying safety factor F_S. However, even though such a claim is commonly encountered in the literature, there is no rigorous connection between GCI-based bounds of solution accuracy and statistical confidence intervals. Such rigorous connection has not, to the best of our understanding, been demonstrated. The eighth challenge, which remains a topic of open research to a great extent, would be to develop rigorous bounds of numerical uncertainty in the context of self-convergence and solution (or calculation) verification.

This list of challenges, which we do not claim to be exhaustive, depicts a state-of-the-practice that seems to be a form of skillful art rather than well-established science. It is our belief, however, that addressing these issues could significantly improve our ability to verify the convergence of discrete solutions and quantify calculation uncertainty. It may also demonstrate that the regimes of convergence of many of our predictive codes are broader than previously thought.

7 Conclusion

Verifying that an analysis code is free of errors and the numerical methods it implements reach rates-of-convergence that match the theory are essential to demonstrate the numerical quality of discrete solutions. Simply speaking, it is the first "V" of the Verification and Validation (V&V) of predictions obtained from numerical simulations. The discipline of code and solution verification in computational engineering and physics is developing tools to assess the source and severity of numerical solution error. The publication outlines a simple procedure to evaluate the asymptotic regime of convergence; estimate a rate-of-convergence based on mesh or grid refinement; extrapolate the discrete solutions; and develop bounds of solution error due to the discretization. Much, however, remains to be done. A non-exhaustive list of topics that warrant further research includes: extending the state-of-the-practice to non-scalar quantities (curves, multiple-dimensional fields); studying the coupling between space and time discretizations; defining a reference mesh for the estimation of solution error; and developing technology to verify adaptive mesh refinement calculations in computational engineering and physics. It is our hope that engineers and physicists, while increasingly relying on modeling and simulation, will understand the importance that numerical issues play in model validation and become an advocate for best practices and more research and development in the discipline of code and solution verification.

Acknowledgments

This work is performed under the auspices of the Advanced Scientific Computing (ASC) verification project at the Los Alamos National Laboratory (LANL). The authors are grateful to Jerry S. Brock, verification project leader, for his support. LANL is operated by the Los Alamos National Security, LLC for the National Nuclear Security Administration of the U.S. Department of Energy under contract DE-AC52-06NA25396.

References

1. Hemez, F.M., Doebling, S.W., Anderson, M.C., A Brief Tutorial on Verification and Validation, *22nd SEM International Modal Analysis Conference*, Dearborn, Michigan, January 26–29, 2004.
2. Brock, J.S., ASC Level-2 Milestone Plan: Code Verification, Calculation Verification, Solution-error Analysis, and Test-problem Development for LANL Physics-simulation Codes, *Technical Report of the ASC Program and Code Verification Project*, Los Alamos National Laboratory, Los Alamos, New Mexico, May 2005. LA-UR-05-4212.
3. Kamm, J.R., Rider, W.J., Brock, J.S., Consistent Metrics for Code Verification, *Technical Report*, Los Alamos National Laboratory, Los Alamos, New Mexico, June 2002. LA-UR-02-3794.
4. Kamm, J.R., Rider, W.J., Brock, J.S., Combined Space and Time Convergence Analyses of a Compressible Flow Algorithm, *16th AIAA Computational Fluid Dynamics Conference*, Orlando, Florida, July 2003. LA-UR-03-2628.
5. Knoll, D.A., Chacon, L., Margolin, L.G., Mousseau, V.A., On Balanced Approximations for Time Integration of Multiple Time Scale Systems, *Journal of Computational Physics*, **185**(2), 2003, 583–611.
6. Buechler, M., McCarty, A., Reding, D., Maupin, R.D., Explicit Finite Element Code Verification Problems, *22nd SEM International Modal Analysis Conference*, Dearborn, Michigan, January 2004. LA-UR-03-4603.

7. Li, S., Rider, W.J., Shashkov, M.J., Two-dimensional Convergence Study for Problems with Exact Solution: Uniform and Adaptive Grids, *Technical Report of the ASC Program and Code Verification Project*, Los Alamos National Laboratory, Los Alamos, New Mexico, October 2005. LA-UR-05-7985.
8. Smitherman, D.P., Kamm, J.R., Brock, J.S., Calculation Verification: Pointwise Estimation of Solutions and Their Method-Associated Numerical Error, *Journal of Aerospace Computing, Information and Communication*, Vol. 4, 2007, pp. 676–692.
9. Hemez, F.M., Brock, J.S., Kamm, J.R., Non-linear Error Ansatz Models in Space and Time for Solution Verification, *1st Non-deterministic Approaches (NDA) Conference and 47th AIAA/ASME/ASCE/AHS/ASC Structures, Structural Dynamics, and Materials (SDM) Conference*, Newport, Rhode Island, May 1–4, 2006. LA-UR-06-3705.
10. Roache, P.J., *Verification in Computational Science and Engineering*, Hermosa Publishers, Albuquerque, New Mexico, 1998.
11. Hemez, F.M., Non-linear Error Ansatz Models for Solution Verification in Computational Physics, *Technical Report of the ASC Program and Code Verification Project*, Los Alamos National Laboratory, Los Alamos, New Mexico, October 2005. LA-UR-05-8228.
12. LeVeque, R.J., *Numerical Methods for Conservation Laws*, Birkhauser, 1990.
13. LeVeque, R.J., *Finite Volume Methods for Hyperbolic Problems*, Cambridge University Press, Cambridge, 2002.
14. Roache, P.J., Perspective: A Method for Uniform Reporting of Grid Refinement Studies, *ASME Journal of Fluids Engineering*, **116**, 1994, 405–413.
15. Stern, F., Wilson, R., Shao, J., Quantitative V&V of Computational Fluid Dynamics (CFD) Simulations and Certification of CFD Codes with Examples, *2004 ICHMT International Symposium on Advances in Computational Heat Transfer*, Norway, April 19–24, 2004. Paper CHT-04-V1.
16. Knupp, P., Salari, K., **Verification of Computer Codes in Computational Science and Engineering**, CRC Press, Boca Raton, Florida, 1st edition, 2002.
17. Roy, C., Review of Code and Solution Verification Procedures for Computational Simulation, *Journal of Computational Physics*, Vol. 205, No. 1, May 2005, pp. 131–156.
18. Hemez, F.M., Functional Data Analysis of Solution Convergence, *Technical Report, Contribution to the Fiscal Year 2007 ASC Verification and Validation Project "Code Verification,"* Los Alamos National Laboratory, Los Alamos, New Mexico, August 2007. LA-UR-07-5758.
19. Salari, K., Knupp, P., Code Verification by the Method of Manufactured Solutions, *Technical Report*, Sandia National Laboratories, Albuquerque, New Mexico, 2000. SAND-2000-1444.
20. Warming, R., Hyett, B., The Modified Equation Approach to the Stability and Accuracy Analysis of Finite Difference Methods, *Journal of Computational Physics*, Vol. 14, 1974, pp. 159–179.
21. Ainsworth, M., Oden, J.T., *A Posterior Error Estimation in Finite Element Analysis*, Wiley Inter-science Series in Pure and Applied Mathematics, Wiley, New York, pp. 16–23, 2000.

Expert Panel Opinion and Global Sensitivity Analysis for Composite Indicators

M. Saisana and A. Saltelli

Summary. Composite indicators aggregate multi-dimensional processes into simplified concepts often aiming at underpinning the development of data-driven narratives for policy consumption. Due to methodological issues, doubts are often raised about the robustness of the composite indicators and the significance of the associated policy messages. In this paper we use expert panel information (derived from budget allocation and analytic hierarchy process) on the relative importance of the underlying indicators included in a composite indicator and run in tandem uncertainty and sensitivity analysis to gain useful insights during the process of composite indicators building. We discuss the extent to which variance-based sensitivity analysis may increase transparency or make policy inference more defensible by using the United Nation's Technology Achievement Index as an illustration.

1 Introduction

Composite indicators enter into public debate once signals of society's performance in complex fields, such as economy, environment or technological achievement are sought. The Consumer Price Index, for example, that considers the costs of about 60 goods and services purchased monthly by a typical household, offers a more complete picture of the relative cost of living in different countries than would the price of a single item (e.g. bread or fuel). The OECD-JRC Handbook on Composite Indicators [22, 23] offers a review and guidelines on constructing such aggregate measures of countries performance. A recent compilation of existing composite indicators lists over 130 such measures from the fields of economy, society, environment, globalization, and technology [1] and an information server on the topic is available at a European Commission site [8].

In mathematical terms, a composite indicator score (or rank) is a function of indicators and weights. Weights may represent the relative importance of each indicator or be implied by the data. The function may involve linear or geometric averaging or use of outscoring matrix in a multi-criteria setting [20].

In their simplest and most frequent form, composite indicators are built by averaging country ranks (or normalised values) across a set of indicators.

There are two polarised schools of thought regarding the uses and misuses of composite indicators. Andrew Sharpe [38] summarises: "The aggregators believe there are two major reasons that there is value in combining indicators in some manner to produce a bottom line. They believe that such a summary statistic can indeed capture reality and is meaningful, and that stressing the bottom line is extremely useful in garnering media interest and hence the attention of policy makers. The second school, the non-aggregators, believe one should stop once an appropriate set of indicators has been created and not go the further step of producing a composite index. Their key objection to aggregation is what they see as the arbitrary nature of the weighting process by which the variables are combined." Along the same lines, Saisana et al. [30] argue that the perpetual debate on the use of composite indicators finds official statisticians, on one side, unwilling to waste or hide information behind a single number of dubious significance, and stakeholders, on the other, tempted to summarise complex and at times elusive processes (e.g. sustainability, welfare) into easy-to-communicate numbers.

This controversy can be put into context if one considers that composite indicators are mathematical models, and as such they represent encoded formalisations of natural systems. Rosen [25] noted that while a causality entailment structure defines the natural system, and a formal causality system entails the formal system, no rule was ever agreed on how to move from perceived reality to a model. In fact, a model will describe some features of the real system but also scientists' choices on how to observe the reality. When building a model or a composite indicator to describe a real-world phenomenon, formal coherence is a necessary but not sufficient property. The choices during the development phase of the composite indicator (e.g. choice of indicators, normalization method, set of weights, aggregation method) will stem from a certain perspective on the issue to be modelled. Reflexivity is thus inherent in a model since the observer and the observation are not separated and the way the human mind approaches the problem is part of the problem itself [14].

What is evident in the lines above is that the painful part in modelling lays in the uncertainty, which according to Höeg [15] is not an accident of the scientific method but its substance. Most modellers today would probably agree that a model cannot be validated in the sense of be proven true [24]. Rather, it is more defensible and correct to say that a model can be corroborated if it passes tests that assess the model's capacity to explain or predict the 'system' in a convincing and parsimonious way.

No matter how subjective and imprecise the description of the system may be, it implies the recognition of the multidimensional nature of a phenomenon to be measured and the effort of specifying the single aspects and their interrelation. Phenomena described by composite indicators are complex, think of concepts such as welfare, quality of education, or sustainability. The aim is to reduce complexity down to a measurable form by replacing non-measurable phenomena with intermediate objectives whose achievement can be observed and measured. The reduction into parts has limits when crucial components of the system are not considered. Although we cannot tackle here the vast issue of quality of statistical information, there is one fact that we find essential for their use. This is the existence of a community of peers (be these individuals, regions, countries) willing to accept composite indicators as their common yardstick for identifying limitations and promoting good practices. In fact, acceptance of a composite indicator relies on both strong scientific basis for its development and, maybe more important, on negotiation [13].

The development of a composite indicator is not straightforward, but it involves theoretical and methodological assumptions that need to be assessed carefully to avoid producing results of dubious analytic rigour [30]. Furthermore, the fact that a composite indicator is likely to be received by a polarized audience calls for stringent standards of rigour and robustness [32]. To maximize their utility and minimize their misuse, composite indicators need to be developed using the best available evidence, documented transparently, and validated using appropriate uncertainty and sensitivity analyses.

In this paper, we address two main questions that a developer of a composite indicator may be asked regarding the reliability of the results: Given the uncertainties/assumptions during the development of the composite indicator, which of them have a strong impact on:

- The difference between composite scores of two countries with similar performance?
- The overall country ranking?

The first question reflects a stakeholders' concern about a suspected bias in certain country scores. The second question pertains to the overall 'plausibility' or robustness of the composite indicator.

To answer these questions, we describe in Sect. 2.2 an easy-to-code procedure for global sensitivity analysis, which assess the influence of input factors on the output by looking at the entire space of assumptions, instead of just a part of it (e.g. derivatives), as recent modelling guidelines recommend (e.g. [9,37]). To exemplify the approach, we focus on uncertainties in the composite indicator building related to the weighting and normalisation issues and run in tandem Uncertainty Analysis (UA) and Sensitivity Analysis (SA). Using UA, we study how uncertainty in the input factors propagates through the structure of the composite indicator and affects the countries composite scores (or ranks). The SA is next applied to assess the impact of each source of uncertainty on the results. Section 3 presents the test case, the Technology Achievement Index (TAI), which together with the Human Development Index, was developed by the United Nations and included in the 2001 Human Development Report [43]. To account for the uncertainties, two normalisation methods and two participatory approaches (budget allocation and analytic hierarchy process) for assigning weights to the TAI indicators were selected.

Section 4 discusses the results and answers the two key questions raised previously on the quality of a composite indicator. Section 5 summarises our conclusions on the role of the combined use of UA and SA in producing more robust or parsimonious composite indicators and in rendering them more defensible in the face of scientific or technical controversy.

2 Methodological Issues and Uncertainties in Building a Composite Indicator

Several methods for calculating composite indicators from a set of underlying indicators are described in the literature and reviewed in [22, 23, 29]. The most frequent methods to render the underlying indicators comparable are either based on re-scaled values with a fixed range but with different mean and standard deviation (1) or on standardised values with fixed mean and standard deviation but different ranges (2). The standardisation method is less sensitive to outliers or extreme values. A composite score Y_c for a c is usually calculated as the weighted average of a total of Q normalised indicators $I_{q,c}$ with the respective weights w_q

$$Y_c = \sum_{q=1}^{Q} I_{q,c} \cdot w_q, \text{where} \begin{cases} I_{q,c} = \dfrac{x_{q,c} - \min(x_q)}{range(x_q)} & (1) \\ I_{q,c} = \dfrac{x_{q,c} - mean(x_q)}{std(x_q)} & (2) \end{cases}$$

where $x_{q,c}$ represents the raw indicator value x_q for country c.

The difference in the composite scores between two countries a and b (3) will be one of the output variables of interest in our uncertainty – sensitivity analysis.

$$D_{a,b} = \sum_{q=1}^{Q} (I_{q,a} - I_{q,b}) \cdot w_q \tag{3}$$

Additionally, the average shift in countries' rank will be explored. This statistic captures in a single number the relative shift in the position of the entire system of countries. It can be quantified as the average of the absolute differences in countries' rank with respect to a reference ranking over the M countries (4).

$$\bar{R}_S = \frac{1}{M} \sum_{c=1}^{M} |Rank_{ref}(Y_c) - Rank(Y_c)| \tag{4}$$

The investigation of the $Y_c, D_{a,b}$ and $\bar{R}_S$ will be the scope of the uncertainty and sensitivity analysis, targeting to provide answers to the issue of quality of a composite indicator.

2.1 Uncertainty Analysis

In the first stage, the uncertain input factors in the estimation of the outputs Y_c, $D_{a,b}$ and $\bar{R}_S$ have to be acknowledged. Generally, the uncertainties in the development of a composite indicator can be attributed to a number of factors induced by modelling choices [23]:

I. The measurement error in the data
II. The mechanism for including or excluding indicators
III. The transformation and/or trimming of the indicators (e.g. taking logarithms of indicators)
IV. The amount of missing data and the imputation algorithm for estimating missing data
V. The normalisation method applied to render the indicators comparable;
VI. The weighting method (e.g. equal weights, weights derived from statistical approaches or expert opinion)
VII. The weights values to be attached to the indicators (e.g. different experts would provide different weights)
VIII. The level of aggregation, if more than one level is used
IX. The aggregation method (e.g. additive, multiplicative, non-compensatory)

All these choices may influence the countries' composite scores and should be taken into account before any further interpretation of the results is attempted. Undoubtedly, the most debated problem in building composite indicators is the difficulty in assessing the relative importance (weight) of the indicators. Cox et al. [7] summarise these difficulties and conclude that many published weighting schemes are either based upon too complex multivariate methods or have little meaning to society. Cherchye et al. [5] observe that the 'lack of consensus' on the relative weights is a defining property of composite indicators, and that while one may hypothesise consensus on the indicators to be included in an index, the weights to be assigned to them will most likely remain controversial. The point of these considerations is that subjectivity and fitness need not be antithetical. They are in fact both at play when constructing a composite indicator [22, 23]. These, only apparently conflicting, properties underpin composite indicators' suitability for advocacy [32].

For these reasons, participatory approaches, such as budget allocation or analytic hierarchy process, are often employed to assign weights to a set of indicators in a way that reflects societal preference. In the budget allocation (BA) method, experts are invited to distribute a budget of points over a set of indicators (usually 3–15 indicators), paying more for those indicators whose importance they want to emphasize [19]. The Analytic Hierarchy Process (AHP) is another widely used technique for multi-attribute decision-making [26, 27]. AHP is based on ordinal pair-wise comparisons of indicators, in which the strength of preference is expressed on a semantic scale from 1 (equality) to 9. The relative weights of the indicators are then calculated using an eigenvector technique [26] that serves to resolve inconsistencies (e.g. loops such as a better than b better than c better than a).

We introduce this subjectivity in our case study, and in particular we focus on three points of the (V) to (VII) steps in composite indicator building: the type of normalisation (re-scaling or standardisation), the weighting scheme (BA or AHP) and the weights attached to the indicators. We anticipate here that we shall translate all these uncertainties into a set of scalar input factors to be sampled from their distributions. As a result, all output variables of interest, namely Y_c, $D_{a,b}$ and $\bar{R}_S$ are non-linear functions of the uncertain input factors, and the estimation of the outputs' probability distribution functions (pdf) is the purpose of the uncertainty analysis. The UA procedure is essentially based on simulations that are carried on (1)–(4), termed henceforth the *model*.

Various methods are available for evaluating output uncertainty. We employ the Monte Carlo approach (for reasons to be explained in the next section) that is based on performing multiple simulations of a model with k input factors. The procedure is as follows:

(a) Assign a pdf to each input factor $X_i (i = 1, \dots, k)$.
(b) Generate randomly N combinations of independent factors $\mathbf{X}^l$ $\left(= X_1^l, \dots, X_k^l\right)$, where $l = 1, \dots, N$.
(c) Evaluate the model, i.e. by computing the output value $\mathbf{Y}^l$ that is either Y_c, the value of the composite indicator for each country, or $D_{a,b}$, the difference between two countries composite scores, or $\bar{R}_S$, the averaged shift in countries' rank.
(d) Close the loop over l, and analyse the pdf of $\mathbf{Y}^l$. The characteristics of the pdf (e.g. variance, higher order moments) can be estimated with an arbitrary level of precision that is related to the size of the simulation N.

These steps constitute our uncertainty analysis. The generation of samples can be performed using various procedures, such as simple random sampling, stratified sampling, quasi-random sampling or others [34].

Once a composite indicator has been produced, modellers may hesitate to submit it to a revision via sensitivity analysis by a third party. This was not the case, however, of recent indices developed by the academia or non-governmental organisations, such as the 2005 Environmental Sustainability Index [10] and the 2006 Environmental Performance Index [11] developed by the Yale and Columbia Universities, the Alcohol Policy Index [2] developed by the New York Medical College, or the 2006 Composite Learning Index [3, 28] developed by the Canadian Council on Learning. In these studies, external review via sensitivity analysis was conducted. This anticipation of criticism by sensitivity analysis is also one of the ten commandments of applied econometrics according to Kennedy [17] who supports Leamer's [18] advocation that when reporting a sensitivity analysis, researchers should explain fully their specification search so that the readers can judge for themselves how the results may have been affected. This is basically an 'honesty is the best policy' approach.

2.2 Sensitivity Analysis Using Variance-Based Techniques

A necessary step when designing a sensitivity analysis is to identify a few summary model outputs that describe concisely, yet exhaustively, the message provided by a model [35]. In the present application for instance, although an uncertainty analysis of all countries scores Y_c and respective ranks was undertaken, sensitivity analysis is only applied to two model outputs, $D_{a,b}$ and $\bar{R}_S$, as they are relevant to the quality assessment of a composite indicator.

For non-linear models, such as our composite indicator happens to be when different normalisation methods, weighting schemes and weights are let to vary, variance-based techniques for sensitivity analysis are recommended [35]. Interesting features of variance-based methods include:

- Model independence (appropriate for non-linear and non-additive models)
- Exploration of the whole range of variation of the input factors
- Assessment of interaction effects among input factors
- Treatment of groups of input factors: uncertain factors might pertain to different logical types and it might be desirable to decompose the uncertainty according to these types

The impact of a given input factor X_i on the output variable can be measured via the so-called *sensitivity measure*, which is defined as the fractional contribution to the model output variance that derives from the uncertainty in X_i. For k independent input factors, the sensitivity measures can be computed using a decomposition formula (5)–(7) for the total output variance $V(Y)$ of the output Y.

$$V(Y) = \sum_i V_i + \sum_i \sum_{j>i} V_{ij} + \cdots + V_{12\ldots k} \tag{5}$$

where

$$V_i = V_{X_i}\left(E_{\mathbf{X}_{-i}}(Y\,|X_i)\right) \tag{6}$$

$$V_{ij} = V_{X_i X_j}\left(E_{\mathbf{X}_{-ij}}(Y\,|X_i\,, X_j)\right) - V_{X_i}\left(E_{\mathbf{X}_{-i}}(Y\,|X_i)\right) - V_{X_j}\left(E_{\mathbf{X}_{-j}}(Y\,|X_j)\right)$$
$$\ldots \tag{7}$$

Note that in computing $V_{X_i}\left(E_{\mathbf{X}_{-i}}(Y\,|X_i)\right)$, the expectation $E_{\mathbf{X}_{-i}}$ would call for an integral over $\mathbf{X}_{-i}$, i.e. over all factors but X_i, including the marginal distributions of these factors. The variance V_{X_i} would imply a further integral over X_i and its marginal distribution. A first measure of the fraction of the unconditional output variance $V(Y)$ that is accounted for by the uncertainty in X_i, is the first order sensitivity measure for the factor X_i (8).

$$S_i = V_i / V \tag{8}$$

Higher terms (above the first-order) in (5) are known as interactions and are present in a non-additive model. Interactions may imply, for instance, that extreme values of the output Y are uniquely associated with certain combinations of model inputs, in a way that is not described by the first-order effects S_i. Interactions may represent an important feature of a model and are more difficult to detect than first-order effects. For example, by using (squared) standardised regression coefficients it is fairly direct to estimate first-order effects, but not interactions. However, higher order sensitivity measures are usually not estimated, as in a model with k factors the total number of sensitivity measures (including the S_i's) that should be estimated is as high as $2^k - 1$. To circumvent this so-called 'curse of dimensionality', the total effect sensitivity measure is computed, which concentrates in a single term all the interactions (including the first-order effect) involving factor X_i. To exemplify, for a model of $k = 3$ independent factors, the three total sensitivity measures would be:

$$S_{T1} = \frac{V(Y) - V_{X_2X_3}(E_{X_1}(Y\,|X_2, X_3))}{V(Y)} = S_1 + S_{12} + S_{13} + S_{123} \qquad (9)$$

And analogously:

$$\begin{aligned} S_{T2} &= S_2 + S_{12} + S_{23} + S_{123} \\ S_{T3} &= S_3 + S_{13} + S_{23} + S_{123} \end{aligned} \qquad (10)$$

The conditional variance in (9) can be written generally as $V_{\mathbf{X}_{-i}}(E_{X_i}(Y|\mathbf{X}_{-i}))$ [16]. It expresses the total contribution to the variance of Y due to non-X_i i.e. to the $k - 1$ remaining factors, so that $V(Y) - V_{\mathbf{X}_{-i}}(E_{X_i}(Y\,|\mathbf{X}_{-i}))$ includes all terms, i.e. a first-order as well as interactions in (5), that involve factor X_i.

An intuitive computational strategy for the estimation of the conditional variances (6), (7), (9) would be the brute-force computation of the multidimensional integrals in the space of the input factors. To obtain $V_i = V_{X_i}(E_{\mathbf{X}_{-i}}(Y\,|X_i))$ for example one would use a set of Monte Carlo points, say 1,000, to estimate the inner expectation for a fixed value of X_i, and then repeat the procedure, say 1,000 times, for different X_i values to estimate the outer variance. In this example, one would need 10^6 points for a single sensitivity measure. This is in fact not necessary, as the computation can be accelerated via shortcuts. Estimators for both (S_i, S_{Ti}) are provided by a variety of methods reviewed in [4, 36]. We describe in the following an easy-to-code and most efficient Monte Carlo numerical procedure that was proposed by Saltelli [31] and improves the original approach introduced by Sobol' [39].

Step 1. Generate a Monte Carlo sample of the k input factors ($2N \times k$, where $N \approx 100 - 1000$) and split it in two matrices of data, $\mathbf{A}$ and $\mathbf{B}$, of size $N \times k$ (11). The recommended sampling strategy for generating the combinations of input factors values is LP_τ, in which quasi-random sequences of vectors are used to generate points uniformly distributed in a unit cube [40, 41]. Alternatively, pure random numbers can also be used to obtain $\mathbf{A}$ and $\mathbf{B}$.

$$A = \begin{bmatrix} x_1^{(1)} & x_2^{(1)} & \dots & x_i^{(1)} & \dots & x_k^{(1)} \\ x_1^{(2)} & x_2^{(2)} & \dots & x_i^{(2)} & \dots & x_k^{(2)} \\ \dots \\ x_1^{(N-1)} & x_2^{(N-1)} & \dots & x_i^{(N-1)} & \dots & x_k^{(N-1)} \\ x_1^{(N)} & x_2^{(N)} & \dots & x_i^{(N)} & \dots & x_k^{(N)} \end{bmatrix}$$

$$B = \begin{bmatrix} x_1^{(N+1)} & x_2^{(N+1)} & \dots & x_i^{(N+1)} & \dots & x_k^{(N+1)} \\ x_1^{(N+2)} & x_2^{(N+2)} & \dots & x_i^{(N+2)} & \dots & x_k^{(N+2)} \\ \dots \\ x_1^{(2N-1)} & x_2^{(2N-1)} & \dots & x_i^{(2N-1)} & \dots & x_k^{(2N-1)} \\ x_1^{(2N)} & x_2^{(2N)} & \dots & x_i^{(2N)} & \dots & x_k^{(2N)} \end{bmatrix} \tag{11}$$

Step 2. Define a matrix C_i formed by all columns of **B** except the ith column, which is taken from **A**(2)

$$C_i = \begin{bmatrix} x_1^{(N+1)} & x_2^{(N+1)} & \dots & x_i^{(1)} & \dots & x_k^{(N+1)} \\ x_1^{(N+2)} & x_2^{(N+2)} & \dots & x_i^{(2)} & \dots & x_k^{(N+2)} \\ \dots \\ x_1^{(2N-1)} & x_2^{(2N-1)} & \dots & x_i^{(N-1)} & \dots & x_k^{(2N-1)} \\ x_1^{(2N)} & x_2^{(2N)} & \dots & x_i^{(N)} & \dots & x_k^{(2N)} \end{bmatrix} \tag{12}$$

Step 3. Compute the model output for all the input values in the sample matrices **A**, **B** and C_i, obtaining three vectors $(N \times 1)$ of model outputs: $y_A = f(A)$, $y_B = f(B)$, $y_{C_i} = f(C_i)$. The three vectors y_A, y_B, y_{C_i} are all we need to compute the first-and total effect sensitivity measures S_i, S_{Ti}, for factor X_i. The computational cost is $2N$ (for matrices **A**, **B**) plus kN (for matrices C_i). The total cost is hence $N(k+2)$, much lower than the N^2 simulations needed in the brute-force method.

Step 4. The first-order sensitivity measures are estimated as follows:

$$S_i = \frac{V_{X_i}\left[E(Y|X_i)\right]}{V} = \frac{y_A \cdot y_{C_i} - f_0^2}{y_A \cdot y_A - f_0^2} = \frac{\frac{1}{N}\sum_{j=1}^{N} y_A^{(j)} \cdot y_{C_i}^{(j)} - f_0^2}{\frac{1}{N}\sum_{j=1}^{N}\left(y_A^{(j)}\right)^2 - f_0^2} \tag{13}$$

where $f_0^2 = \left(\frac{1}{N}\sum_{j=1}^{N} y_A^{(j)}\right)^2$ is the mean, and the symbol $(\cdot)$ denotes the scalar product of two vectors.

Step 5. Similarly, the total-effect sensitivity measures are estimated as follows:

$$S_{T_i} = 1 - \frac{V_{X_{-i}}\left[E_{X_i}(Y|X_{-i})\right]}{V} = 1 - \frac{y_B \cdot y_{C_i} - f_0^2}{y_A \cdot y_A - f_0^2} = 1 - \frac{\frac{1}{N}\sum_{j=1}^{N} y_B^{(j)} \cdot y_{C_i}^{(j)} - f_0^2}{\frac{1}{N}\sum_{j=1}^{N} \left(y_A^{(j)}\right)^2 - f_0^2} \tag{14}$$

A simple explanation of why these formulas work is offered next. In the scalar product $y_A \cdot y_{C_i}$ values of Y computed from **A** are multiplied by values of Y computed from C_i, for which all factors but X_i are re-sampled while the values of X_i remain fixed. If X_i is non-influential, then high and low values of y_A and y_{C_i} are randomly associated. If X_i is influential, then high (or low) values of y_A will be preferentially multiplied by high (or low) values of y_{C_i}, increasing the value of the scalar product. Similarly for the total effect sensitivity measures. The accuracy of both f_0 and $V(Y)$ can be improved by using both y_A and y_B points rather than just y_A in (13) and (14). Consequently, the accuracy of the estimates for S_i, S_{Ti} will be improved although the factors' ranking will not change.

Before using the first-order and total effect sensitivity measures in our case study, we summarise some of their properties that are useful in interpreting the results.

- The first-order sensitivity measure S_i indicates by how much one could reduce the output variance by fixing X_i. It is used for identifying important factors. However, there exists no established threshold. Saisana et al. [30] empirically consider as important an input factor that explains more than $1/k$ of the output variance (i.e. $S_i > 1/k$).
- The total effect sensitivity measure S_{Ti} is used for identifying non-influential input factors. In this case, low S_{Ti} values are sought among the input factors, as these imply that X_i is non-influential and it can, thereafter, be fixed anywhere in its distribution without affecting the output variance.
- By definition S_{Ti} is greater than S_i, or equal to S_i in case X_i is not interacting with other input factors. The difference $S_{T_i} - S_i$ is a measure of how much X_i is involved in interactions.
- The sum of all S_i's is always lower than (or equal to) 1. It is equal to 1 if the model is perfectly additive. The difference $1 - \sum_{i=1}^{k} S_i$ captures the interactions (non-linearity) present in the model.
- The sum of all S_{Ti}'s is always greater than (or equal to) 1. It is equal to 1 if the model is perfectly additive.

The extended variance-based methods, including the method just described (for independent input factors) are implemented in the freely distributed software SIMLAB [36]. Note than when the uncertain input factors X_i are dependent, the output variance cannot be decomposed as in (5). The

S_i, S_{Ti} are still valid sensitivity measures for X_i, though their interpretation changes as, e.g. S_i carries over also the effects of other factors that can be positively or negatively correlated to X_i (see [33]).

3 Case Study: Technology Achievement Index and Expert Opinion

The TAI is a composite indicator developed by the United Nations and described in the Human Development Report [43]. While acknowledging that many elements make up a country's technological achievement, the UN authors suggest that an overall assessment can be facilitated using a composite indicator. Like other composite indicators in the Human Development Report series (e.g. the more familiar Human Development Index), the TAI is suggested for summary purposes, to be complemented by consideration of the underlying indicators. The design of the index reflects two particular concerns. The first is to compare policy effectiveness across countries, regardless of the level of technological development. The second is to identify, and send message to, developing countries. To accomplish this, the index must be able to discriminate between countries at the lower end of the range. The technology achievement concept, as framed by the UN is captured by eight indicators: the number of *patents* granted per 1,000,000 people, the receipt of *royalties* (in US$, per 1,000 inhabitants), the number of *internet* hosts per 1,000 people, *exports* of high and medium technology products (as a share of total goods exports), the number of *telephone* lines per 1,000 people (in logs), *electricity* consumption per capita (in logged kWh), the mean years of *schooling*, and the gross *enrolment* ratio of tertiary students in science, mathematics and engineering. Statistics for the raw data of the indicators for 72 countries are given in Table 1. The data for *patents*, *royalties* and *internet*, exhibit the highest coefficient of variation. The distributions of data for *patents* and *electricity* are the most skewed (positively).

Correlation analysis reveals that the eight indicators have an average bivariate correlation of 0.55 and that six pairs of indicators have a correlation coefficient greater than 0.70. Covariance is further investigated via factor analysis. In order to account for at least 90% of the variance in the full data set, five Principal Components are needed. This result indicates that the phenomenon described by the eight TAI indicators is multidimensional. A higher correlation between the indicators would have resulted in fewer principal components.

Note that, depending on a school of thought, one may see a high correlation among indicators as something to correct for, e.g. by assigning lower weight to correlated indicators. Practitioners of multi-criteria decision analysis [20] would instead tend to consider that correlations need not be corrected, since correlated indicators may indeed reflect different and non-compensable features of the problem.

Table 1. Statistical properties across 72 countries of the eight indicators that compose the technology achievement index

Indicator	Observed minimum value	Observed maximum value	Mean	Std	Skewness
Patents granted to residents (per million people)	0	994	109	199	3.4
Royalties and license fees received (US$ per 1,000 people)	0.0	272.6	29.3	46.1	1.8
Internet hosts (per 1,000 people)	0.0	232.4	32.3	51.1	2.0
High-and medium-technology *exports* (% of total goods exports)	0.0	80.8	30.1	23.0	0.4
Telephones, mainline and cellular (per 1,000 people)	1	901	443	402	0.6
Electricity consumption (kilowatt-hours per capita)	22	6,969	3,496	4,274	2.5
Mean years of *schooling* (age 15 and above)	0.8	12.0	7.2	2.5	−0.1
Gross tertiary science *enrolment* ratio (%)	0.1	27.4	8.1	6.2	1.0

The list of TAI indicators exhibits a typical feature of most composite indicators, i.e. that the underlying indicators are expressed in various measurement units. The TAI authors deal with this problem by *normalising* the original data on a 0–1 scale (1). Prior to this stage, logarithms of the raw data for *telephone* and *electricity* are taken, as these indicators are considered by the TAI authors as important at the earlier stages of technological advance but not at the most advanced stages. Expressing these indicators in logarithms ensures that as the level increases, the indicator contributes less to the overall score. The normalised indicators are next *averaged.*

We depart from the UN deterministic formulation of the index, and allow both the normalisation and weighting procedures to vary. Henceforth, we indicate the extended composite indicator as MC-TAI (for Monte Carlo), which incorporates the three sources of uncertainty (V) to (VII) listed above.

In our case study, the weights for the eight TAI indicators derived from a pilot survey carried out across 20 informed interviewees at the authors' institute, using the two participatory approaches BA and AHP described in Sect. 2.1. Note that for the sake of the present exercise, the interviewees had no bearing on the issue being measured. In principle, interviewees should be part of a community of stakeholders with a legitimate stake. To make an example, in the Internal Market Index for the European Union, the experts were members of a committee representing the EU member states [42].

Results from the pilot surveys on the weights are reported in Fig. 1. The suggested weights for each indicator elicited by BA are less spread than those

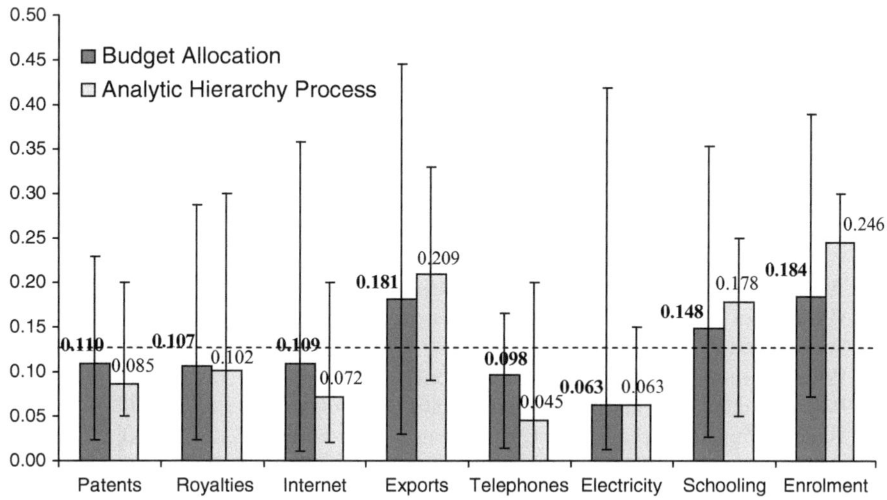

Fig. 1. Statistics on the weights assigned to the eight indicators of MC-TAI. The weights derived from pilot surveys of 20 informed interviewees using budget allocation and analytic hierarchy process. Numbers report mean values, the bars indicate minimum or maximum values. (See Plate 38 on page 324)

by AHP and the variance of the weights across the eight indicators is smaller for BA than for AHP. However, AHP is based on pair-wise comparisons where perception is high enough to distinguish between indicators. In BA all indicators are compared at a glance, and this might lead to circular thinking across indicators, creating difficulties in the assignment of weights, particularly when the number of indicators is high. The different results (per interviewee) between the two participatory approaches reveal the dependence of the human mind on the formulation of a question. Furthermore, there are considerable differences in the weights proposed by the interviewees for the same indicator (e.g., there is not a single pair of individuals who suggest similar sets of weights). Interestingly, unanimity is achieved in judging that the *telephone* and *electricity* indicators are the least important in estimating a country's technological achievement level. Furthermore, although equal weights (=1/8) fall within the upper and lower bounds over the sample of experts, no panel member proposed to weigh all indicators equally, in contrast with the original TAI approach. If a single set of weights were to be used to represent the panel in its entity, then the mean (or median) weight value could be a proper choice. However, much information on the weights would be left unused. To this end, we preferred to incorporate this type of (disagreement) information in the calculation of the composite indicator, and assess its impact on the composite indicator results.

Before doing so, one more remark is in order. Weights in a linear aggregation $\sum I_{q,c} \cdot w_q$ (as in the present study) have the meaning of trade-offs. Hence, what matters in a linear aggregation are the relative weights, which directly

Table 2. The ten uncertain input factors for the analysis

Input factor	Definition	pdf	Range
X_1: Re/St	Normalisation method	Binary	$0 \equiv$ Re-scaled values $1 \equiv$ Standardised values
X_2: BA/AHP	Weighting scheme	Binary	$0 \equiv$ Budget allocation $1 \equiv$ Analytic hierarchy process
X_3: W-patents	Weights' list, patents	Discrete	[1, 2, ..., 20]
X_4: W-royalties	Weights' list, royalties	Discrete	[1, 2, ..., 20]
X_5: W-internet	Weights' list, internet	Discrete	[1, 2, ..., 20]
X_6: W-exports	Weights' list, exports	Discrete	[1, 2, ..., 20]
X_7: W-telephone	Weights' list, telephone	Discrete	[1, 2, ..., 20]
X_8: W-electricity	Weights' list, electricity	Discrete	[1, 2, ..., 20]
X_9: W-schooling	Weights' list, schooling	Discrete	[1, 2, ..., 20]
X_{10}: W-enrolment	Weights' list, enrolment	Discrete	[1, 2, ..., 20]

refer to the substitutability of the different dimensions, instead of the absolute weights. It has, however, been observed (e.g. by Munda and Nardo [21]) that experts usually interpret weights, such as those stemming from a Budget Allocation method, as 'importance coefficients' (cf. [13, p. 10]: "Greater weight should be given to components which are considered to be more significant in the context of the particular composite indicator"). In fact, the panel was literally asked to assign more points to an indicator 'the more important this indicator is'. We will consequently adhere to such an interpretation of the weights derived from BA or AHPP.

The uncertainties we acknowledge in our analysis are captured by ten input factors (Table 2). Input factor X_1 (binary) decides the type of normalisation method to be used, either re-scaling (1) or standardisation (2). Input factor X_2 (binary) guides the selection of the weighting scheme, either BA or AHP. Then factors X_3 to X_{10} represent the reference number of the interviewee, whose assigned weight for a given indicator is taken for the simulation. Factors X_3 to X_{10} are thus sampled from a discrete distribution [?, 1–19], independently of each other. Weights are then scaled to a unity sum. The MC-TAI score is calculated as the weighted average of the normalised indicators. Finally, linear scaling in the range $[0, 100]$ is applied to the composite indicator scores per type of normalisation method, so as to render the results comparable between methods.

Note that in our $k = 10$ dimensional space of uncertainties there are $2 \times 2 \times 20^8 = 1.024 \times 10^{11}$ possible combinations of the input factor values. Given that we cannot afford a full design with so many simulations, we need a representative sampling of the space of uncertainties. For efficiency purposes, we have chosen to use an $LP_{\mathcal{T}}$ (quasi random) sampling scheme [39], with a base sample size $N = 512$. Thereafter, a total of $N(k + 2) = 6,144$ composite scores per country are calculated. These composite scores are non-linear functions of the uncertain input factors, and the estimation of the composite scores pdf is the purpose of our uncertainty analysis.

The uncertainties we assess cover points (V) to (VII) in the uncertainty propagation chain (I to IX) described in Sect. 2.1. The remaining types of uncertainty listed are either non applicable, non relevant, or have been studied elsewhere. There is no information about the measurement error in the data (point [I]), and the UN authors selected one level of aggregation for the indicators in TAI (point [VIII]). The inclusion/exclusion of an indicator (point [II]) is already inherent in our uncertainty, since several interviewees have suggested eliminating some indicators from the dataset (Fig. 1). Finally the transformation of two indicators (e.g. taking logarithms for *telephone* and *electricity*) (point [III]), the imputation of missing data in the dataset (point [IV]) and the choice of the aggregation scheme (e.g. application of multiplicative aggregation or non-compensatory multicriteria analysis) have been studied elsewhere [6, 22].

4 Results

4.1 Uncertainty Analysis on the Countries Composite Scores

The MC-TAI scores (and respective ranks) for the 72 countries were calculated for the 6,144 combinations of normalisation method, weighting scheme and sets of weights. Figure 2 displays the median (black mark) and the corresponding fifth and 95th percentiles of the empirical distribution of the MC-TAI ranks per country. Countries are ordered according to the original TAI ranking. Overlapping whisker plots indicate that the corresponding countries have similar performance. Conversely, if two countries' ranks are not overlapping, the policy inference (i.e. 'one country performs better than another') is robust and independent of the uncertainties. From this analysis one can conclude that:

- For most countries, the original TAI rank is very close to the MC-TAI median rank ($r_{spearman} = .994$, $p < 0.000$) that acknowledges three sources of uncertainty. This implies that the original TAI, although developed using one normalisation method and one set of weights (equal weights), provides a picture of the countries' technological achievement that is not generally biased.
- There is, however, one country, the Netherlands, that is strongly favoured in the original TAI. Originally, the performance of the Netherlands (sixth position) follows the one of the five best countries, including Finland, the United States, Sweden, Japan and Korea. However, when accounting for changes in the normalisation method, the weighting scheme and the set of weights, its performance falls at the ninth position, behind Singapore, United Kingdom, and Australia. Singapore, on the other hand, is ranked sixth in the MC-TAI, instead of tenth in the original TAI.

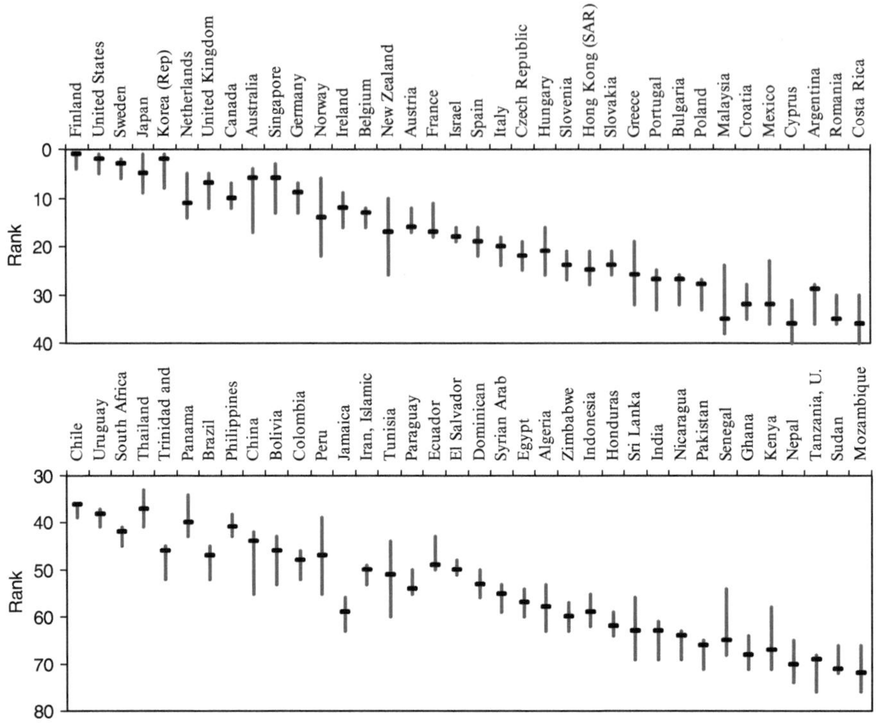

Fig. 2. Results of UA showing the median (black mark) and the corresponding fifth and 95th percentiles of the empirical distribution of the ranks in MC-TAI for 72 countries. Uncertain input factors: normalisation method, weighting scheme, weights for the eight indicators. Countries are ordered according to the original TAI rank

- The United Nations' TAI is not intended to be a measure of leaders in global technology development but to focus on how well a country participates in creating and using technology. Take for example Finland ($\text{TAI}_{\text{original}} = 74.4$) and the United States ($\text{TAI}_{\text{original}} = 73.3$). The United States, a global technology powerhouse, has far more inventions and internet hosts in total than Finland does, while in Finland more is being done to develop a technological skill base throughout the population. When accounting for the uncertainties in the input factors, the message is reinforced, as the distance between the two countries is more evident ($\text{MC-TAI}_{\text{median, Finland}} = 83.1$; $\text{MC-TAI}_{\text{median, USA}} = 78.1$).

The difference between the composite scores of the two countries discussed before, the Netherlands and Singapore (sixth and tenth position in the original TAI, ninth and sixth position in the MC-TAI, respectively) is further studied and depicted in Fig. 3. The left-hand region, where Singapore performs better than the Netherlands, covers about 65% of the total area and this shows

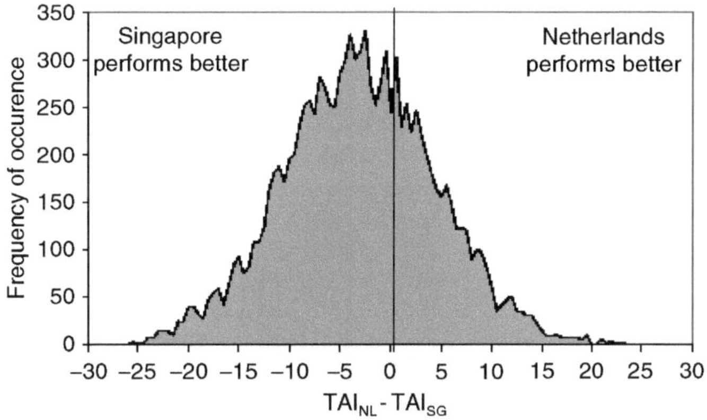

Fig. 3. Result of UA for the output variable $D_{NL,SG}$, difference in the composite scores between the Netherlands and Singapore. Uncertain input factors: normalisation method, weighting scheme, weights for the eight TAI indicators

that Singapore participates in creating and using technology more than the Netherlands does, contrarily to the message conveyed by the original TAI.

We next complement our uncertainty analysis with a sensitivity analysis that employs the variance-based technique described in Sect. 2.2 to apportion the variance in the countries' scores (or ranks) to the ten input factors.

4.2 Sensitivity Analysis Using First-Order and Total Effect Sensitivity Measures

The previous analysis motivates a quantitative estimation of the impact of the ten uncertain input factors on two outputs of interest:

(a) the difference $D_{a,b}$ (3) in the composite scores between two countries. The Netherlands (NL) and Singapore (SG) are selected due to the discussion above.
(b) the average shift in countries' rank, $\bar{R}_S$ (4), taking the original TAI ranking as reference.

The pair of first-order and total effect sensitivity measures (S_i, S_{Ti}), described previously in Sect. 2.2, have been computed using the same set of 6,144 simulations for both outputs of interest.

Output 1: $D_{NL,SG}$, Difference in performance between the Netherlands and Singapore

The sensitivity measures S_i and S_{Ti} for $D_{NL,SG}$ are given in Fig. 4. As summarised in Sect. 2.2, high S_i values indicate important input factors. The most important input factors are the weight for *electricity* ($S_i = .17$), the weight for *enrolment* ($S_i = .12$) and the weighting scheme ($S_i = .14$). All ten input factors, taken singularly, explain 52% of the variance of $D_{NL,SG}$. The remaining 48% is attributed to the interactions among the factors. The

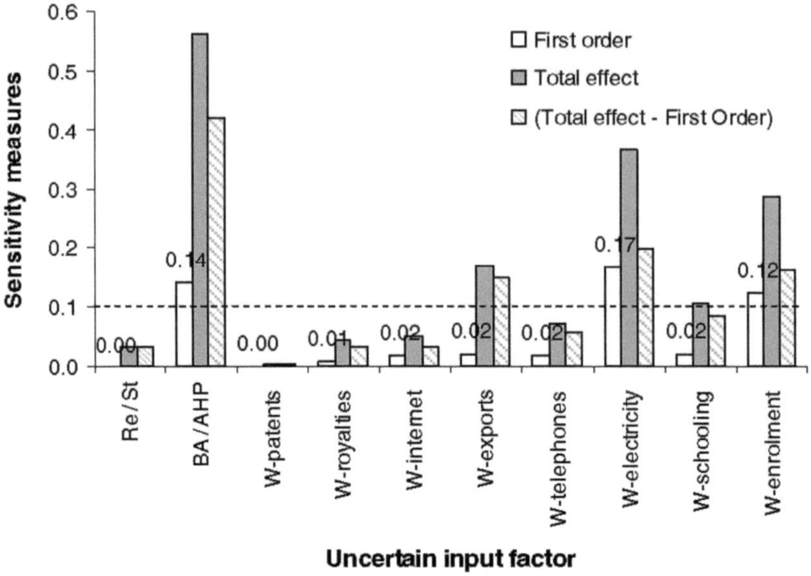

Fig. 4. First-order S_i and total effect S_{Ti} sensitivity measures for the difference between the composite scores of the Netherlands and Singapore $D_{NL,SG}$ for ten ($k = 10$) input factors. Dotted line (at $1/k$) indicates a rule-of-thumb threshold for important factors based on S_i, $S_{Ti} - S_i$

strongest interactions involve the trigger for the weighting scheme ($S_{Ti} - S_i = .42$), mainly with the weight for *electricity*, the weight for *enrolment* and the weight for *exports*. Of the ten input factors, six are non-influential (low S_{Ti} values) to the variance of $D_{NL,SG}$, namely the normalisation method, the weight for *patents*, for *royalties*, for *internet*, for *telephones* and for *schooling*. Interestingly, the weight of *exports* contributes to the output variance more through interactions than as a single factor ($S_{Ti} = .17$). Thus, despite its low S_i value, the weight for exports cannot be fixed anywhere in its possible range of values as this would have an impact on the variance of $D_{NL,SG}$.

The information provided by sensitivity analysis on the importance of the three factors, weighting scheme, weight of *electricity*, weight of *enrolment* is further investigated. By plotting the $D_{NL,SG}$ values versus the 'weighting scheme' and the 'weight of electricity' (Fig. 5a) or the $D_{NL,SG}$ values versus the 'weighting scheme' and the 'weight of enrolment' (Fig. 5b) it is shown that with respect to the original TAI, the MC-TAI favours Singapore when a high weight is assigned to *enrolment* (where Singapore outperforms the Netherlands) or to *electricity* (where Singapore is marginally better than the Netherlands). The relative performance between the two countries displayed on the 'weight of enrolment – weight of electricity' plane (not shown here) confirms that Singapore is better than the Netherlands mostly when the weight for *enrolment* is high, while the weight for *electricity* can only make a difference when extremely high values are assigned to it during the AHP survey.

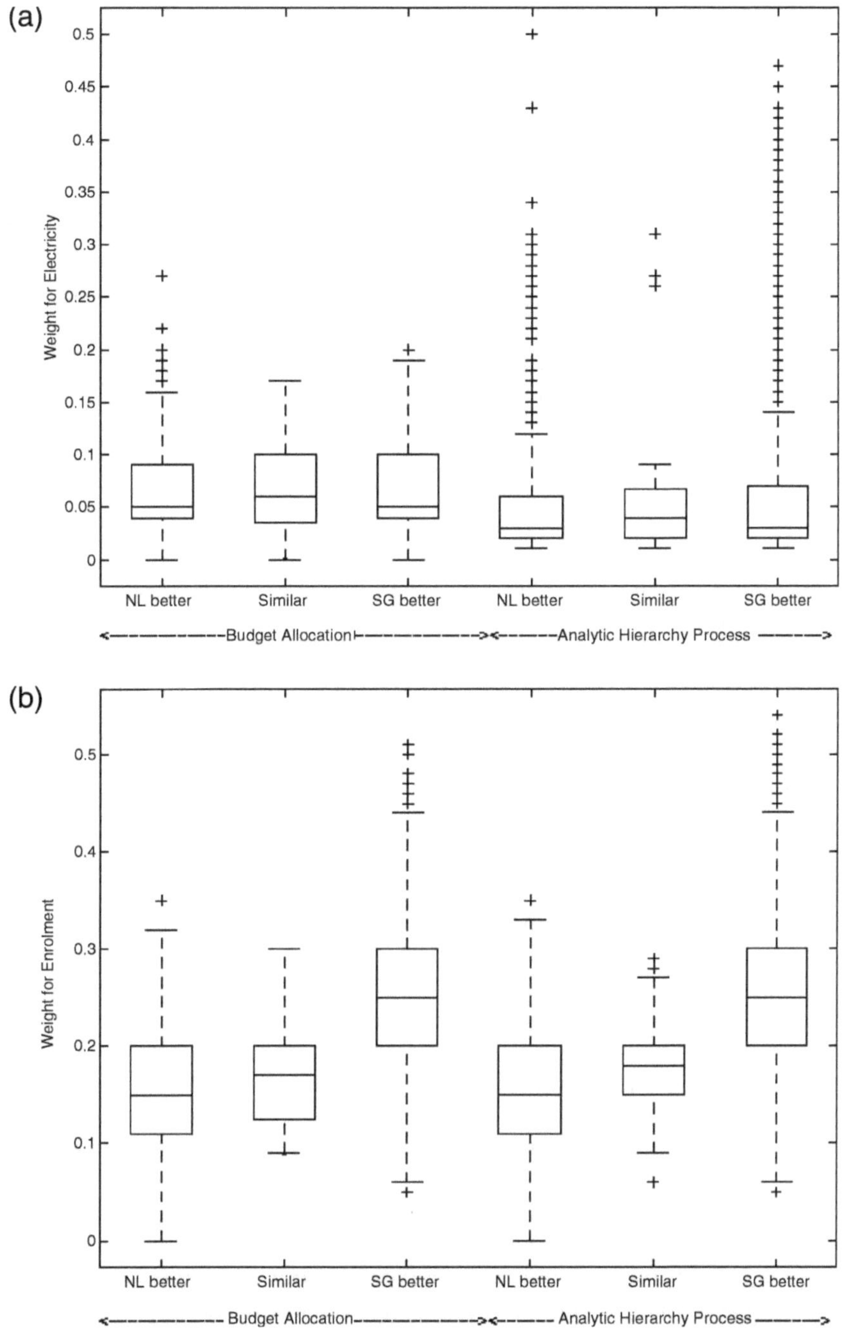

Fig. 5. Boxplots of 6,144 values for the (**a**) weight for electricity and (**b**) weight of enrolment with respect to the two weighting schemes (Budget Allocation, Analytic Hierarchy Process) and the relative performance between the Netherlands (NL) and Singapore (SG). '+' signs indicate outliers

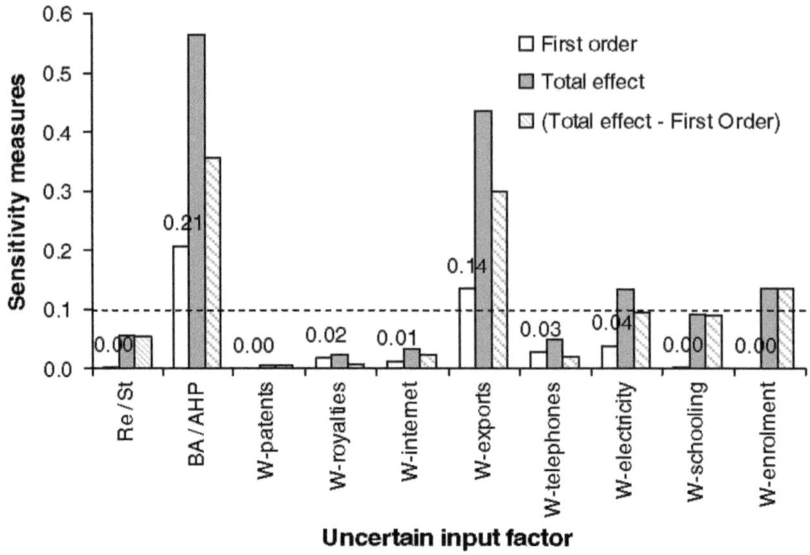

Fig. 6. First-order and total effect sensitivity measures for the output $\overline{R}_S$ average shift in countries' rank with respect to the original TAI for ten ($k = 10$) input factors. Dotted line (at $1/k$) indicates a rule-of-thumb threshold for important factors

Output 2: $\bar{R}_S$, Average shift in countries' rank with respect to the original TAI

The sensitivity measures for our second output of interest, $\bar{R}_S$, are shown in Fig. 6. The weighting scheme and the weight of *exports* are the most important factors, explaining 35% of the variance in $\bar{R}_S$, while all ten input factors taken singularly account for 44% of the variance. The high influence of the weight of *exports* on the variance of $\bar{R}_S$ is due to countries, such as Australia, Singapore, Norway, New Zealand, Greece, Mexico, Chile and Malaysia, for which *exports* values are either lower than the 20th or higher than the 92nd percentiles. The remaining 56% of the variance in $\bar{R}_S$ is due to interactions among input factors. The highest interaction involves the weighting scheme and the weight of *exports*, similarly as for the $D_{NL,SG}$ output. Note that the weight of *exports* contributes to the output variance more through interactions than singularly. Again, the normalisation method has no influence on the $\bar{R}_S$ variance.

This analysis quantified the impact of the input factors on the average shift in countries' rank with respect to the original TAI and has indicated that it is the weighting scheme interacting mainly with the weight of *exports* that can influence country ranking in the MC-TAI with respect to the original TAI. This conclusion could guide the developers of the composite indicator to an optimal resource allocation in terms of time and money spent to converge (e.g. using an iterative process such as the Delphi method) on the weights for just few indicators that are crucial to the policy message conveyed by the countries composite scores or ranks.

The main drawback of variance-based methods is the cost of the analysis, which, in the case of computationally demanding models, can become prohibitive even when using the approach described above. However, this was not the case in the present study since a few thousands of simulations require trivial computational time. The majority, if not all, of composite indicators found in the literature (see list), require negligible computational time and therefore variance-based methods are recommended for composite indicators' case studies.

5 Conclusions and Future Work

Media and policy-makers look with increasing interest at composite indicators as appealing tools to attract the attention of the community, build narratives and help focusing policy debates. Methodological gaps or fragilities in the design of composite indicators may invite politicians to draw simplistic conclusions or the press to communicate misleading information. That is why national and international organisations need to focus on methodological issues before disseminating the results of a composite indicator.

This work has shown how expert panel information for the selection of weights can be plugged into an uncertainty analysis framework and has analysed its impact on the composite indicator's policy message using a variance-based sensitivity analysis method. The focus of our sensitivity analysis on the United Nations' TAI was on quantifying the extent to which countries scores and ranks are influenced by expert opinion for the importance of the underlying indicators (and other 'modelling options'). This is not a coincidence, as this source of uncertainty often drives most of the debate on the composite indicator's ultimate intrinsic value. The analysis was useful in showing that:

- The original version of the TAI provides a picture of the countries relative performance that is similar to the one where several types of uncertainty are considered.
- There are cases where countries ranks are affected by the modelling choices. Such cases were further studied with a view to identify the reason and the regions in the space of the weights that favour one country with respect to another.
- The message of what the TAI aims at measuring (i.e. a country's participation in creating and using technology and not which country is leading in global technology development) can be better communicated to the public when the uncertainties in the input factors are taken into consideration.
- The weighting approach (through interaction with a few weights for the indicators) affects the countries' ranks (useful to focus efforts towards reducing the uncertainty bounds for the MC-TAI), whilst the normalisation method has no influence on the ranks.

As a general recommendation, we would propose that uncertainty and sensitivity analysis are not employed after the composite indicator has been developed, but instead during the construction phase, since the results of these analyses could contribute to the well-structuring of the composite indicator, provide information on whether the countries' ranks are meaningful and could reduce the possibility that the composite indicator may send misleading or biased policy messages.

The verification offered in the present work is nevertheless partial. We did not propagate the uncertainty in the underlying indicators, as error in the data were not available. Furthermore we have implicitly assumed that the debate is entirely captured by the variability in the weights, be it with the budget allocation or analytic hierarchy process alternatives and the normalisation method. While this latter assumption is not far fetched, and one sees in practice examples where this happens, e.g. EU internal market scoreboard, EU e-business readiness index (for a review, see [29]), one might have as well situations where the very concept of composite indicator is rejected by some of the stakeholders, or where the model underlying the weighing or the aggregation is called into question. To make an example, Munda and Nardo [21] have argued that even if weights are customarily assigned as measure of relative importance when using liner aggregation, they have in fact a meaning as substitution rate, whereby e.g. an equal weight for two indicators would mean that we are willing to trade one unit down in one indicator for one unit up in another. Even if we have not propagated these other categories of uncertainty (data uncertainty, aggregation method) in our example, it should be clear to the reader that this can be done in principle without difficulties, following a similar approach such as the one presented herein.

A final remark for future work should be made. Having propagated several uncertainties through the development process of a composite indicator, the overall score is no longer a deterministic value but an empirical distribution. As a result, the composite indicator might seem to lose relevance if a high fraction of countries overlap. In fact, if the purpose of an index in a given policy context were to be of a 'name and shame' nature (e.g. to flog laggard countries into action), then again including layers of uncertainty may be seen as counter productive. A trade-off exists between the level of uncertainty included in the composite indicator and its worthiness, which is herein considered as the capacity of the same index to effectively discriminate among countries. Yet, such a trade-off can only be established by comparing results for several composite indicators for which the same sources of uncertainty have been considered and their impact on the countries ranks assessed. Future work will focus along this line of research.

References

1. Bandura R. (2005) Measuring country performance and state behavior: a survey of composite indices, Background paper prepared for "The new public finance: responding to global challenges", United Nations Development Programme, http://www.thenewpublicfinance.org
2. Brand D. A., M. Saisana, L. A. Rynn, F. Pennoni, A. B. Lowenfels (2007) Comparative analysis of alcohol control policies in 30 countries, *PLoS Medicine*, 4, 752–759, www.plosmedicine.org.
3. Canadian Council on Learning (2007) The 2007 Composite Learning Index: Helping Communities Improve their Quality of Life, Ottawa, pp. 40, www.ccl-cca.ca.
4. Chan K., S. Tarantola, A. Saltelli, I. M. Sobol' (2000) Variance based methods. In *Sensitivity Analysis* (eds. A. Saltelli, K. Chan, M. Scott), pp. 167–197. New York: Wiley.
5. Cherchye L, K. Lovell, W. Moesen, T. Van Puyenbroeck (2007) One market, one number? A composite indicator assessment of EU internal market dynamics, *European Economic Review*, 51, 749–779.
6. Cherchye L., W. Moesen, N. Rogge, T. Van Puyenbroeck, M. Saisana, A. Saltelli, R. Liska, S. Tarantola (2007) Creating composite indicators with data envelopment analysis and robustness analysis: the case of the technology achievement index, *Journal of Operational Research Society*, online 27/06/07, doi: 10.1057/palgrave.jors.2602445
7. Cox D., R. Fitzpatrick, A. Fletcher, S. Gore, D. Spiegelhalter, D. Jones (1992) Quality-of-life assessment: can we keep it simple? *J.R. Statist. Soc.* 155(3), 353–393.
8. EC-JRC (2007) Information server on composite indicators, http://farmweb.jrc.cec.eu.int/ci/ provided by the Joint Research Centre of the European Commission.
9. EPA (2001) Draft guidance on the development, evaluation, and application of regulatory environmental models. Council for regulatory environmental modeling (CREM), http://cfpub.epa.gov/crem/cremlib.cfm whitepapers.
10. Esty D. C., M. A. Levy, T. Srebotnjak, A. de Sherbinin, C. H. Kim, B. Anderson (2006) *Pilot 2006 Environmental Performance Index.* New Haven: Yale Center for Environmental Law & Policy.
11. Esty D. C., M. Levy, T. Srebotnjak, A. de Sherbinin (2005) *2005 Environmental Sustainability Index: Benchmarking National Environmental Stewardship.* New Haven: Yale Center for Environmental Law & Policy.
12. Freudenberg M. (2003) Composite indicators of country performance: a critical assessment. Report DSTI/IND(2003)5, OECD, Paris.
13. Funtowicz, S. O., J. R. Ravetz (1990) *Uncertainty and Quality in Science for Policy.* Dordrecht, NL: Kluwer Academic Publishers.
14. Gough C., N. Castells, S. Funtowicz (1998) Integrated Assessment: an emerging methodology for complex issues, *Journal Environmental Modeling and Assessment*, 3, 19–29.
15. Höeg P. (1995) *Borderliners.* Seal Books Publisher.
16. Homma T., A. Saltelli (1996) Importance measures in global sensitivity analysis of model output. *Reliability Engineering and System Safety*, 52(1), 1–17.
17. Kennedy P. (2007) *A Guide to Econometrics*, Fifth edition. Blackwell Publishing.

18. Leamer E. (1978) *Specification Searches: Ad hoc Inferences with Nonexperimental Data.* New York: Wiley.
19. Moldan B., S. Billharz, R. Matravers (1997) *Sustainability Indicators: Report of the Project on Indicators of Sustainable Development.* SCOPE 58. Chichester and New York: Wiley.
20. Munda G. (2004) Social multi-criteria evaluation (SMCE): methodological foundations and operational consequences, *European Journal of Operational Research,* 158/3, 662–677.
21. Munda G., M. Nardo (2005) Non-compensatory composite indicators for ranking Countries: A defensible setting, European Commission, Joint Research Centre, IPSC, Italy, EUR 21833 EN.
22. Nardo M., M. Saisana, A. Saltelli, S. Tarantola, A. Hoffman, E. Giovannini (2005) Handbook on constructing composite indicators: methodology and users guide, OECD-JRC joint publication, OECD Statistics Working Paper, STD/DOC(2005)3, JT00188147, pp. 108.
23. Nardo M., M. Saisana, A. Saltelli, S. Tarantola (2005) *Tools for Composite Indicators Building, European Commission*, Joint Research Centre, IPSC, Italy, EUR 21682 EN, pp. 131.
24. Oreskes N., K. Shrader-Frechette, K. Belitz (1994) Verification, validation, and confirmation of numerical models in the earth sciences. *Science* 263, 641–646.
25. Rosen, R. (1991) *Life Itself*, Columbia University Press, New York.
26. Saaty T. L. (1980) *The Analytic Hierarchy Process*, New York: McGraw-Hill.
27. Saaty R. W. (1987) The analytic hierarchy process – what it is and how it is used. *Mathematical Modelling*, 9, 161–176.
28. Saisana M. (2007) Robustness issues & Critical Assessment of the 2007 Composite Learning Index, Working paper, at http://composite-indicators.jrc.ec.europa.eu/Document/CLI2007%20Report%20on%20Validation_v1.0.pdf
29. Saisana M., S. Tarantola (2002) State-of-the-art report on current methodologies and practices for composite indicator development, European Commission, Joint Research Centre, IPSC, Italy, EUR 20408 EN.
30. Saisana M., A. Saltelli, S. Tarantola (2005) Uncertainty and sensitivity analysis techniques as tools for the analysis and validation of composite indicators, *Journal of the Royal Statistical Society A*, 168(2), 307–323.
31. Saltelli A. (2002) Making best use of model valuations to compute sensitivity indices. *Computer Physics Communications*, 145, 280–297.
32. Saltelli A. (2006) Composite indicators between analysis and advocacy. *Social Indicators Research*, doi 10.1007/s11205-006-0024-9.
33. Saltelli A., S. Tarantola (2002) On the relative importance of input factors in mathematical models: safety assessment for nuclear waste disposal, *Journal of American Statistical Association*, 97(459), 702–709.
34. Saltelli A., K. Chan, M. Scott (2000) *Sensitivity Analysis*, Probability and Statistics series, New York: Wiley.
35. Saltelli A., S. Tarantola, F. Campolongo (2000) Sensitivity analysis as an ingredient of modelling. *Statistical Science*, 15, 377–395.
36. Saltelli A., S. Tarantola, F. Campolongo, M. Ratto (2004) *Sensitivity Analysis in Practice, a Guide to Assessing Scientific Models.* New York: Wiley. SIMLAB: software for sensitivity analysis at http://www.jrc.cec.eu.int/uasa/prj-sa-soft.asp.

37. Saltelli A., M. Ratto, F. Campolongo, J. Carboni, D. Gabelli, M. Saisana, S. Tarantola, T. Andres (2007) *Global Sensitivity Analysis: Gauging the Worth of Scientific Models*, New York: Wiley.
38. Sharpe A. (2004) *Literature Review of Frameworks for Macro-indicators*, Centre for the Study of Living Standards, Ottawa, CAN.
39. Sobol' I. M. (1967) On the distribution of points in a cube and the approximate evaluation of integrals. *USSR Computational Mathematics and Physics*, 7, 86–112.
40. Sobol' I. M. (1976) Uniformly distributed sequences with an additional uniform property. Zh. v&ymacr;chisl. Mat. mat. Fiz. 16, 1332–1337. English translation: U.S.S.R. Comput. Maths. Math. Phys. 16, 236–242.
41. Sobol' I. M. (1993) Sensitivity analysis for non-linear mathematical models. *Mathematical Modelling & Computational Experiment* 1, 407–414.
42. Tarantola S., M. Saisana, A. Saltelli, F. Schmiedel, N. Leapman (2002) Statistical techniques and participatory approaches for the composition of the European Internal Market Index 1992–2001, European Commission, Joint Research Centre, IPSC, Italy, EUR 20547 EN.
43. United Nations (2001) *Human Development Report*, United Kingdom: Oxford University Press.

A Practical Global Sensitivity Analysis Methodology for Multi-Physics Applications

C. Tong and F. Graziani

Summary. This paper describes a global sensitivity analysis methodology for general multi-physics applications that are characterized by strong nonlinearities and interactions in their input–output relationships, expensive simulation runs, and large number of input parameters. We present a four-step approach consisting of (1) prescription of credible input ranges, (2) parameter screening, (3) construction of response surfaces, and (4) quantitative sensitivity analysis on the reduced set of parameters. Details of each step will be given using simple examples. Numerical results on real applications are available in another paper. Motivated by our experience with some large-scale multi-physics applications, we also propose enhancements to the various steps in the methodology for improving its robustness. The essential computational techniques targeted for this methodology have been implemented in a software package called PSUADE.

1 Introduction

Sensitivity analysis (SA) studies the impact of changes in various model parameters on the model outputs. It is commonly used to examine model behavior for various purposes such as model exploration, model improvement and model validation. The general procedure is to define a model output that represents an important aspect of the model behavior. The values of various model inputs are then varied and the resultant changes in the output variable are analyzed.

Local sensitivity analysis computes or approximates partial derivatives of the outputs with respect to individual input variables at some nominal coordinate. Since the local sensitivity measure may vary with the location of the nominal coordinate, the nominal coordinate should be judiciously chosen consistent with the goal of the sensitivity analysis.

Global sensitivity analysis, on the other hand, studies the output variability when the inputs are varied in their entire allowable ranges. Saltelli et al. [2, 13, 14] have defined global methods by two properties:

1. The inclusion of influence of scales and shapes of the probability density functions for all inputs
2. The sensitivity estimates of individual inputs are evaluated while varying all other inputs (multi-dimensional averaging)

In this paper we consider global sensitivity analysis on simulation models arising from multi-physics applications characterized by some or all of the following features:

- The phenomena (outputs) may be highly nonlinear (with respect to the inputs)
- The simulation model may be expensive to evaluate
- There are many (tens to a hundred) uncertain (input) variables or factors
- The uncertain variables may be correlated

To enable feasible global sensitivity analysis for this type of applications, we use a four-step approach. We will give an overview of this approach in Sect. 2. Sections 3–5 provide the background for the second and fourth steps of the methodology. Section 6 suggests validation strategies for several steps of the methodology. Section 7 describes an implementation framework to facilitate the use of this methodology. Finally, a brief summary will be given in Sect. 8.

2 A Global Sensitivity Analysis Methodology

The sensitivity analysis methodology described in this report is sampling-based ("non-intrusive") in which the model is executed repeatedly for combinations of values sampled from the distribution (assumed known) of the input variables. The essence of this methodology is to use a two-stage (the two stages correspond to step (2) and (3) of the overall methodology) divide-and-conquer approach to deal with high dimensionality (large number of input parameters or factors). This two-stage approach has been advocated in several other applications [13, 16]. We have enhanced this methodology for use in our applications by adding rigor to Step 1 of the overall methodology (namely, prescribing credible input ranges), incorporating an improved screening method, and providing an efficient computational technique for two-way (two-input) interaction studies.

Prior to performing sensitivity analysis, much diligence is required to compile a detailed specification of the model to be studied, since the results concluded from the analysis are valid only with respect to the given specification. The specification includes the simulation model, input parameters that are to be varied (for example, material strength parameters); parameters which will be fixed (for example, the grid resolution or algorithmic parameters such as convergence tolerances) in the current study but which may affect the outcome of the analysis if they are allowed to vary; the simulation output responses; and any other assumptions about the model.

After a detailed specification of the model has been compiled, we are now ready to apply the following four-step methodology:

1. Construct a complete description of the input parameters (that is, the ranges and forms of the distributions) from
   a. Experimental data (from resident physicists)
   b. Literature search (data from published papers elsewhere)
   c. Expert judgment (resident physicists and code developers) and/or
   d. Results of applying the 4-Step methodology recursively to individual physics modules
2. If the number of uncertain input parameters is relatively small (say, less than 10), skip to Step (4). Otherwise, perform a down-select screening analysis on all uncertain parameters. There are several alternatives as to which screening method is best. In the absence of sufficient knowledge about the functional relationship between the input and output variables, we recommend the model-independent Morris-one-at-a-time (MOAT) design [7]: given the ranges and forms of the input distributions
   a. Generate an MOAT design matrix
   b. Perform the simulation experiments (which consists of preprocessing, running multiple simulations, and post-processing)
   c. Compute the MOAT measures and create screening and scatter plots
   d. Analyze the result and identify a subset of input parameters for Step (3)
3. If the simulation is computationally intensive and the output response is a "relatively" smooth function of the uncertain inputs, we should consider using response surfaces (other names are: surrogate functions and emulators) to construct approximate models. For the response surface approach to be effective, a space-filling sampling design is desired. Some useful designs for this step are: quasi-Monte Carlo design, minimax Latin hypercube design, and many more. The sample points as well as the corresponding outputs are then fed into some function approximators such as polynomial regressors, multivariate splines [3], neural network, or

support vector machines. Future sample points can be evaluated cheaply by launching the function approximators.

4. Perform a quantitative sensitivity analysis via variance decomposition techniques
   a. Direct variance decomposition can use sampling design such as the replicated Latin hypercube design for first order sensitivity indices, the replicated orthogonal array for second order sensitivity indices, and the extended Fourier amplitude sampling test (FAST) or direct integration methods for total sensitivity indices
   b. Perform simulation experiments on the selected design
   c. Analyze the results using statistical and graphical analysis techniques such as sensitivity indices, scatter plots and main effect plots

The most important yet challenging task in this methodology is the first step. It is important because proper prescription of the ranges and shapes of the input distributions can dramatically alter the outcome of the analysis. It is challenging because oftentimes these ranges are obtained by carefully analyzing data from physical experiments. For large scale multi-physics applications, physical experiments for the entire multi-physics processes may not be feasible, but smaller physical experiments on individual physics may be viable (For example, by observing the actual bubble growth rates of the Rayleigh–Taylor experiments, one can prescribe realistic ranges for some hydrodynamics parameters). For this reason, we emphasize the benefit of applying this methodology recursively for multi-physics applications (meaning applying this same methodology to individual physics for the purpose of determining credible ranges for some of the input parameters). Validating simulation data against experimental data for individual physics may provide an indispensable tool for making the result of Step (1) more defensible.

For multi-physics applications which are plagued by complex interactions and nonlinearities, it is important that the techniques in Step (2) and (4) above are model-independent, that is, they do not make too many assumptions about the model behaviors. In addition, the techniques should include rigorous analysis of parameter interactions and correlations.

Finally, we would like to to emphasize on the iterative nature of this methodology. After important parameters are identified, their effects can be reduced by tightening the parameter ranges through better understanding of the process. Once accomplished, other input parameters may emerge as important in the next iteration. Also, the iteration will safeguard against the Type II errors (when important parameters are treated as unimportant) in the previous iteration. Furthermore, for very large number of uncertain parameters (say several hundreds), the number of parameters identified as important in the screening analysis may still be too large for the subsequent quantitative step. In this case, additional divide-and-conquer steps coupled with the iterative loop will make the task more manageable.

3 Parameter Screening

There are a number of approaches to perform sensitivity analysis when the number of uncertain parameters is large. One approach is to perform quantitative sensitivity analysis on all input parameters simultaneously using Bayesian techniques. Another approach is to put all the uncertain parameters into a small number of groups and then perform group sensitivity analysis. Still another approach, which is the one explored in this report, uses a two-phase divide-and-conquer strategy [13, 16]. The first phase, to be discussed in this section, consists of employing a low-cost ($O(M)$ simulation runs where M is the number of uncertain input parameters) coarse sampling and the corresponding analysis method to identify a subset of parameters which are candidates for more detailed analysis in Step (3) and (4). Screening methods generally perform well when the number of important inputs is small compared to the total number of inputs under investigation.

For the choice of screening methods, model independence is an important consideration. If it is known that the output variable behaves approximately linearly or monotonically with all the inputs and there is little parameter interactions, we can use a relatively inexpensive screening design such as the Plackett–Burman design [10], which requires only $M + 1$ simulation runs (again, M is the number of inputs). However, if no such knowledge is available, the Morris one-at-a-time (MOAT) method [7] is recommended. The MOAT method has been demonstrated to be effective in capturing parameters that have direct impact on the output of interest throughout the parameter space (that is, when local features are not dominant). Screening methods such as MOAT can also be used to provide an independent verification of prior knowledge about the model. For example, if prior experience tells us that there is no interaction between parameter X and Y, screening provides another venue to confirm this.

3.1 The MOAT Designs

The Morris method is called one-at-a-time (OAT) method because the design varies one input at a time. To generate the Morris design, a base sample point $X^{(1)}$ is created such that each of the M (M is the number of input parameters) components of $X^{(1)}$ is randomly selected from the set $\{0, 1/(p-1), 2/(p-1), \ldots, 1\}$, where p is a pre-selected integer (for example, $p = 4$). The second sample point $X^{(2)}$ is created from $X^{(1)}$ by perturbing one of its inputs by Δ which is a pre-selected multiple of $1/(p-1)$ (an even p and $\Delta = p/(2p-2)$ are recommended to preserve equal probabilities). Subsequent sample points $X^{(i)}, i = 3, \ldots, M+1$ are created in a similar way, namely from $X^{(i-1)}$ by perturbing one of its inputs which has not been perturbed before. After perturbing all inputs, we have $M + 1$ sample points. A necessary condition to satisfy in generating the sample points is that all $X^{(i)}$ lie inside the design space (for example, $[0, 1]^M$).

This process is captured in the following mathematical form:

$$B^* = (J_{M+1,1}X^{(1)} + (\Delta/2)[(2B - J_{M+1,M})D + J_{M+1,M}])P \quad (1)$$

where B^* is an $(M+1) \times M$ normalized design matrix (each row is one normalized sample point), $J_{m,n}$ is an $m \times n$ matrix of all 1's, D is an $M \times M$ diagonal matrix in which each diagonal element is either $+1$ or -1 with equal probability, $X^{(1)}$ is a row vector having the base sample, P is an $M \times M$ permutation matrix, and

$$B = \begin{bmatrix} 0 & 0 & 0 & \cdots & 0 \\ 1 & 0 & 0 & \cdots & 0 \\ 1 & 1 & 0 & \cdots & 0 \\ 1 & 1 & 1 & \cdots & 0 \\ \cdots & \cdots & \cdots & \cdots & \cdots \\ 1 & 1 & 1 & 1 & 1 \end{bmatrix}.$$

To insure this design creation process is a random selection from the distribution of elementary effect (to be defined later), the process uses three randomizations: (1) the base sample point is randomly selected, (2) the direction of the perturbation is random (that is, the creation of D), and (3) the choice of which input to perturb next is a randomized process (reflected by P). After B^* has been generated, this normalized sample has to be mapped onto the actual parameter ranges and distributions before running them through the simulations. This procedure is repeated $R-1$ times (for example, $R = 10$) to ensure enough regions in the design space has been explored.

3.2 The MOAT Analysis

After these sample points have been evaluated, the elementary effects for the inputs can then be calculated by:

$$d_{c(k)} = \frac{f(X^k) - f(X^{k-1})}{\Delta}, k = 2, \ldots, M+1 \quad (2)$$

where $d_{c(k)}$ is defined as the elementary effect for input $c(k)$ (c maps k to its true input number). Note that since we randomly select which input to perturb next, the elementary effect computed using the kth and $(k+1)$th points in general is not the elementary effect for the kth input which explains the $c(k)$ mapping.

Now that we have R elementary effects for each input (let us label them d_i^r for input i and replication r.) Morris proposed two sensitivity measures to analyze the data: μ which estimates the overall effect of each input on the output, and σ which estimates the higher order effects such as nonlinearity and interactions between inputs. The formulas for them are:

$$\mu_i = \frac{1}{R}\sum_{r=1}^{R} d_i^r, \quad \text{and} \quad \sigma_i = \sqrt{\frac{\sum_{r=1}^{R}(d_i^r - \mu_i)^2}{R}}, \quad (3)$$

respectively. Campolongo et al. [13] proposed an improved measure μ^* in place of μ where

$$\mu_i^* = \frac{1}{R}\sum_{r=1}^{R} |d_i^r|. \tag{4}$$

If μ_i^* is substantially different from zero, it indicates that input i has an important "overall" influence on the output. A large σ_i implies that input i has a nonlinear effect on the output, or there are interactions between input i and the other inputs.

The MOAT measures can also be shown graphically by screening plots, which have the x- and y-axes the modified means and standard deviations, respectively. Each input is represented on the screening plot by a point with coordinate (μ_i^*, σ_i). The plot can then be divided into four quadrants with the two quadrants on the right containing sensitive inputs. Expert judgment should be used when defining the four quadrants through intense discussion between physicists and analysts.

An example screening plot is given in Fig. 1. This example is taken from Morris' original paper:

$$y = \beta_0 + \sum_{i=1}^{20} \beta_i w_i + \sum_{i<j}^{20} \beta_{ij} w_i w_j + \sum_{i<j<k}^{20} \beta_{ijk} w_i w_j w_k + \sum_{i<j<k<l}^{20} \beta_{ijkl} w_i w_j w_k w_l$$

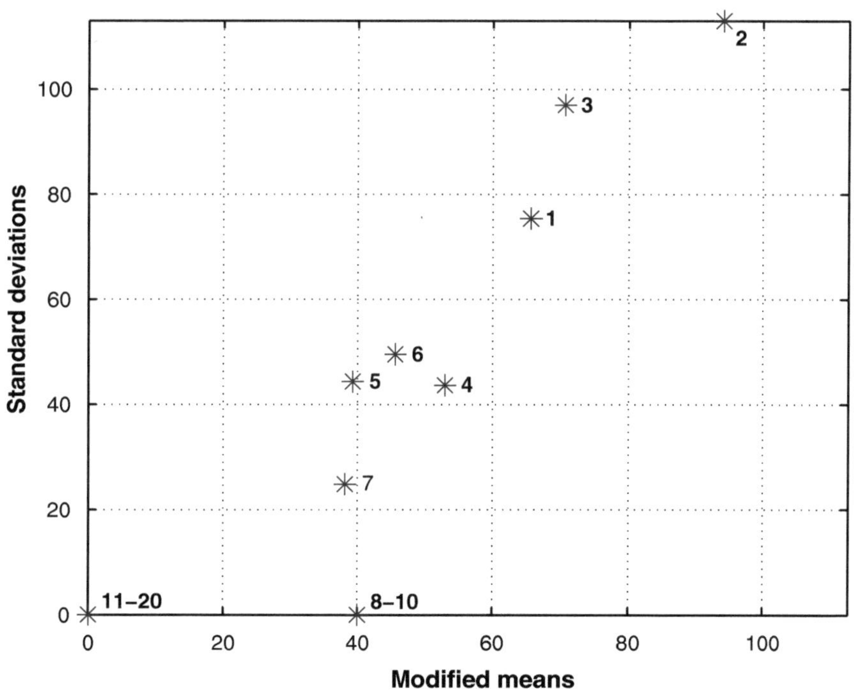

Fig. 1. An example screening plot. (See Plate 39 on page 325)

where $w_i = 2(X_i - 0.5)$ except for $i = 3, 5, 7$, where $w_i = 2(1.1X_i/(X_i + 0.1) - 0.5)$. The other coefficients are assigned as:

$$\begin{aligned} \beta_i &= 20, & i &= 1, \ldots, 10, \\ \beta_{ij} &= -15, & i, j &= 1, \ldots, 6, \\ \beta_{ijk} &= -10, & i, j, k &= 1, \ldots, 5, \\ \beta_{ijkl} &= 5, & i, j, k, l &= 1, \ldots, 4, \end{aligned}$$

and the remainder of first and second order coefficients were generated independently from a normal distribution with zero mean and unit standard deviation, and the remainder of third and fourth order coefficients were set to 0. Here we observe that parameter 1 through 10 are on the right half plane, demonstrating their importance. Parameter 1 through 7, in addition, are located on the upper right quadrant, demonstrating strong nonlinearities and/or interactions.

3.3 Useful Enhancement to Screening Studies

Through our experience with working with a large-scale multi-physics code, we have accumulated a few suggestions for making this screening process more informative and productive. We discuss some of them in the following.

More Graphical Analyses

In addition to the screening plots, scatter plots are also very helpful. The scatter plot for the Morris example is given in Fig. 2. Each triangular symbol represents, for the corresponding input, an elementary effect. The orientation and color of the triangle indicate which level the input (value) was at when the elementary effect was calculated. The scatter plots, in addition to giving detailed information on the standard deviations of the elementary effects, also help identify anomalies as well as separate nonlinearities from interactions.

In addition to scatter plots, bar graphs showing the modified means and standard deviations can also be revealing.

Variable and Multiple Resolutions

The Morris method uses coarse sampling to explore input sensitivities. For example, when $p = 4$, each input is assigned 4 levels. The total number of possible elementary effects for each input is $p^{M-1}[p - \Delta(p-1)]$. A large measure of central tendency for the distribution indicates an input with an important "overall" influence on the output on this particular p scale. If, however, we have the knowledge that the output varies more quickly with a particular input, then it will be more sensible to use a large p for this input.

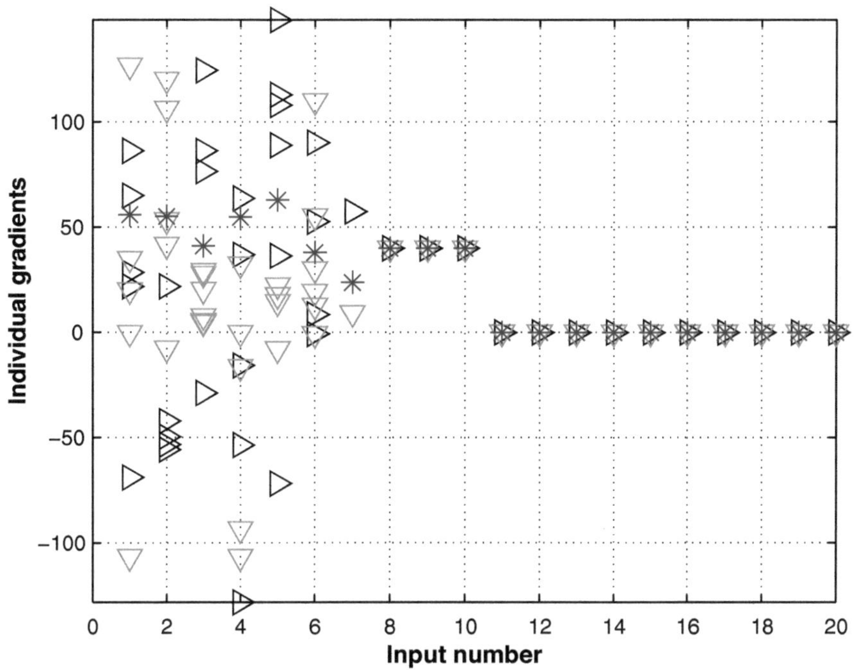

Fig. 2. An example scatter plot. (See Plate 40 on page 325)

Hence, additional knowledge on the model behavior may be helpful in gauging good values of p for each input. This modification can be easily modified from the original Morris method.

Furthermore, if the value of p for a given input is not known and additional resources are available for parameter screening, then a second pass with finer resolution (larger p) can be used to check the earlier assumptions.

Selective Replication

Multi-physics applications often encounter outliers in computing the elementary effects of some inputs. After checking that these outliers are not due to code or other errors, additional runs may be needed to further study whether these outliers are indeed isolated ones. The idea of selective replication is to collect the subset of inputs which we want to study more and create a reduced Morris design for these inputs. The number of replications used for this second step can be determined adaptively when coupled with hypothesis testing of the elementary effects.

Local Refinement

A simple yet powerful idea to study anomalies is the use of local refinement. Suppose for input i, one of its r elementary effect is an outlier. After careful examination, the large gradient computed for the ith input at $X_i^{(1)}$ and $X_i^{(2)}$ is due to one of its end points (say, at $X_i^{(2)}$) appearing to be an anomaly. Focus should then be on investigating whether the offending sample point ($X_i^{(2)}$) is set up and run correctly. In the absence of any evidence of gross errors, one may want to refine the region between $X_i^{(1)}$ and $X_i^{(2)}$ (while keeping all other inputs the same) and study what happens when X_i approaches from $X_i^{(1)}$ to $X_i^{(2)}$. It should be noted that this simple idea is feasible only for methods that varies one input at a time. We learned that simplicity in the sampling design is important to tackle complex applications plagued by frequent anomalies.

Screening for Parameter Interactions

The computational requirement of the MOAT method is relatively inexpensive ($r(M+1)$ where $r \approx 10$ in general). However, one shortcoming of this method is that nonlinearities and parameter interactions are confounded. It is highly desirable to separate nonlinearities from interactions, since negligible interaction between a given parameter with the remaining ones implies that this parameter can be studied independently (that is, decoupled from the rest) in Step (4) due to the additive nature of variance decomposition in this case. To qualitatively study two-parameter interactions, Campolongo and Braddock [1] proposed the extended Morris method. However, the number of simulation runs for this new method is now $O(M^2)$.

In our method implementation, we also provide a qualitative assessment of parameter interaction without additional simulation cost, taking advantage of MOAT's multiple visits to individual input levels. Specifically, let $X_i^{(j)}, j = 1, \ldots, p$ (say, $p = 4$) be the p levels used for input i. If the number of replications r is 10, then it is likely that each of the p levels is used more than once. By comparing the elementary effects on the same levels (again computing standard deviations) a rough assessment of interactions can be made. If a more accurate assessment is needed, one can use more replications or employ the extended Morris method.

3.4 Dealing with Correlated Inputs

The original MOAT method can only handle models with non-correlated or orthogonal inputs. Correlations or non-orthogonalities, however, are common in multi-physics applications exhibiting themselves in the form of:

1. Joint probability density functions
2. Inequality constraints in some analytical forms
3. Constraints imposed by results of related physics simulation

If most or all uncertain inputs are interacting with each other through correlations, the screening process may be greatly complicated. The complication arises because input correlations often disguise as parameter interactions.

In many scientific applications, however, input correlations usually appear in groups pertaining to individual physics. In this case, a natural modification is to put all inputs into uncorrelated groups and then analyze qualitative importance based on group contributions. This is reasonable since some inputs become important only because they are correlated with other inputs. However, if the relationships within groups are also complex, robust group screening may not be possible. We are currently exploring modified screening methods to handle inequality constraints.

3.5 An Enhanced MOAT Method

Using some of the enhancements described above, we have implemented an enhanced MOAT design and analysis tool with the following features:

Variable resolution: In order to support this feature, the original Morris equation (1) has been modified to become:

$$B^* = J_{M+1,1}X^{(1)} + 1/2[(2BP - J_{M+1,M})D + J_{M+1,M}]G. \quad (5)$$

Here, in addition to the notations given before, we have a new diagonal matrix G which has diagonal elements $\{\delta_i = 1/(p_i - 1)\}$ where p_i is the number of levels for parameter i. In addition, the initial seed points are stratified to allow a more even exploration of the parameter space.

Selective replications: This feature creates reduced Morris designs for a small subset of input parameters that have been identified for further screening.

Interaction screening: A simple analysis is added to perform a crude analysis of parameter screening. The results can be used to confirm previous knowledge, or to decouple the uncertain parameters for independent studies.

Visual analysis: This feature creates scatter plots and bar graphs for examining individual sample points and individual gradients.

3.6 Additional Notes

The MOAT method has been successfully used across many applications. There are, however, concerns about its efficacy. Specifically, because of the curse of dimensionality (first used by Bellman to describe the exponential increase in volume associated with adding extra dimensions to a mathematical space), any inexpensive attempts to thoroughly explore this high dimensional uncertain parameter space inevitably incur error of judgment. There are two types of possible errors in the screening exercise. The first type (Type I error), in which unimportant parameters are mistaken as being important, is benign in nature, causing only minor nuisance in Step (3) and (4) of the aforementioned methodology. The second type (Type II error), in which important parameters are mistaken as being unimportant, is more serious.

To address this concern, we first quote Campolongo et al. [2] "Previous exercises, ..., seem to indicate that Morris does not make Type II errors. This is reasonable, since the influence of factors in models follows, according to our experience, a Pareto-like distribution, with few factors accounting for most of variance, and most of the inputs taking up the remaining bit. One has to build a model artificially, in order to get a case where the influence of factors is more uniformly balanced." We also quote Morris: "If an input has a direct impact on the outputs throughout the experimental region, even the nature of that impact changes with location, it takes only a few replications to see it. The danger would come with an input that is important only in a small subregion. Then the probability of stumbling across it would be related to the volume of the subregion in which it is active."

We are currently investigating the use of probabilistic reasoning to characterize Type II error through using hypothesis testing on the elementary effects. The hypothesis testing requires physicists and analysts working together to formulate the hypothesis. This work will be described in more details in an upcoming report.

4 Response Surface Analysis

Response surface methodology (RSM) is a collection of statistical and mathematical techniques useful for developing, improving, and optimizing processes [8]. The original definition of RSM pertains to linear and polynomial regression analyses. We broaden the term RSM to include all function approximation methods. Provided that the model response Y is smooth enough, RSM can be very helpful in reducing the computational cost for global sensitivity analysis. The idea is to create a set of sample points that are space-filling and use them to train a response surface model (or a function approximator). The choice of response surface methods for a given simulation model depends on the knowledge about the simulation model itself. If the simulation model is known to be a linear mapping of its uncertain inputs, then a first-order regression model suffices. If no such knowledge is available about the mapping, more general nonparametric models such as splines, neural network, or Gaussian process models may be more appropriate. An example of a response surface is given in Fig. 3.

Once a good response surface model has been constructed with sufficient accuracy, subsequent analysis can rely on this response surface model which is inexpensive to evaluate. This will facilitate the efficacy of the more quantitative analyses that require a large number of evaluations. Details of how to create response surfaces will not be covered in this paper. We have developed many software tools to create and validate response surfaces in our PSUADE software package.

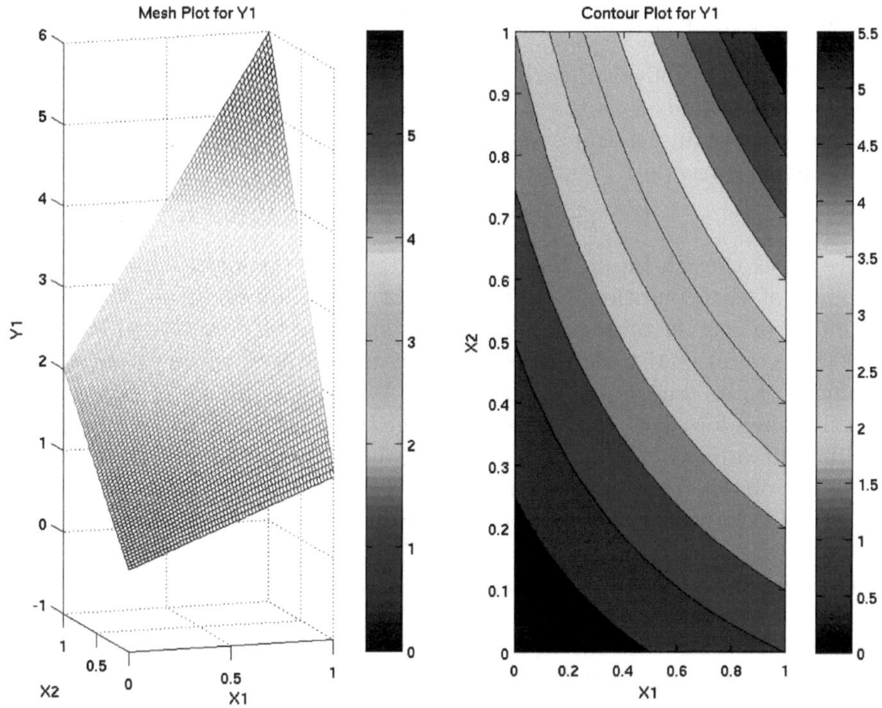

Fig. 3. An example response surface plot. (See Plate 41 on page 326)

5 Variance Decomposition

This section focuses on using variance as an indicator of input importance. The analysis techniques based on regression coefficients, correlation coefficients and their variants have limited applicability since they assume that the model is nearly linear or monotonic. Our attention is mainly on variance-based measures such as correlation ratio and Sobol'/Saltelli sensitivity indices which are less restrictive on the models. For applications with uncorrelated inputs, it is based on the full decomposition of the model variance $V(Y)$ (another notation for σ^2) into

$$V(Y) = \sum_i V_i + \sum_{i<j} V_{ij} + \sum_{i<j<k} V_{ijk} + \cdots + V_{12\cdots M} \tag{6}$$

where $V_i = V(E(Y|X_i))$, $V_{ij} = V(E(Y|X_i, X_j)) - V_i - V_j$, and so on. The total number of terms for M inputs is thus $2^M - 1$. In the present context, we are interested primarily in the main effects (V_i) and two-way interactions (V_{ij}). In the event that the inputs are correlated, the above relationship no longer holds. However, variance-based measures are still useful sensitivity indicators. In this section we describe the computational techniques for main effect and two-way interaction calculations.

5.1 Main Effect Study

The main effect analysis described in this section is due to McKay [6]. The essence of this analysis is the statistical measure called variance of condition expectation. Again, let $E(Y)$ and $V(Y)$ be the prediction mean and variance of an output variable Y, statisticians can tell us that

$$V(Y) = V(E(Y|X_i)) + E(V(Y|X_i)) \tag{7}$$

where X_i is the ith input. Here the first term on the right hand side is the variance of the conditional expectation of Y, conditioned on X_i; and the second term is an error or residual term. A simple interpretation of $V(Y|X_i = X_i^*)$ (the conditional variance) is that it is the reduced variance obtained by setting X_i to its true value X_i^*. In reality, X_i is uncertain so X_i^* is not known exactly. Therefore, it makes sense to average the conditional variances sampled at different values of X_i in its variability range, giving rise to $E(V(Y|X_i))$. It should be obvious now that the difference $V(Y) - E(V(Y|X_i)) = V(E(Y|X_i))$ is a sensible indicator of the importance of input i. Another interpretation of the above relationship is that $V(E(Y|X_i))$ measures the variability in the conditional expected value of Y as the input X_i takes on different values. The residual term represents the variability in Y not accounted for by the input subset X_i.

A related measure is the correlation ratio η^2, which is defined as

$$\eta^2(X_i) = V(E(Y|X_i))/V(Y). \tag{8}$$

A high correlation ratio shows that X_i is important in the variability of the output. If all uncertain inputs are uncorrelated and there are no multi-way interactions, then the sum of the correlation ratios is 1.

There are many ways to compute the correlation ratios. A popular method is to use replicated Latin hypercube sampling together with the corresponding main effect equations. Using Latin hypercube designs, X_i takes on values $X_{ij}, j = 1, \ldots, S$ where S is the number of symbols (or levels). This is to be replicated $R-1$ times so that the total number of simulations (or emulations) required is $N = SR$. Based on Latin hypercube designs, the estimator for the total variance is approximated by (for any $i = 1, \ldots, M$)

$$V(Y) \approx \frac{1}{SR} \sum_{j=1}^{S} \sum_{r=1}^{R} [Y^{(r)}(X_i = X_{ij}) - \bar{Y}]^2 \tag{9}$$

where $Y^{(r)}(*)$ is the output corresponding to the rth replication. The estimator for the variance of conditional expectation for X_i is

$$V(E(Y|X_i)) \approx \frac{1}{S} \sum_{j=1}^{S} \left[\bar{Y}(X_i = X_{ij}) - \bar{Y}\right]^2$$
$$-\frac{1}{SR^2} \sum_{j=1}^{S} \sum_{r=1}^{R} [Y^{(r)}(X_i = X_{ij}) - \bar{Y}(X_i = X_{ij})]^2, \tag{10}$$

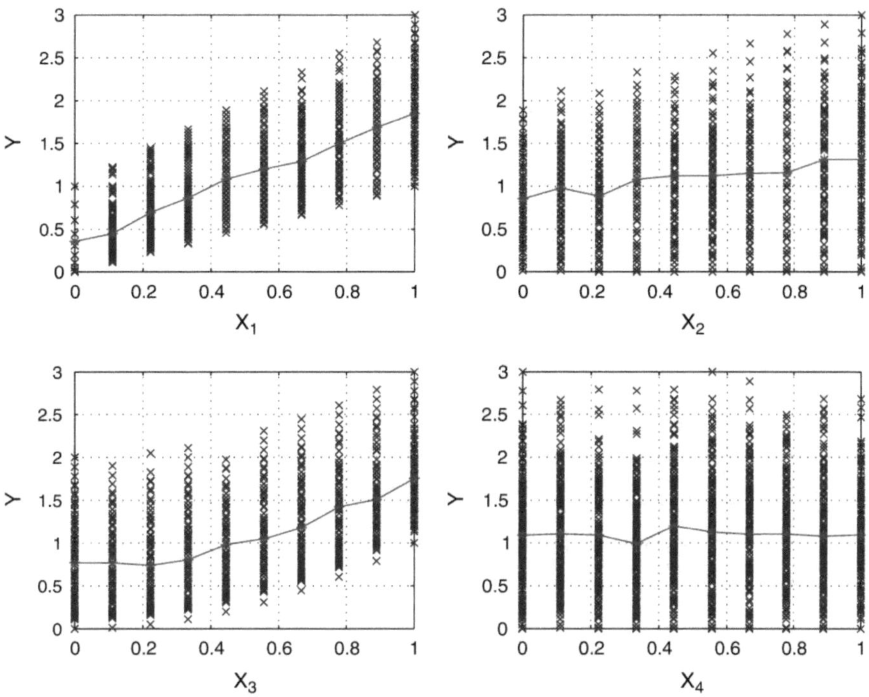

Fig. 4. An example scatter plot from replicated Latin hypercube. (See Plate 42 on page 326)

and the corresponding correlation ratio is computed by dividing VCE(X_i) by $V(Y)$. This correlation ratio is a quantitative measure for the relative importance of X_i. In addition, the scatter plot from the replicated Latin hypercube also gives the trend as X_i takes on different values X_{ij} (by inspecting the line joining the means of all groups, namely X_{ij}'s.) In addition, when R is large enough (say, >50), the parameter space is sufficiently explored such that the difference between the group maxima and minima also indicates interaction effect (that X_i is interacting with some other inputs) although it will take more studies to identify which other inputs it interacts with. An example is given in Fig. 4 where $Y = X_1 + X_1 * X_2 + X_3^2$ for four inputs. Here a close examination of the scatter plots reveals that X_1 and X_2 are involved in some interactions, X_3 is nonlinear, and X_4 has negligible effect.

5.2 Interaction Study

We have extended the idea for main effect analysis to two-way interaction studies (or second-order sensitivities) for uncorrelated inputs. In this case, we employ the following relationship

$$V(Y) = V(E(Y|X_i, X_k)) + E(V(Y|X_i, X_k)) \tag{11}$$

where X_i and X_k are two distinct inputs under consideration. The first term on the right hand side is the variance of the conditional expectation of Y, conditioned on X_i and X_k. Again, the second term is the error or residual term measuring the estimated variance of Y by fixing X_i and X_k. Also, the correlation ratio for the input pair (X_i, X_k) is

$$\eta^2(X_i, X_k) = V(E(Y|X_i, X_k))/V(Y). \tag{12}$$

A high correlation ratio shows that X_i and X_k taken together are important contributors to the output variability. The variance due to the interaction term alone is defined as

$$\hat{V}(E(Y|X_i, X_k)) = V(E(Y|X_i, X_k)) - V(E(Y|X_i)) - V(E(Y|X_k)). \tag{13}$$

$V(E(Y|X_i, X_k))$ can be approximated using many different techniques, for example, by directly evaluating the corresponding integral. Here we advocate the use of replicated orthogonal array (OA) designs with strength 2. Using the OA design, X_i and X_k take on values $X_{ij}, j = 1, \ldots, S$ and $X_{kl}, l = 1, \ldots, S$ where S is the number of levels. The two-way interaction between X_i and $_k$ is obtained by:

$$V(E(Y|X_i, X_k)) \approx \frac{1}{S^2} \sum_{j=1}^{S} \sum_{l=1}^{S} \left[\bar{Y}(X_i = X_{ij}, X_k = X_{kl}) - \bar{Y}\right]^2 \tag{14}$$
$$- \frac{1}{S^2 R^2} \sum_{j=1}^{S} \sum_{l=1}^{S} \sum_{r=1}^{R} [Y^{(r)}(X_i = X_{ij}, X_k = X_{kl})$$
$$- \bar{Y}(X_i = X_{ij}, X_k = X_{kl})]^2,$$

where $Y^{(r)}(X_i = X_{ij}, X_k = X_{kl})$ is the output corresponding to $X_i = X_{ij}$ and $X_k = X_{kl}$ in the rth replication (that is, keeping the two inputs at some fixed values and varying all others); and

$$\bar{Y}(X_i = X_{ij}, X_k = X_{kl}) \approx \frac{1}{R} \sum_{r=1}^{R} Y^{(r)}(X_i = X_{ij}, X_k = X_{kl}). \tag{15}$$

The total variance can be estimated by

$$V(Y) \approx \frac{1}{S^2 R} \sum_{j=1}^{S} \sum_{l=1}^{S} \sum_{r=1}^{R} \left[Y^{(r)}(X_i = X_{ij}, X_k = X_{kl}) - \bar{Y}\right]^2, \tag{16}$$

and finally $\eta^2(X - i, X_k)$ is approximated using 12. Again, the pure interaction term $\hat{V}(E(Y|X_i, X_k))$ can be approximating by subtracting the main effects of X_i and X_k from $V(E(Y|X_i, X_k))$.

This same idea can be applied to the analysis of higher order interaction. For example, to analyze 3-way interaction, a replicated orthogonal array design of strength 3 can be used together with the corresponding formulas for computing the variance of conditional expectations.

5.3 Dealing with Correlated Inputs

If the inputs are correlated, the variance decomposition (6) is no longer valid. The main effect and interaction terms $V(X_i)$ and $V(X_i, X_k)$ can still be computed. However, pure interaction terms can no longer be written as in (13). There are several approaches for handling the case with correlated inputs.

Main effect analysis: The main effect analysis presented previously can be used in this case. However, the first order sensitivity indices will be confounded with interactions due to parameter correlations. If the correlation is in the form of joint probability density functions, the replicated Latin hypercube methods due to Iman and Conover [4] or Stein [15] are useful to produce suitable sample designs for main effect analysis. Direct integration can be used for correlations due to inequality constraints. Ratto, Tarantola, and Saltelli compared several different sampling techniques for estimating the main effect [11].

Group analysis: Jacques and Lavergne [5] proposed a modified approach by grouping the correlated input variables into the same group. Sensitivity indices are then defined for each group (instead of individual inputs) and these multi-dimensional indices are computed based on the Sobol' estimation method using Monte Carlo simulations. If the goal is to reduce the output uncertainty by a certain amount, Tarantola's [16] modified approach can be used to identify a minimal subset of inputs such that fixing these inputs will reduce the variance of Y to a targeted variance V_T.

Total sensitivity indices: Another measure that may be useful in the case of correlated inputs is the total sensitivity index S_{T_i} for input i which is defined as

$$S_{T_i} = V_i + \sum_{i \neq j} V_{ij} + \sum_{i \neq j, i \neq k} V_{ijk} + \cdots + V_{12 \cdots M}. \tag{17}$$

To compute this quantity, we first recall the identity from statistics

$$V(Y) = V(E(Y|X_{\sim i})) + E(V(Y|X_{\sim i})) \tag{18}$$

where ~i means all inputs except input i. For uncorrelated inputs, the total sensitivity index for input i is defined as:

$$S_{T_i} = E(V(Y|X_{\sim i}))/V(Y) \tag{19}$$

which can be computed by the Sobol' method or by using the extended FAST [13].

For correlated inputs, (18) does not hold. However, (19) is still a valid measure for total sensitivity. In this case, (19) can be computed by evaluating the integrals corresponding to the partial variances of all inputs. Since this process is very expensive, it only makes sense to perform this analysis on the response surface models. Both the FAST and Sobol' method for total sensitivity analysis has been implemented in our software package.

To illustrate the case of correlated inputs, we use the following example:

$$Y = F(X_1, X_2, X_3) = X_2 + X_3 \tag{20}$$

where $X_1, X_2, X_3 \in [0, 1]$ and $X_1 + X_2 \leq 1$. The sample mean and variance for this example are 5/6 and 5/36, respectively. The unnormalized main effects of all inputs are $V_1 = 1/48, V_2 = 1/12$ and $V_3 = 1/12$ respectively. However, we also note that

$$V_1 + V_2 + V_3 = 1/48 + 1/12 + 1/12 = 3/16 > V(Y) = 5/36.$$

which is typical in models with correlated inputs. Even though X_1 has no contribution to the output Y, the correlation due to the inequality constraint has induced a small variance of conditional expectation for X_1. In more details, we have

$$\begin{aligned}
V(E(Y|X_1)) + E(V(Y|X_1)) &= 1/48 + 1/9 = 19/144, \\
V(E(Y|X_2)) + E(V(Y|X_2)) &= 1/12 + 1/12 = 1/6, \\
V(E(Y|X_3)) + E(V(Y|X_2)) &= 1/12 + 1/18 = 5/36,
\end{aligned}$$

which do not obey the relationship in (7).
Similarly, we can evaluate the variance integrals to compute the 2-input interaction effects as $V_{12} = 1/18$, $V_{13} = 5/48$, and $V_{23} = 1/6$, and

$$\begin{aligned}
V(E(Y|X_1, X_2)) + E(V(Y|X_1, X_2)) &= 1/18 + 1/12 = 5/36, \\
V(E(Y|X_1, X_3)) + E(V(Y|X_1, X_3)) &= 5/48 + 1/36 = 19/144, \\
V(E(Y|X_2, X_3)) + E(V(Y|X_2, X_3)) &= 1/6 + 0 = 1/6.
\end{aligned}$$

Finally, the total sensitivity indices (unnormalized) for all inputs are $0.0, 1/36$ and $1/12$, respectively. More specifically,

$$\begin{aligned}
V(E(Y|X_{\sim 1})) + E(V(Y|X_{\sim 1})) &= 1/6 + 0 = 1/6, \\
V(E(Y|X_{\sim 2})) + E(V(Y|X_{\sim 2})) &= 5/48 + 1/36 = 19/144, \\
V(E(Y|X_{\sim 3})) + E(V(Y|X_{\sim 3})) &= 1/18 + 1/12 = 5/36.
\end{aligned}$$

For this example, based on the total sensitivity indices we conclude that X_1 is not sensitive at all, which agrees with the the original function (20).

6 Validating our Sensitivity Analysis Methodology

So far we have presented a global sensitivity analysis methodology suitable for large scale multi-physics simulations with large number of uncertain parameters and complex constraints. During the course of applying this methodology to real applications, we have encountered many doubts and skepticisms on its efficacy and accuracy. The question often being asked is: how reliable and accurate are the results from applying this methodology? In this section we take a critical look at this methodology by closely examining the following three major steps of the methodology:

1. The screening analysis
2. The response surface analysis
3. The quantitative sensitivity analysis

We will point out the assumptions behind these steps and then briefly describe a few strategies currently under investigation to address the question. More detailed description of these strategies will be in several other reports which are in preparation.

6.1 Self-Validation in the Screening Analysis

First of all, as discussed in Sect. 3.6, we would emphasize on the importance of understanding when the screening analysis can be most effective. The two desirable conditions are (1) screening methods generally perform well when the number of important inputs is small compared to the total number of inputs under investigation (also called Pareto effect, which states that roughly 20% of the inputs account for 80% of the output variability); and (2) the MOAT method has been demonstrated to be effective in capturing parameters that have direct impact on the output of interest throughout the parameter space. Due to the coarse sampling in the screening procedure, Type II errors (when important parameters are treated as unimportant) may occur. Many practitioners reported [12,13] that Type II errors are unlikely for real applications, and our physicists are less skeptical when they found that the screening results were acceptable or even revealing. In spite of these positive feedbacks, we are convinced of the need to validate the screening step with more rigorous analysis. We thereby advocate three strategies to safeguard against Type II errors: (1) append to the screening procedure with a multi-resolution step which adds replications using finer resolution (that is, larger p_i as described in Sect. 3.3.2); (2) add a rigorous hypothesis testing on the Morris elementary effects to give probabilistic bounds on the believability of the results (briefly described in Sect. 3.6); and (3) provide an iterative feedback loop to capture the errors missed in the last iteration (performed after Step 4 of the methodology presented in Sect. 2). The first strategy helps to verify that the coarseness of the screening sample is adequate, at the expense of additional simulation runs. For the second strategy to be effective, some assumptions such as small asymmetry of the distribution of the elementary effects may be needed. Having additional information about the output response such as smoothness and upper/lower bounds will help to detect the degree of asymmetry and improve the robustness of the hypothesis testing. Strategy 2 can also be part of an adaptive procedure to satisfy the hypothesis by launching more simulation runs.

6.2 Cross-Validation in the Response Surface Analysis

As stated previously, if the simulation runs are very expensive, it may make more sense to create a response surface model first before performing quantitative sensitivity analyses. To construct a reliable and accurate response surface model for general applications, we need a suitable sampling method, a good response surface method, and effective validation tools. A good sampling method should be space-filling and should provide adequate boundary coverage (to avoid extrapolation in the analysis stage). Furthermore, adequate sample size is an important concern without any smoothness assumptions about the output responses. Adaptive response surface methodologies together with good error estimation tools will greatly reduce the computational costs. A few response surface methods have been mentioned in Section 4.1. Cross validation tools such as bootstrapping and jackknifing are useful to validate the response surface models. Many of these features are available from our PSUADE software package.

6.3 Self-Validation in Variance Decomposition

Variance decomposition has a very strong mathematical foundation. The only concern we have is the accuracy of the estimates which depends largely on the sample size. In our variance decomposition we provide a multi-resolution approach to systematically study the accuracy of the estimates by observing the convergence of the estimates using different combinations for sample sizes. These techniques have been implemented in our PSUADE software package and details are described our next paper [18].

7 PSUADE

To be able to perform designed simulation experiments like the ones used in our global sensitivity analysis methodology, many detailed tasks such as setting up the test problem, negotiating the computational resources, extracting and processing the appropriate outputs for examination, packaging the outputs for visualization, applying the analysis tools, optimizing the responses, interpreting the results, etc. are involved. To simplify these tedious bookkeeping chores, we need an integrated environment for such an analysis. There are many important issues in designing the architecture of such an integrated environment:

- A user-friendly interface for setting up the problem and for analyzing the results
- An automated job execution environment suitable for different computer platforms
- Design and analysis tools for this methodology

- Graphics tools for visualizing the results
- A convenient scheme for storing and retrieving the results

Many of the above considerations have been captured in a framework called PSUADE, which is a software package for sensitivity and uncertainty analysis. PSUADE supports a user-friendly interface via input and output filters. Once PSUADE has created a particular sampling design, it systematically feeds the design points into parameter files and calls a user handler which absorbs the sample data and inserts them into the model input files. The user handler then requests computational resource for the evaluation, waits for its completion, and extracts the output data from the model output files into PSUADE's output files. This protocol allows all information exchange to be done via a user-written handler. There is no need to change the simulation source code to accommodate the analysis, and it is sometimes called a "non-intrusive" interface.

To allow multiple evaluations of the model, PSUADE allows function evaluations to be launched asynchronously. PSUADE then monitors the existence of the corresponding output files to determine that an evaluation has been completed. One can prescribe the degree of parallelism so that PSUADE can adequately request resources. PSUADE also has a rich collection of sampling methods such as Monte Carlo, full and fractional factorial, Latin hypercube, orthogonal arrays, and one-at-a-time methods. In addition, some of these methods have been equipped with refinement strategies [17] so that users can start with a small sample size and add more sample points as needed using refinements. Furthermore, uniform, normal and lognormal distributions are available for the inputs.

PSUADE provides a rich set of statistical tools such as the different moments, Pearson correlation coefficients, variance decomposition using main effect analysis, and others. It also provides a few optimization methods via public domain optimization tools.

PSUADE provides graphics support primarily through Matlab scripts. Screening plots, scatter plots, and response surface plots are created in Matlab format so that users can launch Matlab to view them.

PSUADE was written in C++ and runs on many different Unix-based operating systems.

8 Summary

In this report we describe a global sensitivity analysis methodology suitable for large scale multi-physics applications. Various computational and analysis aspects of this methodology from screening to main effect and interaction studies have been discussed in detail. Our methodology is distinguished by an enhanced screening phase, an efficient analysis for two-way interactions, and a recursive mechanism suitable for multi-physics applications. We have applied this methodology to several applications with satisfactory results. Specifically, there are a few lessons learned from our experience with the screening exercises:

1. The screening process is manageable (it is easy to understand, yet effective)
2. The process of applying the screening is as important as getting the results. The process promotes useful communications between scientists and analysts to verify that the observed phenomena make sense
3. Since the screening design exhibits a pattern, anomalies can easily be spotted and investigated

The variance decomposition exercises likewise are revealing, especially with the help of graphical analysis tools. While PSUADE has been able to provide most of the functionalities of this methodology, a better user interface and more graphical tools will be more beneficial.

Acknowledgment

This work was performed under the auspices of the U.S. Energy by Lawrence Livermore National Laboratory under Contract No. DE-AC52-07NA27344. The author would like to thank Dana Goto and Henry Hsieh for providing helpful feedbacks to enhance this methodology.

References

1. F. Campolongo, R. Braddock, The use of graph theory in the sensitivity analysis of the model output: a new screening method, Reliability Engineering and System Safety 64 (1997), 1–12.
2. F. Campolongo, S. Tarantola, A. Saltelli, Tackling quantitatively large dimensionality problems, Computer Physics Communications 117 (1999), 75–85.
3. J. H. Friedman, Multivariate adaptive regression splines, Annals of Statistics 19.1 (1991), 1–141.
4. R. L. Iman, W. J. Conover, A distribution-free approach to inducing rank correlation among input variables, Communication in Statistics Simulation and Computaion, 11 (1982), 311–334.
5. J. Jacques, C. Lavergne, Sensitivity analysis for model with correlated inputs, Reliability Engineering and System Safety, 91(10–11) (2006), 1126–1134.
6. M. D. McKay, *Evaluating Prediction Uncertainty*, Los Alamos National Laboratory Technical Report NUREG/CR-6311, LA-12915-MS.
7. M. D. Morris, Factorial sampling plans for preliminary computational experiments, Technometrics, 21(2) (1991), 239–245.
8. R. H. Myers, D. C. Montgomery, *Response Surface Methodology*, 2nd edn., Wiley Series in Probability and Statistics, 2002.
9. A. B. Owen, Orthogonal arrays for computer experiments, integration, and visualization, Statistics Sinica 2 (1992), 439–452.

10. R. L. Plackett, J. P. Burman, The design of optimum multifactorial experiments, Biometrika, 33 (1946), 305–325.
11. M. Ratto, S. Tarantola, A. Saltelli, Estimation of importance indicators for correlated inputs, *Proceedings of the European Conference on Safety and Reliability, ESREL 2001*, E. Zio, M. Demichela, N. Piccinini (eds.), Torino, Vol. 1, pp. 157–164, 16–20 September 2001.
12. A. Saltelli, S. Tarantola, K. Chan, A Quantitative model-independent method for global sensitivity analysis of model output, Technometrics, 41(1) (1999), 39–55.
13. A. Saltelli, K. Chan, E. M. Scott (eds.), *Sensitivity Analysis*, Wiley Series in Probability and Statistics, 2000.
14. A. Saltelli, S. Tarantola, F. Campolongo, M. Ratto, *Sensitivity Analysis in Practice*, Wiley, 2004.
15. M. Stein, Large sample properties of simulations using latin hypercube sampling, Technometrics, 29(2) (1987), 143–151.
16. S. Tarantola, Quantifying uncertainty importance when inputs are correlated, *Foresight and Precaution*, Cottam, Harvey, Pape, Tait (eds.), pp. 1115–1120, 2000.
17. C. Tong, Refinement strategies for stratified sampling methods, *Reliability Engineering and System Safety,* 91(10–11) (2006), 1257–1265.
18. C. Tong, Toward a more robust variance-based global sensitivity analysis of model outputs, Lawrence Livermore National Laboratory report UCRL-TR-235561, Oct 2007.

Color Plates

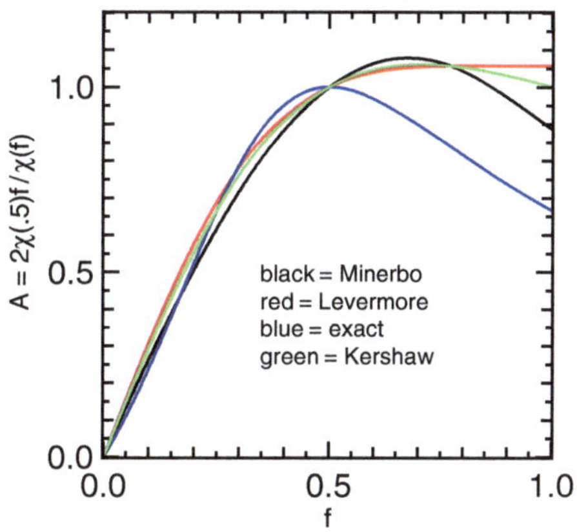

Plate 1: The quantity A defined in the text is shown vs. the scaled flux, f, for four different Eddington factor prescriptions, as indicated. (See Fig. 5 on page 7)

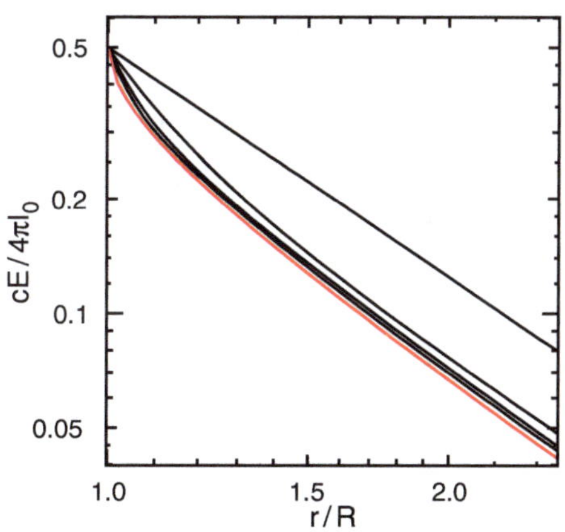

Plate 2: Star-in-space solutions for the normalized energy density vs. radius are shown for S_n methods of different order in comparison with the exact result. The curves are for S_2 (*top*), S_4, S_6, S_8 and the exact solution (*bottom, in red*). (See Fig. 6 on page 8)

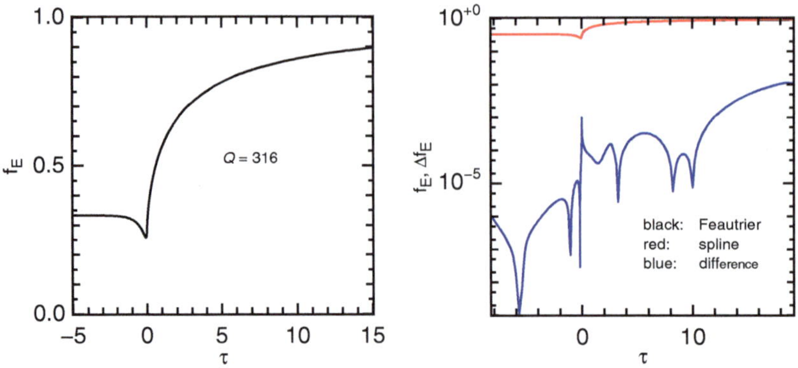

Plate 3: The self-consistent Eddington factor is shown vs. optical depth τ measured from the shock front for a particular radiative shock, $Q = 316$. The flow goes from right to left in the figure. *Left*: values of f_E; *right*: difference between f_E computed from Feautrier and from spline-based quadrature. (See Fig. 9 on page 10)

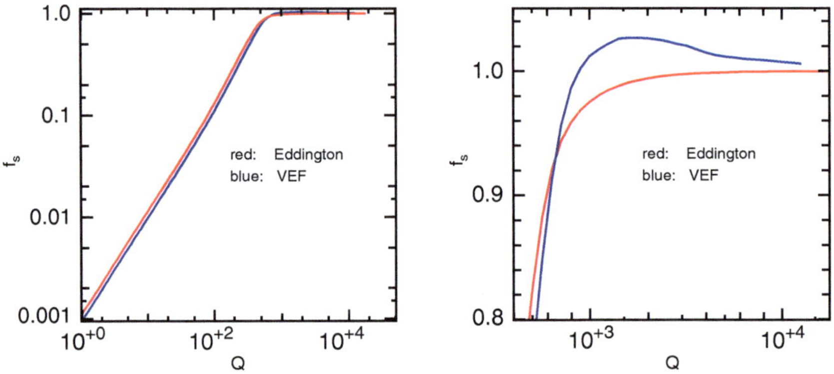

Plate 4: As Fig. 8, but comparing Eddington approximation (*red*) with accurate transport (*blue*). (See Fig. 10 on page 11)

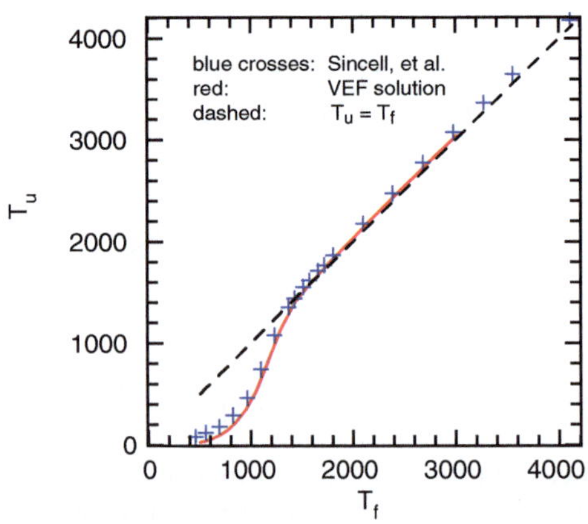

Plate 5: Pre-shock temperature T_u is shown vs. final temperature T_f for the present VEF calculations (*red*) and the VEF results of [SGM] (*blue crosses*). The *dashed line* is $T_f = T_u$. (See Fig. 11 on page 12)

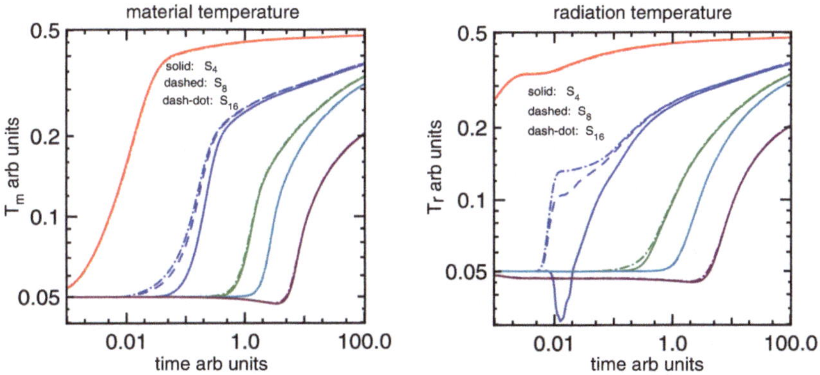

Plate 6: Crooked-pipe results with S_4, S_8 and S_{16} are shown for the time histories of material temperature T_m *(left)* and radiation temperature T_r *(right)* at the five fiducial points in Fig. 12. Data from [PN06]. (See Fig. 13 on page 13)

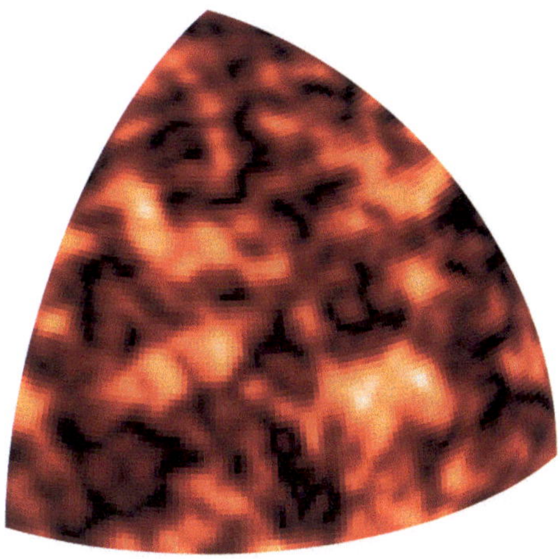

Plate 7: Realization in one octant of the intensity field at a point in an infinite universe filled with uniform opacity and random emissivity, as described in the text. (See Fig. 14 on page 14)

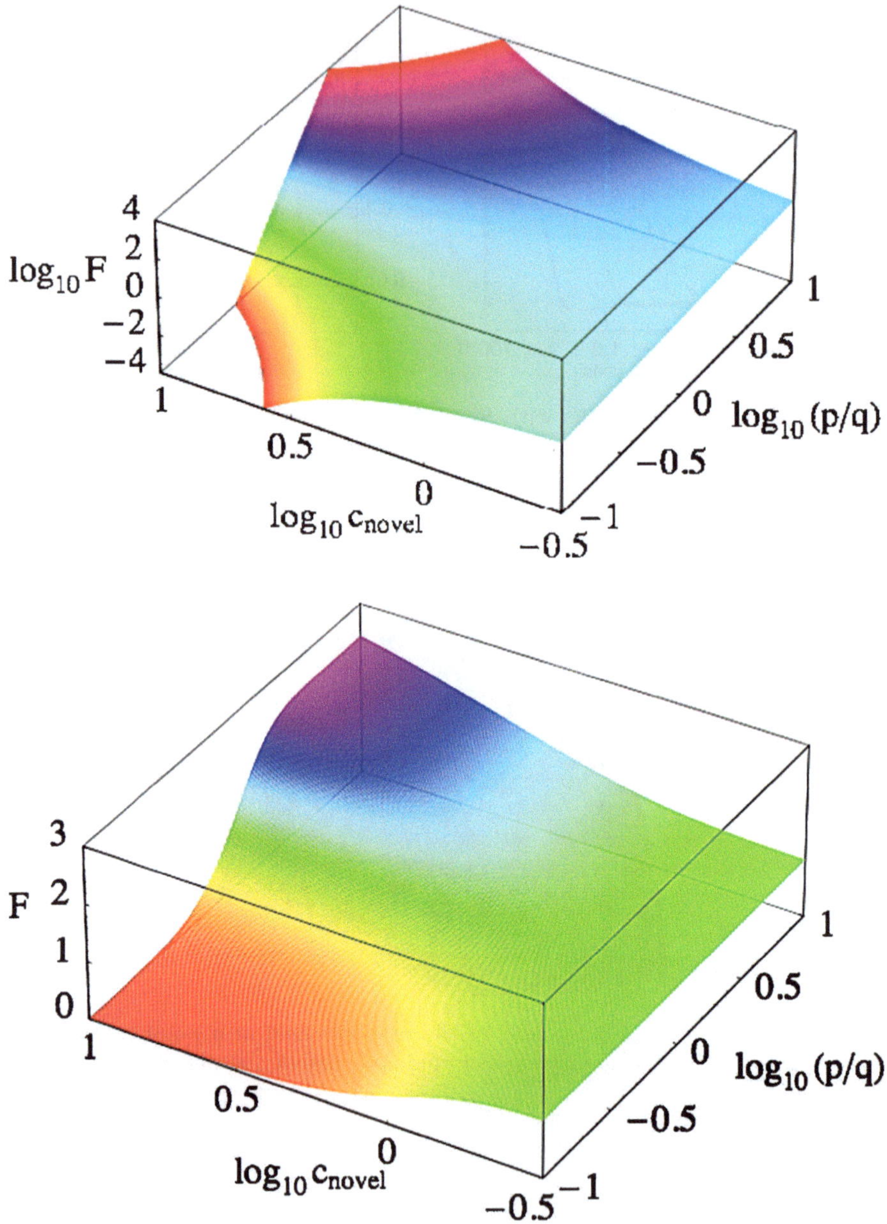

Plate 8: The multipliers defined by (6) and (7) are plotted as functions of p/q and c_{novel} in the upper and lower panels respectively. Note the vertical log scale used for the multiplier (6) in the top panel. (See Fig. 2 on page 39)

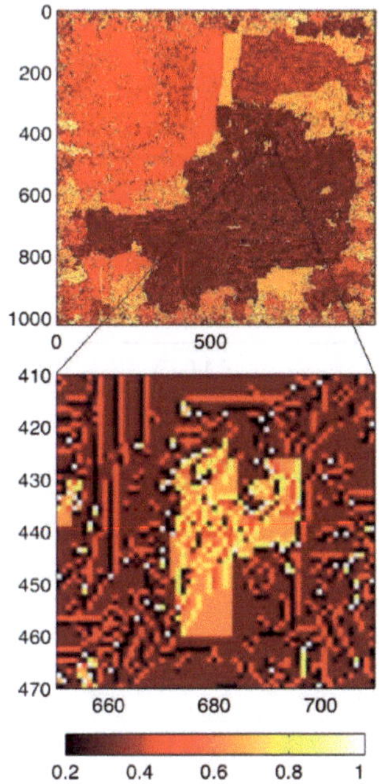

Plate 9: Map of the "stress" field generated by the OFC model immediately after a large avalanche (main shock) at two magnifications. The *upper panel* shows the whole grid of size 1,024 and the *lower plot* represents a subset of the grid delineated by the square in the upper plot. Adapted from [90]. (See Fig. 3 on page 46)

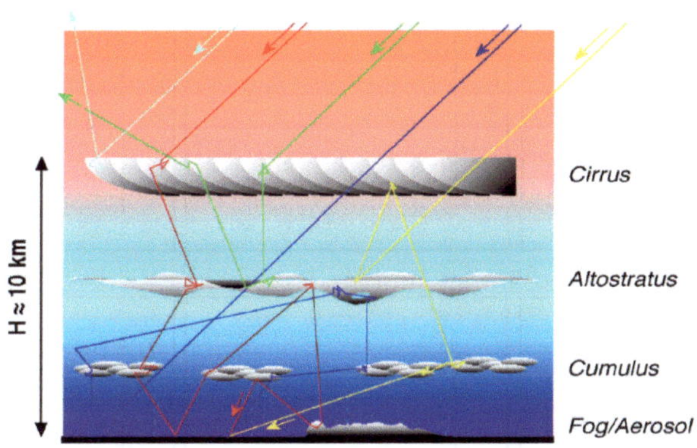

Plate 10: The anomalous diffusion model of solar photon transport at non-absorbing wavelengths in the cloudy atmosphere. In this model, solar beams follow convoluted Lévy walks, which are characterized by frequent small steps (inside clouds) and occasional large jumps (between clouds or between clouds and the surface). The partition between small and large jumps is controlled by the Lévy index α (the PDF of the jump sizes ℓ has a tail decaying as a power law $\sim 1/\ell^{1+\alpha}$). Reproduced from [95]. (See Fig. 4 on page 48)

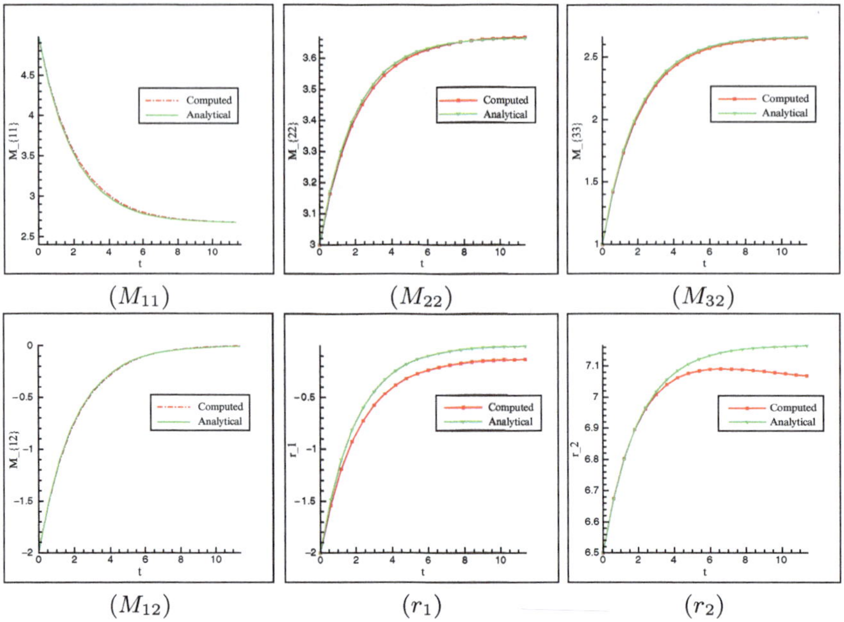

Plate 11: Maxwell type of elastic interactions (number of modes N = 40): *Top row*: momentum flow M_{11}, M_{22}, M_{33}. *Bottom row*: momentum flow M_{12} and energy flow r_1, r_2. (See Fig. 1 on page 93)

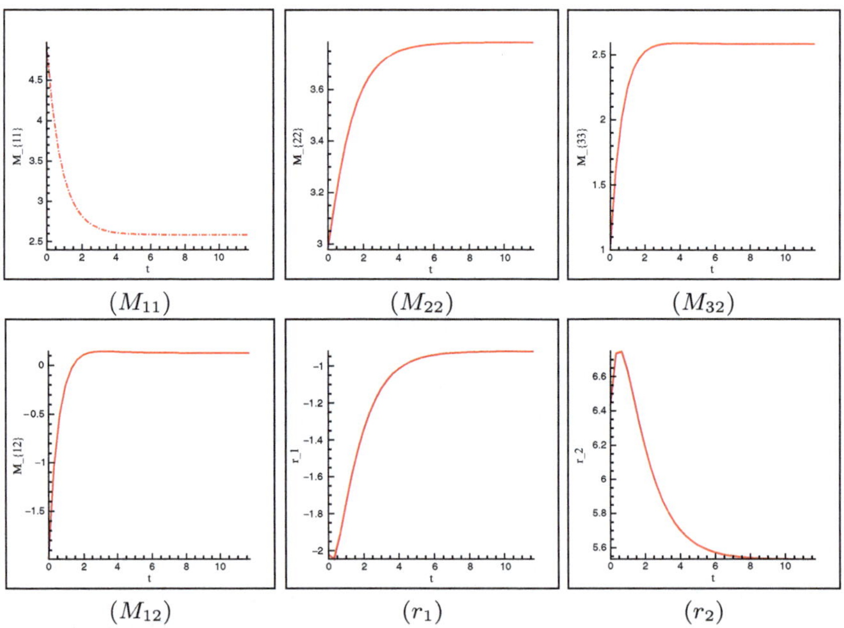

Plate 12: Hard spheres elastic interactions in 3-dimensions (number of modes N = 40): *Top row*: momentum flow M_{11}, M_{22}, M_{33}. *Bottom row*: momentum flow M_{12} and energy flow r_1, r_2. (See Fig. 2 on page 93)

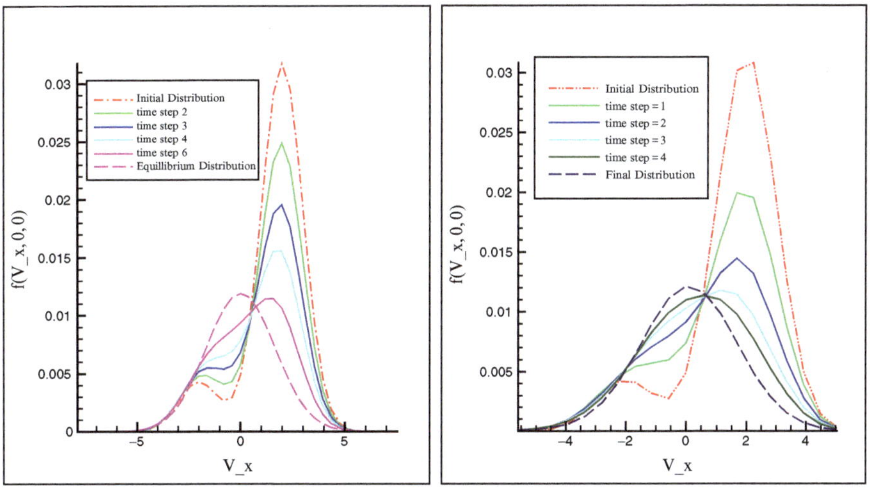

Plate 13: Evolution of the distribution function for elastic interactions. *Left*: Maxwell type interactions, N = 40. *Right*: Hard-sphere interactions, N = 32. (See Fig. 3 on page 94)

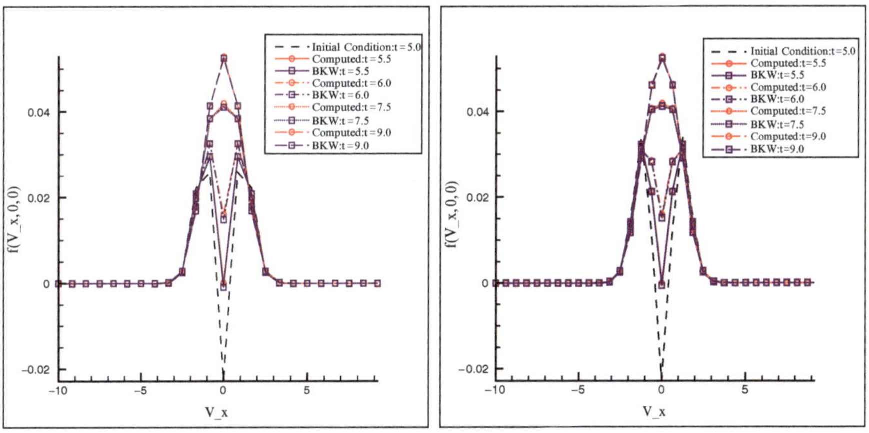

($N = 24$) ($N = 32$)

Plate 14: BKW, $\rho, E(t)$ conserved. (See Fig. 4 on page 94)

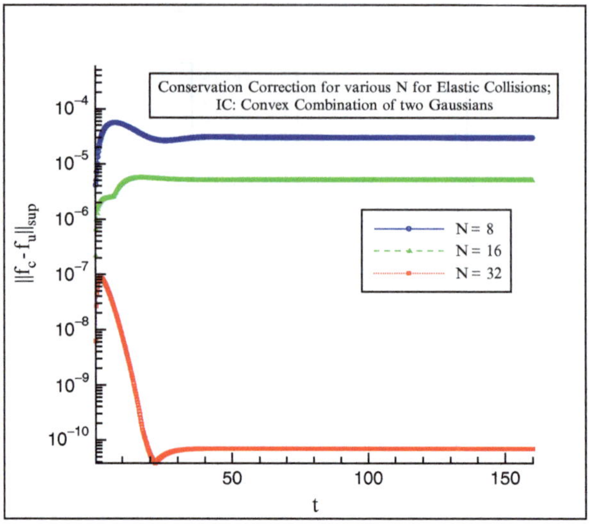

Plate 15: Maxwell type elastic collisions: conservation correction for elastic collisions. (See Fig. 5 on page 95)

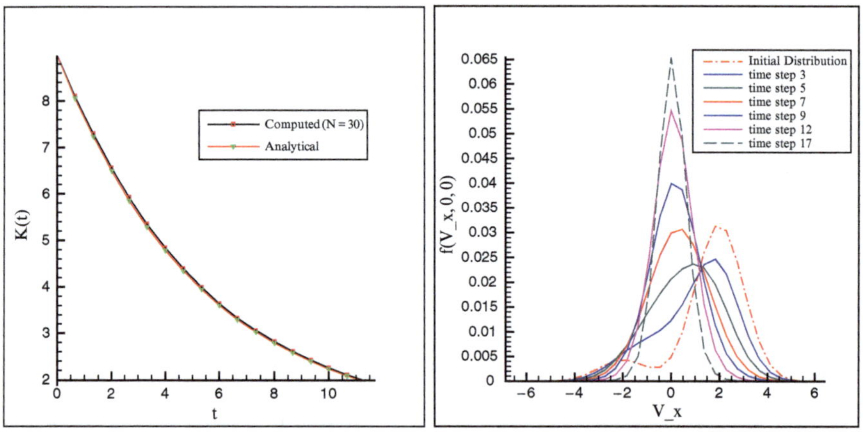

Plate 16: Inelastic: kinetic energy (*left*) and $f(v,t)$ (*right*). (See Fig. 6 on page 96)

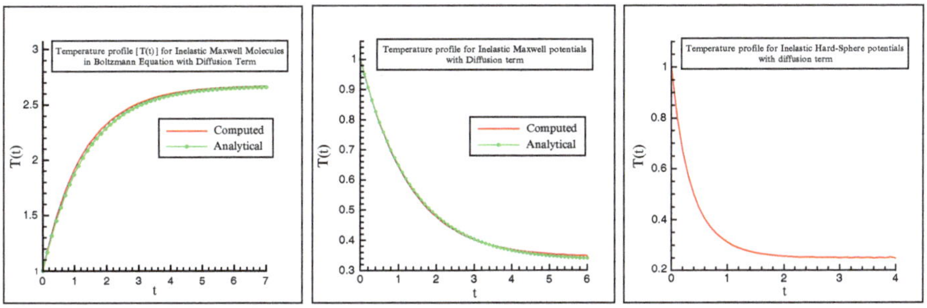

$(T_\infty^{MM} > T_0)$ $(T_\infty^{MM} < T_0)$ $(T_\infty^{HS} < T_0)$

Plate 17: Energy decay for inelastic collisions with a diffusion term, N = 16. *Left and center*: Maxwell type interactions, *right*: hard-sphere interactions.(See Fig. 7 on page 97)

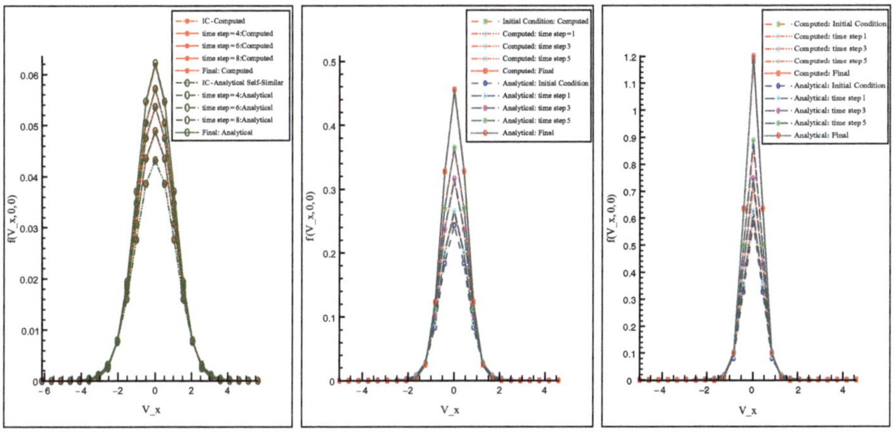

$\mathcal{T} = 1$ $\mathcal{T} = 0.25$ $\mathcal{T} = 0.125$

Plate 18: Maxwell type collisions, slow down process with $\theta = 4/3, \mu = 2/3, N = 24$ evolving computed $f(v,t)$ vs. $f_{\mathcal{T}}^{ss}$ at different values of $\mathcal{T}$. (See Fig. 8 on page 98)

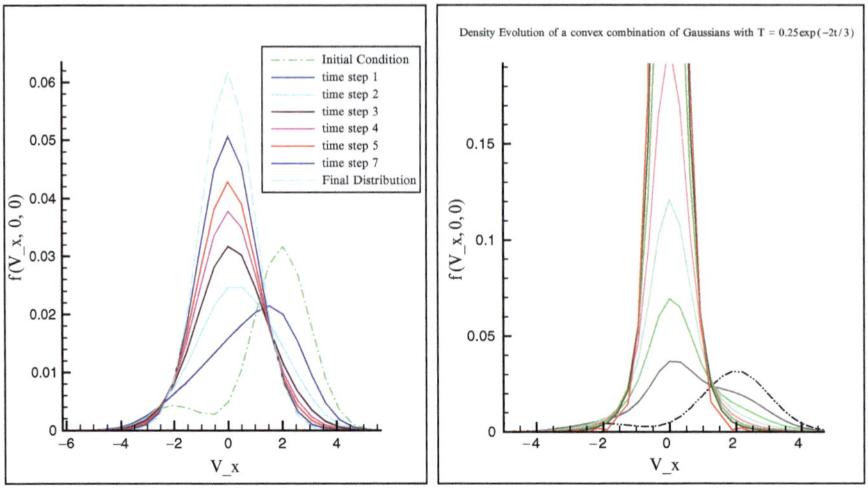

Plate 19: Maxwell type collisions slow down process. Evolving of the pdf $f(v,t)$. *Left*: $\mathcal{T} = 1$, $N = 24$, *right* $\mathcal{T} = 0.25e^{-2t/3}$, $N = 32$. (See Fig. 9 on page 99)

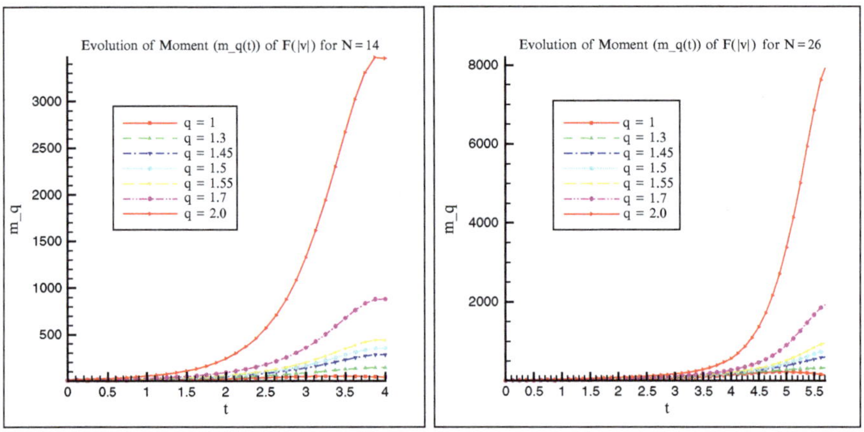

$N = 14$ $\qquad$ $N = 26$

Plate 20: Rescaled moments evolutions $m_q(t)$ for $T = e^{-2t/3}$. (See Fig. 10 on page 100)

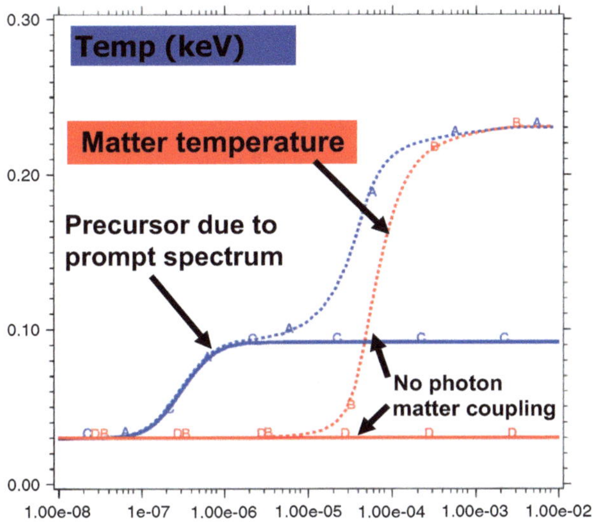

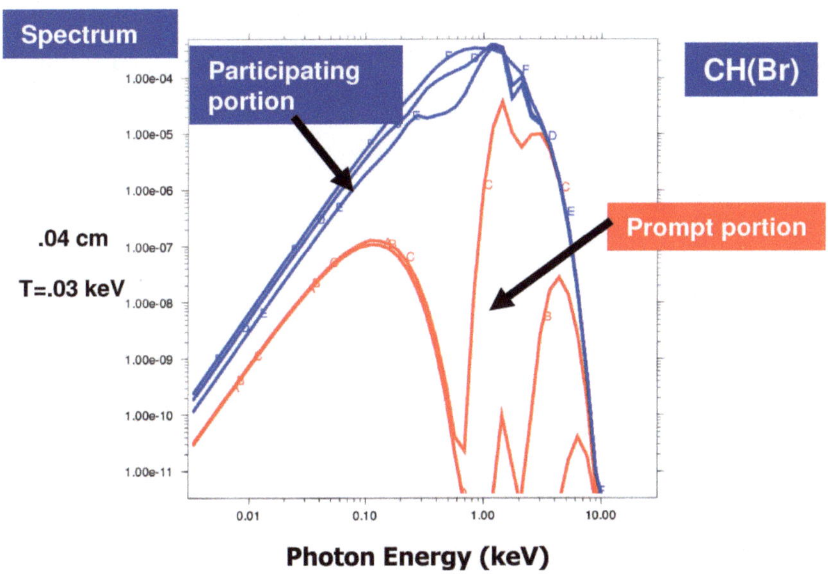

Plate 21: Figure (1a) [top], CH material and radiation temperatures as a function of time, Figure (1b) [bottom] time-snapshots of the radiation spectrum in the CH, showing the prompt portion. (See Fig. 1 on page 155)

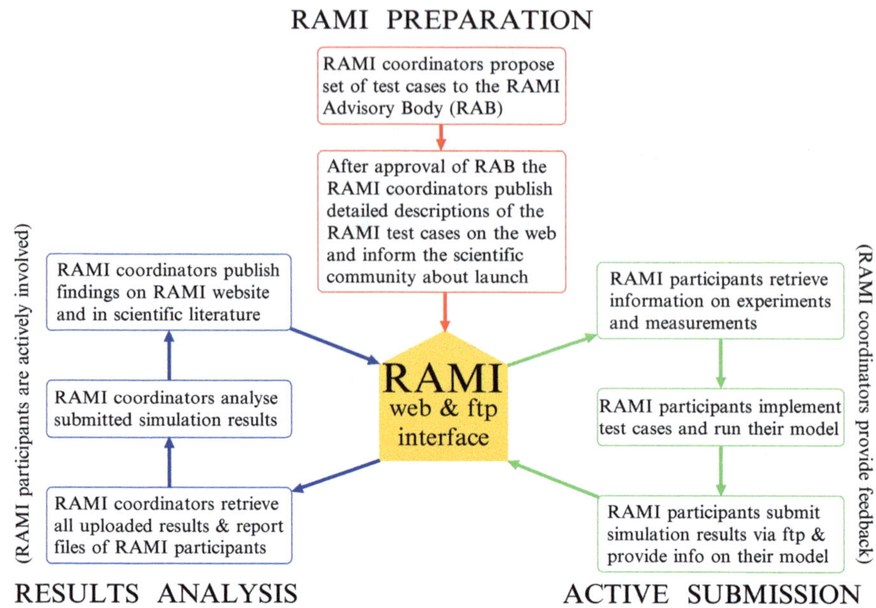

Plate 22: The three main parts of a RAMI phase: RAMI phase preparation (*red*), RAMI active submission (*green*), and RAMI results analysis (*blue*). (See Fig. 1 on page 179)

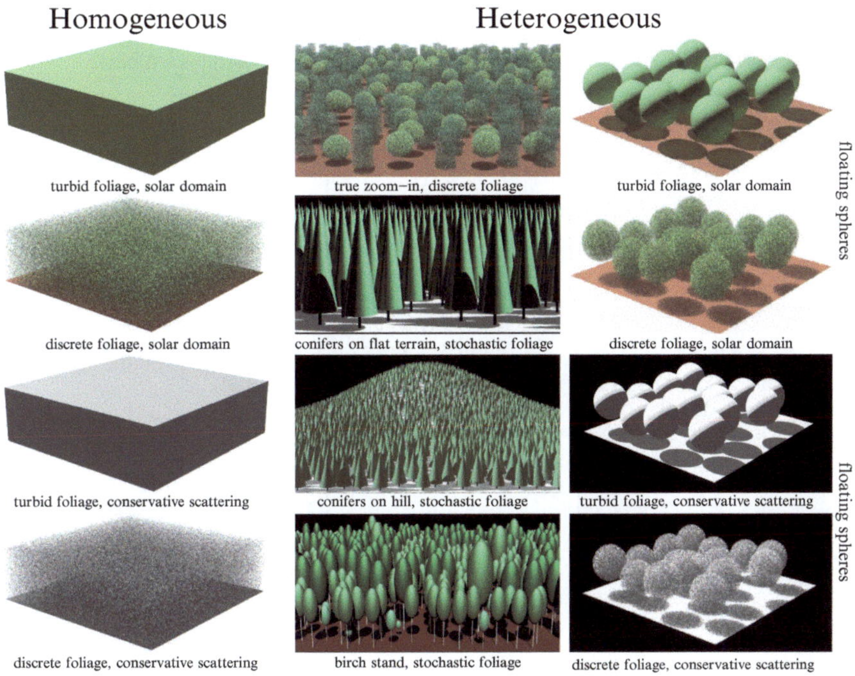

Plate 23: Overview of the structurally homogeneous (*left panels*) and heterogeneous (*middle and right panels*) test cases featuring in the third phase of RAMI. Canopy foliage may be represented through infinitesimally small scatterers (turbid medium), or, through finite sized scatterers having deterministic (discrete) or statistic (stochastic) locations. The spectral properties of both foliage and background material may reflect either values typical for the red and near-infrared spectral region (*colour panels*) or for purely conservative scattering conditions (*grey panels*). (See Fig. 2 on page 181)

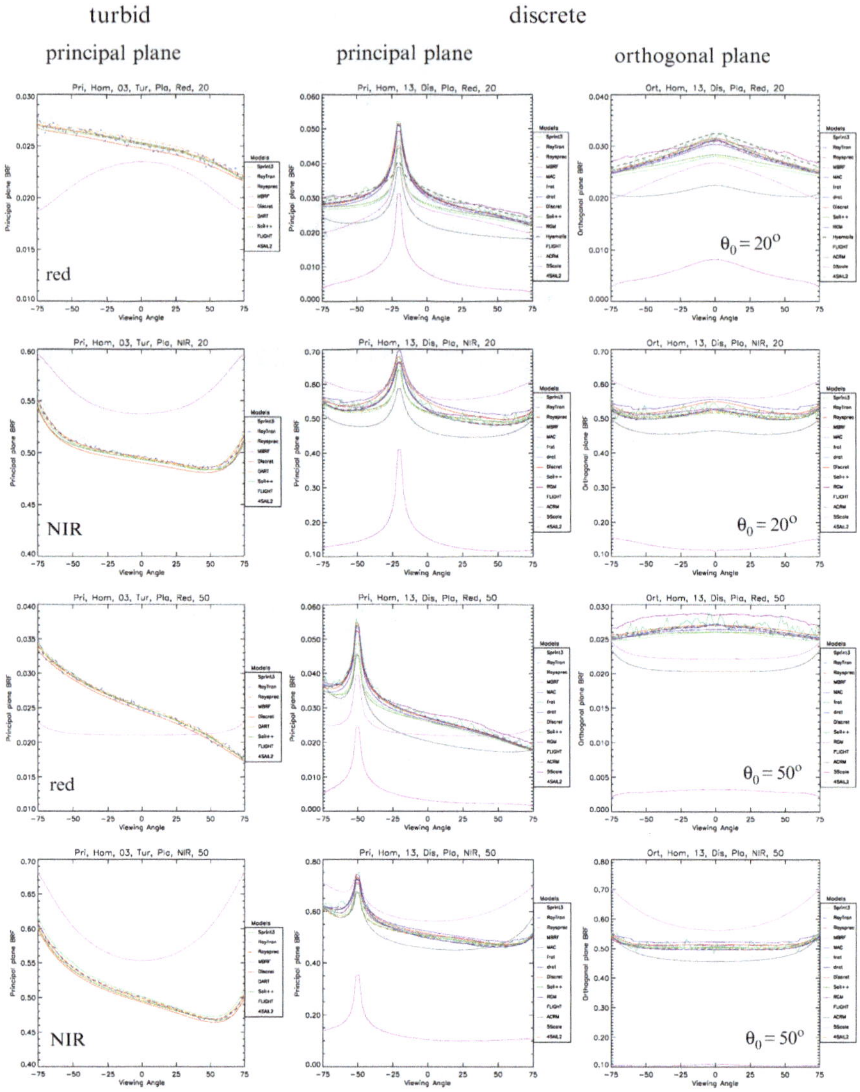

Plate 24: BRF simulations in the red (*first and third row*) and near-infrared (*second and fourth row*) spectral band for structurally homogeneous canopies composed of infinitesimally small (*left column*) and finite-sized (*middle and right column*) scatterers. Simulations pertain to the principal plane (*left and middle column*) and to the orthogonal plane (*right column*) for solar zenith angles (θ_0) of 20° (*top two rows*) and 50°(*bottom two rows*). Note the close agreement between the 3-D Monte Carlo models: `FLIGHT`, `DART`, `drat Rayspread`, `raytran`, and `Sprint3`, respectively. (See Fig. 3 on page 182)

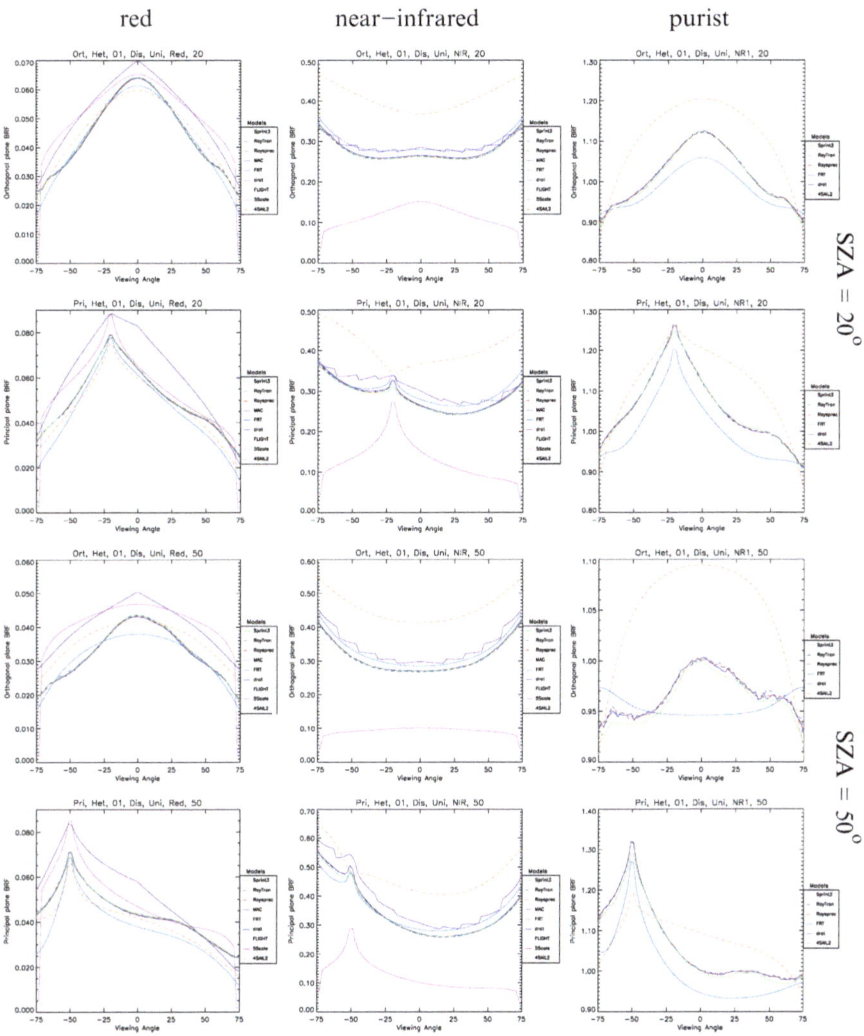

Plate 25: BRF simulations in the orthogonal (*first and third row*) and principal (*second and fourth row*) plane over discrete "floating spheres" test cases with optical properties typical for the red (*left column*) and near-infrared (*middle column*) spectral regimes, as well as, for conservative scattering conditions (*right column*). The top (*bottom*) two rows relate to a solar zenith angle (SZA) of 20° (50°). Note the close agreement among the 3-D Monte Carlo models: `FLIGHT`, `drat`, `Rayspread`, `raytran`, and `Sprint3`, respectively. (See Fig. 4 on page 184)

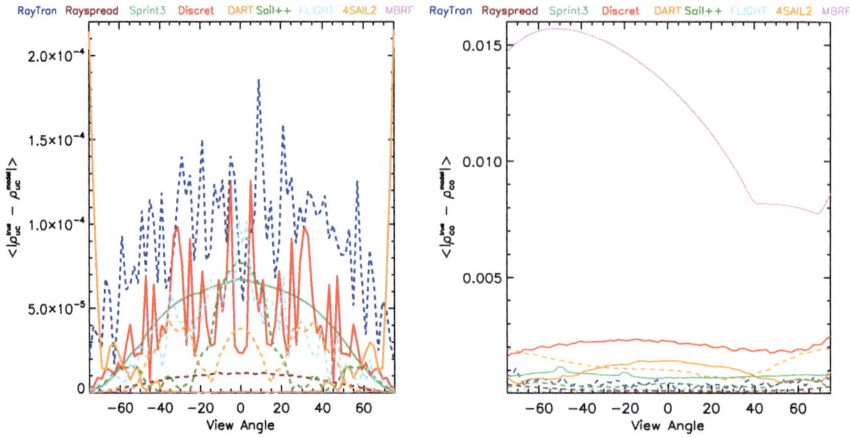

Plate 26: The mean absolute error between model simulations and the analytical formulation of the single-collided, ρ_{co} (*right panel*) and the single-uncollided, ρ_{uc} (*left panel*) BRF components of a homogeneous turbid medium canopy with uniform LND and Lambertian scattering laws. For any view zenith angle the averaging was performed over the principal and orthogonal plane, as well as, for illumination zenith angles of 20° and 50°. (See Fig. 5 on page 189)

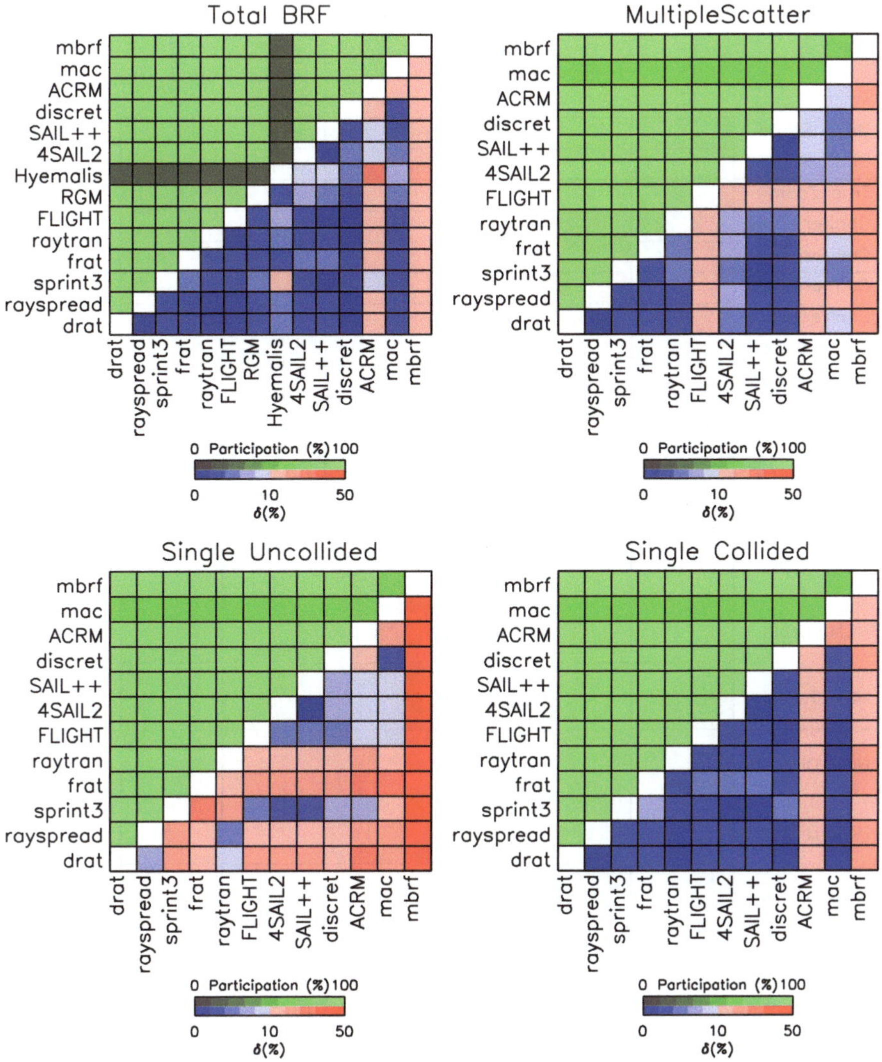

Plate 27: Model-to-model differences $\delta_{m\leftrightarrow c}$ of the total (*top left*), multiple-collided (*top right*), single-uncollided (*bottom left*), and single-collided (*bottom right*) BRF simulations for structurally homogeneous canopies with finite-sized foliage representations in the solar domain. The lower right half of every panel indicates $\delta_{m\leftrightarrow c}$ in (%) (*blue–red colour scheme*), whereas the *top left* half indicates the percentage of available test cases that pairs of models performed together (*black–green colour scheme*). The green colour scale increments in steps of 10%, the blue in steps of 2% (up to $\delta_{m\leftrightarrow c} = 10\%$), and the red in steps of 10% (with a *bright red colour* indicating $\delta_{m\leftrightarrow c} > 50\%$). (See Fig. 6 on page 191)

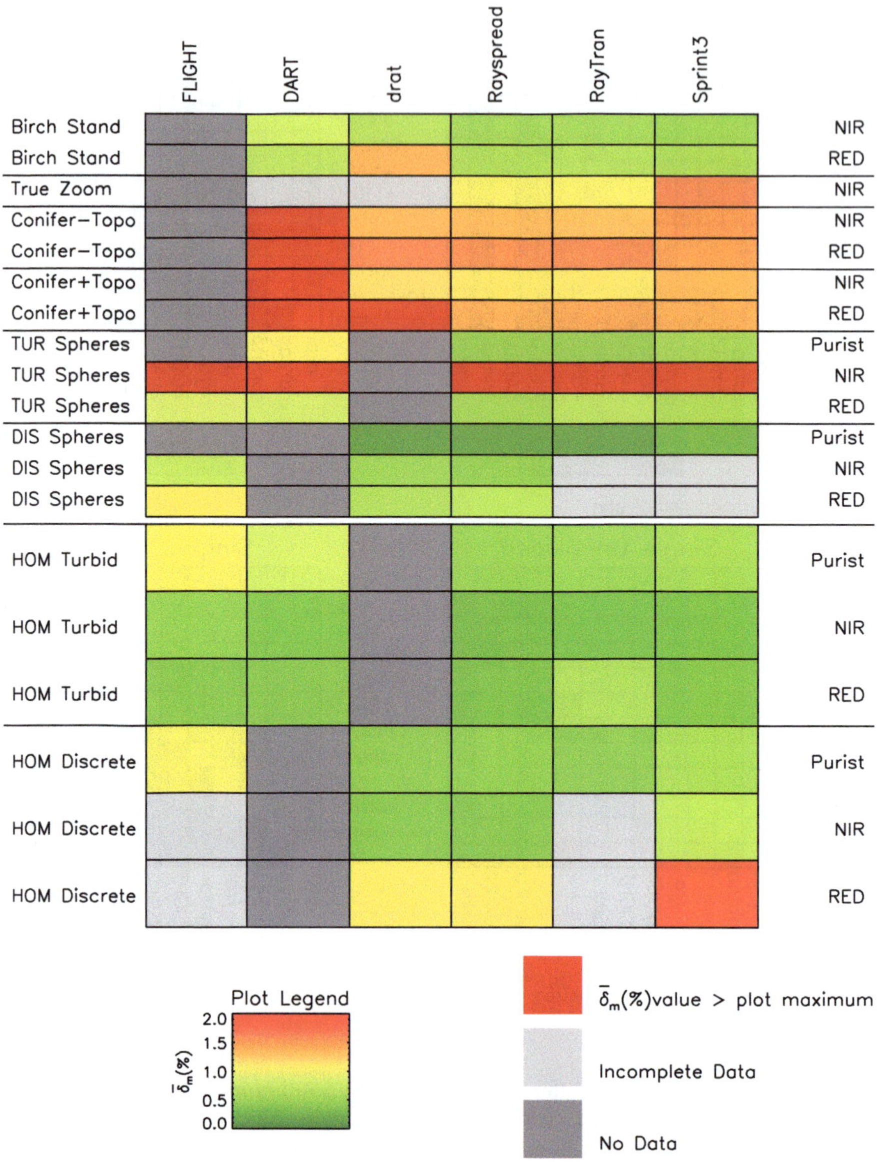

Plate 28: Performance and participation of the "credible" 3-D Monte Carlo models during RAMI-3 for structurally heterogeneous (*top table*) and homogeneous (*bottom table*) discrete canopy representation. Model names are listed on the top of the table (one per column). The experiment identifier is provided to the left, the spectral regime to the right of each table column. *Light* (*dark*) *grey fields* indicate incomplete (no) data submission. The *green–yellow–red colour scheme* represent the integrated model-to-ensemble difference, $\bar{\delta}$ (%) obtained with respect to the mean of all models that have performed the complete set of prescribed total BRF simulations for any given experiment/spectral regime combination. (See Fig. 9 on page 194)

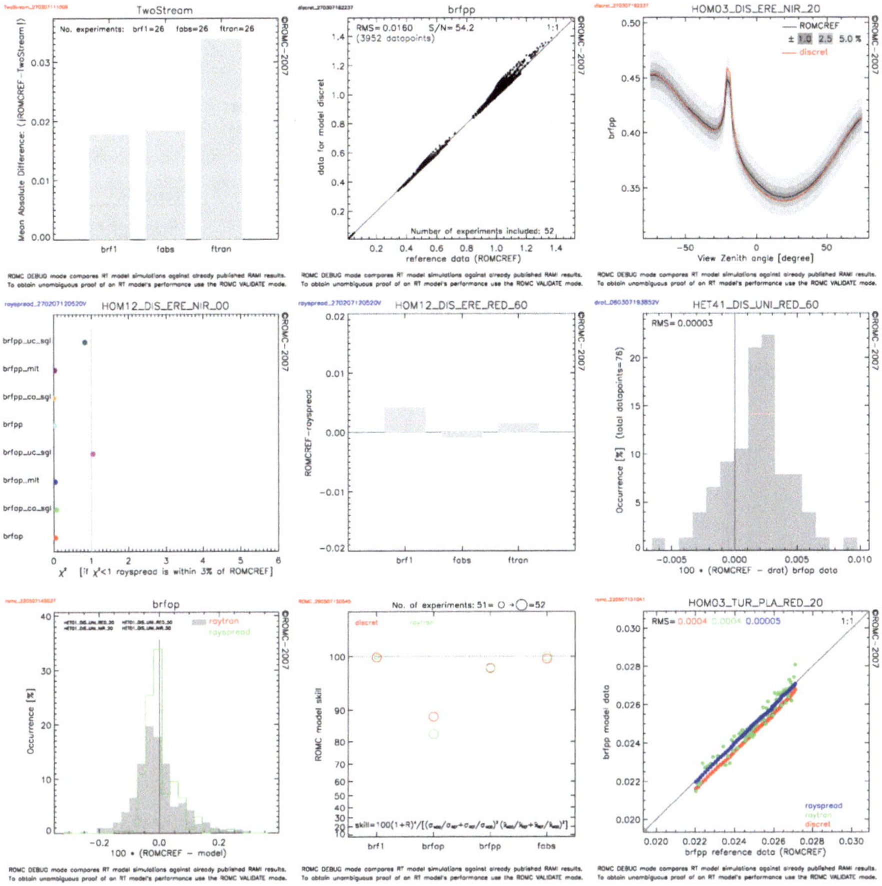

Plate 29: Examples of currently generated ROMC graph types generated in "debug mode" (*top row*), "validate mode" (*middle row*), and using the "interactive" ROMC capabilities when analysing simulation results generated in "debug mode" (*bottom row*). For more information see text. (See Fig. 10 on page 199)

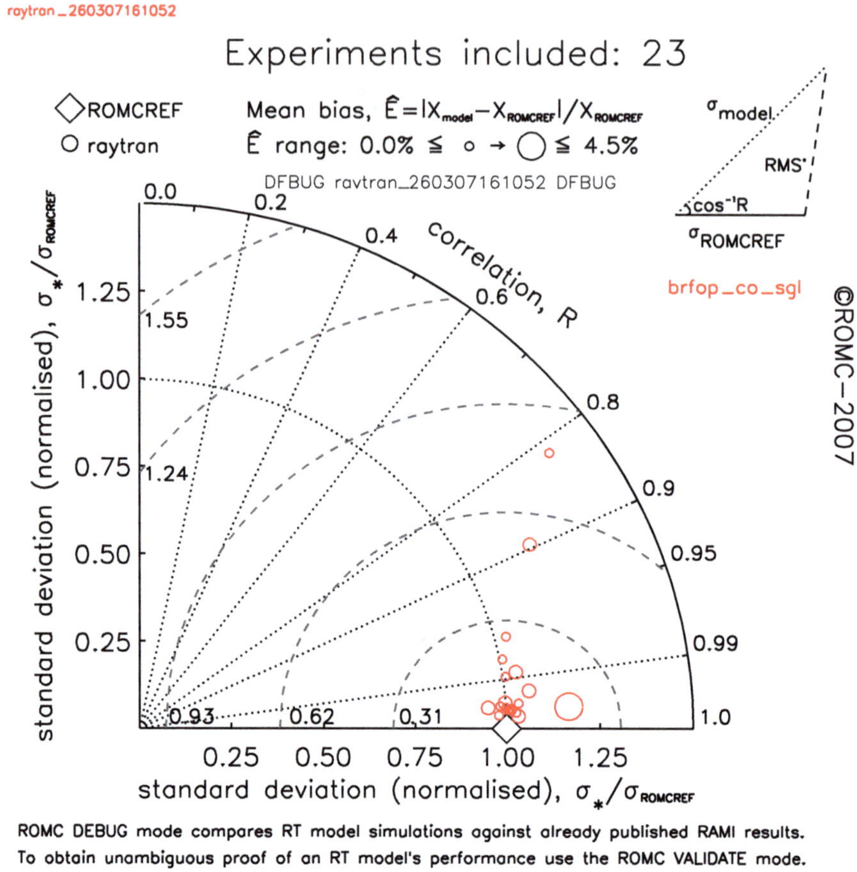

Plate 30: Example of a ROMC generated Taylor diagram in "debug mode" depicting the performance of the **raytran** model in terms of its ability to match the pattern (correlation R) and variability (normalised standard deviation) of the single-collided BRF anisotropy of the reference solution. The normalised RMS difference error (*dashed circular arcs*) and the mean bias (*symbol size*) are also indicated. (See Fig. 11 on page 201)

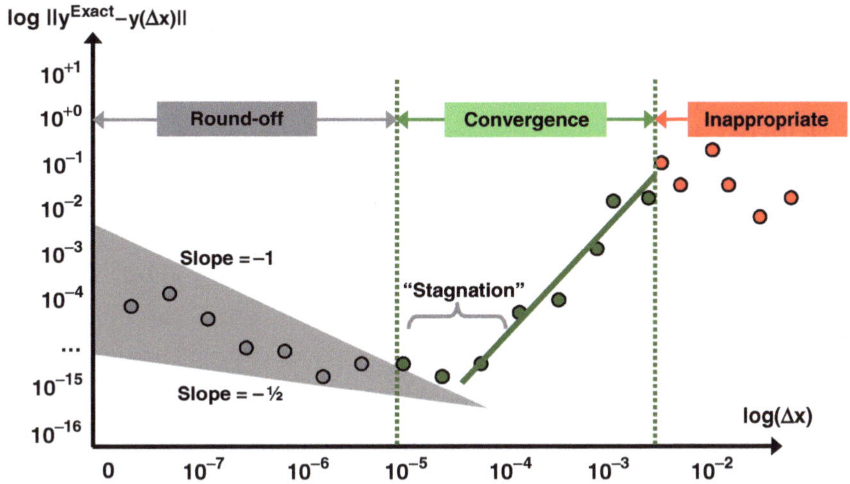

Plate 31: Illustration of the asymptotic regime of convergence. (See Fig. 1 on page 232)

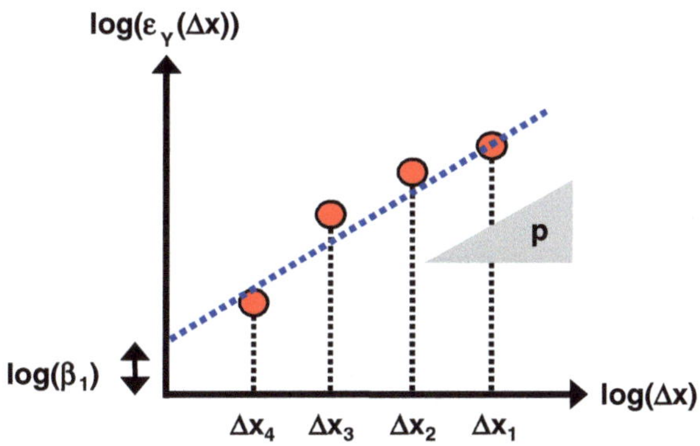

Plate 32: Illustration of solution error $\varpi_Y(\Delta x) = \|y^{\text{Exact}} - y(\Delta x)\|$ vs. mesh size Δx. (See Fig. 2 on page 233)

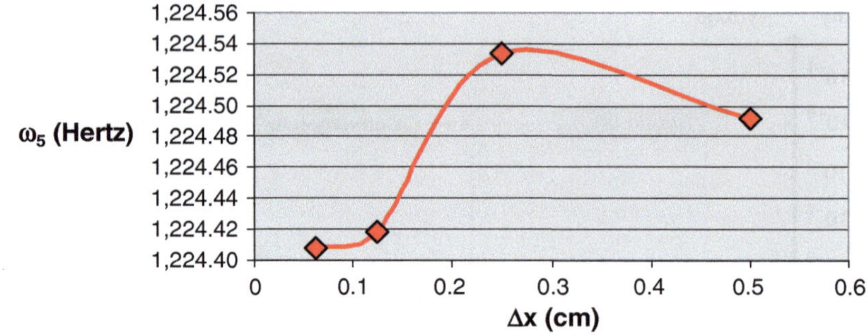

Plate 33: Convergence of a resonant frequency versus element size. (See Fig. 3 on page 236)

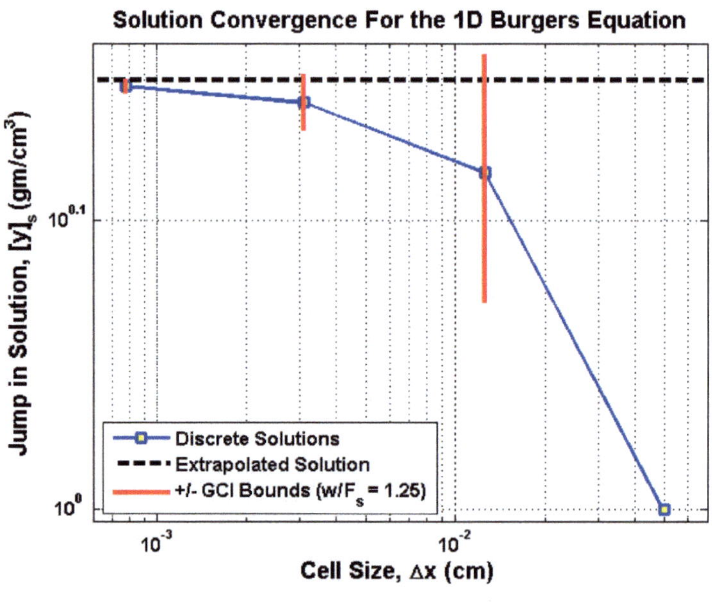

Plate 34: Convergence of a discontinuity $[y]_S$ as a function of cell size Δx. (See Fig. 4 on page 238)

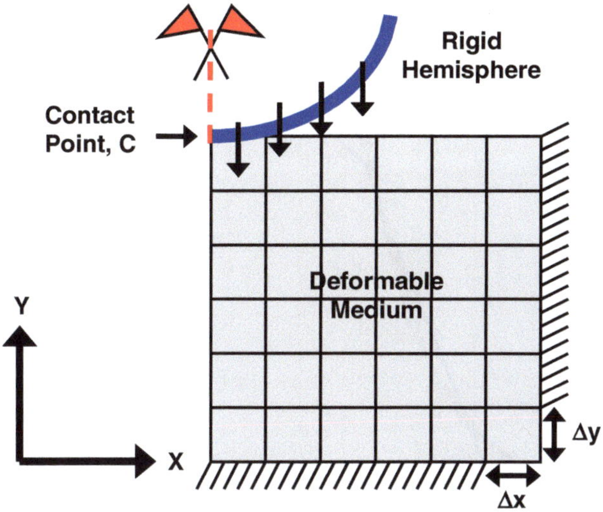

Plate 35: Illustration of the Hertz contact problem. (See Fig. 5 on page 241)

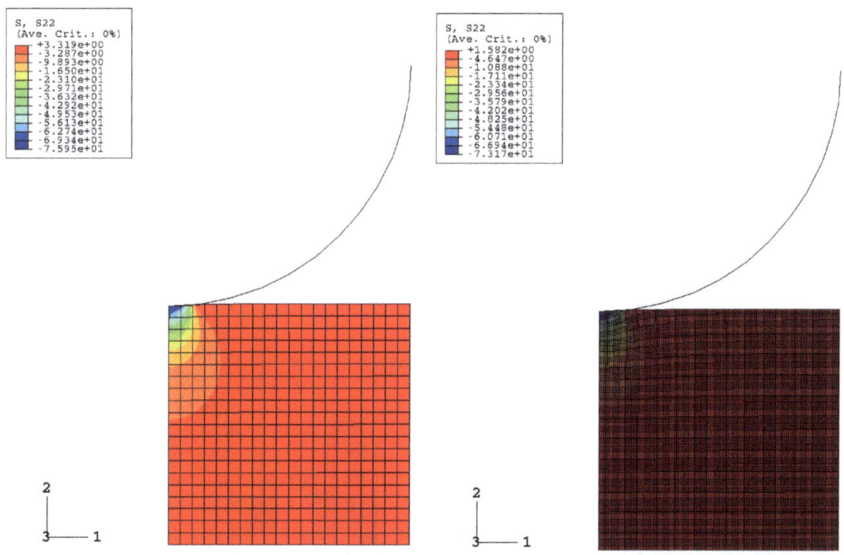

(**a**) Extra-Coarse mesh with 20 elements (**b**) Fine mesh with 160 elements

Plate 36: Stress fields σ_{YY} for the Hertz contact problem. (See Fig. 6 on page 242)

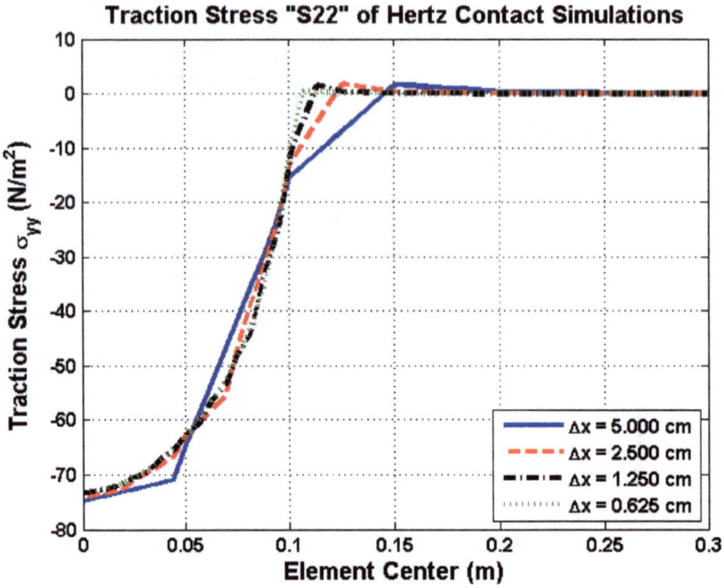

Plate 37: Stress values (σ_{YY}) vs. location (x) of top-row finite elements. (See Fig. 7 on page 243)

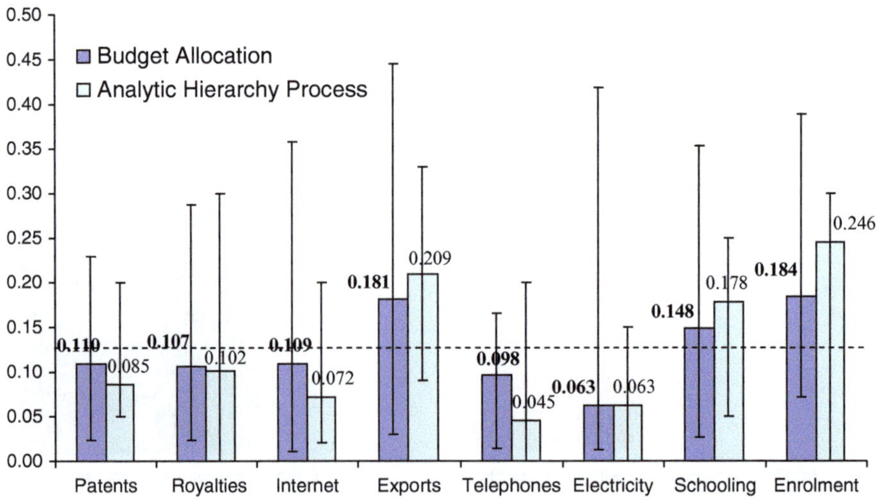

Plate 38: Statistics on the weights assigned to the eight indicators of MC-TAI. The weights derived from pilot surveys of 20 informed interviewees using budget allocation and analytic hierarchy process. Numbers report mean values, the bars indicate minimum or maximum values. (See Fig. 1 on page 263)

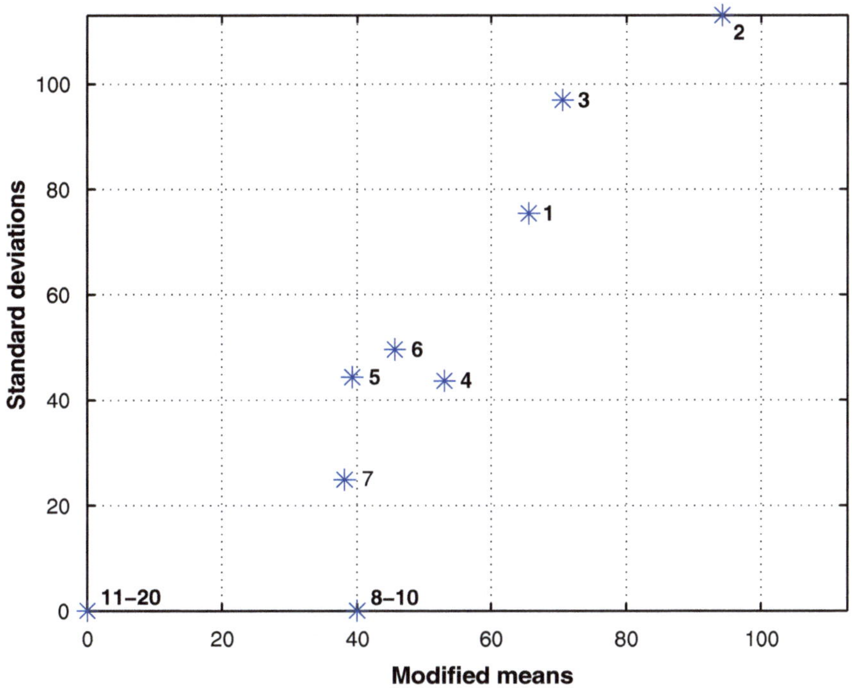

Plate 39: An example screening plot. (See Fig. 1 on page 283)

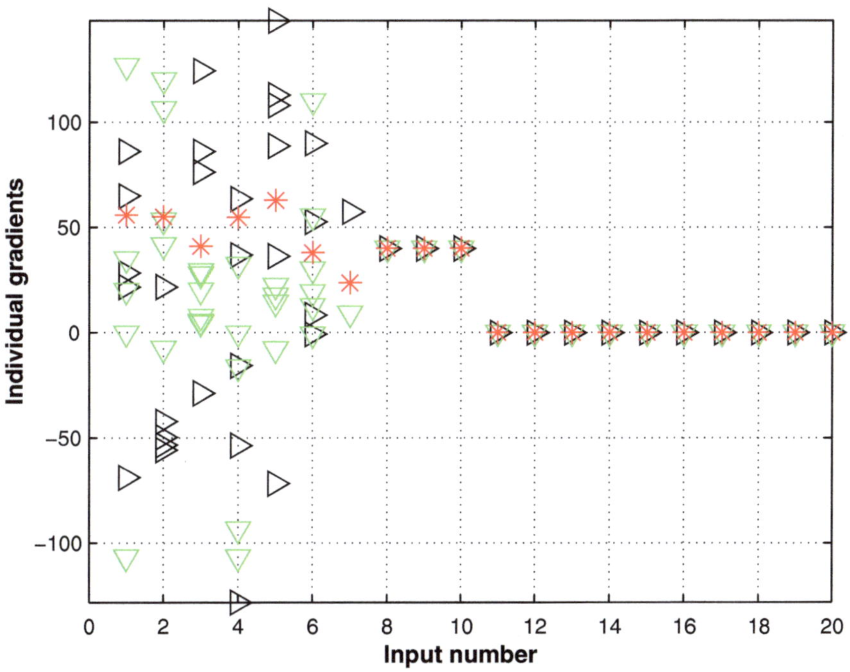

Plate 40: An example scatter plot. (See Fig. 2 on page 285)

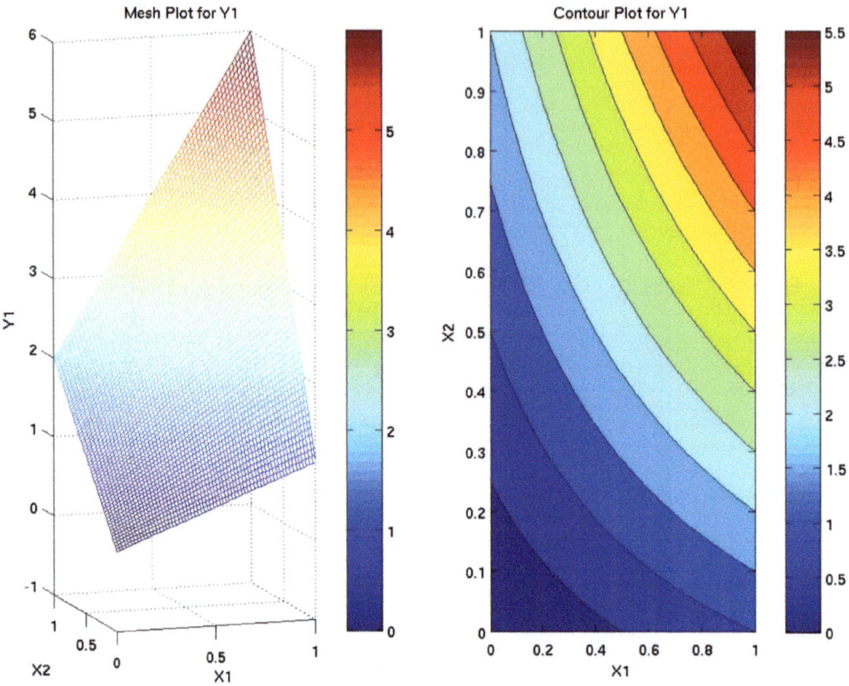

Plate 41: An example response surface plot. (See Fig. 3 on page 289)

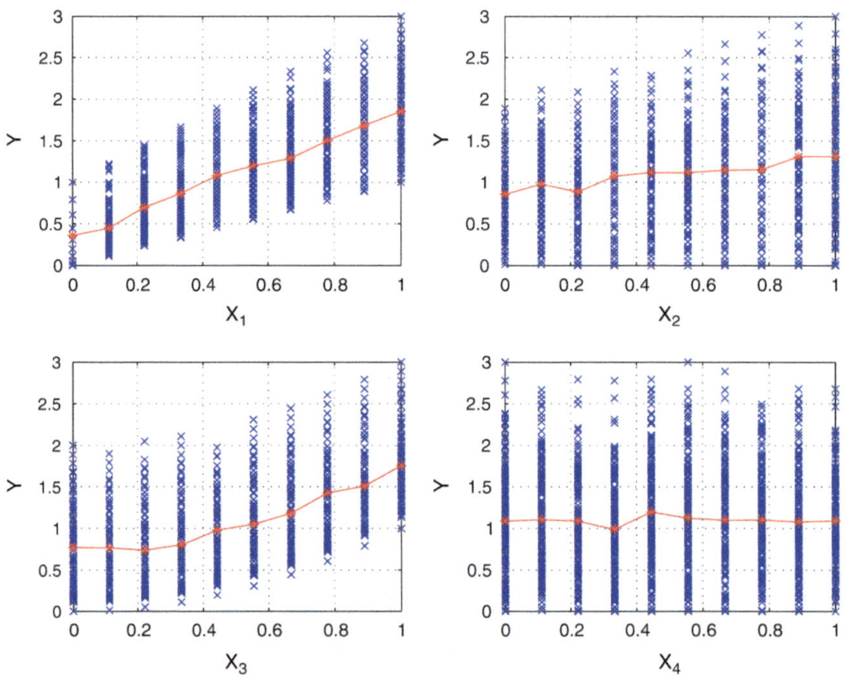

Plate 42: An example scatter plot from replicated Latin hypercube. (See Fig. 4 on page 291)

Editorial Policy

1. Volumes in the following three categories will be published in LNCSE:

i) Research monographs
ii) Lecture and seminar notes
iii) Conference proceedings

Those considering a book which might be suitable for the series are strongly advised to contact the publisher or the series editors at an early stage.

2. Categories i) and ii). These categories will be emphasized by Lecture Notes in Computational Science and Engineering. **Submissions by interdisciplinary teams of authors are encouraged.** The goal is to report new developments – quickly, informally, and in a way that will make them accessible to non-specialists. In the evaluation of submissions timeliness of the work is an important criterion. Texts should be well-rounded, well-written and reasonably self-contained. In most cases the work will contain results of others as well as those of the author(s). In each case the author(s) should provide sufficient motivation, examples, and applications. In this respect, Ph.D. theses will usually be deemed unsuitable for the Lecture Notes series. Proposals for volumes in these categories should be submitted either to one of the series editors or to Springer-Verlag, Heidelberg, and will be refereed. A provisional judgment on the acceptability of a project can be based on partial information about the work: a detailed outline describing the contents of each chapter, the estimated length, a bibliography, and one or two sample chapters – or a first draft. A final decision whether to accept will rest on an evaluation of the completed work which should include

- at least 100 pages of text;
- a table of contents;
- an informative introduction perhaps with some historical remarks which should be accessible to readers unfamiliar with the topic treated;
- a subject index.

3. Category iii). Conference proceedings will be considered for publication provided that they are both of exceptional interest and devoted to a single topic. One (or more) expert participants will act as the scientific editor(s) of the volume. They select the papers which are suitable for inclusion and have them individually refereed as for a journal. Papers not closely related to the central topic are to be excluded. Organizers should contact Lecture Notes in Computational Science and Engineering at the planning stage.

In exceptional cases some other multi-author-volumes may be considered in this category.

4. Format. Only works in English are considered. They should be submitted in camera-ready form according to Springer-Verlag's specifications.
Electronic material can be included if appropriate. Please contact the publisher.
Technical instructions and/or LaTeX macros are available via http://www.springer.com/authors/book+authors?SGWID=0-154102-12-417900-0. The macros can also be sent on request.

General Remarks

Lecture Notes are printed by photo-offset from the master-copy delivered in camera-ready form by the authors. For this purpose Springer-Verlag provides technical instructions for the preparation of manuscripts. See also *Editorial Policy.*

Careful preparation of manuscripts will help keep production time short and ensure a satisfactory appearance of the finished book.

The following terms and conditions hold:

Categories i), ii), and iii):
Authors receive 50 free copies of their book. No royalty is paid. Commitment to publish is made by letter of intent rather than by signing a formal contract. Springer- Verlag secures the copyright for each volume.

For conference proceedings, editors receive a total of 50 free copies of their volume for distribution to the contributing authors.

All categories:
Authors are entitled to purchase further copies of their book and other Springer mathematics books for their personal use, at a discount of 33.3% directly from Springer-Verlag.

Addresses:

Timothy J. Barth
NASA Ames Research Center
NAS Division
Moffett Field, CA 94035, USA
e-mail: barth@nas.nasa.gov

Michael Griebel
Institut für Numerische Simulation
der Universität Bonn
Wegelerstr. 6
53115 Bonn, Germany
e-mail: griebel@ins.uni-bonn.de

David E. Keyes
Department of Applied Physics
and Applied Mathematics
Columbia University
200 S. W. Mudd Building
500 W. 120th Street
New York, NY 10027, USA
e-mail: david.keyes@columbia.edu

Risto M. Nieminen
Laboratory of Physics
Helsinki University of Technology
02150 Espoo, Finland
e-mail: rni@fyslab.hut.fi

Dirk Roose
Department of Computer Science
Katholieke Universiteit Leuven
Celestijnenlaan 200A
3001 Leuven-Heverlee, Belgium
e-mail: dirk.roose@cs.kuleuven.ac.be

Tamar Schlick
Department of Chemistry
Courant Institute of Mathematical
Sciences
New York University
and Howard Hughes Medical Institute
251 Mercer Street
New York, NY 10012, USA
e-mail: schlick@nyu.edu

Mathematics Editor at Springer:
Martin Peters
Springer-Verlag
Mathematics Editorial IV
Tiergartenstrasse 17
D-69121 Heidelberg, Germany
Tel.: *49 (6221) 487-8185
Fax: *49 (6221) 487-8355
e-mail: martin.peters@springer.com

Lecture Notes in Computational Science and Engineering

1. D. Funaro, *Spectral Elements for Transport-Dominated Equations.*
2. H. P. Langtangen, *Computational Partial Differential Equations.* Numerical Methods and Diffpack Programming.
3. W. Hackbusch, G. Wittum (eds.), *Multigrid Methods V.*
4. P. Deuflhard, J. Hermans, B. Leimkuhler, A. E. Mark, S. Reich, R. D. Skeel (eds.), *Computational Molecular Dynamics: Challenges, Methods, Ideas.*
5. D. Kröner, M. Ohlberger, C. Rohde (eds.), *An Introduction to Recent Developments in Theory and Numerics for Conservation Laws.*
6. S. Turek, *Efficient Solvers for Incompressible Flow Problems.* An Algorithmic and Computational Approach.
7. R. von Schwerin, *__Multi Body System SIM__ulation.* Numerical Methods, Algorithms, and Software.
8. H.-J. Bungartz, F. Durst, C. Zenger (eds.), *High Performance Scientific and Engineering Computing.*
9. T. J. Barth, H. Deconinck (eds.), *High-Order Methods for Computational Physics.*
10. H. P. Langtangen, A. M. Bruaset, E. Quak (eds.), *Advances in Software Tools for Scientific Computing.*
11. B. Cockburn, G. E. Karniadakis, C.-W. Shu (eds.), *Discontinuous Galerkin Methods.* Theory, Computation and Applications.
12. U. van Rienen, *Numerical Methods in Computational Electrodynamics.* Linear Systems in Practical Applications.
13. B. Engquist, L. Johnsson, M. Hammill, F. Short (eds.), *Simulation and Visualization on the Grid.*
14. E. Dick, K. Riemslagh, J. Vierendeels (eds.), *Multigrid Methods VI.*
15. A. Frommer, T. Lippert, B. Medeke, K. Schilling (eds.), *Numerical Challenges in Lattice Quantum Chromodynamics.*
16. J. Lang, *Adaptive Multilevel Solution of Nonlinear Parabolic PDE Systems.* Theory, Algorithm, and Applications.
17. B. I. Wohlmuth, *Discretization Methods and Iterative Solvers Based on Domain Decomposition.*
18. U. van Rienen, M. Günther, D. Hecht (eds.), *Scientific Computing in Electrical Engineering.*
19. I. Babuška, P. G. Ciarlet, T. Miyoshi (eds.), *Mathematical Modeling and Numerical Simulation in Continuum Mechanics.*
20. T. J. Barth, T. Chan, R. Haimes (eds.), *Multiscale and Multiresolution Methods.* Theory and Applications.
21. M. Breuer, F. Durst, C. Zenger (eds.), *High Performance Scientific and Engineering Computing.*
22. K. Urban, *Wavelets in Numerical Simulation.* Problem Adapted Construction and Applications.

23. L. F. Pavarino, A. Toselli (eds.), *Recent Developments in Domain Decomposition Methods.*

24. T. Schlick, H. H. Gan (eds.), *Computational Methods for Macromolecules: Challenges and Applications.*

25. T. J. Barth, H. Deconinck (eds.), *Error Estimation and Adaptive Discretization Methods in Computational Fluid Dynamics.*

26. M. Griebel, M. A. Schweitzer (eds.), *Meshfree Methods for Partial Differential Equations.*

27. S. Müller, *Adaptive Multiscale Schemes for Conservation Laws.*

28. C. Carstensen, S. Funken, W. Hackbusch, R. H. W. Hoppe, P. Monk (eds.), *Computational Electromagnetics.*

29. M. A. Schweitzer, *A Parallel Multilevel Partition of Unity Method for Elliptic Partial Differential Equations.*

30. T. Biegler, O. Ghattas, M. Heinkenschloss, B. van Bloemen Waanders (eds.), *Large-Scale PDE-Constrained Optimization.*

31. M. Ainsworth, P. Davies, D. Duncan, P. Martin, B. Rynne (eds.), *Topics in Computational Wave Propagation.* Direct and Inverse Problems.

32. H. Emmerich, B. Nestler, M. Schreckenberg (eds.), *Interface and Transport Dynamics.* Computational Modelling.

33. H. P. Langtangen, A. Tveito (eds.), *Advanced Topics in Computational Partial Differential Equations.* Numerical Methods and Diffpack Programming.

34. V. John, *Large Eddy Simulation of Turbulent Incompressible Flows.* Analytical and Numerical Results for a Class of LES Models.

35. E. Bänsch (ed.), *Challenges in Scientific Computing - CISC 2002.*

36. B. N. Khoromskij, G. Wittum, *Numerical Solution of Elliptic Differential Equations by Reduction to the Interface.*

37. A. Iske, *Multiresolution Methods in Scattered Data Modelling.*

38. S.-I. Niculescu, K. Gu (eds.), *Advances in Time-Delay Systems.*

39. S. Attinger, P. Koumoutsakos (eds.), *Multiscale Modelling and Simulation.*

40. R. Kornhuber, R. Hoppe, J. Périaux, O. Pironneau, O. Wildlund, J. Xu (eds.), *Domain Decomposition Methods in Science and Engineering.*

41. T. Plewa, T. Linde, V.G. Weirs (eds.), *Adaptive Mesh Refinement – Theory and Applications.*

42. A. Schmidt, K.G. Siebert, *Design of Adaptive Finite Element Software.* The Finite Element Toolbox ALBERTA.

43. M. Griebel, M.A. Schweitzer (eds.), *Meshfree Methods for Partial Differential Equations II.*

44. B. Engquist, P. Lötstedt, O. Runborg (eds.), *Multiscale Methods in Science and Engineering.*

45. P. Benner, V. Mehrmann, D.C. Sorensen (eds.), *Dimension Reduction of Large-Scale Systems.*

46. D. Kressner, *Numerical Methods for General and Structured Eigenvalue Problems.*

47. A. Boriçi, A. Frommer, B. Joó, A. Kennedy, B. Pendleton (eds.), *QCD and Numerical Analysis III.*

48. F. Graziani (ed.), *Computational Methods in Transport.*

49. B. Leimkuhler, C. Chipot, R. Elber, A. Laaksonen, A. Mark, T. Schlick, C. Schütte, R. Skeel (eds.), *New Algorithms for Macromolecular Simulation.*

50. M. Bücker, G. Corliss, P. Hovland, U. Naumann, B. Norris (eds.), *Automatic Differentiation: Applications, Theory, and Implementations.*

51. A.M. Bruaset, A. Tveito (eds.), *Numerical Solution of Partial Differential Equations on Parallel Computers.*

52. K.H. Hoffmann, A. Meyer (eds.), *Parallel Algorithms and Cluster Computing.*

53. H.-J. Bungartz, M. Schäfer (eds.), *Fluid-Structure Interaction.*

54. J. Behrens, *Adaptive Atmospheric Modeling.*

55. O. Widlund, D. Keyes (eds.), *Domain Decomposition Methods in Science and Engineering XVI.*

56. S. Kassinos, C. Langer, G. Iaccarino, P. Moin (eds.), *Complex Effects in Large Eddy Simulations.*

57. M. Griebel, M.A Schweitzer (eds.), *Meshfree Methods for Partial Differential Equations III.*

58. A.N. Gorban, B. Kégl, D.C. Wunsch, A. Zinovyev (eds.), *Principal Manifolds for Data Visualization and Dimension Reduction.*

59. H. Ammari (ed.), *Modeling and Computations in Electromagnetics: A Volume Dedicated to Jean-Claude Nédélec.*

60. U. Langer, M. Discacciati, D. Keyes, O. Widlund, W. Zulehner (eds.), *Domain Decomposition Methods in Science and Engineering XVII.*

61. T. Mathew, *Domain Decomposition Methods for the Numerical Solution of Partial Differential Equations.*

62. F. Graziani (ed.), *Computational Methods in Transport: Verification and Validation.*

For further information on these books please have a look at our mathematics catalogue at the following URL: `www.springer.com/series/3527`

Monographs in Computational Science and Engineering

1. J. Sundnes, G.T. Lines, X. Cai, B.F. Nielsen, K.-A. Mardal, A. Tveito, *Computing the Electrical Activity in the Heart.*

For further information on this book, please have a look at our mathematics catalogue at the following URL: `www.springer.com/series/7417`

Texts in Computational Science and Engineering

1. H. P. Langtangen, *Computational Partial Differential Equations.* Numerical Methods and Diffpack Programming. 2nd Edition
2. A. Quarteroni, F. Saleri, *Scientific Computing with MATLAB and Octave.* 2nd Edition
3. H. P. Langtangen, *Python Scripting for Computational Science.* 3rd Edition
4. H. Gardner, G. Manduchi, *Design Patterns for e-Science.*
5. M. Griebel, S. Knapek, G. Zumbusch, *Numerical Simulation in Molecular Dynamics.*

For further information on these books please have a look at our mathematics catalogue at the following URL: `www.springer.com/series/5151`

MIX
Papier aus verantwortungsvollen Quellen
Paper from responsible sources
FSC® C105338

If you have any concerns about our products,
you can contact us on
ProductSafety@springernature.com

In case Publisher is established outside the EU,
the EU authorized representative is:
Springer Nature Customer Service Center GmbH
Europaplatz 3, 69115 Heidelberg, Germany

Printed by Libri Plureos GmbH
in Hamburg, Germany

LINKING PEOPLE, PLACE, AND POLICY

A GIScience Approach

LINKING PEOPLE, PLACE, AND POLICY

A GIScience Approach

edited by

Stephen J. Walsh
University of North Carolina
Chapel Hill, U.S.A.

Kelley A. Crews-Meyer
University of Texas
Austin, U.S.A.

SPRINGER SCIENCE+BUSINESS MEDIA, LLC

Library of Congress Cataloging-in-Publication Data

Linking people, place, and policy : a GIScience approach / edited by Stephen J. Walsh, Kelley A. Crews-Meyer.

p. cm.

Includes bibliographical references and index.

Additional material to this book can be downloaded from http://extras.springer.com.

ISBN 978-1-4613-5337-9 ISBN 978-1-4615-0985-1 (eBook)

DOI 10.1007/978-1-4615-0985-1

1. Nature—Effect of human beings on – Case studies. 2. Natural areas—Management—Case studies. 3. Environmental policy—Case studies. 4. Geographic information systems. I. Walsh, Stephen Joseph, 1951- II. Crews-Meyer, Kelley A.

GF75 .L56 2002
333. 7'2—dc21

2002016133

Printed on acid-free paper.

TABLE OF CONTENTS

CO-EDITORS: ABBREVIATED PROFILES

Stephen J. Walsh is a Professor of Geography, Director of the Landscape Characterisation & Spatial Analysis Lab (LC&SAL), and Research Fellow of the Carolina Population Centre at the University of North Carolina, Chapel Hill, USA. He is the former Amos H. Hawley Professor of Geography (1993-1996) and Director of the Spatial Analysis Unit at the Carolina Population Centre (1992-1997). He is the past Chair of the GIS (1997-1999) and the Remote Sensing (1995-1997) Specialty Groups of the Association of American Geographers (AAG). In 1997, Professor Walsh was the recipient of the Outstanding Contribution Award and Medal from the Remote Sensing Specialty Group of the AAG, in 1999 was awarded Research Honours from the Southeastern Division of the AAG, and in 2001 was awarded Research Scholarship Honours from the Association of American Geographers. He is on the editorial boards of *Plant Ecology* (1996-present), *Journal of Geography* (1995-present), *The Professional Geographer* (1997-2001), and the *Southeastern Geographer* (1992-2000). He also has recently co-authored special remote sensing and GIS application issues in the *Journal of Vegetation Science* (1994) and *Geomorphology* (1998). Professor Walsh's research interests are in GIS, remote sensing, spatial analysis, physical geography and population-environment interactions. Specific research foci are in pattern and process at the alpine treeline ecotone, biodiversity and river dynamics, scale dependence and information scaling, and landuse/landcover change modelling. Major studies underway in Ecuador, Thailand, and North Carolina and Montana, USA.

Kelley A. Crews-Meyer is an Assistant Professor of Geography at the University of Texas – Austin where she teaches remote sensing, GIS, and policy ecology. She is the Director of the department's Geographic Computing Systems as well as an Associate Director of the university's Environmental Science Institute. She is formally affiliated with the Center for Space Research, Population Research Center, and Lazano Long Institute for Latin American Studies. She holds a Ph.D. in Geography from the University of North Carolina and a post-graduate certificate from UNC in Public Policy Analysis with emphasis on environmental policy analysis. Her M.A. and B.S. degrees are from the University of South Carolina in Government and International Studies and Marine Science respectively. Her research centers upon the intersection of GIScience/remote sensing, policy analysis, and quantitative landscape ecology for assessing human-environment interactions across space and time. Her fieldwork and research experiences have concentrated primarily on tropical environments (Amazon, Southeast Asia), with new projects developing in the Choco of Ecuador, the

Peruvian highlands, and the Texas coastal plain and borderlands. Recent publications have appeared in *Geocarto International* and *Agriculture Ecosystems and Environment*, and focused upon longitudinal analysis of products derived from satellite imagery for assessing scale, pattern, and process. Recent awards include those from the American Society of Photogrammetry and Remote Sensing, the GIS Specialty Group of the Association of American Geographers, Sigma Xi Scientific Research Society, and the Open Meeting of the Human Dimensions of Global Change 2001 (UCAR – Union of Concerned Atmospheric Researchers).

Acknowledgements

Cover:
Original watercolor by Esca H. Crews, Jr.;
digital design by Philip H. Page, University of North Carolina.

CD-ROM: Produced by Philip H. Page, University of North Carolina.

Index creation and editorial assistance:
Derek M. Burgess, University of Texas.

Reviewers:
Thomas Bassett
Ling Bian
Michael Binford
Michael Bishop
Eduardo Brondizio
Tom P. Evans
Martha Geores
Myron Gutman
Andrew Hansen
Philip Keating
Nina M. Kelly
Rheyna Laney
Jack Liu
Diana Liverman
C.P. Lo
William McConnell
Robert McMaster
Stephen Perz
Gil Pontius
Basil Savitsky
Rodrigo Sierra
Randolph Wynne

Contributors

Luca Alinovi
Richard E. Bilsborrow
Bruce Boucek
Daniel G. Brown
DongMei Chen
Walter Chomentowski
David Craven
Thomas W. Crawford
Kelley A. Crews-Meyer
Free De Koning
Antonio DiGregorio
Scott A. Drzyzga
Barbara Entwisle
Jefferson Fox
Arthur Getis
Barry Haack
Changchui He
Susan Jampoler
Zdenek Kalensky
Robert K. Kaufmann
Kasper Kok
Eric F. Lambin
John S. Latham
Stephen D. McCracken
Joseph P. Messina
Emilio F. Moran
William K.Y. Pan
Pramote Prasartkul
Julia Reade
Ronald R. Rindfuss
Yothin Sawangdee
Suzanne Serneels
Karen C. Seto
David Skole
Welmoed Soepboer
Elizabeth Solomon
Douglas Stow
D. Michael Thompson
Tom Veldkamp
Peter H. Verburg
Robert Walker
Stephen J. Walsh
Charles H. Wood
Curtis E. Woodcock

Chapter 1

CHALLENGES FOR GISCIENCE: ASSESSMENT OF POLICY RELEVANT HUMAN-ENVIRONMENT INTERACTIONS

Kelley A. Crews-Meyer

Department of Geography, University of Texas, Austin, USA
kacm@uts.cc.utexas.edu

In 2001, National Academy Press published *Grand Challenges in Environmental Sciences* in response to National Science Foundation's request of the National Research Council to identify the most important environmental research agenda items for the next decade that would have the greatest practical bearing on science and policy. Not long afterward, the 2001 Open Meeting of the Humans Dimensions of Global Change community was convened in Rio de Janeiro to ascertain their own priorities for the global agenda. Not coincidentally, a number of emergent themes voiced in these two organizations have also increasingly been discussed by researchers studying human-environment interactions as pressing agenda items for further understanding the interactions between people and the landscape as influencing and being influenced by policy. The NAP 2001 report, for instance, cited land-use dynamics as not only one of the eight (out of 200) most pressing issues facing environmental scientists and policymakers, but moreover recommended land-use dynamics as one of the four areas most likely to achieve significant and practical gains over the next decade given adequate scholarly attention and support. Most important for this work was the continued development of a suite of techniques that have come to be referred to as GIScience. Geographic Information Science (GISc) refers to the integration of two components: scientific theory and information systems. The information systems approach synthesized through the evolution of Geographic Information Systems (GIS) is united with the explicit understanding of the importance of the substantive or thematic nature of the subject of analysis. That is, any application must be informed by understanding both the phenomenon under study as well as the software and methodologies used to assess that phenomenon. GISc unites a variety of spatial digital technologies including but not limited to Geographic Information Systems

(GIS), remote sensing, spatial analysis and statistics, spatial simulations and modeling, data visualization, and global positioning systems (GPS).

In this book, Stephen J. Walsh (Department of Geography, University of North Carolina – Chapel Hill, USA) and Kelley A. Crews-Meyer (Department of Geography, University of Texas – Austin, USA) present a collection of works from both prominent and emerging authors utilizing a GIScience approach to assess the complex interactions among people, place, and policy. More specifically, this book describes a latitude of research centered on the study of human-environment interactions with particular emphasis on land use and land cover dynamics. Concepts, tools, and techniques within GISc serve as the unifying methodological framework for the study of landscapes representing North, South, and Central America as well as Africa and Asia. Analyses conducted through quantitative, qualitative, and image-based techniques are presented to illustrate the methodological breadth and depth of GIScience research assessing the linkages between humans and the environment. The long list of contributors to this collection represents an array of disciplinary and other affiliations recognized as necessary to perform the type of research requiring not a *multi*disciplinary approach but an *inter*disciplinary one. Academicians from geography, sociology, biostatistics, geology, environmental science, and regional studies are represented, as are researchers from both governmental and private sectors.

Increasingly, policymakers and researchers are confronted with problems at multiple spatial and temporal scales or extents, such as global climatic change, land use/land cover change (LULCC) at regional to local settings, and the introduction of new land use policies that have implications across socioeconomic, biophysical, and geographical domains. The interchanges between humans and the environment are played out through land development scenarios and land transformation activities that often create feedbacks and thresholds among people, place, and policy. Yet a lack of adequate tools and theoretical understanding across the social, natural, and spatial sciences has traditionally led researchers to focus on relatively coarse grains of countries and/or counties (or similar units) where aggregate data are available. But it is at the finer social (e.g., villages and households), biophysical (e.g., watersheds, and terrain strata), or spatial (e.g., pixels and facets) scales where spatially explicit information may be more appropriately collected, derived, and applied. Furthermore, even where adequate data have been collected at an appropriate spatial scale in a spatially explicit manner, data analyses have focused on either people or the environment, without suitable integration.

The chapters contained in this book take a step along the path of integration by assimilating thematic domains and space-time scales in the study of land use and land cover dynamics and the drivers of change. A key element of this discourse is the implication of policy and how results are interpreted within a policy-relevant context. The purpose of this book is to address a need for a comprehensive and rigorous treatment of GISc within the context of human-environment interactions as a means of directing areas upon which future research and study should focus to achieve maximal results. Many research and policy communities (including but not limited to human dimensions of global change, LULCC, landscape ecology, human-environment interactions) are collectively viewing the landscape within a spatially explicit perspective, where people are viewed as agents of landscape change that shape and are shaped by the landscape, and where landscape form and function are assessed within a space-time context.

Some challenges and opportunities being addressed include, for instance, work linking specific households and associated demographics with local landuse decision-making as well as satellite-based measures of LULC. Experimenting on how best to link disparate data types has become an integral part of the larger question of forces driving landuse decision-making, and at what levels those forces are more or less understandable and associated with different explanatory factors. The question is not if the processes themselves are scale dependent but if the ability to detect patterns representing those processes is scale dependent. Are forces driving LULC decisions different at household versus community versus regional levels? How can "exogenous" or external effects operating at more macro spatial or temporal scales (such as national policies or climate change) be explicitly incorporated into the understanding of those forces and concomitant changes? What are the spatio-temporal relationships between endogenous and exogenous effects on the linkages between people, place, and policy? Are there significant variations in local to regional to continental patterns of landscape change and the communities and cultures residing in those lands?

From a methodological standpoint, what are the techniques for not only conducting these analyses, but moreover for validating analyses conducted with varying availability of fieldwork and remotely sensed information? How do those techniques change with the advancement of sensor systems—is there a best sensor to use for this vein of research, or is a combination of data sources and platforms critical for successful evaluation? And what mechanisms exist (or should be developed) to ease comparison of studies conducted across scales, geographic areas and cultures, and methodological approaches to distill and diffuse findings and policy relevant information among the stakeholder communities?

One of the problems associated with such a rich selection of themes and chapters is the organizational challenge of bringing together a variety of applications, study areas, and thematic pursuits. Since the book follows first from an emphasis on methodological developments in remote sensing and GISc applied to human-environment interactions, a decision was made to orient the chapters by methodological approach. The first set of chapters outline the operational difficulties associated with bringing together discrete and continuous data, and ultimately for linking people and land. In Chapter 2, Rindfuss et al. test field strategies designed to support the development of a spatially explicit database of people and the land they use. Thompson et al., in Chapter 3, continue the discussion of data integration through the statistical analysis of socio-economic data tied to landscape conditions. In Chapter 4, Seto et al. also explore how best to link remotely sensed imagery and socio-economic data through several statistical approaches. Crawford, in Chapter 5, then specifically addresses point-to-area transformations through Voronoi diagrams and fuzzy set theory as a means of understanding the impacts of village boundaries on functional space.

The concept of boundaries is important for linking people to the places they both impact and are impacted by, often through a variety of processes and policies operating at different space-time scales. Boundaries may be "artificial" (e.g., political or administrative units) or "natural" (e.g., social units such as neighborhood networks or natural units such as watersheds), and impact human-environment research in data collection and in the assessment of policy impacts on land and people. The clearest example of boundary effects involves land ownership or cadastral information. Yet the impacts of land tenure as linked to people and place are contested in different areas of the world. In chapter 6, Fox launches the next theme of the book, land tenure as a key component of both measuring and managing the feedbacks between landusers and the landscape. Fox uses a suite of aerial photography to assess the changing rural landscape in light of evolving national land tenure policies set within a developing country. Walker et al. (Chapter 7) employ satellite imagery in Brazil to assess similar questions with specific regard to forest resources and land titling, and Drzyzga and Brown (Chapter 8) examine spatio-temporal linkages between land ownership, forest use, and landscape fragmentation in the US described across space and time.

In Chapter 9, Walsh et al. continue the exploration of deforestation and fragmentation using a different set of techniques that includes cellular automaton modeling and longitudinal pattern metric analysis. In this chapter, the importance of classification techniques for assessing landuse/landcover change (LULCC) is underscored and acts as a transition to the next set of chapters focusing explicitly on classification analysis. In Chapter 10, McCracken et al. also examine Amazonian frontier deforestation using classification to assess landscape trajectories. Chen et al. (Chapter 11) utilize a variety of source products to examine the potential for improving LULC assessment through multi-resolution (and multi-product) classification in an urban environment. Haack et al. (Chapter 12) also deal with issues of urbanization that add complexity to remotely sensed data analysis, and provide a bridge to the last set of chapters that deal with some larger issues that emerge from incorporating GISc and remote sensing applications. Haack et al. detail the technical side of building a database designed to support mapping, analysis, and prediction, whereas Latham et al. (Chapter 13) follow with a framework describing the United Nation's FAO methodologies used to support an array of development projects. Lastly, in Chapter 14,Veldkamp et al. describe the implications of remote sensing and GIS applications for policy, providing examples of the types of scenarios that open dialogue between analyst and policymaker, achieved through regional-scale simulations of LULC dynamics.

A number of ecological themes run throughout the book, with perhaps the most common being deforestation studies (especially Chapters 8, 9, and 10). In some chapters (see e.g. 7, 9, and 10), deforestation is linked specifically to agricultural extensification, but the understanding of rural/agricultural systems and concomitant land management is a common undercurrent in the text (see Chapters 3 and 6). But both urban settings and urbanization are covered as well, such as in Chapters 4, 8, 11, and 12. Lastly, the chapters provide a wealth of study sites from around the globe representing different areas, different cultures, and different management systems. Four chapters are set in Latin America: three focus on sections of the Amazon (Brazil in 7 and 10, Ecuador in 9) with Chapter 14 including applications in Central America (Costa Rica). There are also two chapters with analysis performed in the US (8 and 11). Further, two chapters detail work performed in Africa (Kenya in Chapter 3 and the continent in Chapter 13). Finally, six chapters represent projects integrating GIS and remote sensing in Asia: China (Chapter 4), Nepal (Chapter 12), Cambodia (Chapter 6), Thailand (Chapters 2 and 5), and the Philippines (Chapter 14). Collectively, the chapters offer a survey of methods, context, theory, and practice conducted within a spatially explicit framework that examines human-environment interactions and assesses LULC characteristics and dynamics by linking people, place, and policy.

Throughout the book graphics depicting remote sensing, GIS, and spatial/statistical analysis; these graphics describe data collection protocols, image processing results, spatial relationships across space and time, research context, and interpretive findings. Within each chapter, figures are presented only in black and white. But in the attached CD-ROM, color graphics are presented. A user interface was developed by Philip H. Page, UNC-Geography, to navigate to chapter-associated graphics by selecting from a list of figure captions per chapter. The color graphics further enhance each chapter, and readers are encouraged to view them as an integral component of the book's overall presentation.

Chapter 2

CONTINUOUS AND DISCRETE: WHERE THEY HAVE MET IN NANG RONG, THAILAND

Ronald R. Rindfuss[1], Barbara Entwisle[1], Stephen J. Walsh[2], Pramote Prasartkul[3], Yothin Sawangdee[3], Thomas W. Crawford[4], and Julia Reade[5]

[1]Department of Sociology & Carolina Population Center, University of North Carolina, USA ron_rindfuss@unc.edu
[2]Department of Geography & Carolina Population Center, University of North Carolina, USA
[3]Institute for Population and Social Research, Mahidol University, Thailand
[4]Environmental Studies Program, Gettysburg College, USA
[5]Joint Center for Housing Studies, Harvard University, USA

Keywords: linking social-environmental data, village territories, spatial transformations, household and community surveys, remote sensing, GIS.

Abstract A major challenge facing the scientific community is to link people to the land so that social behavior can be studied meaningfully in relation to changes in land use and land cover types, landscape conditions, and associated ecosystem processes. One aspect of the challenge is that the linking must take place at appropriate spatial and temporal scales. Another aspect of the challenge is that social data are typically discrete, referring to point locations, whereas data on land cover and land use are typically continuous. In this paper, we examine the linking of people to the land through data transformations that move from the discrete to the continuous and from the continuous to the discrete, with particular reference to our research in Nang Rong, Thailand. We describe approaches used at the village level, which involve setting village territories, use of population distributions models, and integrating social survey data to describe land activities with satellite-based land cover classifications to describe land composition and spatial organization. These approaches are assessed relative to their underlying assumptions, their advantages and disadvantages, and their limitations. We also discuss our experience measuring links at a more micro level, between households and the plots they use. Although our discussion is with reference to a specific setting and site, many of the general points apply to other complex rural environments in which villages consist of clusters of dwelling units surrounded by agricultural lands, and in which households may farm multiple, non-contiguous parcels.

INTRODUCTION

Increasingly, the broad scientific and policy communities concerned with global and environmental change have been requesting more data to obtain a better understanding of land use (as opposed to land cover) dynamics (e.g., Committee on Global Change Research 1999, Lambin et al. 1999, National Research Council 2000). As soon as we consider land use, as opposed to land cover, consideration of human behavior necessarily enters the discourse. Human behavior and social institutions make a hunting preserve of grassland, distinguish an orchard from a forest, and make what appears to be a grassy knoll the tail end of a landfill. It is only by considering the role of human decision-makers that we can begin to consider likely future changes in land use and land cover.

The reasons for interest in land use dynamics are broad, and, if anything, getting broader. At the global level, there are concerns with a) climate change, b) carbon cycling, and c) land degradation, sustainability, and resilience (e.g., Liverman et al. 1998). For example, with respect to carbon cycling, reforestation (either because of deliberate human actions or through abandonment of farm lands) can serve as an important carbon sink. At the local level, land use is directly related to the health of wetlands and estuaries. For example, in eastern North Carolina, increased run-off rates associated with human development (specifically increased levels of impervious surface coverage from the construction of roofs, roads, driveways, sidewalks, and parking lots) are strongly correlated with elevated levels of fecal coliform bacteria and *Escherichia coli* in tidal creeks and estuaries (Mallin et al. 2000).

Remotely sensed data can be used to gain insight into land use. For example, paddy rice fields do not occur naturally; rather, they are evidence of human activity. However, most research questions involving land use require data beyond satellite imagery or aerial photography. Typically, we want to know something about people: their numbers, their characteristics, their social arrangements, their culture and beliefs, and their actions. Further, we frequently want to know how these various human factors are related to land use dynamics, and how land use dynamics, in turn, might affect human activity. To engage this class of questions requires linking land use data with social data, and doing so frequently involves numerous assumptions.

To date, the most common linkage between social data and land use data has tended to favor high levels of aggregation: countries, provinces, districts or counties. Allen and Barnes (1985), for example, examine the causes of deforestation in a sample of 39 developing countries (also see Bilsborrow 1992, Bilsborrow and Geores 1994). Wood and Skole (1998) merge satellite data with census data for *municipios* (similar to U.S. counties) to study patterns of population density and deforestation in the Brazilian Amazon. Such studies can shed light on population processes and land use dynamics, but inferences are limited by the fact that they aggregate and average over important decision-making units on the social side.

In our work in Nang Rong, Thailand, we have established the link at a more local level, the village. In Nang Rong, nucleated villages, consisting of a cluster of dwelling units surrounded by agricultural land, are the dominant settlement pattern. This paper reviews some of the techniques we have used to link data for discrete households and villages collected in social surveys to spatially continuous information on land cover and land use derived from satellite imagery, maps, and other products. However, while the village has served as a useful link in our research, it is not necessarily the only link or the best link given our substantive research questions. We therefore critically examine the reasonableness of the assumptions inherently involved from the perspective of analytical cartography and of the social, biophysical, and geographical context in which the techniques are applied. Advantages of moving the scale at which the links are established from villages to households are reviewed. In Nang Rong, as in many settings, decision-making about land use often occurs at the household level. However, it is not easy to establish links between households and the plots they use. Not only are dwelling units not generally sited on agricultural land, households often cultivate multiple non-contiguous land parcels. The paper concludes with a description of approaches we have considered as well as successfully used in recently completed fieldwork.

THE SETTING: NANG RONG

Our study site is Nang Rong district, Buriram province, northeast Thailand (Figure 1). The district is an agriculturally marginalized environment of approximately 1300 sq. km in size, positioned within the Korat Plateau, characterized by relatively infertile soils, poor drainage, and inconsistent precipitation levels caused by a highly variable monsoonal rainfall pattern (Kaida and Surarerks 1984, Rigg 1987). Over 80 percent of the average annual precipitation occurs between April and November.

Northeast Thailand (or Isaan) contains about one-third of the country's area, has about a quarter of the country's population, and generates about one-fifth of the GNP. The dominant occupation in the region is farming and the majority of farm households own an average of three hectares of land (Ghassemi et al. 1995). Per capita income is the lowest in the country, largely because of low and unstable agricultural production resulting from erratic rainfall and generally poor soils (Parnwell 1988, Arbhabhirama et al. 1988). Approximately one-third of Isaan's land area is unsuitable for successful cropping due either to steepness or laterization (Parnwell 1988). Another one-third of the land is suitable for rice cultivation, though yields are relatively moderate. In many places the soils are very infertile, including areas with high levels of salinity or acidity (Ghassemi et al. 1995). Soils found around the margins of hills and on uplands are often highly susceptible to erosion.

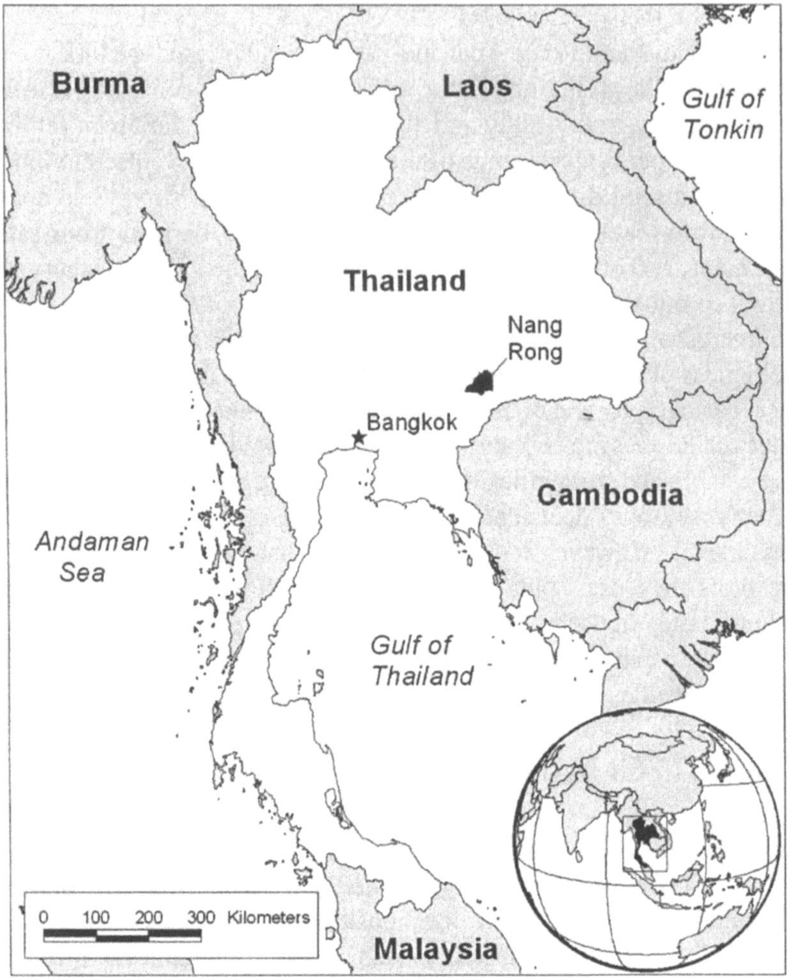

Figure 1. Study area location, Nang Rong District, northeast Thailand.

Deforestation associated with agricultural expansion has been underway for a century or more (Feeny 1988). Prior to World War II, this was associated with increased production of paddy rice, largely for subsistence. More recently, there has also been extensive deforestation in the upland areas of Nang Rong district associated with the cultivation of cash crops such as cassava, in response to shifting world demand. Figure 2 shows land cover classifications for Nang Rong district and a 10 km buffer around the district for 1972/3 and 1997. Massive changes are evident. Over time, the lowlands have been transformed into a landscape matrix dominated by rice paddies, isolated trees in and around the paddies, riparian forests, and forests retained near village compounds (Walsh 1999). In the uplands, forests still are significant in terms of area covered, but cash crops now comprise substantial proportions of the upland area (Walsh

1999). Reforestation is also occurring as a consequence of secondary plant succession, government reforestation programs in conservation forests, and agro-forestry operations primarily involving fruit, rubber, and Eucalyptus tree plantations. The natural vegetation of the district consists of a dry monsoon forest predominated by dwarf dipterocarp trees, and containing areas of grassland, thorny shrubs, and bamboo thickets (Parnwell 1988). The vegetation phenology is largely drought-controlled (Ghassemi et al. 1995, Rundel and Boonpragop 1995).

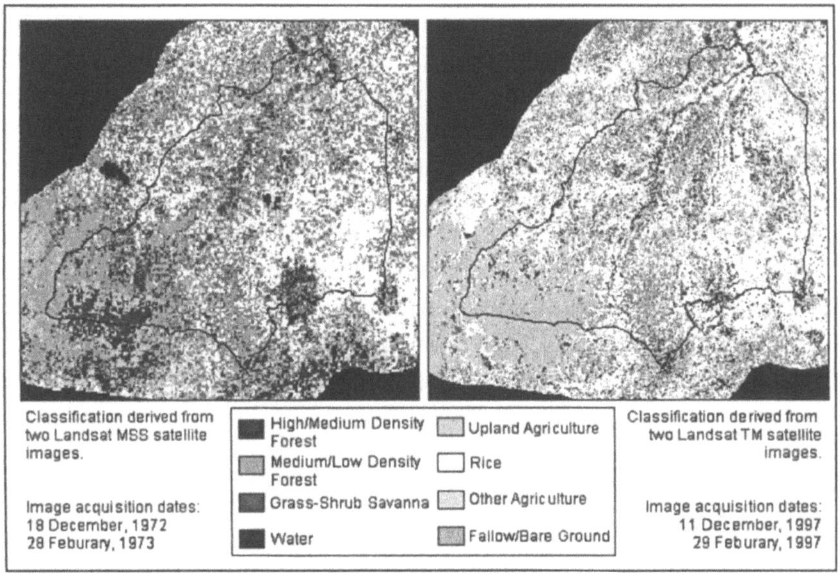

Figure 2. Landsat Thematic Mapper classification of the study area for 1976 and 1997. Changes between the two dates are evident throughout, but most conspicuous in the upland southwest where forests have been replaced by field crops, particularly cassava and to a lesser extent, sugar cane.

Nang Rong has become our laboratory for the study of land cover and land use change, particularly the human dimensions of this change (Entwisle et al. 1998). For example, patterns of land use change depend on the location of villages. In Nang Rong, settlement occurred in several waves, the largest beginning in the 1950s and extending through the 1970s. Given a set of site conditions defined by topography, hydrography, and soils as well as by climate and rainfall, where will households moving into an undeveloped or underdeveloped area choose to cluster their dwellings? Road development is another factor. Are roads the cause, consequence, or corollary of settlement patterns? The highway (referred to as the "Chokechai – Dech-udom Highway") linking Nang Rong to Korat (the third largest city in Thailand and a regional node of the Northeast), and ultimately to Bangkok, was built by the military during the late 1960s, and surely this facilitated migration into the region, but building and upgrading of other roads in the district appears to have been a more local matter (Crews-Meyer 2000). In Nang Rong, the frontier closed in the 1970s. Subsequently, the dominant direction of migration shifted from in to out, in response to the rapid growth of urban job opportunities in construction, manufacturing, and services, especially in Bangkok and periphery. Our research has documented that the decisions made by young people about whether to leave or stay in the rural villages is central to land use and availability (Rindfuss et al. 1996, Entwisle et al. 1998), especially to the conversion of forest to cash crop production. These are only a few of the ways that we are exploring connections between population process and land use change.

The data for our research are described elsewhere (www.cpc.unc.edu/projects/nangrong; Entwisle et al. 1998, Walsh et al. 1999a and b, Rindfuss et al. 2000). Here, our concern is with the linking of people to land through approaches that move from the continuous to the discrete, and from the discrete to the continuous. On the spatial side, thematic GIS coverages (e.g., aerial photographs beginning in 1954, and Landsat, SPOT, and NOAA AVHRR satellite data from the 1970s to the 1990s) have been assembled. On the social side, community and household interviews were conducted in 1984, 1994, and 2000. Research completed to date joins the 1984 and 1994 social data with spatial data from the 1970s to the 1990s. Village serves as the linking unit in this research. With the 2000 data, linking will be possible at the household level as well.

DISTRIBUTING POPULATION ACROSS THE NANG RONG LANDSCAPE

Until recently, our population data have been linked to a point representing the location of the center of the village. Because data on land cover and land use are spatially continuous, we have had to relate the village (point) information to a broader spatial area. If we are interested in how various social variables might be related to land cover and land use, then it is necessary to specify how far out the village's relationships with the land might go, and in what directions. If we are asking questions about the effects of local aspects of the environment on the behavior of individuals or households, then we still need to specify the extent and direction of the village's spatial territory. Either way, we need allocation rules and procedures to distribute the social and demographic information linked to the landscape at the centroid of the residential village outward to the areas that might be influenced by village members, and we need to know the dimensions of the "village-sheds" or territories to capture land use and land cover information for surrounding lands.

Entwisle et al. (1998), Evans (1998), Rindfuss and Stern (1998), Walsh et al. (1998), Wood and Skole (1998), and Crawford (2000) have addressed the issue of setting village territories, or other small areas of use or ownership, in developing countries where boundary information is generally lacking. In the

Thailand setting, formal administrative boundaries of villages are often absent, but even if available, such boundaries provide little meaning in this agrarian landscape. In Nang Rong district, most of Thailand, as well as in other similar agriculturally dominated environments, "functional" boundaries more explicitly describe village territories and their areal extents. These functional boundaries are less arbitrary than administrative boundaries and reflect agriculturally related land use and ownership patterns by households contained within a specified village. Land title in Nang Rong can be vague and/or absent, and land ownership and use patterns of households are generally discontinuous across the landscape. Functional village boundaries are set through formal (e.g., rent) and informal (e.g., kinship ties) use arrangements; land titles are specified through a complex system involving multiple levels of land ownership. With transportation alternatives historically limited and natural boundaries imposed as areal constraints, functional village territories can follow an inverse distance weighting function that is biased towards lands nearby to village centers. The more formal administrative boundaries are often susceptible to the modifiable areal unit problem that has been described by Openshaw (1984) and Bailey (1994).

Spread Functions

Our initial and typical approach has been to use radial buffers around the village centroid. For example, to examine the effect of available, ambiguously titled, forest cover on the propensity of young adults to migrate out of their home village, we used a 2.0 km radial buffer around each village centroid (Rindfuss et al. 1996). The argument is that ambiguously titled forest cover could be converted into crop land by young adults and that they would be more likely to stay in their home village if such land were available. We used a 2.0 km radial buffer because that is approximately the distance that someone could walk in a half-hour in the hot tropical sun, and for the time period under investigation, walking was the dominant mode of transportation to one's fields. To take another example, we examined the effect of population growth on the extensification of agriculture (Entwisle et al. 1998). The goal was to assess the effects of village size and growth, and the density of village settlement, on the percent of land in upland crops. We experimented with 1.5, 2.0, 2.5, and 3.0 km radial buffers to obtain a sense of where effects might be most pronounced and found that population effects are felt more at the margins of village settlements.

Using a simple radial buffer around a village makes a number of assumptions that might or might not be appropriate. The first assumption is how far from the village might household members be willing to go for daily activities, in this case, to farm. Clearly, this will depend on a number of things including mode of transportation, availability of land, and the extent to which available land is flood prone. In the early part of our time-series, for example, farmers would walk to their fields. How far were they willing to walk, carrying their tools, under a hot tropical sun? Two kilometers seems a reasonable guess. Figure 3 shows all villages in Nang Rong in 1994 with a 2.0 km radial buffer around them. Note that there is some land that does not fall within the buffer area of any village. Yet, when we drape this figure on top of a land cover classification for 1994, it is clear that there is land in rice cultivation that is not within the 2.0 km buffer of any village. From the perspective of the land, 2.0 km is not sufficient. Further, farmers nowadays travel to their fields in trucks as well as on foot. Yet for many villages, as discussed below, going out 2.0 km is too far; 1.5 or 1.0 km would be more appropriate. Villages likely vary in the extent of their geographic reach, over time and over space. Assigning the same buffer to every village in every time point misses this variation.

Figure 3. Nang Rong District, Thailand: 2-km radial buffers around the 1994 survey villages.

Variable buffer radii might be assigned to each village depending upon population size or some other demographic, biophysical, or geographical characteristic. Villages in Nang Rong are more homogeneous with respect to population size than in many parts of the world, but there is variation nonetheless. Additionally, for villages in close proximity to one another, and within some prescribed distance threshold, a joint village centroid could be designated, based on distance or weighted by a population characteristic, and a radial buffer drawn from that point instead of from the centroid of an individual village's nuclear settlement pattern. Also, village territories might be set relative to the total land under cultivation as reported in the social survey and spatially defined through a remote sensing land use and land cover classification. In this way, we might enhance or "value-add" to the concept of radial buffers around villages by reflecting characteristics of individual or joint villages. However, this enhancement needs to be based on knowledge of where households in that village actually have their land, but such information is not available for 1984 or 1994. Does it matter if we fail to take into account variation in village territory? It may, but empirically we will not know until we validate ways of measuring the extent of village territories. It will be possible to do this with the 2000 data, and will be reported in subsequent papers.

Figure 3 makes evident another problem with making a 2.0 km assumption: the radial buffer of any given village is likely to overlap with the buffer of multiple other villages. Extending out to 3.0 km exacerbates this problem as Figure 4 amply illustrates. With a 3.0 km radius, there are some land pixels that are within the buffer of 23 villages. With this level of competition from near-by villages, the analyst either has to control for the number of villages nearby or make an assumption that most likely is untenable, namely that nearby villages do not affect the population-environment relationship that one is investigating.

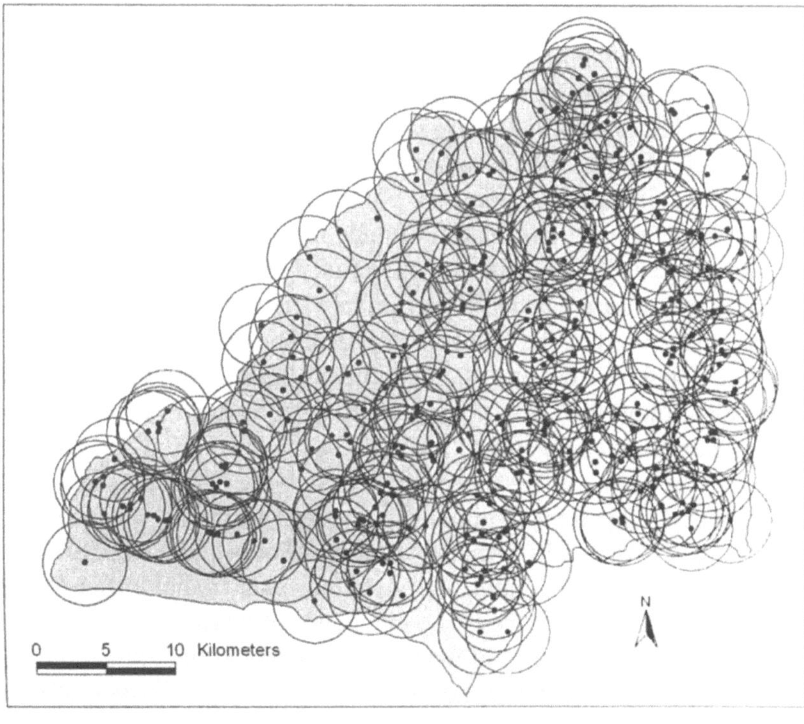

Figure 4. Nang Rong District, Thailand: 3-km radial buffers around the 1994 survey villages.

Another assumption with using a radial buffer approach is that the land associated with the village is circular in shape (although elliptical or other geometries could be alternately used). Clearly, there will be numerous situations where this is not the case. Natural barriers, such as rivers or very steep terrain, likely lead to an irregular shape because villagers would be reluctant to cross the barrier. Man-made barriers such as roads might have the same effect (Crews-Meyer 1999). Soil suitability would also affect where villagers farm. As Crawford (2000) has shown, soil suitability for rice farming, the dominant crop, varies widely across Nang Rong district; significant areal extents of rice production occur on marginally suitable and unsuitable soils (Welsh 2001). One way around these difficulties is a region growing approach after Evans (1998), further examined by Crawford (2000). Region growing within the Thai village setting used Fuzzy Set Theory (Zadeh 1965, Burrough and McDonnell 1998, Crawford 1999) to address uncertainty and imprecision in the placement of village territories by considering transitional borders between village territories where shared use occurs. Because of the pattern of villages and the varying density of their point locations, village territories may overlap and grade into territories of other villages. Using the shortest-path distance between multiple villages and the concept of a least cost surface, the core and periphery of village territories were defined using a fuzzy definition of boundary. In this volume, Crawford discusses extending the Ordinary Voronoi diagram by altering or relaxing the underlying assumptions to define village territories in Nang Rong in the absence of cadastral and village boundary data.

Finally, the radial buffer approach assumes a fairly crisp edge to the boundary of the land used by the village. Evans (1998) tested this assumption in his fieldwork. He selected 18 villages and followed eight transects from the village center recording, with the aid of local informants, information on village ownership. The transects ended at the point where villagers from the sample village no longer owned any land. Connecting the end points of the eight transects provided a rough estimate of "functional" village boundaries based on ownership (but not use). His work definitively showed that the shape of lands around the village used by the village is irregular, and that boundaries where one village's functional territory ended and another's began were fuzzy rather than crisp. He also found that, on average, most villagers travel less than 2.0 km from the nuclear residential area to their fields.

Given that the radial buffer approach has a number of dubious assumptions associated with it, why use it? For many purposes, one would want to know the location of the lands actually farmed by the households in the village, a topic to which we return below. Without such data, however, radial buffers have the advantage of being straightforward to calculate. Further, they start from the perspective of human activity and the location of residential units. The choice of the radius length was determined by our knowledge of the type of farming being done in the district, mode of transportation between dwelling unit and farm plot, and our general knowledge of the organization of daily activities in agricultural Nang Rong. Thus, even though the radial buffer approach was not designed with the intent of distributing the population across the landscape, it can be straightforwardly adjusted to accommodate general information on the activities of those populating the landscape. As previously suggested, buffer dimensions can be variably assigned to villages based upon social (e.g., total population), biophysical (e.g., total land under cultivation), and geographical (e.g., distance to nearest set of neighboring villages) factors. In this way, village territories can be set through radial buffer dimensions linked to social survey data and to data derived though remote sensing and/or GIS techniques.

A drawback to the radial buffer approach, as noted above, is that there is a reasonable probability that some land will not be included in the buffer of any village, unless the radius is made very large, which in turn likely negates their advantage from a social-behavioral perspective. Figure 5 shows the areas in the district not contained in any of the 2.0 km buffers around the 310 survey villages in 1994. Note the substantial areas of cultivated agriculture in these unassigned areas as defined through a LULC (land use/land cover) classification of Landsat

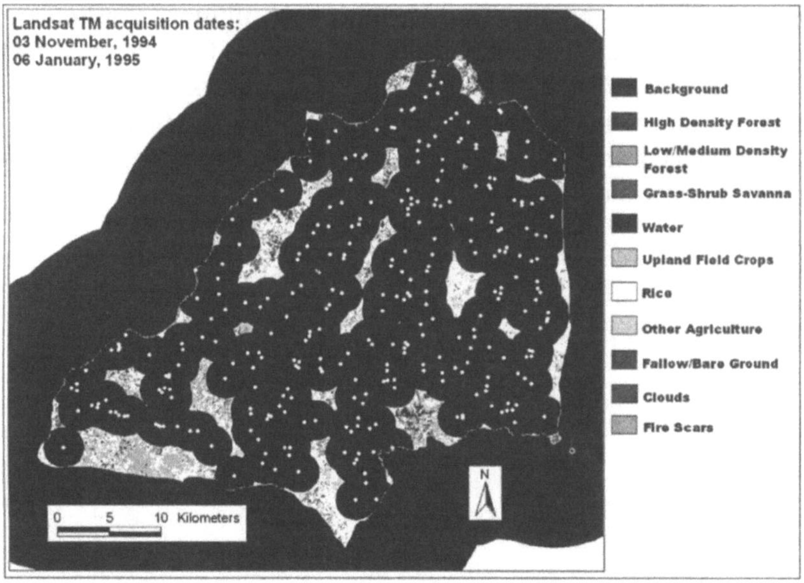

Figure 5. Unassigned areas in the study area for village territories defined through a 2-km radial buffer; classified LULC mapped through 1994/95 Landsat satellite data.

Thematic Mapper satellite data. Note that the lands located along the outer margins of the district may be within the functional village territories of villages outside of Nang Rong district. That is less likely the case for lands located more towards the interior of the district. To eliminate such problems we considered alternative approaches that would insure that all land was allocated. Perhaps, Thiessen polygons or Ordinary Voronoi diagrams are the best of these alternatives (Boots 1980 1987, Boots and Murdoch 1983). A Thiessen polygon approach is generally applied in continuous space (Okabe et al. 1994). Here, the approach was used to assign all the land in Nang Rong district to a polygon, seeded by the placement of the district villages, such that no overlap of village territories occurred, all land within the district was accounted for in one of the defined polygons, bounded by the district outline, and the polygons varied in size and shape as determined by the areal distribution of the villages. Figure 6 shows the land in Nang Rong divided by a Thiessen polygon approach. Note that here we did not take into account the population size or composition of the villages in Nang Rong, but such an extension is certainly possible. An implicit assumption of the Thiessen polygon approach is that crisp boundaries between polygons exist. Also, small distances between villages imply a territory that is areally restricted because of the bounding polygons.

Consider villages labeled A and B in Figure 6. The territory associated with village A is over 40 times smaller than the territory associated with village B. It is unlikely that this reflects where households in those villages farm, but rather is a consequence of settlement timing, with the earliest villages established near the main rivers and the most fertile lands. Figure 7, by overlaying the main hydrographic features on the Thiessen polygons, illustrate this point. A relationship between the location of major hydrographic features and density of villages is generally visible. The location of villages likely reflects the soil suitability of nearby land, location of transportation facilities (initially rivers and railroads, and increasingly roads), and the flooding patterns during the monsoon, and access to water for the cultivation of paddy rice.

Figure 6. Nang Rong District, Thailand: Thiessen polygons around 1994 survey villages; small (A) and large (B) polygons identified.

Returning to Figure 6, the assumption of crisp boundaries is also questionable, especially in parts of the district where the density of village settlement is high. Households in Village B likely compete with households from nearby villages for available farmland, and as a result, boundaries are likely to be fuzzy. This is difficult to capture with the Thiessen polygon approach. To test the assumptions behind the Thiessen polygon approach again requires information about where specific households in each village actually farm. Crawford (2000) developed expansion models of the Ordinary Voronoi diagram to examine the influence of land use and land cover, transportation network, distance between the nearest and second nearest villages to the focal village, and demographic characteristics of villages in defining territories. Findings suggest the importance of a graded membership function for lands located adjacent to multiple villages.

Population Distribution Model

Another set of substantive and methodological questions that we have addressed involves scale and pattern issues. Landscape ecology theory holds that all landscapes contain structure, function, and change (Forman and Godron 1986, Urban et al. 1987, Turner 1990). To what extent do relationships hold as one moves across a series of spatial and/or temporal scales? There has been some research on scale dependencies associated with biophysical processes and natural landscapes (e.g., Bian and Walsh 1993, Brown et al. 1993, Walsh et al. 1998). We were interested in extending this research to examine the nature of patterns across spatial and temporal scales for a landscape that has a pronounced social or human imprint. To do so, we needed a mechanism for assigning people and related social variables to each pixel in the landscape (Walsh et al. 1999b, Walsh et al. 2001).

We used a simple population distribution model (Walsh et al. 1999b). In the agricultural areas, the population variables were distributed on a per pixel basis using an equal-value spread function. Initially, a 1993 Landsat Thematic Mapper digital data set was classified into land cover types and subsequently recoded into agriculture and non-agriculture. For the recoded classification, a mask was generated of all pixels throughout the study area that were classed as agriculture, primarily rice in the lowlands and field crops in the uplands. Using the 1994 reported population of the 310 survey villages in the district, a 5 km radial buffer for defining village territories (so that nearly all land within the district was assigned to one or more village territories), and the agriculture mask previously generated, selected population variables were distributed on a per-pixel basis for each village territory using an equal-value spread function. The population distribution was not biased by geographic position on the landscape or any other site characteristic.

Figure 7. Nang Rong District, Thailand: Thiessen polygons around 1994 survey villages and hydrography superimposed.

This work makes a simplifying assumption that people are spread evenly throughout the village territory, at least the agricultural territory. Clearly, other assumptions could have been used. It would have been possible to weight cells to bias the distribution of people to upland or lowland crops. However, to do so would require using land cover data to distribute people, and then the social data would no longer be independent of the spatial characteristics of the cover data, making analyses of the relationship between the two problematic. Other possibilities include assigning impedance values to land cover types so that some cover types, such as water, would impede the flow of people. Similarly, we could have routed people to fields using existing road structure or some type of inverse weighting scheme from village centroids of the nuclear settlement pattern. The problem with these variations is that we needed more information on which fields households in any given village actually farm. Basic information that we collected in our 2000 round of data collection, but which had heretofore been unavailable, will allow us to make such connections of people to place.

We have also considered using distance decay coefficients to construct a population potential surface (Pooler 1987) so that we could improve on the equal-value spread function previously discussed. Walsh et al. (2001) assessed the sensitivity of landscape greenness, measured through the Normalized

Difference Vegetation Index, to population distribution parameters, defined through population potential surfaces. Population potential is a normative measure of a pixel's accessibility to nearby village populations and was computed after Yeates (1968). Population potential was derived by considering the population of each village (secured from our social surveys), the geographic position of each village (digitized from our study area base-maps), land cover and landscape greenness at the pixel level (mapped from classified satellite data), distances of pixels to villages (computed from a GIS), and a distance decay exponent (derived from hypotheses about travel distances to farming sites). A related variable, population density (or population pressure) was also computed by summing, for each pixel, agricultural land and total population amounts for every village within a 5 km radius. The total amount of agricultural land used by all villages within a 5 km distance of each pixel was divided by total population to create the population pressure variable.

LINKING LAND AND PEOPLE: IN-SITU APPROACHES

There has been relatively little work linking people and agricultural plots at the household level. An exception includes Kikby (1973) who mapped plot data in Mexico, but for a relative limited number of parcels compared to the work described here. This is unfortunate because important decisions about land use are made at the household level. In many parts of the world, households make decisions about what to plant, how much fertilizer and pesticide to use, what trees to cut, whether to build a structure, how to dispose of waste, and whether to convert land to urban uses. A comprehensive understanding of the forces that drive land use and changes in land cover requires studies of household decisions, the determinants and consequences of these decisions, and of policy and environmental processes at higher levels.

Why have there not been more studies aimed at the household and plot levels? The few studies that have done household-to-land linking have focused mainly on Amazonia, where pieces of land owned and used by households tend to be quite large, extending back from roads, and the household responsible for that land lives in a dwelling unit located within the parcel's boundaries (e.g., Moran et al. 1994). Figure 8a illustrates this "fishbone" pattern. In Amazonia, by "seeing" the parcel of land on a remotely sensed image, calibrating its location through ground control using a GPS unit, and finding and interviewing members of the household, researchers have been about to link household-level social data to agricultural plots and measurements based on remotely sensed data.

The pattern shown in Figure 8a does not apply to many areas around the world, however. Another common pattern is for agriculturalists to reside in clusters of dwelling units (villages) at a distance from their land, rather than directly adjacent to or on that land. It is also common for households to own or use multiple parcels of land. There need not be one-to-one correspondence between the social and spatial units. This complexity is illustrated in Figure 8b, which provides, at a glance, an intuitive understanding of the difficulty involved in linking social data about households with remotely sensed land cover data.

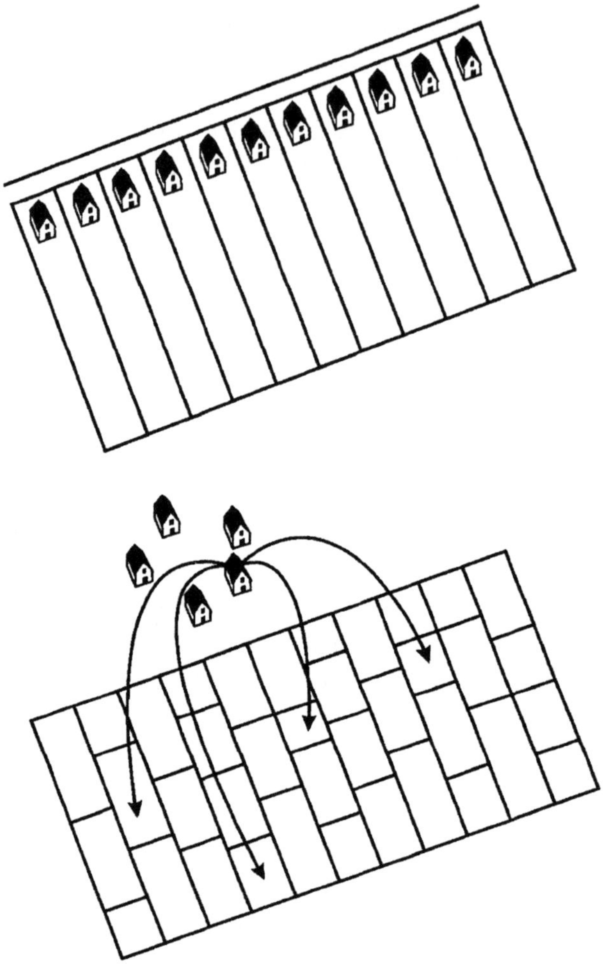

Figure 8. Fishbone land-ownership pattern characteristic of portions of Amazonia (top-a), and the distributed pattern characteristic of northeast Thailand (bottom-b), after Liverman et al. (1998).

In some research settings, cadastral maps showing plot boundaries and ownership may be readily available. Assuming that the maps are spatially referenced, complete, up-to-date and accurate, they provide valuable information with which to link households and plots. Cadastral maps may not be available, however, especially in developing countries. Or, they may be incomplete, as is the case in Nang Rong. As part of the 2000 data collection, we obtained all available cadastral maps. Figure 9 shows the areas covered and not covered by these maps. Whereas coverage is fairly complete in the eastern-half of the district, it is substantially incomplete in the western-half, where some of the most dramatic changes in land cover and land use have occurred. Further, cadastral maps are unlikely to include land over which households only have "use rights," including recently converted forest lands important to the story of land use change in many places.

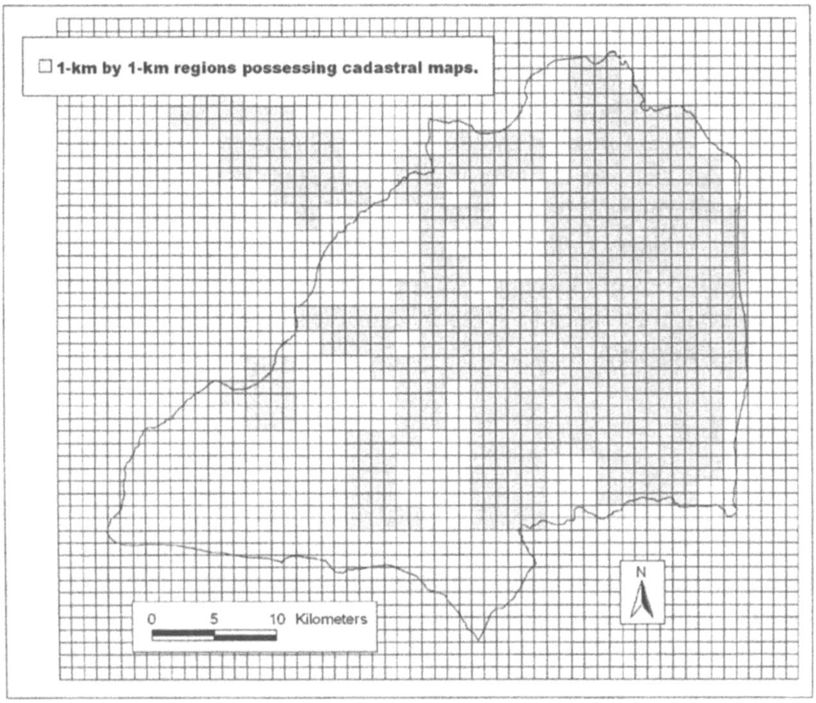

Figure 9. Cadastral map coverage of Nang Rong District, Thailand in 1999.

Over the past several years, we have explored various approaches to linking households and agricultural plots in the complex socio-spatial environment of Nang Rong. The goal was to develop approaches appropriate for relatively large samples of households and plots (1,000's), able to handle multiple ties, that can be implemented with or without cadastral maps, and flexible as to whether the focus is ownership, use or both.

Our initial approaches focused on methods involving travel to individual field plots and the use of GPS devices to collect locational data, which was then linked, to household identifiers in a variety of ways. Two of these approaches required household informants to identify land owned or used by household members; the third approach relied instead on village informants. Subsequently, we developed approaches that relied on maps as data collection devices. In household interviews, land owned and used by household members was indicated directly on maps prepared for that purpose and locational identifiers recorded in the household questionnaire booklet. At the village level, a similar exercise was performed by a group of informants, covering land owned or used by households in the village. The approaches we have tried can be arrayed across two dimensions: (1) data collection through interviews of *household* informants, or *village* informants; (2) locational information collected through *direct* observation, or *indirectly* through the use of maps.

The first field test, which occurred in December 1997, evaluated and compared three approaches: flag method; household interview method; roving group interview method. This field test provided an important baseline for the subsequent development of alternative approaches. Believing that our experiences can be useful to researchers in other settings, we describe them here. Our initial test was followed by additional field tests, which culminated in the set of methods that we actually used in the 2000 data collection: a hybrid approach combining multiple elements. The intellectual odyssey we traveled from the first field test to the final data collection would be worth documenting, but for the purposes of this paper, we focus on the endpoints of our experience, with occasional references to some of the other approaches tried and lessons learned along the way.

Flag Method

The flag method combines the detailed knowledge that household members presumably have about the location of land owned or used, with the efficiency of collecting information about plot location at the village level. The method consists of household representatives placing numbered flags in their land parcels. At the time of distribution, the numbers on the flags assigned to each household would be recorded along with identifying information for the household. After the flags are placed, a data collection team would travel from plot to plot over predefined spatial areas, measuring each plot's location using a GPS and recording the plot identification numbers on the flags. The plot identification numbers -- i.e., the numbers on the flags -- would serve as the link between plot and household. Prior to field-testing, there was a lot of enthusiasm for this method among the UNC investigators. However, it failed in the field for a number of reasons -- in fact, the method was never implemented.

First, in the Nang Rong setting, renters and other users felt it would be necessary to get explicit permission from the legal owner to place a flag on the land they use. In the village, one man explained the situation by saying, "It is like a house. If you don't own the house, you are afraid to go in until the owner gives you permission. People who own the land may be happy to help. People who don't own land will not take a flag." Asking all renters to gain appropriate permissions seemed like an unfeasible prospect.

Second, villagers told us that finding the flags after they had been placed would be a problem. For areas planted in rice, it would be possible, in principle, only when the crop was not at a mature stage. Flags were mounted on one-meter rods and would be difficult to see amidst a mature crop. For other areas, especially upland areas, finding the flags would be extremely difficult at many times during the year due to the taller vegetation. Sugar cane, an important upland crop grown in Nang Rong, can grow to a height of two meters. When asked their opinion of the different methods, several villagers commented that their biggest worry was they we would be unable to find all the flags. Half-joking, one person said we would need a helicopter to fly around and look for the flags. In addition to concerns about being able to see the flags, villagers were worried that children, water buffalo, or pranksters would pick up or trample the flags.

Household Interview Method

This method collects information on plot location, neighbor adjacencies, and other attributes within the context of a household interview. We tried two approaches to collecting plot location. In the first pretests, interviewers traveled with a household representative (by foot or other means as available) to each plot within some pre-defined spatial areas, at which point plot locations were measured using a GPS. In later pretests, interviewers brought map products into the household, asking a household representative to identify the location of plots owned or used by the household directly onto the map.

Although we had a lot of confidence in the quality of the data collected by actually traveling to the field with a household representative, this approach did not generalize well to the sample of 8,500 households we hoped to include in actual data collection. Time and money costs could not be justified within our budget constraints. Nevertheless, we did use this strategy to check the quality of data collected by other means for a sample of areas and households in the main data collection.

In contrast, the use of map products within the household interview was problematic. Respondents were quite heterogeneous with respect to their ability to orient to and "read" the maps. This may be a basic literacy issue, especially for older respondents, who are the ones likely to be at home and available to participate in an interview. According to the 1994 household survey data, more than a quarter of villagers aged 50 and older had less than a primary education, which consisted of four years of formal schooling. Alternatively, it may be that some people are not very good at reading maps -- irrespective of their level of formal education. But in addition, household interviewers did not themselves always feel comfortable with the maps. Although all of the interviewers had a college education, only some had a well-developed spatial ability. More up-to-date maps and more training might help to alleviate this problem, but it would not solve it entirely. The maps were also awkward to carry from house to house. Because of their own lack of comfort with the maps, and the difficulties experienced by some of the respondents, interviewers perceived the maps to be a burden. We decided against using map products within household interviews because of these problems.

Nevertheless, in the household interviews conducted as part of the pretests, we found that members could easily identify their plot neighbors. When asked if others could identify their neighbors, one man responded, "Yes, because they often talk together in the fields about cultivation. It is like a big chain–if you keep going from neighbor to neighbor, you will know the whole country! Farmers like to chat on the way to and from the fields. They might need labor and talk to people on the way. People in rural areas know each other very well, not like in the big cities." It was clear that social networks in Nang Rong are at least partially structured by the spatial topology of land use and ownership. We used this result in the design of the household portion of the data collection method used in 2000.

Roving Group Interview Method

This method relies on a small number of informants knowledgeable about the ownership and use of specific parcels to identify and link plots and households. Along with a data collection team, the informants systematically rove the landscape, identifying the owners and users of each plot within a predefined area. Plot locations are measured using a GPS and then linked to a household. Group data collection works well in the Nang Rong setting. One informant said that people would be "more comfortable" in a group, and that an individual "might be more 'lonely' being isolated with the research team." Data quality is also improved as each informant considers and reacts to the information provided by the others.

The success of the roving group depends on selecting knowledgeable informants. In the first field test, informants were more successful identifying plots and households in areas closer to the village. One set of sample plots was directly south of the residential cluster, running parallel to a stream, and within 1.5 km of the village center. The other set of sample plots was located in an upland area about 6 km from the village center. Complicating data collection in the latter area was the taller and denser upland vegetation that obscured visual observation and made passage more difficult. There appeared to be variation related to experience. For example, older officials, who had lived in the village a long time, tended to be better informants. One of our most knowledgeable informants was a trapper of frogs and other small animals, who needed permission from local landowners to traverse the fields as part of his work activities.

The drawback to the roving group interview method was its extensive time demands. All of the methods tried initially involved direct observation of field plots and their locations. Some of these were at a considerable distance from the village, and from each other. Further, traversing the fields took a lot of time, even when plots were concentrated and located close to the village.

Hybrid Approach Used in 2000

As we refined our methodology, increasingly many of the elements of different approaches were melded together and used in data collection that began in February 2000 and completed in November 2000. The precise indicators of linking will be further analyzed in a subsequent paper. We have built in a number of checks and redundancies so that it will be possible to evaluate this component of our fieldwork. However, all indications from the field (such as cooperation from households and village leaders, reports from field staff, and hand counts of matching people and plots) suggest that for the 8,500 households of interest, we were successful in linking them to the land that they use.

Since households live in nucleated villages and then farm parcels of land that primarily do not include their dwelling unit, we needed procedures for both the parcels of land that contain their dwelling unit and other parcels of land that they use. The protocols were very different for these two distinct land parcels, and we describe both below. For ease of exposition, we refer to two different types of field workers: household interviewers and the spatial team. The spatial team had specialized training, which included interpreting maps and aerial photographs, the use of GPS devices, and storing and downloading locational digital files.

Dwelling Unit Plots

Our overall goal was to link households to all parcels of land relevant to their activities. Parcels containing the dwelling unit were the easiest to identify and link. The work was done entirely by the spatial team.

The first step was a visit to the village, meeting with the village headman and describing the purpose of the entire study. The village headmen keep a list of all households in the village. This list has the name (first and last) of the head of the household along with a unique number identifying the household. In a typical village, this number was essentially a running number, beginning with 1 for the first household on the list and going up to the total number of households in the village for the last household listed. In a situation like this, we used these numbers as IDs for the household. If this was not the case, we would assign equivalent numbers.

Then, with the village headman or an assistant headman accompanying the spatial team, they would go from dwelling unit to dwelling unit, recording the name and ID number of the household, taking a GPS reading of the location of the dwelling unit, and recording ID information from the file that was stored on the GPS receiver. Occasionally, but rarely, the team would come to a dwelling unit that contained a household that was not on the list of households provided by the village headman. In such a situation, they would simply record the name of the household head and give the household the next available ID number.

In general, this part of the fieldwork was straightforward and unobtrusive. Respondent burden was low, primarily involving the village headman. If others in the village asked what was going on, an uncomplicated description of the purpose of the study was provided. The fact that villages are nucleated and relatively small (on average about 100 households) made the task manageable. It also helped that it is part of the responsibility of village headmen to know every household in their village.

In the household interview component of the study, there was a section about the parcel of land on which the household's dwelling unit stood. Questions about the land parcel included size, ownership, and whether any agricultural crops were grown on this land parcel. Since the household ID number on the list used by the spatial team was an ID number also used by the household interviewing team, linkage of these two files is possible.

Further, we had obtained and digitized cadastral maps for all areas of Nang Rong that had clear titles recorded at the district land office. These land parcels have unique ID numbers, precise boundary lines, and an accompanying list that contains both the ID number and the name of the person who owns the land parcel. From the household interview we have the name of every member of that household. Thus we will be able to match the household to the cadastral information by matching on names. While this is more difficult than matching on ID numbers, the uniqueness of most Thai names[1] (given and surname) makes this a feasible task. Matching to the cadastral files provides the shape of the dwelling unit parcel as well as an independent estimate of its size.

Farm Plots

Linking households to their agricultural plots (that is plots they farm but not containing their dwelling unit) is far more difficult than for plots containing their dwelling unit. Our approach involved two stages of data collection, and then hand matching between them. Our approach also employed insights from the social network literature. Finally, our approach involved letting people tell us the information they knew best, and we subsequently put this information together for the needed household-to-plot linkage.

The first stage of the data collection process took place in household interviews. Given the ambiguity in the meaning of a "farm plot," during the process of constructing the questionnaire, we did some qualitative, cognitive lab work (Tanur 1992) to determine the best term to use to refer to farm plots. Basically, this involved talking with small groups of individuals from Nang Rong, and asking them about the different Thai terms that could possibly be used to refer to a parcel of land. We went both forward and backwards: asking them to tell us what certain terms meant to them and asking them what term would be best to describe our concept of farm plot. The Thai word "plang" emerged as being closest to our concept, but it also was clear that using this term, *plang*, still left some ambiguity. So, in the questionnaire we use the term *plang* and explain that we mean by this a single contiguous piece of land. We then go on to say that we want to ask about the plangs that the household uses, defining use as follows:

"By uses, I mean the plangs that your household uses to grow crops, livestock, fruit trees, eucalyptus trees, fish ponds or anything else to benefit your household. This includes joint use. By joint use, I mean your household and one or more other households cultivate this land jointly or together."

After fixing terminology, we asked respondents how many farm plots they used, followed by a series of detailed questions about each plot.

Two types of information collected in the household interview help match individual plots to other parts of the data collection. First is the size of the parcel, in whatever units of measurement are relevant and known by the household (e.g., rai, wa: 400 wa = 1 rai; 1 wa = 4 square meters; 1 rai = 1,600 square meters). Second are the names and village of residence of those who work the fields neighboring the plot in question. If the village of residence is the same as the village where the household resides, then the interviewer checks against a list of households in that village and writes down the ID number for the household being considered.

The second component of the process involves the spatial team. For each village we created a map that started with a 1994 aerial photograph, 1:50,000 scale, centered on the village, and covering 36 square kilometers. An overlay coverage was developed and registered on the aerial photograph, where we indicated significant features like rivers and roads. In addition, for any portion of the 36 square kilometers for which we had cadastral lines, we layered those cadastral lines on top of the aerial photograph. We then cut this map into nine tiles, printed each tile on 36 x 36 inch paper, and laminated them (Figure 10). A map of the entire district was also available, showing, among other things, the location of all villages.

The spatial team went back to each village after all the household interviews were completed and met with the village headman, describing the purpose of this phase of the data collection process. The spatial team would ask the village headman to arrange for a group of 4-6 knowledgeable individuals to meet with the spatial team. When this group was assembled, the nine tiles would be assembled as in Figure 11, and the informants would be asked to imagine they were in an airplane flying over the village. Significant features were pointed out on the maps. Once they were oriented to the nine tiles, the informants would be asked to point out areas where village households farmed. At this point, there were some elaborations and discussions. The practical process of the spatial data collection can be divided into two stages. The first stage was to divide the area in the set of nine tiles into 4-5 sub-areas, and to select 2-3 farmers who were using the land in each sub-area. The second stage was to invite those selected farmers in each sub-area to participate in the group discussion. Thus, in each village, the spatial team arranged 4-5 groups, corresponding to the defined sub-areas. Each group was composed of at least two standing key informants, and 2-3 farmers who were using land in that sub-area. The idea was that we assumed it was impossible for one group of farmers (2-3 persons) to know the details of land use in that vast area. Protocols were established so that the first group of farmers and standing key informants in a village would identify users in sub-areas to be invited to join the group discussions.

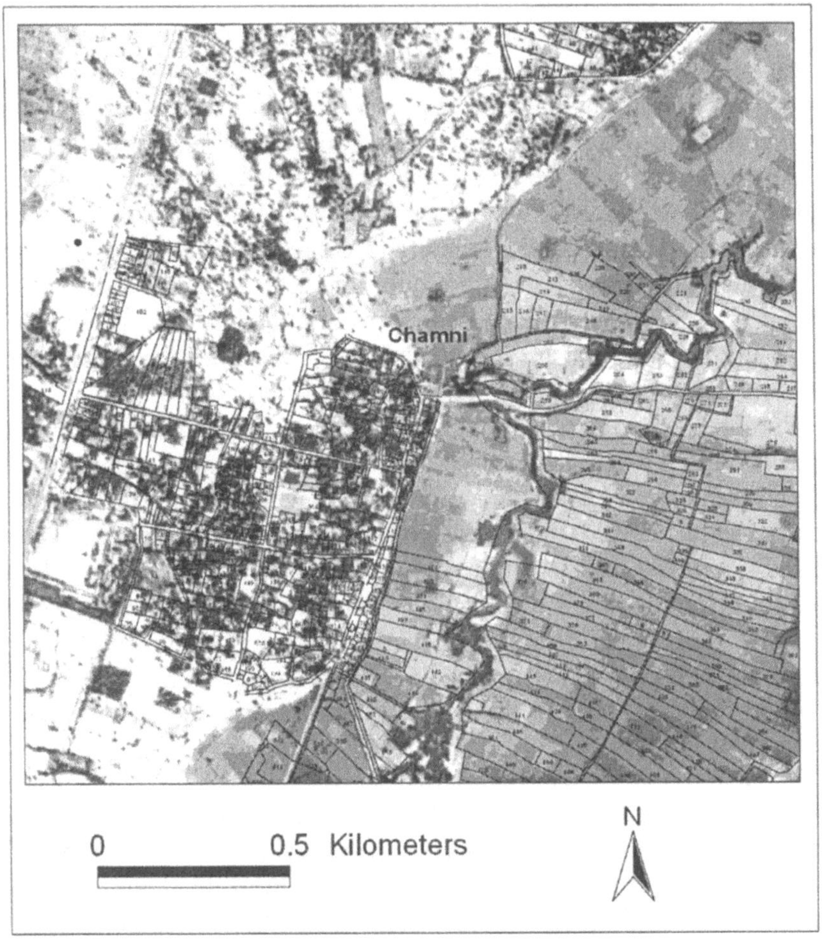

Figure 10. Field maps used by the spatial team for year 2000 survey – image composite of aerial photography, cadastral lines, village point location, and selected image annotation.

Operationally, one of the defined sub-areas would be chosen, and the assembled informants would be asked to identify specific parcels used by village households. This would proceed, parcel by parcel, with the spatial team recording parcel ID numbers, name of the household using the parcel, and household IDs. When a section was finished, if some neighboring parcels were not farmed by households in the village, informants were asked to provide the name and village of those parcel users. Then, the spatial team asked about the next area in which village households farmed, and so on until all of the areas were enumerated.

Informants were also asked about parcels used by village households outside the nine-tile area. When there were such parcels, the spatial team would record the name and ID number of the household, as well as the name and ID number of the village that was the closest to the parcel. Overall, this group interviewing process went well. Typically there was agreement among group members. When there was not, friendly discussion took place to resolve inconsistent views across group members.

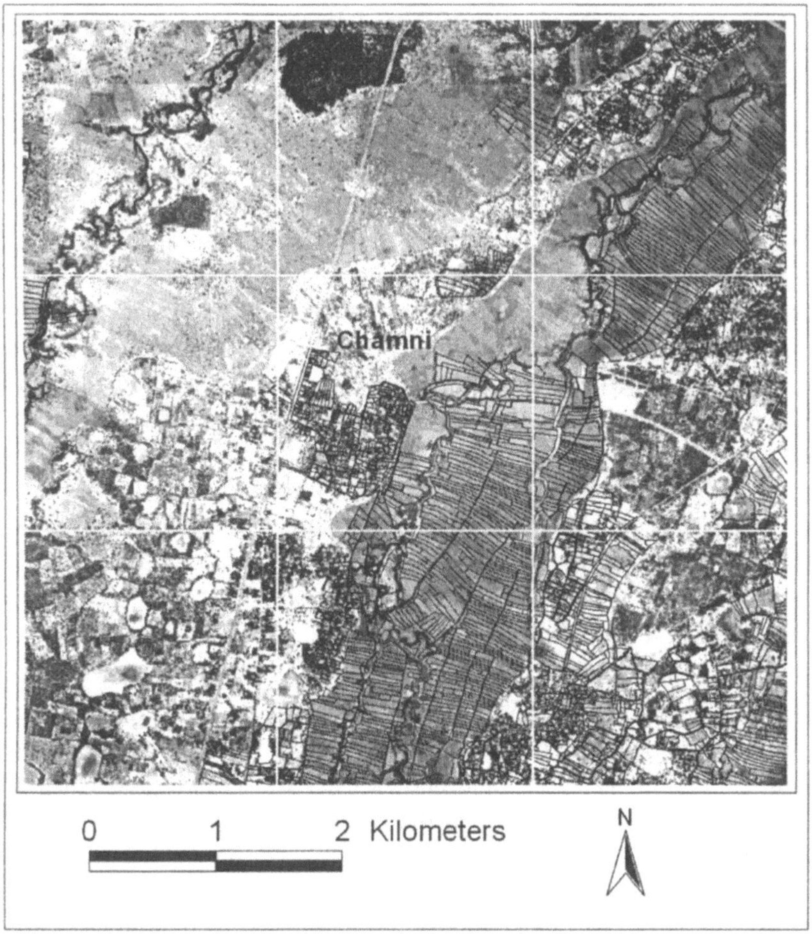

Figure 11. Field maps arranged in a 3 x 3 map matrix covering 36 square kilometers and centered on the selected survey village.

After the second component of this process was finished, the spatial team matched the information from the two components -- the household interviews and the village informant interviews -- a tedious but crucial element of the design. The team started with the information provided in household interviews, and then looked to match with information from group interviews. In the matching, a match was recorded even if the directionality was off, that is it was considered a match if the household said their neighbor to the north was Mr. V and the group interview said Mr. V was the neighbor to the east.

We recorded numerous details of the matching process. After the data entry is complete, high on our agenda will be analyses of the quality of the match. For now we can report that field counts indicate that just over 80% of the parcels mentioned by the household were matched to the data provided by the group interviews. To put this in perspective, in countries in Western Europe or North America, response rates above 70% in household surveys are considered acceptable. In the household phase of the fieldwork, only six households refused to participate. Of the rest, it appears that we have linked 80 percent of the parcels households use to specific locations in Nang Rong, providing the opportunity to subsequently link to remotely sensed data, as well as other coverages in our GIS.

Finally, field parcels were randomly selected and visited in the field by an adult member of the corresponding household and the spatial team. Six parcels per village were randomly chosen. The spatial team asked the village headman or senior person to draw two bottle caps that were numbered 1-9 on the facedown side of the caps. The numbers represented the nine map tiles. Then, a key informant would be asked to toss a coin on each chosen map tile. The area where the coin fell was marked. Three parcels around the marked location were selected for spot-checking in the field. Members of the households using those parcels were invited to lead the spatial team to those land parcels. The team collected GPS coordinates at the approximate center of the parcels. The GPS coordinates will be used for spot-checking the link between the household and spatial data in subsequent analyses.

CONCLUSIONS

Recent interest in changes in land use, as opposed to land cover, along with advances in geographic information systems, the improved quality and availability of remotely sensed data, and the availability of the global positioning system (GPS) challenge the scientific community to develop better links between the continuous data typically available on the spatial side with the spatially discrete data typically available on the social side. This paper has described the nature of this challenge as manifested in ongoing research on relationships between population process and landscape dynamics in Nang Rong, Thailand. Initially, we linked our population data to a point representing the location of the village, and the challenge was how to relate this point information to a broader spatial area partitioned through pixels. Linking at the village level provides important insight not possible at higher levels of aggregation such as the district, province, or country level. Nevertheless, in making a village level link, we had to make a set of assumptions about the location of the agricultural fields farmed by village residents and the nature of boundaries with other nearby villages -- assumptions, as we argue, that are not completely consistent with what we know about land use in Nang Rong.

Linking at a lower level of aggregation, households to plots, will both allow us to examine these assumptions and to address at a host of questions associated with household decision-making, how characteristics of the land influence these decisions, and what the impacts are of the decisions made. This paper has described how such links can be established, even in complex socio-spatial environments such as Nang Rong. Given household-plot links, and with the GIS tools now available, we can then link to various GIS coverages containing attributes of those parcels such as soil characteristics, topographic position, use/ownership history and physical accessibilities to residences, roads, water, or market centers. With strongly linked household-plot data, it would be possible to model risk abatement strategies directly, examining the distribution of plots over a risk surface and household decisions about land use in relation to the location of plots as well as household characteristics. Household decisions about whether to farm the land they own, rent some of it out, or rent additional lands can be assessed relative to labor supply and migration behavior. Linking households to land farmed is also be a powerful generator of social network data, assuming that neighboring farmers talk to one another about a variety of matters. This is turn makes it possible to model the diffusion of new technologies as a function of detailed social and spatial location as well as the characteristics of potential adopters. Relationships between social and biophysical drivers of land use/land cover change can be scale dependent, and indeed, are scale dependent in our setting (Walsh et al. 1999). Strongly linked household-plot data will make it possible to extend this insight to the lowest socially meaningful level of aggregation, providing a better anchor for the large-scale models that influence national and international policy.

NOTES

[1]There are a number of reasons why Thai names are much more likely to be unique than American, European, Korean or Chinese names. Put differently, in Thailand the John Smith problem is less likely to arise. Surnames were not used in Thailand until King Rama VI decreed in 1920 that they should be used. Further, when families were registering their surnames, there were checks to make sure that there were no duplicates. Thus in principle in the 1920s each family had a unique surname. While the checking for duplicates at registration time was by hand, and hence not perfect, the overall effect was to have the vast majority of surnames be unique. Unlike many Western countries, there is no tendency for fathers to give their first names to their son or mothers to give their first name to their daughter. Instead, sometimes a syllable from the mother's name and a syllable from the father's name might be joined to create a name for

the child. In short, frequently first names are made up, and thus have a higher probability of being unique.

The Thai alphabet has 44 consonants, 21 vowels and 5 tonations. Hence, there is a greater possibility for creating a wide range of unique names than in the Western alphabet.

A computerized file exists containing the names of approximately 40,000 individuals who were listed in the household rosters of the 1994 data collection. In this file, there were only 130 cases where more that one person had the same first and last name. For these reasons, matching by name should be less problematic than in most other countries.

ACKNOWLEDGEMENTS

The work reported in this paper is supported by the National Aeronautics and Space Administration under its Land-Cover and Land-Use Change program (NAG5-6002). This paper is part of a larger set of interrelated projects funded by the National Institute of Child Health and Human Development (R01-HD33570 and R01-HD25482), the National Science Foundation (SBR 93-10366), the EVALUATION Project (USAID Contract #DPE-3060-C-00-1054), MacArthur Foundation (95-31576A-POP), and the Mellon Foundation. The larger set of projects involves various collaborations among investigators at the University of North Carolina, Carolina Population Center and the Departments of Sociology and Geography, and investigators at the Institute for Population and Social Research (IPSR), Mahidol University, Bangkok, Thailand.

REFERENCES

Allen, J.C. and Barnes, D.F., 1985. The cause of deforestation in developing countries. *Annals of the Association of American Geographers* 75(2): 163-184.

Arbhabhirama, A., Phantumvanit, D., Elkington, J., and Ingkasuwan, P.. 1988. *Thailand Natural Resources Profile.* Singapore: Oxford University Press.

Bailey, T.C., 1994. A Review of statistical spatial analysis in geographical information systems. In, *Spatial Analysis and GIS*, (Fotheringham, S. and Rogerson, P. eds.), London: Taylor & Francis, 13-44.

Bian, L. and Walsh, S.J., 1993. Scale dependencies of vegetation and topography in a mountainous environment of Montana. *The Professional Geographer* 45(1): 1-11.

Bilsborrow, R.E., 1992. *Rural Population Dynamics and Agricultural Development: Issues and Consequences Observed in Latin America.* New York: CIIFAD.

Bilsborrow, R.E. and Geores, M., 1994. Population, land-use, and the environment in developing countries: what can we learn from cross-national data? In: *The Causes of Tropical Deforestation: The Economic and Statistical Analysis of Factors Giving Rise to the Loss of the Tropical Forests* (Brown, K. and Pearce, D., eds.). London: University College Press.

Boots, B.N. 1987. Modifying Thiessen Polygons. *The Canadian Geographer* 31(2): 160-169.

Boots, B.N., 1980. Weighting Thiessen Polygons. *Economic Geography* 56(3): 248-257.

Boots, B.N. and Murdoch, D.J., 1983. The spatial arrangement of random Voronoi Polygons. *Computers & Geosciences* 9(3): 351-365.

Brown, D.G., Bian, L., and Walsh, S.J., 1993. Response of a distributed watershed erosion model to variations in input data aggregation levels. *Computers & Geosciences* 19(4): 499-509.

Burrough, P.A. and McDonnell, J., 1998. *Principles of Geographical Information Systems*. Oxford: Oxford University Press.

Crawford, T.W. 1999. A comparison of region building methods used to examine human-environmental interactions in Nang Rong district, northeast, Thailand. *Proceedings, Applied Geography Conference* (Schoolmaster, F.A., editor), 22: 366-373.

Crawford, T.W., 2000. *Human-Environment Interactions and Regional Change in Northeast Thailand: Relationships between Socio-Economic, Environment, and Geographic Patterns*. Doctoral Dissertation, Department of Geography, University of North Carolina – Chapel Hill.

Crews-Meyer, K.A., 2000. *Integrated Landscape Characterization Via Landscape Ecology and GIScience: A Policy Ecology of Northeast Thailand*. Doctoral Dissertation, Department of Geography, University of North Carolina – Chapel Hill.

Crews-Meyer, K.A. 1999. Modeling landcover change associated with road corridors in northeast Thailand: integrating Normalized Difference Vegetation indices and accessibility Surfaces. *Proceedings, Applied Geography Conference*, (Schoolmaster, F.A., editor), 22: 407-416.

Committee on Global Change Research. 1999. *Global Environmental Change: Research Pathways for the Next Decade*. Washington: National Academy Press.

Entwisle, B., Walsh, S.J., Rindfuss, R.R., and Chamratrithirong, A., 1998. Land use/land-cover and population dynamics, Nang Rong, Thailand. *People and Pixels* (Liverman, D., Moran, E.F., Rindfuss, R.R., and Stern, P.C., editors), National Academy Press, Washington, D.C., 121-144.

Evans, T.P., 1998. *Integration of Community-Level Social and Environmental Data: Spatial Modeling of Community Boundaries in Northeast Thailand*. Doctoral Dissertation, Department of Geography, University of North Carolina, Chapel Hill.

Feeny, D., 1988. Agricultural expansion and forest depletion in Thailand, 1900-1975. In J.F. Richards and R.P. Tucker (eds.), *World Deforestation in the Twentieth Century*, Durham, NC: Duke University Press, 112-143.

Forman, R.T.T. and M. Godron. 1986. *Landscape Ecology*. New York: John Wiley & Sons, Inc.

Ghassemi, F., Jakeman, A.J., and Nix, H.A. 1995. *Salinisation of Land and Water Resources: Human Causes, Extent, Management and Case Studies*. Sydney: University of New South Wales Press Ltd.

Kaida, Y. and Surarerks, V. 1984. Climate and agricultural land use in Thailand. In: M.M. Yoshino (ed.), *Climate and Agricultural Land Use in Monsoon Asia*. Tokyo: University of Tokyo Press, 231-254.

Kirkby, A.W., 1973. The use of land and water resources in the past and present valley of Oaxaca, Mexico. *Prehistory and Human Ecology of the Valley of Oaxaca* (K.V. Flannery, ed.). Ann Arbor: University of Michigan (Museum of Anthropology).

Lambin, E. F., X. Baulies, N. Bockstael, G. Fischer, T. Krug, R. Leemans, E. F. Moran, R.R. Rindfuss, Y. Sata, D. Skole, B.L. Turner II, C.Vogel. 1999. *Land-Use and Land-Cover Change (LUCC) Implementation Strategy*. Stockholm: IGBP Secretariat.

Liverman, D., Moran, E.F., Rindfuss, R.R., and Sterns, P.C., 1998. *People and Pixels: Linking Remote Sensing and Social Science*. Washington, DC: National Academy Press.

Mallin, M. A, K. E. Williams, E.C. Esham, and R.P. Lowe. 2000. An effect of human development on bacteriological water quality in coastal watersheds. *Ecological Applications* 10(4): 1047-1056.

Moran, E.F., Brondizio, E., Mausel, P., and Wu, Y. 1994. Integrating Amazonian vegetation, land-use, and satellite data. *BioScience* 44(5): 329-338.

National Research Council. 2000. *Grand Challenges in Environmental Sciences*. National Academy Press.

Okabe, A., Boots, B.N, and Sugihara, K. 1994. Nearest neighbor operations with generalized Voronoi diagrams: A Review. *International Journal of Geographical Information Systems* 8(1): 43-71.

Openshaw, S., 1984. *The Modifiable Areal Unit Problem, Concepts and Techniques in Modern Geography*. Norwich: Geo Books.

Parnell, M.J.G., 1988. Rural poverty, development and the environment: the case of Northeast Thailand. *Journal of Biogeography* 15: 199-208.

Pooler, J. 1987. Measuring geographical accessibility: a review of current approaches and problems in the use of population potentials. *Geoforum* 18(3): 269-289.

Rigg, J., 1987. Forces and influences behind the development of upland cash cropping in North-East Thailand. *Geographical Journal* 153(3): 370-382.

Rigg, J., 1991. Homogeneity and heterogeneity: an analysis of the nature of variation in Northeastern Thailand. *Malaysian Journal of Tropical Geography* 22(1): 63-72.

Rindfuss, R.R. and Stern, P.C. Linking remote sensing and social science: the need and the challenges. In, *People and Pixels: Linking Remote Sensing and Social Science*, (Liverman, D., Moran, E.F., Rindfuss, R.R., Stern, P.C., eds.), Washington, DC: National Academy Press, 1-27.

Rindfuss, R.R., Walsh, S.J., and Entwisle, B., 1996. Landuse, competition, and migration. Paper Presented at the *Population Association of America Annual Meeting*, New Orleans, LA.

Rindfuss, R.R., Chattopadhyay, A., Toshiko, K., and Sethaput, C., 2000. Migration and Longitudinal Data Analysis: Implications of individual and family processes. Paper presented at the Population Association of American, Los Angeles.

Rundel, P. W. and Boonpragop, K. 1995. Dry forest ecosystems of Thailand. In: Bullock, S. H., Mooney, H.A., and Medina, E. (eds.). *Seasonally Dry Tropical Forests*. New York: Cambridge University Press, 93-123.

Tanur, J. M. (ed.) 1992. *Questions About Questions: Inquiries into the Cognitive Bases of Surveys*. New York: Russell Sage Foundation.

Turner, M.G., 1990. Spatial and temporal analysis of landscape patterns. *Landscape Ecology* 4(1): 21-30.

Urban, D.L., O'Neill, R.V., and Shugart, H.H., 1987. Landscape ecology: a hierarchical approach can help scientists understand spatial patterns. *Bioscience* 37(2): 119-127.

Walsh, S.J. 1999. Deforestation and agricultural extensification in Northeast Thailand: a remote sensing and GIS study of landscape structure and scale. *Proceedings, Applied Geography Conference*, (F.A. Schoolmaster, editor), 22: 223-232.

Walsh, S.J., Butler, D.R., and Malanson, G.P., 1998. An overview of scale, pattern, and process relationships in geomorphology: a remote sensing and GIS perspective. *Geomorphology* 21(3-4): 183-205.

Walsh, S.J., Crawford, T.W., Welsh, W.F., and Crews-Meyer, K.A., 2001. A multiscale analysis of LULC and NDVI variation in Nang Rong District, Northeast Thailand. *Agriculture, Ecosystems, & Environment,* 85: 47-64.

Walsh, S.J., Entwisle, B., and Rindfuss, R.R., 1999a. Landscape characterization through remote sensing, GIS, and population surveys. In, S. Morain (ed.), *GIS Solutions in Natural Resource Management: Balancing the Technical-Political Equation*, Sante Fe: OnWard Press, 251-265.

Walsh, S.J., Evans, T.P., Welsh, W.F., Entwisle, B., and Rindfuss, R.R., 1999b. Scale dependent relationships between population and environment in northeast Thailand. *Photogrammetric Engineering and Remote Sensing* 65(1): 97-105.

Welsh, W.F., 2001. *Agro-Ecological Sustainability and Land Degradation Potential in Nang Rong, Thailand.* Doctoral Dissertation, Department of Geography, University of North Carolina, Chapel Hill.

Wood, C. and Skole, D., 1998. Linking satellite, census, and survey data to study deforestation in the Brazilian Amazon. In, *People and Pixels: Linking Remote Sensing and Social Science*

(Liverman, D., Moran, E.F., Rindfuss, R.R., and Stern, P.C., eds.), Washington, DC: National Academy Press.

Yeates, M. 1968. *An Introduction to Quantitative Analysis in Human Geography.* New York: McGraw-Hill, Inc.

Zadeh, L.A., 1965. Fuzzy sets. *Information and Control* 8(3): 338-353.

Chapter 3

LAND USE STRATEGIES IN THE MARA ECOSYSTEM: A SPATIAL ANALYSIS LINKING SOCIO-ECONOMIC DATA WITH LANDSCAPE VARIABLES

D. Michael Thompson[1], Suzanne Serneels[2], and Eric F. Lambin[2]

[1]*16 Park Road, Woking, Surrey, GU22 7BW, United Kingdom*
mthomp@net2000ke.com
[2]*Department of Geography, Universite Catholique de Louvain, 3 Place Louis Pasteur, B-1348 Louvain-la-Neuve, Belgium*

Keywords: land use strategies, spatial modeling, Masai Mara, pastoralists, logistic regression, land cover changes.

Abstract This paper presents an analysis of the determinants of shifts in production among a Maasai pastoral population adjacent to the Masai Mara National Reserve in Kenya. The Maasai have traditionally followed a pastoral livestock land use system, which has been largely compatible with wildlife. More recently Maasai land use systems have diversified, with large scale mechanized cultivation and tourism now providing considerable incomes. Large-scale cultivation has in particular been seen to compete with both traditional pastoral livestock keeping and with wildlife-based tourism. We used a set of landscape-, accessibility-, and socio-economic-based variables to test assumptions regarding the forces seen to be driving production system and land use change. We first analyzed which land use strategies are followed by the pastoralists in the study area, and then used generalized logit models to assess the determinants of these strategies. We then tested whether groups of households diversifying from a traditional cattle-keeping economy can be related to patterns of land cover change around their homesteads. This is highly relevant to unfolding policy in the field of environment and development in East Africa where there is widespread concern in policy circles over the expansion of mechanized cultivation and its impacts for wildlife and biodiversity conservation.

Analysis of land use strategies of pastoralists show the emergence of clearly distinguishable groups diversifying into small-scale cultivation, large-scale mechanized cultivation, tourism, and a mixture of small-scale cultivation/tourism. In all of these diversifying groups, livestock keeping remains a central component of the land use strategies chosen. The generalized logit model results indicate that distance to the Maasai Mara National Reserve (MMNR) as the variable with most explanatory power, linked to choice of land use strategy, followed by landscape attributes (elevation and slope). These variables can be seen as influencing the economic returns possible from chosen different strategies. Distance to MMNR can also be seen as inverse distance to the nearest market center (Narok), and therefore linked to distance-economic return models. Specifically, among the groups diversifying into mechanized cultivation, distance to the MMNR followed by agro-ecological zone and slope provided the greatest explanatory power among households choosing this land use. The emergence of these variables, all of which affect the economic returns achievable from large-scale cultivation, supports economic-based models that explain the process of conversion of rangeland to cultivation. However, these variables do not fully explain land use strategies. On the reduced sample for which full socio-economic variables were available, leadership, education, and wealth (as measured in livestock holdings) were all significantly linked to mechanized cultivation as a choice of land use. Although place matters, socio-economic factors also intervene in the decision-making process on land use. Households in the cattle keeping baseline group are less likely to have a leadership position than households in any of the other land use strategy groups. Leadership status and networks can be used to obtain lands at a location that is favorable for development, or will allow revenues from distant sources.

Households are often involved in several land use activities that have an impact on the environment in a wide range around the homestead, so linking households to changes in land cover is difficult. Only mechanized farming as a land use strategy is strongly spatially linked to participating households. The modeling techniques described here are useful to shed light on the underlying factors linked to land conversion. Our findings suggest that while explanatory variables based on economic models are most significant, patterns of decision-making rooted in social and economic structures of pastoral communities are important in understanding the land use production strategies.

INTRODUCTION

In the last few decades, conversion of grassland, woodland, and forest into cropland and pasture has risen dramatically in the tropics (Houghton 1994). These changes in land cover are the result of land use decisions made at the household level. Much literature assumes that these decisions are based on a strategic response to macro-economic or local economic conditions (e.g., Walker and Homma 1996). Modeling of these economic conditions leads to assumptions of land use change based on land potential (Norton-Griffiths 1996). These analyses relegate other factors to a subsidiary role.

This paper seeks to increase our understanding of the relative importance of economic versus landscape and social factors that shape land use/land cover changes. Typically, remote sensing data are used to investigate land cover and land use changes at the landscape level (Liverman et al. 1998). However, at this level, investigations of patterns of land use and land cover changes may miss much of the underlying dynamics and mechanisms of change that take place at the household level (McCracken et al. 1999). An alternative approach is to understand change as the result of actions of particular user groups and individual households (Turner et al. 1994, Liverman et al. 1998). Although linking remote sensing with socio-economic household data – linking people to pixels -- seems intuitive and conceptually straightforward, some major challenges have to be addressed (Liverman et al. 1998). An appropriate level of aggregation of information derived from the social data and the natural environment has to be identified. In most studies, remote sensing observations and socio-economic data are linked at the scale of the administrative units used for the collection of socio-economic data (Green and Susman 1990, Skole et al. 1994). Examples include Mertens et al. (2000), who integrated household survey data with remotely sensed data at the village level to study the impact of macro-economic change on deforestation in Southern Cameroon; and Entwisle et al. (1998) who linked population dynamics at the village level with patterns of land use/land cover change in Thailand. In these studies, remote sensing data could not be linked to field observations directly, as people live in nucleated villages, away from the multiple, non-contiguous fields they cultivate. Elsewhere, McCracken et al. (1999) successfully analyzed land cover changes at the property level, because each household lived on a well-defined plot and had little impact on adjacent land. Few studies have attempted to link socio-economic data on pastoralist land use to remote sensing data, although Boone et al. (2000) used interviews of Maasai head-of-households to characterize the effect of droughts and El Niño weather patterns on their livestock through changes in vegetation biomass. In this study, household-interview results were extrapolated based on patterns of vegetation change derived from NOAA AVHRR-NDVI data.

The Maasai living in our study area are primarily pastoralists, who use very diverse areas away from their permanent settlements to graze their livestock at certain periods of the year (Homewood and Rodgers 1991, Potkanski 1994a). The system of grazing rights is complex. Animals graze on communal land, so a given area is used by different families, often coming from different regions. Linking livestock-related land cover change data to people is thus difficult. Other patterns of resource use such as cultivation and income from tourism may take place closer to the homestead and may be more easily related to households.

Our interest in this paper is to analyze which of a series of possible explanatory variables best explain the observed diversification of land use strategies of Maasai pastoralists around the Masai Mara National Reserve in Kenya, and to link this to data on land cover change. Existing literature suggests that economic factors (returns to land) based on land potential will increasingly drive Maasai decisions to change land use in the Mara ecosystem (Norton-Griffiths 1996, Aligula 1997, Ottichilo 2000). To test these assumptions, we first analyzed which land use strategies are followed by the pastoralists in the study area, then assessed the determinants of these strategies, based on variables chosen to describe the physical and socio-economic environment in which the households operate. Subsequently, we tested whether groups of households following specific land use strategies are related to patterns of land cover change around their homesteads.

BACKGROUND

Study Area

The study area (10,694 km²) is located in Narok District, in southwestern Kenya, between 34°45'E - 36°00'E at the Kenya-Tanzania border and adjacent to the Serengeti National Park in Tanzania (Figure 1 – on the CD-ROM). The rangelands surrounding the Masai Mara National Reserve can be divided into three range units based on biogeographic and climatic differences. The western unit consists mainly of grasslands and comprises the Masai Mara National Reserve. The Loita Plains stretch out in the northeastern part of the study area and are covered by dwarf shrub and whistling thorn (*Acacia* drepanolobium) grasslands. The eastern area, with the Siana Hills and Plains, supports *Croton* bush and several other woody species interspersed with grasslands (Stelfox et al. 1986). The Masai Mara National Reserve was established in 1961. The 1,523 km^2 reserve is a formal conservation estate, where land use is restricted to wildlife viewing tourism. The rangelands surrounding this conservation area act as a buffer zone between the Masai Mara National Reserve and the cultivated "high potential lands" in the northern part of Narok District. These rangelands contain year-round communities of resident wildlife, and migratory wildlife spill out onto them in the dry season. As such, the grazing resources are important to the Maasai pastoralists and their livestock, and the wildlife alike.

Historically the study area has been inhabited since the colonial period by Maasai of the Purko section (Hollis 1905, Lamprey 1984). Colonial policy largely focussed on the control of the "Maasai" as a distinct ethnic identity in a process linked to the appropriation of land away from Maasai use (Waller 1988). The creation of the Southern Maasai Reserve in 1911 allowed European ranchers to settle on some of the best Maasai grazing land in the central rift valley around Naivasha and Nakuru (Waller 1976). This also resulted in the removal of Purko Maasai from more northern areas and their movement southwards to join Purko Maasai families already established in our study area. The creation of this Southern Maasai Reserve forced the bulk of the Maasai into a single reserve measuring some 12,000 square miles, and excluding much of the best grazing lands (Waller 1976, Kerven 1992). Thereafter, further piecemeal appropriation of land has occurred, for example, through the creation of national parks and national game reserves (Nairobi National Park, 1946; Tsavo National Park, 1948; Maasai Mara, 1963; Amboseli, 1974), and for experimental government cattle ranches and cattle holding grounds[1].

Since Independence in 1963, the Government of Kenya has pursued a policy of privatization of rangelands. The Land Group Representatives and Land

Adjudication Act of 1968 enabled land previously held in trust for its inhabitants by the Government to be adjudicated into Group Ranches (GR).

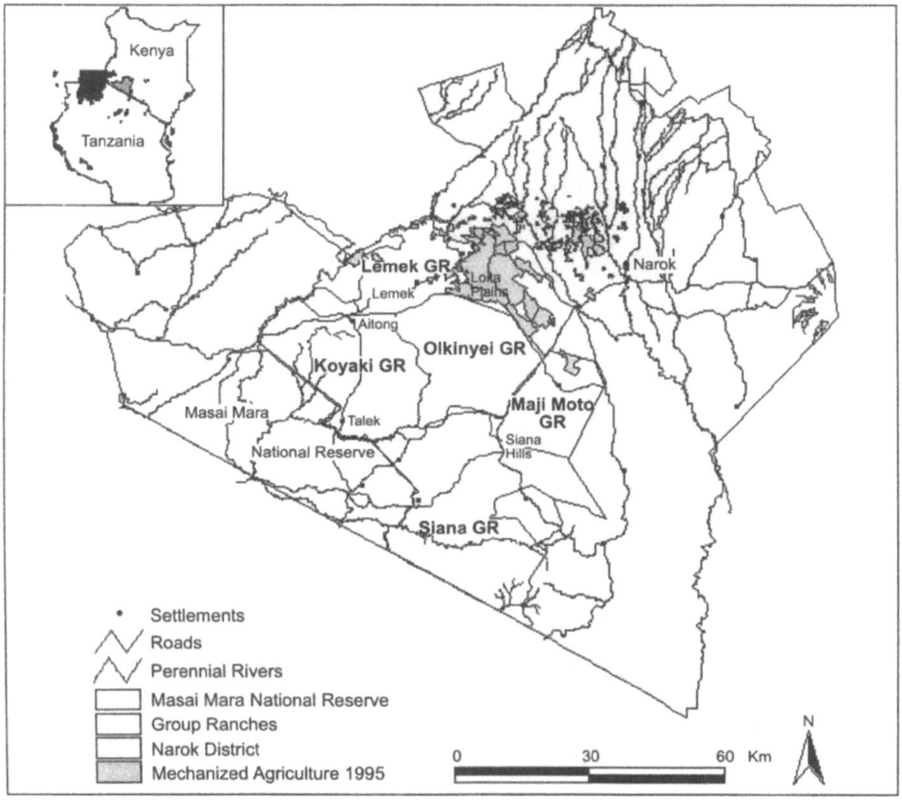

Figure 1. Study area location.

Important changes in land use have taken place in the study area during this period. The major change has been due to large-scale wheat farming, which was initiated in the early seventies, and rapidly expanded from the eighties onward (Serneels et al. in press). In Narok District, an area of 4,875 ha in the northern part of the Loita eco-unit was cultivated in 1975 (Karime 1990). Between 1975 and 1995, an additional 44,000 ha were converted to large-scale wheat farming in the Loita Plains (Figure 2). Elsewhere other changes included rangeland modification by pastoralists and smallholder agriculture. Together, they account for, in total, 8.4% of the study area that has been subject to a decrease in vegetation cover over the 20-year study period.

Livestock keeping forms the basis of the economy of our study population. This has proved a remarkably resilient system, enabling households to take advantage of seasonally available range and water in an otherwise highly variable semi-arid environment. Communal land management systems have been governed by negotiable access described by Potkanski (1994) as a series

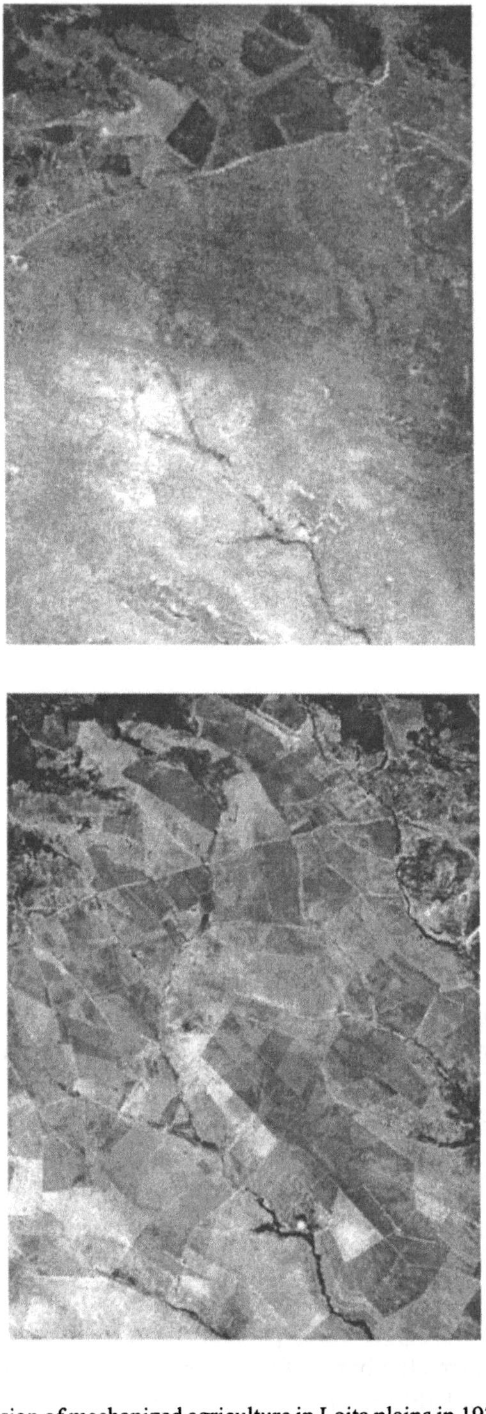

Figure 2. Expansion of mechanized agriculture in Loita plains in 1975 (top) and 1995 (bottom).

of primary and secondary rights to grazing and water. Systems of homestead organization, clan and section have allowed these negotiable systems of access

to operate across vast geographical areas (Spencer 1988, Homewood and Rodgers 1991). These systems have proved adept in facilitating household's entry into diversified productions systems. Thus, in our study area, a married man might have access to labor, livestock, and income from three or more homesteads at a distance of 20-kilometers or more from each other. These homesteads will often be located on different group ranches and may sometimes be located in different Districts or even spread between Kenya and Tanzania. Thus, the permanent homestead, comprising most of the livestock may be in one location, children attending primary school may stay with another wife and live in a separate homestead made up of similar neighbors in another location, while an elder son or another wife may live in another homestead near to a tourist lodge and generate an income from selling curios. Cattle controlled by the married man may be distributed among these three different homesteads, and income from each homestead in the form of money from wage labor or sales from tourism curios may be returned to him. This system also operates to allow a permanent homestead located in an area unsuitable for cultivation to benefit from cultivation at a different site (personal observation).

In addition to allowing simultaneous diversification into different production systems, this system has ensured that livestock keeping has run alongside even areas of large-scale mechanized cultivation (Thompson, forthcoming). In addition there are opportunities to generate considerable incomes from the leasing of large areas of land to outside entrepreneurs (usually Kenyans of European or Asian origin). These leases have generally been for large-scale cultivation or for tourism operations. This has generated a contest among Maasai to arrange leases over wider areas of Maasai lands (Thompson and Homewood, submitted). Maasai that have leadership responsibilities, and are able to speak for larger areas of land, are in a strong position to withhold or manipulate information on how these leases are arranged. Levels of education, experience in dealing with outside systems, and leadership have thus become important attributes by which the contest for the considerable incomes available have been played out (Thompson, forthcoming).

CONCEPTUAL FRAMEWORK

While the transition between forest/woodlands and grasslands is largely driven by edaphic factors and disturbances such as fires, heavy browsing by elephants, and natural succession (Ruess and Halte 1990, Dublin et al. 1990, Sprugel 1991, Turner et al. 1995, Sinclair and Arcese 1995), our interest is in factors driving pastoral decisions over production systems and associated changes in land cover (Figure 3). Present pastoral land use decisions are derived from a complex interplay of factors encompassing historical land use patterns, diversifying opportunities for income generation and unfolding rules governing access to land, capital and technology to exploit diversification.

Existing literature suggests a number of possible explanatory variables underlying changes in natural vegetation to cultivation. The expansion of large-scale mechanized farming, for example, may not only be linked to areas of suitable agroclimatic potential, but also to the availability of land for rent or sale, and to economic factors such as cereal and input prices, access to the market and transportation costs (Brush and Turner 1987, Serneels and Lambin 2001). Changes in land tenure may increase opportunities for the expansion of mechanized farming (Norton-Griffith 1996, Aligula et al. 1997, Ottichillo 2000) or conversely, the confused nature of land privatization programs may inhibit large scale agricultural expansion (Thompson, forthcoming).

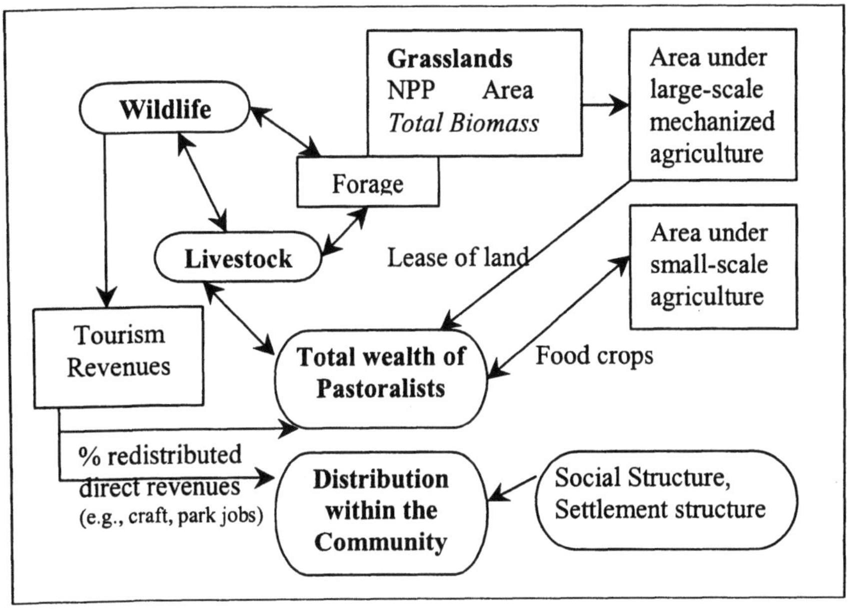

Figure 3. Conceptual model.

Cultivation of small plots, mostly maize, is becoming increasingly common among Maasai, as is the development of permanent settlements (Lamprey and Walle 1990). Existing literature variously suggests that the expansion of small-scale agriculture is driven by an increasing human population (Jewell 1980, Lamprey 1984, Prins 1992, McCabe et al. 1997), by a drive to maximize/diversify production (Sikana et al. 1993, Southgate and Hulm 1996), or as being influenced by other factors such as increasing education/exposure to outside agricultural systems, promotion of Maasai individuals to positions of leadership, and so on (Spencer 1988, Galaty 1992). Similarly, individual pastoralist livestock holdings reflect pastoralist wealth, and are based at least in part on decisions by pastoralists on the investment of household resources in herds (Dahl and Hjort 1976, Spencer 1988). As production systems diversify, there may be links between income and individual livestock holdings

(particularly as cash incomes are obtained from farming and, from wildlife-related tourism activities).

Finally, policies at national or sub-national level have a bearing on land use. These are often difficult to include in a fine-scale spatial analysis, for lack of appropriate ways to represent the impact of policies in a spatially explicit way. However, where policies have spatial manifestations (for example, in privatization of land, construction of market centers, creation of conservation areas), distance to these centers can be used as a proxy. Three major categories of factors can therefore be used to model income composition and links to social and land-use/spatial characteristics. These are: (i) who decides on and who benefits from different land use activities (socio-economic status of the household); (ii) the natural landscape attributes of different locations, which influence land use (slope, elevation and agro-climatic potential); and (iii) macro-policy and accessibility factors which are reflected in land tenure, access to markets, roads and water, creation of conservation areas. Policy instruments in particular can be manipulated to influence the decision-making process of agro-pastoralists, and therefore modify land use changes and their impacts on the ecosystem. The three factors above may lead to considerable spatial variability in preferred land use options, as manifested both on the remote sensing images an by the fine-scale socio-economic surveys.

DATA

Socio-Economic Data

Data were gathered by semi-structured interview over a period of eighteen months (May 1998 to November 1999) in single-round broad-scale and detailed multi-round surveys at five study sites in Narok District, Kenya (Thompson, forthcoming). The unit of study is the household (*Olmarei* - singular), which corresponds to the gate cut in the stockade around the *boma* (homestead) for a married man and his household. Bomas are classically made up of groupings of *Ilmareta* (plural of *Olmarei*), from 1 in our sample to as many as 10.

We interviewed 288 heads of household on details of household composition, education level (years of schooling), livestock holdings (cattle and small stock), cultivation of maize and/or wheat, and other types of production/sources of income (shares in tourism facilities, dividends from wildlife associations, wage employment and remittance from wage earners working away from the household). Data collected by survey was cross-referenced by participant observation, livestock counts and field measurement during the multi-round survey. At a later stage of the survey, information on the leadership status of the head of household (no, minor or senior leadership position) was also collected. For the latter, information is available for only 132 households. If a household had been involved in agriculture, the area cultivated was calculated as the average of the cultivated area in the last three years[2]. The homestead of each household was located with a global positioning system (GPS), which enabled us to represent each household in a geographical information system (GIS). When multiple households in one boma were interviewed, these households were represented by one set of GPS coordinates (the boma).

To reduce the variance in the socio-economic data, the data were categorized. The variables concerning cultivation or leasing of land were transformed to binary variables, indicating presence of cultivation or leasing of land, or absence of cultivation or leasing for a given household. Education level, household size, and livestock equivalents per reference adult were grouped in four categories, representing the four quartiles of the data range. The other income variables were collected as categorical information.

GIS Database

The GIS database was created by digitizing the topographic and thematic maps. The vector data were rasterized to a spatial resolution of 100 m. The roads network, rivers and villages were manually digitized from the 1:250,000 scale Survey of Kenya Topographic maps (1980), and updated with a Landsat Themtic Mapper (TM) image (21 January 1995) through on-screen digitisation. As was outlined in the conceptual framework, three major categories of factors are used to link the income composition with social and land use/spatial characteristics. Besides the factors describing the socio-economic status of the household, these are the accessibility factors and natural landscape attributes.

Accessibility was represented by a set of distance variables, created as a series of buffers expanding from each arc or point of the network. The layers were generated as isotropic surfaces. We assumed that the accessibility is not hindered by natural barriers such as hills or dense vegetation. Based on former work (Serneels and Lambin 2001), distance to roads suitable for wheeled traffic, distance to village centers or tourist lodges, distance to permanent water sources, distance to the Masai Mara National Reserve, distance to the district capital and distance to the area of mechanized agriculture (i.e., distance to the nearest tractor ploughed field larger than 5 ha) were selected as important accessibility variables. Roads suitable for wheeled traffic are considered important for import or export of goods and produce. Village centers provide administrative, health and educational services and shops, while tourist lodges can provide tourism-related employment. Maasai pastoralists value the proximity of permanent water for their livestock. Perennial rivers, permanent springs, and artificial water holes with permanent water supply were used in the analysis. Distance to the Masai Mara National Reserve and to the district capital allow for testing the influence of conservation areas *versus* regional market agriculture. Such variables were introduced to test the spatial relationship between the location of the wheat farms and the homestead of the people involved in mechanized farming.

The contour lines were extracted from the Arc/Info Digital Chart of the World (ESRI 1993), at the 1:1,000,000 scale, with intervals of 1,000 feet. A Digital Elevation Model (DEM) was constructed from the contour lines and the resulting elevation map was reclassified in units of 100 m intervals. The DEM allowed for the calculation of the slopes (%). The map of agro-ecological zones represents the potential agricultural suitability of a region, based on a combined interpretation of soil quality, rainfall, and elevation for each location (Jaeholtz 1982). Agro-ecological zones were introduced as a binary variable, indicating land suitable for rainfed agriculture (maize, wheat, or sorghum) and drylands suitable as rangelands only.

Land tenure was included as a binary variable, indicating private or communal land. In our study area, actual land privatization is a very recent phenomenon, that took place only last year, thus after the collection of socio-economic and land cover change data. However, the process of subdivision has been going on for several years before being formalized and we assume that many group ranch members anticipated the privatization in their land use decisions. Hence, the variable "land tenure" was included as a binary, independent variable, indicating which group ranches were subdivided in individual land titles, without indicating individual plot boundaries.

Land Cover Changes

A time-series of three Landsat images was acquired for the years 1975 (Landsat MSS), 1985, and 1995 (Landsat TM). Change-detection techniques were applied to image pairs and nine trajectories of change were mapped (Serneels et al. in press). The change trajectories are based on changes in the normalized difference vegetation index (NDVI) of a given pixel. NDVI-image pairs of consecutive dates were subtracted. A pixel in the difference image was considered as having been converted from one vegetation type to another if its difference in NDVI between the two dates was larger than a given threshold value. The thresholds were chosen conservatively as to exclude the more subtle changes within vegetation types (Serneels et al. in press). The resulting land cover change map (with a pixel resolution of 100 m) was validated in the field. A sample of changed and unchanged areas was selected and visited in the field. The actual state of the vegetation was analyzed and, where possible, the former vegetation cover was checked with the aid of aerial photographs or through interviews with local people. The field visits also allowed us to link the land cover changes to broad categories of land use. The land cover changes were grouped into changes related to: (i) the expansion of mechanized farming, (ii) subsistence agriculture and development of permanent settlements (smallholder impact), and (iii) vegetation regrowth.

Table 1. Spatial variables used in the study.

Spatial Variables	*Type*	*Unit*	*Mean*	*Std*	*Min*	*Max*
distance to Masai Mara National Reserve	continuous	km	16.1	16.4	0	58
distance to Narok	continuous	km	67.7	13.5	33	84
Distance to mechanized farming	continuous	km	25.2	16.2	0	50
distance to roads	continuous	hm	14.6	22.4	0	112
distance to villages	continuous	hm	50.9	35.6	3	203
distance to water	continuous	hm	20.6	15.1	0	119
elevation	continuous	hm	17.8	1.2	16	21
slope	continuous	%	1.1	1.7	0	7.4
agro-ecological zone suitable for rainfed agr.	binary				1	2
land tenure	binary				1	2

METHODS

Clustering of Land Use Strategies

The households were clustered according to land use strategies, based on the following variables: (i) revenues from shares in tourist facilities and/or from dividends in wildlife associations, (ii) acreage under maize cultivation, and (iii) acreage under wheat cultivation and/or income from leasing land to contractors for wheat cultivation, and (iv) wage labor (Table 2). All households practiced livestock keeping, which is the basic land use strategy. The two binary variables indicating revenues from tourist facilities and wildlife associations were added, resulting in the following classes: no revenues, revenues from one source, and revenues from both sources. Furthermore, the two variables concerning wheat farming and land leasing were added, resulting in three classes: no cultivation/no leasing, cultivation or leasing, and cultivation and leasing.

We used the K-means clustering technique with Euclidean distance to cluster the households. This algorithm splits a set of objects into a selected number of groups by maximizing between-cluster variation relative to within-cluster variation. K-means starts with one cluster and splits it into two clusters by picking the cases furthest from the center as a seed, one for each cluster, and assigning each case to the nearest seed. It continues splitting one of the clusters into two (and reassigning cases) until a specified number of clusters is formed.

K-means reassigns cases until the within-groups sum of squares can no longer be reduced.

The households belonging to the different land use strategy clusters were represented on a map. The spatial distribution of the different clusters was studied and the degree of spatial overlap between the different land-use strategies was assessed.

Statistical Modeling of Land Use Strategies

To understand the explanatory factors of the different land use strategies, we used the following possible determinants of land use production choices: socio-economic status (household size, education level, leadership status, wealth in livestock equivalents per reference adult[3]); landscape attributes (elevation, slope, agro-ecological zone); and accessibility factors (distance to Masai Mara National Reserve, Narok, mechanized farming, roads, villages and water).

These explanatory variables were tested in a multivariate model with the different land use strategies as dependent variable. Independent variables were chosen to represent socio-economic characteristics of households most commonly collected in surveys of this nature (e.g., Homewood and Rodgers 1991, McCabe et al. 1997a and 1997b). Landscape factors reflect constraints to cultivation. Distance to the Masai Mara National Reserve was included, as it indicates the potential for development of tourist-related activities. Distance to the district capital was included as a proxy for market accessibility. As discussed in the introduction, livestock keeping strategies are remarkably flexible. Livestock keeping was found to be universally present and to run alongside other production choices. Strictly speaking, livestock holding is a land use strategy and thus part of the dependent variable. However, we used it as an independent variable because, in this pastoralist society, livestock are frequently used as an indicator of wealth (Grandin 1988, Bekure et al. 1991, Homewood 1992). For the leadership status of the head of household, data were collected for only 132 households, so this variable was included in a second model, based on a reduced sample, to test its importance.

As the dependent variable is nominal and the independent variables are either categorical or continuous, we used the generalized logit model. The statistical procedure specifies that the response functions are generalized logits of the marginal probabilities for the dependent variable. The generalized logits are obtained by taking the logarithm of the ratio of two probabilities. The denominator of each ratio is the marginal probability corresponding to the last observed level of the variable – the *baseline category* (Agresti 1990) or the *reference group* (Hosmer and Lemeshow 1989), noted as P_b. The numerator is the marginal probability corresponding to the land-use strategy under consideration, noted as P_c, with α being the intercept and β_i the parameter estimate for the independent variable X_i.

$$logit\ P_c = log\ (P_c/P_b) = \alpha + \beta X_1 + \beta X_2 + \ldots + \beta X_n \qquad (1)$$

If r is the number of response levels of the dependent variable, the procedure calculates r-1 equations, or conditional models, each giving the probability of observing a considered land use strategy rather than the baseline land use strategy for the set of independent variables. The logit for comparing two non-baseline land use strategies is the difference between the logit of these two land use strategies with the baseline strategy.

Generalized logit models are an extension of the logistic multiple regression (LMR) techniques (Hosmer and Lemeshow 1989, Agresti 1990). Logistic regression has become a common tool in landscape analyses (Pereira and Itami 1991, Vega Garcia et al. 1995, Narumalani et al. 1997, Bian and West 1997), but few authors have yet applied generalized logit models to land use studies (Chomitz and Gray 1996, Nelson et al. 1999, Mertens and Lambin 2000).

Model Building

Prior to the construction of the generalized logit models, all independent variables were tested for collinearity. Variables were excluded when a multi-collinearity with the other independent variables of $R^2 > 0.8$ was found. Distance to the district capital, distance to the mechanized farming area, distance to the Masai Mara National Reserve, and land tenure were all strongly related to each other ($R^2 > 0.8$). Similar models were constructed with each of the variables and the model with the highest explanatory power was chosen. Some authors (Chomitz and Gray 1995, Nelson et al. 1999) included spatial lag parameters to control for spatial autocorrelation in their models. Spatial autocorrelation in the error term causes inefficient but asymptoticallyunbiased estimates of β. In our data set, we found a very strong multi-collinearity between the X and Y coordinates introduced as spatial lag variables, and some of the distance variables. Distance to the Masai Mara National Reserve and to the district capital were strongly correlated with the Y-coordinate, as these variables represent a north-south gradient. The X-coordinate correlated strongly with elevation, as there is a gradient in the elevation in the study area from east to west. The distance to park and elevation variables were thus also used as spatial autocorrelation indicators.

The initial models were constructed with a full set of uncorrelated independent variables. Subsequently, variables were eliminated from the model, if they did not significantly contribute to its overall explanatory power, and if they did not have a confounding effect on the remaining variables. Once a reduced model was built, each of the remaining variables was tested for interaction with other variables. When necessary, an interaction term was added to the model. All continuous and ordinal categorical variables were tested for linear behavior. A second set of generalized logit models was then constructed following the above procedures, but based on a reduced sample of households, for which information on the leadership status of the head of household was available.

Impact on Land Cover

We tested whether land use strategies in a pastoralist community are related to changes in land cover. In the study area, there is an increasing tendency towards permanent settlement of pastoralist households. We therefore examined a possible spatial link between households following a given land use strategy and land cover changes in the area surrounding their homesteads. We formed buffers of 5-km around each group of households and calculated the percentage of land cover change, as detected on the satellite images, within each area, distinguishing between changes related to the development of mechanized agriculture, subsistence agriculture, development of permanent settlements (referred to below as smallholder impact), and vegetation regrowth (Serneels et al. in press). These categories of land cover changes were analyzed for each group of households, following the same land use strategy, and for the period 1985-1995, as it overlaps with the time-frame of the household interviews.

Results

Descriptive Statistics

All the interviewed households owned livestock, with large variance in the number of animals kept per reference adult (Table 2). Most households were involved in other economic activities as well. Among these alternative sources of income, cultivation (maize), and tourism related activities (receiving income from shares in tourist facilities or dividends from wildlife associations) are the most important. Few households generate income through wage employment and remittance from household members working elsewhere. Although very few people are involved in mechanized agriculture, either directly through cultivation of wheat, or indirectly through the leasing of land to external wheat farmers, this is a very important activity in terms of income generation. The areas cultivated for wheat are significantly larger than those for maize and the number of bags harvested on average is larger.

Cluster Analysis

Four land use strategies were identified (Table 3). The households of the first group complement their income from livestock primarily with revenues from small-scale subsistence maize cultivation (Table 3). This group was used as the baseline group in the generalized logit models, as it represents the households that have diversified least (avoiding a shift to mechanized cultivation or income from tourism). Households of the second group have turned to tourist-related activities as an additional source of income. Most households get income from dividends from the wildlife associations (n=121), of which 62 households also receive income from shares in tourist facilities. Others (n=9) only get income from shares in tourist facilities. Members of the third group combine livestock herding with mechanized farming, or the leasing of land to wheat farmers. The households of the fourth group get additional revenues from maize cultivation and from dividends from wildlife associations (n=24), shares in tourist facilities (n=7), or both (n=24).

Table 2. Socio-economic data used in the study.

Socio-Economic Data	*Mean*	*STD*	*Min.*	*Max.*
Maize acreage (for those cultivating: n=118)	2.7	2.8	0.25	20
Maize harvest (# 90 kg bags/acre – for those cultivating)	3.14	4.3	0	25
Wheat acreage (for those cultivating: n=28)	11.6	8.6	3	36
Wheat harvest (90kg bags/acre – for those cultivating)	8.13	6.18	0	20
Wheat leasing income in US$ (for those leasing land: n=17)	1497	640	308	2769
Livestock equivalents per reference Adult	12.2	11.1	0.6	83.7
Household size (reference adults)	9.5	5.7	1	35
Education level of head of household (no. years schooling)	1.4	3.4	0	15

Households of the first group cultivate the smallest maize fields (1.4 acres ±1.2, n=33). Acreages in the third (3.6 acres ± 1.9, n=26) and fourth (2.9 acres ± 3.4, n=59) groups are not statistically different (probability level (p) = 0.28), but both groups have high variances, with cultivated surfaces ranging from 1 to 12 acres. There is also a large variance in the number of acres allocated to mechanized wheat farming (11.5 acres ± 8.7, n=28), as there is in the income generated from leasing land to external farmers for wheat cultivation (1,437 US$/year ± 609, n=17).

The wage-earning variable turned out to have little discriminatory power in the model. Only in the third group of households were there a few households getting revenues through wage earning. Differences between the other groups were not significant.

Spatial Distribution of Land Use Strategies

Although the clustering is entirely based on income sources, a spatial pattern of land use strategies emerged (Figure 4). The households belonging to the first group and the households getting additional revenues from tourism (group 2) cover the same area (69% spatial overlap), but the households are distributed differently. The gravity center for group 2 is in Talek, at the border of the Masai Mara National Reserve, while most households belonging to group 1 live farther away from the Masai Mara National Reserve, south of Lemek. All households involved in mechanized wheat cultivation are located in the Loita Plains, in the northernmost part of the study area, close to the wheat farming area. The households getting revenue from tourism and subsistence cultivation are mainly located around the village of Aitong and south of Lemek. The land use strategy chosen by a household is thus at least in part determined by place-specific variables.

Table 3. Final clusters of land-use strategies.

Land Use Strategy (LUS)	*Maize*	*Wheat*	*Tourism*	*Wage*
1: livestock herding with some subsistence cultivation (N=54)	Some	No	No	Some
2: livestock herding with income from wildlife tourism (N=136)	No	No	Yes	Some
3: livestock herding with mechanized agriculture (N=29)	Yes	Yes	No	Few
4: livestock herding with income from tourism and subsistence cultivation (N=59)	Yes	No	Yes	Some

Statistical Modeling of Land Use Strategies

Full Sample Model

The multinomial logistic model explaining groups of land use strategies consists of three equations, each calculating the logit for a non-baseline, land-use strategy with reference to the baseline strategy (group 1). We tested alternative models with the different distance variables that were highly correlated. The model with the "distance to the Masai Mara National Reserve" variable had the highest explanatory power (McFadden's ρ^2=0.44) and is analyzed below. We first discuss the overall model performance and the contribution of the variables to the explanatory power of the entire model. Subsequently, the three logits that comprise the model are discussed separately.

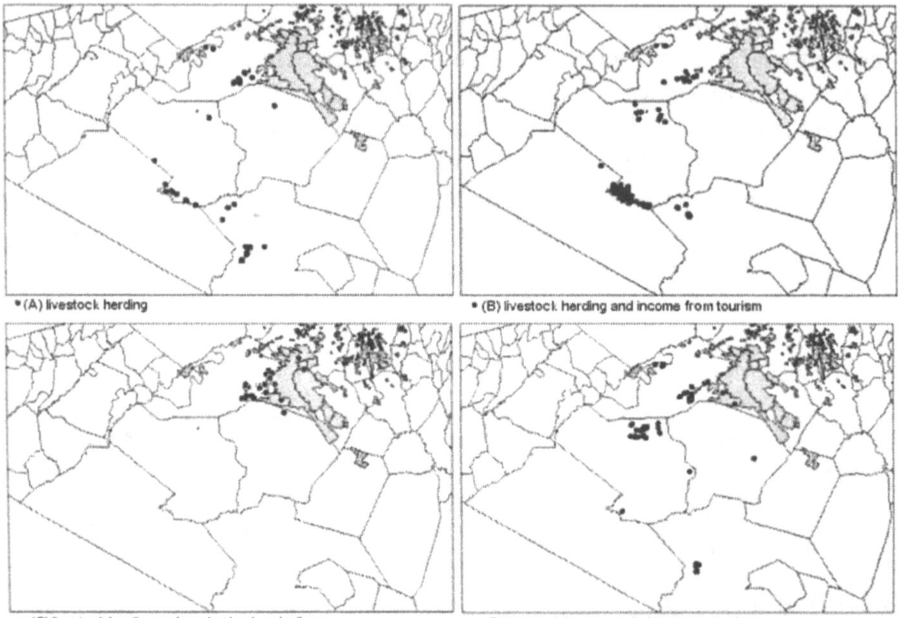

Figure 4. Spatial distribution of land use strategies.

The final generalized logit model has eight explanatory variables, of which the education level of the head of household and the number of livestock equivalents per reference adult have a non-significant Wald χ^2 statistic (χ^2=4.8, p=0.18; χ^2=3.8, p=0.27). However, both variables contributed significantly to the model, the difference in –2log-likelihood with and without the variable being significant (p<0.05). Distance to park is the variable with the highest overall explanatory power in the model (χ^2=44.7, p=0.00001), followed by elevation (χ^2=32.6, p=0.0001), slope (χ^2=19.8, p=0.0001), and distance to water (χ^2=13.1, p=0.0001). The importance of distance to park in the model indicates the presence of spatial autocorrelation. Households living at similar distances to the park are more likely to have similar land use strategies.

The explanatory power of the variables in each of the logits is given in Table 4. In the first logit, the landscape variables (slope and elevation) have the highest explanatory power. Households receiving income from tourism to supplement their income from livestock herding are more likely to be found at

Table 4. Full sample generalized logits model for non-baseline land use strategies.

Land use strategy	***Independent variables***	***Param. Est.***	***Stand. Error***	***Chi-Square***	***Prob***	***Odds Ratio***
INCOME FROM TOURISM VS. CATTLE FARMING	intercept	24.4658	4.3935	31.01	0	
	distance to park	0.0218	0.019	1.31	0.2522	1.022
	distance to water	-0.00081	0.016	0	0.9593	1
	distance to village	-0.00441	0.00555	0.63	0.4266	1
	elevation	-1.3638	0.258	27.95	0.0001	0.25
	slope	-0.4301	0.1401	9.43	0.0021	0.65
	agroecol. zone, rainfed agric.	0.7011	0.5734	1.49	0.2215	2.01
	livestock equivalents	0.195	0.1855	1.1	0.2933	1.21
	education level	0.6811	0.392	3.02	0.0823	1.97
INCOME FROM MECHANIZED FARMING VS. CATTLE FARMING	intercept	-27.6972	13.3855	4.28	0.0385	
	distance to park	0.3385	0.062	29.81	0.0001	1.4
	distance to water	-0.0272	0.0321	0.72	0.3964	0.97
	distance to village	-0.0357	0.0128	7.83	0.0051	0.96
	elevation	1.0707	0.687	2.43	0.1191	2.92
	slope	-0.5166	0.1891	7.46	0.0063	0.59
	agroecol. zone rainfed agric.	-2.1646	1.1303	3.67	0.0555	0.11
	livestock equivalents	0.6212	0.3357	3.42	0.0643	1.86
	education level	0.4742	0.4948	0.92	0.3378	1.61
INCOME FROM TOURISM AND SUBSISTENCE CULTIVATION VS. CATTLE FARMING	intercept	-0.9301	4.4922	0.04	0.836	
	distance to park	0.1333	0.0256	27.06	0.0001	1.14
	distance to water	-0.0782	0.0236	11.01	0.0009	0.92
	distance to village	-0.0148	0.00773	3.64	0.0565	0.98
	elevation	0.0559	0.2291	0.06	0.8073	1.05
	slope	-0.6301	0.1452	18.82	0.0001	0.53
	agroecol. zone rainfed agric.	-0.7396	0.6826	1.17	0.2786	0.47
	livestock equivalents	0.3182	0.2243	2.01	0.156	1.37
	education level	0.7881	0.3729	4.47	0.0346	2.19

lower altitudes and on flatter land than the reference households. Lower elevation is linked to lower local agro-ecological potential, making cultivation more difficult and thus possibly influencing households to rather diversify into tourism. This variable probably needs to be interpreted in terms of spatial autocorrelation as well. Households living close to each other (in areas with the same altitude or slope) are more likely to pursue the same land use strategy. Accessibility variables have little power in the first logit, given the large spatial overlap between the baseline group and the pastoralists profiting from tourism. None of the socio-economic variables contributes significantly to the model.

Accessibility variables are important in the second logit. The homesteads of households involved in mechanized wheat farming or leasing land to outsiders for wheat farming are more likely to be found away from the Masai Mara National Reserve, on flatter land and closer to village centers than the households of the reference group. Distance to the Masai Mara National Reserve is inversely correlated to the distance to the district capital. Thus easier access to the main regional market is an important spatial determinant of the land use strategy associated with mechanized farming. Landscape variables including agro-ecological zone are of secondary importance.

In the third logit, distance to park again explain best the difference between households generating income from a mixture of tourist-related activities, subsistence farming and livestock herding, and the households in the baseline group. The former households are more likely to be found in the vicinity of permanent water sources and villages, on flatter land and farther away from the Masai Mara National Reserve than households of the reference group. Of secondary importance are factors of socio-economic status. The probability of belonging to this fourth group of households rather than the reference group increases with the education level of the head of household. Although this variable did not significantly contribute to the overall model explanation, it is significant in this particular logit. The households are not settled close to the Masai Mara National Reserve. They get dividends from the wildlife association that is located outside the park borders.

Overall, the landscape and accessibility variables have a higher explanatory power than the social variables in predicting the probability of belonging to one of the land use strategy groups other than the baseline. The overall prediction success of the full sample model is 70%. Land use strategies of tourism with livestock holdings and mechanized farming with livestock holdings are best predicted, with 83% and 79% of the cases correctly predicted. Prediction success is lower for the tourism with subsistence farming and livestock group (54%) and the reference group (49%). But, for each of the land use strategy classes, the model improves the prediction success, compared to a purely random model.

Because of strong collinearity between the distance to park and land tenure variables, we could not include both variables in one model. The model where the "distance to park" variable was substituted by "private land tenure" had elevation and private land tenure as most important explanatory variables (ρ^2=0.33). Households involved in cultivation were more likely to live on private land than on communal land. However, conclusions on the strength of association between mechanized farming and privatization of land are not drawn here, as there are confounding factors. Land tenure at the time of rapid expansion of mechanized farming (1975-95) on the group ranch affected by this expansion (Lemek) was communally held, with the same land tenure system as neighbouring group ranches. Although land tenure has been privatized on Lemek since 1995 - maybe in part due to the increased value of land resulting from mechanized farming - it could equally be argued that distance to park (or to the market) is the significant factor explaining the presence of mechanized farming rather than land tenure. Privatization of land tenure on Lemek could have been coincidental, or a consequence rather than a cause of the expansion of mechanized farming.

Reduced Sample Model

The model developed for the sample of households for which information on the leadership status of the head of household was available also has a high overall explanatory power (ρ^2=0.48). In this model, landscape variables have a lower explanatory power compared to the full sample model, while the importance of the socio-economic variables has increased. Household size contributed significantly to the model and was included, while the agro-ecological zones variable was dropped from this model, as it did not contribute significantly. The explanatory power of the variables for each of the three logits comprising the model is given in Table 5.

All logits have positive parameter estimates for livestock equivalents. Thus, the likelihood of being involved in other income generating activities than livestock herding increases with an increasing number of livestock equivalents. This supports more recent literature suggesting a shift in cultivation strategies of pastoralists away from meeting subsistence and towards generating wealth (Sikana 1993, Southgate and Hulme 1996, Homewood et al. 2000). Leadership position and education level are important explanatory variables in the second and third logit (mechanized agriculture vs. cattle farming and subsistence agriculture with tourism vs. cattle farming). The likelihood of being involved in mechanized farming increases with increasing education level of the head of household, or if the head of household holds a leadership position in the community. However, the interaction between education level and leadership position is negative, thus leaders are less likely to have received a formal education. Compared to the full sample model, the percentage of correctly predicted outcomes is similar for the second and the fourth land-use strategy groups. The reduced sample model better predicts the basic land use strategy (group 1), but performs less well for mechanized farming/livestock class (group 3).

Table 5. Generalized logit model for land use strategy based on a reduced sample.

LAS	*Indep. variable*	*Param. Est.*	*Std. Error*	*Chi-Square*	*Prob.*	*Odds Ratio*
	intercept	28.2089	7.5543	13.94	0.0002	
INCOME	dist. to park	0.0566	0.031	3.33	0.0682	1.06
FROM	dist. to water	-0.00142	0.026	0	0.9564	1
TOURISM	dist. to village	-0.0176	0.0122	2.07	0.1506	0.98
VS.	elevation	-1.6352	0.4325	14.29	0.0002	0.19
CATTLE	slope	-0.5168	0.2555	4.09	0.0431	0.61
FARMING	livestock equivalents	0.834	0.3322	6.3	0.0121	2.29
	education level	1.3537	0.7218	3.52	0.0607	3.02
	household size	0.0374	0.3103	0.01	0.904	1.01
	leadership	2.4368	1.3093	3.46	0.0627	12.3
	educ*lead	-2.0456	1.0654	3.69	0.0548	0.17
	intercept	-6.7528	19.4949	0.12	0.7291	
INCOME	dist. to park	0.2944	0.0738	15.91	0.0001	1.34
FROM	dist. to water	0.0665	0.044	2.28	0.1308	1.07
MECHANIZED	dist. to village	-0.0527	0.0181	8.43	0.0037	0.94
FARMING	elevation	-0.2251	1.0675	0.04	0.833	0.79
VS.	slope	-0.6086	0.3468	3.08	0.0793	0.55
CATTLE	livestock equivalents	1.3275	0.5238	6.42	0.0113	3.72
FARMING	education level	1.9917	0.8613	5.35	0.0207	5.64
	household size	-0.8492	0.5174	2.69	0.1007	0.41
	leadership	6.0628	1.7864	11.52	0.0007	467.2
	educ*lead	-5.7152	1.9764	8.36	0.0038	0.0046
	intercept	2.7842	7.6284	0.13	0.7151	
INCOME	dist. To park	0.1326	0.037	12.84	0.0003	1.14
FROM	dist. To water	-0.0695	0.0345	4.06	0.044	0.93
TOURISM AND	dist. To village	-0.0203	0.0138	2.18	0.14	0.97
SUBSISTENCE	elevation	-0.2548	0.3994	0.41	0.5235	0.75
CULTIVATION	Slope	-0.7962	0.2742	8.43	0.0037	0.46
VS.	livestock equivalents	1.1059	0.3713	8.87	0.0029	3.01
CATTLE	education level	1.5915	0.717	4.93	0.0264	3.67
FARMING	household size	-0.2565	0.3397	0.57	0.4502	0.74
	leadership	3.6237	1.2956	7.82	0.0052	41.3
	educ*lead	-2.5336	1.0034	6.38	0.0116	0.11

Impact on Land Cover

Given the spatial overlap between the first two land use strategy groups there is little difference in land cover changes measured for these groups (Table 6). Changes between 1985 and 1995 in the area occupied by those groups can largely be attributed to the development of permanent settlements (smallholder impact). Localized degradation around permanent water holes is also present. Of the buffer zones around the households engaged in mechanized farming, 22.3% was subject to a change in vegetation cover between 1985 and 1995. This includes 14.8% change due to land conversion to mechanized farming and 7.5% to subsistence farming or settlement impacts. The percentage of land cover change in the areas around the households involved in subsistence agriculture is slightly higher than for the households not involved in agriculture. As some of the households practicing subsistence agriculture live close to the mechanized farming area, part of the land in their immediate surroundings (3.7%) was converted to mechanized farming.

Table 6. Land cover changes in areas with different land use strategies.

% of area corresponding to a LAS	*livestock*	*inc. from tourism*	*mechanized farming*	*tourism+ subsistence farming*
clouds/no data	5.6	6.8	0.0	6.5
Change in land cover	**5.0**	**4.4**	**22.3**	**9.1**
due to smallholder impact	*4.1*	*4.1*	*7.5*	*5.5*
due to mechanized farming	*0.9*	*0.3*	*14.8*	*3.7*
no change	89.2	88.6	75.9	84.3
Vegetation regrowth	0.2	0.2	1.8	0.1
total area (ha)	100,050	95,880	45,915	96,775

DISCUSSION

Three possible sets of explanatory variables underlying production choices have been analyzed: socio-economic, landscape attributes, and accessibility. The full sample model, when all households were taken into account, was dominated by the accessibility and landscape variables (distance to the Masai

Mara National Reserve, elevation, slope, and distance to water and villages). The socio-economic variables had a lower explanatory power. This might be due to the high variance in the socio-economic data. Distance to the Masai Mara National Reserve also indicates the presence of spatial autocorrelation. Neighboring households are more likely to follow the same land use practice. For example, all households involved in mechanized agriculture are located close to each other, at a large distance from the Masai Mara National Reserve. Surprisingly, mechanized farming is less likely to be found in agro-ecological zones more suitable for rainfed agriculture. Slope seems to be the more significant factor here. Cultivation is practiced on flat land, despite unreliable rainfall. The input of nitrogen fertilizer and the use of new drought-resistant wheat varieties may be able to offset the disadvantages of lower agro-ecological zone. Mechanized agriculture is the most profitable land use strategy (Thompson and Homewood, forthcoming), with high returns in case of a successful harvest. Farmers might thus take higher risks, cultivating marginal lands.

Although place matters, other factors also intervene in decisions about land use strategies, as shown in the results of the reduced sample model. In this model, accessibility variables still predominate, but the socio-economic variables (leadership position and livestock equivalents) have gained in importance. Households in the reference group, the traditional pastoralists, are less likely to have a leadership position than households in any of the other land use strategy groups. Thus, if the social position of the head of household improves, more opportunities to generate income from other sources than those associated with traditional land uses are created. For example, influential people can get revenues from dividends in a wildlife association, even if they live far from this wildlife association. Without a leadership position, they would not be entitled to receive dividends. Educated people are also more likely to benefit from income sources other than those generated from traditional land uses. However, leaders are less likely to have received a formal education. Most leaders are village elders, while most educated people are young people who have not yet obtained a leadership position.

A shortcoming of the models developed in this study is the low quality of some of the independent variables used. The spatial resolution of 100 m at which the analyses were performed has been dictated by the size of the small-scale land cover changes. At lower resolution, many of the changes would have been either ignored, or largely overestimated. However, some of the independent variables (e.g. elevation and agro-ecological zones) were used here at a higher resolution than the implied precision of the base products. The use of more detailed information on endogenous variables such as soil fertility, rangeland quality and rainfall variability would improve the model. It might also give more information on the appropriateness of the followed land use strategies, provided that the land-use activities take place near the homestead.

While livestock raising is the basic land use strategy in the study area, this study did not address the linkages between spatial and temporal movements of livestock and land cover changes. Livestock movements are largely determined by the social networks of the family to which the animals belong and differ considerably from one family to another, with one household having access to different grazing areas for their livestock and several families using a given grazing area. This many-to-many relationship in space and time is not easily represented in a spatially explicit way.

To gain a better understanding of the impact of the land use policy concerning the subdivision of group ranches in private land titles, a more detailed approach than the one followed in this study is necessary. Feedback mechanisms possibly exist between the different stages of implementation of this policy and land use decisions made by the Maasai in our study area. These mechanisms have not been represented in our models, due to lack of information. Likewise, mapping of the plot boundaries and linking them to households might reveal interesting patterns.

Linking changes in land cover in a given area to the land use strategy of the households from that area is often difficult. Households are often involved in several land-use activities that have an impact on the environment in a wide range around the homestead. Hence the small difference in land cover changes in the areas around the first, second and fourth land use strategy groups. Mechanized farming as a land use strategy is strongly spatially linked to participating households.

CONCLUSION

Land use strategies followed by pastoralists were analysed in a spatially explicit way. Although distance to the Masai Mara National Reserve was found to be the explanatory variable with the greatest significance, social variables are also important to explain the land use strategies followed by the households. Of the social variables introduced, leadership position and livestock wealth proved to be the most important explanatory variables. Distance to the Masai Mara National Reserve is inversely correlated to the distance to Narok, thus suggesting that both market factors and conservation policies are influencing the location of mechanized agriculture. Leadership status and networks can be used to obtain lands at a location favorable for development, or will allow revenues from distant sources. Links between leadership position, livestock wealth and cultivation (both mechanized and small scale) support literature, which describes the diversification and intensification of production systems in pursuit of wealth accumulation.

Our findings are consistent with studies that suggest a continued expansion of mechanized cultivation from the existing concentration of farming areas (Norton-Griffiths, 1996). Our results suggest that the factors of greatest importance for processes of intensification of land use are accessibility and socio-economic factors. This suggests continued pressure on natural vegetation, which will be at its greatest in areas furthest from the Masai Mara National Reserve and closest to the existing mechanized farming belt.

Although it is possible to describe land use strategies with spatial and socio-economic variables, analyzing the environmental consequences of these land-use strategies is more complicated. The Maasai are a pastoralist society with most of the land being managed and used communally. Hence the difficulties in attributing the changes in land cover to certain households. We showed here that there certainly is a strong link between the location of the homestead of people involved in mechanized farming and the land cover changes that this entails. However, the link with land cover changes was less obvious for the other land use strategies. Changes in these areas were also less important, with on average 5% of the area subjected to land cover change between 1985 and 1995.

Despite these difficulties, the modeling techniques described here are useful in shedding light on the underlying factors linked to land conversion. These findings suggest that greatest significance linked to choice of land use strategy is provided by variables linked to economic factors, and landscape attributes. Within this, patterns of decision-making rooted in social and economic structures of pastoral communities are important in understanding land use production strategies.

NOTES

[1]The Olchoro-Oiroua area in our study area was given a land title as an experimental government cattle ranch managed by the veterinary department, while land at Nkorinkori on Lemek group ranch was held under government land titles as livestock holding grounds. The existence of these titles facilitated their appropriation by Maasai connected with the newly independent administration during the late 1960s.

[2]Cultivation history was collected and found to have commenced in all cases within the last fifteen years. Data was collected on field sizes during the three years in which fieldwork was carried out. Small-scale expansion and contraction of field sizes was recorded during this time, therefore an average field size over this period was used.

[3]Livestock Equivalents are used as a basis for comparison between herds, and are based on 250 kg live weight and herd and flock characteristics of Kajiado Maasai (Bekure et al. 1990). The Reference Adult measure is a basis for comparing food energy requirements of the human population as calculated by Little (1980).

ACKNOWLEDGEMENTS

This work was partly funded by the European Commission under the INCO-DC programme (contract ERBIC18CT960070). Additional support came from DFID-ESCOR R6828, the Parkes Foundation, University of London Central Research Fund and University College London Graduate School. Special thanks go to Prof. K. Homewood, who commented on earlier versions of this manuscript. We would like to thank the Department of Resource Surveys and Remote Sensing (DRSRS), and particularly M. Said, for granting us access to its GIS database. We are grateful to the Government of Kenya, the Kenya Wildlife Service, the management of Lemek, Koiyaki and Siana GRes, and to the many Maasai households and individuals who gave hospitality, information and insights. Field work was in particular only possible with the help of M. Ololoigero, F. Naimodo, and J. Mpoe, who were particularly instrumental in pursuing the aims of this research, amongst the many others too numerous to mention. Two anonymous reviewers helped to further improve the manuscript.

REFERENCES

Agresti, A. 1990. Categorical data analysis. New York: John Wiley & Sons.

Aligula, H.M., Siad, M., and Njuguna, E.C., 1997. Geo-information for sustainable rangeland management: A case study of the arid and semi-arid lands in Kenya. Paper presented to a conference on Geo-information for sustainable and management, International Institute of Aerospace Surveys and Earth Sciences, International Technology Centre (ITC), Enschede, Netherlands.

Behnke, R.H. and Scoones, I., 1993, Rethinking range ecology: Implications for Rangeland Management in Africa, ch 1 in Behnke, Scoones and Kerven, C., 1993 (eds) Range Ecology at Disequilibrium, Overseas Development Institute, London.

Bekure, S., de Leuuw, P.N., Grandin, B.E., Neate, P.J.H., 1991. Maasai herding. an analysis of the livestock production system of Maasai pastoralists in eastern Kajiado District, Kenya. International Livestock Centre for Africa, Nairobi.

Bian, L. and West, E. 1997. GIS modelling of elk calving habitat in a prairie environment with statistics. Photogrammetic Engineering and Remote Sensing 63 (2): 161-167.

Boone, R.B., Galvin, K.A., Smith, N.M. & Lynn, S.J. 2000. Generalizing el Niño effects upon Maasai livestock using hierarchical clusters of vegetation patterns. – Photogrammetic Engineering & Remote Sensing, 66: 737-744..

Brush, S.B. and Turner, B.L., 1987. The nature of farming systems and views of their change. In: Turener, B.L. and Brush, S.B., (Eds.) Comparative farming systems.- The Guilford Press, New York, 11-48.

Chomitz, K.M. and Gray, D.A. 1996. Roads, land use, and deforestation: a spatial model applied to Belize. - The World Bank Economic Review 10 (3): 487-512.

Dahl, G., and Hjort, A. 1976, Having Herds, Pastoral growth and Household Economy. Studies in social anthropology. Department of Social Anthropology, University of Stockholm.

Dublin, H. T., Sinclair, A. R. E., and McGlade, J., 1990. Elephants and fire as causes of multiple stable states in Serengeti-Mara woodlands. - Journal of Animal Ecology, 59, pp. 1147-1164.

Entwistle, B., Walsh, S.J., Rindfuss, R.R., and Chamratrithirong, A., 1998. Land-use/land-cover and population dynamics, Nang Rong, Thailand. – In: Liverman D, Moran E F, Rindfuss R

R, Stern P C (eds.), People and Pixels. National Academy Press, Washington, D.C.: pp. 121-144.

ESRI, 1993. Digital chart of the world. Environmental Systems Research Institute Inc. Redlands, California.

Galaty, J.G., 1992. "The land is yours": Social and economic factors in the privatisation, sub-division and sale of Maasai ranches. - Nomadic Peoples 30:26-40.

Grandin, B.E., 1988. Wealth and pastoral dairy production: a case study from Maasailand. Human Ecology 16, 1-21.

Green, G.M. & Sussman, R.W. 1990. Deforestation history of the eastern rain forests of Madagascar from satellite images. – Science, 2: 212-215.

Hollis, A.C., 1905, The Masai, their language and folklore. Clarendon Press, Oxford.

Homewood K., E.F. Lambin, E.Coast, A. Kariuki, I. Kikula, J. Kivelia, M. Said, S. Serneels, M. Thompson (submitted) Long term changes in Serengeti-Mara wildlife and land cover: pastoralists or policies? Proceedings of the National Academy of Science.

Homewood, K.M., 1992, Development and the Ecology of Maasai Pastoralist Food and Nutrition. Ecology of Food and Nutrition, Vol 29, pp 61-80.

Homewood, K.M. and Rodgers, W.A. 1991. Maasailand Ecology: pastoralist development and wildlife conservation in Ngorongoro, Tanzania. - Cambridge University Press, Cambridge, pp 298.

Hosmer, D.W. and Lemeshow S. 1989. Applied regression analysis. New York: John Wiley & Sons.

Houghton, R.A. 1994. The worldwide extent of land-use change. – Bioscience, 44:305-313.

Jatzold R. 1982. Explanation of the evaluation of the natural potential, method of the agro-ecological zonation – In: Ministry of Agriculture Farm Management Handbook of Kenya, Vol II/B, Central Kenya, Government of Kenya.

Jewell, P., 1980, Human ecology in savanna environments in D.R. Harris (ed) Human ecology in savanna environments. Academic Press, London.

Karime, O. N. I., 1990. The changing land use patterns in the Ngorengore Loita plains. Department of Resource Surveys and Remote Sensing, Ministry of Planning and National Development, Kenya, pp. 32.

Kerven, C., 1992. Customary commerce: A historical reassessment of pastoral livestock marketing in Africa. ODI, London.

Lamprey, R. H., 1984. Maasai impact on Kenya savanna vegetation: a remote sensing approach. - PhD thesis, University of Aston in Birmingham, U.K., pp. 349.

Lamprey, R., and Waller, R., 1990. The Loita-Mara region in historical times: patterns of subsistence, settlement and ecological change. - In: P. Robertshaw (ed.) Early pastoralists of South-western Kenya. Memoirs of the British Institute in Eastern Africa, Nairobi.

Lamprey, R.H., 1984, Maasai Impact on Kenya Savannah Vegetation: A Remote Sensing Approach, PhD Thesis, University of Aston.

Little, M., 1980, Designs for human-biological research among savanna pastoralists. In: Human ecology in savanna environments. D.R. Harris (Ed.). Academic Press, London.

Liverman D, Moran E F, Rindfuss R R, Stern PC (ed.) 1998. People and pixels: linking remote sensing and social science. National Acadamy Press, Washington D.C.

McCabe, J. T., Mollel, N., and Tumaini, A., 1997a, Food security and Nutritional Status, ch II.14 in Thompson, D.M., (ed) Multiple Land Use: the experience of the Ngorongoro Conservation Area, Tanzania, IUCN Nairobi.

McCabe, J.T., Schofield, E. C., and Nygaard Pederson.G, 1997b, Food security and the Role of Cultivation in the Ngorongoro conservation Area, ch II.14 in Thompson, D.M., (ed) Multiple Land Use: the experience of the Ngorongoro Conservation Area, Tanzania, IUCN Nairobi.

McCracken, S.D., Brondizio, E.S., Nelson, D., Moran, E.F., Siqueira, A.D. & Rodriguez-Pedraza, C. 1999. Remote sensing and GIS at farm property level: demography and deforestation in the Brazilian Amazon. – Photogrammetric Engineering & Remote Sensing, 65:1311-1320.

Mertens, B. and Lambin E.F., 2000. A spatial model of land-cover change trajectories in a frontier region in southern Cameroon, Annals of the Association of American Geographers, 90 (3): 467-494.

Mertens, B., Sunderlin, W.D., Ndoye, O. & Lambin, E.F. 2000. Impact of macroeconomic change on deforestation in South Cameroon: integration of household survey and remotely-sensed data. – World Development, 28:983-999.

Narumalani, S., Jensen, J.R., Althausen, J.D., Burkhalter, S. and Mackey H.E. Jr., 1997. Aquatic macrophyte modelling using GIS and logistic multiple regression. Photogrammetric Engineering & Remote Sensing, 63 (1): 41-49.

Nelson, G.C., Harris, V. & Stone, S.W. 1999. Spatial econometric analysis and project evaluation: modeling land use change in the Darién. Washington, D.C.

Norton Griffiths, M., 1996. Property rights and the marginal wildebeest: an economic analysis of wildlife conservation options in Kenya, Biodiversity and Conservation 5, 1557-1577

Ottichilo, W.K., Said, M.Y., Koster, H.N., Prins, H.H.T., J de Leeuw, I.M.A. Heikonig, 2000. The dynamics of migratory wildebeest in the Maasai Mara Ecosystem, Doctoral Thesis (2000) ISBN 90-5808-197-4, Wageningen University & Research Centre, the Netherlands.

Pereira, J.M.C., Itami, R.M. 1991. GIS-based habitat modelling using logistic multiple regression: a study of the Mt. Graham red squirrel. - Photogrammetric Engineering and Remote Sensing 57 (11): 1475-1486.

Potkanski, T., 1994a, Pastoral economy, property rights and traditional mutual assistance mechanisms amongst the Ngorongoro and Salei Maasai of Tanzania. IIED Pastoral Land Tenure Monograph Series n°2, London.

Potkanski, T. 1994b. Property concepts, herding patterns and managemant of natural resources among the Ngorongoro and Salei Maasai of Tanzania. - IIED Pastoral Land Tenure Series n° 6. pp. 74.

Prins, H.T., 1992, The Pastoral road to extinction: Competition between wildlife and traditional pastoralism in East Africa, - Environmental conservation.

Ruess, R.W. and Halter, F.L., 1990. The impact of large herbivores on the Seronera woodlands, Serengeti National Park, Tanzania. - African Journal of Ecology, 28, 259-275.

Serneels, S. and Lambin, E.F. 2001. Proximate causes of land-use change in Narok District, Kenya: a spatial statistical model. - Agriculture, Ecosystems and Environment, 85: 65-81.

Serneels, S. and Lambin, E.F 2001. Impact of land-use changes on the wildebeest migration in the northern part of the Serengeti-Mara ecosystem. - Journal of Biogeography, 28 (3): 391-408.

Serneels, S., Said, M.Y. and Lambin, E.F., 2001. Land-cover changes around a major East-African wildlife reserve: the Mara Ecosystem (Kenya). - International Journal of Remote Sensing, *in press*.

Sikana, P.M., Kerven, C.K., and Behnke, R.H., 1993, From subsistence to specialised commodity production: commercialisation and pastoral diarying in Africa. ODI Pastoral Development Network.

Sinclair, A.R.E. and Arcese, P. 1995. Serengeti in the context of worldwide conservation efforts. - In: Serengeti II. Dynamics, management and conservation of an ecosystem. (eds Sinclair, A.R.E and Arcese, P.) University of Chicago Press, Chicago, 31-46. Singida, I., 1984. Land and population problems in Kajiado and Narok, Kenya. - African Studies Review, 27, 1, pp.23-39.

Skole, D.L., Chomentowski, W.A., Sallas, W.A. & Nobre, A.D. 1994. Physical and human dimensions of deforestation in Amazonia. – Bioscience, 44: 314-322.

Southgate, C. and Hulme, D., 1996, Land, water and local governance in a Kenyan wetland in a dryland: The Kimana Group Ranch and its environs, Rural resources and rural livelihoods paper 4, Institute for Development Policy and Management, Manchester University

Spencer, P., 1988, The Maasai of Matapato, A study of rituals of rebellion. Manchester University Press.

Sprugel, D.G., 1991. Disturbance, equilibrium and environmental variability: what is 'natural' vegetation in a changing environment? - Biological Conservation, 58,1-18.

Stelfox, J. G., Peden, D. G., Epp, H., Hudson, R. J., Mbugua, S. W., Agatsiva, J. L., and Amuyunzu, C. L., 1986, Herbivore dynamics in southern Narok, Kenya, - J. wildl. Management 50, 2 pp. 339-347.

Thompson, D.M., 1998, The role of cultivation in Maasai Production Systems on three GRes adjacent to the Maasai Mara National Reserve in Kenya, Department of Anthropology, First Year Report/PhD Upgrading proposal, University College London.

Thompson, M. & Homewood, K. Entrepreneurs, elites and exclusion in Maasailand – Human Ecology (submitted).

Thompson, M. forthcoming. Livestock, Cultivation and Tourism; Livelihood Choices and Conservation in Maasai Mara Buffer Zones, Kenya, PhD thesis in prep, Department of Anthropology, UCL, London.

Turner, B.L., Meyer, W. & Skole, D. 1994. Global land-use/land-cover change: towards an integrated study. – Ambio, 23:91-95.

Turner, M.G., Gardner, R.H. and O'Neill, R.V., 1995. Ecological dynamics at broad scales: ecosystems and landscapes. - BioScience Supplement 1995, 29-35.

Vega Garcia, C., Woodard, P.M., Titus, S.J., Adamowicz, W.L., and Lee, B.S. 1995. A logit model for predicting the daily occurrence of human caused forest fires. - Int. J. Wildland Fire 5 (2): 101-111.

Walker, R. & Homma, A. 1996. Land use and land cover dynamics in the Brazilian Amazon: an overview. – Ecological Economics, 18:67-80.

Waller, R., 1976, The Maasai and the British, 1895-1905: The origins of an Alliance. - Journal of African History 17.

Waller, R., 1988, Emutai: crisis and response in Maasailand 1883-1902, in Johnson and Anderson (eds) The ecology of survival.

Walsh, S.J., Evans, T.P., Welsh, W.F., Entwisle, B. & Rindfuss, R.R. 1999. Scale-dependent relationships between population and environment in Northeastern Thailand. – Photogrammetric Engineering & Remotes Sensing, 65:97-105.

Chapter 4

MONITORING LAND USE CHANGE IN THE PEARL RIVER DELTA, CHINA

Karen C. Seto[1], Robert K. Kaufmann[2], and Curtis E. Woodcock[3]

[1]*Center for Environmental Science and Policy, Institute for International Studies, Stanford University, Stanford, CA 94035-605, USA*
kseto@stanford.edu
[2]*Department of Geography & Center for Environmental Science and Policy, Boston University, Boston, MA 02215, USA*
[3]*Department of Geography & Center for Remote Sensing, Boston University, Boston, MA 02215, USA*

Keywords: remote sensing, land use models, China.

Abstract The research presented in this chapter focuses on the Pearl River Delta, a rapidly developing region in southern China. We combine high-resolution satellite imagery and socioeconomic data to address the following questions: How rapidly is agricultural land being lost? What is the rate at which urban areas are expanding? We highlight some of the methodological challenges encountered in our research to link remotely sensed imagery with socioeconomic data. Specifically, we discuss the effects of classification accuracy and bias on land use models, present several statistical approaches to integrating time series satellite data with socioeconomic data to model land use, and discuss some of the patterns of land use change in the Pearl River Delta. Our analysis indicates that urbanization increased by more than 300% during the period 1988-1996. Most of the new urban areas were converted from agricultural land. The geographic pattern of land use change follows the Hong Kong-Guangzhou corridor and mirrors foreign direct investment flows.

INTRODUCTION

Remote sensing has revolutionized the way in which scientists observe and study landscapes. More than just pictures, satellite images provide information on multiple levels and scales from which conclusions about the earth's land surface can be drawn. Historically, remote sensing has been used by natural scientists to collect data on terrestrial land cover features, but its use by social scientists to understand the human processes that drive observed spatial patterns is relatively new. Yet, remote sensing, with its synoptic survey of the land, offers a unique opportunity to study the history and characteristics of human interactions with the environment at a geographical coverage not previously possible through conventional ground observations. Currently there exists a 25-year archive of high-resolution Landsat data. This archive is being continued by Landsat 7 and other earth observing satellites that will allow studies of land cover change and the fundamental social and economic processes that drive them.

Monitoring land cover is only the first step in addressing the issues surrounding human dimensions of global environmental change. Equally important, the drivers that cause land cover change must be understood. There are two major causes of land cover conversions: those that occur naturally (e.g., forest fires, pests), and those that are driven by human actions (deforestation, agriculture). To assess land cover conversions caused by natural forces, the physical conditions that induce change, such as climate variability, and the relation between them, must be understood. Alternatively, to assess land cover changes caused by human actions, researchers must assess the social, political, economic, and environmental factors that motivate human behavior and affect land use. Therefore, understanding the causes of land use change is a two-fold exercise. First, a history that provides information on types of land use, land use change, as well as rates and magnitude of change, is required. Second, the socioeconomic, political, and environmental factors that affect how people use natural resources must be identified and the relations among them understood.

Our research combines the strengths of the remote sensing observational record with time-series modeling techniques to evaluate the socioeconomic factors that drive urbanization in a rapidly developing region of China, the Pearl River (*Zhujiang*) Delta. The project brings together expertise in remote sensing, statistical modeling of economic systems, and the Chinese economy to provide a more comprehensive understanding of urbanization and the effects of market liberalization policies on land use. Our research goals are to: (1) monitor the rate and patterns of land use change over a 9-year period, from 1988 to 1996; (2) develop a statistically rigorous model of land conversion to evaluate the causes of urbanization; and (3) forecast future rates of land conversion based on assumptions about the directions of the Chinese economy. The underlying research questions are: What are the socioeconomic drivers of land use change? How have the numerous changes in the economy affected land use, and in particular, urbanization? What drives the conversion of land to more intensive uses, such as the conversion of farmland to urban areas?

This chapter highlights our efforts to pursue the goals stated above and discusses some of the larger issues we faced in this endeavor. In particular, we address the following issues: (1) statistical approaches to integrate socioeconomic and remote sensing data to model land use; (2) the effects of classification accuracy and bias on land use models; and (3) patterns and drivers of urbanization in the region. The chapter is organized as follows: Section 2 provides a description of the Pearl River Delta and changes in policies that have led to the opening of the Chinese economy. A description of the remote sensing and socioeconomic data used in the study, and a discussion on our approach to image classification are provided in Section 3. The effects of classification accuracy and bias on land use models are discussed in Section 4. Section 5 describes statistical methods for integrating remote sensing data and socioeconomic data in land use models. Results from the land use change analysis for the Pearl River Delta are provided in Section 6, and concluding remarks are presented in Section 7.

THE PEARL RIVER DELTA

Our study area is the Pearl River (*Zhujiang*) Delta (PRD), one of the fastest growing regions in China, located in the southern province of Guangdong (Figure 1). The PRD includes several basins drained by the Xi (west), Dong (east), and Bei (north) rivers. The area of the PRD that this study includes is defined by one Landsat TM scene, and covers 26,000 km^2. Located between 21°N and 23°N Latitude and crossed by the Tropic of Cancer, the Delta is characterized by a dry season, November through April, and a rainy season, May through October, during which typhoons are common. The population of the Delta is more than 21 million, representing one-third of the total population of

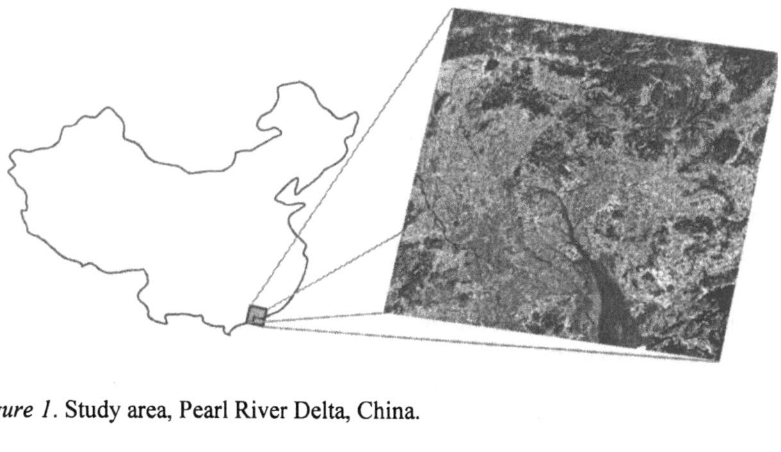

Figure 1. Study area, Pearl River Delta, China.

the province. The region is fertile, has a strong agricultural history, and can support two to three crops per year. The semi-tropical monsoon climate combined with fecund alluvial soils makes the Delta a national leader in the production of rice, lychees, bananas, pond fish, and sugar cane. In addition to its natural resources, the Delta benefits from its geographic proximity to Hong Kong and cultural ties to overseas Chinese investors, who provide the region with access to capital, technological innovations, and managerial expertise.

The unparalleled economic growth in the PRD started in 1979, when the central government gave the Guangdong Province preferential treatment through two main mechanisms: decentralization policies and market-oriented reforms. These include the abolition of collective farming in favor of household and individual farming, or the "household responsibility system,"[1] and agricultural price reforms[2]. The household responsibility system introduced subcontracting and allowed communes to divide their land among farmers, who could sell a portion of their crops on the free market. The price reforms led to an increase in prices for grain and other agricultural outputs. In response, farmers chose their crops based on market conditions and increased agricultural production dramatically (Lin 1988). Shortly after these reforms, the central government established special economic zones (SEZs) for the purposes of attracting foreign investment, promoting exports, and evaluating different economic policies (Chu 1998). In essence, the SEZs are economic laboratories that test various free-market reforms. The PRD is home to Shenzhen, arguably the most influential and prosperous SEZ in the country. The economic policies tested in Shenzhen reverberate throughout the country. To compliment the SEZs, various coastal regions were granted SEZ-like status and designated as open cities and economic zones. In 1985, the Pearl River Delta Open Economic Region was designated and given special tax and trade status (Lin 1997). In addition, the privatization of some government and communally-held enterprises shifted economic control from top-down state planning towards free market principles.

The social and environmental impacts of these policies and reforms have been and continue to be debated (Muldavin 1996), but the economic impacts are self-evident: between 1979 and 1999, China's GDP grew at an average annual rate of 10.4 percent (World Bank 2000). The World Bank estimates that since the advent of the reforms, more than 200 million people have been lifted out of poverty (World Bank 1996) and the OECD projects China to be the world's largest economy by 2020 (Maddison 1998). The Delta exemplifies China's economic growth. From 1988 to 1996, real GDP for counties in the Delta increased by 350 to 550% (Statistical Bureau of Guangdong, various years).

Rapid economic development has affected the economic welfare of the Chinese people in numerous ways. Most apparently, it has raised per capita income as indicated by the rise in the ownership of private vehicles, increases in both the quality and quantity of living space per capita, and improved dietary standards. The reforms also have important implications for urbanization. Prior to the reforms, China had a stringent *hukou*, or household registration system, which limited people's residential mobility; migration from rural to urban areas was nearly impossible. Additionally, the *danwei*, or work unit, provided basic goods and services such as housing, health care, food ration tickets, and education. Together, the *hukou* and danwei controlled geographic mobility and curbed urbanization. Reforms have relaxed the *hukou*, diminished the importance of the danwei, and increased the number of "floating" workers created by gains in agricultural efficiency. Combined with economic opportunities in the coastal regions and increased industrial and commercial activities, there have been massive migrations from rural to urban regions and unparalleled rates of urban growth. Compounding these pressures on land use are changes in the Chinese diet. In China, as in many developing countries, higher income tends to increase meat consumption. Due to trophic inefficiencies, a meat-based diet requires more land than a grain-based diet. This has profound implications for land use in China,

According to Article 10 of the Constitution of the People's Republic of China, all land belongs to the state, and individual farmers have no property rights. During the pre-reform period, land use was determined by the state. Land was considered to have neither value nor price, since output from the land also had no price. Although land was officially the property of the state, it was managed by production brigades through collective land ownership. In 1988, Article 10 was amended to allow the transfer of land use rights (Sharkawy et al. 1995). Ownership remains in state hands, but land use rights are available by negotiation, bid, or auction. As a result, farmers and collectives can rent their land to foreign and local ventures, and large areas of farmland have been converted to urban uses.

DATA AND IMAGE CLASSIFICATION

Remote Sensing Data

Our satellite data consist of nine predominantly cloud-free Landsat TM images (WRS2 path 122, row 44) acquired between October and March (Table 1). Ideally, anniversary images would have been chosen, but the Delta is located in the tropics, and clouds during much of the rainy season obscure the area. The images are georeferenced to a Universal Transverse Mercator map projection provided by the Institute for Remote Sensing Application in Beijing, and resampled to 30 x 30m pixels using a nearest neighbor resampling algorithm with a first order polynomial. The number of ground control points (gcp) used for registration varies by image, and in all but one scene, the root mean square error (RMSE) of the registration process is less than one-third of a pixel (Table 1).

To correct for changes in atmospheric conditions, illumination angles, and seasonal variation across the images, we used a relative radiometric normalization method. This empirical technique employs a two-date density plot of digital number (DN) values in which stable features in the images form a natural "ridge" in the plot. A regression line is fit to the ridge, which represents the linear relationship between the two images. Using the 1988 image as the master, two-date density plots were constructed for each spectral band of each scene, and the regression line was used to match each date to the 1988 scene. This approach has limitations if change constitutes a large portion of the image, or if there are significant seasonal differences among images. However, in this case it was as effective as a variety of other more complicated algorithms, including absolute atmospheric corrections where DNs are converted to surface reflectances (Song et al. 2001).

Table 1. Characteristics of Landsat images used in the study.

Acquisition date	*Number of gcp's*[1] *for registration*	*Geometric Registration RMSE*[2]
10 December 1988	25	± 0.2893
13 December 1989	23	± 0.2914
30 October 1990	18	± 0.2364
02 February 1991	18	± 0.2412
20 January 1992	16	master image
24 December 1993	20	± 0.2618
08 November 1994	17	± 0.3162
30 December 1995	24	± 0.2970
03 March 1996	25	± 0.2776

[1]Ground control points
[2]Root mean square error of x,y coordinate

Classification Approach

During the years of a state-planned economy, farmers had no choice in how to use their land. Our hypothesis is that the move to a market economy allows farmers and land users to allocate resources based on economic return. We assume that the main factors that drive land conversion is the economic return to capital, land, or labor associated with a particular land use. Given the choice between two parcels of land, the one with the lower opportunity cost will be converted first. In this regard, we separate land based on its value, land with an economic return, and land with no economic return, and define four land use categories: urban and agriculture (land uses with economic return) and natural vegetation and water (land without economic return). In this analysis, it is assumed that water that is converted to urban areas was not used previously for economic-related purposes; they were not reservoirs for drinking water or irrigation, but rather naturally occurring water bodies. Ideally, these types of water uses would have been differentiated. However, it is difficult to distinguish reservoirs from lakes in imagery without ancillary information. From these four categories, we develop two land use change classes: *agriculture→urban* and *natural→urban*, where *natural→urban* is an aggregation of change classes *forest→urban*, *shrub→urban,* and *water→urban*. The rationale behind combining these three classes into a single change class is that they all represent conversions from non-economic uses to an economic use of the land – that is, these conversions have little or no opportunity cost. Therefore, we do not differentiate between conversion from water, forest, or shrub to urban areas. The two change categories, *agriculture→urban* and *natural→urban*, are the dependent variables in our land use models.

One of the challenges in this study is the use of only remote sensing data to identify changes in land use, not just changes in land cover. For example, the conversion of field crops to fish ponds is a change in land cover, but does not represent a change in the general land use class, agriculture. Alternatively, some changes in land use do not manifest themselves as changes in land cover, such as the conversion of unmanaged water to fish ponds. This lack of a direct relation between land cover and land use constitutes a technical difficulty for assessments of land use change that use remote sensing alone. Particularly for applications that link remote sensing with human activity, the land cover versus land use differentiation is important, because the latter emphasizes the functional role of land in economic activities (Campbell 1983), while the former does not. Therefore, confounding the two may over- or under-estimate the relation between economic activity and land use. For example, misclassifying all water bodies as fish ponds would inflate the amount of agricultural land and may over-estimate the impact of economic reforms on farm output. To overcome these measurement differences, we develop a conceptual model that links land covers with land uses. Based on the idea that land use classes are composed of component land-covers, land cover is used as a "bridge" to link the images and the land use classes. In essence, we map the component land covers and then aggregate them into land use and land use change classes. In this context, the land use classification process can be viewed as an exercise in disaggregation. First, the land use and land use change classes, which are constructs that relate to human activity or "use" of land, are disaggregated into their key biophysical manifestations, or land covers. Next, the land covers are disaggregated into their fundamental spectral manifestations.

Twenty-three stable and change classes were identified in the images (Table 2). The seven stable classes include: *water, forest, shrub, fish pond, agriculture, transition,* and *urban.* The *agriculture* class includes rice paddies, fruit orchards, and field crops, which are disaggregated into their spectral components during image classification in order to generate better classification results. The stable urban class is represented by two main points along the development continuum: 1) areas that have been cleared and have not had extensive construction (*transition*), and 2) new urban areas, roads, and older sections of cities and towns (*urban*). These seven stable classes are used to generate sixteen change classes, which are consolidated into the two classes for the land use models.

In our initial step, we generate a two-date change map with the 1988 and 1996 images using a multidate Tasseled Cap transformation and a Bayesian Maximum Likelihood classification. Other techniques are used to monitor change in the PRD, including stacked Principal Components (Li and Yeh 1998) and entropy models (Yeh and Li 2001). We select a multidate Tasseled Cap transformation, because it is scene independent, directly associated with physical scene features that affords intuitive interpretation, and has been used successfully to monitor change (Fung 1990, Collins and Woodcock 1996). The Tasseled Cap transformation rotates TM data and creates three new planes: Brightness (B), Greenness (G), and Wetness (W), which are directly associated with physical scene attributes (Crist and Cicone 1984). The Brightness band is a weighted sum of all six reflective bands and can be interpreted as the overall brightness or albedo at the surface. The Greenness band primarily measures the contrast between the visible bands and near-infrared bands and is similar to a vegetation index. The Wetness band measures the difference between the weighted sum of the visible and near-infrared bands and the mid-infrared bands and can be interpreted as a measure of soil and plant moisture. The six-dimensional multidate, multispectral space consisting of BGW values for date 1 and difference BGW values (□B, □G, □W) between two dates that captures the original spectral location for the land covers and the magnitude and direction of change.

A total of 809 training sites (7807 pixels) was selected for the twenty-three stable and change classes. The distribution of sites for each class is shown in Table 2. The sites are selected based on visual interpretation of the images in

Table 2. Stable and change classes, number of training sites for each class, and classes for land use model.

Stable Classes	*Sites*	*Change Classes*	*Sites*	*Classes for Land use Model*
water	34	water to fish pond	22	
		water to agriculture	48	
		water to transition	36	natural to urban
		water to urban	26	natural to urban
forest	31	forest to water	21	
		forest to transition	36	natural to urban
		forest to urban	18	natural to urban
shrub	35	shrub to water	14	
		shrub to transition	33	natural to urban
		shrub to urban	23	natural to urban
fish pond	22	fish pond to transition	12	agriculture to urban
agriculture	117	agriculture to water	26	
		agriculture to fish pond	34	
		agriculture to transition	83	agriculture to urban
		agriculture to urban	47	agriculture to urban
transition	28	transition to urban	24	
urban	39			

the laboratory and the field. In the laboratory, an analyst familiar with the region selects large areas (10-30 pixels), which are representative of the classes of interest. After this initial area selection, training sites within the areas are selected during fieldwork in China in February, 1998. In the field, sites are georeferenced using a global positioning system (GPS) and digitized onto the image. The inputs to the classifier are six bands of data for the 809 training sites: BGW calculated for 1988, and changes in BGW (□B, □G, □W) between 1996 and 1988.

This change-detection technique produces a reliable change map with a high overall accuracy of 93.5% as confirmed by a field-based accuracy assessment in March and April, 1999 (Seto et al. in press). However, we require remote sensing data compatible with the socioeconomic data, which are available at annual intervals at the county level. Rather than utilizing only the two end dates, 1988 and 1996, we use the sequential time-series remote sensing data to generate annual estimates of land use change. We employ two methods to produce annual estimates of land use change. The first is an extension of the multidate Tasseled Cap transformation, where we apply the 1988-1996 training statistics on intervening pairs of images, 1988-1989, 1988-1990, etc. By using the 1988-1996 change map as a mask for all other images, we process only those pixels which change during the nine year period. The mask procedure obviates the problem of identifying change pixels in excess of those identified in the change map.

We develop the second method as a new approach to change-detection using time-series econometric techniques. The techniques are described in detail elsewhere (Kaufmann and Seto 2001), and only the most noteworthy points are described here. The method is a two-step change-detection procedure where in the first step, a fixed effects estimator is used to parameterize models of the stable land cover classes. In the second step, the models for each class are combined to simulate changes between land cover classes at all possible dates. Competing models are evaluated with a test for predictive accuracy and the date of change is determined by the model that predicts the observed land conversions most accurately. The technique is not intended to be a stand-alone change-detection methodology, but rather, given a two-date change map and a sequential time-series of images for the intervening years, the methodology assesses the year in which land use changes occur. Both the multidate Tasseled Cap and econometric techniques produce favorable results with the training data, but only the Tasseled Cap approach is adopted for all the scenes as it is easier to implement.

Socioeconomic Data and Scale of Analysis

One motivation for our research is to develop land use models that provide a better understanding of how policy levers may affect land use. Therefore, our unit of analysis is a scale where policy can influence land-use: the county. In

China, the county government has the power to plan and oversee a substantial part of economic, urban, and rural development. Counties use different strategies to attract foreign investment with different levels of effectiveness. Towns, the next lower administrative level, do not have the same authority to execute far-reaching economic policies. Twenty-five counties are included in the Landsat scene, but only twelve counties are included in this study due to data limitations – the other thirteen counties are only partially included in the image (Figure 2).

Figure 2. Counties included in the image. The twelve counties included in the study are: Nanhai, Dongguan, Panyu, Foshan, Shunde, Shenzhen, Zengcheng, Huadu, Sanshui, Guangzhou, Baoan, and Jiangmen.

Demographic and economic data are obtained from the Guangdong Statistical Yearbooks and include eighty-three variables, such as: agricultural population, urban population, number of town/village enterprises, output of grain, gross output value of agriculture, gross output value of industry, output in primary sector, which comprises mainly of agriculture (farming, forestry, animal husbandry, and aquaculture), output in the secondary sector, which is mainly industry, and output in tertiary sector, which includes all industries not included in the agriculture or industrial sectors.

The short temporal sample of the land use change data, eight years, cannot be used to estimate regression results for individual counties that can be interpreted in a statistically meaningful manner. Instead, the land use change and socioeconomic data are pooled to form a set of panel data that includes eight years of observations for all of the twelve counties. Panel data have several advantages over a cross-sectional or time series data set. These include the account and control for heterogeneity across samples, the ability to develop more complex models, the increase in data points which results in more degrees of freedom and hence reduces collinearity among explanatory variables, gains in statistical efficiency, and reduction in omitted variable bias (Hsiao, 1986; Baltagi, 1995). The socioeconomic data are collected at the county level at annual intervals, and the remote sensing estimates are imported into a GIS and county boundaries delineated to generate county level, yearly estimates of land use change. County boundaries are provided by the Institute of Remote Sensing Application (IRSA) in Beijing and the Center for International Earth Science Information Network (CIESIN) at Columbia University.

CLASSIFICATION ACCURACY, BIAS AND THEIR EFFECTS ON MODEL RESULTS

In most land use change models, the dependent variable is land use change, or some proxy for land use change. When satellite images are used to estimate land use and land use change, analysts tend to emphasize the accuracy of the classification. This emphasis implies that poorly classified imagery will have adverse effects on land use models. There exists a variety of metrics to assess the accuracy of classification results, including producer and user accuracies, the Kappa statistic, and overall accuracy. However, the literature includes little or no discussion on the effects of biased classification results on land use models. Regardless of the type of land use model employed, inputs to statistical models usually are assumed to be distributed normally about a mean of zero. A mean of zero indicates that the data are unbiased – that is, there is no systematic error in the data. In terms of satellite data for land use models, bias can be evaluated by looking at differences between the attributed date and type of land conversion and the actual date and type of land conversion. A large bias, either negative or positive, indicates a strong proclivity to methodically misidentify the date or type of land use change, and consequently, any observed relation may not be true (Pindyck and Rubinfeld 1991).

If the date of land use change is systematically assigned before or after the actual date of land conversion, the area estimates for land use change will be biased. This bias can make it difficult for statistical models to link remotely sensed data on land use changes with the demographic and economic variables that cause these changes. Making this link depends on the timing of changes in land use and socioeconomic conditions. For example, a model that specifies land use change as a function of current and lagged rates of economic activity will indicate no relation between these variables if the change-detection technique consistently identifies a date of change prior to the actual date of land conversion. This implies that when data obtained from satellite imagery are used to estimate statistical models of the relation between land use and its socioeconomic drivers, the importance of bias may be equal to that of accuracy and must be tested explicitly when evaluating the results from change-detection techniques.

There is an inverse relation between the percentage of an image classified and classification accuracy (Richards and Jia 1999), and the same may be true for the percentage of an image classified and classification bias (Kaufmann and Seto 2001). Not surprisingly, with a stringent classification criterion, only the pixels with a significantly distinct spectral signature will be classified, which results in a smaller percentage of the image classified, but with higher accuracy. With increasingly less stringent classification criteria, more of the image will be classified, including pixels with more "gray area" and overlapping spectral signatures. An issue then is whether to classify fewer pixels in an image, but with greater confidence and hence less bias and higher accuracy, or whether to classify all the pixels in an image, but at the risk of greater bias and lower accuracy. Traditionally, remote sensing has been used by the natural science communities to measure quantitative and qualitative land cover characteristics, and the production of a land cover map is essential to these goals. However, if satellite data are used as input to statistical models of land use, then perhaps we need to rethink how the data can be adapted to suit these needs. While we do not have the answer to these questions, these issues highlight new challenges that need to be addressed by the social science and remote sensing communities.

INTEGRATING REMOTE SENSING DATA WITH SOCIOECONOMIC DATA

We integrated the remote sensing data with socioeconomic data by: (1) generating land use classes that can be used as dependent variables in land use models; (2) extracting annual, county-level land use estimates that can be linked with the annual socioeconomic data; and 3) developing land use models based on panel econometric techniques to understand the long-run relation between land use change and key demographic and economic variables. Using rates of urbanization from the *agriculture→urban* and *natural→urban* classes as dependent variables, the long-run relation can be understood by estimating land conversion factors which translate socioeconomic activity into a demand for land. A simple model of the relation between land use change and socioeconomic variables can be specified as follows:

$$\text{Agriculture}\rightarrow\text{Urban} = \alpha + \beta_1 * \frac{\text{GDP}}{\text{Population}} + \beta_2 * \frac{\text{Value of Agricultural Output}}{\text{Agricultural Area}} + \varepsilon \qquad (1)$$

where α, β_1, and β_2 are estimated from the panel data. This relation can be estimated from the panel using a variety of statistical techniques (Hsiao, 1986). The simplest method is to estimate Equation (1) using ordinary least squares (OLS). Despite its seeming simplicity, this specification imposes a major assumption on the regression coefficients: that α, β_1, and β_2 are homogenous, i.e., the same for each county and period. For Equation (1), this means that a unit increase in per capita GDP has the same effect on the conversion of agricultural land to urban uses for each county for each time period.

A less restrictive model of land use change that allows parameter estimates to vary across counties (but to remain constant through time) can be specified as follows:

$$y_{it} = \alpha_i + \beta_i x_{it} + \varepsilon_{it}, \qquad i = 1,\ldots,N \quad t = 1,\ldots,T \tag{2}$$

where α_i is a 1 x N vector and β_i is a K x N matrix of regression coefficients, x_{it} is a T x K matrix of socioeconomic variables, and ε_{it} is a disturbance term with a mean of zero and a constant variance, σ_u^2. Equation (2) can be viewed as an unrestricted model of land use change in that it allows coefficients to vary among counties. For the example given by Equation (1), the effect of per capita GDP on the conversion of agricultural land can vary by county. The hypothesis that the slopes (β_i) and intercepts (α_i) are the same across counties can be tested explicitly. This test compares the residual sum of squares of an unrestricted model (RSS_u) (Equation (2)), which allows α_i and β_I to vary by county, with the residual sum of squares of a restricted model (RSS_r) (Equation (1)), which assumes α_i and β_i to be equal across counties (Hsiao 1986). The following test statistic, which is evaluated against an F distribution, can be used to evaluate the null hypothesis that slope coefficients do not vary among counties:

$$F_1 = \frac{\dfrac{(S_r - S_u)}{(N-1)(K+1)}}{\dfrac{S_u}{NT - N(K+1)}} \tag{3}$$

where S_r-S_u measures the increase in the unexplained sum of squares caused by restricting the values of the α_i and β_i to be equal across counties, N is the number of individual counties, T is the number of observations, and K is the number of regressors. The null hypothesis is that the slope and intercept coefficients are the same across individual counties – that is, $\alpha_i = \alpha_2 = \ldots = \alpha_N$ and $\beta_1 = \beta_2 = \ldots = \beta_N$. If this null cannot be rejected, the test statistic, F_1 with (N-1)(K+1) and NT-N(K+1) degrees of freedom will not exceed the value of the F distribution at some predetermined level of significance ($p<0.05$) and

Equation (1) can be estimated using OLS. If the test statistic exceeds the threshold for statistical significance, the null hypothesis is rejected, and OLS estimates for Equation (1) will be statistically inefficient. To alleviate this source of inefficiency, we must determine whether the slopes and/or intercepts vary among counties.

To test whether the slopes are the same across counties,

$$y_{it} = \alpha_i + \beta x_{it} + \varepsilon_{it} \qquad \begin{array}{l} i = 1,\ldots,N \\ t = 1,\ldots,T \end{array} \tag{4}$$

its residual sum of squares is compared to Equation (2) using:

$$F_2 = \frac{\dfrac{(S_r - S_u)}{(N-1)K}}{\dfrac{S_u}{NT - N(K+1)}} \tag{5}$$

where S_u is the residual sum of squares from the unrestricted model (Equation (2)) and S_r is the residual sum of squares from the restricted model (Equation (4)). The test is similar to Equation (3) in that an F-test can be used to evaluate whether the increase in residual sum of squares for the restricted model (Equation (4)) is significant relative to the residual sum of squares from the unrestricted model (Equation (2)). The test evaluates the null hypothesis that $\beta_1 = \beta_2 = \ldots = \beta_N$ which is imposed on the restricted model (Equation (2)). If F_2 with (N-1)K and NT-N(K+1) degrees of freedom exceeds the critical threshold, then the slope coefficients are not identical, but rather vary across individuals. In this case, the slope and intercept coefficients for individual counties must be allowed to vary across counties, and a random coefficient model must be used. If F_2 does not exceed the critical threshold, i.e., it is not statistically significant, then the slope coefficients do not vary across individual counties and either a fixed or random effects estimator can be used to test the models.

The fixed and random effects estimator allows the intercepts (β) to vary across individual counties through a variable intercept model. These models assume that the effects of the excluded variables either: 1) vary across time, but remain constant across cross-sectional units; 2) vary across cross-sectional units, but remain constant over time; or 3) vary across both time and individual. There are two main methods to estimate a fixed effects estimator, which treat unobserved individual effects as fixed over time. The first is a dummy variable approach in which individual counties are assigned dummy variables to represent the effects of omitted variables. These omitted variables can represent anything not included explicitly in the model, such as the effect of access to capital. These effects are assumed to be heterogeneous across individual counties but remain constant through time or are the same for all counties but are different for each time period. This method becomes computationally overwhelming for samples with a large value for N or T. A second method involves subtracting mean values of the time series observations from each individual. This transformation is given as:

$$X'_{it} = X_{it} - \overline{X}_i \tag{6}$$

where X'_{it} is the transformed value for variable X, X_{it} is the value of X for individual i at time t, and $\overline{X}_i$ is the mean value for the variable X for individual i. The transformed variables, X'_{it}, are then used to estimate the regression equation with no intercept.

The random effects model is sometimes referred to as an error components model and treats the effects of the unobserved omitted variables as random variables. The random effects estimator treats the disturbance term in Equation (4), ε_{it}, as having three components:

$$\varepsilon_{it} = \varepsilon_i + \varepsilon_r + \varepsilon_t \tag{7}$$

where ε_i is the error of the individual component, ε_r is the error of the random component, and ε_t is the error of the time component. The random effects model calculates a separate variance component (individual, random, and time) for each individual county.

If the test statistic given in Equations (3) and (5) exceeds the critical threshold – i.e., reject the null hypothesis of equal slopes – then unique regressions must be calculated for each county. To determine whether the coefficients from these individual regressions originate from some common data generating process, a random coefficient model is used. The random coefficient model assumes that the coefficients are realizations of a random process. The assumption is that there is a stochastic process that generates the coefficients, and that this stochastic process is stationary. The objective is to estimate the common mean and appropriate variances and covariances. The Swamy random coefficient model is presented as follows (Swamy 1970, Swamy 1971):

$$y_i = X_i\beta_i + \varepsilon_i \tag{8}$$

where $\beta_i = \overline{\beta} + \mathbf{v}_i$, $\mathbf{v}_i \sim (0, \mathbf{V})$. Each county has a separate coefficient vector, but each county's response through time remains constant. The vectors have a common mean $\overline{\beta}$. To estimate the random coefficient model, $\overline{\beta}$ and V must be estimated. After estimating V, the hypothesis that $V = 0$ can be tested and predictions can be made. For counties i = 1,…,N and T observations, $\overline{\beta}$ can be estimated by writing all NT observations as:

$$y = \tilde{X}\bar{\beta} + Xv + \varepsilon \tag{9}$$

The matrix $\tilde{X}$ is a matrix of stacked X_is and X is a block diagonal matrix. The covariance matrix of the composite error term, $Xv + \varepsilon$, is Ω. The generalized least squares (GLS) estimator can then be used to estimate the random coefficient estimator. Although the random coefficient model allows the β to vary by individual counties, there is still the assumption that there is a set of constant parameters which describes the process, such as the unvarying functional form, the constant zero mean and fixed variance of the distribution of the error term, and the constant absence of serial dependence in the error term (Hendry 1997). Numerous land use models and specifications are tested, and in all cases, the null hypothesis that the slopes and intercepts are equal across individual counties is strongly rejected ($p<0.01$). This eliminates the possibility of using pooled OLS techniques. This result is not surprising because there is no *a priori* reason to assume that the coefficients of the independent variables and intercept terms are constant across the twelve individual counties. Similarly, the null hypothesis that the slopes are equal across individual counties also is rejected strongly ($p<0.01$). Because the hypothesis of common slopes is rejected, a random coefficient model is used to estimate the land use models. The two most important driving forces in urbanization are access to capital, usually in the form of foreign direct investment, and differences in wages between agricultural and industrial sectors (Seto 2000).

LAND USE CHANGE IN THE PEARL RIVER DELTA

Between 1988 and 1996, the Pearl River Delta experienced a scale of land conversion unprecedented in the history of the country; approximately 1905 km^2 of land were converted to urban uses, an increase in urban area of 364%. From an estimated 720 km^2 of urban area in 1988, or 2.67% of the study area, the Delta's urban land increased to over 2625 km^2 by 1996. By 1996, urban areas occupied almost 10% of the study area. While roughly a quarter of the new urban areas were previously forest, shrub, or water, most were converted from farmland, approximately 1376 km^2 (Table 3).

One interesting and unexpected type of land use change is the conversion of farmland and other areas to reservoirs. With the increase in population and industrial activity, reservoirs and dikes have been constructed to provide the booming region with an adequate water supply. Perhaps more surprising is that 16 of the 21 km^2 of new reservoirs were previously agriculture, and only 5 km^2 were converted from forests and shrubs. This indicates that the conversion of agricultural land for reservoirs bears a lower opportunity cost than the conversion of other land. One explanation may be that agricultural areas are closer to the cities, while forested regions are located further and the increase in distance carries higher transportation costs.

Table 3. Area estimates of land uses and land use change, 1988-1996.

Class	km^2	*% of study area*
water	2110	7.82
natural vegetation	11906	44.13
agriculture	10166	37.68
urban	720	2.67
agriculture to water	16	0.06
natural vegetation to water	5	0.02
water to agriculture	151	0.56
natural vegetation to urban	529	1.94
agriculture to urban	1376	5.09
new urban, 1996	2625	9.7

The pattern of urbanization is not surprising and follows a corridor between Hong Kong to Guangzhou (For an illustration of the rate and pattern of land use change, refer to Animation 1 on the CD-ROM and Figure 2). This development corridor mirrors two aspects of economics in the PRD: the construction of the new Shenzhen-Guangzhou highway, and the migration of many Hong Kong manufacturing firms to nearby counties.

The central and provincial governments recognize the threat of declining agricultural land, and have supported initiatives to reclaim part of the Delta for agriculture. During the study period, approximately 151 km^2 of the Delta's water areas were converted to farmland, mainly in the western region of the Delta (Figure 3). This increase offsets about 11 percent of the loss in agricultural land converted for urban uses.

A comparison of remotely sensed and official estimates indicates that the rates of agricultural land conversion are considerably greater than those reported in official statistics (Seto et al. 2000). The remote sensing estimates of total agricultural land loss for ten counties[3] is 11 percent greater than the 789 km^2 reported in statistical yearbooks (Figure 4). However, these differences between satellite-derived versus official estimates are blurred when the figures from the counties are aggregated. Similarly, official estimates for agricultural land are consistently smaller than remote sensing-derived estimates (Figure 5). Remote sensing estimates of total agricultural area in ten counties for 1990 is 6724 km^2, 115 percent greater than the 3119 km^2 reported in government yearbooks.

Figure 3. Example of land use change in the Pearl River Delta. The white areas in the western part of the Delta represent new agricultural land. The white areas in the remainder of the figure are urban areas, converted from either agriculture or natural vegetation.

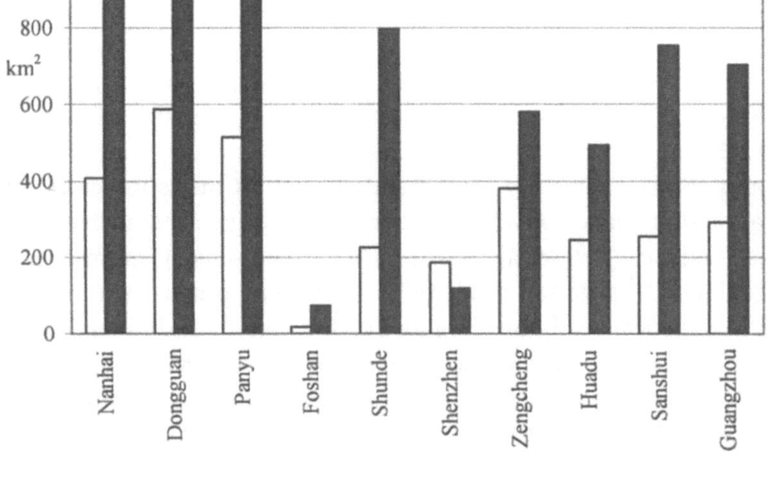

Figure 4. Agricultural land for ten counties in the Pearl River Delta, 1990. (From Seto et al. 2000; reprinted with permission from *Nature.*)

Institutional forces probably are largely responsible for the apparent bias in government statistics. During the Great Leap Forward period (1958-1961), farmers consistently underreported the size of their agricultural land, because grain quotas were based on agricultural land size. Therefore, understating agricultural land area reduced a farmer's quota. This bias has not been eliminated completely by market liberalizations. The tendency to understate the size of agricultural land is reinforced by a moratorium on the conversion of agricultural land. To hide the lease of land to commercial developments, farmers underreport the size of their agricultural land holdings. Similarly, official figures of total land conversion is less than estimates derived from remote sensing because farmers and landholders seek to avoid paying taxes which accrue when land is leased for commercial purposes. The difference in wages between the agricultural and industrial sectors is an important component of land conversion. As farm wages decline relative to industrial wages, farmers abandon agricultural work and seek employment in the industrial sector. If economic returns to agricultural work are higher than returns to industrial or manufacturing work, farmers will have fewer incentives to convert their land to urban uses or lease their land to nonfarm uses.

Foreign direct investment (FDI) has and continues to play a crucial role in the development and urbanization of the area. FDI is used mostly to finance residential and transportation projects. In the PRD, myriad residential construction projects are styled after the "Southern California suburban" motif. These tract-home style houses and condominiums have a single design layout, but are usually available in four shades of pastel pink. Like "master-planned communities" in the United States, some of these development projects are accessible only through gates, and often, armed guards. In the transportation sector, FDI finances the construction of super-highways and high-quality roads that can support heavy cargo trucks and other large vehicles. Transportation investments are particularly lucrative for investors because they generate revenues via tolls for use. The Shenzhen-Guangzhou super-highway is a 6-lane road which costs approximately US$8 in tolls from end to end. Although this is a relatively high price, buses, construction trucks, and cargo vehicles easily recoup the fee. What once was a journey which took more than four hours through small dirt roads and low-grade streets is now an easy one and a half hour drive with frequent toll booths. As toll roads increasingly become the transportation routes of choice, the high volumes of traffic on these roads will allow investors to continue earning a high return on their investment over the long term.

Patterns of urbanization and land use change follow not only the road network and the corridor of towns with overseas connections. Most of the FDI in the region comes from overseas Chinese investors in Hong Kong, Taiwan, and the United States (Statistical Bureau of Guangdong, various years). The investments often are directed toward villages with familial and professional contacts (Xu and Li 1990, Eng 1997). Thus, *guanxi*, the Chinese word describing connections, or networks, is fundamental to the location of FDI. With the opening of the economy and the government's warm welcome of FDI, overseas Chinese have invested heavily in their hometowns and villages. Increasingly, libraries, concert halls, housing tracts, and schools are built and named after overseas Chinese donors. In exchange, investors obtain prominence, name recognition, and respect from local residents and officials. The latter is particularly important as it builds *guanxi* – important ties for future land lease negotiations.

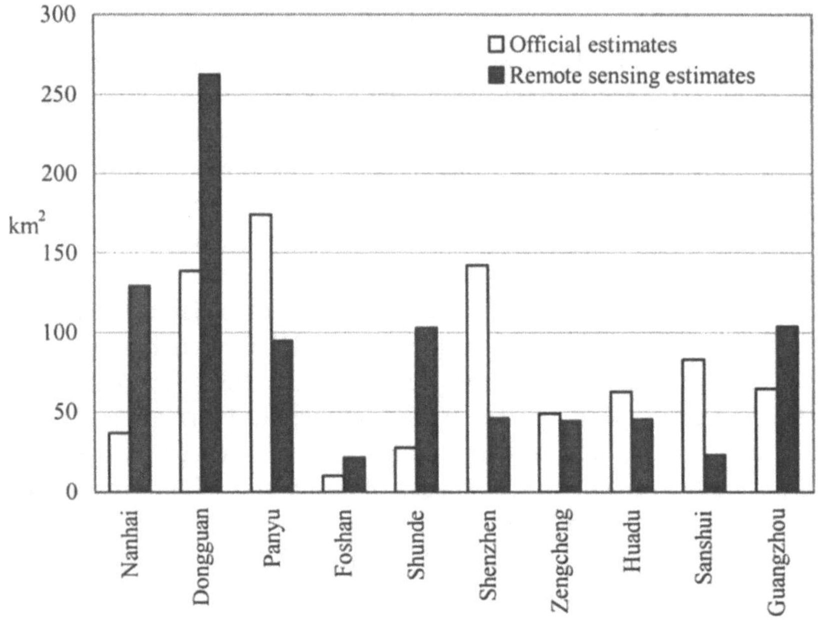

Figure 5. Agricultural land loss for ten counties in the Pearl River Delta, 1990-1996. (From Seto et al. 2000; reprinted with permission from *Nature*.)

CONCLUSION

One of the methodological challenges in land cover and land use change research is to couple remote sensing data with socioeconomic analysis. Challenges remain to develop methodologies that fully utilize the sequential time series of Landsat observations and that link the images with economic data at appropriate spatial and temporal scales. Information on human and economic activity is clearly more difficult to detect than land cover characteristics because it requires assumptions about how land use is manifest in land cover. Our results indicate that a conceptual model of land use classes comprised of spectrally diverse land covers provides a good framework to infer land use change from land cover characteristics.

Panel econometric techniques offer a wealth of methods to analyze satellite imagery and land use change. The techniques may be most suitable for study areas with few sample periods, but multiple individual observations. The various approaches discussed in this chapter illustrate how different specifications allow testing of theories about the relation between economic variables and land use change.

The rate of urbanization and land conversion in the Pearl River Delta is unprecedented. Historically, the Pearl River Delta has been a region with relatively little land conversion and population growth. Land use patterns and agricultural practices in much of the Delta have remained constant for hundreds of years. Land conversion is largely driven by wage differentials between the agricultural and industrial sectors, and foreign direct investment. Given that economic growth has directly improved the living standards of most citizens in the Delta, urbanization rates of over 300% between 1988 and 1996 are not unexpected, but still impressive.

NOTES

[1]The ten counties are: Nanhai, Dongguan, Panyu, Foshan, Shunde, Shenzhen, Zengcheng, Huadu, Sanshui, and Guangzhou.

[2]For more information on the agricultural reforms, readers should refer to Garnaut et al. (1996).

[3]The ten counties are : Nanhai, Dongguan, Panyu, Foshan, Shunde, Shenzhen, Zengcheng, Huadu, Sanshui, and Guangzhou.

REFERENCES

Baltagi, B. H. (1995). *Econometric Analysis of Panel Data*. New York, John Wiley and Sons.

Campbell, J. B. (1983). *Mapping the land; aerial imagery for land use information*. Washington, DC, Association of American Geographers.

Chu, D. K. Y. (1998). "Synthesis of economic reforms and open policy," in *Guangdong: Survey of a province undergoing rapid change*. Y. M. Yeung and D. K. Y. Chu, Eds. Hong Kong, Chinese University Press: 485-504.

Collins, J. B. and Woodcock, C. E. (1996). "An assessment of several linear change detection techniques for mapping forest mortality using multitemporal Landsat TM data." *Remote Sensing of Environment* 56(1): 66-77.

Crist, E. P. and Cicone, R. C. (1984). "A physically-based transformation of Thematic Mapper data - the TM tasseled cap." *IEEE Transactions on Geoscience and Remote Sensing* 22: 256-263.

Eng, I. (1997). "The rise of manufacturing towns: externally driven industrialization and urban development in the Pearl River Delta of China." *International Journal of Urban and Regional Research* 21: 554-568.

Fung, T. (1990). "An assessment of TM imagery for land cover change detection." *IEEE Transactions on Geoscience and Remote Sensing* 28 (681-684).

Garnaut, R., Guo, S. and Ma, G., Eds. (1996). *The Third Revolution in the Chinese Countryside*. New York, Cambridge University Press.

Hendry, D. F. (1997). *Dynamic Econometrics: Advanced Texts in Econometrics*. New York, Oxford University Press.

Hsiao, C. (1986). *Analysis of Panel Data*. New York, Cambridge University Press.

Kaufmann, R. K. and Seto, K. C. (2001). "Change detection, accuracy, and bias in a sequential analysis of Landsat imagery in the Pearl River Delta, China: econometric techniques." *Agriculture, Ecosystems and Environment* 85(1-3): 95-105.

Li, X. and Yeh, A. G. O. (1998). "Principal component analysis of stacked multi-temporal images for the monitoring of rapid urban expansion in the Pearl River Delta." *International Journal of Remote Sensing* 19: 1501-1518.

Lin, G. C. S. (1997). *Red Capitalism in South China: Growth and Development of the Pearl River Delta*. Vancouver, UBC Press.

Lin, J. Y. (1988). "The household responsibility system in China's agricultural reform: a theoretical and empirical study." *Economic Development and Cultural Change* 36: S199-S224.

Maddison, A. (1998). *Chinese Economic Performance in the Long Run*. Paris, OECD.

McMillan, J., Whalley, J. and Zhu, L. (1989). "The impact of China's economic reforms on agricultural productivity growth." *Journal of Political Economy* 4: 781-807.

Muldavin, J. S. S. (1996). "The impact of reform on environmental sustainability in rural China." *Journal of Contemporary Asia* 26(3): 289-322.

Pindyck, R. S. and Rubinfeld, D. L. (1991). *Econometric Models and Economic Forecasts*. New York, McGraw-Hill, Inc.

Richards, J. A. and Jia, X. (1999). *Remote Sensing Digital Image Analysis : An Introduction*, Springer Verlag.

Seto, K. C. (2000). "Monitoring and modeling land use change in the Pearl River Delta, China, using satellite imagery and socioeconomic data." Boston University, Department of Geography, Ph.D. Dissertation, Boston.

Seto, K. C., Kaufmann, R. K. and Woodcock, C. E. (2000). "Landsat reveals China's farmland reserves, but they're vanishing fast." *Nature* 406: 121.

Seto, K. C., Woodcock, C. E., Song, C., Huang , X., Lu, J. and Kaufmann, R. K. (in press). "Monitoring Land use Change in the Pearl River Delta Using Landsat TM." *International Journal of Remote Sensing.*

Sharkawy, M. A., Chen, X. and Pretorius, F. (1995). "Spatial trends of urban development in China." *Journal of Real Estate Literature* 3: 47-59.

Song, C., Woodcock, C. E., Seto, K. C., Pax Lenney, M. and Macomber, S. A. (2001). "Classification and change detection using Landsat TM data: When and how to correct atmospheric effects?" *Remote Sensing of Environment* 75: 230-244.

Statistical Bureau of Guangdong (various years). *Statistical Yearbook of Guangdong*. Beijing, China Statistical Publishing House.

Swamy, P. A. V. B. (1970). "Efficient inference in a random coefficient regression model." *Econometrica* 38: 311-323.

Swamy, P. A. V. B. (1971). *Statistical Inference in Random Coefficient Regression Models*. New York, Springer-Verlag.

World Bank (1996). *Poverty in China: what do the numbers say?* Washington, D.C., World Bank.

World Bank (2000). *World Development Report*. Washington, D. C., World Bank.

Xu, X.-Q. and Li, S.-M. (1990). "China's open door policy and urbanization in the Pearl River Delta region." *International Journal of Urban and Regional Research* 14: 49-69.

Yeh, A. G.-O. and Li, X. (2001). "Measurement and monitoring of urban sprawl in a rapidly growing region using entropy." *Photogrammetric Engineering and Remote Sensing* 67(1): 83-90.

Chapter 5

SPATIAL MODELING OF VILLAGE FUNCTIONAL TERRITORIES TO SUPPORT POPULATION-ENVIRONMENT LINKAGES

Thomas W. Crawford

Environmental Studies Program, Gettysburg College, Gettysburg, PA 17325
tcrawfor@gettysburg.edu

Keywords: population-environment linkage, functional region, Voronoi diagram, fuzzy set

Abstract Geographic information science plays a key role in linking population and environmental data to support a variety of substantive questions. A common linkage approach is the polygon overlay procedure that relates administrative boundary files and their associated social data with environmental data layers to derive relevant variables. Formal administrative boundaries are lacking, both in the real and digital senses, in many parts of the developing world as one moves to fine scales of analysis. Because population-environment research often focuses on developing regions, it is necessary to develop linkage approaches that can circumvent these problems. The key task of this research is to enable linkages between population and environmental data by transforming discrete village points, with associated demographic and economic attribute data, into functional regions that partition the landscape into village territories. Resulting village regions can be stored in a GIS database and overlaid with environmental layers to derive village-level environmental descriptors. The ordinary Voronoi diagram is a classic point-to-area transformation. There are several extensions to the Ordinary Voronoi diagram. Operating within a raster environment, this paper presents a new extension called the kth nearest-point, shortest-path (KSP) Voronoi diagram. Additionally, Fuzzy Set Theory is applied to the extension to represent the spatial fuzziness in village land ownership patterns within the study area. This approach can be applied to other regions similarly characterized by a nuclear settlement pattern. Results are presented, and issues regarding accuracy assessment, model adjustments, and use of the model to derive linked village-level population and environmental descriptors are discussed.

INTRODUCTION

Developing spatially linked data and approaches that extend across the social, biophysical, and geographical domains is a basic and applied problem that needs to be addressed as part of the human-environment emphasis in geographical research. This emphasis examines the integration of social, spatial, and biophysical patterns and processes through the spatial representation and characterization of social variables and landscape features operating within a geographic information science (GIScience) framework.

A common approach for linking across thematic domains is the polygon overlay procedure that integrates administrative boundary files with associated social, environmental, and geographical data layers to derive relevant, spatially explicit measures for social units at defined hierarchical levels. During the past few decades, mapping efforts by national governments and international agencies have developed digital boundary files of administrative units at multiple hierarchical levels for many of the world's regions. For example, near global coverage of national and provincial boundaries exist in public use GIS datasets, and finer scale boundaries are available for practically all "more developed" nations. Unfortunately, fine scale data below the provincial or district level are lacking for many parts of the world.

Of particular need is the generation of theory and practice in estimating "village territories" or other small areas in developing countries where boundary information is lacking (Entwisle et al. 1998, Evans 1998, Rindfuss and Stern 1998, Walsh et al. 1998, Wood and Skole 1998). A major consideration when defining or estimating territories at the village level is that administrative boundaries are *formal* regions, as opposed to *functional* regions, that are often arbitrary and may not represent actual functional linkages between human populations and the areal features that serve as their resource base, in addition to being the areas which they impact through human modifications. A second consideration is that administrative boundaries often do not exist in many parts of the developing world, particularly as one moves to finer scales of analysis (Rindfuss and Stern 1998). Because researchers who study human-environment relationships often focus on developing regions at relatively fine spatial scales such as the village, it is necessary to develop methods to effectively represent functional village territories that can act as a bridge across disparate data types.

This paper develops and tests a novel approach for linking data across thematic domains, and spatially representing functional village territories by introducing a hybrid extension of the Ordinary Voronoi diagram (e.g., Thiessen Polygons) that uses non-Euclidean distance and incorporates concepts from Fuzzy Set Theory. The approach is applied to a set of villages in Nang Rong District, Buriram Province, northeast Thailand (Figure 1), a poor, rural district that has experienced significant social and biophysical transformations during the past few decades associated with deforestation, agricultural extensification, and population migration.

Figure 1. Nang Rong study area.

Problem Statement

Two basic questions regarding human-environmental processes in Nang Rong, Thailand stimulated this research: (1) What is the nature of inter-village variation in resource endowments such as soil quality, accessibility to water and roads, and topographic characteristics? (2) Is this variation related to village economic outcomes such as agricultural yields or asset holdings and demographic behavior such as out-migration? Within a broader context, how might deforestation and landuse/landcover change be linked to social and demographic characteristics of households and villages? What are the drivers of landscape change, and how might people, land parcels, and village territories/regions be spatially linked and examined within an automated, digital environment where location, topology, and thematic and temporal relationships are central to understanding landscape form and function?

To answer these questions, it is essential to know the spatial extent of each village territory. With such knowledge, and assuming the existence of a spatial database containing relevant thematic layers, one can perform standard GIS overlay procedures to calculate areal estimates of various biophysical factors (e.g., soils, elevation, landuse/landcover, hydrography) for use as village-level environmental profiles that characterize landuse, land suitability, and resource endowments.

In northeast Thailand, villages are organized in nuclear patterns (Figure 2) with agricultural lands arrayed outward from the village settlement and spatially constrained by the position of other villages, natural barriers (e.g., rivers), and existing transportation networks and modes of travel. Evans (1998) reported that most village territories within Nang Rong, based upon observed land ownership patterns, were limited to a 1.0 km radial distance around the nuclear village center, but ownership patterns could extend up to 10 km from the village compound. A village's agricultural lands generally adhere to an inverse distance weighting function such that nearby lands are more strongly associated with respective proximal villages. In this environment, administrative boundaries (e.g., political delineations) are less important than functional boundaries (e.g., delineations of activity spaces) that are influenced by road and river positions, land use and land cover mosaics, watershed boundaries for management of water and other natural resources, cultural and ethnic associations with other villages, labor sharing agreements, and kinship ties.

In Nang Rong, locations of village central points (defined as the location of a village's central street intersection) are known, but village boundaries are not defined. How does one characterize village environmental descriptors when its landscape features occur over an undefined area? One approach is to perform overlay procedures within a GIS using village point locations as one of the input layers. However, this approach excludes from consideration the vast majority of a village's area by focusing on a single dimensionless point. Another approach is to recruit village informants to aid in a map survey of village boundaries using either a global positioning system (GPS) or other field measurement methods. These approaches are appropriate for a single village case study (Reenburg 1995), but may be infeasible when applied to a larger number of villages. For example, the Nang Rong study area examined in this research includes over 300 administratively defined villages that form a complete coverage of the district area as defined by the Thai government in the early 1990s. Another approach, which is the core of this paper, is to transform discrete village points into functional regions that partition the landscape into estimated village territories based on plausible methods and assumptions. Thus, this paper's theoretical focus is on point-to-area spatial data transformations that maintain a sensitivity to the distance decay properties of transitional boundaries such as those associated with Nang Rong villages.

Figure 2. Nuclear village settlement pattern with a residential core and agricultural fields along the periphery.

A classic point-to-area transformation is the *Ordinary* Voronoi diagram in which regions are built from a finite set of discrete points in a planar space by assigning every location in the space to the nearest discrete point (Okabe et al. 1992). The Ordinary Voronoi diagram is defined using Euclidean distance. Several extensions of the ordinary diagram have been proposed or developed by researchers from various disciplines. Two such extensions are the kth *nearest-point* and *shortest-path* Voronoi diagrams. The kth nearest-point diagram is similar to the Ordinary Voronoi diagram, but differs by assigning every location in the planar space to the kth nearest discrete point rather than to the nearest point. The shortest-path diagram is conceptually the same as the Ordinary diagram, except that it is defined using shortest-paths (accumulated travel costs) rather than Euclidean distances. In this paper, these two extensions are combined into a kth *nearest-point, shortest path digital* diagram to model village territories in Nang Rong. For simplicity, this hybrid extension is referred to as the *KSP* diagram. The term "digital" is used to connote the fact that this extension operates in a raster environment.[1]

In Nang Rong, the nuclear village structure, lack of formal boundaries, and mixed village land ownership at increasing distances from village centers make the problem of defining village territories amenable to the concept of fuzzy sets (Evans 1998). With fuzzy sets, a membership function quantifies a grade-of-membership (GOM) that ranges continuously from 0 to 1 for every object under consideration (Bonham-Carter 1994). Objects with values close to 1 are core objects, and, conversely, peripheral objects have low values closer to 0. In this context, defining village regions is essentially a classification problem where land parcels are the objects and village units are the sets.

The Ordinary Voronoi diagram and its various extensions yield regions with *crisp* boundaries. In other words, discrete lines partition the entire space into distinct regions according to the rules of the particular Voronoi diagram. In the present application, crisp village boundaries are not realistic representations because of the mixed ownership of land parcels extending from multiple villages. To more realistically represent the fuzziness of functional regional boundaries, a membership function for land parcel membership in village territories is defined from the KSP diagram. The resulting normative model described here is called a digital *Fuzzy KSP Voronoi diagram*. The method can be generalized for many other applications where the objective is to link population data to environmental data using overlays with areal representations of nodal (nuclear) regions having transitional boundaries.

In short, the goal of this chapter is to model functional village regions where villages are characterized by a nuclear settlement pattern. By establishing the spaces of the village regions it is possible to subsequently link socio-economic data collected through intensive field surveys at the village level to enable rich analyses of reciprocal human-environment relationships and processes. These functional regions inherently have indeterminate boundaries, and therefore a fuzzy set approach is an appropriate framework (Burrough 1996). The kth nearest point Voronoi diagram for $k = 1$ and $k = 2$ is essential to this approach because it enables one to demarcate the indeterminate boundary zone between proximal village regions. Using this approach, one may create fuzzy functional regions where each grid cell has a grade of membership score for "belonging" to a certain village. Alternatively, and if desired, one can establish crisp regions assigning grid cells wholly to the village with the highest grade of membership score. Taking the crisp approach is akin to establishing formal regions, and runs the risk of masking the mixed nature of village landuse and ownership along interstitial regions. Towards the end of the chapter, the accuracy of resulting fuzzy functional regions are compared to circular radial buffer regions around each village points to assess the utility of this approach.

Within the context of geographic information science, this chapter blends together several concepts and methods in a new way to address the problem of linking spatially discrete population data to spatially continuous areal data using Nang Rong district in northeast Thailand as a case study. The concepts and methods that are employed and integrated include: functional and formal regions, spatial data transformations, Voronoi diagrams, shortest-path distances, Fuzzy Set Theory and fuzzy object representation, accuracy assessment, and derivative thematic layers via overlay procedures. The approach described extends beyond the Nang Rong research context. Many regions of the world are characterized by nuclear settlement patterns that exhibit core and peripheral areas. Information regarding the space or spaces of human interaction with the surrounding environment is often lacking and may need to be estimated using a novel approach. For example, many parts of Africa and South and Southeast Asia exhibit nuclear settlement patterns. Finally, the ability to link human populations with their environments using regional and fuzzy set concepts such as core and periphery may be of use within agricultural landuse contexts such as

Von Thunen's theory of varying landuse intensity along the spectrum of core to periphery.

FUZZY SET THEORY AND REGIONALIZATION

Fuzzy Set Theory has emerged in the spatial sciences as a way to deal with the uncertainty and imprecision inherent in many spatial phenomena (Leung 1983, Leung 1985, Banai 1993, Altman 1994, Edwards 1994). A concise treatment of fuzzy sets and their relevance to geographic applications can be found in Burrough and McDonnell (1998). Fuzzy Set Theory has been applied in many geographic contexts including but not limited to soil classification and mapping (McBratney and Odeh 1997), landuse/landcover classification (Fisher and Pathirana 1990, Foody 1992, Fisher and Pathirana 1993, Atkinson et al. 1997, Brown 1998a 1998b), land suitability and evaluation (Sui 1992, Groenemans et al. 1997), regionalization (Ponsard and Tranqui 1985, Harris et al. 1993), defining spatial relationships (Robinson 1990), and indeterminate boundary issues (Leung 1987, Burrough and Frank 1996, Evans 1998, and Brown 1998b).

Traditionally, researchers represent regions as *crisp* sets using Boolean logic. A crisp set allows only for a binary membership function such that an individual element either is or is not a member of a set. For example, a land parcel may either belong wholly to county A or wholly to county B, but it can not belong partially to both or either. For formal administrative regions like counties or land ownership parcels in developed countries, crisp sets are entirely plausible. However, for functional (nodal) regions, which by definition have core and peripheral areas, crisp sets introduce a sense of discreteness and precision that may not match reality.

For both crisp and fuzzy sets, the degree to which an observation z is a member of set A is expressed by a membership function. For crisp sets the value of the membership function can either be 0 or 1. More formally, membership in a crisp set A is defined as:

$$\mathrm{MF}(z) = 1 \qquad \text{if } b_1 \leq z \leq b_2 \tag{1}$$

$$\mathrm{MF}(z) = 0 \qquad \text{if } z < b_1 \text{ or } z > b_2 \tag{2}$$

where b_1 and b_2 define exactly the boundaries of set A.

With fuzzy sets, the membership function expresses the grade of membership scores that typically ranges continuously from 0 to 1. Observations close to the core concept of a particular set have values close to or equal to 1, which indicates a strong degree of membership in the set. Conversely, observations further away have smaller values. Burrough and McDonnell (1998, p. 269) define Fuzzy Sets formally as follows:

If Z denotes a space of objects, then the fuzzy set A in Z is the set of ordered pairs:

$$A = (z, \mathrm{MF}_A(z)) \quad \text{for all } z \in Z \tag{3}$$

where the membership function $\mathrm{MF}_A(z)$ is known as the "grade of membership of z in A."

Thinking in terms of modeling Nang Rong village regions leads to the following translation of the previous definition applied in a raster environment: Z denotes all land parcels (grid cells) in the study area. Set A is the functional region (fuzzy set) for village A and consists of all land parcels in the study area for which the grade of membership for belonging to village A is greater than zero. Land parcels with values close to 1 have a strong degree of membership in village A's functional region and vice versa. The next section describes how this translation was implemented.

DATA AND METHODS

A variety of spatial data were used to model Nang Rong village regions. Discrete data included point features representing all district villages, line features of district roads, and polygon/line features of district hydrography. The discrete features had been previously digitized from Thai government topographic maps (1:50,000) and stored in a GIS database to support related projects within the study area (Entwisle et al. 1998). Village point locations were validated using a GPS receiver during fieldwork. All GPS coordinates acquired in the field were differentially corrected through use of base-station files collected at a nearby facility. In some cases multiple villages were located very close together (within 500 meters) in village clusters. In these instances, clustered villages were merged into a single point to represent a settlement of multiple villages with a lineage of recent administrative splits of a formerly unified village[2]. The final derived village location data contained points for all 240 villages (or village clusters) within Nang Rong district. All of the discrete digital features described above were converted into raster format at a resolution of 60 x 60 m (3600 square meters)[3]. Other raster data included a level-1 landuse/landcover (LULC) classification derived from a 1994 Landsat Thematic Mapper (TM) satellite digital data set. Raster roads and hydrography features were merged with the LULC coverage to produce a final grid-surface that contained agriculture, forest, roads, and water.

A Digital KSP Voronoi Diagram

Consider first a standard cost allocation procedure. This procedure assigns locations to the nearest source site where "nearness" is defined as shortest-path distances. If the source sites and locations completely fill a planar surface, then the cost allocation result is also a *shortest-path (SP)* diagram. The cost allocation procedure is also a kth *shortest-path (KSP)* diagram for $k = 1$. However, for $k > 1$, computing a *KSP* diagram becomes more difficult. This section presents an algorithm to compute a *KSP* diagram using the 240 Nang Rong village points and a friction surface derived from a satellite-based landuse/landcover (LULC) grid[4].

First, 10 km radial buffer grids were generated for all 240 points. A 10 km distance was selected because it is unusual that village territories in Nang Rong extend outward further than 10 km from its center point. In fact, previous research (Evans 1998) suggests that most villagers travel less than 2 km from the nuclear center to their agricultural land, and most travel 1-km from the village centroid.

Second, for each of the 240 grids, distances from each grid cell to the village center was calculated as least accumulated cost (shortest-path) distances across a frictional surface[5]. The friction surface had a cell size of 60 x 60 m and contained approximately 850,000 cells that were weighted differentially by LULC type so that barriers and conduits to movement could be simulated. Roads were assigned a low weight to simulate their role as movement conduits. Agriculture had a weight twice that of roads to indicate a higher difficulty of movement across certain crop types. The dominant agriculture in the district is rice grown in paddies surrounded by raised bunds, slightly elevated retaining walls designed to impound water. Villagers commonly use the bunds to travel to and from their households and fields. Forests were assigned a moderately high frictional weight to reflect more difficult movement through their denser cover. Water had a very high weight to indicate the relative difficulty of crossing a river in the absence of a bridge.[6]

Third, 240 grids were stacked and queries issued to select from the assembled stack the 1st, 2nd, 3rd, ..., or k^{th} ranked cost distance for each grid cell location, and also to select the village that was associated with each of the selected cost distances (Figure 3). It was possible for ties to occur though the probability was very small. To further decrease the chance of a tie, small random integers were added to every grid cell in each of the 240 grids prior to ranking. The random integers were set small enough to break ties without altering the original rank orders in non-tie situations. The results of each query yielded KSP Voronoi diagrams for $k = 1,2,3, ..., n$.

Figure 4 shows results of the algorithm for $k = 1$ and $k = 2$ for a selected portion of the study area. To show polygons for all 240 villages simultaneously would create a fairly confusing map, so a sample is presented here. For every location x in the striped region, Village A is the second closest village point when distance is measured as a shortest-path distance across the friction surface. Note that the region is not contiguous, though it is possible in principle. Also notice that a river located northwest of Village A acts as a barrier such that very few cells across the river are part of its $k = 2$ region. However, a road (not shown) running northwest from Village A causes a small sliver on the opposite side of the river to be part of its $k = 2$ region. The "$k = 1$ region" consists of the grid cells for which Village A is the closest village using a shortest-path distance definition of closeness. Similarly, the "$k = 2$ region" consists of the grid cells for which Village A is the second closest.

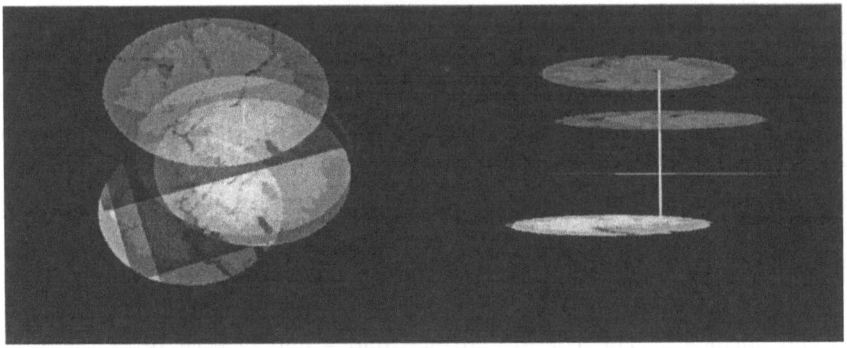

Figure 3. Vertical and horizontal views of stacking grids and selecting the kth closest village per cell. Each village had a circular grid that showed shortest-path distances to the centrally located village point. Distances are in grayscale with darker grays being more distant. For simplicity, only four grids are shown. Visually, "shadows" appear in the vertical view due to the stacked nature of the grids.

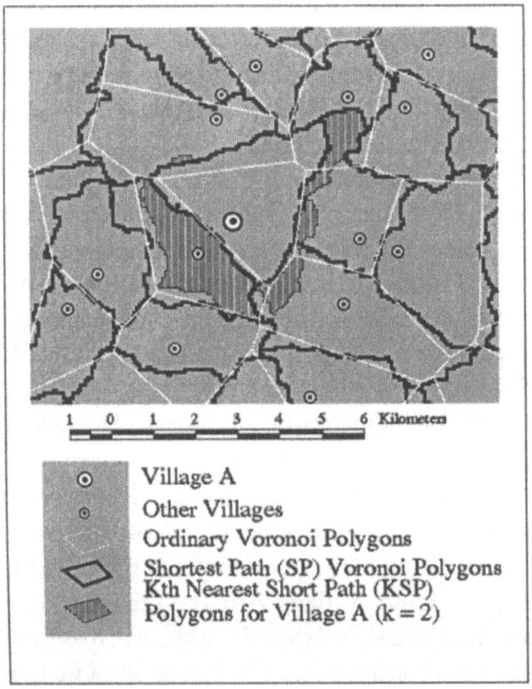

Figure 4. KSP diagram for Village A.

Again, each grid cell contains attribute information that identifies the first, second, third, ... k^{th} closest village and the respective shortest-path distances. This information alone may be useful in research that models resource or network flows. However, in this research, the information is useful in that it enables one to model fuzzy village functional regions and link these regional spaces to village-level socio-economic data collected in the field.

The previous section produced crisp KSP Voronoi polygons for $k = 1$ and K = 2. This section presents a method to transform the crisp polygons into fuzzy regions. The rationale behind this step is that nuclear villages in Nang Rong exhibit decreasing land ownership in a distance-decay fashion away from the nuclear center. To transform the $k = 1$ and k = 2 polygons (which are really discrete raster zones at this point) from crisp to fuzzy regions requires specification of a membership function. From above we know the shortest-path distances, $d_{1,s}$ and $d_{2,s}$, and their associated village points, p_i, for every grid cell in the $k = 1$ and k = 2 KSP diagrams. In other words, for every grid cell we know the shortest-path distance to the closest and second closest villages. We also know for each cell which villages are the first and second closest.

A Fuzzy KSP Voronoi Diagram

Each grid cell observation therefore must have an attribute, z, that serves as input for the membership function. Here z is defined as:

$$z_{p_1} = d_{1,s} / (d_{1,s} + d_{2,s}) \tag{4}$$

$$z_{p_2} = 1 - z_{p_1} \tag{5}$$

where,
p_1 = the first closest village
p_2 = the second closest village
$d_{1,s}$ = the shortest-path distance to the closest village, p_1
$d_{2,s}$ = the shortest-path distance to the second closest village, p_2

[1]Note that z can range from 0 to 0.50. To explain, in a first example, assume that a planar space contains only two village points, points p1 and p2. A grid cell located exactly at p1 will have d1,s= 0, d2,s> 0, and z = 0 for p1 and z = 1 for p2. Thus, z = 0 represents the core concept of a grid cell belonging to a point pi. In a second example, a grid cell located approximately one-half way between village p1 and p2 will have d1,s ≅ d2,s causing z ≅ 0.50 for both p1 and p2. In a third example, a grid cell located very close to village p1 will have a very small z-value, perhaps z = 0.01, which is very close to the core concept, z = 0, of "belonging to village p1."

If the membership function is simply defined as MF(z) = 1 – z, then in the first example from above, the grid cell receives a grade of membership of 1.00

for belonging to the closest village, p_1. This is logical because the cell is located exactly at the location of village point p_1. For the third example where the grid cell is located very close to village point p_1, the grid cell will receive a grade of membership of approximately 0.99 for belonging to the closest village p_1 and approximately 0.01 for belonging to the second closest village p_2. In reality, the grid cell most likely belongs to village p_1 even though village p_2 admits a small possibility of claiming the cell. In other words, if MF(z) = 1 – z, then exactly one cell will have a grade of membership equal to 1.00 -- the cell located exactly at the location of the village point p_1. Each village would have one and only one "core" cell – not a very realistic situation.

To account for this situation, the membership function was specified as a modification of a sinusoidal function. There are several commonly used non-linear as well as linear functions that might be appropriate (Burrough and McDonnell 1998). The specific function used for this research is defined as:

$$\mathrm{MF}(z_{p_i}) = 1 \qquad \text{if } z_{p_i} < .5 - f \qquad (6)$$

$$\mathrm{MF}(z_{p_i}) = 0 \qquad \text{if } z_{p_i} > .5 + f \qquad (7)$$

$$\mathrm{MF}(z_{p_i}) = [1 + ((z_{p_i} - .5 + f)/f)^2]^{-1} \qquad \text{if } .5 - f \le z_{p_i} \le .5 \qquad (8)$$

$$\mathrm{MF}(z_{p_i}) = 1 - [1 + ((.5 - z_{p_i} + f)/f)^2]^{-1} \qquad \text{if } .5 + f \ge z_{p_i} > .5 \qquad (9)$$

Where, f = a fuzzy parameter set exogenously by the analyst to take possible values [0,.5]; p_i = the first or second closest villages.

Note that altering the f parameter alters the amount of fuzziness. Low f values are less fuzzy, whereas high f values are more fuzzy. The "correct" f value to use is an empirical question whose answer will vary by application. Values can be selected by trial and error such that results "look right" or can be selected based on field data used to calibrate the model.

The membership function in Equations 6-9 was applied to the Nang Rong data. Results are shown for a selected portion of Nang Rong in Figure 5. The black regions are village "core" areas with MF(z) = 1.00. Higher f values, cause the level fuzziness to increase as is indicated by the wider bands of light values which are hypothesized to demarcate territorial boundaries or transition zones. The river in the southeast portion of the Figure 5 stands out clearly as a boundary for all three f values.

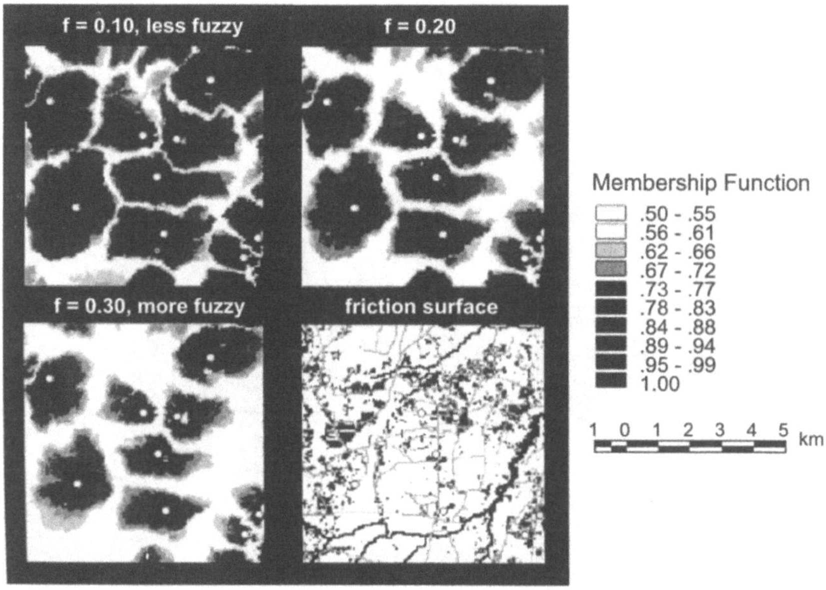

Figure 5. Fuzzy functional village regions.

ACCURACY ASSESSMENT

Accuracy of fuzzy KSP models were assessed by comparing resulting regional boundaries with estimates of actual village boundaries based on field data for a subset of villages. Additionally, the performance of fuzzy KSP models were compared to results for the Ordinary Voronoi diagram and radial buffer models using radii of varying sizes.

Analyses utilized transect data collected by Evans (1998) for several villages in Nang Rong designed to assess the nature of village boundaries. Evans started at the village center and collected data along transects emanating in the eight cardinal directions. He recorded, with the aid of local informants, information on village ownership for each land parcel that intersected the transects. The transects ended at the point where villagers from the source village no longer owned any land. Connecting the end points of the eight transects provides a rough estimate of "true" village boundaries that can be used to assess model results.

Figure 6 shows the "true" crisp boundaries along with the fuzzy regions for selected villages. The regions were all developed with $f = 0.20$ based on trial and error and field familiarity. In image panel 1 (upper left of Figure 6), the model boundary between the two villages agrees with the "true" boundaries derived from transect data, and the shape of the northern village generally parallels the boundary line. Visual inspection of the other panels suggests mixed, though encouraging results. Village 35's fuzzy region (lower right of Figure 6) captures its transitional boundary zone with its nearest neighbors to the north. It also has a long "tail" extending to the south that is evident in the transect data though slightly misaligned. Village 217's boundary (lower left of Figure 6) generally follows the transect boundary with fuzziness being more prevalent to the northwest.

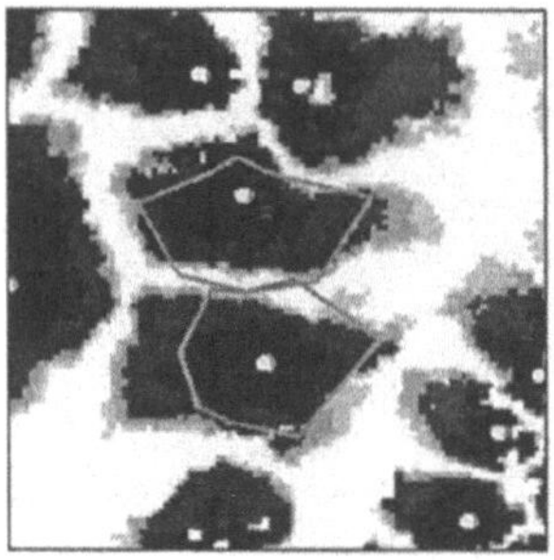

1. Villages 127 and 152

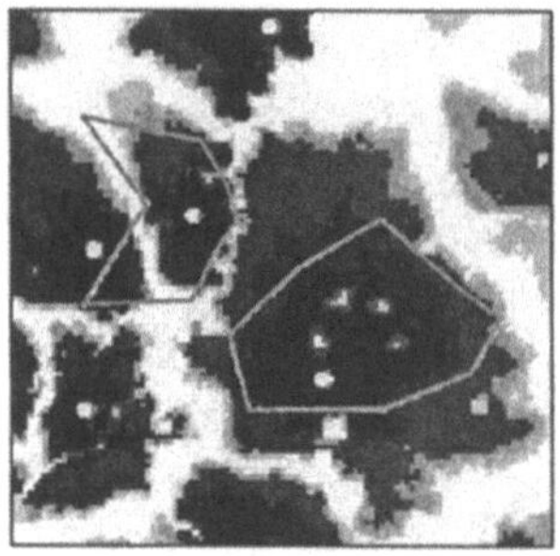

2. Villages 42 and 117

3. Village 217

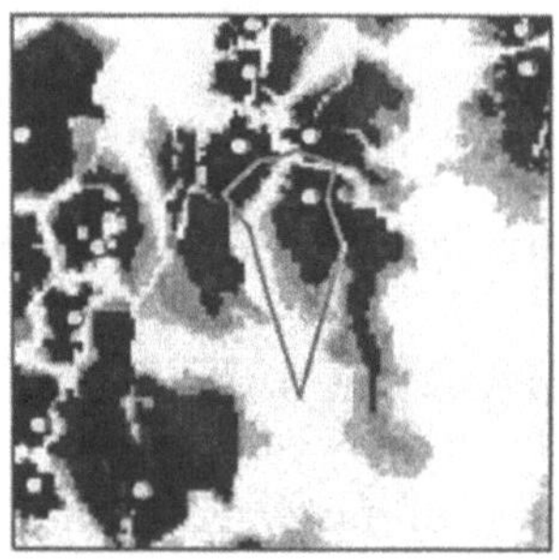

4. Village 35

Figure 6. True village territories based on transect data (Evans 1998) are shown in gray lines – fuzzy functional village regions are shown in grayscale.

Mean *absolute* deviations were calculated as the absolute distance between transect endpoints and model boundaries along the respective transects' cardinal directions. Using a sample of 13 villages each with 8 transects yielded a sample size of 102 (two unreliable transects were omitted). Figure 7 shows results for fuzzy KSP models (*f*=2) with varying grade-of-membership (*GOM*) thresholds. To explain, multiple village regions were iteratively defined as those grid cells for which the observed grade of membership, *GOM*, was greater than the defined grade of membership thresholds, *GOM*, which ranged from 0 to 1.00 at increments of 0.10. Figure 7 also shows results for various radial buffer models. Radial buffers were generated around each village central point using radii from 500 to 3000 m at 250 m increments. Resulting circular village regions had a uniform shape and size. For the fuzzy models, deviation reached a minimum value at *GOM* = 0.90 for model *FZ90*. The nomenclature *FZ90* means that this is a fuzzy model with a grade of membership threshold of 0.90. For the radial models, deviation reached a minimum using a radius of 750 m for model *RB750*. The nomenclature *RB750* means that this is a radial buffer model with a radius of 750 m. Thus, models *FZ90* and *RB750* were the top performers of these respective approaches.

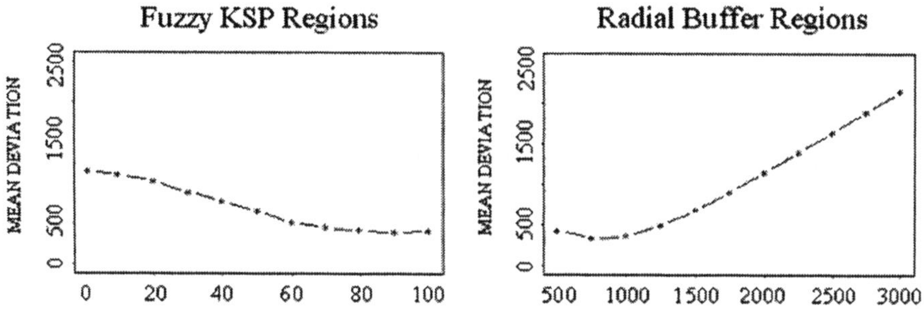

Figure 7. Mean deviation between modeled boundaries and transect endpoints.

Table 1 shows descriptive statistics of absolute deviations for *FZ90*, *RB750*, and the Ordinary Voronoi diagram (OV). *FZ90* had a mean value almost 300 m less than that of OV. *FZ90*'s mean value was about 50 m larger than that of *RB750*. This was likely due to *FZ90*'s relatively high maximum value because *FZ90* and *RB750* had identical medians. Because computing the fuzzy KSP diagram requires significantly more effort than the other two approaches, it is worthwhile to determine if it provides significantly better results. To approximate normality, a square root transformation was applied to all observations, and paired t-tests were used to determine if mean absolute deviations differed among models (Table 2). Both *FZ90* and *RB750* performed significantly better than OV ($t = -5.44$ and -5.62, $p = .00$); there was no significant difference between *FZ90* and *RB750* ($t = 0.42$, $p = .68$).

Table 1. Absolute deviation (meters) between model and transect endpoints.

	FZ90	*RB750*	*OV*
min	4	3	16
1st q.	94	150	296
median	269	269	537
mean	378	325	697
3rd q.	543	440	100
			9
max	2015	1170	303
			6

Table 2. Paired t-tests of absolute deviations; t-statistic shown with p-value in parentheses; *One-tailed test; ** Two-tailed test.

	FZ90	*RB750*	*OV*
FZ90	-	0.42 (.68**)	-5.44 (.00*)
RB750		-	-5.62 (.00*)
OV			-

Table 3 shows descriptive statistics of deviations (*not absolute*) for *FZ90*, *RB750*, and *OV*. The high positive value for *OV* (562 m) means that, on average, the *OV* model substantially over-estimated true boundary extents. The small negative values for both *FZ90* (-54 m) and *RB750* (-98 m) mean that these models slightly under-estimated true boundary extents. Paired t-tests were used to determine if deviations differed among models (Table 4). Paired differences should be interpreted here as differences in modeled transect lengths since the actual transect length (from field data) is included in deviation calculations for each observation and therefore will cancel out. Thus, test results will determine if the modeled transects are significantly different, and, extending the interpretation, will suggest if the different model approaches yield significantly different village regions. Both *FZ90* and *RB750* were significantly different than *OV* ($t = -11.11$ and -9.51, $p = .00$); *FZ90* appears to be substantially different than *RB750,* but the difference is not significant at traditional levels ($t = 0.99$, $p = 0.16$).

Table 3. Deviation (meters) between model and transect endpoints.

	FZ90	*RB750*	*OV*
min	-1369	-1170	-1154
1st q.	-317	-340	126
median	-32	-100	491
mean	-54	-98	562
3rd q.	172	221	810
max	2015	639	3036

Table 4. Paired t-tests of deviations; t-statistic shown with p-value in parentheses; *One-tailed test.

	FZ90	*RB750*	*OV*
FZ90	-	0.99 (.16*)	-11.11 (.00*)
RB750		-	-9.51 (.00*)
OV			-

To summarize the analysis results, the Ordinary Voronoi diagram performed the worst of the three village territory models in terms of mean absolute deviation (an error measure). The fuzzy KSP and radial buffer models were clearly better than the Ordinary diagram, however, there was no significant difference between absolute deviations for the fuzzy and radial models. This suggests that the fuzzy model is no better than the radial buffer model and vice versa. Given that radial buffers are much simpler to derive than the fuzzy territories, a possible conclusion, in this case, is that radial buffers should be preferred over the fuzzy territories. However, drawbacks to the radial buffers include their unrealistic uniformities of shape and size throughout the study area. Visual inspection and analysis of transect deviations (non-absolute) for each model suggested that territory shapes and sizes were substantially (though not significant statistically) different. A middle ground between choosing one model over the other is to retain both models for use in generating two distinct sets of village environmental descriptors, one pertaining to each model approach, via GIS overlay procedures. Sensitivity analyses can then be performed on subsequent analytical models that use the environmental descriptors to assess the robustness of models (beta coefficients, R-squared, etc.) using the different village territory approaches (Crawford 1999).

One of the aims of this research is to integrate social and environmental data into a spatially explicit format for the generation of relevant variables to support substantive analyses of human-environment relationships. Using the estimated village regions yielded by the discrete region-building models described previously, it is fairly straightforward to derive village-level environmental variables from relevant GIS data layers via overlay procedures. For interval and ratio data, descriptive statistics (e.g., central tendency and/or dispersion) can be generated for aggregations of grid cells falling within respective village regions. For example, one could examine the central tendency and distribution of elevation or slope values for individual villages using a gridded digital elevation model. The same general reasoning applies to the generation of percentages for categorical data layers. For example, one could examine the percentage of village space classified as having high or low land suitability ratings for particular agricultural types using a soil or land suitability layers.

CONCLUSIONS

In the Nang Rong study area, but also elsewhere in developing countries, representing settlement patterns as a discrete point location is common. Detailed maps of actual household locations are generally lacking and information that describes the geographic position of farm plots associated with households and/or villages, such as cadastral maps, are often times non-existent or inaccessible. Therefore, an approach needed to be developed and evaluated that is capable of representing and estimating the areal extent of land used by population groups at defined scales. This chapter has been primarily concerned with defining village regions for integration of social and environmental data at relatively fine scales, in this case, at the village level for the Nang Rong study area in northeast Thailand. It has demonstrated a novel method to estimate functional village territories for linkage with population data to support spatially explicit analyses of human-environment interactions. While the empirical work in this chapter is specific to the Nang Rong study area, the conceptual issues raised here may be germane to many other settings similarly characterized by a nuclear settlement pattern and lacking functional region definitions for small villages or settlements.

Previous work (Walsh et al. 1999) has attempted to integrate social and environmental data, but in a different way by using the pixel (or grid cell) as the unit of analysis. In the previous work (Walsh et al. 1999), population counts were distributed from village points in the Nang Rong study area to individual cells using a population distribution model. This former approach can be called a "people to pixels" linkage with the name indicating the linkage direction. With the people to pixels approach, village population data were disaggregated and distributed throughout the landscape to "populate" pixels. In contrast, the work presented here operates in the opposite direction by aggregating grid cells and linking them to the village unit of observation for village level analyses. The approach demonstrated in this chapter can be called a "pixels to people" linkage since it requires aggregation of estimated village pixels and then the aggregation to respective villages. A drawback with this approach when considering issues of statistical power in subsequent analyses is that the number of observations may be fairly low depending on the size of the study area and the number of villages or settlements. The former approach, people to pixels linkage, allows for a larger number of observations (e.g., hundreds of thousands of study area pixels in the case of Nang Rong using 60 x 60 m pixels), however, at present much validation work remains to be done for population distribution models.

A fundamental question, then, is what linkage direction is most appropriate for spatially oriented human-environment research? This question remains to be answered. A speculative answer is that it depends on what variable is the main outcome of interest. If a social variable is the main interest, and it is reported at the village level in census or survey data, then the approach advocated in this chapter may be more appropriate. If a spatially varying environmental variable is the main interest, and it is reported at the pixel level in remotely sensed imagery or GIS data layers, then the former approach may be more valid. To more completely understand the nature of human-environment interactions, researchers must come to grips with these issues at multiple scales of analysis.

NOTES

[1]The rationale for using a raster environment is that resulting regions will be used in conjunction with environmental coverages (e.g., satellite imagery and

digital elevation models) that are commonly in a raster format. As such, the precision of resulting regions is linked to the resolution of the raster environment. At successively smaller resolutions, results approximate the precision that is obtained using continuous space. Choice of resolution should be governed by the specific research context. The goal in this paper is to obtain plausible village territories that exhibit fuzzy boundaries based on non-Euclidean distances. Operating in the raster environment is a convenient approach to operationalize these goals or similar goals for other applications.

[2]Villages units in Thailand are defined formally for administrative purposes. Each village has a headman and a village council. Occasionally, a village grows in size to a point where it is necessary to split the former village into two separate villages even though they reside in a common nuclear cluster. Thus, two nearby "split" villages were formerly a single unified village or settlement cluster and are re-aggregated into a single village for this analysis.

[3]Other applications may use different cell resolutions depending on the specific research context. A 60 x 60 m resolution was selected to enable resulting village territories to be overlaid efficiently with environmental data (e.g., satellite imagery and digital elevation models) that were originally stored or resampled at that resolution.

[4]Using the Nang Rong data, KSP polygons for $k = 1$ through $k = 6$ were computed. The algorithm can also be used for greater k values. In this paper, results are used for $k = 1$ and $k = 2$. Thus, we only two "competing" villages were considered in this modeling scenario. The approach can be extended to incorporate more than two competing villages ($k > 2$), however, at added computational complexity.

[5]10 kilometer radial buffer grids were used to reduce computational time and space costs. For example, creating 240 least cost grids for the entire district (approximately 50 x 50 km) was not necessary since practically all villagers travel less than 10 km to their fields. Creating a 10 km buffer and masking out the remaining area was more efficient. Other applications may require different buffer sizes or may not need to buffer at all, if the number of points and the size of the study area is small.

[6]Friction weights were estimated based on the field experiences in the study area. Related research (Crawford 1999) using different region defining methods slightly altered the friction weights and found that resulting regional boundaries were fairly similar.

ACKNOWLEDGEMENTS

Gratitude is expressed to Barbara Entwisle, Ronald R. Rindfuss, Stephen J. Walsh, and the Landscape Characterization & Spatial Analysis Laboratory, Department of Geography of the University of North Carolina-Chapel Hill for their various efforts in support of the research. Additionally, field data and insights from Tom Evans of Indiana University-Bloomington aided this study.

REFERENCES

Altman, D. 1994. Fuzzy set theoretic approaches for handling imprecision in spatial analysis. *International Journal of Geographical Information Systems* 8(3): 271-289.

Atkinson, P. M., M. Cutler, and H. Lewis, 1997. Mapping sub-pixel proportional land cover with AVHRR imagery. *International Journal of Remote Sensing* 18(4): 917-935.

Banai, R. 1993. Fuzziness in geographical information systems: Contributions from the analytic hierarchy process. *International Journal of Geographical Information Systems* 7(4): 315-329.

Bonham-Carter, G.F. 1994. *Geographic Information Systems For Geoscientists: Modelling With GIS.* New York: Elsevier Science Inc. (Pergamon Press).

Brown, D. G. 1998a. Mapping historical forest types in Baraga County, Michigan as fuzzy sets. *Plant Ecology* 134(1): 97-111.

Brown, D. G. 1998b. Classification and boundary vagueness in mapping presettlement forest types. *International Journal of Geographical Information Science* 12(2): 105-129.

Burrough, P.A. 1996. Natural objects with indeterminate boundaries, in, P.A. Burrough and A.U. Frank (eds.), *Geographic Objects with Intdeterminate Boundaries*, Bristol, PA: Taylor & Francis, pp. 3-28.

Burrough, P.A. and A.U. Frank, eds., 1996. *Geographical Objects With Indeterminate Boundaries.* London: Taylor & Francis.

Burrough, P.A. and J. McDonnell, 1998. *Principles Of Geographical Information Systems.* Oxford: Oxford University Press.

Crawford, T.W. 1999. A comparison of region building methods used to examine human environment interactions in northeast Thailand. *Papers and Proceedings of the Applied Geography Conference*, 22: 366-373.

Edwards, G. 1994. Characterising and maintaining polygons with fuzzy boundaries in geographic information systems. *Advances in GIS Research* 1: 223-229.

Edwards, G. and K. E. Lowell, 1996. Modeling uncertainty in photointerpreted boundaries. *Photogrammetric Engineering & Remote Sensing* 62(4): 377-391.

Entwisle, B., S. J. Walsh, R. R. Rindfuss, and A. Chamratrithirong, 1998. Land-use/land-cover and population dynamics, Nang Rong Thailand, in D. Liverman et al. (eds.), *People and Pixels*, Washington: National Academy Press, pp. 121-144.

Evans, Tom P. 1998. *Integration Of Community-Level Social And Environmental Data: Spatial Modeling Of Community Boundaries In Northeast Thailand.* Ph.D. dissertation, Department of Geography, University of North Carolina.

Fisher, P.F. and S. Pathirana 1990. The evaluation of fuzzy membership of land cover classes in the suburban zone. *Remote Sensing of Environment* 34(2): 121-132.

Fisher, P.F. and S. Pathirana 1993. The ordering of multitemporal land-cover information derived from Landsat MSS data. *Geocarto International* 8(3): 5-14.

Foody, G. M. 1992. A fuzzy sets approach to the representation of vegetation continua from remotely sensed data: An example from lowland heath. *Photogrammetric Engineering & Remote Sensing* 58(2): 221-225.

Groenemans, R., E. Van Ranst, and E. Kerre, 1997. Fuzzy relational calculus in land evaluation. *Geoderma* 77(2-4): 283-298.

Harris, T. R., S. W. Stoddard, and J. C. Bezdek, 1993. Application of fuzzy set clustering for regional typologies. *Growth & Change* 24(2): 155-165.

Leung, Y. 1985. Basic issues of fuzzy set theoretic spatial analysis. *Papers – Regional Science Association* 58: 35-46.

Leung, Y. 1987. On the imprecision of boundaries. *Geographical Analysis* 19(2): 125-151.

McBratney, A B. and I. Odeh, 1997. Application of fuzzy sets in soil science: fuzzy logic, fuzzy measurements and fuzzy decisions. *Geoderma* 77(2-4): 85-113.

Okabe, A., B. Boots, and K. Sugihara, 1992. *Spatial Tessellations: Concepts And Applications Of Voronoi Diagrams*. New York: John Wiley and Sons.

Ponsard, C. and P. Tranqui, 1985. Fuzzy economic regions in Europe. *Environment and Planning A 17*(7): 873-887.

Reenburg, A. (1995). The spatial pattern and dynamics of a Sahelian agro-ecosystem. *GeoJournal* 37(4): 489-499.

Rindfuss, R. R. and P. C. Stern, 1998. Linking remote sensing and social science: The need and the challenges. In D. Liverman, E. F. Moran, R. R. Rindfuss, and P. C. Stern (eds.), *People and Pixels: Linking Remote Sensing and Social Science*, Washington: National Academy Press, p. 1-27.

Robinson, V.B. 1990. Interactive machine acquisition of a fuzzy spatial relation. *Computers & Geosciences* 16(6): 857-872.

Sui, D. Z. 1992. A fuzzy GIS modeling approach for urban land evaluation. *Computers, Environment & Urban Systems* 16(2): 101-115.

Walsh, S.J., T.P. Evans, W. F. Welsh, R. R. Rindfuss, and B. Entwisle, 1998. Population and environment characteristics associated with village boundaries and landuse/landcover patterns in Nang Rong district, Thailand. *Proceedings of Pecora 13, Symposium on Human Interactions With the Environment: Perspectives From Space.* Bethesda, MD: American Society for Photogrammetry and Remote Sensing (ISBN-1-57083-055-X [CD Publication]), 395-404.

Walsh, S.J., T.P. Evans, W.F. Welsh, B. Entwisle, and R.R. Rindfuss, 1999. Scale dependent relationships between population and environment in northeast Thailand. *Photogrammetric Engineering and Remote Sensing 65(1): 97-105.*

Wood, C.H. and D. Skole, 1998. Linking satellite, census, and survey data to study deforestation in Amazonia, In D. Liverman et al. (eds.), *People and Pixels*, Washington: National Academy Press, pp. 70-93.

Chapter 6

UNDERSTANDING A DYNAMIC LANDSCAPE: LAND USE, LAND COVER, AND RESOURCE TENURE IN NORTHEASTERN CAMBODIA

Jefferson Fox

East-West Center, Honolulu, Hawaii, USA
foxj@eastwestcenter.org

Keywords: participatory assessment, swidden agriculture, land use transitions, aerial photography.

Abstract This paper seeks to describe changes in land cover, land use practices, and tenure systems in several villages in northeast Cambodia over the last 50 years. The project integrated the development of a spatial database (based on 1953 and 1996 aerial photographs and 1:50,000 base maps) with socioeconomic information collected for a development project, and an analysis of relevant socioeconomic policies. Over the last half-century, land use and total tree cover have remained stable, but fragmentation of the tree cover has increased extensively. Land use has begun to change recently as both Cambodian and foreign investors invest in industrial agricultural crops such as palm oil, rubber, cassava, and kapok. In the past, farmers had a clear sense of village lands, but specific boundaries between villages were not traditionally required. This study suggests that national land tenure policies are making it increasingly difficult for farmers to maintain their traditional swidden land use practices. Simultaneously, market pressures -- the commercialization of subsistence resources and the substitution of commercial crops for subsistence crops -- are encouraging farmers to engage in new and different forms of commercial agriculture. Combined, these forces will eventually cause a major change in land use practices from swidden agriculture to commercial crops, and a change in land cover from secondary vegetation to monocultural agriculture. These changes have significant implications for biodiversity, watershed hydrology, and carbon sequestration, as well as the lives and livelihoods of local people.

INTRODUCTION

Scientists, resource managers, policy-makers, and planners increasingly recognize that land-use change is a major driver of global change, through its impacts on climate, ecosystem processes, biogeochemical cycles, biodiversity and, even more importantly, human activities (Nunes and Auge 1999). To understand land-use changes, scientists and resource managers must consider the roles played by individual farmers, decision-makers, institutions—including land tenure, and the inter-level integration of processes at one level with those at other levels of aggregation. As Nunes and Auge (1999) point out, a village connected by paved roads to world markets feels the pressure of international commodity price shifts a great deal more than communities with poor road infrastructure, and are likely to make very different decision about land use. Without understanding the human dynamics behind land use change, we cannot understand changes in land cover, nor predict the outcomes of policy intervention

The world knows little about land cover in northeast Cambodia or about the customary land use practices and tenure systems of the highland people, who have inhabited this area for generations. The little evidence available suggests that land cover in the area is being transformed by rapid and extensive political, economic, and ecological changes (Global Witness 1998). If trends from elsewhere in Southeast Asia hold in Cambodia, then much of this change is being driven by various types of market pressures: the commercialization of subsistence resources, the substitution of commercial crops for subsistence crops, and the growth of outside control over the production or extraction and marketing of local resources (Young 1999). It is important to generate baseline data on the effects of commodification on local resource management systems to understand the impact of these changes on land cover, sustainable resource use, and landscape transformation (Nunes and Auge 1999). There is also an urgent need for a comprehensive study of traditional land use practices and systems of customary tenure to recognize and protect the rights of the indigenous peoples (Peluso 1992).

Recently, a few research teams have begun linking social science, remote sensing, and other spatial data at the community and household levels to understand land use and land cover change (e.g., Guyer and Lambin 1993, Skole et al. 1994, Moran et al. 1994, Fox et al. 1995, Walsh et al. 1999). These researchers come from diverse research communities and have diverse theories and substantive interests; yet they all recognize the need to link specific remotely sensed pixels and spatial coordinates to specific decision-makers (households and communities). Among these teams at least two types of social science methodologies are used. One group collects quantitative social science data at the household level using formal questionnaires (Walsh et al. 1999). The power of longitudinal designs is well established in the social sciences, providing numerous analytical and statistical advantages to cross-sectional designs. The other group uses participatory assessment techniques to solicit non-quantitative information in a manner that is more relevant, timely, usable, and inexpensive than sample surveys (Reid 2000, Fox et al. 2000). Development workers have adopted participatory assessment techniques to enable local communities to speak for themselves, to organize themselves, and to identify and solve their own problems (Chambers 1994). Scientists use participatory assessment techniques, such as semi-informal and group interviews to develop a set of hypotheses about the causes and consequences of changes in land use and land cover.

This project integrated the development of a spatial database with information collected through participatory assessment techniques. This paper describes changes in land cover, land use practices, and tenure systems in several villages of Ratanakiri province over the last 50 years. The paper concludes with two policy recommendations for preserving ecological diversity in the region and protecting the stability of the human-forest relationship.

GEOGRAPHY OF RATANAKIRI

The Cambodian province of Ratanakiri, "the mountain of precious stones," lies about 600 km northeast of Phnom Penh. It is bordered by Vietnam on the east and Laos on the north, and covers approximately 12,500 km^2 (Figure 1). Due to its distance from major regional centers and high prevalence of malaria, the province remained remote and isolated from Western influences until recently (Bourdier 1995). With the exception of two ethnographies (Fontanel 1967, Matras-Troubetzkoy 1967-1968, published in 1983), no study of human geography or anthropology had been undertaken in the province until the 1990s.

The Sesan and Srepok Rivers cross the province flowing west from Vietnam to the Sekong River, a tributary of the Mekong (Figure 1). The northern portion of the province, between the Sesan River and the Laotian border, is covered with broadleaf evergreen forest. Approximately 12,600 people, 18% of the province's population, live there. South of the Srepok River, the province is covered with a tropical deciduous forest. Approximately 7,000 people, 10% of the population, live there (Bourdier 1995). The remaining area, between the two rivers, is composed of red basaltic soils on a high plateau (300 m in elevation) and is covered with tropical secondary forests, forests "formed as a consequence of human impact" (Brown and Lugo 1990:3). This area includes the provincial capital, Ban Lung. Approximately 51,000 people, over 70% of the population of the province, live there. Ethnic communities in Ratanakiri include the Brao, Jarai, Kachah, Kraveth, Krung, and Tampuen (Lebar et al. 1964).

This study was conducted in the high plateau area in *Khum* (commune) Poey of O Chum district, which consists of 10 villages (*phum*) covering approximately 100 square kilometers. Poey is located on the unimproved road that runs north from Ban Lung to the Vietnamese border (Figure 1). Elevation ranges from 100 to 400 m. The region has a monsoonal climate, with a rainy season beginning in May or June and lasting until October or November. Annual rainfall is always above 2,000 millimeters, and can reach 2,950 millimeters in Ban Lung (Bourdier 1995). Almost no rain falls between December and April. Vegetation is composed primarily of broadleaf evergreen and deciduous forests (Bourdier 1995). People of Krung ethnicity populate the 10 villages of Khum Poey.

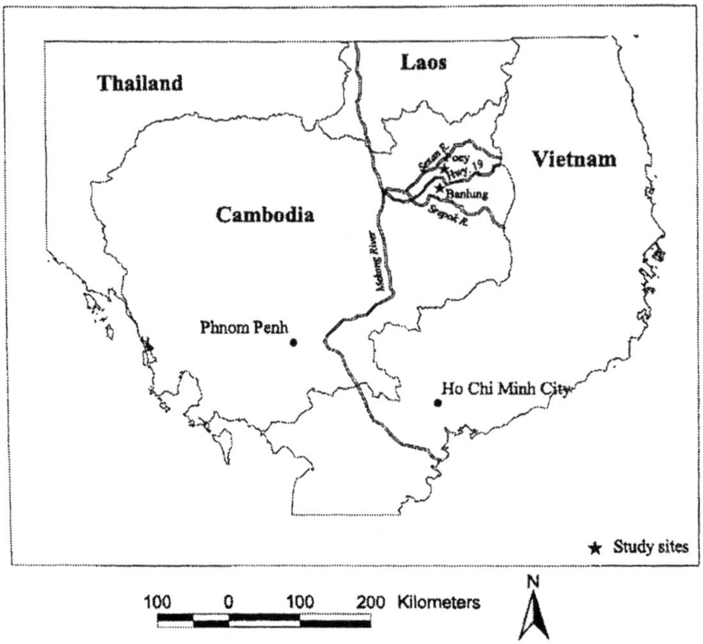

Figure 1. Study sites in Ratanakiri Province, northeast Cambodia.

Swiddening, called *chamkar* by the Krung, and also known as slash-and-burn, or shifting cultivation, is the major land use practice in the village and is practiced by households in all ten villages. Swiddening involves cutting living vegetation in the dry season, letting it dry, burning it late in the dry season, and then planting a crop in the ashes early in the wet season. *Chamkar* fields are cleared in January; burning commences in March and continues through April; the first corn is planted in May and rice seeds are sown in June or July; corn is harvested in August and rice in October and November. The different components of the agricultural season (clearing, burning, planting, weeding/growing, and harvesting) are noted with small ceremonies or agricultural rituals. All livestock raised in the village, (i.e., buffalo, pigs, and chickens) are used for these ceremonies and for appeasing forest spirits when people fall sick or have bad dreams (Matras-Troubetzkoy 1983).

In a swidden agricultural system the perceived dichotomy between agriculture and forest is for the most part artificial. Swidden fields, secondary forests, and mature forests are all part of the same agroecosystem. Farmers manage vegetation to optimize the production of useful products (edibles, medicinals, thatch, lumber, etc.), while minimizing labor inputs. In many swidden systems a portion of the landscape is managed as a forest reserve for the production of materials not generally available in secondary forests (i.e., wildlife, building timber, rattan, certain medicinals, mushrooms, etc.; see Brookfield 1988, National Research Council 1992, Kellman and Tackaberry 1997).

METHODS

The project integrated the development of a spatial database with information collected through semi-informal interviews with farmers and other key informants and analysis of relevant socioeconomic policies. Interviews were conducted in both Khmer and Krung languages and translated into English by experienced field workers. This database served as a framework for analyzing changes in land cover and forest patterns, as well as a tool for analyzing the information and insights collected in interviews and policy analysis (Fox et al. 1995).

Approximately 40 semi-structured interviews were conducted with people from four villages: Kres, Kralaa, Tannich, and Ganchueng (Figure 2) during five field visits between June 1995 and October 1998. People interviewed included local key informants and farmers including traditional village chiefs, community leaders, and older people who could provide holistic histories of land use dynamics in the study area. Household interviews were conducted with individual farmers to collect data on how families arrange their swidden fallow plots across the landscape, how they collaborate with neighboring farmers and communities for allocating and zoning land use, and how they control fires.

Semi-structured interviewing (Chambers 1985, Grandstaff and Grandstaff 1987) is a methodology developed to enable outsiders to learn about rural conditions in a more cost-effective manner than formal survey questionnaires. Semi-structured interviewing uses no question list, printed questionnaire, or other means of strictly controlling the interview. Informants are selected opportunistically rather than randomly. The interviewer starts with the guiding issues in mind, curiosity, and willingness to analyze issues and responses as the interviews develop. Interviewer and informant interact and explore issues using specific questions until a logical and internally consistent picture is pieced together. The interview depends upon mutual rapport and the interviewer's ability to evaluate responses and to respond with relevant follow-on questions. For example, interviewer and informant can talk about farming practices, including various costs, the returns to the different enterprises, related problems, and other income sources. By the end of the interview, income can be calculated without the interviewer ever asking the usually inappropriate question "What is your income?" Farmer identified problems and practices and perceptions new to the interviewer are also explored. Semi-structured interviewing is not designed to solicit data that can be analyzed in a quantitative manner; rather it is an exploratory and iterative tool that allows a process of rapid and progressive learning to take place.

Recent aerial photographs (1996) were used in the field as tools for soliciting discussion from local leaders and farmers on past land use practices, present land use conflicts, and plans for the future. Local people made sketch maps delineating customary land use boundaries, tenure boundaries, sites of conflicts, and the location of valuable resources. Important boundaries demarcating features such as community-protected forests, and mining roads and pits were mapped using GPS technology and the guidance of local people.

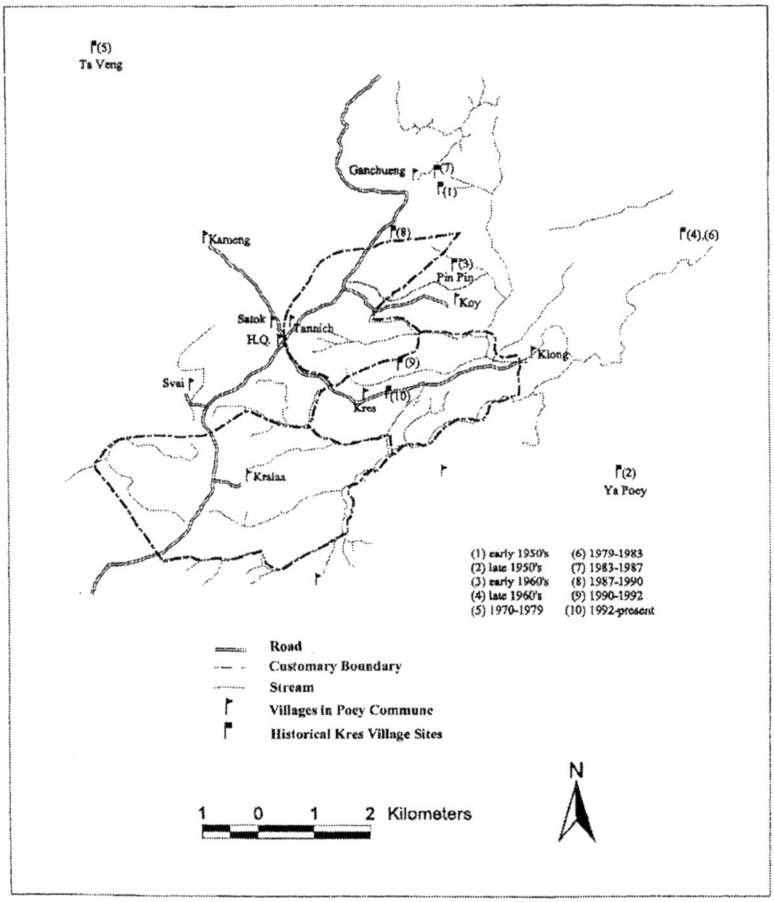

Figure 2. Spatial history of Kres village between the early 1950's and today.

A spatial database of the study site was developed using 1996 color aerial photographs (nominal scale: 1:25,000—FinnMapp). Aerial photo interpreters at the Cambodian Integrated Resource Information Center in Phnom Penh manually interpreted the aerial photographs (with the aid of a stereoscope), classified land cover into seventeen classes, and digitized the maps. These classes were then consolidated into six types to compare with similar studies in other parts of mainland Southeast Asia (Fox et al. 1995, Long-Chun-lin et al. 1999, Xu et al. 1999, Fox et al. 2000). The six types included mature forests (broadleaf evergreen, broadleaf deciduous, and mixed deciduous/evergreen), secondary forest, swidden, paddy, plantations, and village/other lands. The secondary forest or successional vegetation was subdivided into three groups—closed-canopy forest; open-canopy forest; and grass, bamboo, and bushes (recent fallow) (Brown and Lugo 1990). Differences observed in color/tone, pattern, and texture on the aerial photographs were used to differentiate mature forest from secondary forest (and among the different types of secondary forest). The closed-canopy forests may or may not have been primary forest, but fieldwork in the region suggests that even areas usually classified as forest may be more accurately described as intermediate or advanced secondary vegetation (Potter et al. 1994).

A spatial database was also developed for an area adjacent to the southern boundary of the study site using panchromatic 1953 (nominal scale: 1:40,000) aerial photographs taken as part of a comprehensive mapping of Indochina undertaken by the French military in 1952-53; 1996 aerial photographs were also available, as mentioned above. Aerial images from 1952 were not available for the study area because the French government classified images of the Cambodian-Laos border. Land cover on the 1953 photographs was manually interpreted (using a stereoscope) and classified into nine classes (fewer classes were delineated on the 1953 photographs because some classes, such as lowland rice, did not exist in 1953, and also because of differences in scale and color of the two sets of photographs). These classes were then grouped into the same types as the 1996 images. All photographs were registered to a map base at a scale of 1:50,000. Land-cover categories were digitized and entered into a geographic information system (GIS) database (Arc/Info software on a Sun Sparc workstation).

Transition matrices showing changes in land cover through time were developed by overlaying grid-cell data for a given period with grid-cell data for different time periods. A grid program within the GIS was used to transform the classified polygonal data for 1953 and 1996 into a set of grid cells, each measuring 30 m by 30 m. These data were used to calculate the transition matrix for the 1953-1996 period.

RESULTS

Changes in Land Cover

In 1996 approximately 4% of the total area of Poey commune was devoted to active swidden (Table 1). Mature forest cover (primarily broadleaf evergreen) covered 50% of the landscape. Various types of secondary forests accounted for another 46% of the landscape. Less than 1% of the landscape was devoted to villages and permanent agricultural fields.

Table 1. Land Cover in Poey in 1996 (interpreted from aerial photographs).

Land cover	Ha	%	No. of Fragments	Mean size (ha)
Forest	4,663	50	52	90
Secondary forest	4,255	46	164	26
• Closed-canopy forest	3,536	38	53	67
• Open-canopy forest	476	5	56	9
• Grass, bamboo, and bushes	243	3	55	4
Swidden	419	4	143	3
Paddy	0	0	0	0
Plantations	3	<1	2	1
Village and other	37	<1	42	1
Total	9,377	100	401	23

Land cover change in the area immediately to the south of Poey has been extensive over the last 43 years (Table 2). This area, which includes the provincial capital, Ban Lung, is more densely populated than Poey. In 1953, the Ban Lung area had only 18% forest cover (broadleaf evergreen), 60% secondary forest, and 21% swidden cover. Little information on land use in the area exists for 1953, but these data suggest a fairly dense population actively involved in swidden agriculture. By 1996, forest cover had increased to 26% of the landscape, secondary forest had decreased to 51%, and swidden agriculture had decreased to 14% (Table 2). These data suggest a less densely populated landscape than in 1953 (as expected from the disruptions caused by the Khmer Rouge), but a landscape that is still more densely populated than is Khum Poey to the immediate north. Closed-canopy secondary forests decreased from 45% to 30%, while open-canopy secondary forests increased from 6% to 18%. Overall tree cover (forest and secondary forest) in Ban Lung remained constant during the 43-year period (Table 2). The greatest change during this period was a decrease in swidden (21% to 14%) and an increase in land devoted to villages and plantations, primarily rubber and palm oil (1% to 8%). In Poey, however, plantations still account for less than 1% of the landscape (Table 1).

Changes also occurred in the number and size of forest fragments. In 1953 there were 20 fragments of forest with an average size of 166 ha (total area was 3,319 ha). By 1996 this number had grown to 85 fragments with an average size of only 56 ha (total area was 4,769 ha). The number of secondary forest fragments grew from 93 (11,234 ha) to 444 (9,622 ha) and the mean size decreased from 121 ha to 22 ha (Table 2). Mean fragment size for forest and closed-canopy secondary forest categories has larger in Poey (90 ha and 67 ha respectively, Table 1) than in Ban Lung (56 ha and 34 ha, Table 2).

Table 2. Land cover in Ban Lung in 1953 and 1996 (interpreted from aerial photos).

Land cover	1953		1996		1953		1996	
	Ha	%	Ha	%	No. of Frag-ments	Mean size (ha)	No. of Frag-ments	Mean size (ha)
Forest	3,319	18	4,769	26	20	166	85	56
Secondary forest	11,234	60	9,622	51	93	121	444	22
• Closed-canopy • forest	8,415	45	5,646	30	27	312	167	34
• Open-canopy • forest	1,177	6	3,326	18	31	38	186	18
• Grass, bamboo and bushes	1,642	9	650	3	35	47	91	7
Swidden	3,951	21	2,638	14	175	23	706	4
Paddy	0	0	88	<1	0	0	22	4
Plantation	0	0	733	4	0	0	56	13
Village and other	148	1	808	4	35	4	51	16
Total	18,652	10	18,658	100	323	58	1,364	14

Table 3 is a transition matrix showing historical changes in land cover in the Ban Lung area. During the period 1953-1996, 43% of the mature forest cover that existed in 1953 was lost. Forest was converted primarily to closed-canopy secondary forest (17%), active swidden (11%), and open-canopy secondary forest (8%). On the other hand, 24% of the closed-canopy secondary forest, 14% of the active swidden, and 13% of land around villages, streams, and roads reverted to mixed broadleaf evergreen and deciduous forest (Table 3). The 57%

Table 3. Transition matrix of land-cover classes in Ban Lung.

1996	*Forest*	Secondary forest			Swid.	Paddy	Plan-tation	Village & other
1953	%	*Closed*	Open canopy %	Grass %	%	%	%	%
Forest	57	17	8	3	11	<1	3	<1
Secondary forest	46	92	74	11	45	<4	12	17
• Closed-canopy	24	34	15	4	15	<1	4	4
• Open-canopy	11	34	22	3	19	1	4	6
Grass, bamboo bushes	11	24	37	4	11	2	4	7
Swidden	14	36	23	3	15	<1	4	5
Village and other	13	16	8	4	12	<1	3	44

broadleaf evergreen forest that remained stable between 1953 and 1996 (Table 3) represent approximately 10% of the landscape of the study area. Likewise, 34% of the closed-canopy secondary forest also remained stable throughout this period; this represents 15% of the landscape. In 1996, at least 25% of the landscape was covered with closed-canopy mixed broadleaf evergreen and deciduous forest that was more than 43 years old.

Shifting Villages and Customary Tenure

Using aerial photographs and sketch mapping techniques, we were able to trace the spatial history of what is now called Kres village through at least nine voluntary movements and one forced movement over the last 50 years (Figure 2). The oldest known location of Kres is reported as a village called Ban Poey located near present day Ganchueng (Figure 2). The village headman or elder at that time was Ta Van Poey. Ban Poey was inhabited in the late French and early Sihanouk period (early 1950s). The village then relocated to the southeast in what today is La Ak commune and became known as Ya Poey village led by Ya Nanch in the Sihanouk period (late 1950s). The ancestors of residents from present day Kres, Ganchueng, and Satok villages lived in both Ban Poey and Ya Poey. In the early 1960s, the village moved close to the site of present day Koy village. At that time the village was called Pin Pin and the village elder was Ya Bot. During the Lon Nol period in the late 1960s (a period of disruption), the village moved deeper into the forest to the east near the O Touk River (Figure 2). This is the first time that the village was known as Kres and it was led by Kan Thaw and Chanly.

A grid program within the GIS was used to transform the classified polygonal data into a set of grid cells, each measuring 30 x 30 m. This table was developed by overlaying grid-cell data for 1953 with grid-cell data for 1996. In the row labeled forest, 57% of what was classified as forest in 1953 was still classified as forest in 1996, 17% was classified as closed-canopy secondary forest in 1996, and so on.

It was from this site that the Khmer Rouge forcibly relocated farmers to Taveng District (north of their traditional homelands) to grow lowland irrigated rice (Figure 2). Farmers remained there until the fall of the Khmer Rouge in 1979 when they returned to Kres village. They stayed there, along with current day residents of Koy village, until 1983. In 1984 they and the Koy residents moved to a site near present day Ganchueng (Figure 2). They stayed in at least two different sites around Ganchueng until approximately 1990. They then moved to a site near the current Kres village for two years before moving to their present site.

In interviews, farmers in Poey Commune identified a clear sense of village lands. Specific boundaries between villages were not traditionally required unless cultivation areas of two villages met one another. Farmers expressed the belief that if they farm on the other side of another village's *swidden* fields and have to cross that village's fields frequently, the spirits will be unhappy and cause misfortune or death. When a village's *swidden* fields are adjacent to those of another village, village elders may meet to decide the boundaries. In most cases the physical location of swiddens and the taboos against crossing each other's field for cultivation define the limits of cultivation.

Alternatively, from interviews it was also clear that village boundaries have become more firmly set in recent years. In 1997 the commune leader in Poey convened village headmen to discuss and delineate village boundaries. Between village boundaries, members of Poey Commune have started prohibiting the use of their lands for swidden by people from other villages (even in areas where other farmers do not have to cross their fields). Population data collected by village headmen suggest an average population density of approximately 30 people per km^2 in the five villages we mapped.

Farmers of Poey Commune reported that traditionally they would clear their swidden fields, use them, and then abandon them. Once a field was abandoned, a farmer had no future claim to that piece of land. Today, however, evidence from interviews suggests that many farmers in Poey Commune are beginning to consider their current swidden fields as private property. Neither customary nor government institutions currently exist for protecting this concept of private claims by swidden farmers of their agricultural lands.

Farmers in Poey reported limited interactions with government officials. While they report no agricultural extension or forest extension services, they are aware that the government prohibits them from cutting the forest for new swidden fields without government permission. Farmers reported that they do not have legal papers for any of their lands, although they would like to have legal title. They claim that they have managed their lands for several generations and that in their opinion it is their land. They claim they share their forest resources with neighboring villages and that they have few land conflicts within or between villages.

Land Use Practices

In interviews, farmers reported that they generally used their swidden fields for approximately three years and then abandoned them. They reported that they preferred to use their old fields as long as possible because they require less labor than opening new fields. Farmers reported that they initially choose their fields so as to be able to enlarge and expand around the sides of the originally cultivated fields and then only after three to five years shift to another place.

Farmers reported having two to four different sets of fields that they use for three to five years before fallowing. The number of fields a household manages depends on the amount of labor available within the household. The length of period a household uses a field depends on a number of variables, including rate of weed infestation, proximity to trails and passing livestock (a nuisance), and opening of new fields elsewhere. Fallow periods may vary from 3 to 30 years depending on soil fertility and land availability, but most people reported leaving their fields fallow for approximately 7 years. Households reported having approximately one to three ha at each swiddening site. They reported that they preferred to locate their fields within a one-hour walk of the village as they usually go back and forth every day. They also prefer fields close to water.

Farmers reported they burned their *chamkar* when they perceive that the fields are ready and there is sufficient wind for a good burn. Each family decides when and where to burn. While farmers in general do not discuss burning with each other, a farmer should try to inform his or her neighbors before burning. If a fire spreads to neighbors' fields and the farmer has not informed them first, then he must help them prepare their fields to burn again.

The main crop is swidden rice, but farmers reported more than 20 to 30 varieties of vegetables, green leaves, and fruits are grown in the *chamkar*. Other crops include cassava, taro, sugarcane, maize, sweet potatoes, yams, gourds, beans, peppers, sesame, tobacco, pineapples, eggplants, tomatoes, pumpkins, and cucumbers. Fruit trees are grown in the villages and *chamkar*, bearing bananas, jackfruit, cashews, papaya, and mangos.

Farmers in Poey reported that they distinguish between nearby secondary forests and swidden fields that are part of the long-agricultural rotation system, and more distant forests with older growth. The more distant forest areas are less clearly delineated and are shared with other communities. Farmers reported different rules and taboos governing the use of shared old growth forests (Colm 1997, Poffenberger 1999). People believe that in some forests resident spirits forbid cutting and other activities. Different spirits, each with their own taboos or spiritual regulations that effectively provide for forest and wildlife conservation (Colm 1997, Poffenberger 1999), are reported to rule different forests. A careful inventory of Phnom Tapin Forest, one of four forests managed by the Ya Poey Forest Conservation Association, found close to 200 tree species and over 3,000 species of flora and saplings (Bahn 1997).

Farmers in Poey report that they subsist self sufficiently by swidden farming and collecting forest products. If the previous harvest is good they claim they can eat rice all year long. Non-timber forest products collected in the village include bamboo and bamboo shoots, rattan, forest resins, tubers, wild taro (used as a pig forage), wild pigs, deer, and free roaming chickens. In addition to using forest products as famine food, farmers sell forest products in the market. In interviews, farmers suggested that they earn their cash income from collecting and selling forest products (resin, fruit, leaves), and not from their agricultural fields.

Since 1993, when Cambodia opened up for international investment, over a dozen coffee, rubber, cashew and other estate crop concessions of 100 to 20,000 hectares have been granted in Ratanakiri (Poffenberger 1999). In addition, close to one-half of the 1.16 million hectares in the province have been set aside as protected areas. Finally, six dams are planned and, in some cases, flood areas could stretch across nearly two-thirds of the province from Vonsai to the Vietnamese border. By one calculation (Butterfield 1997), 130% of the province's land area has either been granted as concessions or placed in protected status. This does not take into consideration the 72,000 people who live in the province and earn their living by farming, hunting, and gathering. These developments are happening in an unplanned fashion and without reference to land zoning or customary land use. The process of approving contracts and concessions is not clear. Prior consultation with existing communities (or even with the Provincial Government in the case of concessions) is poor or non-existent. In this study, farmers reported that they were unhappy with the "land grab" by commercial companies as well as government agencies. They see these developments as threatening their customary domain, their land use practices, as well as their remaining farmland and ancestral landscape (Colm 1997, Poffenberger 1999).

DISCUSSION

Land cover in the region has been both stable and dynamic. The GIS analysis revealed that between 1953 and 1996, 77% of the Ban Lung landscape remained under forest or some type of secondary forest, and in Khum Poey fully 96% of the landscape was still under forest or secondary forest as recently as 1996. The rise of the Khmer Rouge to power and the subsequent forced removal of farmers from the region in the 1970s had a significant but temporary impact on land use and land cover. The expulsion of farmers from their homelands and the prohibition of swidden agriculture are the most likely causes for the increase in forest cover observed in the Ban Lung area. When freed from enforced labor camps in the lowlands, farmers returned to their ancestral homelands and re-engaged in their traditional swidden agricultural practices. Since the 1960s, farmers have also devoted a small but growing portion of the landscape to growing cash crops, primarily rubber, palm-oil, and cashews. In the long run, the major change in land use occurring in this region is the slow but steady decrease in swidden agriculture and the concomitant growth in the area of cash crops.

When we look at any individual plot, however, land cover may have changed several times during this period. As a land use practice, swiddening results in relatively small patches of disturbance that, after they are abandoned, regenerate into secondary forest. The GIS analysis showed these patches to be dispersed non-uniformly across the landscape, and with few exceptions, such as the Khmer Rouge period, disturbed regularly through time. Swiddening thus causes fragmentation of forest cover as formerly homogeneous patches of forests are converted into a mosaic of tree cover in different stages of regeneration.

Government officials and policy makers in Phnom Penh, as well as at the provincial level in Ban Lung, generally regard the ethnic groups that inhabit this region as constantly moving their villages across the landscape. This study suggests that during the last 50 years farmers in northeast Cambodia did not move as widely as generally perceived. When villages did move, it tended to be within the village's ancestral cultivation area and often only a few kilometers. Moves were generally for specific reasons: political upheaval, government relocation, bad omens, excessive illness, or other hardships at the current site. If part of a village decided to separate from the original group because of population increase or conflicts, the breakaway group looked for new, available land. Such occurrences generally happened only once in two or three generations. Kralaa village in Poey commune stayed in the same location from most of the last 100 years, while neighboring Kres village moved more than 10 times in the last 40 years.

In terms of land tenure, farmers in Khum Poey have always had a clear sense of village lands. However, specific boundaries between villages were not traditionally required unless the cultivated areas from two villages met one another. In response to pressures from government agents, land speculators, and others interested in acquiring or using land in the area, farmers have begun to demarcate village boundaries since 1998. While people of Krung ethnicity did not traditionally view land as a private resource, many farmers in Poey are beginning to consider their current swidden fields as their private property. This is probably because land is becoming scarcer. Neither customary nor government institutions currently exist for protecting the private claims of swidden farmers to their agricultural lands.

In terms of land use practices, the granting of the forest concession as well as the palm oil estates have negated the customary land tenure rights of most highland communities and hence put these communities into a position of inevitable conflict with the state. The national government is unofficially encouraging lowlanders to immigrate into this area from other provinces and to encroach on the lands that uplanders claim through customary tenure systems by allowing lowland settlers (but not uplanders) to apply for land certificates. In addition, commercial investments, supported by both Cambodian and foreign investors, are increasing, especially for industrial agricultural crops such as palm oil, rubber, cassava, and kapok (Colm 1997, Poffenberger 1999).

There is an urgent need to develop legal mechanisms for recognizing traditional land use practices and systems of customary tenure to protect the rights of the indigenous peoples. Ya Poey Community Forest is an excellent example of the positive type of community/government interaction that can be promoted through policies that recognize the land use practices and customary rights of the highlanders.

CONCLUSIONS

Findings from this study suggest that the practice of swidden cultivation and the extent of forest cover (including both mature and secondary forests) have remained stable in northeastern Cambodia over the last 50 years. This despite the fact that national and provincial planners have attempted to "control" swidden cultivation through polices that ban shifting cultivation, declare forest reserves from which people are excluded, resettle people into the lowlands, and introduce settled agriculture either *in situ* or in a new location. The minor changes in land use that have occurred have involved a shift from swidden to monocultural tree crops — rubber and palm oil.

This suggests that two forces will increasingly determine land-use systems. First, national land tenure policies — the nationalization of forest lands and efforts to increase control over upland resources by the central and provincial governments — will make it increasingly difficult for farmers to maintain traditional swidden land-use practices. Second, market pressures—the commercialization of subsistence resources and the substitution of commercial crops for subsistence crops — will encourage farmers to engage in new and different forms of commercial agriculture and forest exploitation (Robinson and Bennett 2000). Combined, these "pushing" and "pulling" forces will eventually cause major shifts in land-use practices from swidden agriculture to commercial crops, and a change in land cover from secondary forest to commercial agriculture, most likely a monoculture.

This study also suggests that although swidden cultivation is responsible for maintaining a large portion of the landscape under some type of secondary vegetation, swiddening has also led to the fragmentation of forest cover, producing a highly heterogeneous and fragmented cover of secondary vegetation. This research suggests that the negative effects of swiddening on land cover (changes from homogeneous forest to highly heterogeneous cover of secondary vegetation) may be offset by the stability of swidden agriculture as the primary land use system in the region and the fact that swidden systems protect secondary forest cover. It supports the argument that true deforestation will occur through the replacement of swidden by permanent agriculture, if current land tenure and economic policies are not modified (de Jong 1997).

Chazdon (1998) suggests that tropical biodiversity conservation is undergoing a conceptual transition: the true value of isolated forest fragments, logged forests, and secondary forests is gradually being recognized. This study reinforces this new conservation paradigm that encompasses human-impacted lands (Schelhas and Greenberg 1996). Government officials and resource managers in Cambodia should respond in two ways. First, they should work to improve swidden systems through greater investments in research on maintaining the biodiversity of fallows, while also increasing their productivity and soil-sustaining properties. Second, they should facilitate the design and implementation of appropriate land tenure and economic policies that will empower local people to manage and utilize their own land and forest resources. In this way they can work toward preserving the ecological diversity and stability of human-forest relationships.

Finally, this study demonstrates the value of using participatory assessment techniques in collecting information in less developed countries. The social economic information collected in Khum Poey was not collected in a quantifiable manner nor from a random sample of farmers, yet the reiterative nature of the exercise as well as the logical and internally consistent story that has been pieced together assures us of the accuracy of the information. This information was collected in an on-going project to enable farmers in Khum Poey to make legal claim to the forestlands they have traditionally managed. This effort resulted in the formalization by community members of rules for using their communal forestlands. These rules were later endorsed by the heads of the District Provincial Forestry, Agriculture, Environment and Land Titles offices in the province. According to the Governor, this process represents a *de facto* recognition at the provincial and local government level for Ya Poey Community Forest. This contract approved a ninety-nine year forest protection concession for the Ya Poey Community Forest Association (Paterson 1997).

ACKNOWLEDGEMENTS

I would like to acknowledge financial support for my fieldwork in Ratanakiri from the Rockefeller Brothers Fund; The Ford Foundation; the International Development Research Center, Canada; and Associates in Rural Development, World Bank funded Forest Policy Reform Process Project. I am indebted to numerous people with whom I have worked in the field. In particular I am indebted to Gordon Paterson, the Non Timber Forest Project Coordinator for Novib and Oxfam (United Kingdom) in Ratanakiri, who organized many of the field trips and provided valuable insights into resource management in Ratanakiri. Fieldwork in Khum Poey was conducted with among others Vann Piseth, Sovanna, and Dom Taylor Hunt, of IDRC. Mr. Hor Hong of the Provincial Environment Department made his time and car readily available. Finally, I would like to thank the residents and commune and village chiefs in Khum Poey for their hospitality and guidance.

REFERENCES

Bahn, C. 1997. An economic analysis of tropical forestland use options, Ratanakiri Province, Cambodia. Singapore: EEPSEA

Bourdier, F. 1995. Health, women and environment in a marginal region of northeastern Cambodia. Cuban National Committee of the IGU: Environment, Society, and Development, Cuba, July 31-August 5, 1995.

Brookfield, H. 1988. The new great age of clearance and beyond. In J. Denslow and C. Padoch (editors). People of the Tropical Rain Forest. Berkeley: University of California Press.

Brown, S. and A. Lugo. 1990. Tropical secondary forests. Journal of Tropical Ecology 6:1-32.

Butterfield, R. 1997. Land use allocations and community livelihoods: Policy issues. Associates in Rural Development, Forest Policy Reform Process Project, Phonm Penh.

Chambers, R. 1994. The origin and practice of participatory rural appraisal. World Development 22(7): 953-969.

Chambers, R. 1985. Shortcut methods in social information gathering for rural development projects. In Putting People First: Sociological Variables in Rural Development, ed., Michael M. Cernea. Washington, D.C.: Oxford University Press, World Bank.
Chazdon, R. 1998. Tropical Forests—Log'Em or Leave'Em. Science 285:1295-1296.
Colm, S. 1997. Options for land security among indigenous communities in Ratanakiri, Cambodia. Non-timber Forest Products Project, Banlung, Cambodia.
de Jong,W. 1997. Developing swidden agriculture and the threat of biodiversity loss. Agriculture, ecosystems, and environment 62:187-197.
Fox, J., J. Krummel, S. Yarnasarn, M. Ekasingh, and N. Podger. 1995. Land use and landscape dynamics in northern Thailand: Assessing change in three upland watersheds. Ambio 24:328-334.
Fox, J., T. M. Dao, A.T. Rambo, P. T. Nghiem, T.C. Le, and S. Leisz. 2000. Shifting cultivation: A new old paradigm for managing tropical forests. BioScience 50(6): 521-528.
Global Witness. 1998. Going places: Cambodia's future on the move. March 1998. http://www.oneworld.org/globalwitness/reports/GoingPlaces/index.htm
Grandstaff, S. and T. Grandstaff. 1987. Semi-structured interviewing by multidisciplinary teams in RRA. In Proceedings of the International Conference on Rapid Rural Appraisal edited by Term Charoenwatana. Rural Systems Research and Farming Systems: Khon Kaen University, Khon Kaen, Thailand.
Guyer, J. and E. Lambin. 1993. Land Use in an Urban Hinterland: Ethnography and Remote Sensing in the Study of African Intensification. American Anthropologist 95(4): 839-859.
Kellman, M. and R. Tackaberry. 1997. Tropical Environments: The functioning and management of tropical ecosystems. London: Routledge Press.
Lebar, F., G. Hickey and J. Musgrave. 1964. Ethnic groups of Mainland Southeast Asia. New Haven: Human Relations Area Files Press.
Long Chun-lin, Jefferson Fox, Lu Xing, Gao Lihong, Cai Kui, and Wang Jieru. 1999. State Policies, Markets, Land-Use Practices, and Common Property: Fifty Years of Change in Yunnan, China. Mountain Research and Development 19(2): 133-139.
Matras-Troubetzkoy, J. 1983. Un Village en Foret: L'essartage chez les Brou du Cambodge. Paris: SELAF.
Moran, E., D. Brondizio, P. Mausel and You Wu. 1994. Integrating Amazonian vegetation, land-use and satellite data. BioScience 44(5): 329-339.
National Research Council. 1992. Sustainable Agriculture and the Environment in the Humid Tropics. Washington, D.C.: National Academy Press.
Nunes, C. and J. Auge, editors. 1999. Land-use and land-cover change (LUCC): Implementation Strategy. IGBP Report 48 and IHDP Report 10. Stockholm: IGBP Secretariat, Royal Swedish Academy of Science.
Paterson, G. 1997. Protection and management of forests by indigenous communities: The example of Ya Poey Forest, O Chum District, Ratanakiri, Cambodia, Paper presented at the IMC Workshop, Phonm Penh, 15-16 September.
Peluso, N. 1992. The political ecology of extraction and extractive reserves in East Kalimantan, Indonesia. Development and Change 23:49-60.
Poffenberger, M., editor. 1999. Communities and Forest Management in Southeast Asia. Berkeley, Ca. and Gland, Switzerland: World Conservation Union-IUCN.
Potter, L., H. Brookfield, Y. Byron (1994). The Sundaland region of Southeast Asia. In J. Kasperson, R. Kasperson, B. Turner II (Eds.) Regions at Risk: Comparisons on Threatened Environments. Tokyo: United Nations University Press
Robinson, J. and E. Bennett, editors. 2000. Hunting for sustainability in tropical forests. New York: Columbia University Press.
Reid, R. 2000. Land Use and Land Cover Dynamics in Response to Changes in Climatic, Biological and Socio-Political Forces: The Case of Southwestern Ethiopia. Landscape Ecology 15: 339-355.
Schelhas, J. and R. Greenberg, editors. 1996. Forest Patches in Tropical Landscapes. Washington, D.C.: Island Press.

Skole, D., W. Chomentowski, W. Salas, and A. Nobre. 1994. Physical and human dimensions of deforestation in Amazonia. BioScience 44(5): 323-328.

Walsh, S., T. Evans, W. Welsh, B. Entwisle, and R. Rindfuss. 1999. Scale-dependent relationships between population and environment in northeastern Thailand. Photogrammetric Engineering and Remote Sensing 65 (1): 97-105.

Young, O. 1999. IDGEC (International Dimensions Global Environmental Change) Science Plan: Institutional dimensions of global environmental change. International Human Dimensions Programme on Global Environmental change (IHDP) Report No. 9. Bonn Germany.

Xu Jianchu, Jefferson Fox, Lu Xing, Nancy Podger, Stephen Leisz & Ai Xihui. 1999. Effects of Swidden Cultivation, Population Growth, and State Policies on Land Cover in Yunnan, China. Mountain Research and Development 19(2):123-132.

Chapter 7

THE IMPACT OF LAND TITLING ON TROPICAL FOREST RESOURCES

Robert Walker[1], Charles H. Wood[2], David Skole[1], Walter Chomentowski[1]

[1]*Department of Geography & Basic Science and Remote Sensing Initiative*
Michigan State University, East Lansing, MI, USA
rwalker@bsrsi.msu.edu
[2]*Center for Latin American Studies, University of Florida, Gainesville, FL, USA*

Keywords: household surveys, land tenure, tropical forests, Brazil.

Abstract The impact of land tenure security on tropical deforestation and land management, more generally, is an issue that has received a great deal of discussion, and which possesses significant policy implications. If land tenure encourages planning over the long term, then the awarding of titles and the enforcement of private property rights may provide a decentralized approach to controlling rapid loss of forest. The economic theory on the role of tenure security on the management of assets such as land is well articulated, and in the case of tropical forest predicts a reduction in forest exploitation and destruction. Despite the conceptual appeal of the theoretical argument, little empirical work has been accomplished addressing specifically the role of secure land tenure on rates of forest clearance in tropical frontiers. The present chapter seeks to rectify this situation by combining data from a detailed survey of households with a time series of classified satellite images. In the summer of 1996, the authors conducted a survey along the Transamazon Highway in the state of Pará, Brazil, compiling information from 347 lots belonging to small-holders. In addition to the survey, they compiled a six year time-series of classified Landsat TM images for an 11-year period between 1986 and 1997 (1986, 1987, 1988, 1991, 1992, 1997). These images cover the survey area and provide a dynamic account of the land cover changes for a sub-sample of the lots queried in the original survey. The chapter gives an account of the survey sample, and results of a statistical analysis addressing the impact of title on the commercial exploitation of hardwoods and on rates of deforestation. Given knowledge of the year of titling and a series of images bracketing 11-years, the research considers, on a lot-by-lot basis, rates of deforestation before and after the awarding of title. The paper is concludes with a discussion of the implications of findings for policy.

INTRODUCTION

Tropical deforestation represents a major threat to local, regional, and global environments given the widespread ecological changes that occur with forest clearance. Such changes often represent biomass and species reductions, particularly when human-dominated landscapes replace natural systems. Even when forest lands are sustained under low intensity fallow, with long-term regrowth of secondary vegetation, biomass and diversity of tree species are reduced from levels found in old growth and primary forest (Uhl et al. 1988, Viera et al. 1996). Nowhere is the problem of deforestation more serious than in the Amazon basin, which suffers the largest forest losses worldwide and is the focus of the present chapter. The Brazilian portion of the basin alone has experienced substantial deforestation over the past several decades, losing between 10,000 and 20,000 km^2 of tropical moist forest annually (e.g., Skole and Tucker 1993, INPE 2000).

One key to controlling environmental impacts associated with tropical deforestation is to reduce its rate through government policy. Such policies tend to be enforcement-based, as with state protection of natural areas, or they seek to promote incentives leading to environmentally sustainable behavior. This later category includes investments in agricultural intensification and the identification of low impact agriculture and extraction activities, as well as the construction of carbon sequestration markets.

The present chapter addresses one such incentive-oriented approach that has been advanced as a general solution to a wide range of local-level environmental problems, and in particular resource depletion, namely the provision of secure land tenure to resource-dependent farmers (Mueller et al. 1994). The theoretical arguments that have been offered in support of this position are cogent, but little empirical work has been done to test the idea, which we refer to as the Land Tenure Security (LTS) hypothesis. Thus, the primary goal of this chapter is to assess elements of the LTS hypothesis within the context of tropical forest conservation, and to this end it describes the results of an analysis of survey data taken from small producers in an active frontier area of the Brazilian Amazon.

A secondary goal, however, is to illustrate a methodological approach to conducting research in tropical forest regions where the necessary data may not be available for the asking. In such a situation, the task of obtaining and organizing information requires as much energy and ingenuity as conducting the final analysis and interpreting the results. As shall become apparent in this chapter, attempting to answer the question of the role of land tenure in tropical forest change required an unusual combination of data sources. In particular, we found it necessary to rely on survey research methods from the social sciences, as well as on a geographic information science (GISc) application involving classification of remotely sensed images and data base functions linking survey and satellite information. Only in this way were we able to produce data needed for the analysis. Such integrated approaches may hold the methodological key to answering some of our most challenging questions about global environmental change.

The chapter is organized as follows. The introduction is followed by a discussion of the main theoretical issues involving the link between land tenure security and resource use behavior. Previous research on the subject is discussed, both from theoretical and empirical perspectives. In addition, shortcomings of the LST hypothesis are considered with respect to protection of the environment. The methodology describes "what" and "how" things were implemented in the research, which combined a household survey with data from remotely sensed satellite images. The sample and results from the statistical analysis are described; conclusions complete the chapter.

Tenure security can affect many farm decisions as shall become apparent in the discussion, and it is not our intention to provide an exhaustive empirical test on all counts. Rather, we consider two issues commonly mentioned in this regard, namely the exploitation of hardwood by the rural poor, and deforestation resulting from land occupation. The stylized fact to be addressed is that land tenure security reduces impulses to both log hardwoods and to clear forest at an excessive rate.

PROPERTY RIGHTS AND ENVIRONMENTAL CONSERVATION

Land Tenure Security: Incentives

Interest in the link between property rights and resource conservation has a long history in resource and environmental economics. In early work on open access resources, Gordon (1954) developed a model for the fisheries case that demonstrated complete rent dissipation over the long run, with total costs exhausting total revenues at an industry level due to the lack of barriers to entry. Later, Clark (1976 and 1985) showed how rent dissipation could produce catastrophic impacts on fisheries, with ultimate collapse of the resource base. The free entry problem of open access was transformed into powerful metaphor for the land case by Hardin's "tragedy of the commons," which provided an ideological argument for the utility of private property in solving problems of resource depletion and environmental degradation (Hardin 1969).

Hardin's formulation has sparked widespread reaction and theoretical revision (e.g., Ellickson 1991, Ostrom et al. 1991). Indeed, anthropologists have long observed non-privatized property institutions that do not deplete resources (Hames and Vicker 1983, Dove 1986, Posey and Balee 1989, Moran 1990). In particular, "common" property allows for exclusionary use by well-defined groups, and thereby overcomes the rent dissipation problem of open access, even with low productivity due to the absence of land markets (Feder and Feeny 1991, Beaumont and Walker 1996).

Despite such empirical and theoretical observations, consolidation of private property in land has been viewed as a vehicle to economic development due to the salubrious effect it has on landholders, at least theoretically (Feder and Feeny 1991). With well-defined rights, often expressed in the legal document of the *title*, a farmer -- free from the threat of eviction -- can derive income from land, and can sell, mortgage, or use it to obtain credit. For reasons such as these, owners are generally more likely to make investments than they would be in the absence of security in their holdings (Feder et al.1988, Eggertsson 1990, Mueller et al. 1994). An additional impact of secure land tenure is resource conservation (Mueller et al. 1994). Resource conservation in this context necessarily refers to renewable resources, since the optimal use of non-renewables dictates hasty and complete exploitation, in the absence of price increases for the resource commodity (Hotelling 1931). For small holders in tropical frontiers, the resources in question are mainly soils and forest cover, both of which are potentially renewable if properly managed.

Degradation of soil resources is often cited as an impetus to tropical deforestation through the process of *invasive forest mobility* (Leslie 1980, Meyer 1980, Schmithusen 1980, Office of Technology Assessment 1984, Ross 1984, Walker 1987, Repetto and Gillis 1988). By this story, lack of land ownership -- interacting with poverty -- encourages the *mining* of soil nutrients with attendant fertility decline and loss of soil structure. Presumably, secure land tenure mitigates this impact by inducing the use of modern inputs to prolong soil productivity, thereby reducing destructive land uses and resulting mobility. A secure property right could also bring about land improvements or switches in the farming system to perennials or tree crops that are less consumptive of soil nutrients (Arnold 1983).

Forest resources and their management present a more complicated picture given that forest cover represents more than one value to the farmer. As a source of fuel wood, game habitat, and microclimate control, forests are important to the household economy of the farm, and the welfare of its occupants. At the same time, continuous tree cover inhibits the use of land, the primary agricultural production factor. Trees must be cleared to enable farming, and in many cases to secure a holding through the demonstration of *productive* land use. Moreover, the high *market* value of hardwoods may represent an irresistible temptation to a poor farming household in need of liquid capital for farm investments. The exploitation of tropical hardwoods, much of it from smallholder properties, has become a major issue in discussions on deforestation in the Amazon basin, the focus of the present chapter (Nepstad et al. 1999).

From a pure profit maximization perspective, the physical growth rates of tropical hardwoods are probably too low to protect them from the liquidation incentives associated with non-renewable resources[1]. But it must not be forgotten that smallholder households are family farms, not corporations, and their optimization behavior includes consideration of welfare dimensions that would be of little import to a commercial venture (Nakajima 1969, Barnum and

Squire 1979, Thorner, Kerblay, and Smith 1986). The desire of small holders to leave a piece of land, with a valuable forest bequest, must be factored into the equation when assessing the impact of land tenure security. Only through defendable private property rights can such a desire be fulfilled.

Hardwood conservation on the one hand, and maintenance of forest cover on the other, do not represent the same resource management impulse. Nor do they necessarily lead to a consistent environmental outcome. Conserving valuable hardwoods need not translate directly into a reduced amount of deforestation, and the conservation of forest values such as maintenance of microclimate and game habitat. On the other hand, valuable trees are often taken selectively given their low density, spatial distributions, at least in the Amazon basin (Verissimo et al. 1992). As a consequence, hardwood depletion (of marketable species) may leave behind a fairly intact forest, at least from a gross measurement perspective. Unfortunately, much forest cover is "valueless" in a narrow commercial sense. Since game species tend to vanish early in the colonization process due to over-hunting, and since fuel wood can be taken from secondary regrowth, incentives to preserve the spatial extent of forest may be quite weak, and insufficient to overcome the agricultural necessity for productive, treeless, land. The analysis that follows addresses the role of land tenure security on the maintenance of both productive hardwoods and the *extent* of forest cover.

Land Tenure Security: Outcomes and Contradictions

While the hypothesized impacts of secure land tenure on resource conservation make good theoretical sense, there are several ways in which contradictory outcomes might be expected to occur from the perspective of environmental protection. First to note is the long-standing requirement for land occupation in the Brazilian Amazon that "productive use" be demonstrated prior to the emission of title. In some respects, this institutional quirk has turned the theoretical reasoning of the LST hypothesis on its head, by making the acquisition of title casual to initial deforestation, at least during the initial phase of land occupation in *terra devoluta*, federal lands not marked for colonization but "occupiable" under the principle of *usocapião*. The demonstration of productive use was originally intended to reduce the speculative acquisition of unused land. Despite efforts to change elements of the law in Brazil, we continue to observe such demonstrations in land occupation. Indeed, colonists have a word for it, the *marcação*, a large patch of deforestation meant to show that a piece of land has been spoken for. Under the principle of usocapião, an individual who occupies an unused parcel of government land for a year and a day can apply for land title.

Other ways in which land tenure security and environmental outcomes might stand in contradiction relate to issues of scale, in both spatial and technical dimensions. The LTS hypothesis focuses on individual land managers, and necessarily is framed within a partial equilibrium context. Should the institutional environmental evolve on a forest frontier to one ensuring secure property rights, the gains in forest conservation on individual holdings under the LST could be wiped-out in the aggregate due to the general equilibrium effect on factor movements and, in particular, migration. To the extent that securing ownership increases the rent potential of individual sites through investments and reduced costs of claim enforcement, the extensive margin of agriculture is likely to advance into undeveloped lands according to the dictates of von Thünen (Nerlove and Sadka 1991). Adjustment to a general equilibrium at regional or national scale would presumably induce new migrations to frontier areas, and more land could be cleared in the aggregate even with reduced deforestation at local level on individual parcels.

A second spatial issue is linked to the trans-boundary externalities generated by certain forms of resource conservation. Land tenure security might well encourage soil maintenance through the use of modern inputs such as fertilizers and pesticides, thereby reducing rates of migration to forested areas. However, the positive effect for soils could be bought at the price of environmental pollution should the farm chemicals find their way into local streams and groundwater.

Another possible contradiction between forest conservation and secure land tenure in the Amazonian case results from the scale economies of production. Although poor farmers without title are often portrayed as shifting cultivators penetrating and degrading very large tracts of land in an endless cycle of forest loss (Meyers 1980), Walker (1999) has shown that the land demands of such systems are not that great. In the Amazonian setting, more heavily capitalized operations make a much more significant deforestation impact, per enterprise (Walker et al. 2000), particularly in an agro-ecological environment providing strong scale economies in ranching (Mattos and Uhl 1994). To the extent that secure land tenure enables the development of such enterprises through the protection of investments, it contributes to deforestation.

The present chapter presents results of what is best regarded as a partial equilibrium analysis. In particular, we address the management of individual farm properties, and do not consider the impact of institutional change on the advance of the frontier. Moreover, we focus on small holders, and do not address the role played by secure land tenure in promoting the creation of large ranches through property aggregations. We argue that a partial equilibrium approach is necessary given the spatial scale issue of adjustments to general equilibrium, and that a focus on small holders removes confounding effects possibly associated with the wealth endowment of the agents, which might affect discount rates.

As for the various elements of the LTS hypothesis, we limit ourselves to assessing only the claim regarding a link between land tenure security and resource conservation (Mueller et al. 1994). In this regard, we focus on the conservation of forest resources, both the maintenance of commercial hardwood and the extent of forest *cover*. Prior research has addressed soil conservation, showing that tenure security influences use of modern inputs and land improvements (Feder et al. 1988, Place and Hazell 1993).

METHODOLOGY

The methodology employed was statistical in nature, using data generated from a social science survey and an application of Geographic Information Science to remotely sensed imagery. The survey addressed the activities of small producers and was undertaken in the summer of 1996, in the municipio, Uruará (1996 population: 11,194), State of Pará, Brazil. Three teams of two individuals each administered a 20-page survey during the month of July. In all, 261 proprietors were interviewed, covering 347 individual lots. The lot is the physical entity of approximately 100 hectares adopted by the Brazilian colonization institute, INCRA, in the days of land distribution, to be allocated to individual households. Since the 1970s, some land concentration has occurred, as reflected in our data containing information on 261 properties, or economic entities under the control of one owner, who altogether possess 347 physical lots of about 100 hectares each. In the analysis and data presentation to follow, we identify the unit of observation as a lot or property.

Uruará is located on the Transamazon Highway between Altamira and Ruropolis, an old settlement frontier originally opened in the early 1970s with road building. The survey obtained information on ownership status, household demography (including family size, age composition, and migration behavior), durable goods possession, farming system components, and agricultural technology. The sample was collected from properties on the main highway and from travessões, crossroads running north and south off the main axis of the Transamazon to distances of nearly 100 kilometers in some places. The zone of survey activity covered about 100 kilometers, stretching east and west of Uruará, and may be observed in Figure 1, a mosaic of images covering the central part of the State of Pará, in the eastern Amazon basin.

The sample possesses a large random component, but does not completely exclude certain systematic biases inherent in the sampling process. We base this observation on an assessment of the sample composition process. Efforts at random sampling of small producers by distance and by household count proved intractable. Counting houses is difficult given that proprietors are often absent; moreover, structures may be hidden from view. Distance sampling is complicated by the fact that a residence may not appear at the distance specified for a sampling stop. In practice, we sampled according to a principle of "first opportunity," together with an attempt at geographic representation. To this end, we consulted maps to identify under-sampled areas, then travelled to those areas, conducting interviews when we could find a proprietor or knowledgeable resident on a lot. Daily updates of the geographic coverage of the interviews enabled us to obtain a widely dispersed sample across the municipio. Houses with absent proprietors or knowledgeable individuals, or difficult of access, are mainly omitted from the sample.

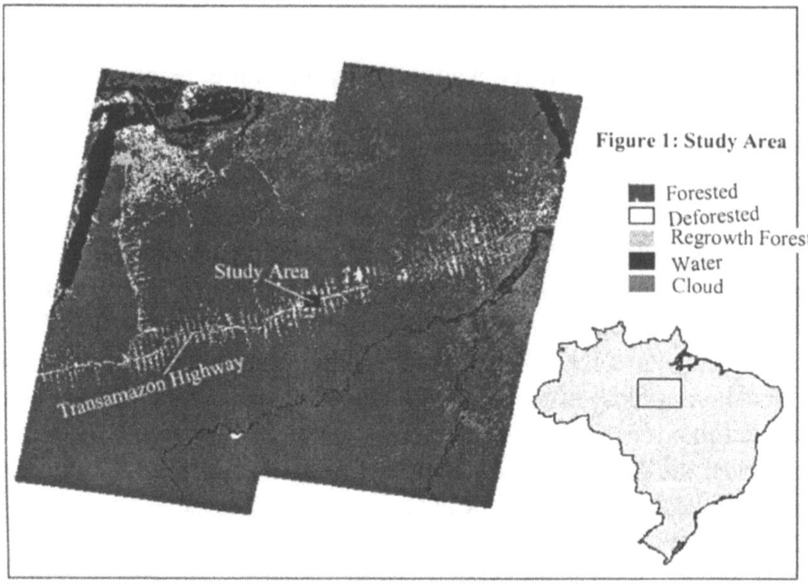

Figure 1. Image mosaic of survey study area.

In Uruará, an absent proprietor is not to be confused with an absentee landlord, in which case we do not expect a systemic difference between the sample and those excluded due to "absence," at least along dimensions of household wealth. A typical absence was for personal family reasons, or to head to town for a purchase or sale. On the other hand, "difficult of access" properties would probably show a systematic difference with the collected sample, since they are typically occupied by extremely poor families. Thus, the sample is reasonably random within a population excluding those that are at the very margins of existence and food security. As indicated, large land holdings have been excluded. This was because such properties tend to be titled. The study area is mainly composed of colonists with properties consisting of the 100 hectare lots made available by INCRA. Nevertheless, six aggregated holdings in excess of about 1500 hectares are apparent on satellite images covering the area (Walker et al. 2000).

General demographic characteristics of the sample are as follows. Average family size is 7.12 individuals, with 2.85 children. The age of the household head is 47 years, and he or she possesses 2.02 years of schooling. Households consume most of their corn (60 %) and bean (70 %) production, and use very little hired labor (28 person-days per year). These statistics suggest the sample is drawn from subsistence households with imperfect market attachments, consistent with the notion of a peasantry (Ellis 1993, Walker et al. forthcoming).

Data from the Geographic Information Science Application

The survey results were integrated with land cover data derived from satellite images. This was accomplished as follows. First, a map of the property boundaries was obtained from the Brazilian colonization agency (INCRA). This was digitized into a cadastral layer, and entered in vector format into GIS software. The map contained property numbers associated with colonization, which were also identifiable on the survey responses. In this way, it proved unnecessary to use GPS to identify the location of the properties on the cadastral layer. The map was field verified, and geo-referenced by direct observation of features in the landscape, such as road intersections.

We next collected a time-series of Landsat TM imagery from path 226/row 63 and path 227/row 63, covering most of the area of the survey activity. These images were geo-referenced and co-registered, then used to classify land cover, using an unsupervised algorithm, into five classes: forest, regrowth, deforested land, clouds, and water. The satellite-based overages, generated in raster format, were then converted into vector format for compatibility with the digital property map.

The GIS overlay function was used to produce the cover classes at the lot level. To this end, the cadastral layer was overlain on the classified images. The data on the class covers for individual lots were output to files readable by the statistical software. For the statistical approach, the cover and survey data were combined. It proved unnecessary to mosaic the images, given a 10 percent coverage overlap and the relatively small size of the properties. Nor was it necessary to use GPS in delimiting property boundaries, given the regular settlement geometry of the lots, readily discernible on the satellite images. Ultimately, excessive cloud cover restricted the analysis to 1986, 1987, 1988, 1991, 1992, and 1997.

DESCRIPTIVE STATISTICS AND FINDINGS

Descriptive Statistics

Table 1a and 1b gives the basic information on the titling status for the full sample. Land title is used in the analysis as a proxy for land tenure security, which is mainly unobservable as it reflects the psychological state of an owner. The unit of observation given in the table is for individual lots, and the data show title type, occupation times, and time to title emission. A title classification can be *definitive*, meaning the proprietor has in his or her possession the formal land title document conferring possession; or, it can be *provisional,* which in Uruará is typically an authorization to occupy land, or a recognition that land has been occupied (See Alston et al. 1995 for a discussion on the titling process in contemporary Brazil). Fully 202 lots, or 58 percent of those sampled, were in possession of definitive or provisional title, and of the titled lots, a majority (67%) possessed definitive documentation. The time to receive a definitive title averages 2.18 years in the sample, and occupation times are 12.29 years for lots with definitive titles and 6.97 years for all others, including lots with provisional titles or no documentation whatsoever.

Table 1a. Titling status of survey lots: titling institution[a]; provisional title type[b].

	Universe	Titled Lots	INCRA	ITERA	Authoriza-tion	Recogni-tion
UN-TITLED	145 (42%)					
TITLED	202 (58%)		154	1		
Defin-itive title		135 (67%)				
Provisio-nal title		67 (33%)			45	21

[a]Of the 202 titled lots, 47 do not report titling institution. [b] Of the 67 provisional titles, 1 does not report provisional title type.

Table 1b. Titling and occupation times.

	Occupation Time	Time to Receive Title
TITLED LOTS (definitive, with documents)	12.29 years (N = 200)	2.18 years (N = 141)
PROVISIONAL & UNTITLED LOTS	6.97 years (N= 138)	not observed

Tables 2 and 3 characterize the households in the sample by type of farming system, and demonstrate that the Uruará site is quite typical of the Amazon basin. Table 2 gives physical breakdowns in harvest volumes and areas planted for the main regional crops. Among annuals, rice production dominates, with a gross production of about 3000 kilograms per property (2266 per lot). Given average rice areas of a little over three hectares, this indicates low productivity yields on the order of 1000 kilograms per hectares. Cocoa production accounts for the greatest volume of perennials production, with about 800 kilograms produced per property (613 per lot), followed by pepper (658 kg/property; 495 kg/lot) and coffee (394 kg/property; 296 kg/lot). Pepper appears to be in decline, while coffee is growing in importance[2]. The sample also reveals an impulse to pasture conversion. The average amount of pasture per property is 29 hectares, supporting 24 animals (22 hectares per lot, with 18 animals on average). These data appear broadly consistent with the diversified farming systems found throughout the Amazon basin associated with small producers. Such systems typically include annuals, perennials, and cattle, and the dominant land cover component is pasture (Dale et al. 1993, Pichón 1997, Vosti et al. 1998).

Table 2. Production systems by component: averages for properties and lots.

COMPONENT	Properties (N=261)	Lots (N=347)
Rice Production (kg)	3007 (3.06 ha)	2266 (2.31 ha)
Bean Production (kg)	284 (.68 ha)	214 (.51 ha)
Cocoa Production (kg)	816	613
Coffee Production (kg)	394	296
Black Pepper Production (kg)	658	495
Total Pasture (ha)	29	22
Cattle	24	18

Table 3 further characterizes the sample by presenting values associated with system components (annuals, perennials, and cattle), aggregated to property level for the multiple-lot holdings. Evidently, annuals production generates the most income across the sample of properties, 2900 reais on average (1$ = about 2 reais), out of a total yearly revenue of 5417 reais. The average of the

Table 3. Revenue components: averages over properties and lots, in Reais.

	Properties	Lots
Annuals[1]	2900 (.57)[2]	2186 (.53)
Perennials[3]	2063 (.30)	1549 (.32)
Cattle[4]	517 (.13)	388 (.15)
Total	5417	4070
Land Value[5]	27657	20907

[1]Summing annuals revenue over production of rice (.70 reais per kg), beans (.75 reais per kg), corn (.30 reais per kg); prices are producer prices for 1996 provided by EMATER, State of Pará.
[2] Number in parentheses is average for proportion of total revenue provided by the particular activity, annuals production, perennials, or ranching.
[3] Summing perennials revenue over production of cocoa (.85 reais per kg), coffee (1.12 reais per kg), and black pepper (1.40 reais per kg); same source and year as annuals, except for black pepper calculation, which uses 1995 price.
[4] Assumes young bull weight of 200 kg, a 20 percent offtake rate for herds, and a price of .55 reais per kg carcass.
[5] Land values are self-reported estimates of sales price.

percentage of total revenue accounted for by annuals is 57 percent[3]. Self-reported land values for properties are 27,657 reais on average, or 20,907 per lot, yielding an imputed land price of about 209 reais per hectare. The fact that land has value indicates that incentives exist to establish property rights, with land tenure security. The low value is consistent with conditions of an active frontier, where variation in the strength of the property institution can be expected at individual property level. Such a setting is necessary to the research design, which requires variation in this measure.

Links Between Titling and Logging

Table 4 gives information on the resource use behaviors considered in the analysis pertaining to logging, namely hardwood conservation (depletion) and timber market participation (non-participation). The analysis on deforestation, or extent of forest cover, follows in the next section. The data in Table 4 are frequency counts of lots by category, as well as mean calculations of individual trees sold (self-reported); thus, individual properties including multiple lots are counted more than once. The data presented are for cells reflecting two, two-way classifications, one for each of the resource use behaviors. The individual classifications themselves are based on the resource use categories and on titling status, which is addressed using two definitions. In the first instance, title is taken as the presence of a definitive title, and lack of title is lack of definitive title; secondly, title is defined more broadly to include both definitive and provisional titles, and "no title" refers to the condition of absence of both definitive and provisional title (See Table 1).

Table 4. Observed frequencies.

	Definitive Title	No Definitive Title	Title[1]	No Title
Forest Conservation	122 (95%)[2]	176 (94%)	186 (96%)	112 (93%)
Forest Depletion	6	11	8	9
Market Participation	13 (4.25/2.54)[3]	28 (2.98)	24 (4.16/3.0)	17 (2.46)
Non Participation	73 (85%)	96 (77%)	110 (82%)	59 (78%)

Title/No Title indicates response to question, *does lot possess title*, as perceived by owner. As such, this category distinguishes between individuals with definitive or provisional title, and all others.

[2] Percentages in parentheses are based on frequencies showing behavior within title category; for example, 95 percent of lots with definitive titles evidence "forest conservation," defined as the current possession of valuable tree specimens on the lot.

[3] The parenthetic numbers for the market participation row show average number of trees sold last year, by title category. The two numbers for the definitive title and title category reflect averages calculated with and without the outlier of 150 trees sold from a titled lot.

A forest conservation outcome is defined as the current existence of valuable trees on a lot. Counts are conditioned on the original status of tree stands, namely the early existence of valuable timber. Any lot depleted prior to arrival was excluded in order to isolate the use activities and outcomes of current owners. The top two rows show strong forest conservation impulses across all categories of titling, with slight variation (between one and three percent) attributable to actual titling status. Among those lots with definitive title, 95 percent continue to possess valuable trees, and the percentage climbs to 96 percent when provisional titles are included. Alternatively, lots lacking definitive title show a conservation rate of 94 percent; those lacking both definitive and provisional title conserve at a rate of 93 percent. In general, relatively few lots in any of the title categories report having depleted their forest reserves since arrival.

Non-participation in recent timber sales (the preceding year) also shows relatively high frequencies overall, although percentages across the various title categories are less than observed in the forest conservation case. The same pattern observed for forest conservation is evident here, with relatively more titled lots not engaged in market activities than untitled lots. Nevertheless, the separation between the categories in relative frequency is considerably stronger, between five and ten percent depending on the title definition. Eighty-five percent of the lots with definitive titles did not participate, while the comparable relative frequency for lots without definitive title is 77 percent. For the broadly conceived title definition, the relative frequencies are 82 percent for titled lots, and 78 percent for untitled lots. Data on average number of trees sold per lot, by title category, is presented in the market participation row. The averages range between about 2.5 and 4.0 across the titling categories. When an outlier of 150 trees sold from a titled lot is included in the calculations, the average on titled lots tends to be larger than untitled lots, for both definitions of titling. Exclusion lowers the average taken from definitely titled lots below the sales amount from untitled lots.

Table 5 extends the analysis of Table 4 with the results of binary logistic regression. As before, individual lots are the unit of observation, and two behaviors are considered, namely conservation of valuable trees and non-participation in timber markets (during the previous year). The logistic model is implemented by taking the resource use behavior as the dependent variable, which is dichotomous for both conservation (verses depletion) and non- market participation (verses participation). Independent variables were selected to control for household circumstances that might affect the resource use outcome such as length of residence, distance from the main highway, age of the

Table 5. Logistic results[1].

		Coefficient	Significance
Forest[2] Conservation	Definitive Title (Δ = .09)	.00118 (1.001)	.9976
	Any Title (Δ = .11)	.9338 (2.544)	.0178
Market[3] Participation	Definitive Title (Δ = .03)	.6636 (1.942)	.1281
	Any Title (Δ = .02	.5337 (1.705)	.2215

[1]Odds ratio is in parentheses under coefficient value; Δ is under title definition.
[2]Control variables include length of residence, distance from main highway, age of household head and years of schooling, and initial wealth of family.
[3]Control variables same as conservation case, with wealth variables omitted due to estimation sample size.

household head, years of schooling, and wealth status of the household defined on durable goods possession. Results are presented for two dummy variables indicating two types of land titling status. A lot achieves definitive title status when it possesses a definitive title. Alternatively, any title status is achieved when the lot possesses either definitive or provisional title.

The results are consistent with the LTS hypothesis in that the coefficients for the title variables are all greater than zero, leading to odds ratios that area greater than one. This indicates that titled properties are more likely to engage in conservation behaviors than untitled properties. Significance, however, only occurs when the titling definition is sufficiently broad to include those with provisional authorization to occupy land, given in the table as any title. Here, the result is reasonably strong for long-term conservation, but only suggestive for non-market participation behavior. The significance levels, however, encompass a rejection region with two-tails (Hosmer and Lemeshow 1989), in which case they may be halved if the prior hypothesis of a positive impact is assumed. This strengthens the findings, although the ρ values for the individual regressions are somewhat low. Such values are not expected to be very high (Mertens and Lambin 2000).

Links Between Titling and Deforestation

Tables 6, 7, and 8 address the issue of *deforestation* as opposed to logging. Here, the approach taken is different given measurement issues. In particular, it was necessary link lot level data with classified Landsat TM satellite images, since self-reported magnitudes of deforestation may be understated (Walker et al. 2000). The same may be said for the logging data, but these would be impossible to obtain through remote sensing.

The analytical design is that of a panel analysis for individual lots, for which we have both title information and a set of classified images. We attempt to measure the rates of deforestation before and after title acquisition, *on the same lot*. Although it would have been preferable to use a complete data series, with an image for each year, cloud cover prevented this. Due to the resulting data limitation, the deforestation analysis must be interpreted in a cautionary light. Although there are 347 lots in the actual survey, a fairly large number were located outside the bounds of the INCRA map used to locate them on the satellite images. Moreover, the nature of the analysis required that each lot be cloud-free for each year in the series. One hundred ninety-one lots meet this analysis criterion.

Table 6 shows deforestation for the analysis periods, in both absolute and yearly magnitudes. Deforestation was measured as the change in *Forest* cover between successive years, and rates were calculated by dividing this amount by the number of intervening years. In the upper panel are average yearly rates, which vary between 1.14 and 2.49 hectares per year for the 191 lots in the sample. The lower panel shows the total deforestation for all pairwise combinations of the available satellite data. As can be observed, the average amount of deforestation is 17.72 hectares for the entire analysis period covering eleven years.

Table 6. Average deforestation and magnitudes.

Average Deforestation on Survey Lots

Period	Yearly Average
1986-87	2.22 ha
1987-88	1.46 ha
1988-91	1.14 ha
1991-92	2.49 ha
1992-97	1.62 ha

Deforestation Magnitudes of Specified Intervals

Period	Absolute Magnitude
1986-87	2.27 ha
1986-88	3.68 ha
1986-91	7.13 ha
1986-92	9.62 ha
1986-97	17.72 ha
1987-88	1.46 ha
1987-91	4.90 ha
1987-92	7.40 ha
1987-97	15.49 ha
1988-91	3.44 ha
1988-92	5.93 ha
1988-97	14.03 ha
1991-92	2.49 ha
1991-97	10.59 ha
1992-97	8.09 ha

Ideally, we would like to measure deforestation both before and after the award of a title on the same lot. This requires that (1) we can measure the rate of deforestation before title; and (2) that the title, in fact, be definitive. Otherwise, we cannot measure an "after title" rate of deforestation. An ideal panel sample would consist of lots both acquired and titled after 1986, because only then can we measure rates before and after the titling event with any degree of exactitude for the lots in question. This requirement, however, further restricts sample size, since many of our sample households have been in the region for quite a while. The upper panel of Table 7 shows only 14 lots meet these various criteria, and that the yearly rate of deforestation is actually greater *after* title acquisition (1.91) than *before* (1.52).

Table 7. Deforesation with and without definitive title.

Deforestation on Lots with Definitive Title	
(yearly average hectares)	
Before Title Acquisition [1]	1.52 (n = 14)
After Title Acquisition (lots occupied 1986 and after)	1.91 (n = 14)
After Title Acquisition [2] (lots occupied prior to 1986)	1.48 (n = 48)
Deforestation on Lots without Definitive Title	
Before Title Acquisition[3]	1.82 (n = 86)

[1]To calculate "before title acquisition" deforestation on titled lots, it is necessary to limit the sample to those acquiring lots in 1986 or after, the period covered by satellite images. Deforestation before 1986 is not observed.
[2]Deforestation is yearly average after 1986. Titles may have been obtained a number of years prior to 1986, in which case the repeated value only reflects part of the post title acquisition period.
[3]Deforestation on lots without title by definition is "before title acquisition."

This restricted sample can be augmented by adding two type of lots, those that received title prior to 1986, for which deforestation *after* title reception can be computed in the post-1986 period, and lots **without** title, for which deforestation *before* title acquisition can be computed, also in the post-1986 period. Note, however, that given the start date of the image series, the deforestation rates so calculated are yearly averages **only** for the period, 1986-1997. These samples show deforestation to be somewhat higher on the untitled lots, although the average rate of deforestation (1.82 hectares per year) is less than what is observed *after* title acquisition in the limited panel (1.91 hectares per year; n = 14). Overall, these data do not suggest that land titling, in and of itself, reduces rates of deforestation.

The deforestation magnitudes were further investigating inferentially by T-test and regression analysis, as presented in Table 8. Both titled and untitled lots were included, acquired before and after 1986. The T-test on the difference of mean annual deforestation rates does not reveal any difference whatsoever, and the rates are quite similar, between 1.70 and 1.80 hectares of deforestation annually. The lower panel of the table further examines the impact of title by controlling for other key variables known to affect deforestation, such as period of ownership, distance from highway (The Transamazon Highway), family size, and use of hired labor. The regression analysis focuses on the definition of a dummy variable for *before* (=0) and *after* (=1) title acquisition (also using titled lots acquired prior to 1986 as well as those that remained untitled as of the time of survey, 1996). Note for the analysis data base, only 14 lots generated observations *before* and *after* title acquisition. The titled lots acquired prior to the analysis period contributed only *after* title observations (dummy = 1) for the period 1986-97, and the untitled lots gave only *before* title observations, for the same period. The results of the analysis are presented in the lower panel of Table 8.

Table 8. Regression results.

T-Test on Average Deforestation Rates Before and After Title

	Average Rate	Standard Deviation
Before Title	1.78 (n = 100)	1.66
After Title	1.70 (n = 62)	1.43

T-values (probability)	Variance Assumptions
.319 (.749)	Unequal
.308 (.758)	Equal

Regression of Yearly Average Deforestation

Variable	Beta Coefficient (α-probability)
"After Title" Dummy	.285 (.338)
Years owned	-.083 (.002)
Distance from Highway	-.011 (.394)
Family Size	.0091 (.736)
Hired Work	.002 (.493)
$R^2 = .07$	F = 2.014 (.067)

As with the T-test, these show no significant effect for title on rates of deforestation. The fit of the model is low overall, and only the length of the period of ownership is significant, suggesting a slowing down in deforestation with time, as would be expected under the life-cycle dynamics of a household (Walker et al. 2000, Walker et al. forthcoming). The other control variables all function as would be expected, although they are insignificant. The rates of deforestation decrease with distance from the main highway, but increase with available workforce.

DISCUSSION AND CONCLUSION

The results suggest that a link exists between land titling and resource conservation behavior, and that this link is consistent with theory. In particular, land tenure security is associated with forest conservation and non-participation in timber markets. It is important to keep in mind that in the stronger set of findings, land tenure security is defined broadly as the possession of a definitive or provisional title. Evidently, security emerges once some documentation exists for the land parcel. Unfortunately, the results do not isolate the effects of land tenure security associated with freedom from eviction and freedom to sell or mortgage land. Nevertheless, the broadly conceived titling definition includes those who cannot legally sell, but are probably free from eviction fears, namely lots with some provisional form of title. Given that significant findings emerge only under this definition, this would seem to suggest that resource conservation arises on the basis of long-run attachment to the land.

The results on hardwood management do not translate into rates of deforestation. As we have argued, there are a number of reasons why such an outcome might be expected, and each of these reasons could affect our analysis. On the other hand, the results may also reflect the inherent limitations of sample size, in which case our statistical tests do not possess sufficient power to be able to reject the null hypothesis of no effect.

In addition, we cannot rule out the strong possibility of errors arising from measurement. Lack of a sequential series means we have had to infer rates for lots whose titling and acquisition dates to not conform exactly to the time points in the series (1986, 1987, 1988, 1991, 1992, and 1997). For example, consider a lot acquired in 1982 and titled in 1989. We estimate *before* title deforestation as the rate calculated from 1986 to 1988, and the *after* title deforestation as the rate calculated from 1988 to 1997. Unfortunately, our image series does not possess the 1982, 1989, or 1996. Although we have no way of gauging the magnitude of errors associated with such a temporal mismatch, we suspect they are important and that the LTS hypothesis requires further testing for the case of deforestation.

But we must also consider the possibility that secure land tenure has little impact on deforestation -- as opposed to logging -- among small holders. This is not an inconsistent outcome, given tree distributions are often patchy, in which case hardwoods may be preserved at the same time as extent of forest is reduced. Indeed, it would seem that small holders have a strong economic imperative to create a land *factor* upon arrival on a forested lot, given subsistence requirements of the household. Moreover, land tenure security could trigger excessive rates of clearance by the paradigm itself, since investments are greatly facilitated with land title. Given the initial expense and widespread preference for ranching even among small producers, land tenure security may actually be implicated in the current expansion of pasture in the Amazon basin. If so, then land titling does not hold out much hope as a solution to the problem of forest loss in the Amazon frontier. Of course, reduced rates of logging are highly desirable, especially with the observed link between selective logging and fire contagion (Nepstad et al. 1999).

Several important issues remain to be addressed. Foremost among these is the existence of an effect of land titling on forest clearance. Our analysis leaves this question unanswered, but does suggest that increased sample size could resolve the issue one way or the other. The spatial concentration of the sample over a 100 kilometer stretch of highway meant that years with clouds affected relativity high numbers of households. For the case of the Transamazon Highway, one approach would be to implement this same approach for a larger stretch of highway, thereby keeping large numbers of households in the estimation sample even during relatively cloudy years.

The second issue concerns the relative effect of land tenure security on logging and deforestation. As noted, land tenure security does reduce incentives to exploit hardwoods and, by implication, to selectively log. We have also argued the contrary point -- that the awarding of land title may in fact *promote* deforestation. Thus, a key policy question that remains is whether the reduction in fire damages resulting from land tenure security compensates any potential deforestation, *resulting from land tenure security*. If the answer is yes, then it might be reasonably argued that land titling programs are good for the Amazon forest. If the answer is no, then perhaps policies based on the protection of natural areas are preferable. Hopefully, the methodology elaborated in this chapter will prove useful to those who engage these important issues in the population-environment discourse.

NOTES

[1]Renewable resources with maximal growth rates less than the interest rate are "inferior;" profit maximization dictates immediate liquidation and investment in financial assets generating the interest rate (Clark 1973).

[2]Other perennials currently grown in the region include coconut palm trees and cupuaçu. Thus far, production from these alternatives remains low.

[3]It is important to note that many household consume substantial portions of their annuals production. In particular, 44 household out of 261 consume 100 percent of their rice and beans production. Hence, the term revenue, is misused in the strictly economic sense. A better terminology is probably 'value' of household production.

ACKNOWLEDGMENTS

This research was supported by (1) the Geography and Regional Science Program of the National Science Foundation, under grant NSF-SBR-95-11965, by (2) the National Geographic Society, and by (3) The National Aeronautics and Space Administration's LBA-Ecology Program (NCC5-291). We are grateful to Stephen Perz and Marcellus Caldas for their technical support, and we would like to thank those who helped us collect the data in the field, Celio

Armando Palheta Ferreira, Rui de Almorim Carvalho, Luiz Guilherme Texeira Silva, Andre Caetano, Roberto Porro, and Fabiano Ton. The Basic Science and Remote Sensing Initiative of the Michigan State University provided much necessary infrastructure and data.

REFERENCES

Alston, L.J. 1996. Empirical work in institutional economics: An overview. In Empirical Studies in Institutional Change. (Eds.) L.J. Alston; T. Eggertsson, and D.C. North. pp. 25-58. Cambridge: Cambridge University Press.

Alston, L, J.; Libecap, Gary, D., and Schneider, Robert. 1995, Property rights and the preconditions for markets: The case of the Amazon frontier. *Journal of International and Theoretical Economics.* 151(1): 89-107

Alston, L, J.; Libecap, Gary, D.; Mueller, Bernardo. 1997. Violence and the development of property rights to land in the Brazilian Amazon. In The Frontiers of the New Institutional Economics. John, N. Drobak and John V.C. Nye (eds.) Pp. 145-163. Academic Press.

Arnold, J.E.M. 1983. Economic considerations in agroforestry projects. *Agroforestry Systems* 1: 299-311.

Barnum, H.N. and Squire, L. 1979. A Model of an Agricultural Household: Theory and Evidence. Baltimore, MD: The John Hopkins University Press.

Beaumont, P. And Walker, R. 1996. Land degradation and property regimes *Ecological Economics* 18: 55-66.

Clark, C.W. 1973. Profit maximization and the extinction of animals. *Journal of Political Economy* 81: 950-961.

Clark, C. W. 1976. Mathematical Bioeconomics: The Optimal Management of Renewable Resources. New York: John Wiley & Sons.

Clark, C. W. 1985. Bioeconomic Modelling and Fisheries Management. New York: John Wiley & Sons.

Dale, Virginia, H., O'Neill, Robert V., Pedlowski, Marcos, and Southworth, Frank. 1993. Causes and effects of land-use change in central Rondonia, Brazil. *Photogrammetric Engineering and Remote Sensing* 59(6): 997-1005.

Dove, M. 1986. The ideology of agricultural development in Indonesia. In *Central government and Local Development in Indonesia.* (Ed.) McAndrews. pp. 221-247. Singapore: Oxford University Press.

Ellickson, Robert, C. 1991. Order without law: How neighbors settle disputes. Cambridge, MA: Harvard University Press.

Ellis, F. 1993. *Peasant economics.* Cambridge: Cambridge University Press.

Dove, M. 1986. The ideology of agricultural development in Indonesia. In Central government and Local Development in Indonesia. (Ed.) McAndrews. pp. 221-247. Singapore: Oxford University Press.

Eggertsson, T. 1990. Economic Behavior and Institutions. Cambridge: Cambridge University Press.

Feder, G. and Feeny, D. 1991. Land tenure and property rights: Theory and implications for development policy, *The World Bank Economic Review* 5(1): 135-153.

Feder, G., Onchan, T., Chalamwong, Y., Hongladarom, C. 1988. Land Policies and Farm Productivity in Thailand. Baltimore: the Johns Hopkins University Press.

Gordon, H.S 1954. The economic theory of a common property resource: The fishery. Journal of Political Economy 62 124-142.

Hames, R. And Vickers, W. (Eds.) 1983. *Adaptive Response of Native Amazonians.* New York: Academic Press.

Hosmer, D.W. and Lemeshow, S. 1989. *Applied Logistic Regression.* NY: John Wiley & Sons.

Hotelling, H. 1931. The Economics of Exhaustible Resources, *Journal of Political Economy* 39: 137-175.

Instituto Nacional de Pesquisas Espaciais (INPE). 2000. Desmatamento da Amazônia Brasleira,1998-1999. Available at INPE,http ://www.inpe.br/Informacoes_Eventos/ amz1998_1999 /pagina7.htm, May.

Leslie, Alf J. 1980. "The Government-TNC Relationship in Tropical Timber Concession Contracts," Paper delivered at the Asia and Pacific Regional Workshop on Negotiations with Transnational Corportions in the Tropical Hardwoods Sector, Pattaya, Thailand.

Libecap, G.D. 1989. Contracting for Property Rights. Cambridge: Cambridge University Press.

Mattos, M. and Uhl, C. 1994. Economic and ecological perspectives on ranching in the Eastern Amazon, *World Development* 22(2): 145-158.

Mertens, B. and Lambin, E. 2000. Land-Cover-Change Trajectories in Southern Cameroon, *Annals of the Association of American Geographers* 90(3): 467-494.

Moran, E. 1990. The Ecosystem Concept in Anthropology. Washington, DC. American Association for the Advancement of Science.

Mueller, B., Alston, L., Libecap, G. D., Schneider, R. 1994. Land, property rights and privatization in Brazil, *The Quarterly review of Economics and Finance* 34: 261-280.

Nakajima, C. 1969. "Subsistence and Commercial famliy farms: Some Theoretical Models of Subjective Equilibrium." Pp. 165-185 in Subsistence Agriculture and Economic Devlopment, ed. C.R. WhartonChicago: Aldine Publishing Company.

Nepstad, D.C., Veríssimo, A., Alencar, A., Nobre, C., Lima, E., Lefebvre, P., Schlesingr, P.,Potter, C., Moutinho, P., Mendoza E., Cochrane, M.A., Brooks, V. Large-scale impoverishment of Amazonian forest by logging and fire. Nature 398, 505-508 (1999).

Nerlove, M. and Sadka, E. 1991. Von Thunen's Model of the Dual Economy. *Journal of Economics* 54: 97-123.

Moran, E. 1990. *The Ecosystem Concept in Anthropology*. Washington, DC. American Association for the Advancement of Science.

Myers, Norman. 1980. "The Present Status and Future Prospects of Tropical Moist Forest," *Environmental Conservation*, 7, 101-14.

Office of Technology Assessment. 1984. Technologies to sustain Tropical Forest Resources. Washington, D.C.: U.S. G.P.O.

Ostrom, E., Walker, J. and Gardner, R. 1991. Covenants with and without the sword. Unpublished manuscript. Workshop in political theory and policy analysis. Bloomington: Indiana University.

Pichón, Francisco, J. 1997. Settler Households and Land-Use Patterns in the Amazon Frontier: Farm-Level Evidence from Ecuador. *World Development* 25(1): pp. 67-91

Place, F. And Hazell, P. 1993. Productivity effects of indigenous land tenure systems in Sub-Saharan Africa. *American Journal of Agricultural Economics* 75: 10-19.

Posey, D. And Balee, W. (Eds.) 1989. Resource Management in Amazonia: Indigenous and Folk strategies. *Advances in Economic Botany Monograph*, No. 7. New York: New York Botanical Gardens.

Repetto, Robert and Malcolm Gillis, eds. 1988. Public Policies and the Misuse of Forest Resources. Cambridge; Cambridge University Press.

Ross, Michael S. 1984. "Forestry in Land use Policy for Indonesia," unpublished Ph.D. dissertation, University of Oxford, Green College.

Sanchez, P.A. 1994. "Alternatives to Slash-and-Burn: A Pragmatic Approach for Mitigating Tropical Deforestation." Pp. 451-479 in Agricultural Technology: Policy Issues for the International Community, ed. J. R. Anderson. Cambridge: CAB International and the World Bank.

Schmithusen, Franz. 1980. "Forest Utilization Contracts—A Key Issue in Forest Policy and in the Development of the Tropical Hardwoods Sector," FAO Paper 5, Pattaya, Thailand.

Serrão, E.A.S. and A.K.O. Homma. 1993. "Brazil." Pp. 265-351 in Sustainable Agriculture and the Environment in the Humid Tropics. Washington, DC: National Academy Press.

Skole, D.L. and C.J. Tucker. 1993. "Tropical Deforestation and Habitat Fragmentation in the Amazon: Satellite Data from 1978 to 1988." *Science* 260:1905-1910.

Thorner, D., B. Kerblay and R.E.F. Smith. 1986. A.V. Chayanov on the Theory of Peasant Economy. Madison: University of Wisconsin Press.

Uhl, C. Buschbacher, R. And Serrao, E.A.S. 1988. Abandoned pastures in Eastern Amazonia. I. Patterns of Plant Succession. *Journal of Ecology* 76: 663-681.

Verissimo, A., Barreto, P., Mattos, M., Tarifa, R., and Uhl, C. 1992. Logging Impacts and Prospects for Sustainable Forest Management in an old Amazonian Frontier: The Case of Paragominas. *Forest Ecology and Management* 55: 169-199.

Viera, I.C.G., Salomão, R. Rosa, N. Nepstad, D.C., and Roma, J.C. 1996. *Ciencia Hoje* 20 (119): 38-44.

Vosti, Stephen, A., Witcover, Julie, and Carpentier, Chantal Line. 1998a. *Arresting Deforestation and Resource Degradation in the Forest Margins of the Humid tropics: Policy, Technology, and Institutional Options for western Brazil.* Final Report. Washington DC: International Food Policy Research Institute.

Walker, R.T. 1999. The Structure of Uncultivated Wilderness: Land Use beyond the Extensive Margin. Journal of Regional Science. Vol. 39(2): 387-410.

Walker, R.T., Moran, E., and Anselin, L. 2000. Deforestation and Cattle Ranching in the Brazilian Amazon: External Capital and Household Processes. *World Development.* Vol. 28(4): 683-699.

Walker, R.T., Perz, S. Caldas, M. and Texeira da Silva, L.G. (forthcoming). Land Use and Land Cover Change in Forest Frontiers: The Role of Household Life Cycles. *International Regional Science Review.*

Walker, R.T. and Smith, T. E., 1993. Tropical deforestation and forest management under the system of concession logging: a decision-theoretic analysis. *J. Region. Science*, 33(3): 387-419.

Chapter 8

SPATIAL AND TEMPORAL DYNAMICS OF OWNERSHIP PARCELS AND FOREST COVER IN THREE COUNTIES OF NORTHERN LOWER MICHIGAN USA, CA. 1970 TO 1990

Scott A. Drzyzga[1] and Daniel G. Brown[2]

[1]*Department of Geography, Michigan State University, East Lansing, MI 488, USA2 drzyzgas@msu.edu4*
[2]*School of Natural Resources & Environment, University of Michigan, Ann Arbor, MI 48109, USA*

Keywords: land ownership, forest fragmentation, spatial pattern analysis.

Abstract This chapter explores the relationship between land ownership subdivision and forest fragmentation to better understand the ways humans and forests have interacted in a temperate forest region in the North Central USA. Geographic information technologies were used to assess the dynamics of landscape configuration of private land ownerships and forest habitats in three forested counties over a 20-year period. Spatial and temporal patterns of land parcel subdivision were analyzed using digitized plat maps, while patterns of forest cover and forest cover change were assessed using digital remotely sensed data. Changes in land parcel subdivision and forest cover were observed in an area surrounding a small and growing central place, and their coincident patterns were used to evaluate the degree to which these phenomena are linked. The results show no direct connection between fragmentation of ownerships and that of forests. This finding highlights the complex interactions that govern landscape change, and indicate a need for new models that incorporate growing "footloose" populations, that is populations whose livelihoods are not directly tied to the land on which they reside.

INTRODUCTION

Historical observations of social and landscape change have provided important evidence of the interaction between society and environment (Blaikie and Brookfield 1987, Naveh 1991, Weins 1992, Blaikie 1995). The need for improved understanding of these interactions is true in tropical regions (Walker and Homma 1996, Faminow 1998, Walker et al. 2000), where deforestation is an important source of global atmospheric carbon and is implicated in substantial biodiversity loss. A similar need also exists in forested temperate landscapes of the Northern Hemisphere, where land use changes associated with demographic changes and economic shifts (e.g., abandonment of agricultural lands) have been driving increases in forest cover (Casperson et al. 2000).

Many rural areas in the U.S. are experiencing large increases in both residential (U.S. Department of Commerce 1972, 1982, and 1992) and seasonal populations (Stewart 1994, Stynes et al. 1997). At least some increases are driven by an increasingly "footloose" population, which, because of improved transportation, telecommunications infrastructure, and transformations in demographic and economic situations, is not as dependent on a specific land base or urban system for its livelihood. These population increases are feared by many who believe natural resources and aesthetic qualities associated with regenerating temperate forests will subsequently be destroyed. Such a concern raises a general need to better understand the interaction between humans and forests and a specific need to reveal how recent historical development processes have influenced contemporary landscape patterns in temperate forests. Potential impacts include forest fragmentation and consequent wildlife habitat effects, and more complicated management strategies for forest resources.

Historical and contemporary processes affect landscapes over different spatial and temporal scales, but all can leave indelible patterns. These patterns are interesting, because they have implications for landscape organization and structure. For example, in the midwestern USA, many roads and land parcel boundaries coincide with public land survey system[1] (PLSS) section lines and intersect at section corners. These regular patterns influence patterns of land ownership subdivision and, in turn, the shape and ownership of forest patches. In essence, spatially structured land use decisions have produced conspicuously organized and structured land cover patterns. Therefore, a spatially explicit understanding of historical land uses is necessary for explaining contemporary land cover patterns.

Land use and land cover are related, but they are not always equivalent. The association between them is neither the same for all locations nor constant in any one location over time. For example, in frontier regions where economic growth depends on an extractive economic base (e.g., the Brazilian Amazon), land use and land cover are often semantically equivalent. Land-use activity associated with logging leads to a deforested land cover (Lambin 1997). Therefore, satellite images can often be used to detect land-use change through observations of the biophysical characteristics of the land. However, in regions with a service or information driven economy, like most of the contemporary United States and Europe, land use and land cover are less likely to be equivalent. Similarly, observable land cover patterns may reflect not only current land use, but may also be the result of a particular historical trajectory of use.

Spatial technologies associated with geographic information science (GISc), which include aerial photography, satellite remote sensing, geographic information systems, and spatial analysis, can make an important contribution to the study of landscape phenomena, because they allow spatially explicit data to be stored, integrated, analyzed and visualized. Spatial technologies make possible detailed mapping and analysis of areas with large extents and numerous observations. Land cover, ownership, and use are examples of complex spatio-temporal phenomena that coexist over very large areas. GISc offers the unique ability to map, integrate and query, effectively and efficiently, observed coincidences between these landscape phenomena over multiple scales.

In brief, our purpose is to assess and characterize historical dynamics of, and interactions between, private land ownership and forest cover changes in a forested region. We start by defining terms, establishing the setting, and describing the spatial model and technologies used to characterize the relationship between the two phenomena. Next, we examine our results and assumptions. Last, we identify a need for new models that should incorporate growing "footloose" populations, i.e., populations whose livelihoods are not directly tied to the land on which they reside. Our specific objectives are to:

- quantitatively describe the spatial and temporal patterns of land ownership subdivision in a forested region in Northern Lower Michigan, USA between 1970 and 1990;
- quantitatively describe the spatial and temporal patterns of forest cover and fragmentation during the same period;
- characterize the relationship between changes in parcels, forests cover, and forest fragmentation; and
- use these results to improve our understanding of the link between population development processes and contemporary landscape patterns.

DEFINITION OF TERMS

The terms fragmentation (Brown and Vasievich 1996, Theobald et al. 1996) and parcelization (Barlowe 1978, Erickson 1995) have been used synonymously in past research to mean the division or subdivision of land ownership in a landscape. However, their meanings are different and future application of the results of this study requires a careful definition of terms.

Land ownership *parcelization* refers to the subdivision of land under a single ownership into smaller parcels under a diverse ownership. The Michigan Society of Planning Officials (MSPO) described forest *fragmentation* as the result of cutting forests into configurations where "edge species of plants and animals have proliferated" and deep forest habitats and resident species have declined (Warbach and Norberg 1995:iv). This definition implies physical landscape change. However, the MSPO also describes forest fragmentation as a process of subdividing large forested patches among many parcels under diverse ownership. This process does not necessarily effect a loss of habitat or scenic area, but is an impediment to forest management. Forest management strategies are difficult to construct and implement when many landowner viewpoints must be considered. The second definition does not imply physical landscape change. The former, physical definition of forest fragmentation is employed in this study, recognizing that the political connotation of the term forest fragmentation is very important to some planners and planning agencies.

STUDY AREA AND BACKGROUND

Grand Traverse, Kalkaska, and Crawford Counties (Figure 1 and Figure 2) were selected as suitable areas for study for several reasons. The three counties are representative of extreme cases in Northern Lower Michigan USA, as each experienced substantial population growth over the 20-years between 1970 and 1990, although not concurrently. The three counties are also expected to incur continued population growth through the next 20-years (Grand Traverse County Planning Commission, 1997; Warbach and Norberg, 1995). Table 1 lists population figures for the three-county region, Northern Lower Michigan USA (NLM), and the entire State. The %CHANGE figures in Table 1 demonstrate that population growth occurred more rapidly in NLM than in the state. In turn, the three selected counties grew more rapidly than all of NLM.

Brown and Vasievich (1996) documented in their sample study of the Upper Midwest, USA that Grand Traverse and Crawford Counties underwent parcelization from 1960 through 1990. Their study examined 136 sample sites across seventeen counties. Each county, including Grand Traverse and Crawford, was characterized by decreasing average parcel sizes. Grand Traverse, Kalkaska, and Crawford Counties were selected also because the three counties constitute one contiguous region, and land cover change phenomena may be studied at scales larger and smaller than the county scale.

Historical land use practices fragmented the forest resource in this region. An initial public land survey, a booming late nineteenth century timber industry, agricultural development, and establishment of public land areas affected forest distributions, compositions, and structures directly and indirectly. Growing human populations and a variety of land use choices continue to influence remnant forests and forest patches. Yet, many areas are exhibiting signs of forest regrowth and succession. Recent forest inventories and analyses of the Northern Lower Peninsula have indicated a six percent increase in total forest area between 1980 and 1993 (Leatherberry 1994).

Figure 1. The State of Michigan, its regions, the study area, and large cities mentioned in the text.

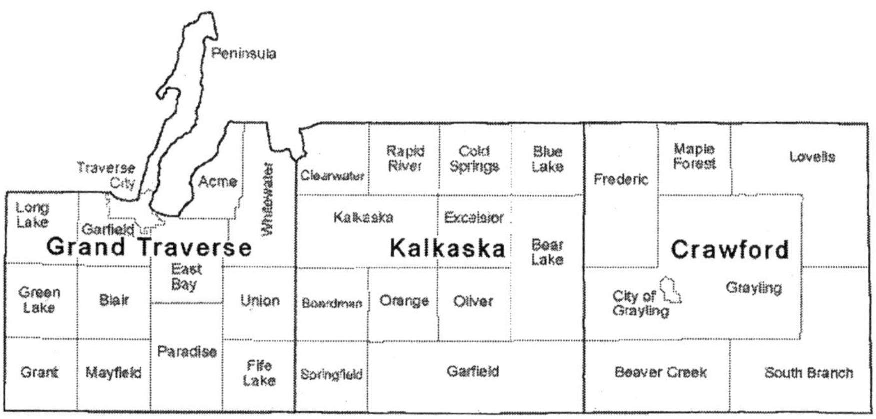

Figure 2. Political entities in the study region.

Traverse City, which is a relatively small city (population was 14,532 in 2000) located in Grand Traverse County, is the economic and retail center of northwest Lower Michigan. In it, the heaviest concentration of buildings and economic activity is located in the central business district (CBD), where many commercial, retail and service establishments are located. The most important land use outside the center is housing. Just outside the CBD, apartment buildings and single family residences are distributed evenly along a rectangular street network. Beyond the regular network, low-density single-family lots extend into the countryside with parcel sizes generally increasing with distance from the CBD. Beyond the urban extent are regions dedicated to agricultural and forest uses. In summary, the general structure of services, employment, and housing forms a discernable annular pattern of land uses around Traverse City.

Table 1. Population counts and changes.

County	*1970*	*1990*	*Change*	*%CHANGE*
Grand Traverse	39175	64273	25098	64
Kalkaska	5272	13497	8225	156
Crawford	6482	12260	5778	89
Region	50929	90030	39101	77
NL Michigan	383965	538534	154569	40
State	8881826	9295287	413461	5

Source: U.S. Department of Commerce: 1972, 1992.

The area around Traverse City is expected to continue a positive population growth trend into the year 2020 (Warbach and Norberg 1995). The Traverse City region attracts recreational development because of existing access to lake front areas, access to large population centers (e.g., Grand Rapids is 225 km, Detroit is 410 km, and Chicago is 510 km) gained from new highway construction, and proximity to public forests. Manson and Groop (1988) have characterized the area as a "retirement region" because primary population increases are associated with immigrating older persons; secondary increases are associated with local employment that serve the older community. Yet, residential population growth represents only a portion of total population growth in this region. Stewart (1994) and Stynes et al. (1997) have demonstrated that seasonal populations contribute substantially to rural landscape development in the Great Lakes region. Stynes et al. (1997) also demonstrated that the population of a Great Lakes county can be six or seven times the official resident population, because official census counts do not include seasonal homes or seasonal home owners in population totals.

The period from 1967 to 1995 is unique to Michigan's landscape evolution, because the Subdivision Control Act (SCA) of 1967, a statewide minimum lot-size zoning law, was in effect and enforced during this period. In brief, the SCA provided for an expensive platting process to be imposed on landowners that subdivided land into parcels smaller than 4.05 ha (10 acres). Norgaard (1984) found that the statewide response was a proliferation of parcels just larger than 4.05 ha in size. Negative by-products of this proliferation are higher infrastructure-per-capita costs and land cover fragmentation. This legislation affected both residential and seasonal home development. Therefore, the period between 1970 and 1990 should also be useful for studying landscape changes influenced by this piece of legislation.

Similarities exist between the land cover characteristics and land use activities found in Lower Michigan forests and those found elsewhere. For example, Michigan forests are influenced by minimum lot-size zoning legislation, as are forests in other areas of the USA. Theobald et al. (1996) describe how existing minimum lot-size zoning ordinances in Colorado pave the way for creation of large lots with few houses, high infrastructure costs, and increased land cover fragmentation. Also, Michigan forests have been regenerating (or replaced with plantations) since the 1900s, but are now subject to the complex effects of multiple uses. Similar afforestation processes have been observed in the temperate forests of Central Europe; Hüttl et al. (2000) indicates the rigor of both the current regrowth trend and the set of demands placed on these forests by modern society (e.g., labor, primary production, and recreation). Scarascia-Mugnozza et al. (2000) report that similar demands are being placed on the declining forests in Mediterranean regions. New research (Puric-Mlademovic et al. 2000) also indicates peri-urban development has caused a significant reduction in forest cover and increased forest fragmentation to occur in the Regional Municipality of York, United Kingdom. This kind of cause-and-effect relationship represents one of the principle concerns for Michigan planners (Wyckoff 1995), especially in the Traverse City region, which is the setting for this study. So, the human activities and ecological processes present in the study region share broad commonalities with those found in other temperate regions of the world.

METHODS

Representing Historical Human Influences

Human influence on the landscape can be expressed in terms of parcelization because existing land parcels are subdivided as more new landowners move into an area. Plat map books (Rockford Map Publishers Inc., 1969-1991) were collected for the study area for the approximate years 1970, 1980, and 1990. When plat map books were unavailable for the exact dates, the closest available date was used as a surrogate. County plat map books contain detailed maps that show the spatial location and extent of parcels in geographic townships. Each parcel is labeled with the name of the owner and, in some cases, the approximate size of the parcel. The locations of roads, schools, cemeteries, and administrative boundaries are also identified. Patterns of land ownership and parcel boundaries were digitized into a GIS database using Arc/Info (ESRI Inc., Redlands, CA) software. Each digitized parcel map was registered and georeferenced to the UTM coordinate system given ground control coordinate information obtained from surveyed section corners (MIRIS 1978).

After digitization, all polygons in the parcel database were classified into one of five categories: privately-owned large parcels, privately-owned small tracts, public lands, open water, and other. Privately-owned large parcels are mapped individually in the plat map books. Private ownerships are maintained by either individuals, groups of individuals, or corporations. Privately-owned small tracts are collections of private lots that share a common boundary. Each lot is not mapped individually, rather, all contiguous lots are mapped as a single polygon. Because the distribution of parcel areas within small tract polygons cannot be discerned from the plat maps, we assumed average parcel sizes within each polygon to be equal to 0.20 hectares (0.5 acres). According to Kufour (1984), the average lot size within small tract areas tended between 0.12 and 0.20 ha in size between the years 1970 and 1980. Kufour (1984) also noted that lots tended to be larger, (0.40 ha) during the 1950s and 1960s. Because several older, small tract areas remain throughout the landscape, we used the larger, 0.20 ha estimate. Public land areas are those areas owned and maintained by public schools and government entities. Open water areas are permanent inland water features, such as lakes, reservoirs, and ponds. In Michigan, navigable open water areas are held in the public trust and cannot be privately owned. Any parcel that did not fit directly into any above stated ownership group was classed into the "other" category. These parcels include ownerships maintained by utility companies and all military lands.

Quantifying Human Influence

A useful way to visualize land ownership parcelization is to calculate and compare parcel size frequency distributions for different dates. Changes between measured frequencies can be used to infer temporal changes on the landscape. So, such distributions were calculated and graphed for all three dates. Another useful way to quantify land ownership parcelization is to calculate and compare average parcel sizes for different dates. Average parcel sizes, were calculated using an average parcel size index (APS) similar to that used by Brown and Vasievich (1996). The APS index is a measure of average parcel size for a given area and sensitive to parcel area changes.

The geometric mean was used to calculate the APS index because the distribution of large parcel areas within any sampled area tends to be positively skewed. The APS index is derived using the following equation:

$$APS = \left(e^{\left(\left(\rho_{large} \frac{\Sigma \ln A}{n_{large}}\right) + \left(\rho_{small} \cdot \ln 2000\right)\right)}\right) / 10000$$

where:

ρ_{large} = the proportion of private land area classified as large parcels
ρ_{small} = the proportion of private land area classified as small tracts
$\frac{\Sigma \ln A}{n_{large}}$ = the mean of natural log transformed large parcel areas (m^2)
2000 = the assumed size (m^2) of all lots in small tract areas
10000 = a constant to convert meter^2 to hectares

The calculated geometric mean of large parcels and the assumed geometric mean of small parcels within small tracts are weighted according to the proportion of private land area each class covers within a sample unit and then summed into a single measure.

Representing Historical Landscapes

Three pairs of Landsat Multispectral Scanner (MSS) satellite imagery were collected that cover the study area for the approximate years 1973, 1985, and 1991. Each satellite image was registered, georeferenced, and distributed as part of the North American Landscape Characterization (NALC) data set at a spatial resolution of 60 m (Lunetta et al. 1998). Table 2 lists our image inventory. Slight registration errors were detected and corrected using PCI ImageWorks (PCI, Richmond Hill, Ontario) workstation software and known UTM coordinate system information obtained for regional road networks stored in vector GIS format (MIRIS, 1978).

Table 2. Inventory of NALC imagery used for target dates.

	1970		1980		1990	
	West	East	West	East	West	East
Path/Row	22/29	21/29	22/29	21/29	22/29	21/29
Date acquired	6/09/73	8/01/75	6/28/85	8/08/85	7/15/91	8/09/91
Sun elevation	60	53	59	53	56	52
Sun azimuth	128	127	124	131	133	129

All NALC images are distributed with a 60 meter pixel resolution.

Clouds in the images were identified and removed through on-screen polygon digitizing according to their popcorn shapes and corresponding displaced shadows. The cloud and cloud-shadow polygons were then used to "mask" the images, thus removing pixel values associated with these features.

In an effort to further reduce the amount of non-essential variation between corresponding images, each individual scene was clipped by the areal extent of the study area plus a 1.5 kilometer buffer zone.

The histogram matching method (Jensen 1996, Homer et al. 1997, Vogelmann et al. 1998) was used to improve the comparability of spectral response values in adjacent images for mosaicking purposes. Regions of spatial overlap were selected and the distribution of brightness values in one image was normalized to match the distribution of brightness values in the other image. This normalization, or matching, was carried out for each spectral band and each date.

After each pair of masked and clipped images were histogram matched and mosaicked, a composite cloud mask, which was generated by aggregating all clouded areas within all six images, was used to mask the three resultant image mosaics. This masking was done in order to maintain a consistent area of analysis throughout the experiment.

Our primary remote sensing task was to identify and distinguish all forested areas from non-forested areas. The target classification scheme included five classes: forested areas, not forested areas, open water, clouds, and background. The ISODATA clustering algorithm (Tou and Gonzalez 1977) was used to identify statistically homogenous areas within each landscape mosaic. The four MSS image bands, plus a fifth Normalized Difference Vegetation Index (NDVI) band, were used as inputs to the classification. The NDVI band, a mathematical combination of MSS bands 2 and 4 (Jensen 1996), has been found to be a sensitive indicator of the presence of green vegetation (Lillesand and Kiefer 1994). Vegetated areas, like trees and forest canopies, will generally yield higher NDVI values than bare soils, urban surfaces, water, or clouds. The algorithm converged on a solution set of 72 clusters for each mosaic.

Subsequent to cluster analysis, three sets of 16 aerial photograph mosaics (each set was obtained for the approximate dates 1970, 1980, and 1990) were used to aide cluster identification and accuracy assessment. The photos had already been scanned at 2 m resolution, georeferenced, and mosaicked into images corresponding to the 5 x 5 km sample areas in Grand Traverse and Crawford Counties used by Brown et al. (2000). The temporal sets of photographs were randomly divided into two subsets of eight by maximizing the geographic extents covered by each. The first subset for each date was used to identify and classify clusters and assign them to historic land cover types. The second subset was set aside for use in testing the quality of the final classification maps (Drzyzga 2000).

For accuracy assessment, stratified random sampling was used to select pixel locations within derived land cover areas that were also represented by historical aerial photographs. This process was stratified so that small, but potentially important, areas were not undersampled (Lillesand and Kieffer 1994). Next, all selected pixel locations were attributed with both image and reference class information. Finally, error matrices were calculated to summarize the classification accuracy at each date (Table 3). This method is often preferred to the post classification training set method, because the random method allows for the selection of pixels that represent borders between adjacent land cover categories. In a study of forest fragmentation, boundaries between forested and non-forested land cover categories are important ecological features and should be considered in any accuracy assessment. The post classification training sets method tends to avoid such boundaries as homogeneous areas are preferred. A water mask was created using published 1:100,000 Digital Line Graph hydrography data (U.S. Department of the Interior 1993) to remove pixels associated with water from each classified image. In reality, open water boundaries are rarely static in nature - even over geologically short ten and twenty year periods. However, we assumed constancy of the presence and extent of open water within the study area in order to isolate changes between forested and non-forested land covers.

Table 3. Error matrices and summary statistics for NALC image mosaics.

	Airphoto Reference Data			
1973	NF	FOR	ROW	USER
NF	284	59	343	82.8
FOR	81	440	521	84.5
COL	365	499	864	
PROD	77.8	88.2		
			K_{hat} =	66.5
			PCC =	83.8
1985	NF	FOR	ROW	USER
NF	275	29	304	90.5
FOR	56	499	555	89.9
COL	331	528	859	
PROD	83.1	94.5		
			K_{hat} =	78.8
			PCC =	90.1
1991	NF	FOR	ROW	USER
NF	231	28	259	89.2
FOR	20	529	549	96.4
COL	251	557	808	
PROD	92.0	95.0		
			K_{hat} =	86.3
			PCC =	94.1

FOR = Pixels as land Forested
NF = Pixels as land Not Forested
USER = User's accuracy
PROD = Producer's accuracy
K_{hat} = Estimate of Kappa Statistic
PCC = Percent Classified Correctly

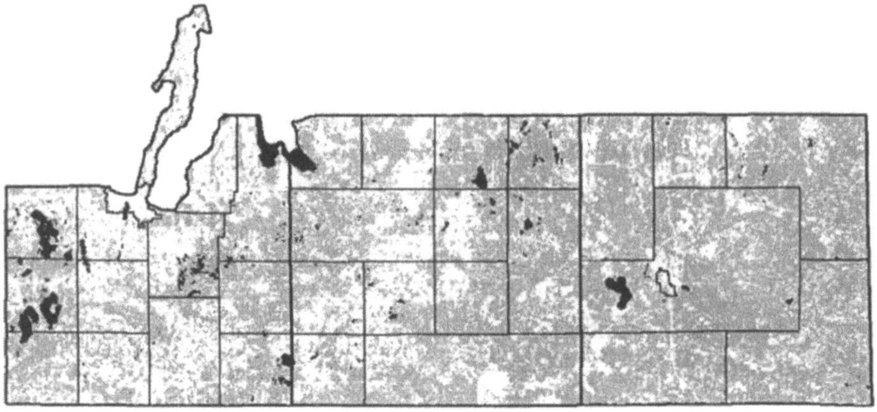

a. 1973

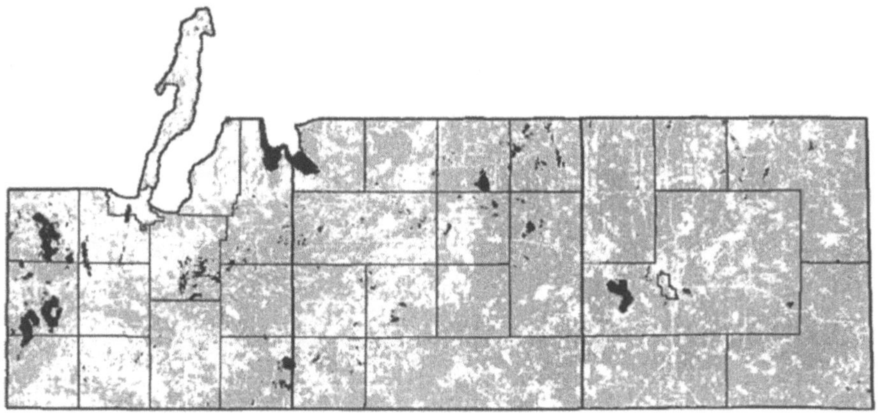

b. 1985

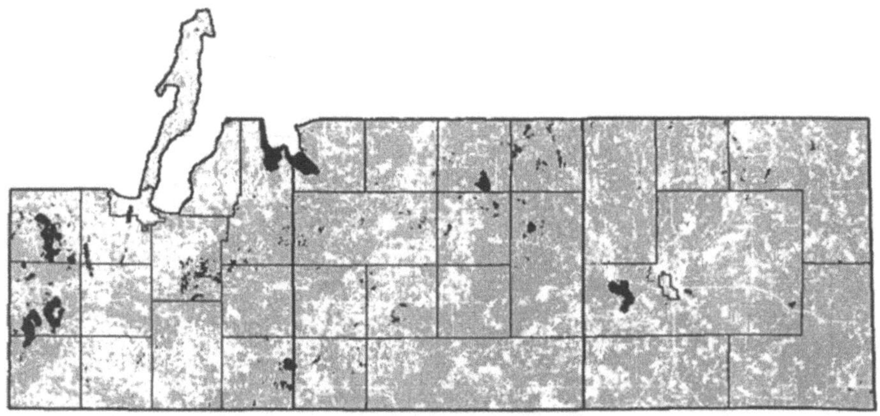

c. 1991

Figure 3. Forest cover data for (*a*) 1973, (*b*) 1985, and (*c*) 1991. Forested areas shown in green hue or gray tone. Inland waters shown in black. Buffer omitted from display.

Finally, the forest cover data were converted into a vector GIS format and overlayed with the parcel data. Spatial overlay allowed further partitioning of the forest data into private and non-private categories. The data were partitioned to ensure changes in privately owned parcels could be compared to changes in privately owned forests. All subsequent references to changes in forest cover and forest fragmentation are limited to changes observed on privately owned lands. Figures 3a, b, and c represent the three final forest cover data sets.

Quantifying Forest Cover

For any given region, the total area as forested was estimated by summing all polygon areas classified as forested. Also, the proportion of land as forested was estimated by dividing the sum of privately owned forest areas by the sum of all non-water and non-cloud, private areas. The amount of land as private remained nearly constant over the twenty-year period. Therefore, calculated changes in the proportion of land as forested can be attributed to observed changes in forested area (the numerator) and not to changes in the amount of land as private (the denominator).

Quantifying Forest Fragmentation

Landscape ecology is based upon the premise that landscape patterns are indicative of ecological processes (Forman and Godron 1986, Turner 1989, Gustafson 1998) and quantitative metrics allow measurable links to ecological processes to be determined (Frohn 1998). In the landscape ecology literature, spatially distinct and contiguous areas within a landscape class are referred to as patches. Patches can be quantitatively described in terms of size, shape, interior core area, and complexity (Forman and Godron 1986, Milne 1991, Turner and Gardner 1991, Baker and Cai 1992, Li and Reynolds 1993, McGarigal and Marks 1995, and Riitters et al. 1995).

Studies of forest fragmentation often use metrics to describe interior core areas and edge effects (Merriam and Wegner 1992, Villard et al. 1992, Fahrig and Merriam 1994). Interior core areas, sometimes referred to as interior habitat areas, are internal patch areas where some natural species and ecosystems can function independently and better sustain themselves. These interior areas contain requisite energy, nutrients, materials, and gene pools necessary for self-perpetuation. Edges are portions of an ecosystem (forest patch) near the perimeter where influences of surrounding areas prevent natural interior environmental conditions (Forman 1995) and can be considered habitat by other species. Those forest areas influenced by the effects of non-forest edges are often called disturbed forest areas. Thus, patch size and shape can influence population dynamics and survival as two patches of the same area but with different amounts of edge may produce core areas of differing quality and species composition (Saunders et al. 1985, Saunders et al. 1991).

Gustafson (1998), however, is critical of the current set of landscape metrics often used in the landscape ecology literature and of the lack of any means to interpret them consistently. He states, "[t]here is seldom a one-to-one relationship between index values and pattern (that is, several configurations may produce the same index value)" (p.150). In the same paper he argues...

> "[v]ery few pattern indices produce values that are useful by themselves. Their most instructive use is in comparing alternative landscape configurations, either the same landscape at different times or under alternative scenarios, or different landscapes represented by using the same mapping scheme and at the same scale" (p.152).

So, we constructed a new landscape index that can be a) useful by itself, b) useful for describing conditions of forest fragmentation, and c) have a direct relationship with degrees of fragmentation. The core fragmentation (CF) index (Drzyzga 2000) can be derived with the following equation, which is expressed here in condensed terms:

$$CF = \frac{CORE_{max} - CORE_{obs}}{CORE_{max}}$$

The CF index is interpreted as the proportion of the maximum potential core area ($CORE_{max}$) of a patch or class that is not core area ($CORE_{max} - CORE_{obs}$) because of shape induced edge effects. The index is constrained between zero and one. Calculated CF values approaching a value of one indicate a highly fragmented landscape with all forested areas being influenced by edge effects. Urban forests, with few core areas would generally fit into this category. Calculated CF values approaching zero indicate a forest unfragmented and subjected to minimal edge effects. Such areas would include interior areas of very large forest stands.

Sampling Landscapes with Political Units

Land use planning in the USA is often based on a local decision process that is directed by local legislation. Under such legislation local planning boards are given the authority to develop ordinances to regulate land uses, protect natural resources, and limit congestion of human populations. Land-use regulation in Michigan occurs at the township level, therefore townships are important organizational units for land and ecological planning. For this reason, each digital landscape was sampled using township units (see Figure 2). Sampling in this way allows for parcel data to be aggregated in meaningful ways and, perhaps, an opportunity to uncover spatial patterns associated with township behaviors.

Sampling Landscapes with Concentric Zones

First-hand observations indicate that the spatial structures of land uses and land consumption around Traverse City appear annular with respect to the CBD. Given this general structure, we sampled the region using concentric zones that radiate from the CBD (Figure 4). Of course, this sampling method borrows from the oldest theories of landscape organization (von Thünen 1842) and urban structure (Burgess 1925), and these theories suffer gross assumptions about the behavior of individuals acting on a landscape. However, given our observations, we proceeded with this basic sampling scheme and the idea of making informed adjustments during future research based on our results.

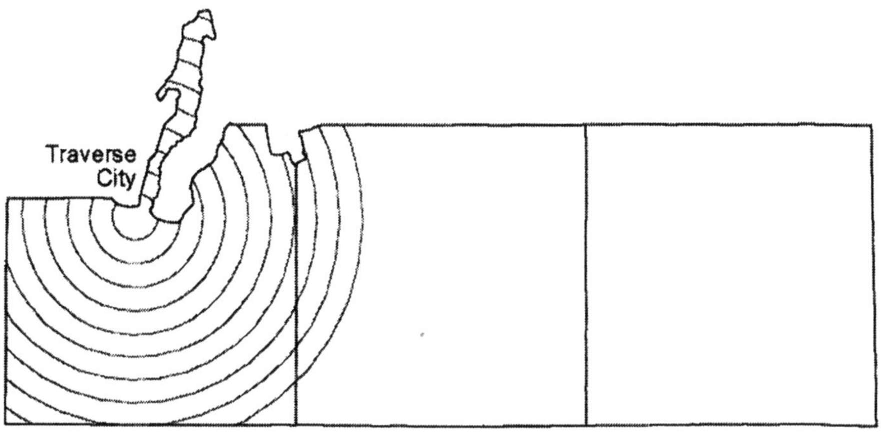

Figure 4. Concentric zones around Traverse City, Michigan.

Land ownership and land cover data were divided into several distance bands of equal width (3 km) and APS, percent forest, and CF index values were calculated for each band to characterize the distributions of parcelization and forest structure around Traverse City. The spatial patterns of land consumption, forest cover, and forest fragmentation radiating from Traverse City can be visualized by plotting calculated index values against the middle distance value of each band and interpolating values between known points.

To estimate changes in the potential extent of urban land uses around Traverse City, we compared each APS value to a threshold value of 4.05 ha. We chose a parcel size threshold value of 4.05 ha, because the equivalent in acres is used by Michigan lawmakers to distinguish lands available for development from open lands unavailable. Starting from the center of the city and moving outward, the urban extent can be marked at the first distance that average parcel size values exceed the threshold value. Movement of the urban extent can be tracked over time by plotting this relationship for successive dates and observing positional changes in the calculated intersection.

Should results indicate the extent of developable areas associated with Traverse City increased over time, those areas that came under urban pressures may have experienced decreases in forest cover and increases in forest fragmentation. Conversely, should the extent of developable areas have decreased over time, those areas relinquished from urban pressures may have experienced increases in forest cover and decreases in forest fragmentation. Population figures reported for the Traverse City region from 1970 to 1990 indicate a positive population growth trend. This trend leads to the hypothesis that the urban extent of Traverse City increased over time and consequently, fragmentation increased in pressured areas within it.

RESULTS

Parcelization

Calculated APS values show that regional average parcel sizes decreased (see Table 4) from 9.6 ha in 1970, to 4.9 ha in 1980, and to 4.1 ha by 1990. Changes between index values also indicate that average annual parcelization rates decreased over time. These trends coincided temporally with residential population growth trends. Care should be taken, however, in interpreting this temporal coincidence as a causal relationship.

Spearman's rank-order correlation coefficients between township population growth rates (average number of persons per year) and parcelization rates (average change in parcel size per year) indicate that township parcelization rankings were statistically unrelated to residential population growth rankings (Table 5). Furthermore, correlation results of lagged rankings suggest that changes in land subdivision were not related statistically to earlier residential changes. These statistically insignificant correlations do not necessarily discount the role new resident populations played in the land subdivision process, but suggest that other factors must have played a large role. Seasonal populations are not accounted for in census counts and were not accounted for in this simple test for correlation. Stewart (1994) and Stynes et al. (1997) have shown how seasonal home development and seasonal populations have contributed to rural landscape development in the Great Lakes region. It is likely that new second or seasonal homeowners also influenced the subdivision process. These results suggest that a measured change in residential population alone may not serve as an adequate indicator of potential land ownership change for this region.

Table 4. Average parcel size (APS, in hectares), percent forest, and core fragmentation (CF) index values for counties and townships by date.

	APS			Percent Forest			CF Index		
	1970	1980	1990	1973	1985	1991	1973	1985	1991
Region totals	*9.6*	*4.9*	*4.1*	*54.0*	*55.7*	*58.7*	*0.662*	*0.524*	*0.543*
Crawford	**5.7**	**3.3**	**3.0**	**74.1**	**77.6**	**75.6**	**0.597**	**0.405**	**0.461**
Beaver Creek	7.8	3.9	3.1	69.5	75.3	70.2	0.620	0.439	0.534
Frederic	6.2	3.4	2.9	73.0	77.5	77.5	0.555	0.376	0.421
C. of Grayling	3.4	7.0	-	28.3	36.4	23.7	0.710	0.600	0.715
Grayling	3.4	2.4	2.2	78.0	75.3	71.9	0.543	0.441	0.509
Lovells	6.8	4.3	3.9	80.7	83.7	81.7	0.636	0.368	0.435
Maple Forest	10.5	5.1	4.3	57.6	63.7	64.4	0.607	0.497	0.544
South Branch	5.6	3.0	3.3	74.3	84.2	83.0	0.631	0.310	0.339
Grand Traverse	**12.2**	**5.8**	**4.3**	**43.8**	**40.2**	**44.2**	**0.728**	**0.662**	**0.669**
Acme	10.8	6.3	4.3	31.6	26.0	34.1	0.792	0.699	0.752
Blair	11.8	5.0	3.7	41.0	36.6	39.6	0.723	0.670	0.654
East Bay	7.1	3.6	2.8	46.9	44.2	49.3	0.724	0.684	0.691
Fife Lake	16.9	5.6	5.4	48.9	55.6	58.2	0.713	0.549	0.563
Garfield	6.8	3.3	2.3	27.7	21.4	25.0	0.862	0.870	0.867
Grant	17.9	9.5	7.6	56.7	44.6	49.0	0.611	0.601	0.607
Green Lake	10.9	4.8	3.7	59.9	59.1	61.9	0.605	0.541	0.552
Long Lake	8.2	3.8	2.5	42.7	40.7	44.6	0.762	0.730	0.715
Mayfield	23.1	10.3	8.1	41.5	27.4	31.1	0.779	0.770	0.784
Paradise	22.5	7.8	6.5	46.2	48.2	49.8	0.710	0.627	0.618
Peninsula	6.9	4.9	3.0	29.2	25.1	27.4	0.848	0.855	0.870
Traverse City	0.7	0.4	-	16.3	27.5	24.6	0.957	0.992	0.978
Union	15.9	8.6	6.7	63.0	72.6	75.4	0.691	0.508	0.506
Whitewater	11.7	5.6	4.5	42.7	42.0	49.0	0.694	0.576	0.607
Kalkaska	**10.7**	**5.4**	**4.8**	**53.8**	**60.9**	**65.9**	**0.647**	**0.501**	**0.492**
Bear Lake	7.3	3.8	3.0	76.6	85.5	87.2	0.601	0.313	0.295
Blue Lake	5.4	4.0	3.0	72.6	76.1	79.9	0.525	0.538	0.506
Boardman	11.6	6.9	5.8	41.1	49.0	54.5	0.805	0.614	0.606
Clearwater	11.6	5.9	4.6	53.0	61.9	66.5	0.591	0.487	0.491
Cold Springs	6.5	3.2	2.9	78.2	81.6	83.8	0.454	0.399	0.367
Excelsior	12.4	6.3	5.3	49.1	58.0	64.3	0.682	0.419	0.439
Garfield	9.5	5.1	5.3	54.7	60.5	64.4	0.690	0.530	0.535
Kalkaska	9.9	5.1	4.5	47.6	52.0	59.6	0.729	0.608	0.578
Oliver	11.3	5.7	5.7	60.9	66.1	72.0	0.567	0.365	0.364
Orange	17.5	6.1	6.0	47.1	55.8	60.9	0.706	0.501	0.502
Rapid River	9.0	4.4	4.2	43.1	52.3	56.4	0.609	0.521	0.512
Springfield	13.0	6.5	5.7	44.0	52.1	59.2	0.713	0.576	0.550

Changes in parcel size frequency distributions were calculated and compared to infer changes that occurred on the landscape (Figure 5). At first glance, one can see each distribution is skewed, is multi-modal, and has peaks occurring in the 4, 8, 16, 32, and 65 hectare bins. These peaks portray the effects of recursive halving or quartering of survey sections - results typical of landscapes surveyed according to the PLSS.

Table 5. Spearman's rank correlation coefficients that characterize the relationships between township population growth and parcelization rates.

Average annual parcelization rates for	*Average annual population growth rates* 1970s	1980s
1970s	0.05 (n=33)	
1980s	0.24 (n=31)	0.21 (n=31)

All correlation coefficients were determined to be not significantly different from zero using a 95 percent confidence interval.

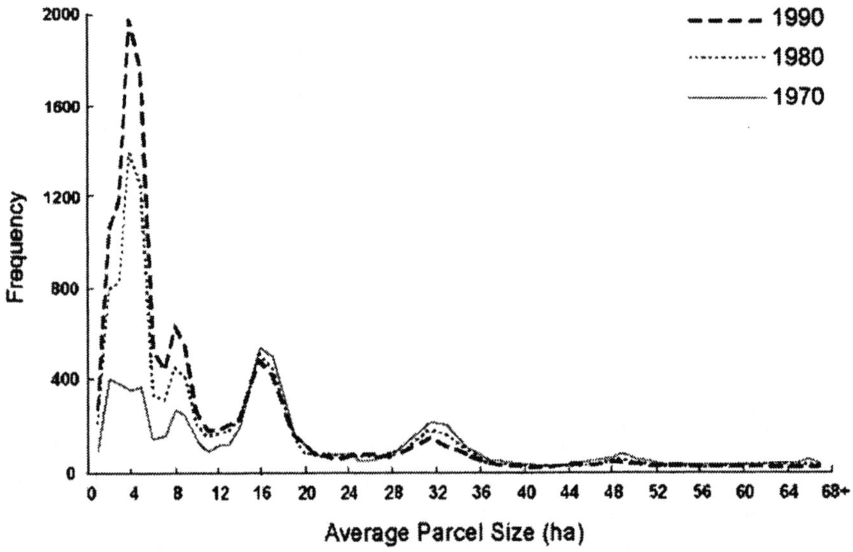

Figure 5. Frequency distributions for private parcels. Numbers of parcels with area values less than 1ha omitted for display. Omitted values are; 55874 (1970), 89697 (1980), and 94936 (1990).

Figure 5 reveals large increases in the number and aggregate extent of small parcels and coincident decreases in the number and aggregate extent of large parcels over time. (Numbers of parcels less than 1 ha in size were calculated by dividing the total area classified as small tracts by 0.2 ha, the assumed average parcel size in small tract areas.) In 1970, parcels smaller than 1 ha, in aggregate, consumed 6% (11415 ha) of private land, while parcels larger than 65 ha consumed more than 19% (43710 ha). By 1990, parcels larger than 65 ha covered only 4% (10620 ha) in aggregate, while parcels smaller than 1 ha consumed 10% (19740 ha). Newly created parcels were carved out of larger parcels given the finite amount of private land. On this landscape however, many newly created parcels seem to have been carved out of the largest large parcels.

Figure 5 also reveals large increases in the number and aggregate extent of parcels 4 to 5 hectares in size. These increases likely reflect the influence the SCA had on the parcelization process during this time. These parcels are important for three reasons. First, many parcels of this size now exist; second, they accounted for a relatively large area in aggregate (5%) by 1990; and third, we suspect that many land cover changes have occurred on these parcels. For example, the possibility exists that large parcels, perhaps once held in agriculture, were subdivided into these 4.05+ ha parcels. If this is true, then the likelihood exists that subsequent residential development was concentrated on only a small portion of each parcel (e.g., a driveway, house and lawn), thereby leaving the balance of each to regenerate from an open field condition into a young forest condition. This scenario might support the forest growth trend reported by Leatherberry (1994).

Regional Forest Ownership Parcelization

Although we defined fragmentation with respect to a type of physical landscape configuration, the ownership connotation is important to some planners and planning agencies. A useful way to recognize and visualize private forest parcelization is to calculate and compare forested parcel size frequency distributions for different dates. For our purposes, a forested parcel is a parcel that holds at least 50 percent of its area as forest. Parcels in small tract areas are not considered because without specific boundary data, the spatial coincidence between individual lots and forest polygons cannot be calculated. Figure 6 illustrates frequency distributions of forested parcel areas.

Each distribution displayed in Figure 6 is positively skewed, is multi-modal, and has peaks occurring in the 4, 8, 16, 32, and 65 hectare bins, like each total private parcel distribution. Forested parcel distributions are similar to total parcel distributions (Figure 5), because many parcels hold an areal majority as forest cover.

Changes between measured parcel size frequencies can be used to infer changes in forest ownership parcelization. For example, Figure 6 reveals large increases in the number and aggregate extent of small, forested parcels and coincident decreases in the number and aggregate extent of large forested parcels over time. Additional spatial queries revealed that in 1970, forested parcels smaller than 5 ha but not including small tracts (n=1300) contained 3%

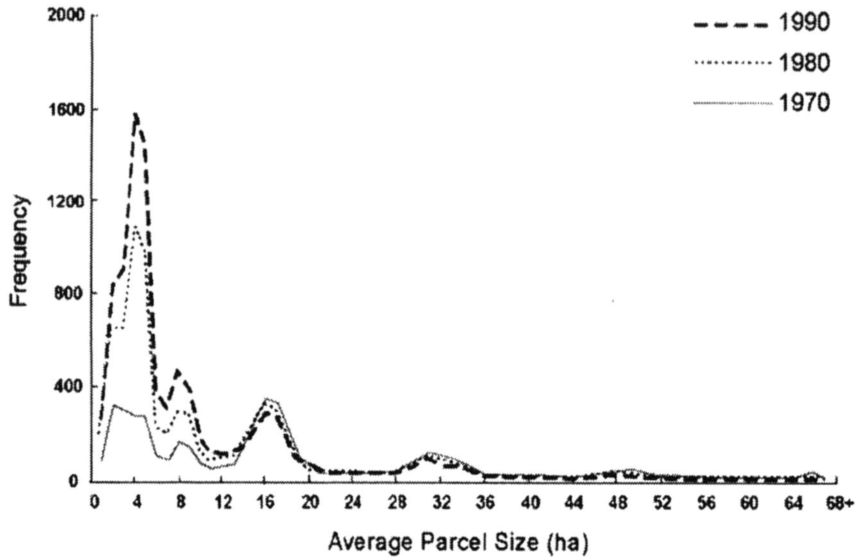

Figure 6. Frequency distributions of forested parcel sizes. Numbers of lots with area values less than 1ha not calculated.

(3100 ha) of privately owned forest, while forested parcels larger than 65 ha (n=287) contained more than 26% (30000 ha). By 1990, forested parcels larger than 65 ha (n=187) contained only 19% (21300 ha) of forest, while forested parcels smaller than 5 ha (n=5190) consumed 13% (19430 ha). Obviously, the number of forest ownerships increased as thousands of new landowners gained control over private forests. It is no wonder that forest managers are concerned about how to best implement new management strategies, especially when so many forest owner viewpoints must be recognized and considered.

Figure 6 also reveals large increases in the number and aggregate extent of parcels 4 to 5 hectares in size. Again, increases in the number of parcels that fall into this category likely reflect the influence that the SCA had on the parcelization process.

Parcelization Results for Townships

The APS index was used to estimate the average size of parcels for each county and township (Table 4). In every case but one (i.e., the City of Grayling), APS values decreased over time. Temporal changes in APS values indicate that average annual parcelization rates were greater during the 1970s than during the 1980s. Although change magnitudes varied over space, parcelization increased almost everywhere.

Almost all small tract areas mapped within the City of Grayling on the 1970 map had, by 1980, been cartographically reclassified by the publisher (Rockford

Map Publishers Inc. 1969-1991) into polygons associated with the incorporated city. Subsequently, no small tract areas were used in the calculation of the 1980 APS index and the resultant value did not measure their influence. No APS values were calculated for the City of Grayling and Traverse City for 1990 because all private parcels had been cartographically reclassified into city-owned parcels by 1990. Hence, each respective value was replaced with a no data entry in Table 4.

Parcelization Results for Concentric Zones

Land consumption patterns were summarized for the area around Traverse City by calculating APS index values for a set of concentric zones radiating from the geometric center of the City (Figure 7). Based on our knowledge of the applicable land use law, we assumed 4.05 ha to be the largest parcel size

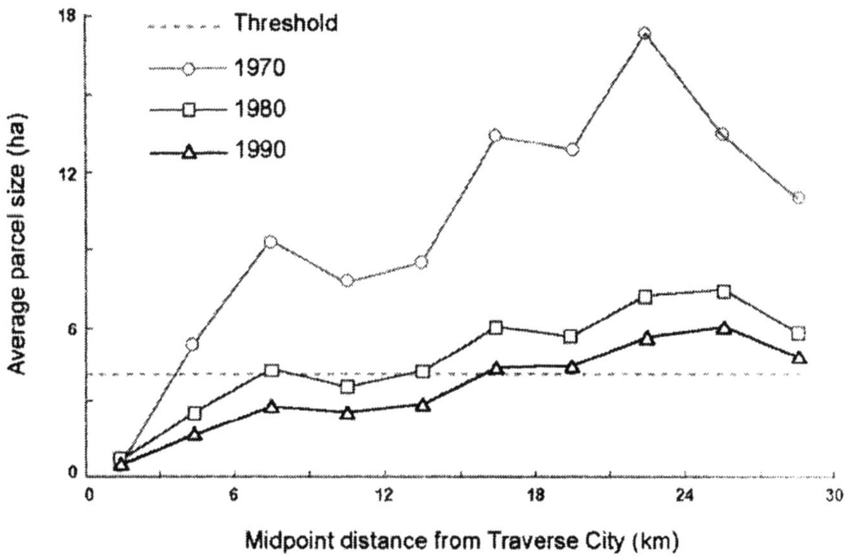

Figure 7. APS values for concentric zones by date.

associated with developable parcels. The approximate extent of potential urban land uses could be estimated for each date by comparing APS values calculated for all 1970, 1980, and 1990 distance bands to the threshold value. Figure 7 clearly shows that the first distance at which average parcel size values exceeded the threshold value increased over time. In 1970, the first intersection occurred between the first and second bands. In 1980, the first intersection occurred between the second and third bands. By 1990, the intersection occurred between the fifth and sixth bands. Movement of this intersection away from the city suggests the approximate urban extent of Traverse City increased over time. This movement can be characterized as a form of urban sprawl.

The parcel size gradient flattened over time. Such flattening indicates locations farther from the city subdivided faster than locations closer to the city. This trend can be interpreted to mean places farther from Traverse City gained better access to the city and its amenities, and therefore attracted new landowners, or proximity to the city became less important over time. Either of these reasons, or a combination of both, may explain why adjacent townships like Acme, East Bay, and Garfield experienced large increases in resident populations during the same period that Traverse City experienced small losses. Whatever the cause, the parcel landscape became more homogenous over time.

Regional Forest Changes

Regional percentages of private land as forested were calculated for each date. The percentage of forested private land increased over time from 54.0 percent in 1970, to 55.7 percent in 1980, and to 56.7 percent by 1990. Because the amount of private land remained almost constant over time, differences between percentages show regional forests increased over time. Also, average annual changes suggest the region experienced higher regrowth rates during the 1970s than during the 1980s. These regrowth trends coincide temporally with residential population growth trends.

CF indices were calculated to estimate fragmentation conditions for each of the three image dates. Regional CF values decreased from 0.66 in 1973 to 0.52 in 1985 but increased slightly to 0.54 by 1991. In general, changes between dates show forest fragmentation decreased between 1973 and 1991. The long-term trend is, however, insensitive to the shift from a period of defragmentation to a period of fragmentation. This shift provides evidence to support Wyckoff's (1995) concern that continued development may ultimately increase fragmentation in this region.

Forest Results for Counties and Townships

In most townships, we observed mixed trends of forest growth and loss (Table 4). Townships such as Bear Lake, Blue Lake, Union, and Fife Lake exhibited 18-year growth trends. Only Grayling Township exhibited an eighteen-year trend of forest loss. These mixed results show the forest recovery trend is an aggregate trend and does not apply to all areas within the region. These results also indicate that forest growth occurred in many townships that exhibited increases in both population and parcelization. Therefore, the association between land ownership change and forest cover change is neither simple nor perfectly indirect.

Grayling Township is the largest township in Crawford County and it contains many public land areas. The large amount of forest loss in a township that contains so many private lands adjacent to protected areas was surprising. Additional research revealed several large forest fires occurred in areas in or around Grayling Township between 1985 and 1991 (U.S. Forest Service, 1997). For example, the Stephan Bridge Road fire of 1990 burned 2450 ha of private and public forest in Crawford County (Winter and Fried 1997). Most of the forest losses associated with the Stephan Bridge Road fire occurred in Grayling Township thus, they contributed to the loss trend observed between the 1985 and 1991 satellite images.

The forest fragmentation index (CF values) was calculated for the region, each county, and each township (Table 4). No dominant fragmentation trend existed over the entire eighteen-year period. Crawford and Grand Traverse Counties exhibited trends of defragmentation from 1973 to 1985 and trends of fragmentation from 1985 to 1991. Most townships within these two counties, with exceptions like Blair, Long Lake, and Union Townships, exhibited similar trends. Kalkaska County demonstrated an eighteen-year defragmentation trend even though townships like Clearwater, Excelsior, and Garfield exhibited increased forest fragmentation between 1985 and 1991. Grayling Township exhibited an increase in fragmentation after 1985, probably due to changes in forest amount and structure caused by fire.

Forest Results for Concentric Zones

The spatial structures of forest cover around Traverse City can be visualized by plotting percent forest values against zone midpoints (Figure 8). In general, percentage values increased with increasing distance from the city. These results are quite reasonable as lesser amounts of forest can be expected to exist in locations closer to the city as more land in these locations was dedicated to anthropogenic land uses. However, some interesting temporal changes can be observed. For example, the spatial structure of forest cover changed between 1973 and 1985 as some areas experienced forest declines while others did not. These spatial differences occurred even though all areas exhibited increases in land ownership subdivision. However, the general structure remained the same between 1985 and 1991.

CF values were calculated and plotted for each concentric band in order to characterize the spatial structure of forest fragmentation (Figure 9). In general, CF values tended to decrease with an increase in distance from the city. This general structure is reasonable because higher degrees of forest fragmentation tend to exist in areas with relatively little forest or many forest edges, like in urban areas. However, specific temporal changes indicate that the spatial structure of forest fragmentation changed between 1973 and 1985. In areas beyond the 1985 urban extent, forest fragmentation decreased during a period of forest loss. Also, it appears fragmentation decreased at an increasing rate with an increase in distance. The general structure remained almost constant between 1985 and 1991, even though all areas exhibited forest growth.

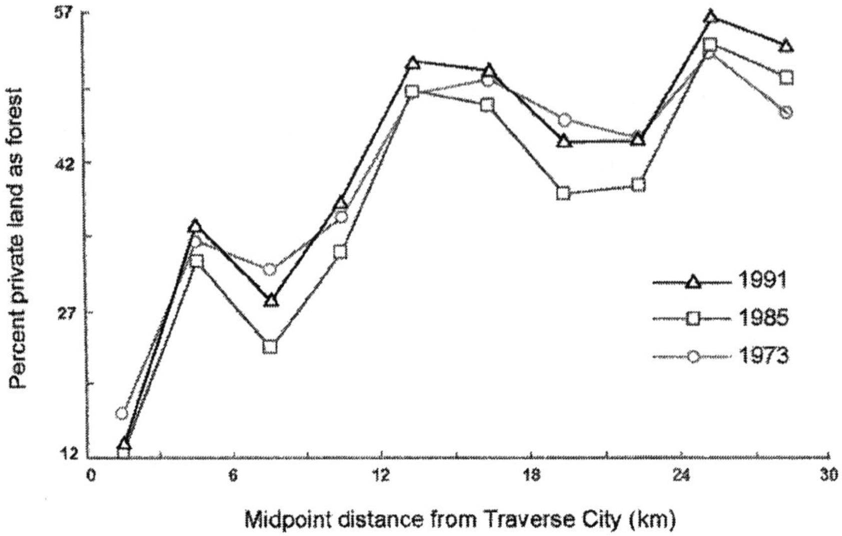

Figure 8. Percent forest values for concentric zones by date.

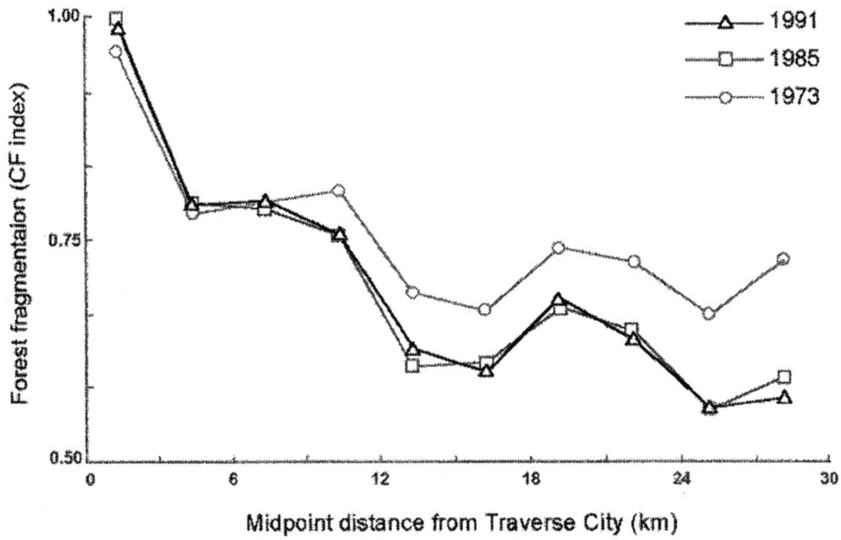

Figure 9. CF index values for concentric zones by date.

DISCUSSION AND CONCLUSIONS

We examined the spatial patterns on a temperate forested landscape in the USA and dynamic changes that occurred on that landscape. The static conditions tend to represent the cumulative effects of past land use and cover decision processes, while the dynamic changes represent current processes. The analysis of static conditions, for example, revealed the effects that the histories of land subdivision regulations have had on the size distribution of parcels (Figure 5). Analysis of changes revealed the effects that land use decisions, in aggregate, had on forest cover during a period of rapid population growth. Though change related information resulting from simple differencing of static information is subject to data uncertainty, it reveals more about recent trends on the landscape.

Average parcel sizes, forest amount, and forest fragmentation indices all show strong spatial gradients outward from the dominant urban center of the region. (Additional inquiry revealed similar gradients, but with lesser areal extents, radiated from smaller places like the City of Grayling and the Village of Kalkaska in Kalkaska Township.) Parcel sizes and forest covers increase while forest fragmentation decreases with increasing distance from Traverse City. This spatial coincidence is reasonable because one would expect areas with many small parcels to be associated with more anthropogenic land covers. Also, one would expect areas with many large parcels to contain lower densities of anthropogenic land covers and open spaces available for natural forest covers. Therefore, the spatial concurrence of the average parcel size gradient with the static forest cover and fragmentation gradients supports a proposed link between parcelization and fragmentation (Brown and Vasievich 1996, Bell 1995, and Saunders et al. 1991).

It is worth noting that the relationship between spatial structures of parcelization and forest cover and fragmentation is actually reversed when examined at the county level (Table 4). Although Grand Traverse County had the highest average parcel size in 1970, it had the lowest average forest cover and highest fragmentation index values. Conversely, Crawford County, with the smallest parcels on average in 1970, had the most, and least-fragmented, forest. The aggregation of diverse land use and cover situations to the county level renders data at that scale inappropriate for analysis of these relationships.

The analysis of parcel and forest dynamics failed to support a link between changes in parcel size and changes in forest cover and fragmentation. In general, areas that exhibited more rapid declines in parcel size tended not to have higher fragmentation rates. In fact, areas that were greater than about 10 km from Traverse City tended to have rather high rates of decline in average parcel size in the 1970s (Figure 7) but also exhibited rather rapid declines in forest fragmentation during roughly the same period (Figure 9). A simple interpretation would suggest that forest defragmentation was associated with land subdivision, at least within this particular study area. However, Spearman's rank-order correlation coefficients between average annual parcelization and fragmentation rates, which were calculated for township entities (see Table 6), indicate no significant relationship existed between parcel and forest dynamics.

Table 6. Spearman's rank correlation coefficients that characterize the relationships between township parcelization and forest fragmentation rates.

Average annual fragmentation rates	*Average annual parcelization rates* 1970s	1980s
1970s	-0.12 (n=33)	
1980s	-0.07 (n=31)	-0.04 (n=31)

All correlation coefficients were determined to be not significantly different from zero using a 95 percent confidence interval.

These findings are important because they suggest that parcelization and fragmentation processes in this region were not directly associated in the same way as static landscape conditions. This means that we cannot use static landscape patterns, which reflect historical landscape evolution, as a proxy for understanding contemporary landscape processes. Furthermore, these findings demonstrate a disconnection between the processes affecting changes in land ownership and the processes affecting land cover. Land subdivision does not necessarily imply the destruction of the forest. In fact, in some areas, land subdivision was associated with forest regrowth. However, it is important to reiterate that although the existence of forests may not be threatened by the parcelization of land ownership, management of forests as a natural resource is much more difficult under diverse ownerships. Therefore parcelization is much more likely to threaten the forest products industry than the forests themselves.

We have based our analysis on a class of models that describe the spatial organization of land uses around an urban center using concentric zones. These models can be rightfully criticized because they usually require gross assumptions be made about a landscape (e.g., a featureless and isotropic plain) and the behavior of individuals or households acting upon it (e.g., each acts with perfect information to maximize welfare). We used concentric zones to sample our landscape data because we observed annular patterns that seemed quite similar to the annular patterns described by von Thünen (1842), Burgess (1925), Alonso (1964) and Muth (1969). This sampling method has proven quite useful for describing static patterns of parcel size, forest cover, and forest fragmentation gradients around Traverse City. Yet, this method, and the class of models associated with it, offers little to explain the processes that are changing these spatial structures.

Models that try to explain land use patterns are often based on the additional assumption that people living on a landscape are coupled with that landscape. For example, the farmers in von Thünen's (1842) isolated landscape live on and work the land, and interact spatially with a central market. Alonso (1964) substituted commuters for farmers and a central business district for the central market, and constructed a model to describe population density functions surrounding a central source of employment. These models and others like them (e.g., Muth 1969) assume that people live, work, and play on the same landscape. Although, such assumptions may be most appropriate for frontier regions where economic growth depends on an extractive economic base, they are less appropriate for regions with service or information driven economies.

Many of the people acting on the landscape do not depend on Traverse City for employment or income, nor do many depend on the land on which they live. Plentiful natural amenities have attracted second homeowners who live and work elsewhere (e.g., Detroit, Flint, Grand Rapids, or Lansing), but come to the region as part-time residents for weekends, weeks, or a season at time (Stewart 1994). For example, it is commonly known in Michigan and other states in the region, that the states becomes "top-heavy" on summer weekends because so many people living in southern and more urbanized areas of the region travel northward to access natural amenities and recreational opportunities. Also, many older persons have moved permanently into the area to retire and tend to prefer environmental qualities to urban amenities (Manson and Groop 1988). Older persons also tend to depend on transfer payments (e.g., social security and pensions) for income. Non-employment income may originate from any location outside the study area and can arrive by mail or electronic transfer. Therefore, these new landowners, either seasonal or resident, are potentially "footloose" and can choose to live anywhere on the landscape because they are free from location-specific employment constraints. Another segment of the population is also becoming "footloose." According to Fuguitt and Beale (1996: p.170), "There is a segment of large-scale urban society, ...that seems increasingly dissatisfied by the stresses and costs of megalopolitan and corporate life [e.g., Detroit, Flint, Grand Rapids or Lansing - authors' inclusion], drawn to outdoor amenities, and fully mindful of space-negating qualities or modern information and communication technologies." It seems likely that such persons have been, and may continue to be, attracted to this region, but do not depend on it or its resources for employment or income.

The increasingly "footloose" populations of northern Lower Michigan and other temperate forest regions do not conform well to the models handed down from the economic geography of agricultural or urban systems. Our work suggests a need for models that can incorporate strong amenity drivers to explain land use decisions and that model land use (e.g., development) and land cover (e.g., forest clearing or regrowth) as separate processes (Brown et al. 2000). Also, our work suggests a need for models that incorporate multiple foci, which must include both urban and natural central places. To predict the influences of population growth in rural environments, we need to understand both the evolving spatial structure and the evolving land management decisions of that population.

NOTES

[1]The Public Land Survey System, or the U.S. Rectangular Land Survey System, is a rectangular land-survey system that uses a series of origins, baselines and meridians to partition land on the ground. Six-mile squares of land are referenced north and south of baselines, and east and west of meridians. Each six-mile square, which is called a township, is divided into 36 one-mile squares, which are called sections. Each section contains approximately 259 ha of land. Most lands in the U.S.A. were surveyed according to this system and many private property boundary locations are documented with respect to section corners and lines.

REFERENCES

Alonso, W. 1964. *Location and Land Use: Toward a General Theory of Land Rent.* Cambridge, MA, Harvard University Press.

Baker, W.L. and Cai, Y. 1992. The r.le programs for multiscale analysis of landscape structure using the GRASS geographical information system. *Landscape Ecology* 7(4): 291-302.

Barlowe, R. 1978. *Land Resource Economics.* Englewood Cliffs, Prentice-Hall, Inc.

Blaikie, P.M. 1995. Political ecology in the 1990s: An evolving view of nature and society. *CASID Distinguished Speaker Series* No. 13, Center for Advanced Study of International Development, Michigan State University, East Lansing, Michigan.

Blaikie, P.M. and Brookfield, H. 1987. *Land degradation and society.* Metheun, London.

Brown, D.G., Pijanowski, B.C., and Duh, D.J. 2000. Modeling the relationships between land use and land cover onprivate lands in the Upper Midwest, USA *Journal of Environmental Management* 59: 247-263.

Brown, D.G. and Vasievich, J.M. 1996. A study of land ownership fragmentation in the upper Midwest. *Proceedings.* GIS/LIS '96 Annual Conference and Exposition. Denver, CO. p. 1179-1186.

Burgess, E.W. 1925. "Growth of the City," in *The City*, R.E. Park, E.W. Burgess and R.D. McKenzie, pp. 47-62. Chicago: University of Chicago Press.

Caspersen, J.P., Pacala, S.W., Jenkins, J.C., Hurtt, G.C., Moorcroft, P.R., Birdsey, R.A. 2000. Contributions of land-use history to carbon accumulation in US forests *Science*, 290: 1148-1151

Drzyzga, S.A. 2000. Land ownership parcelization and forest fragmentation in three northern Michigan counties. *Masters Thesis.* Department of Geography, Michigan State University, East Lansing, MI.

Erickson, D.L. 1995. Rural land use and land cover change: Implications for local planning in the River Raisin watershed. *Land Use Policy* 12(3): 223-236.

ESRI Inc. 1995. Arc/Info v. 7.2. Environmental Systems Research Institute, Redlands.

Fahrig, L. and Merriam, G. 1994. Conservation of fragmented populations. *Conservation Biology.* 8(1): 50-59.

Faminow, M.D. 1998. *Cattle, deforestation and development in the Amazon: an economic, agronomic, and environmental perspective.* Oxford: CAB International.

Forman, R. T. T. 1995. *Land Mosaics: The ecology of landscapes and regions.* Cambridge University Press, Cambridge.

Forman, R. T. T. and Godron. M. 1986. *Landscape ecology.* John Wiley & Sons, New York.

Frohn, R.C. 1998. *Remote sensing for landscape ecology: New metric indicators for monitoring, modeling, and assessment of ecosystems.* Lewis Publishers, Washington, D.C.

Fuguitt, G.V., and Beale, C.L. 1996. Recent trends in nonmetropolitan migration: Toward a new turnaround. *Growth and Change* 27:156-175.

Fuguitt G.V., Brown, D.L., and Beale, C.L. 1989. *The population of rural and small town America*. Russell Sage, New York.

Grand Traverse County Planning Commission. 1997. *Focus 2020: A County Master Plan, 1997*. [Online] Available at: http://www.grandtraverse.org/masterp/. (Viewed on 6 June. 1997).

Gustafson, E.J. 1998. Quantifying landscape spatial pattern: What is the state of the art? *Ecosystems* 1: 143-156.

Homer, C.G., Ramsey, R.D., Edwards Jr., T.C., and Falconer, A. 1997. Landscape cover-type modeling using a multi-scene Thematic Mapper mosaic. *Photogrammetric Engineering and Remote Sensing* 63(1): 59-67.

Hüttl, R. F., Schneider, B.U., and Farrell, E. P. 2000. Forests of the temperate region: gaps in knowledge and research needs. *Forest Ecology and Management* 132: 83-96.

Jensen, J. R. 1996. *Introductory digital image processing: A remote sensing perspective, 2nd Edition*. Prentice-Hall, Inc., Upper Saddle River.

Kufuor, O. K., 1981. An Analysis of Private Land Fragmentation by Land Holdings of less than 11 Acres in Michigan. PhD Dissertation. Department of Resource Development, Michigan State University, East Lansing, MI.

Lambin, E.F. 1997. Modelling and monitoring land-cover change processes in tropical regions. *Progress in Physical Geography* 21:16-27.

Leatherberry, E.C. 1994. Forest statistics for Michigan's Northern Lower Peninsula Unit, 1993. *Resource Bulletin NC-157*. U.S. Department of Agriculture, Forest Service, North Central Forest Experiment Station, St. Paul, MN.

Li, H. and Reynolds, J.F. 1993. A new contagion index to quantify spatial patterns of landscapes. *Landscape Ecology* 8: 155-162.

Lillesand, R.M. and Kiefer, R.W. 1994. *Remote sensing and image interpretation, 3rd Edition*. Wiley & Sons, New York.

Lunetta, R.S., Lyon, J.G., Guindon, B., and Elvidge, C.D. 1998. North American Landscape Characterization dataset development and data fusion issues. *Photogrammeteric Engineering and Remote Sensing*. 64(8): 821-829.

Manson, G.A., and Groop, R.E. 1988. Concentrations of nonemployment income in the United States *The Professional Geographer* 40(4): 444-450.

McGarigal, K. and Marks, B.J. 1993. *Fragstats: Spatial pattern analysis program for quantifying landscape structure*. Corvallis, OR: Oregon State University.

Merriam, G. and Wegner, J. 1992. Local extinctions, habitat fragmentation and ecotones. In Hansen, A.J. and Di Castri, F. (Eds.) *Landscape Boundaries*. Springer Verlag, New York.

Michigan Resource Iinformation System (MIRIS). 1978. *Digital land use and land cover data* [CD-ROM not distributed] State of Michigan, Department of Natural Resources, East Lansing, MI.

Milne, B. T. 1991. Lessons from applying fractal models to landscape patterns. Pages 199-235 in Turner, M. G. and Gardner, R. H. (Eds.). *Quantitative methods in landscape ecology*. Springer-Verlag, New York.

Muth, R.F. 1969. *Cities and Housing: The spatial pattern of urban residential land use*. University of Chicago Press, Chicago.

Naveh, Z. 1991. Some remarks on recent developments in landscape ecology as a transdisciplinary ecological and geographical science. *Landscape Ecology* 5:65-74.

Norgaard, K.J. 1984. Impacts of the Subdivision Control Act of 1967 on Land Fragmentation in Michigan's Townships. Ph.D. Dissertation. Department of Agricultural Economics, Michigan State University, East Lansing, MI.

PCI. 1997. ImageWorks Version 6.2. PCI. Richmond, Ontario.

Puric-Mladenovic, D., Kenny, W.A., and Csillag, F. 2000. Land development pressure on peri-urban forests: A case study in the Regional Municipality of York. Forestry Chronicle 76(2): 247-250.

Riitters, K.H., O'Neill, R.V., Hunsaker, C.T., Wickham, J.D., Yankee, D.H., Timmins, S.P., Jones, K.B., and Jackson, B.L. 1995. A factor analysis of landscape pattern and structure metrics. *Landscape Ecology* 10(1): 23-39.

Rockford Map Publishers, Inc. 1969-1991. *Land Atlas and Plat Map Book.Rockford Map* Publishers, Inc., Rockford, IL.

Saunders, D. A., Hobbs, R.J., and Margules, C.R. 1991. Biological consequences of ecosystem fragmentation: A review. *Conservation Biology* 5: 18-32.

Saunders, D. A., Rowley, I., and Smith, C.T. 1985. The effects of clearing for agriculture on the distribution of cockatoos in the southwest of Australia. Pages 309-321 in Keast, A., Recher, H.D., Ford, H., and Saunders D. (Eds.) *Birds of eucalypt forests and woodlands: Ecology, conservation, and management.* Royal Australian Ornithologists Union, Surrey Beatty and Sons, Chipping Norton, Australia.

Scarascia-Mugnozza, G., Oswald, H., Piussi, P., and Radoglou, K. 2000. Forests of the Mediterranean region: gaps in knowledge and research needs. *Forest Ecology and Management* 132: 97-109.

Stewart, S. I. 1994. The Seasonal Home Location Decision Process: Toward a Dynamic Model. PhD. Dissertation. Department of Parks and Recreation, Michigan State University, East Lansing, MI.

Stynes, D.J., Zheng, J.J., and Stewart, S.I. 1997. Seasonal homes and natural resources: patterns of use and impact in Michigan. *General Technical Report NC-194.* U.S. Department of Agriculture, Forest Service, North Central Research Station. St. Paul, MN.

Theobald, D.M., Gosnell, H., and Riebsame, W.E. 1996. Land use and landscape change in the Colorado Mountains II: A case study of the East River Valley. *Mountain Research and Development* 16(4): 407-418.

Thünen, J.H. von. 1842. Der Isolierte Staat. In Hall, P. (Ed.), *Von Thünen's Isolated State.* Pergamon. London.

Tou, J.T. and Gonzalez, R.C. 1977. *Pattern recognition principles.* Wesley Publishing Co., Addison.

Turner, M.G. 1989. Landscape ecology: The effect of pattern on process. *Ann. Rev. Ecol. Syst.* 20: 171-197.

Turner, M.G., and Gardner, R.H. (Eds.) 1991. *Quantitative methods in landscape ecology.* Springer-Verlag, New York.

U.S. Department of Commerce. 1972. *1970 Census of Housing and Population.* U.S. Government Printing Office, Washington D.C.

U.S. Department of Commerce. 1982. *1980 Census of Housing and Population.* U.S. Government Printing Office, Washington D.C.

U.S. Department of Commerce. 1992. *1990 Census of Housing and Population.* U.S. Government Printing Office, Washington D.C.

U.S. Department of the Interior, U.S. Geological Survey. 1993. *1:100,000-Scale Digital Line Graph (DLG) Data: Hydrography and Transportation - Area 7 Northern Great Lakes States.* [CD-ROM] U.S. Department of the Interior, U.S. Geological Survey, National Mapping Division. Reston, VA.

U.S. Forest Service, Department of Agriculture. 1997. *Great Lakes Fires greater than 10 acres, 1985-1995.* [Online] Available at: http://198.87.4.38/gla/natdist/images/10fireglo.jpg. (Viewed on 12 Jun. 1999).

Villard, M.A., Freemark, K.E., and Merriam, G. 1992. Metapopulation dynamics as a conceptual model for neotropical migrant birds: An empirical investigation. Pages 474-482 in Hagan, J.M. and Johnston, D.W. (Eds.) *Ecology and conservation of neotropical migrant landbirds.* Smithsonian Institution Press, Washington, D.C.

Vogelmann, J.E., Sohl, T. and Howard, S.M. 1998. Regional characterization of land cover using multiple sources of data. *Photogrammetric Engineering & Remote Sensing* 64(1): 45-57.

Walker, R., and Homma, A. 1996. Land use and land cover change dynamics in the Brazilian Amazon: an overview. *Ecological Economics* 18:67-80.

Walker, R., Moran, E., and Anselin, L. 2000. Deforestation and cattle ranching in the Brazilian Amazon: External capital and household processes. *World Development* 28(4): 683-699.

Warbach J.D. and Norberg, D. 1995. Michigan Society of Planning Officials *Trend Future Project: Public Lands and Forestry Trends Working Paper*. Michigan Society of Planning Officials, Rochester, MI.

Weins, J.A. 1992. What is landscape ecology, really? *Landscape Ecology* 7:149-150.

Winter, G.J. and J.S. Fried. 1997. Assessing the Benefits of Wildfire Risk Reduction: A Contingent Valuation Approach (PDF). *Proceedings of the 1996 Society of American Foresters National Convention*. p. 320-325.

Wyckoff, M.A. 1995. *Michigan Society of Planning Officials Trend Future Project: Final Report*. Michigan Society of Planning Officials, Rochester, MI.

Chapter 9

CHARACTERIZING AND MODELING PATTERNS OF DEFORESTATION AND AGRICULTURAL EXTENSIFICATION IN THE ECUADORIAN AMAZON

Stephen J. Walsh[1], Joseph P. Messina[2], Kelley A. Crews-Meyer[3], Richard E. Bilsborrow[4], and William K.Y. Pan[4]

[1] *Department of Geography & Carolina Population Center, University of North Carolina, USA swalsh@email.unc.edu*
[2] *Department of Geography & BSRSI, Michigan State University, USA*
[3] *Department of Geography, University of Texas, Austin, USA*
[4] *Department of Biostatistics & Carolina Population Center, University of North Carolina, USA*

Keywords: land use and land cover dynamics, longitudinal household survey, satellite classification, pattern metrics, cellular automata, landscape simulation.

Abstract We examine human-environment interactions in the Oriente region of the Ecuadorian Amazon through a Geographic Information Science (GISc) perspective. A remote sensing time-series is used to represent LULC (land use/land cover) dynamics, a GIS to assess resource endowments at local and regional settings, a longitudinal household survey to measure socio-economic conditions and changes over time at the farm or household *finca* level, and a community-level survey administered to community leaders, farmers, teachers, women, and health workers in places ranging from tiny communities to the largest city, Lago Agrio to measure infrastructural linkages between households and their communities. Here, we (a) describe the multi-thematic, spatially explicit database assembled to address deforestation and agricultural extensification; (b) indicate changes in *finca* demographic and land use characteristics reported in the 1990 and 1999 household surveys; (c) use a hybrid digital classification approach for repeatable LULC characterizations for selected Landsat Thematic Mapper (TM) and Multispectral Scanner (MSS) images; (d) apply pattern metrics to classified satellite data to assess the composition and spatial organization of the landscape through trajectories of landscape structure; and (e) use a cellular automata (CA) approach to simulate LULC patterns for antecedent and future periods, given historical and current LULC patterns, as defined through the satellite time-series and hypotheses about the importance of geographic access to and spatial diffusion from the region's central city.

Preliminary analyses of the longitudinal household surveys indicate the importance of land subdivisions since 1990 in sample *fincas* located in close proximity to Lago Agrio and other main towns in the region, and as a consequence, a reduction in the mean size of family land holdings. Analyses of the pattern metrics, calculated from the classified satellite time-series, suggest that the Oriente is undergoing a conversion of LULC types from high-density forest to agriculture as well as to low and medium density forest through secondary plant succession. These changes result from (a) initial deforestation by migrant households settling along roads built by petroleum companies in the early 1970s; (b) further clearing of forests to support subsistence agriculture and increasingly commercial cultivation; and (c) secondary plant succession of agricultural lands related to declining soil fertility and pasture formation. Simulation models indicate the important role of the central city, Lago Agrio, and of road networks that connect farms to community services and off-farm employment. CA models indicate that, over time, urbanization will become more significant and land fragmentation through deforestation and agricultural extensification more concentrated in areas currently remote from Lago Agrio. Areas surrounding Lago will become more mature, with a mix of crops, pasture, and residential/commercial development, while high-density forest areas, previously concentrated in large tracts will be converted to heterogeneous low to small medium density forest tracts, on *fincas* intermixed with agriculture and pasture.

INTRODUCTION

In Ecuador's Amazon, a region known as the Oriente, small-scale farmers have been the primary direct agents of deforestation (Bromley 1989, Rudel and Horowitz 1993, Southgate and Whittaker 1994, Pichon and Bilsborrow 1999), although forests were initially cleared to lay oil pipelines and roads. The usual sequence of land clearing by migrant settlers is, first, small areas of food crops for subsistence, then cash crops, and finally pasture as soil fertility declines. Eventually, settlers may abandon land and move on to start new farms elsewhere, as has been common in Brazil, but rare in Ecuador because of better soils and less land available for pioneer activities. Farmers in the Oriente tend to cultivate newly cleared patches of land within their *fincas* (farms of approximately 50 ha in size, 250 m x 2 km) as soil fertility declines, economic opportunities occur, or land is sub-divided through kinship ties and/or formal land sales.

Drawing upon a growing body of human-environment literature (e.g., Bilsborrow 1987, Marquette and Bilsborrow 1994, Jolly 1994, Moran et al. 1994, Wood and Skole 1998, Walsh et al. 1999), the effects of the growing frontier population on the local environment may be seen as occurring through a number of contextual or mediating factors representing political, human, and landscape ecology theory. Political ecology (Blaikie and Brookfield 1987) is concerned with those factors and forces that are imposed upon a system as context and include socio-cultural norms (e.g., the desire to raise cattle, and gender roles), the level of living and technology, local and national government policies, commodity prices at local to global scales, and physical infrastructure (e.g., roads). Human ecology argues that people are active agents on the landscape that shape and are shaped by it (Pattison 1969). Activities such as the gathering of fuelwood, land use conversion, and the out-migration of young adults because of excessive land fragmentation are examples. Landscape ecology points to the interactions between landscape composition, spatial organization, and time as factors affecting human behavior (Walsh et al. 1999).

Based on these theories as well as prior research in the study site, we map, model, and visualize LULC dynamics and suggest several key structural characteristics of the landscape over space and time, which suggest how people are interacting with the land, and how the land may be affecting human behavior within a system influenced by feedbacks and thresholds. Land use changes in the Ecuadorian Oriente depend partly on existing patterns of land use, which may involve critical points where a small amount of land use change significantly alters feedback processes and leads to a new pattern or equilibrium state.

We have assembled a Landsat Multispectral Scanner (MSS) and Thematic Mapper (TM) time-series to characterize LULC dynamics over a nearly quarter century, and also collected household survey data on a representative sample of migrant colonists and a community survey of services and facilities at such locations. DEMs and derived terrain and landscape variables are used to describe resource potentials and geographic accessibilities for the 418 *fincas* first assessed in 1990 and revisited in 1999. The result is a powerful data infrastructure appropriate to explore LULC dynamics through mapping, modeling, and scientific visualization. Finally, sets of spatially explicit tools have been assembled to integrate the data and generate images and interpretations of changes in landscape structure over time and space. LULC patterns and scenarios of the past and present as well as for future periods have been generated for discrete time periods at scales ranging from the local to the regional.

The first objective of this paper is to describe the diverse data sets assembled to examine deforestation and agricultural extensification in the Oriente, and how they have been integrated into a GIS. We then discuss how selected socio-economic conditions at the farm-level have changed over the nearly 10-years between the surveys; how remote sensing digital classifications are applied to categorize LULC; how ecological pattern metrics are used to quantify the composition and spatial organization or structure of LULC so that landscape form and function are linked; and how LULC simulations are derived through cellular automata approaches where transition probabilities are used to generate

LULC scenarios for selected time-frames based on statistical estimates of the drivers of LULC dynamics.

STUDY SITE

Lying in the westernmost part of the huge Amazon basin, the Ecuadorian Oriente serves as a unique and ecologically significant laboratory for research on human-induced Amazon LULC dynamics and change associated with growing population settlements. The site contains several areas of extraordinary biodiversity, making it a priority conservation zone. Indeed, the region is considered one of the 11 ecological "hot spots" in the world (Myers 1988, Myers et al. 2000), so the ongoing deforestation has important implications for biodiversity, resilience, sustainability, and their mediation through government and non-government environmental policies.

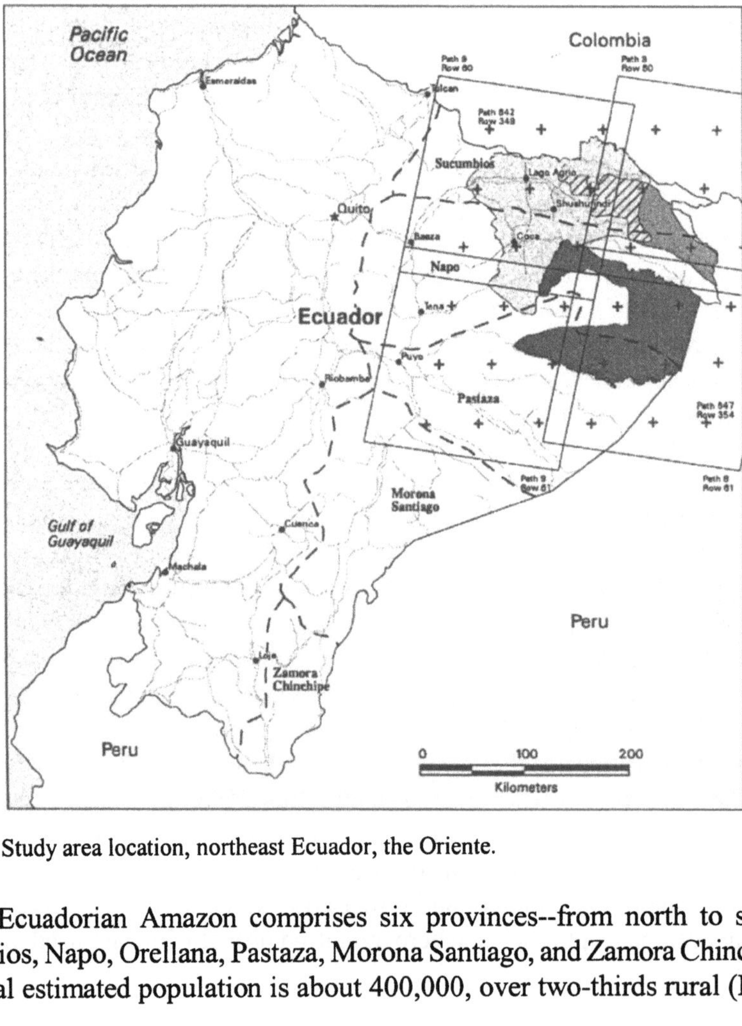

Figure 1. Study area location, northeast Ecuador, the Oriente.

The Ecuadorian Amazon comprises six provinces--from north to south, Sucumbios, Napo, Orellana, Pastaza, Morona Santiago, and Zamora Chinchipe. The total estimated population is about 400,000, over two-thirds rural (INEC

1991). Rapid population growth has occurred since the early 1970s, with almost half its population born outside the region. Government policies encouraged migration, as the Oriente was perceived as an area with almost infinite space and resources and thus an "escape valve" to relieve socio-economic imbalances in other regions, especially pressures on the land in the Sierra or Highlands (Pichon and Bilsborrow 1999). Settlers in the Oriente are small, poor farmers, who have migrated to the region in search of land, to support their families, rather than often being large ranchers, as in Brazil. Most have received little or no benefit from government subsidies, credit or agricultural extension. While there is no large-scale timber extraction, much small-scale cutting of trees occurs, on both settler plots and indigenous lands. Having no distinct dry season, little "slash-and-burn" techniques are used; instead "slash-and-mulch" approaches are more commonly followed.

Since 1990, the Oriente has become more densely populated, due to both population growth and to the fact that many large areas have been set-aside as national parks and conservation areas or have been legally titled to indigenous populations. Thus, since then, colonist families have been sub-dividing plots to hand down parcels to their children as well as selling portions to new migrants coming into the region. With the Ecuadorian economy still generally depressed, since the early 1980's, migrants are still indeed, arriving in significant numbers, in search of work in the burgeoning towns as well as for a piece of land. Population also continues to grow rapidly from natural population growth, or the excess of fertility over mortality of the existing population: thus total fertility rates are much higher than elsewhere in Ecuador, still close to 6 births per woman (vs. 3.3 in the rest of the country, according to the latest national demographic survey in 1999 (CEPAR 2000)).

Access to the Oriente was made possible initially by a petroleum boom, which began, in the early 1970's. Once oil was discovered the oil companies built roads to lay pipelines, which eventually crossed the Andes to the Pacific Coast at Esmeraldas. Since the early 1970's petroleum has generally provided over one-half of both Ecuador's merchandise export earnings, and federal government revenues. Once the roads were built, migrant settlers poured in, mostly from the crowed Sierra, expanding the agricultural frontier eastward. In addition to the individual farm dwellings, clusters of dwellings also evolved around major road intersections and petroleum encampments, and sometimes grew into market towns. Lago Agrio is the largest city in the region, but still with only about 30,000 people in 2000.

The semi-permeable barriers of the major rivers and internal and external political boundaries shape transport, trade, and communications, as the region borders Colombia and Peru. While the Ecuadorian government facilitated the provision of permanent ("escrituras") or provisional ("certificados de posesion") land titles to in-migrant families to encourage settlement in the 1970's and 1980's through the Instituto de Reforma Agraria y Colonizacion (IERAC), IERAC was eliminated and replaced by another far weaker government agency,

INDA, in 1993, so most new arrivals and children inheriting plots do not have legal land titles. The resulting increased insecurity of land titles may be influencing subsequent land use practices in new ways, as well as the permanence of migrant settlement.

Although the Ecuadorian Amazon generally has better soils than the rest of Amazonia (e.g., much of Brazil) -- which is characterized by sedimentary soils of low fertility, high acidity, and aluminium toxicity -- it is still subject to a rapid decline of nutrients from agricultural use, particularly when the ground is not covered continuously by vegetation. Pockets of high fertility volcanic soils and rich alluvial riverine soils alternate with poor quality red soils throughout the study area. Much of the land is also surprisingly hilly and susceptible to soil erosion, and most is easily compacted by cattle.

SPATIALLY EXPLICIT DATA

All socio-economic, biophysical, and geographical data collected to support the goals of this research have been integrated into a GIS database, creating a powerful data storage and retrieval system that is spatially explicit, temporally referenced, and thematically diverse. Below, we describe the central data sets collected for this research with emphasis on the spatial- and temporal-explicitness of the information.

Household Surveys

A unique feature of the project is the implementation of a 1999 socio-economic household survey in the same precise geographic sites or farms as were covered in a 1990 survey. This has produced a longitudinal household-level data set for a statistically representative sample of farm plots covering the entire major area of colonization in Ecuador's Amazon. Thus in 1990, the first University of North Carolina -- Chapel Hill survey in the region was carried out, with support from the U. S. National Science Foundation and the World Wildlife Fund. A two-stage sampling design was used to select migrant farm households, who were virtually all spontaneous settlers. The sample frame used was a list of farm "sectors" (aggregates of farm plots sites) made available for settlement by colonists by IERAC, the Ecuadorian land titling agency. In the first stage, sectors were sampled with probabilities of selection proportional to estimated size, (i.e., proportional to the number of plots in each sector). Thus, 64 sectors were selected from the total of 275 sectors settled by that date in the jurisdiction of the Aguarico and Coca offices of IERAC. In the second stage, *fincas* were selected from each sample sector again roughly in proportion to the size of the sector, with a cluster of 5 to 10 contiguous plots selected randomly from each sector (all farm plots located on a selected *finca* were included in the sample). Once interviews were completed (with only a 1% refusal rate), the final completed sample comprised 419 settler plots located on 405 *fincas*, representing about 6% of all colonist plots in the main lowland settlement areas of the northernmost Amazonian provinces.

In 1999, the same plots were re-visited (except for one sector containing 10 *fincas* on the Colombia border that was excluded for security reasons), and all farms and sub-divisions within the same geographic space were re-visited (and all new households and subdivisions separately interviewed), yielding a sample of 767 farms, plus 109 additional residential urban lots or "*Solares*" created by "parcelization" near several major towns, reflecting incipient urbanization. Detailed questionnaires were again, as in 1990, administered to both the head of the household (almost always male in this frontier region) and the spouse separately, and a community-level survey was also administered (in 2000). The questionnaire of the head included information on plot acquisition, land size, and tenure; agricultural production and inputs; livestock; contact with government agencies and non-government agencies, use of credit and extension services; use of hired labor and work off the farm; migration history, future intentions to migrate or stay, and aspirations for children. The spouse provided information on household composition, including out-migration of members from the farm since they came to the Amazon; housing conditions, water and fuel sources, household possessions; health, fertility and mortality; gender roles and environmental problems. In addition, the spatial location of crops and other forms of land use and changes since 1990 were noted on sketch maps for each farm, including the location of all sub-divisions and houses. The location of dwellings, plots, communities, roads, etc. were also geo-referenced (using hand-held GPS units) to link farm survey data to other information in the GIS database. Community questionnaires were administered to community leaders, farmers, teachers, and groups of women and health workers, in 60 places ranging from tiny communities to the largest city, Lago Agrio.

GIS Coverages

Contour lines and spot elevations on the 1:50,000 scale maps of the study area have been digitized to yield a digital elevation model that has been processed to produce slope angle, slope aspect, slope position, slope convexity/concavity, soil moisture potential, solar radiation potential, and flood inundation potential. Point elevations, terrain sinks, and hydrography have been integrated with contour information to determine drainage potential for a number of terrain derivatives.

Relative Slope Position, a measure of slope position as a percentage distance from a particular point from the nearest valley to the nearest ridge, was defined by surface flow hydrological analysis using a depressionless DEM (O'Callaghan and Mark 1984). Slopes were measured by identifying flow accumulation (the accumulated weight of all cells flowing into each downslope cell) based on flow direction (the direction of steepest descent from one cell to another) (ESRI, 2000). Slope Shape (planform curvature) is the product of three measures of curvature -- planform curvature (curvature perpendicular to slope), profile curvature (curvature parallel to slope), and overall curvature (profile and planform curvature combined). Since curvature is related to a surface's rate and degree of drainage, planform curvature most affects the distribution of downhill flow (McKnab 1989). Using such measures, the Landform Index was computed following the procedure implemented by McKnab (1993). The Topographic Relative Moisture Index (Parker 1982) was also computed to stress the importance of local landform, including exposure to wind, solar radiation, and cumulative surface flow, in inferring local moisture patterns.

Roads and communities were also captured from the 1990, 1:50,000 scale base maps and entered into the database. Roads were categorized by surface type and fieldwork linked estimated travel speeds to each of the road types. Geographic accessibility estimates were generated for an updated road network for each surveyed *finca* to the central town, Lago Agrio, using Euclidean distance, network distance, and accumulated travel time (McDaniel 2000). A shape file and locational coverage of all sample *fincas* and development sectors were generated for linking household survey data to the remotely-sensed images and the derived GIS coverages of a farm's resource endowments. The GIS was also used to calculate the topologic relationships of farms to other farms and to other key landscape features; the measures were also placed within the database.

Linkages between the household survey data and the GIS and remote sensing data sets have been developed within the Arc/Info software systems. Relational joins and key fields were identified so that we could link farm and household attributes from our 1990 and 1999 surveys with LULC dynamics from the satellite time-series (ESRI 2000). Linkages between locational and non-locational information within the database also facilitated the development, calibration, and verification of the LULC simulations for antecedent and subsequent time periods. A geodetic control network was established through GPS technology so that a well populated and spatially-referenced set of control points could be used, for example, to geometrically correct our satellite images and to conflate GIS coverages integrated in the spatially explicit data base. A GPS base-station in Quito operated by Aeromapa was contracted to collect reference coordinates so we could perform differential corrections of our rover files.

Remote Sensing Data

A satellite time-series was assembled that extends from the mid 1980s to the late 1990s. While some synthetic aperture radar (SAR) data sets have been obtained for cloud penetration, soil moisture mapping, and sub-canopy assessments, optical systems have been primarily emphasized. A host of Landsat TM images (and now some 1999 Ikonos data) have been acquired, preprocessed, and classified for Paths 8 & 9/Rows 60 & 61. The currently assembled time-series includes images for 16 December 1984, 23 January 1987,

23 April 1988, 22 January 1989, 22 December 1989, 28 April 1990, 19 February 1991, 14 July 1992, 11 August 1996, 3 September 1996, 21 October 1996, 30 August 1997, 6 September 1997, 15 November 1999, and 9 November 2000. Standard geometric and radiometric corrections have been applied for image rectification, digital enhances to increase the acuity of the data, classifications generated for information extraction, and image change-detections applied for characterizing LULC dynamics (Jensen 1996).

METHODS

Empirical Models of Household Decision-Making

Each land use decision a migrant colonist farmer makes at a given point in time has implications for other forms of land use. More land may be put into pasture, for example, by converting land from perennial crops, annual crops, or fallow land, or by clearing primary forest. Land taken out of crops may be turned over to pasture or left to regenerate through secondary plant succession. Thus, land idle or land in regeneration as secondary forest in 1999 (a small fraction of the total in 1990) was included in pasture land in the 1990 cross-sectional statistical analyses. Below, we present a few tables that compare differences in socio-economic characteristics of farm plots surveyed in 1990 and 1999. An unexpected finding in the 1999 follow-up survey was the widespread subdivision of *fincas* through inheritance and land sales. Informal and formal land transactions occurred throughout the sample sectors even in remote areas, but the creation of *solares* or housing lots was common near towns, especially Lago Agrio and Joya de Las Sachas. Geographical accessibility to Lago Agrio seemed to be one major driver of subdividing land into *solares*. The following tables illustrate elements of the subdivision phenomena as well as differences in demographic and socioeconomic characteristics of settler plots between 1990-1999. A more comprehensive analysis of farm characteristics and their implications for deforestation and agricultural extensification will be presented in subsequent papers; here, the intent is only to provide preliminary findings.

Table 1 shows some of the demographic differences of households on plots surveyed in 1990 and 1999. Two of the most important components are the population and distance to market variables. Total population nearly doubled during the time period. Recall that the 1999 survey revisited the same *fincas* as in 1990 so this increase is due mainly to the subdivision of property. The service sector is expanding at the expense of the primary sector, although small subdivided *solares* or lots generally support small-scale in garden plots. The average distance to market decreased from 28.2 km to 20.4 km. This decrease indicates properties closer to market towns tend to be subdivided more. This provides better access for people working in Lago Agrio and other central market towns associated with the expansion of the service and non-farm employment sectors. It is also evident that electricity is starting to reach more sectors along that main road. The smaller proportion with legal title to their plots is a consequence of subdivision combined with replacement of IERAC by a completely inactive government land reform agency in 1993 called INDA.

Table 1. Comparison of 1990 vs. 1999 household demographics, assets, and tenure characteristics. * Excludes population living in *solares* (housing lots).

	1990 (N = 418 HH's)	**1999 (N = 763 HH's)**
Average people per HH	6.6	5.7
Population (farms only)*	2761	3813
Average distance to market	28.2 km	20.4 km
Road distance	24.2 km (n=394)	19.1 km (n=682)
Walking to road	5.3 km (n=214)	2.6 km (n=333)
Canoe	34.1 km (n=35)	18.7 km (n=30)
Head born in Oriente	4.6 %	8 %
Electricity in dwelling	14 %	32 %
Own chainsaw	30 %	20 %
Have full legal title	50 %	34 %
Gini Coefficients for Landholdings	0.26	0.50

Table 2 illustrates differences in demographic characteristics, land tenure, and credit of long-term owners versus those becoming heads of land parcels since 1990. The data suggest the latter are more likely to engage in off-farm employment, less likely to have an intact farm, a land title, or to have ever received credit than those already living in the region in 1990 (time of the first survey). Legal title facilitated receipt of credit, usually used to purchase cattle on lands converted from forest, to cultivated crops, then to pasture.

Table 2. Differences in Plots of Long Term (>9 yrs) vs. New Owners, 1999.

	Long Term	**New Owner**
Number of households	315	393
Head born in Oriente	6 %	12 %
Off-farm employment	49 %	60 %
Hire day laborers	48 %	40 %
Intact farm (no subdivision since 1990)	40 %	21 %
Hold Full Legal Title	53 %	21 %
Receipt of credits/loans	34 %	12 %
Cattle	75 %	31 %

Table 3 provides summary data on differences in land use. The single most important point illustrated in Table 3 is the dramatic decline in farm size with the introduction of new farms created through land subdivision. With smaller plots, the new owners have less land in forest and pasture and more in the more intensive forms of land use - perennials (mostly coffee) and annual crops.

Table 3. Long Term (>9 yrs.) vs. New Owner Land Use, 1999.

Land Size and Use	Long Term	New Owner
Mean Plot Size	34.2 ha	18.4 ha
% Forest	42 %	36 %
% Pasture	30 %	23 %
% Perennial Crops	21 %	287%
% Annual Crops	9 %	14 %

LULC Classification

A multi-phase hybrid classification approach was used to categorize LULC within the study area for all image dates within the assembled time-series. The approach was designed to be repeatable across all images in the time-series, rely upon image characteristics through derived statistical measures, limit the reliance on in-situ data in the classification process because of the antecedent nature of time-series data and the difficulty of validating historical LULC classification, and strategically integrate acquired aerial photography with expert knowledge of the landscape achieved through local informants and team experiences in the region.

Described by Messina et al. (2001), we used a supervised hybrid classification approach for LULC characterization across the assembled time-series that involved the ISODATA decision-rule to define 100 "naturally" occurring spectral classes. The initially defined classes were subsequently reduced to about 30 classes through the interpretation of the transformed divergence and divergence statistics generated as outputs from the classification process. ERDAS Imagine software was used for the classifications (ERDAS 1997). Following the unsupervised classification, a supervised classification was applied using a maximum likelihood classifier to relate unclassed pixels to the 30 spectral classes (i.e., the training data) defined through the unsupervised classification. This approach allows for the generalization of classes to a few key LULC types as well as the expansion of cover types to additional classes.

Pattern Metrics

Fragstats software (McGarigal and Marks 1993) was used to characterize the composition and spatial organization of LULC patterns represented through the hybrid digital classifications of selected time-frames for the Ecuadorian study site. Pattern metrics were generated at the landscape level for the northern Intensive Study Area (ISA), which is centered on Lago Agrio and measures approximately 90,000 ha (see Figure 2). Theoretically, the metrics are associated with the science of landscape ecology (Forman and Godron 1986), in which scale, pattern, and process are explicitly interrelated and where landscape form is used to examine landscape function.

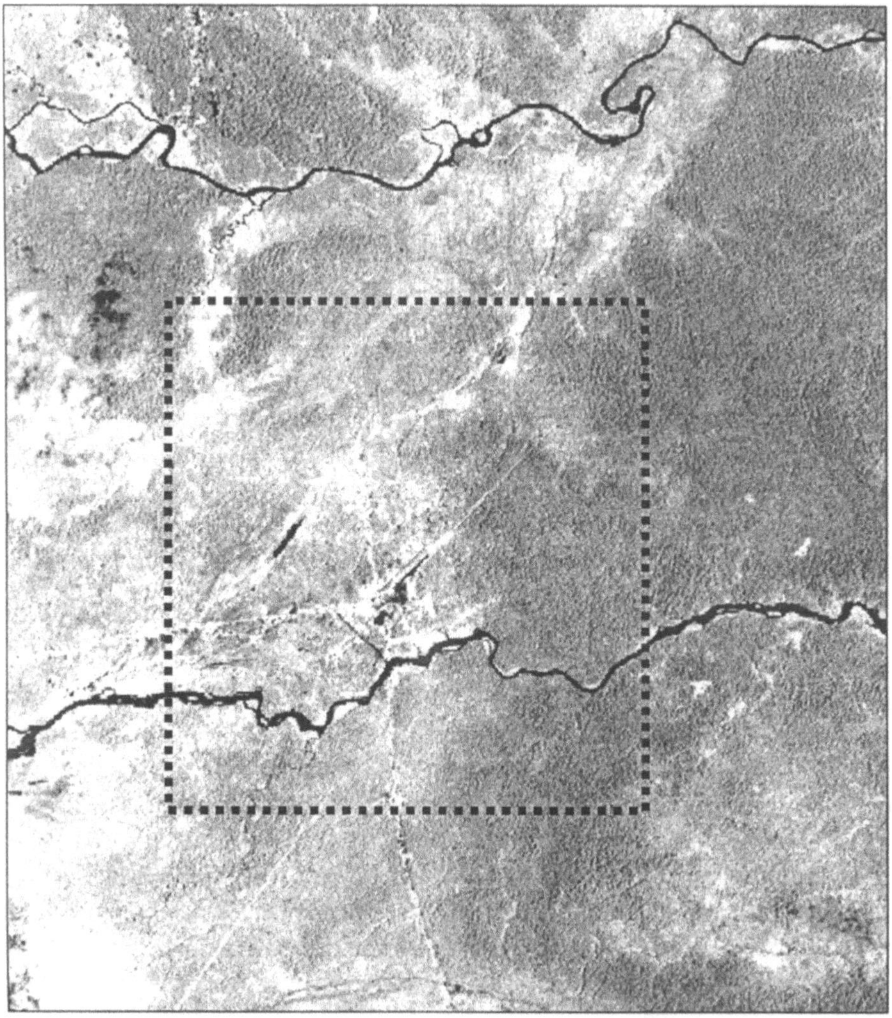

Figure 2. Band 4 Landsat MSS data with the boundary of the Northern ISA identified, and area of 90,000 ha.

ANALYSIS

Statistical Models of Land Use and Land Cover

Pichón (1999) and Pan et al. (2001) conducted cross-sectional multivariate regression analyses based on the 1990 and 1999 household surveys to examine the determinants of LULCC. The models have a similar political-ecology approach, but the statistical techniques and variable definitions are slightly different. Pichón (1999) used multinomial Tobit estimation in which the dependent variables were the *proportions* of land in different forms of land use, while Pan et al. (2001) used a censored General Linear Multivariate Model where the dependent variables were the absolute amounts of land in each form of land use (defined as in Pichon). The presence of the size of plot as an explanatory variable in Pan et al. (2001) is intended to control for size effects, so that the interpretations of the remaining explanatory variables may be compared to the degree definitions are similar. Thus, the dependent variables are the proportions (in Pichon) or areas in ha (in Pan et al.) of land in perennial crops, annual crops, pasture, and forest (primary and secondary combined). While most independent or explanatory variables have the same definitions, to facilitate testing for differences in structural effects over time in the factors linked to variations in land use, there are slight differences, as indicated in Table 4. In addition, a few variables (not shown) were included in the Pichon model which we consider endogenous so they were excluded, so the results of the two models cannot be directly compared. However, they are sufficiently similar that insights regarding changes in the determinants of LULCC can be made.

First, the demographic variables such as region of origin, education, and household size do not have consistent effects over time, with number of consumers in 1990 having a strong negative relationship with forest cover, but not in 1999. The signs of the modest effects of children in 1999 are at least logical, as children eat annual crops but not coffee. The overall weakness of demographic variables in 1999 may be because their effects were already incorporated in other included variables, notably household labor and plot size. Ecological factors such as soil and topography may also have different apparent effects because of flatter areas being closer to roads and far more likely to experience subdivisions. The distance variables continue to be the most powerful factors of all, especially the effects of walking distance on forest cover. Plot size has been discussed above, with its effects in the 1999 model functioning mainly as a control variable, with larger plots having more land in all forms of land use. Duration of residence has fairly consistent effects in the same direction. Owning additional land, better measured in 1999, has fairly consistent effects. Having a legal title to the land appears to have different effects, but this may be due to its being correlated with duration of residence (see Table 2). The active government land titling agency of IERAC was replaced in 1993 by a very weak organization, INDA, essentially ending the possibility of obtaining legal land titles for those acquiring subdivisions. The last three variables measure labor availability on the farm and off. The results for the two time periods are nearly identical, indicating the importance of labor availability in land use decisions of these migrant farmers living in essentially a labor-scarce environment. Both more household labor and hired labor is linked

Table 4. Multivariate analyses of land use patterns (1990 results from Pichón and Bilsborrow 1999; 1999 results from Pan et al. 2001).

1990 vs. 1999 Variable Definition	Perennial Crops	Annual Crops	Pasture	Forest
Non-Native (90) vs. Born in Oriente (99)	+ ** NS	- * + *	NS - *	NS NS
Education of Head	NS NS	+ *** NS	+ *** NS	NS - ***
Consumers (90) vs. Children (99)	+ *** - *	NS + *	+ * NS	- *** NS
Good Soil (90) vs. Black Soil (99)	+ ** NS	+ ** + ***	NS NS	- *** NS
Hilly (90) vs. Flat (99) Terrain	NS NS	- *** + ***	- *** NS	+ *** - ***
Plot Size (ha)	- *** + *	- * + ***	+ * + ***	+ *** + ***
Road Distance to Market (km)	- *** NS	NS NS	NS - ***	+ ** + ***
KM Walking to Road (90) vs. Need to Walk (99)	- *** NS	- ** NS	- ** - ***	+ *** + *
Years Living on the Farm	NS + ***	NS NS	+ *** + ***	- * - ***
Title	- *** NS	- *** NS	NS + ***	+ ** - ***
Own Multiple Plots (90) vs. Ha of Additional Land (99)	NS - ***	NS - ***	- *** NS	+ *** NS
Household Labor	+ *** + ***	+ ** + *	NS + *	- *** - *
Hired Labor	+ *** + ***	+ ** NS	NS + ***	- *** - ***
Off-Farm Employment: Months (90) vs. Earnings (99)	- *** - ***	NS NS	- *** NS	+ *** + ***

Statistical significance: * p<0.05; ** p<0.01; *** p<0.001; NS = Not Significant.

to more land in annual and perennial crops, as is expected given the low labor demands of raising cattle. The more household labor is allocated to off-farm employment, resulting in less perennial crops (coffee requires a lot of labor to pick the beans by hand) and more land in forest. Many surveys conducted in the Amazon have focused only upon social conditions to build LULCC models, but the results of Table 4 show that many factors impact LULC decision-making.

Pattern Metrics and Landscape Structure

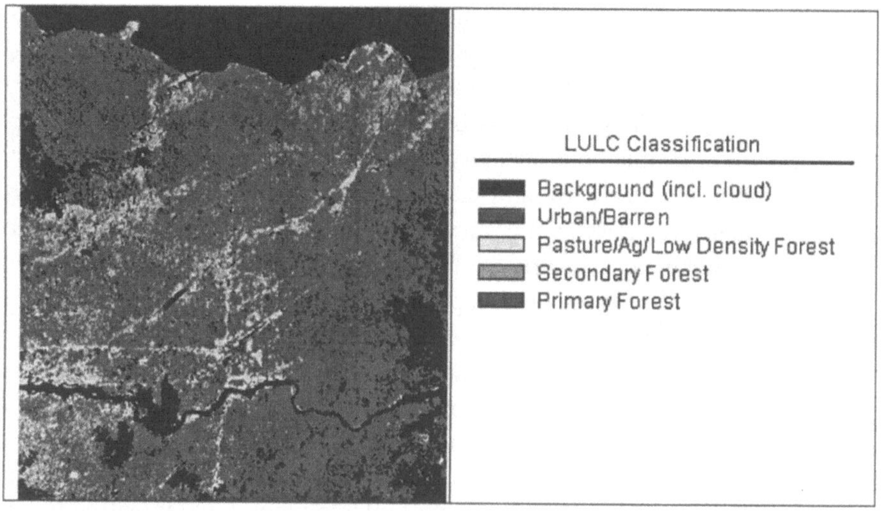

Figure 3. The Ecuadorian Oriente, the Northern ISA classified image for 1973.

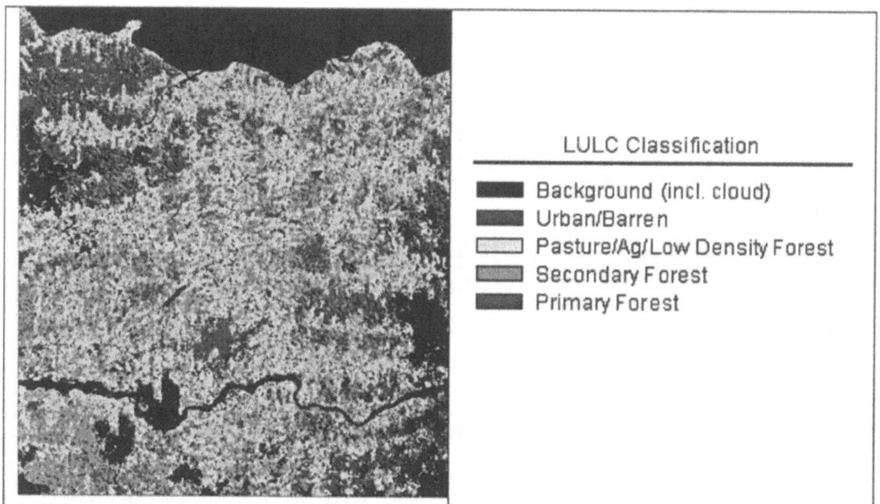

Figure 4. The Ecuadorian Oriente, the Northern ISA classified for 1999.

To assess the changing nature of the Oriente landscape, a time-series of satellite imagery was assembled, comprised of Landsat MSS and TM scenes from 1973 (see Figure 3: our initial conditions for simulation modeling), 1986, 1989, 1996, and 1999 (see Figure 4). After each of the images was resampled to 30 x 30 m for compatibility, Fragstats (McGarigal and Marks 1993) was used to generate pattern metrics for both the landscape as well as individual classes to quantify spatial organization and composition. Table 5 summarizes the selected pattern metrics run for the Northern ISA for 1973, 1986, 1989, 1996, and 1999. Figures 5a - 5h indicate the changes in each metric for the five time periods. Figure 5a is a plot of the number of patches occurring at the landscape level, a measure of fragmentation. Figure 5b is a plot of patch density, a measure of structural fragmentation. Figure 5c is a plot of edge density, a measure of standardized edge on a per unit area basis. Figure 5d is a plot of mean patch size, a measure of spatial heterogeneity. Figure 5e is a plot of mean distance to nearest neighbor, a measure of landscape pattern. Figure 5f is a plot of nearest neighbor standard deviation, a measure of patch dispersion. Figure 5g is a plot of landscape shape, a measure of landscape diversity defined through a perimeter to area ratio for the landscape as a whole; and Figure 5h is a plot of interspersion/juxtaposition, a measure of adjacency.

Comparing the metrics for each year, it is apparent that in 1973 the northern ISA was dominated by high density forest at over 80% of the land cover, and average patch size over 50 ha (as opposed to 1-3 ha for the other three land cover types). High density forest also showed the greatest patch contiguity and closest like-neighbors, as well as the lowest interspersion / juxtaposition index, thereby suggesting a solid matrix of high density forest with little else in terms of significant patches of alternate cover-types.

Table 5. Landscape level pattern metrics by year of analysis.

	1973	**1986**	**1989**	**1996**	**1999**
Number of Patches	6607	9710	9997	11,523	8371
Patch Density (#/100 ha)	9.85	14.47	14.90	17.18	21.48
Mean Patch Size (ha)	10.16	6.91	6.71	5.82	8.02
Edge Density (m/ha)	49.90	96.42	107.75	112.68	98.71
Landscape Shape	56.46	86.59	93.92	97.12	88.07
Double Log Fractal Dimension	1.46	1.45	1.47	1.43	1.44
Mean Nearest Neighbor (m)	93.27	90.57	85.44	85.81	97.17
Nearest Neighbor Stand. Deviation (m)	137.12	101.16	93.08	78.38	109.9
Interspersion/ Juxtaposition (%)	49.23	80.38	87.00	91.14	74.97

In sharp contrast, by 1986 one-half of the landscape was still covered by high density forest, but one-third of that landscape was occupied by pasture/agriculture. However, at this point in time high density forest still had the fewest, but largest patches (and lowest patch density), as well as the lowest interspersion/juxtaposition index, and was effectively still functioning as the background matrix. Three years later, the preponderance of high density forest had again been reduced, with pasture/agriculture increasing to 40% of the total land cover. Meanwhile, mean nearest neighbor and interspersion/juxtaposition index values of high density forest continued to increase, representing greater fragmentation and lesser dominance. By 1996, the conversion from a landscape dominated by high density forest to one dominated by pasture/agriculture was complete. High density forest represented one-third of the landscape, while pasture/agriculture represented 45%. The remaining land cover was evenly distributed between the other two classes (urban/barren and medium density forest), although these classes, by far, had the greatest number of patches and the greatest patch density, indicative of potential future change. By 1996 the largest patches were pasture/agriculture, having replaced high density forest.

The trajectories of increasing fragmentation of high density forest, increasing consolidation of pasture/agriculture, and a transition in the dominance of the former to the latter appear to have tapered off by 1999, when the conversion to pasture/agriculture stabilized at just under 45% of the landscape. An average patch size of 20 ha, lowest number of patches, and lowest patch density characterized pasture/agriculture as the matrix land cover type, although the higher interspersion/juxtaposition rate suggests (1) a recently consolidated and relatively unstable ecological matrix, and (2) a dominant land cover type whose nature is inherently fragmented in its spatial organization. Visual inspection of the classified images reveals this fragmentation in the traditional piano key or fishbone pattern of deforestation noted in other Amazonian areas and related to household LULC dynamics.

Focusing momentarily upon each class as it changed over time, it is apparent that the urban/barren class never had much of an impact on the total land cover, and in fact stabilized relatively early and confined itself to the town of Lago Agrio. However, as land around Lago Agrio becomes less available for settling and the area transitions from a frontier environment to a more market-oriented one, land subdivision and residential development indicates an urban expansion along transportation corridors. As mentioned above, the pasture/agriculture class went from barely existing in 1973 to dominating the landscape by 1996, to becoming a seemingly more stable (from a land cover perspective) matrix by 1999. Concurrently, the vast expanses of high density forest that covered 85% of the landscape in 1973 steadily declined to 20% by 1999, with the vast majority located in the northern part of the study site. Medium density forest showed the least consistent trends among land cover types, likely because it is effectively a successional category, and given the tremendous bio-growth rates in the Amazon, it can appear almost as quickly as it can disappear.

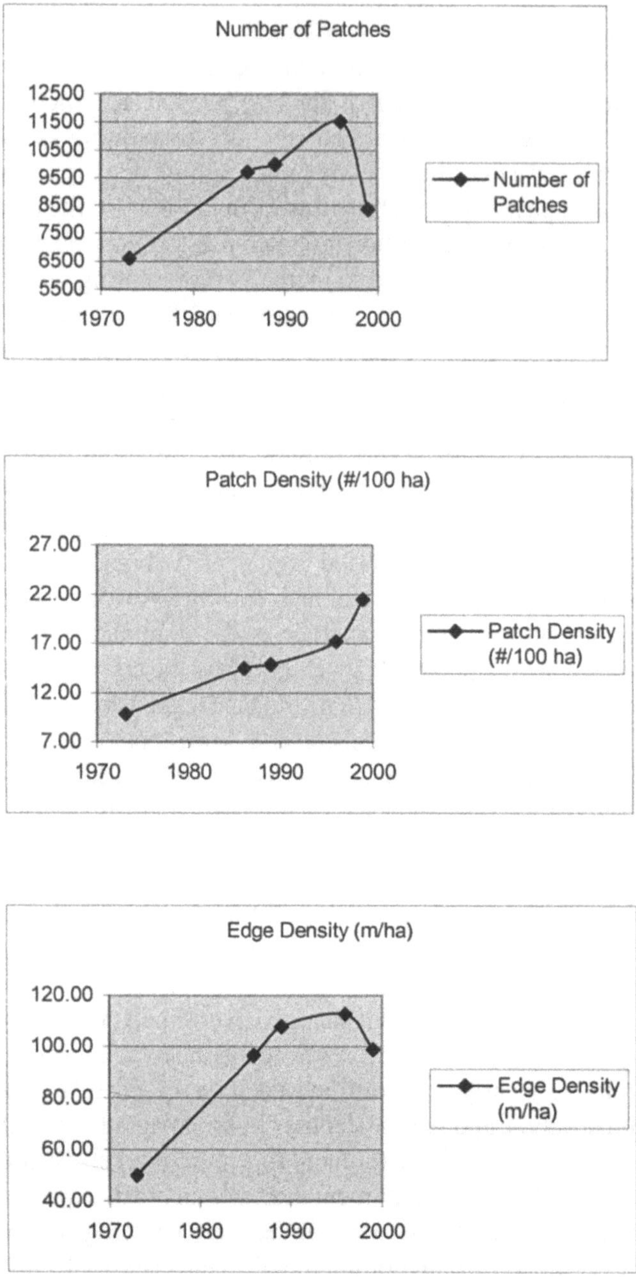

Figures 5a, 5b, and 5c. Changes in metric performance for the five time periods assessed: number of patches, patch density, and edge density.

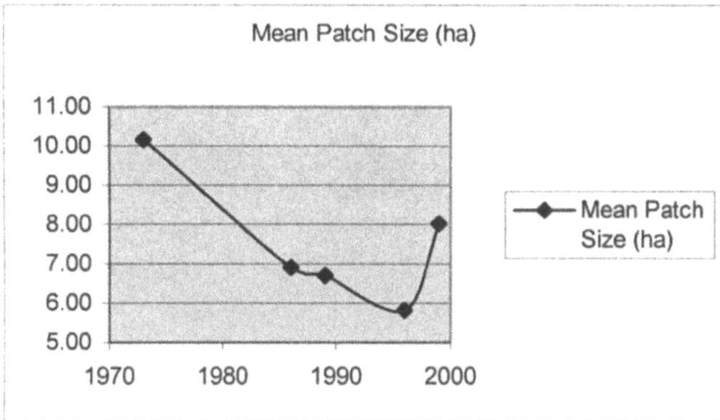

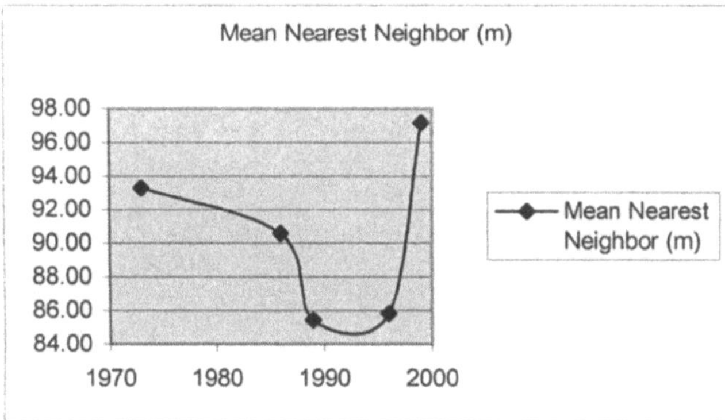

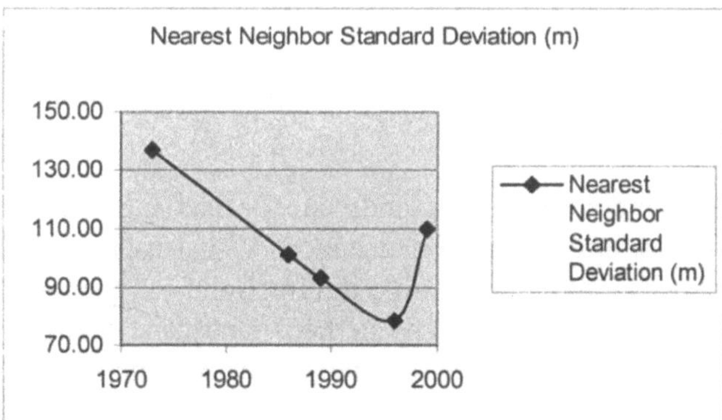

Figures 5d, 5e, and 5f. Changes in metric performance for the five time periods assessed: mean patch size, mean nearest neighbor, and nearest neighbor standard deviation.

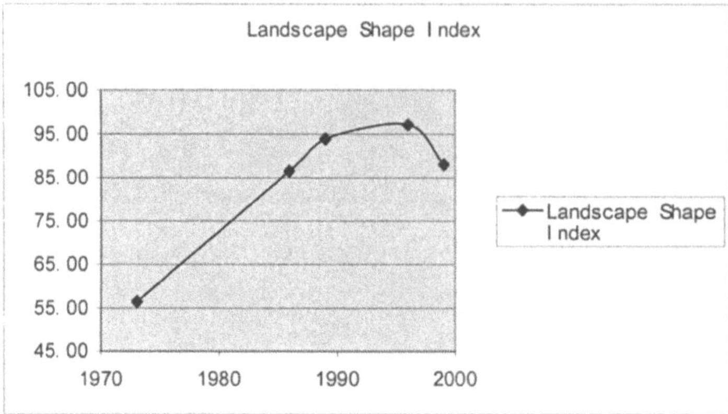

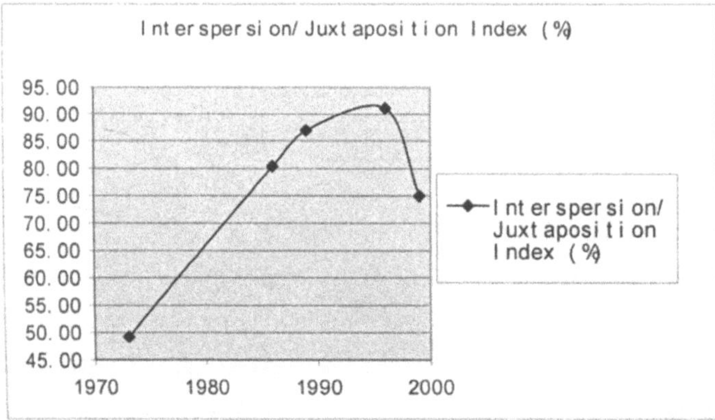

Figures 5g and 5h. Changes in metric performance for the five time periods: landscape shape and interspersion/juxtaposition.

Graphically, the changing metrics point to a landscape that has settled into a new trajectory of land cover composition and organization. Visual inspection of the associated graphics reveals 1996 as a possible turning point for changing landscape composition, ecological function, and perhaps landscape stability. Certainly, the addition of metrics from images acquired in the upcoming years is necessary to verify this hypothesis, and the acquisition of additional images for intervening years in the 1973 – 1999 study period would also increase confidence in this hypothesized trajectory from forest to pasture/agriculture.

Cellular Automata: Modeling in Ecuador

A cellular automaton (CA) system consists of a regular grid of cells, each of which can be in one of a finite number of k possible states, updated synchronously in discrete time steps according to a local interaction rule. The state of a cell is determined by the previous states of a surrounding neighborhood of cells (Wolfram 1984). The rule contained in each cell is essentially a finite state machine, usually specified in the form of a transition function or growth rule that addresses every possible neighborhood configuration of states. The neighborhood of a cell consists of the surrounding (adjacent) cells. For 1-D (one-dimensional) CA models, a cell is connected to r local neighbors (cells) on either side, where r is a parameter referred to as the radius (each cell has 2r+1 neighbors, including itself). The ability of a system to grow and then alter its rate of growth and possibly reverse itself or "die" is a fundamental attribute in biological or human system CA modeling. The systems modeled by Clarke et al. (1996, 1997) and the modeling scheme proposed by Messina and Walsh (2001) both attempt to follow biological patterns. Further, the local operations defined within the Oriente include agent-based functions for change that will be estimated from the socio-economic household survey data collected in 1990 and 1999 as part of this research.

Transition probabilities for the cellular automata model depend on the state of a cell, the state of its surrounding cells, the physical characteristics of the cell (e.g., terrain, soil quality, vegetation, hydrology), and the weights associated with the infrastructure context of the cell (e.g., proximity to roads, markets, schools, and medical facilities, time since settlement). These weights are determined from models of land use based on the household survey data and the GIS database. An important aspect of our CA model is that noncontiguous cells may affect the matrix of transition probabilities for a given cell.

Within the broad category of ecosystem models, the dynamic spatial simulation model, a subset of non-process based models, serves to predict the spatial and temporal patterns of LULC change. As a physical world abstraction, the model reflects the sequence of occupation and agent interaction with the landscape. Specifically, the model describes the location and time of settlement and the attendant agent based decisions. The occupation and interaction change characterization includes the type, magnitude, direction, and rate of changes of various types of anthropogenically-driven activity. Ideally, given the rate and spatial organization of land conversion patterns at different hierarchical scales (the farm, sector, and region), it is possible to determine landscape structure as a measure of land condition through pattern metrics that represent the composition and spatial organization of the local and regional landscape. Within CA and the broader theoretical framework of Complexity Theory, it is expected that there will be both non-linear and hierarchical relationships among the physical, spatial, and socio-economic factors that govern LULC change. Part of the modeling requirement then, is the focus on modeling the highest level of abstraction necessary to capture the details of the system (Dale et al. 1993).

Within the Ecuadorian Amazon, local and regional characteristics are driven by decision-making at the household level and captured in the longitudinal socio-economic surveys. These decisions manifest as changes upon the landscape in both the composition and spatial organization of LULC types. The integration of this household decision and the landscape change is through the assumption of neighborhood interaction. The functional neighborhood defines both the knowledge base diffusion constraints as well as a biophysical control on landscape elements and associated change conditions or possibilities.

Morphological and Cellular Processing

Morphological and cellular processing refers to a class of spatial filtering operations used to facilitate the CA process by changing the shape of the binary "ON" region of a class within a larger class-map. The method of assigning membership to the active community of cells is very specific to cellular processing which uses arithmetic evaluations of pixel values over a specified neighborhood and compares this result with an assigned threshold:

$$\varnothing(i, j, \acute{\eta}) = \sum_{(n,\, m) \equiv \acute{\eta}} I(n, m)$$

where n indexes the neighborhood pixels around the center pixel. The second component calculates the number of eight connected groups of pixels in the neighborhood $x(i, j, \acute{n})$. The results are compared with user set thresholds $T_{\varnothing}$ and T_x respectively. The center pixel is then essentially turned ON or OFF.

Two specific cellular operations are used during the modeling process; Erosion and Dilation. Erosion is used within the context of the modeling environment to remove specific ON pixels along the edges of boundaries and to increase the size of holes or islands, depending upon one's perspective.

Dilation, the more commonly used function within the Ecuador modeling scheme, is used to fill holes or add pixels along a region boundary. Dilation changes an OFF pixel to ON if the number of neighborhood ON pixels exceeds a threshold.

The adjustment of threshold values can create a scenario that closely mimics convolution. By adjusting the threshold values, which may range from 0 to 7, the rate of change is controlled and isotropic effects can be constrained (Seul et al. 2000). Using the morphological methods, the model employs rules defined by real, surveyed, or hypothesized drivers to produce an output image of the anthropomorphized landscape. Physical elements such as hydrology, soils, and slope are incorporated as part of the biophysical database. These layers function as exclusionary devices or promoters of change depending upon the question asked and the modeling strategy.

Growth Rules

There are three types of growth that are directly influenced or modeled by the CA model. The first, organic growth, describes the spreading outward of settlements from existing urban centers. Within the Northern Intensive Study Area, there is only one dominant city, Lago Agrio. The process is controlled by and based upon some of the tenants of central place theory (Hudson 1985). Organic growth represents the tendency of cities to expand. Under morphological constraints, focal filters are used to create a field of neighbor effects to approximate the urban influence and to allow development to expand based upon the location of both existing cells and random cells. The second type, diffusive or dispersive growth, reflects tendencies for a random dispersal of urban centers regardless of fixed proximity functions (Clarke et al. 1996). This type of growth simulates the settlement of a region based upon conditions not necessarily fully modeled by the system.

The local (or neighborhood) operations defined within the Oriente include agent-based functions for change derived from the household survey data collected in 1990 and 1999. The local effect measuring the urban centers as defined within the social gradient surface uses only those areas defined on the image-based input maps. In effect, agricultural expansion is modeled in the CA framework, but the seed input images of initial conditions and their respective updates will control the specific assignment of urban centers for central place purposes. Finally, transportation and accessibility influence growth. Access-based development encourages anthropomorphized cells to manifest along the transportation network. The likelihood of settlement and consequent deforestation along a road is high yet variable depending upon access to markets. Recent work on the effects of distance to markets and agent-based activity has concluded that both Euclidean and network effects exist (McDaniel 2000). The development of rules defining the network effect focuses on the household surveys, fieldwork validation of travel surfaces, and image interpretation.

Random Numbers and Self Modification

The random number generation procedure is considered vital to successful CA modeling within the LULCC context. Thresholding is possible by developing a variable number of random pixels per iteration. Building iterative tests for non-linear thresholds necessitates an additive set of rules, which works in conjunction with the basic iterative tenants of CA modeling. Self-modification is included to better approximate the S-curve population-environment change rate, and necessary, as the model would otherwise produce linear or exponential growth (Clarke et al. 1997).

Operationalizing the CA Model

The output layers of each process are input into the model as a whole. The output from the model is a spatially explicit characterization of the landscape with probabilistic drivers. Model calibration is via comparison of the model's output to a historical data set with respect to key variables including urban and agricultural extensification, deforestation, and reforestation (see Figure 6). As the complexity of the seed image directly influences the shape and pace of growth, special care is taken in the seed creation phase, including sensitivity testing.

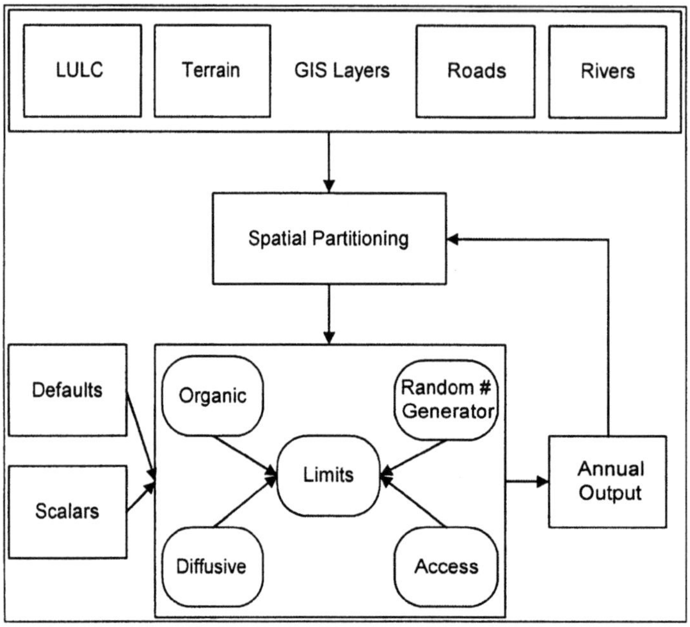

Figure 6. Structural design of the CA model.

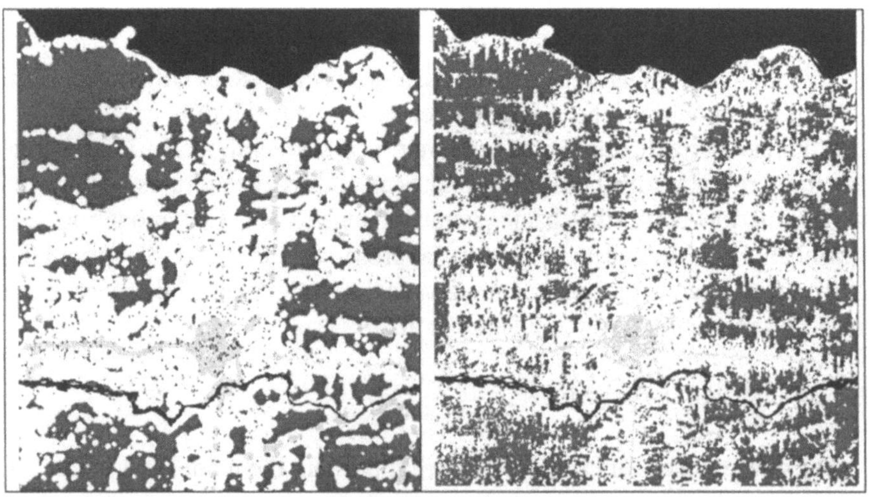

Figure 7. Simulated 1996 vs. Control 1996 output.

The relative regional homogeneity of the biophysical landscape (e.g., climatic and soil regimes), and the importance of local farmers in deforestation, as a response to the geographic accessibility to towns and the expanding transportation infrastructure, permitted the wider application of the model to multiple social and physical scales. While the spatial fit of the model was less than perfect and there are differences between the simulated and the control images (but similar trends are obvious), it is important to realize that no methods exist to quantitatively evaluate the fit. The simulations (Figure 7) suggest a more homogenous landscape with time, a scenario that fits the theoretical understanding of how in-migration of farmers into existing *fincas* through resale, sub-division of *fincas* to those engaged primarily in the burgeoning service sector, and the establishment of new development sectors alters the natural landscape through deforestation and agricultural extensification. However, the selected ISA was already substantially converted to agriculture by the time of imaging, and the *fincas* were sufficiently established and aged for secondary plant succession of crops to pasture and then to forest to occur. Land conversion occurs, particularly given the soil infertility curves that link yields to LULC decisions, and the motivation of farmers to consider opening new areas in their *fincas* for agricultural extensification instead of intensification of existing plots through use of fertilizer and other inputs. The pattern metrics, previously presented, suggest that the landscape is becoming more spatially homogenous with fragmentation increasing slightly for some classes. The contiguity of the landscape is decreasing with time for both the mapped and modeled dates. These non-linear swings in LULC were expected.

CONCLUSIONS

The integration of data characterizing the social, biophysical, and geographical domains, organized within a spatially explicit mapping, modeling, and visualization environment, offers significant power and flexibility for examining the drivers of LULC dynamics. Here, household surveys, GIS coverages, GPS coordinates, and a satellite time-series have been used to assess LULC patterns across space and time through analysis of a longitudinal household survey data, and cellular automata models. While much remains to be done to fully explore the drivers of landscape dynamics in the Oriente of Ecuador, thus far, we have created a powerful data infrastructure that offers tremendous potential for exploring LULC change hypotheses.

The spatially explicit and temporally sensitive nature of the GIS database offers significant opportunities, but also serious challenges for characterizing LULC patterns and assessing scale-dependent drivers of landscape change. On the opportunity-side, we can analyze the assembled data through multivariate models that, among other things, inform us about the important variables and their relative weights for explaining household decisions about LULC change. Such information can be integrated into the CA models to simulate LULC dynamics for selected time frames. Sensitivity analyses allow us to explore alternate variables and weights to examine their relative importance on simulated LULC patterns. Also, "what if" scenarios can be developed from the empirical statistical models and implemented through the CA-based simulations. On the challenge-side, we need to continue to develop approaches for better linking social science data with biophysical and remotely sensed data. Among the questions are how to join longitudinal data to satellite data assembled within an image time-series? What are the appropriate spatial scales for linking data -- the pixel, the household or community, or some terrain unit? How do we validate simulations for future time periods? Finally, how can LULC simulations be related to environmental and other policies that have national, regional, or local implications?

At present, there are more questions than answers, but the development of a spatially explicit data infrastructure and a set of spatial digital tools that can be applied to the study of population-environment relationships provides the potential to discern new and important insights about how people and place are interrelated, how LULC changes as a consequence, and how people and the environment interact through feedback mechanisms and critical thresholds.

ACKNOWLEDGEMENTS

This project is a collaborative project carried out by the Carolina Population Center (CPC), Department of Geography, and the Department of Biostatistics at the University of North Carolina, in conjunction with Ecociencia, a leading non-profit research organization in Quito, Ecuador. Professors Richard E. Bilsborrow and Stephen J. Walsh are the Principal Investigators. We are grateful to the US National Aeronautics and Space Administration (NASA) (NCC5-295) and the Mellon Foundation for supporting all aspects of the project, and to the Summit and Compton Foundations and PROFORS (Proyecto Forestal, in Lago Agrio, Ecuador, funded by GTZ, the German foreign aid program) for supporting the household survey data collection in the Oriente region of the Amazon in 1999. In addition, NASA provided assistance with access to Landsat satellite imagery, the CPC Spatial Analysis Unit (Stephen McGregor and Brian Frizzelle) and the Department of Geography's Landscape Characterization & Spatial Analysis Lab supported spatial and biophysical data collection and processing of remote sensing and GIS data, and the Centro de Estudios de Poblacion y Desarrollo Social (CEPAR) in Quito entered and processed the 1999 household survey data and the 2000 community survey data, which were further processed at CPC by graduate students supported by traineeships funded by the National Institute for Child Health and Human Development (HD07168).

Data from the earlier 1990 Amazon survey were collected and processed with funding provided by the US National Science Foundation, the World Wildlife Foundation, and the CPC, and logistical support was received from several

Ecuadorian government agencies, including CONADE (National Planning Agency), the Ministerio de Agricultura y Ganaderia, and INIAP (Agricultural Research Center, in Francisco de Orrellana, Napo).

REFERENCES

Bilsborrow, R.E. 1987. Population pressure and agricultural development in developing countries: a conceptual framework and recent evidence. *World Development* 15(2): 183-203.

Blaikie, P. and Brookfield, H. eds. 1987. *Land Degradation and Society*. New York: Metheun & Company, Ltd.

Bromley, D. 1989. *Economic Interests and Institutions: The Conceptual Foundations of Public Policy*. Oxford: Blackwell Publishers.

Centro de Estudios sobre Poblacion y Desarrollo Social, Encuesta Demografica y de Salud Materna e Infantil. *ENDEMAIN III*. 2000. Quito: CEPAR and Centers for Disease Control.

Clarke, K.C., Gaydos, L. and Hoppen, S. 1997. A self-modifying cellular automaton model of historical urbanization in the San Francisco Bay area. *Environment and Planning B*, 23: 247-261.

Clarke, K.C., Hoppen, S. and Gaydos, L. 1996. Methods and techniques for rigorous calibration of a cellular automaton model of urban growth. *Third International Conference/Workshop on Integrating GIS and Environmental Modeling*. Santa Barbara: National Center for Geographic Information and Analysis, Santa Fe, New Mexico, 1996.

Dale, V.H., O'Neil, R.V., Pedlowski, M. and Southworth, F. 1993. Causes and effects of land-use change in Central Rondonia, Brazil. *Photogrammetric Engineering and Remote Sensing*, 59(6): 997-1005.

Environmental Systems Research Institute, Inc. 2000, *Arc/Info*.

ERDAS., *IMAGINE 8.4*, 1997. ERDAS, INC. Atlanta, GA.

Forman, R.T.T. and Godron, M. 1986. *Landscape Ecology*, John Wiley & Sons, New York.

Hudson, J.C. 1985. *Plains Country Towns*, University of Minnesota Press, Minneapolis.

INEC - Instituto Nacional de Estadistica y Censos, *V Censo de Poblacion y IV de Vivienda*. 1991. Quito, Ecuador.

Jensen, J.R. 1996. *Introductory Digital Image Processing*, Prentice Hall, Upper Saddle River, New Jersey.

Lambin, E.F. 1994. Modelling deforestation processes. *A review. TREES series B, research report no 1*, European Commission.

Marquette, C. and Bilsborrow, R.E. 1994. *Population and the environment in developing countries: literature survey and research bibliography*. NY: UN Population Division.

McDaniel, P.M. 2000. *Household and Community Drivers of Land Clearing: Human-Environment Interactions in the Ecuadorian Amazon*. Master's Thesis: Department of Geography, University of North Carolina - Chapel Hill.

McGarigal, K. and Marks, B.J. 1993. *Fragstats: Spatial Pattern Analysis Program for Quantifying Landscape Structure*. Forest Science Department, Oregon State University, Corvallis, Oregon.

McNab, W.H. 1989. Terrain shape Index: quantifying effects of minor landforms on tree height. *Forest Science*, 35: 91-104.

McNab, W.H. 1993. A topographic index to quantify the effect of mesoscale landform on site productivity. *Canadian Journal of Forest Research*, 23: 1100-1179.

Messina, J.P. and Walsh, S.J., 2.5D Morphogenesis: Modeling Landuse and Landcover Dynamics in the Ecuadorian Amazon. *Plant Ecology*, in press.

Messina, J.P., Walsh, S.J., Taff, G. and Valdivia, G. 1999. The application of cellular automata modeling for enhanced landcover classification in the Ecuadorian Amazon. Proceedings of 4th International *Conference on GeoComputation*. Mary Washington College, Fredericksburg, VA, gc_081 (CD-ROM).

Myers, N. 1988. Threatened bota: hotspots in tropical forests. *The Environmentalist*, 8(3): 1-20.

Myers, N., Mittermeier, R., da Fonseca, G.A., Kent, J. 2000. Biodiversity hotspots for conservation priorities. *Nature*, 403: 853-858.

Moran, E.F, Brondizio, E., Mausel, P., Wu, Y. 1994. Integrating Amazonian vegetation, land-use, and satellite Data. *BioScience*, 44(5): 329-338.

O'Callaghan, J.F. and Mark, D.M. 1984. The extraction of drainage networks from digital elevation data. *Computer Vision and Image Processing*, 28: 323.

Pan, W.K., Murphy, L., Sullivan, B. and Bilsborrow, R. 2001. *Population and Land Use in Ecuador's Northern Amazon in 1999: Intensification and Growth in the Frontier.* Paper presented at Population Association of America annual meeting, Washington DC.

Parker, A.J. 1982. The topographic relative moisture index: an approach to soil-moisture assessment in mountain terrain. *Physical Geography*, 3(2): 160-168.

Pattison, W.D. 1969. Four Traditions of Geography. *Annals of the Association of American Geographers,* 59.

Pichón, F. and Bilsborrow, R.E. 1999. Land Use Systems, Deforestation, and Demographic Factors in the Humid Tropics: Farm-Level Evidence from Ecuador. In, *Population and Deforestation in the Humid Tropics*, R.E. Bilsborrow, ed. Liège, Belgium: International Union for the Scientific Study of Population.

Rudel, T. and Horowitz, B. 1993. *Tropical Deforestation: Small Farmers and Forest Clearing in the Ecuadorian Amazon.* New York: Columbia University Press.

Seul, M., O'Gorman, L. and Sammon, M.J., 2000. *Practical Algorithms for Image Analysis: Description, Examples, and Code.* Cambridge University Press, Cambridge.

Southgate, D. and Whitaker, M. 1994. *Economic Progress and the Environment: One Developing Country's Policy Crisis.* Oxford: Oxford University Press.

Walsh, S.J. 1999. Deforestation and Agricultural Extensification in Northeast Thailand: a Remote Sensing and GIS Study of Landscape Structure and Scale. In *Proceedings, Applied Geography Conference*, Schoolmaster, F.A. (Ed.), 22: 223-232.

Walsh, S.J., Evans, T.P., Welsh, W.F., Entwisle, B., and Rindfuss, R.R. 1999. Scale-dependent relationships between population and environment in northeastern Thailand. *Photogrammetric Engineering and Remote Sensing*, 65(1): 97-105.

Walsh, S.J., Bilsborrow, R.E., McGregor, S.J., Messina, J.P., Pan, W., Frizzelle, B.G., Crew-Meyer, K.A., Mena, C. 2001. *Landscape dynamics in the Ecuadorian Amazon: characterizing Ddforestation and agricultural extensification through space & time dependent models.* LBA-Ecology Open Meeting, Atlanta.

Wolfram, S. 1984. Cellular automata as models of complexity. Nature, 311: 419-424.

Wood, C. and Skole, D. 1998. Linking Satellite, Census, and Survey Data to Study Deforestation in the Brazilian Amazon. In, *People and Pixels: Linking Remote Sensing and Social Science,* Liverman, D., Moran, E.F., Rindfuss, R.R., and Stern, P.C., editors, Washington, DC: National Academy Press.

Chapter 10

DEFORESTATION TRAJECTORIES IN A FRONTIER REGION OF THE BRAZILIAN AMAZON

Stephen D. McCracken, Bruce Boucek, and Emilio F. Moran

ACT/CIPEC, Student Building 331, Indiana University, Bloomington, Indiana 47405-6101
steve_mccracken@hotmail.com

Keywords: Amazon, Brazil, frontier, colonization, agriculture, deforestation.

Abstract This chapter provides a conceptual framework for a micro-level approach for studying farm-level and landscape change in the context of an agricultural frontier region of the Brazilian Amazon. The development of a property grid overlay (with 3913 properties) facilitates the extraction of data on land cover change derived from the analysis of a time-series of remotely sensed images from 1970 to 1996. Time-segments between remotely sensed images permits the identification of the period of initial clearing and for analyzing the amount of farmland that is cleared annually by approximate age of the farm for the 25-year period. Deforestation is greatest in the beginning of farm settlement and slows down in subsequent years followed by a leveling off around year 18-20. The analysis suggests that by the year 2020, after 50 years of colonization, there will be between 24-32 percent of the original forest remaining.

INTRODUCTION

Over the past 30-years, more than 15 percent of the Amazon Basin's tropical moist forest has been cut down. This has had impacts on atmospheric composition (Andreae and Schimel 1989, Crutzen and Andreae 1990), on regional climate (Dickinson 1987, Gash et al. 1996), on biological diversity (Wilson 1988), and on the lives of native and immigrant human populations (Moran 1988 1993). Deforestation rates have been highest in Ecuador, but total amounts deforested in Brazil dwarf what is happening elsewhere in the tropical world. The total area of deforestation in Brazil is five times that of Mexico, the second most deforested country in Latin America, according to FAO estimates.

Estimates of the nature and extent of deforestation vary (Dickinson 1987, Malingreau and Tucker 1988, Skole and Tucker 1993, INPE 1998 1999). Booth (1989) and Setzer and Pereira (1991) reported that up to 8 million hectares of forest were burned in 1987 – and this peak was surpassed by a substantial margin in 1995 (INPE 1998). Little deforestation had taken place in the Brazilian Amazon before the 1970's, with as little as 30,000 square kilometers deforested, or 0.6 percent of the region. Between 1978 and 1988 deforestation increased to 6 percent, and from 1988 to the end of the century, this amount had surpassed 15 percent.

Deforestation has not proceeded across the Brazilian Amazon in a random fashion. Rather, it has followed the roads associated with national development planning. First, along the Belém-Brasília highway, then along the Transamazonian highway, along the Cuiabá-Santarém highway, and along the Cuiabá-Porto Velho (BR-360) highway connecting Rondônia to the south of Brazil and to Manaus. It is particularly along the Belém-Brasília and along the Cuiabá-Porto Velho highways that the highest rates have been observed along what has come to be called the "arc of deforestation" (see Figure 1 for a Basin-wide analysis that highlights the arc). Deforestation along this arc represents the advance of the cattle pasture frontier from the Center-West of the country into the southern part of the Amazon Basin. Most of the deforestation in the Basin is associated with conversion of forest into extensive cattle ranching on very large properties. Elsewhere we have shown that many of these deforested areas revert to secondary successional vegetation, because of the aggressive invasion of pioneer species (Moran et al. 1994 1996, Tucker et al. 1998). However, where enough labor and capital is available for continued reclearing of these areas, there is floristic impoverishment through repeated burning episodes, and greater probability of persistence in pasture.

Alongside these large cattle ranching projects, there has also been a considerable inflow of immigrants from all over Brazil to claim land offered by colonization agencies (Moran 1981, Smith 1982). This has been particularly true along the Transamazon highway and in Rondônia along the BR-360 highway. Here, too, we find a preponderance of total land deforested in pasture (and degraded pasture) following short cycles of cultivation (Moran 1981, Moran et al. 2000). The relationship between population and the environment is not a direct one. Rather, it is mediated by many other factors such as the land tenure system in place, the economic and environmental uncertainties, the limitations posed by a particular age and gender household structure on the labor and capital available, the past experience of household members in the new habitat, and the role of capital and macroeconomic policies.

This paper will present results of research in one of the regions along the Transamazon Highway in an effort to elucidate the spatial and temporal dynamics of deforestation trajectories as mediated by the demographic characteristics of immigrating households. Unlike much of the literature currently available, the analysis presents not only a landscape analysis of deforestation, but also a property-level analysis. The trajectories of deforestation are based on time-series analyses of Landsat Thematic Mapper (TM) images overlaid on the property grid in a GIS for the 3913 properties in the study area. The demographic analysis is based on a survey of 402 sampled farm households. The final section of the paper makes deforestation projections to 2010 and 2020 for the entire study region based upon the trajectories of deforestation examined at property level.

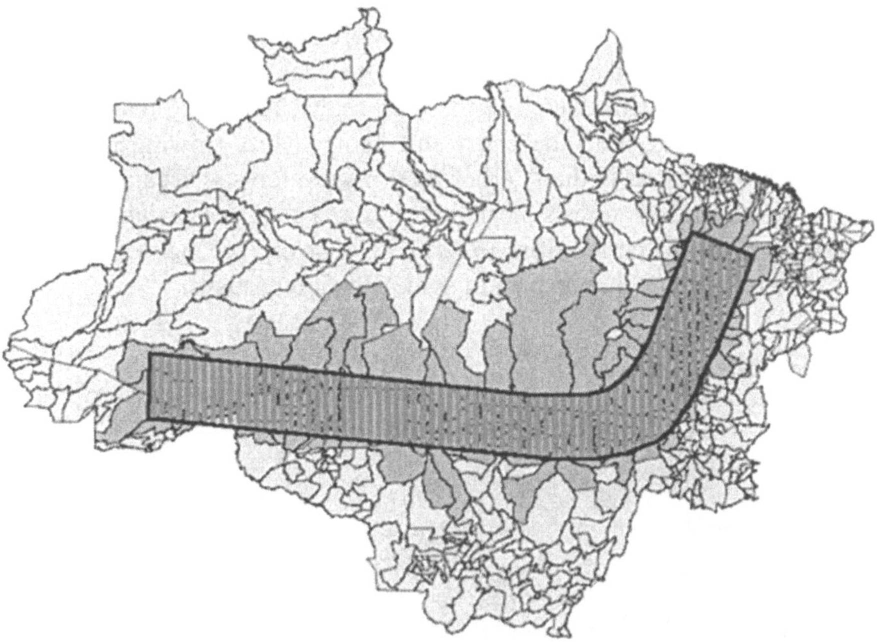

Figure 1. Amazon-wide Arc of Deforestation (Source: Instituto de Pesquisa Ambiental da Amazônia -IPAM).

STUDY AREA

The study area is the region west of the city of Altamira in the lower Xingu Basin of the Brazilian Amazon. The area was viewed as a showcase for colonization in the 1970s when the military began its Program of National Integration. The implementation of these colonization programs attracted researchers interested in monitoring the process of land occupation (Moran 1976 1981, Smith 1982, Fearnside 1987). After the initial four years, the government blamed the farmers for the low grain yields in the first three years, and reduced the subsidies of the settlement scheme. Over the next 25 years, many of the immigrants abandoned or sold their properties, while a steady flow of new immigrants continued to arrive – buying the land of those who desisted, and occupying new properties further on. Settlement has followed a very orderly spatial pattern in large part due to the gridded property demarcation pattern imposed on the settlement scheme; settlement took place first along the highway closest to Altamira and then along nearby side roads. Settlement proceeded in layers extending outward from the main road and from the emerging towns of Brasil Novo and Medicilândia. All properties are approximately 100 hectares in size with predictably similar rectangular layouts of about 500 by 2000 meters (Figure 2). The result is one associated in the literature with what has come to be known as the "fishbone pattern" of Brazilian colonization.

As settlers arrive they begin to clear forest in their property to make a farm. They tend to do so from the front of the lot towards the back. When the project started, farmers were told that they must keep the back half of their 100-hectare lots as forest reserve. However, over the years settlers have interpreted this view in a variety of ways, with the end result being that a growing number of properties have cleared virtually all of their mature forests. Remaining areas tend to be secondary forests near water sources or areas unsuitable for cultivation. Over the period from 1970 to 1996 we can see the steady pace of deforestation in Figure 3 in an area of some 384,000 hectares.

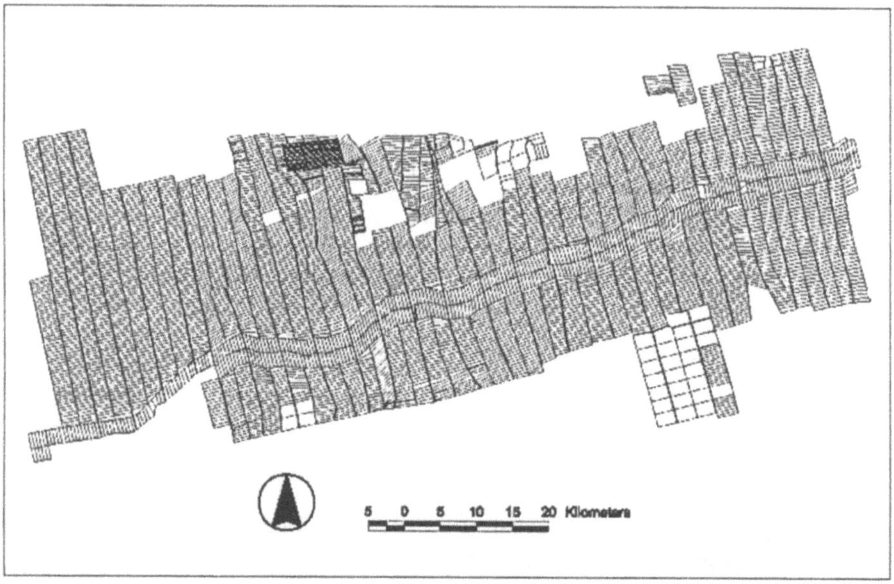

Figure 2. Property grid overlay developed for study area, colonization along the Transamazon Highway, Altamira, Pará.

The area's above-average soils have near neutral pH, and relatively favorable base saturation (i.e., nutrients available to plants). These soils (i.e., alfisols) are interspersed with the proverbial acid low nutrient soils, chiefly oxisols. Portions of the study area also have areas of rugged topography for the Amazon, where steep gradients make transportation and farming more difficult. The road was built so that it cut through some of the best soils and early settlers were able to make claim to these soils. Elsewhere we have shown that farmers who were able to identify or claim the alfisols have been able to stay on their properties far more often than those who selected properties with poor soils, and that land turnover on properties with poor soils is high (Moran et al. 2000). We have also shown that over time farmers have learned that properties with very poor soils can support pasture but little continuous cropping, whereas properties with alfisols have a more diversified farming strategy with equal amounts of land in pasture and in intensive cultivation (mostly cocoa and sugar cane). The persistence of cattle even on the best soils seems to be a risk aversion strategy at the farm level since cattle prices and these two commodities seem to follow opposite trajectories in the economy (Moran et al. 2000).

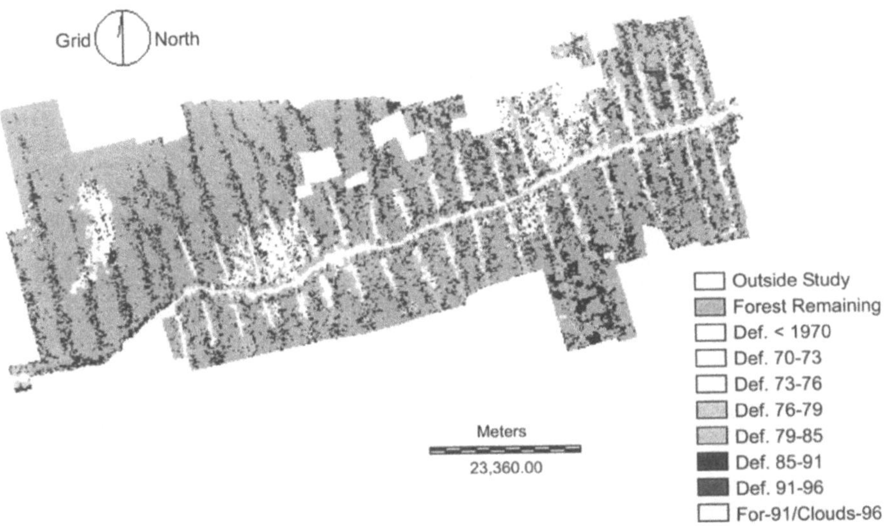

Figure 3. The observed pace of deforestation among frontier farms along the Transamazon Highway, km 20-140, 1970-96.

The population has changed over time in area of origin, economic status, and demographic structure as evidenced from information from our sample survey of 402 households. The aging process of households over time is illustrated in the current age-sex composition of households arriving since 1985 compared to those arriving before 1976. Recent arrivals are typically made up of small nuclear households made up of young and middle-aged adults with a large number of children (Figure 4a). Households that arrived before 1976 are now (in 1998) much older with young adult children and teenagers and an increasing number of adult children having left the farm (Figure 4b). The parents, the initial colonists, are now in their 50s and 60s. As these households age there is a tendency to maintain male children on the farm as a labor-retention strategy while marrying daughters out and sending them to nearby towns for employment or schooling. The changing age-sex composition associated with aging of households can be expected to result in changing labor composition of households and changing agricultural strategies and patterns of deforestation over the domestic life course.

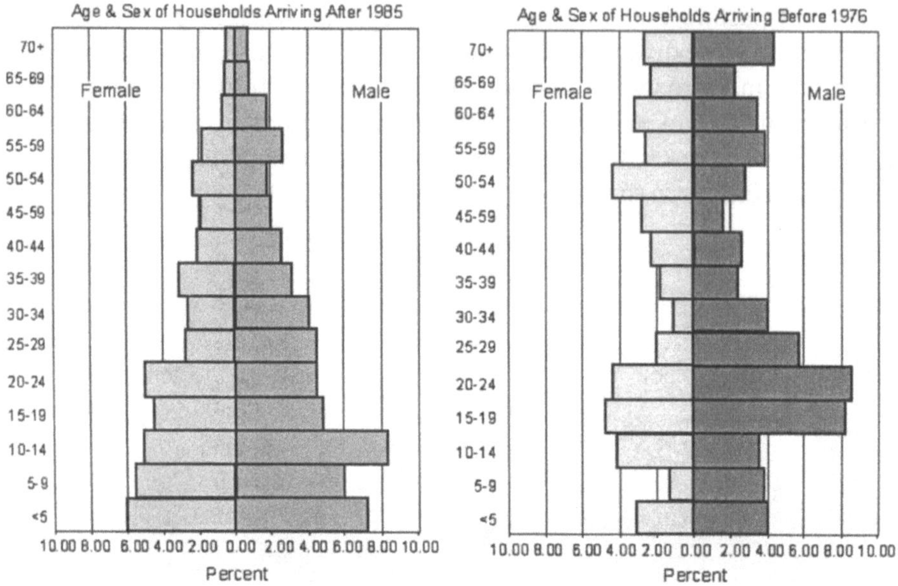

Figure 4a (left). Current composition of households arriving after 1985; *Figure 4b (right),* current composition of households arriving before 1976.

Our methodology reflects a conceptual model underlying our project. (McCracken et al. 1999). In this conceptual model (Figure 5), we propose that households in the frontier follow a trajectory of changing land use that reflects the changing composition of the household and its labor and as their development cycle interacts with features of their social, political, economic and biophysical environment. We suggest that cohorts of farm settlement pass through stages of the domestic life cycles and farm development as their households age over a 25-year or generational time-span. From young adults with young children to older adults beginning to pass their farm property to their children, we hypothesize a steady shift in land use from annual crops favored by young households, followed by strategies made possible by accumulation of capital and more household labor as children are old enough to work and shifts to labor saving land uses as children begin to exit the household through marriage and parents' increasing age. The latter would favor agroforestry, and plantation agriculture (sugar cane, cocoa, and increasingly coffee) and pasture and raising cattle over annual crops. As frontier households develop and consolidate their family farms deforestation subsides; families are increasingly involved in managing fallows, maintaining pastures and developing their agroforestry activities

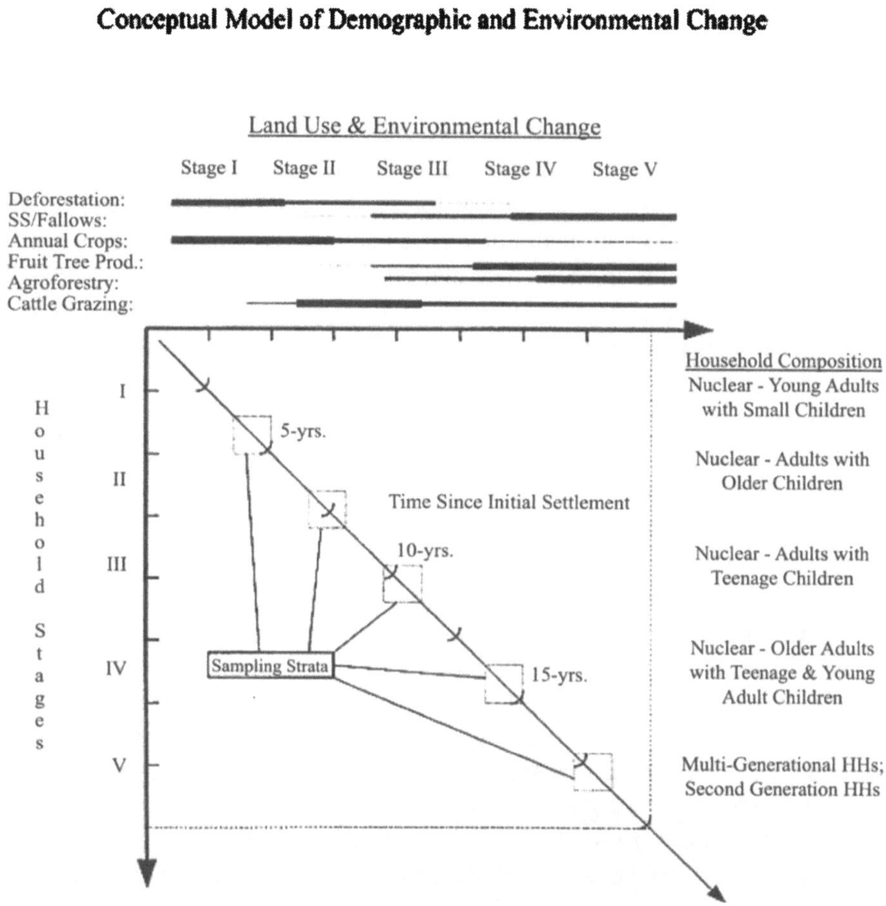

Figure 5. A conceptual model of the domestic life cycle of households, and expected trajectory of land use and environmental change at the farm-household level.

METHODS

For our analysis we have used a set of 10-images at approximately three-to-six-year intervals that cover the period 1970 to 1996. Aerial photographs for 1970 and 1978 were acquired and the cover of forest and non-forest areas was visually interpreted using a mirror stereoscope (Fairchild Type F-71B). A controlled mosaic of the interpreted photos was produced. The controlled mosaic interpretation map was scanned and map development undertaken, including thematic coding, resulting in a georeferenced map in digital form with 4-classes of land cover that distinguished: forest, non-forest, river, and road. These data were then integrated into a GIS. Landsat Multispectral Scanner

(MSS) digital data were acquired for 1973, 1975, 1976, 1978 and 1979. Landsat Thematic Mapper (TM) digital data were acquired for 1985, 1988, 1991 and 1996 for this analysis. Because of incomplete coverage of different image sources the classifications for 1975 and 1976, and 1978 and 1979 were combined to yield a composite of land cover for January 1976 and 1979.

The development of the property grid GIS data layer (see Figure 2) followed a process similar to that used for deriving the 1970 and 1978 forest cover data layers. The property grid used for this analysis was created from a series of scanned property maps provided by the Colonization Agency (INCRA). The original purpose of these grids did not foresee these applications. Rather, they were a rapid and topographically unverified layout of the numbered properties used to know which farmers had claimed what properties. Initially, the surveyors attempted to lay the lot markers along this ideal grid, but over time adjustments were made to take into account topographic and hydrologic features ignored at the outset. These maps were georeferenced to previously geo-corrected spatially referenced scanned topographic maps of the study area. The scanned and geo-corrected property grid was then used as a base on which to manually digitize a vector polygon property grid in Arc/Info GIS.

We found in the field that even the road network from the available maps did not always conform to reality given local adjustments by farmers and road builders to the topography and hydrology. To define the properties, we began by identifying the rear of the properties. This was done by making a mathematical interpolation of distance. Using a slope/aspect algorithm in IDRISI, we identified the lines that defined the midpoints between the side roads, and this was used to represent the rear property lines that were commonly under a thick cover of primary forest. The next step was to ensure that the satellite images were all georeferenced correctly and were registered to each other so that there were no discrepancies between years in the location of the properties. Once this was accomplished we digitized the properties on-screen in ERDAS Imagine. The time-series helped identify most of the properties because of distinctive patterns of land clearing on each property up to the boundary lines. We then assigned a unique ID to each property to assist us in developing our sampling framework. During fieldwork in December of 1998, we collected differentially-corrected GPS points, using a base-station and back-pack Trimble GPS equipment, at property boundaries along every feeder road at intervals of 5-10 kilometers, in the re-georeferencing of the property grid.

This process illustrated many problems with our original mapping of properties based on INCRA colonization maps and greatly improved the overlay of a new property grid for extracting data at the property level. Because the original INCRA maps did not have adequate cartographic information and geographic references, the final fit between the property maps (based on the INCRA land distribution scheme), GPS points, and satellite images was based on an iterative process of successively fitting the property grid to the GPS points collected in the field until an acceptable fit was achieved. Our successive evaluation of the final property grid was based on visual examination of its overlay on multiple image classifications, where property boundaries are often discernable from differences in clearing and land-cover. Getting a perfect fit is extremely difficult given the imperfect geographic features of the original colonization maps and the large area covered in this study. Visual inspection suggests that the majority of properties are within a pixel (30 meters) of the proposed boundaries. Pixel-level accuracy is better than one might obtain from maps generated at 1:100,000 scale, and ideally, better than those produced with maps at a scale 1:25,000. Based on visual inspection of remotely sensed imagery and intimate knowledge of the primary investigators about individual properties and the landscape leads us to suggest a acceptable fit for near 90 percent of farms in the area. Our approach may not work everywhere, given the unusual circumstance. In this colonization scheme, land is divided into equal rectangular plots along a network of roads, and evenly spaced along the highways. These regular patterns make definition of the property boundaries more apparent than in other areas where it may be difficult to ascertain the size and shape of properties without doing property-level topographic surveying. The end result of this two-year effort of GIS layering and field verification is a useful tool for extraction of data at the property level.

We believe that the development of a farm property grid within a GIS and a satellite time-series dataset represents a substantial breakthrough for conceptualizing and interpreting social science and environmental questions about land use and land cover change (Liverman et al. 1998). Previous work for an area this size, 3,840 sq km, typically analyzes and describes land use and land cover change from a landscape perspective at best, without addressing the highly variable land uses that farmers engage in. We have been able to clarify in this manner, evidence for forest degradation from logging and fire invasion, and to understand the impact of soil fertility on crop choice in the study area (McCracken et al. 1999, Moran et al. 2000).

One of the objectives of this project was to develop a robust geographic information system for understanding population-and-environment dynamics. In addition to the property grid layer discussed above, we have developed the previously described GIS data layers for 1970 and 1978 with 2 classes (forest and non-forest); Landsat TM-based classification layers with forest, advanced secondary succession, intermediate secondary succession, initial secondary succession, bare soil, pasture, and cropland. Topographic and hydrographic layers were developed from maps from the Instituto Brasileiro de Geografia e Estatistica (IBGE).

The property grid overlay, combined with remotely sensed land cover classification, provides two unique opportunities for research endeavors. First, in contrast with more traditional approaches to analysis of population-and-environment that focus primarily on landscape or pixel level analyses, a property grid provides an opportunity to analyze changes at a more appropriate level of analysis, at the level of the individual actors. In the context of this research, it is farms and settler-households. Within a GIS with land cover classifications, the property grid can be used to extract information associated with all farm lots for analysis. Data extractions, using ArcView3.2 and an interface with ACCESS, and statistical analysis using STATA have been used in our research activities. Secondly, the property grid also provides advantages for drawing systematic random samples in areas where developing sampling frames can be very difficult. In our particular case, we were interested in drawing a sample of colonist households who had arrived on the frontier over the entire period of colonization for detailed interviews on socio-economic and demographic characteristics of households, agricultural strategies, and land-use history.

A key component of the research project seeks to disentangle the period, cohort and age effects associated with trajectories of deforestation at the farm level. For example, we are interested in whether different government policies or market changes differentially affect agricultural strategies of households on farms in different stages of development or in different stages of the domestic lifecycle. Based on area deforested in a farm lot between satellite images we were able to identify the period of initial occupation of each farm property in our study area. We used evidence of the clearing of 5 hectares as a signal that farming activities had begun, based upon field experience that households cleared a minimum of 3 hectares per year during the beginning of settlement. This allowed us to define cohorts of farm properties and their respective colonist-households based upon when clearing of property began (see Figure 6 for a view of the cohorts that were identified for our study area). These "farm-property cohorts" in turn were used to draw a disproportionate stratified random sample of 402 households for detailed interviews on socio-economic and demographic characteristics of households, agricultural strategies, and land-use history. We used a disproportionate stratified random sample to ensure that relatively equal shares of the surveys were carried out among each cohort of

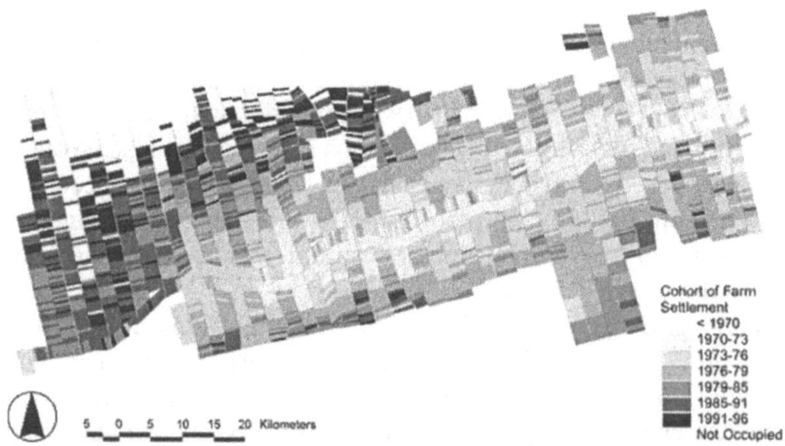

Figure 6. Map of properties by period forest clearing was initiated (Farm Cohorts).

settlers. Equal shares of households among each cohort of farms, and, to a considerable extent, cohorts of settlers, will enable analyses of potential period and cohort effects at different stages (or ages) of farm development.

The analysis for this paper focuses simply on the age effects of farm development on the individual-level patterns of deforestation and predicts individual and landscape level patterns of deforestation from information derived from all the farm properties identified in the property grid. A simply and exploratory way to arrive at an age pattern of deforestation is to regress the number of years of activity on the farm area or percent farm area deforested based on the latest satellite image classification available, in our case, 1996. This approach yields high R-squares, on the order of 60+ percent, and suggests a very linear pattern of deforestation – of about 2.5-3.0 hectares of forest cleared annually. While areas of clearing, on the order of 2.5-3.0 hectares annually, is quite consistent with general information from farmers, this approach is heavily influenced by the accumulated outcome rather the underlying age pattern associated with deforestation over the course of farm development. An alternative approach, used here, takes advantage of the time-series of remotely sensed image classifications and captures the higher levels of deforestation during the initial years of farm occupation and the subsequent declines in deforestation after several years of farming activities, which are consistent with knowledge from many years of field experience with these farming households.

Farm size, and area in forest were extracted for each of the 3,913 farm properties identified in the property grid for each remotely sensed image. The area deforested between dates of each remotely sensed image were used to estimate the percent of farm area deforestation during each interval (between satellite images). These are not "true rates" in that they are based on farm area and not area remaining in forest at the beginning of each period. Since the average size of the farm properties approximates 100 hectares, the percent farm area deforested represents the number of hectares cleared on a farm. This approach avoids the problem of exponential decay associated with a true rate and more appropriately reflects the amount of area a household clears to put into productive activities as they carry out their farming activities. This information was divided by the number of years in the interval to obtain a "mean annual rate of deforestation" for each property for each period of observation (pre-1970, 1970-73, 1973-76, 1976-1979, 1979-85, 1985-91, 1991-96). The "start-date" for each property was established by taking the mid-point for the interval in which they were first noticed as having had farm activities. For example, farm properties settled between July of 1985 and July of 1991 when the satellite images were collected, initiated activities over the period with a mean number of years of activity around three. The mean annual rate of deforestation would be the percent of forest cleared by July 1991 divided by 2.5. In the subsequent interval, between 1991 and 1996, the additional percent farm area deforested is divided by five, and is associated with these approximately eight-year-old farms.

Using the time interval between satellite images, appropriately associated with the approximate age of the farm at the end of the interval, we end-up with 13,383 time intervals for the analysis based on the 3,913 farm properties in the study area. Four hundred and eight farm lots that were not as yet occupied by 1996 are not included in this analysis of the age pattern associated with farm settlement and development.

The objective of this analysis is to obtain an age pattern of deforestation at the farm level over time in much the same way demographers develop mortality curves for predicting age structure in the future or in the development of life tables. The result is the trajectory of land use and land cover change that are the focus of this paper. Figure 7 (and Figure 6) illustrates the number of farm properties occupied by 1996 by the period in which farm activities began, of farm cohorts. The pattern is not uniform – rather it shows the increasing occupation of new farm properties through the end of the 1970s associated, first, with the initial government-directed colonization, and followed by the continued distribution of land through the mid-to-late 1970s. Settlement of new areas declined in the early 1980s and was revived by increases in cocoa prices in the

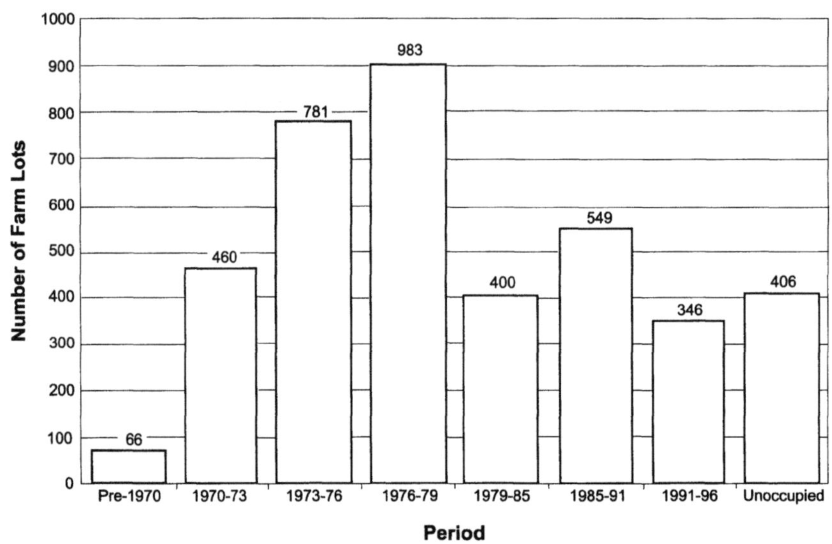

Figure 7. Properties by period forest clearing was initiated (Farm Cohorts).

1980s. Disease and declining cocoa prices resulted in reduced occupation of new farm properties in the first-half of the 1990s. These fluctuations represent important "period" effects on farm occupation and deforestation at the landscape level. In the absence of considerable changes in government policies or market prices for products from the area, we can expect farm occupation to continue slowly as 2nd generation households are formed and move to unoccupied properties in our study area and beyond. We can expect continued occupation in older consolidated frontiers to be driven by the natural increase of the local population as children of the initial colonists come of age and move on to establish their own farms.

RESULTS AND DISCUSSION

To obtain an age pattern of deforestation at the farm level, we fitted our mean annual rates of farm area deforested with ordinary least-squared regression (OLS) with a simple linear model based on age, and curvilinear models with the inclusion of age-squared and a quadratic equation with an age-cubed term. The results are presented in Table 1. In Figure 8, the observations for each interval (between remotely sensed classified images) are presented by approximate age of the farm, and a line from the quadratic equations is used to illustrate the general age pattern associated deforestation on frontier farms. The satellite image for 1996 had several properties with cloud cover in 1996. In this case, we estimated the percent area that was in forest in 1991 that was under cloud cover in 1996. The inclusion of this variable had no considerable impact on the results and was not in included in this presentation. An important observation from both the table and the figure is that there is substantial variation about the regression line; the models explain only 13 to 16 percent of the mean annual age patterns of deforestation. As can be seen in Figure 8, there are substantial variations below and above the line; many individual farms experienced mean annual rates above 10 percent during their initial settlement. These fitted lines simply reflect an average tendency on these farms that translate into highly

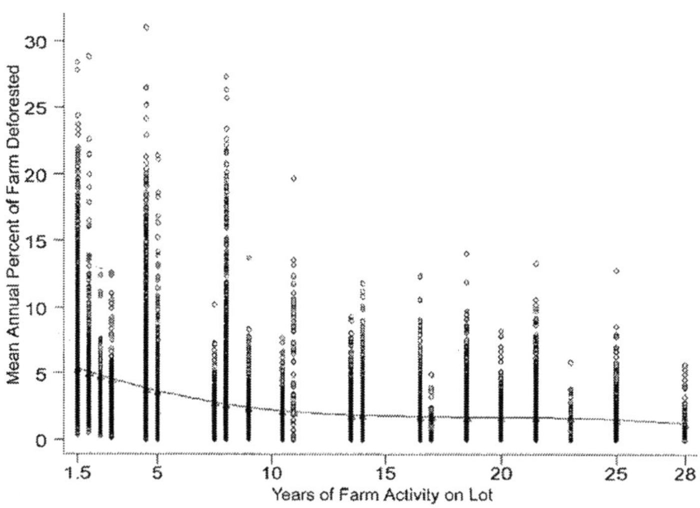

Figure 8. Mean annual percent of farm area deforested by years of farm occupation.

variable rates that occur on any given real farm. What occurs on any given farm is a result of complex dynamics that include: the quality of soils on the property and their location; the amount, age, and gender composition of household labor; and their access to credit or capital – to name but a few of a number of factors that influence the behavior of households. Investigation of these factors is being carried out in separate analyses with information from the 402 surveyed farm households. Other variables for consideration and developed in the GIS are topography and distances to the highway and Altamira. Exploratory analysis of mean slope of the property and various distance measures indicate that they are statistically significant but the associated coefficients are quite small in this analysis of mean annual percent of farm area cleared. Distances to the highway and to Altamira are also highly correlated with age of the farm and importance of distance to farmers, it can be expected, is likely to have changed over the years. Both distance and ruggedness of the farmland are more likely to affect overall levels of deforestation at the farm level rather than specific rates during the course of farm development. This will be discussed further in the analysis of predicted versus actual areas deforested by 1996 in the subsequent section. The fitted regression lines for percent of farm area in forest cleared annually are presented in Figure 9. The figure illustrates the very different implications they have for finding an age pattern of deforestation and for predicting deforestation at the farm level into the future.

Table 1. OLS Regression on mean annual percent of farm area deforested.

Mean Annual % Farm Area Deforested	Linear	Curvilinear-Sq.	Quadratic
Independent Variables			
Years on Lot	-0.1724	-0.4667	-0.7200
Years-squared		0.0125	0.0374
Years-cubed			-0.0007
Constant	4.7712	5.8458	6.3826
Observations	13383	13383	13383
F	2007.24	1269.45	870.43
Prob > F	0.0000	0.0000	0.0000
Adj. R-squared	0.1304	0.1594	0.1631

All results were significant at the 0.01 level

The simple linear regression suggests that farms clear approximately 4.8 percent of the land during the 1st year (year 0) based on the y-intercept. In each subsequent year, the percent of farm area cleared declines by 0.17 (4.8-0.17). By the 12th year, percent of farm annually cleared falls to less than three percent;

by the 20th year, it falls to around 1.5 percent per year. This model suggests that additional clearing stops by the 30th year of farming. Applying these rates to a 100-hectare farm suggests that 32 percent of farmlands would remain in forest.

The subsequent models based on age-squared and age-cubed suggest curvilinear patterns, but with very different consequences for long run expectations about deforestation. Both suggest that deforestation is greater in the beginning and declines to about 2 percent by the 12th year (year 13) and begin to level off by about year 18-20. After the 20th year, the simpler model using age and age-squared, suggest that deforestation would gradually increase by year 25 and then accelerate. The implication is that by the 36th year of farm activities, there would be no more original forest remaining on the farm. The line fitted by the quadratic equation provides a different portrayal of an expected age pattern associated with farm development. It begins with high rates of farm clearing in the initial years, levels off by around years 18-20 at about 2 percent, and then declines slowly after 30 years with no more additional deforestation after 35 years. Applying these rates to a typical farm of 100 hectares suggests that about 24 percent of the farm would continue in forested area.

The fitting of regression lines yields quite different scenarios for the future of farm development and deforestation in agricultural areas on the frontier. The linear regression fits the data well, but understates the amount of deforestation associated with initial settlement of the family farm and overstates the amount of forest remaining in future years. The curvilinear model suggested, by inclusion of a squared-term for age, indicates an increasing deforestation after 20 years that would result in complete deforestation of these farms by year 36. The quadratic model provides a general pattern consistent with field observation and knowledge of the area.

In the initial stages of colonization many families sought to establish their farms quickly and invested the capital they brought with them in clearing large areas of the property. Within a few years many found that they were unable to manage the land they had cleared as weed invasion occurred. Later cohorts of colonist families appear to be taking a more conservative approach, often based on knowledge gained elsewhere in the Amazon. Now many new arrivals purchasing older farm properties and focus their attention on clearing secondary succession left by previous owners. Often these areas are close to the house and more accessible than remaining forested areas in the back of the farm lot. As an increasing share of farmland is placed in productive activities, particularly labor-intensive activities such as cocoa and coffee, or maintaining pasture and caring for cattle, an increasing share of family labor is devoted to these endeavors and less to clearing forested land. As children of initial colonists become adults and begin to form families, we may witness new patterns. Further deforestation on the parental farm may continue, but fragmentation of properties in this area is uncommon. Most adult children who remain in agricultural activities typically purchase neighboring farmland or move to new areas of colonization. During the course of interviews, particularly during conversations with older and 2nd generation colonists, many voiced concerns with retaining some forested areas for future lumber needs, and for conserving areas around sources of water. The maintenance of a near quarter of the land in forest in an older frontier, after the majority of farms have been established, may suggest some optimism. It also must be remembered that these are now highly fragmented areas with substantial loss of biodiversity and have important implications for the local climatic and hydrologic systems.

If we take the regression results in Table 1 and Figures 8 and 9, we can calculate predicted area deforested for the area over the last 26 years for each property based on initial period of occupation to compare our results with observed levels of deforestation. We chose to use the results from the quadratic equation based on their intermediate estimation of the age pattern of deforestation; these are presented in Figure 10. The predicted amount of area remaining in forest is overstated for 1976 and 1979, when compared with the observed levels, and understated in 1991 and 1996. This can be explained in part by the ambitious efforts of the early colonists, and by renewed deforestation on these farms associated with the increase in cocoa prices in the late 1970s and early 1980s. Establishing a very general baseline of deforestation associated with the age pattern of farms now provides us with the opportunity to further explore "period" effects associated with temporal changes in policy and market fluctuations as well as "cohort" effects that distinguish different waves of settlement. By projecting out the percent of area deforested on these farm properties, we anticipate that nearly 37 percent of the area will remain in forest by 2010. If new settlement takes place on the remaining properties (408) during the first part of this decade, we can expect only 31 percent of the area remaining in forest. Barring major changes in policies and prices, this will fall to 24 percent and level out.

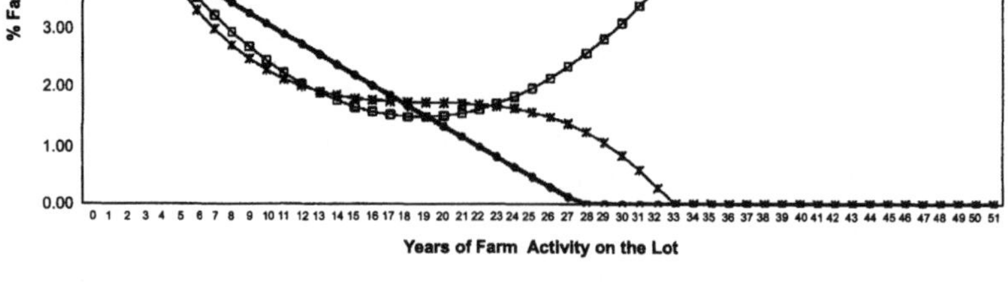

Figure 9a. Estimated percent farm area cleared annually from regression results; various methods.

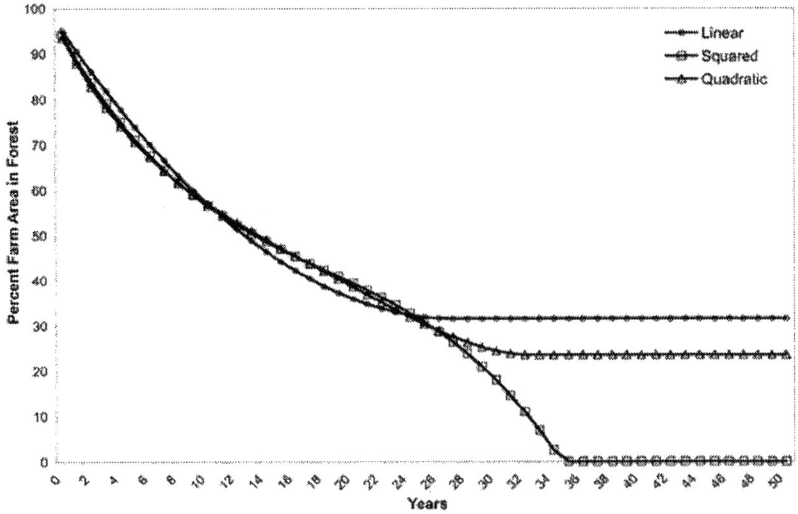

Figure 9b. Predicted forest remaining based on regression results; various methods.

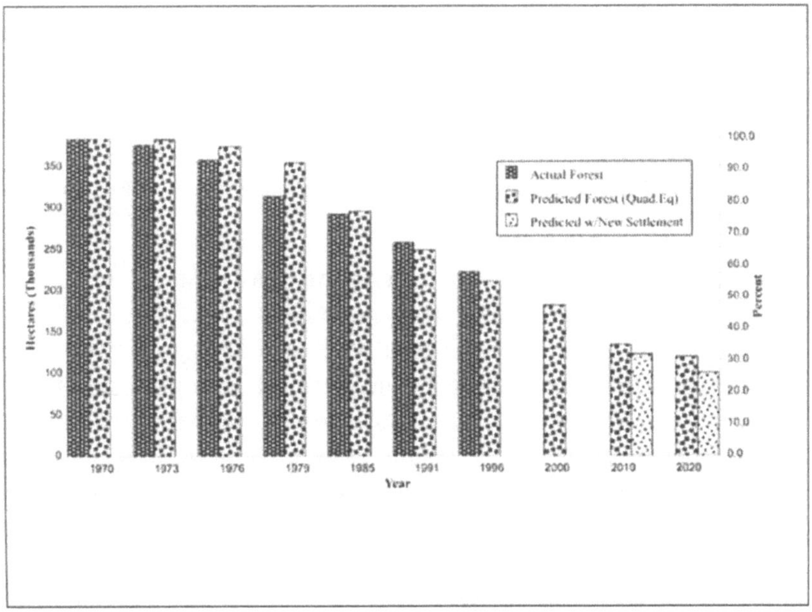

Figure 10. Observed and expected area of remaining forest in the study area, 1970-96, and predicted for 2010 and 2020; based on farm level projections.

Fitting the regression lines and deriving a general pattern of deforestation by age of the farm also provides an opportunity for exploring spatial patterns not directly identifiable when working with tabular survey data. A map of residuals of observed minus predicted deforestation in 1996 (Figure 11) illustrates interesting spatial patterns. Areas with light gray shading represent farms with smaller-than-expected deforestation, given the length of settlement, while dark gray shading represents areas with greater levels of deforestation, whereas white represents areas within one standard deviations of the expected value.

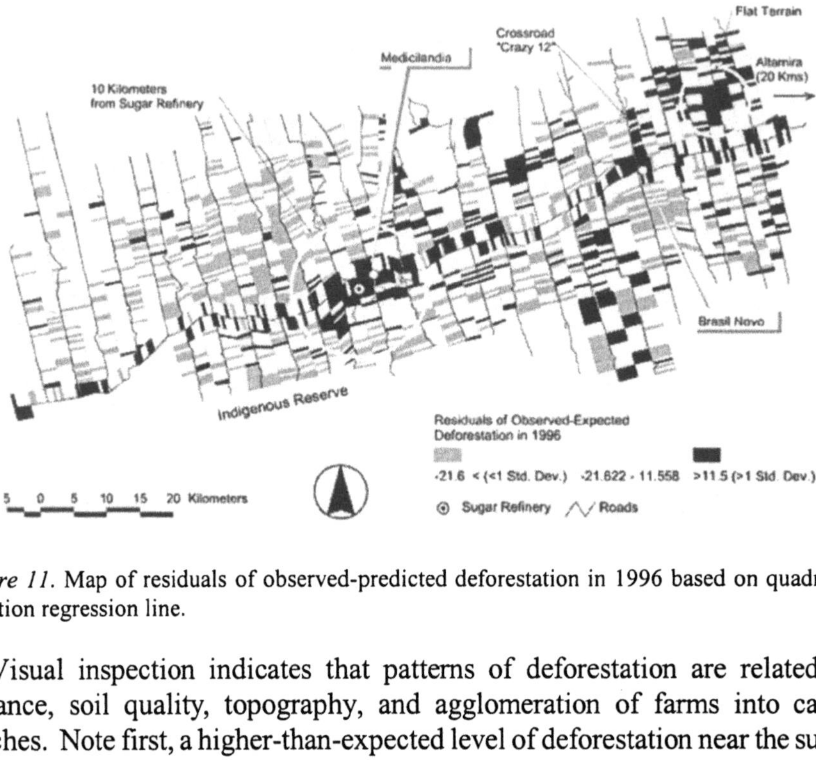

Figure 11. Map of residuals of observed-predicted deforestation in 1996 based on quadratic equation regression line.

Visual inspection indicates that patterns of deforestation are related to distance, soil quality, topography, and agglomeration of farms into cattle ranches. Note first, a higher-than-expected level of deforestation near the sugar cane refinery near Medicilândia. Sugar cane, once cut, has diminished potential for sugar production the longer it is not processed. Areas within 10-kilometers of the sugar refinery are, as a result, associated with higher levels of deforestation. Similarly, we note areas of deforestation greater-than-expected in the northern section of the property grid closest to Altamira. This area is flat, and given its proximity to Altamira, is an area where large-scale cattle ranching activities are increasing on agglomerated farm properties. In contrast the extreme southern fringe of the study area, bordering the indigenous reserve, is very rugged terrain and has more forest given the age of the farms. Topography may be important in the particular farming strategies pursued, and the overall amount of area cleared, but it also affects transportation routes. Properties on side roads leading southward from the town of Brasil Novo, for example, have lower-than-expected deforestation while north of the town one group of properties has higher-than-expected deforestation while another group, on a parallel road, has lower-than-expected deforestation. This area, to the north of Brasil Novo, is known as the "Crazy 12" in reference to the road numbered 12. Variability in road conditions impact farm families differently in addition to issues related to distance and the topography of their farms.

CONCLUSIONS

In this paper, we analyzed the relationship between length of time on the farm and pattern of deforestation. The tightness of the fit illustrates that there is substantial variation among farmers in developing their homesteads overtime, but that a general pattern emerges that is associated with age or length of time on the farm (Brondizio et al. in press, McCracken et al. in press). Future work will combine these estimates with survey data to better understand the complex array of factors at the household level that affect these patterns. Similarly, the base-line estimate of the age pattern of deforestation among frontier farms will serve to disentangle "period and cohort effects" associated with changing policies and market prices and differences among the various waves of colonists' settlement. Future analysis with a property grid will be used to analyze transition matrices of forest-conversion to annual crops, pasture and perennial crop production observable with multiple satellite images. Ultimately we envision analyses of these transitions between various land covers through the construction of multi-state life tables to analyze the flows between different types of land-use. The very general age-pattern of deforestation on farm properties provides the base line for this work.

Our predictive analysis of deforestation suggests that over a 50-year period (i.e., from 1970 to 2020), total original forest cover will decline to 24-32 percent (based on quadratic and linear model fits, respectively). Our analysis based on individual properties, suggests that rates of deforestation slow down considerably over time, but as new roads and communities are created, new pockets of frontier expansion and consolidation are established. As Brazil undertakes in 2001 a new, and ambitious program of road-building (i.e., Avança Brasil), the implications of these findings for future deforestation should give considerable pause to development agencies. The rates for the area studied here could very well become the rates for the Amazon as a whole as more of the region becomes road accessible.

REFERENCES

Andreae, M. O. and D. S. Schimel (eds.) 1989. *Exchange of Trace Gases between Terrestrial Ecosystems and the Atmosphere.* New York: Wiley-Interscience.

Booth, W. 1989. Monitoring the Fate of Forests from Space. *Science* 243:1428-1429.

Brondizio, E., S. McCracken, E.F. Moran, A.D. Siqueira, D. Nelson, and C. Rodriguez-Pedraza. In press. The Colonist Footprint: Towards a conceptual framework of deforestation trajectories among small farmers in Frontier Amazonia. In: *Land Use and Deforestation in the Amazon*. C. Wood and R. Porro (eds.) Gainesville, University of Florida Press.

Crutzen, P. J. and M. O. Andreae. 1990. Biomass Burning in the Tropics: Impacts on Atmospheric Chemistry and Biogeochemical Cycles. *Science* 250:1669-1678.

Dickinson, R. E. (ed.) 1987. *The Geophysiology of Amazonia: Vegetation and Climate Interactions*. New York: Wiley.

Fearnside, P.M. 1987. Causes of deforestation in the Brazilian Amazon. In: *The Geophysiology of Amazonia*. R.E. Dickinson (ed.) Pp. 37-57. New York: Wiley

Fearnside, P. M. 1996. Amazonian Deforestation and Global Warming: Carbon Stocks in Vegetation Replacing Brazil's Amazon Forest. *Forest Ecology and Management* 80(1-3):21-34.

Gash, J. H. C., C. A. Nobre R. L. Victoria, and J. M. Roberts (eds.) 1996. *Amazonian Deforestation and Climate*. New York: Wiley.

Instituto Nacional de Pesquisas Espaciais. http://www.inpe.br/ Website consulted 1998, 1999.

Liverman, D. E.F. Moran, R.R. Rindfuss and P.C. Stern (eds.) 1998. *People and Pixels: Linking Remote Sensing and Social Science*. National Academy Press: Washington D.C.

Malingreau, J. and C. J. Tucker. 1988. Large Scale Deforestation in the Southeastern Amazon Basin of Brazil. *Ambio* 17:49-55.

McCracken, S., E. S. Brondizio, D. Nelson, E. F. Moran, A. D. Siqueira, and C. Rodriguez-Pedraza. 1999. Remote Sensing and GIS at Farm Property Level: Demography and Deforestation in the Brazilian Amazon. *Photogrammetric Engineering and Remote Sensing* 65(11):1311-1320.

McCracken, S., A.D. Siqueira, E.F. Moran, and E.S. Brondizio. In press. Land-use Patterns on an Agricultural Frontier in Brazil: Insights and Examples from a Demographic Perspective. In: *Land Use and Deforestation in the Amazon*. C. Wood and R. Porro (eds.) Gainesville, University of Florida Press.

Moran, E. F. 1976. Agricultural Development in the Transamazon Highway. Bloomington: Indiana University/Latin American Studies Working Papers.

Moran, E. F. 1981. *Developing the Amazon*. Bloomington: Indiana University Press.

Moran, E. F. 1993a. Deforestation and land use in the Brazilian Amazon. *Human Ecology* 21:1-21.

Moran, E. F. 1993b *Through Amazonian Eyes: The Human Ecology of Amazonian Populations*. Iowa City: University of Iowa Press.

Moran, E. F., E. S. Brondizio, P. Mausel and Y. Wu. 1994. Integration of Amazonian vegetation, land use and satellite data. *BioScience* 44:329-338.

Moran, E. F., A. Packer, E. Brondizio and J. Tucker. 1996. Restoration of vegetation cover in the Eastern Amazon. *Ecological Economics* 18:41-54.

Moran, E. F., E. S. Brondizio, J. M. Tucker, M. C. Silva-Forsberg, S. McCracken, I. Falesi. 2000. Effects of Soil Fertility and Land use on Forest succession in Amazonia. *Forest Ecology and Management* 139: 93-108.

Moran, E.F., E.S. Brondizio, and S. McCracken. In press. Trajectories of Land Use: Soils, Succession, and Crop Choice. In: *Land Use and Deforestation in the Amazon*. C. Wood, and R. Porro (eds.) Gainesville: University of Florida Press.

Setzer, A.W. and M.C. Pereira. 1991. Amazonia biomass burnings in 1987 and an estimate of their tropospheric emissions. *Ambio* 20:19-22.

Skole, D. and C. Tucker. 1993. Tropical Deforestation and Habitat Fragmentation in the Amazon: Satellite Data from 1978 to 1988. *Science* 260:1905-1910.

Smith, N. 1982. *Rainforest Corridors*. Berkeley: University of California Press.

Tucker, J., E. Brondizio and E. Moran. 1998. Rates of Forest Regrowth in Eastern Amazonia and a Comparison of Altamira and Bragantina Regions. Pará State, Brazil. *Interciencia* 23(2):64-73.

Wilson, E. O. (ed.) 1988. *Biodiversity*. Washington, DC: National Academy Press.

Chapter 11

MULTI-RESOLUTION CLASSIFICATION FRAMEWORK FOR IMPROVING LAND USE/COVER MAPPING

DongMei Chen, Douglas Stow, and Arthur Getis

Department of Geography, San Diego State University, San Diego, CA 93182
dchen@esri.com

Keywords: Multi-resolution analysis, land use/cover, classification, spatial analysis.

Abstract Airborne and satellite remotely sensed data have been the major source for generating land use/cover maps. With the development of new remote sensing systems, hyper-spectral and very-high spatial resolution images will soon provide a source for continuously sampling of the earth surface from local, regional to global scales. A considerable amount of previous research has been devoted to exploring the magnitude and impact on image analysis when shifting scale from coarse to fine resolutions. This chapter presents a multi-resolution analysis and classification framework for selecting and integrating suitable information from different resolutions and analytical techniques into classification routines. The multi-resolution approaches were tested using simulated multi-resolution images for a portion of the rural-urban fringe of the San Diego Metropolitan area. We demonstrate that the multi-resolution classification approaches developed in this chapter can significantly improve landuse/cover classification accuracy when compared with those from single-resolution approaches.

INTRODUCTION

Human activities such as urban, industrial, energy, or agricultural expansions, driven by increasing resource consumption from increasing affluence and population growth, have dramatically changed land use/cover patterns on the surface of the earth. Concern about the impact of these activities and their immediate, as well as long-term, consequences makes the analysis and prediction of land use/cover a basis for the rational management of our environment.

Airborne and satellite remotely sensed data have been the major source for generating land use/cover maps and GIS layers. There are two major ways of generating land use/cover maps through remotely sensed data. One is by visually interpreting image features and manually digitizing land use/cover unit boundaries in the image. This approach generally can produce accurate land use/cover maps. Unfortunately, it is expensive and usually unaffordable especially when applied to a large area. The other approach is by conducting semi-automated image classification that provides a low-cost approach to generating land use/cover information, but with a loss of precision and accuracy.

Many methods of semi-automated image classification of remotely sensed data have been established for improving the accuracy of land use/cover classification during the past forty years (Jensen 1996). Different algorithms and strategies including supervised, unsupervised, and hybrid training approaches (Richards 1986), parametric and nonparametric classifiers, segmentation (Conner et al. 1984), artificial neural networks (ANN) (Hepner et al. 1990, Civco 1993, Jensen et al. 2001), fuzzy sets (Wang 1990, Fisher and Pathirana 1990), and knowledge-based systems (Wharton 1989, Kontoes and Rokos 1996) have been developed and tested for various environments. However, most of these methods were developed for use with moderate-to-coarse resolution data ($\geq$ 10 m) and were employed in single-resolution image classification. Few studies on semi-automated methods of land use/cover mapping and analysis have been conducted on images with 1 m to 5 m spatial resolution.

With the development of new remote sensing system, hyper-spectral and very-high spatial resolution images will soon provide a set of continuous samples of the earth surface from local, regional to global scales. The spatial resolution of various satellite sensors ranges from 1 to 25,000 m now. Furthermore, high resolution airborne data acquisition technology has developed rapidly in recent years. As an increasing number of high resolution data sets become available such as Earthwatch, Space Imaging (IKONOS), Orbimage, IRS, ADAR, DOQQ, etc., there is an increasing need for more efficient approaches to store, process, and analyse these data sets of varying spatial resolution.

A considerable amount of previous research has been devoted to exploring the magnitude and impact on image analysis when shifting scale from coarse to fine resolutions (Gong and Howarth 1992, Barnsley and Barr 1992 1996, Cihlar 2000). Due to the increasingly heterogeneous spectral-radiometric characteristics within land use/cover units portrayed in high resolution images, many applications of traditional single resolution classification approaches have not led to satisfactory results.

Based on the scene model concept suggested by Strahler et al. (1986) different land use/cover objects have different sizes and patterns, which require different image analysis approaches. Spatial autocorrelation existing in each class is an important factor influencing classification results at each resolution level. From a practical standpoint, building a framework to represent, analyse and classify images represented by multiple resolutions is necessary in order to capture unique information about land use/cover classes that vary as a function of scale.

This chapter discusses the limitations of conventional single-resolution approaches and presents on theoretical background on multi-resolution analysis and classification framework for remotely sensed data. The multi-resolution framework is based on the development of analytical techniques and strategies of selecting and integrating suitable information from different resolution into classification routines.

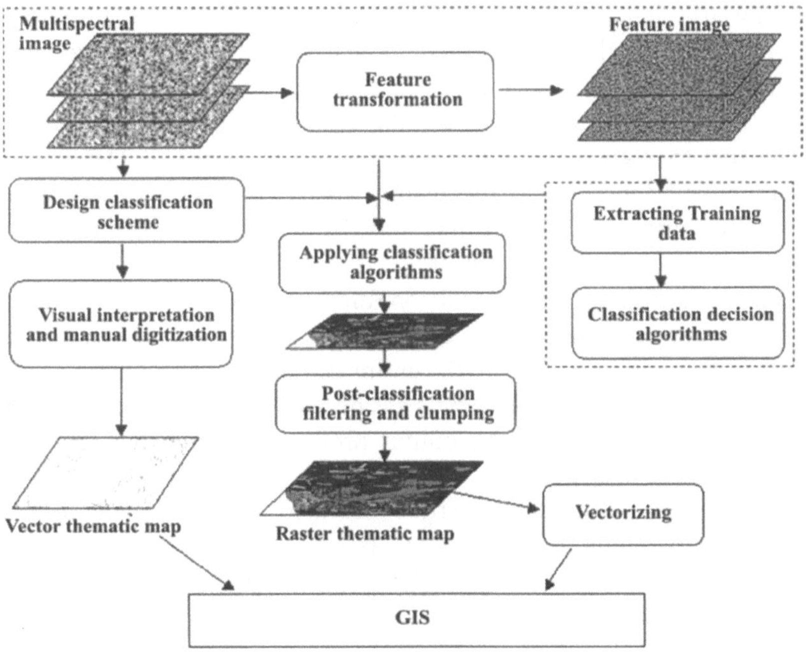

Figure 1. Data flow for generating a GIS layer of land use/cover using remotely sensed data.

Figure 1 shows the basic data flow used in visual interpretation and semi-automated computer approaches for generating GIS layer of land use/cover from remotely sensed images. Computer-assisted classification procedures are the focus of this chapter. In the following the different stages of computer-assisted classification procedures are briefly summarized.

The basic aim or assumption of image classification is that the ground scene represented by image data can be partitioned into spatial units of homogeneous categories or classes. Each category or class is assumed to correspond to a distinct range of spectral reflectance values. The level of detail in the classification process, i.e., the number and types of classes defined, is dependent on the classification purpose, the scale of the study area and spatial resolution of the data. Input for the classification process is in the form of multi-spectral imagery from a satellite or aircraft. Each image pixel contains digital number (DN) values that correspond to reflectance or emittance in one or many spectral wavebands, and/or transformed data derived from DN values. In a typical computer-assisted classification shown in Figure 1, thematic classification involves five major steps:

(1) Feature transformation - Modify a multispectral image by a spatial or spectral transform, such as by selecting a subset of bands, using principal component analysis (PCA) to reduce the data dimensionality, or generating spatial texture/contexture bands, etc. In many cases spectral properties alone are not sufficient for classifying land use/cover features, so the variability of neighboring pixels is quantified to generate spatial texture/contexture features that can be used in classification decisions.
(2) Training - Extract signatures for the pixels to be used for training the classifier to recognize certain categories, or classes. Determine the discriminant functions in the feature space. This step is accomplished through either a) supervised methods, b) unsupervised methods (cluster analysis), or c) a hybrid approach.
(3) Application of classification decision rules – Apply discriminant functions, such as parametric and non-parametric decision algorithms, segmentation, fuzzy sets, neural nets, etc. to the entire feature image and label all pixels. Once each pixel or block of pixels has been assigned to a feature class, the classes are labelled and the output product is a thematic map of the distribution of image feature classes.
(4) Postprocessing – Filter the classified results to generalize the high detailed results, in particular, from per-pixel classification approaches.
(5) Accuracy assessment – Compare the classification with reference data. As discussed by Campell (1996) and Congalton (1991).

The classification scheme should be carefully designed prior to implementation of the classification routine. A successful design for a classification scheme cannot be achieved without a good understanding by an image analyst of the information required, the limitation of the remote sensing data, the techniques available, and most importantly, the information requirements of the user. Very often, a scheme is designed as a compromise between the information required and the data quality and capabilities of classification techniques available. The design of a classification scheme is not a major concern in this chapter.

Table 1 summarizes the main spatial data assumptions for the various types of feature selection, training data and classification algorithms. For each type of classification approaches reviewed, consideration is given to the type of scene model in terms of H (high) or L (low) resolution models suggested by Strahler et

an image (e.g., 30 meters per pixel for TM imagery or 10 meters per pixel for SPOT panchromatic imagery). Clearly, the degree and type of information that can be extracted from digital imagery depends on the spatial resolution. Often the ground sampling distance (*pixel size)* is used to represent the spatial resolution. But ground sampling distance of an image can be different from the spatial resolution of the sensor that records the image after resampling. However, for this chapter spatial resolution is used in both its strict meaning and its broad meaning of ground sampling distance of an image. When the spatial resolution of an image is changed, it means the image is resampled to a different ground sampling size, which means that the ground sampling distance (GSD) is changed.

According to Townshend (1988), there are four criteria for selection of the spatial resolution of an image system: a) the geometrical properties of the system, b) the ability to separate point targets, c) the ability to measure periodicity of repetitive targets and d) the ability to measure spectral properties of small targets.

Previous research on scale dependence in digital images has often focused on changes in the information derived from images as a function of the spatial measurement interval. The measurement interval, or *measurement (analysis) scale* is the distance or unit at which measurement or spatial analysis is taken, such as the lag used in calculating semi-variance, or the sampling distance. It is associated with the spatial resolution of a sensor or image.

Resolution-Related Problems in Image Classification

As an increasing number of sources of multispectral remotely sensed data become available, the selection among different spatial resolutions becomes difficult. To select an image with appropriate spatial resolution for an image classification, one should examine the characteristics of scene content, especially the changing pattern of a scene as a function of changes in analytical scale and resolution. For example, how does the Ground Sampling Distance (GSD) (e.g., 1, 5, 10 or 100 m) affect the statistical parameters used in classifiers and the result of classification in land use/cover classification using remotely sensed imagery?

The spatial resolution of an image substantially affects image classification results. Spatial variation of image brightness values is important in the classification of remotely sensed data since pixels are grouped into classes because of similarities and classes are distinguished because of dissimilarities. Most studies on spatial resolution have examined the accuracy of estimating some property on the ground with remotely sensed imagery of different spatial resolutions (such as Welch 1982, Latty et al. 1985, Irons et al. 1985, Toll 1985, Johnson and Howarth 1987, Marceau et al. 1994, Hlavka and Livingston 1997). These studies are empirical and the choice of spatial resolution depends on the availability of images.

Markham and Townshend (1981) found that image classification accuracy is affected by two factors. The first factor is the influence of boundary pixels on classification results. The second factor which influences classification accuracy is that finer spatial resolution increases the spectral variation of land cover types. Variation within a land cover unit decreases the spectral separability of classes and results in lower classification accuracy.

Based on a forest environment, Marceau et al (1994a, 1994b) evaluated the impact of measurement scale and spatial aggregation on the information context and classification accuracy. Their results revealed that the impact of resolution change is greater than the change of aggregation level. The accuracy on a per-class basis showed that at a particular aggregation level, some classes are better classified at fine spatial resolution while others require coarser resolution.

The advancement of spatial analysis methods and technologies for handling spatial data increases the chance of successively examining spatial pattern and processes from a multiscale and/or multilevel point of view (Cao and Lam 1997). At the same time, inferring the underlying spatial patterns at different scales remains a challenge. Depending upon the observation scale, patterns that appear homogeneous at a coarse scale may be heterogeneous at a fine scale. For example, in a LANDSAT TM image, the urban residential environment is sampled as a relatively homogeneous entity. But when observed at 1 to 5 meter resolution, residential areas mostly consist of individual buildings, roads, sidewalks and vegetation. In this case, neighborhood controls such as spatial/texture methods might be applied, or post-classification spatial processing (e.g., Barnsley and Barr 1996) might be implemented to aggregate land cover components to identify the residential class.

Several techniques have been employed to assess appropriate (or optimal) spatial resolutions. Although a particular classification can achieve the best result from a single resolution appropriate to the class, there is no single resolution that would give the best results from all classes (Marceau et al. 1994b). Clearly landscape objects (e.g. land cover/use polygons) are not the same size, therefore, the analysis scale corresponding to one object may not be applicable to the others.

Several researchers have also emphasized the importance of multiscale approaches. For example, De Cola (1993, 1994 and 1997) proposed multiscale visualization as a component of exploration and analysis of image data. Friedl (1997) introduced a multiscale simulation model of spatial variation and used it in an environmental study using satellite data. Franklin and Woodcock (1997) addressed the advantage of multiscale approaches in vegetation mapping. Couloigner and Ranchin (2000) developed a multiscale approach for automatic road extraction and indicated the potential of the approach. However, there are few of any works on developing multi-scale approaches used in classification.

The question of how one might choose scales that best represent each land use/cover type is the basis for proposing multi-resolution classification approaches. Such approaches should incorporate information obtained from multi-resolution data and exploratory data analysis to improve the accuracy and effectiveness of land use/cover classification procedures. The following illustrates the multi-resolution classification framework.

MULTI-RESOLUTION IMAGE ANALYSIS AND CLASSIFICATION FRAMEWORK

A proposed multi-resolution image analysis and classification framework for integrating information from multiple resolutions is illustrated in Figure 2. Central to the approach are the following assumptions and hypotheses:

- There is no single spatial resolution that is optimal for classifying land use/cover types, but an appropriate spatial resolution range does exist for specific land use/cover types and a given classification routine and mapping approach.
- Information obtained for a range of spatial resolutions is useful for discriminating land cover/use classes, but improvements in accuracy are a function of the complexity of the environment, analysis methods, and land cover/use types.
- Choosing the appropriate resolution range can be aided by information on the spatial characteristics of image brightness using spatial analytical methods.
- Different land use/cover types can be identified based on the their spectral DNs and their spatial features extracted from different resolutions.
- With classification accuracy as the standard for evaluating effectiveness, multiple spatial resolutions and analysis techniques are superior to single-resolution approaches.

The framework proposed here focuses on the examination of image pattern/structure/autocorrelation using different spatial analytical techniques in order to select appropriate methods in different stages of classification such as training strategy, feature extraction, scene models, and classification accuracy. Multi-resolution images can be generated through different aggregation methods. Spatial analysis techniques for measuring the pattern size and degree of autocorrelation are computed for each training class in order to determine whether they can guide selection of training data, high-resolution or low-resolution classification models, and the range of spatial resolutions used for classification. The strategies of incorporating information from resolutions are developed.

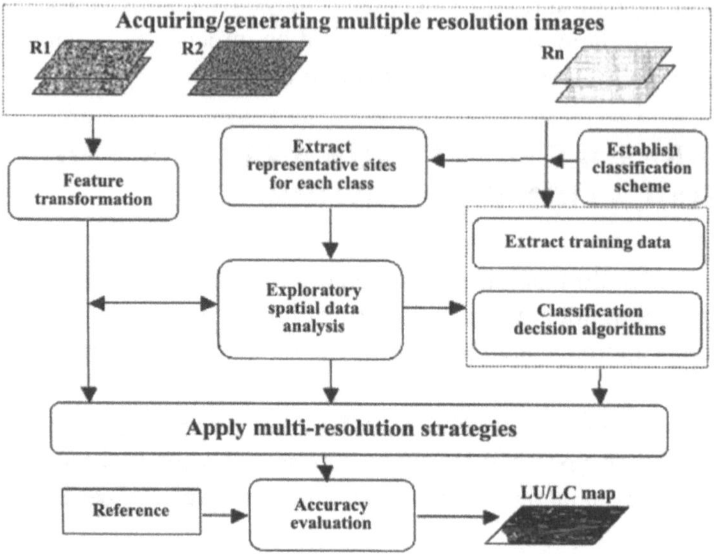

Figure 2. A proposed multi-resolution image analysis and classification framework. Rn stands for variable image spatial resolution.

Generating Image Resolution Pyramids

Multi-resolution images can be created in two ways: (1) by integrating different resolution images acquired by different sensors such as the multi-resolution images shown in Figure 3; and (2) aggregating fine resolution images into different coarse resolution levels (i.e., image pyramids). Obtaining images of different resolutions from different sensors could have advantage of including more spectral information that can be used to identify different objects, but it is expensive. The misregistration between different images also would increase the processing cost and reduce classification accuracy.

It is generally easier to extract spatial information from an image over a range of resolutions. Often an integration of information at different resolutions is required, *upscaling (bottom-up scaling)* to apply information from a fine resolution at a coarse resolution, *downscaling (top-down scaling)* to apply information from a coarse resolution at a fine resolution and *isoscaling* to use information at the same resolution.

The process of changing resolution of data from fine to coarse is often called *aggregation (*Bian and Butler 1999*), degradation* (Justice et al. 1991), or *reduction* (Schowengerdt 1997), and will alter the statistical and spatial characteristic of the data. A series of images at different resolutions are often called image pyramid layers. A pyramid is defined as,

$$X(L) = AGGREGATION\ (X(L-1)) \quad (1)$$

where $X_{(L)}$ is the image at resolution level L and AGGREGATION is any operation on the image at level L-1 that reduces its resolution for level L.

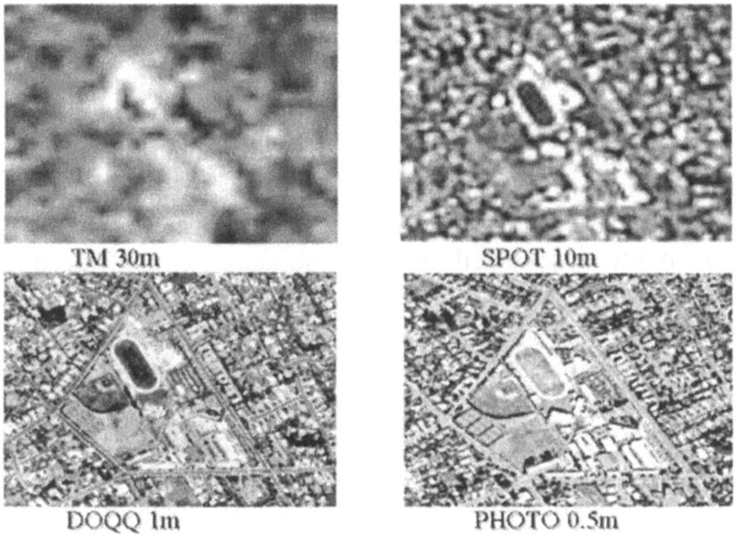

Figure 3. An example of multi-resolution images acquired from different sensors for an subarea near Del Mar, San Diego.

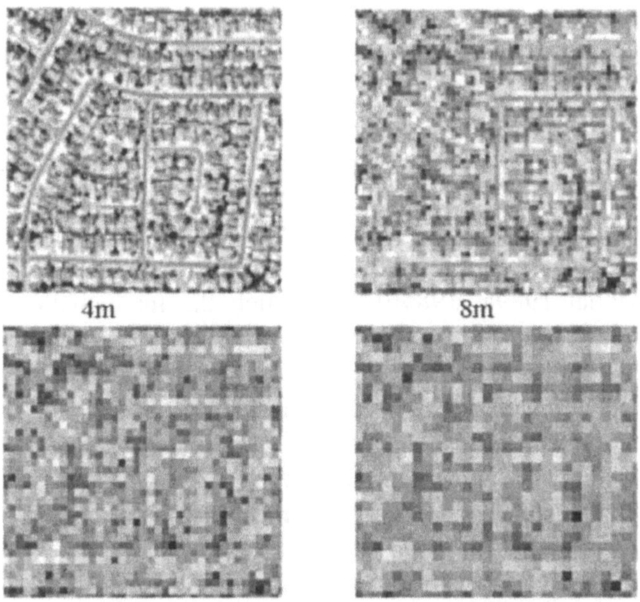

Figure 4. An example of an image pyramid generated from a 1m DOQQ subset of single-residential area.

For example, aggregation can be achieved by computing mean value of 2 x 2 pixel neighborhoods. Figure 4 illustrates one example of this. In remote sensing and digital image processing, generating image resolution pyramid layers is often used to efficiently display, store and analyze image data.

There are several types of methods used in the operation of *AGGREGATION*. The simple types of aggregation methods include averaging, central-pixel, median, sub-sampling, nearest neighbor, cubic convolution, etc. Simple aggregation methods are often used to generate coarser resolution images representing original finer resolution images in "scaling up" analysis. This type of aggregation often loses detail, especially variance for highly heterogeneous regions. Several researchers such as Bian and Butler 1999, Marceau et al. 1994a have compared effects of simple aggregation methods in remote sensing. Central-pixel and median values are not realistic physical models of a sensor's IFOV process. The nearest neighbor and cubic convolution algorithms of interpolating data induce sharpening or smoothing effects that influence the resulting analysis (Marceau et al. 1994a). Collins and Woodcock (1999) suggested using a geostatistical variance estimation algorithm rather than a simple averaging method to better characterize resolution-dependent variance of coarse images from fine-resolution images. However, due to its computational complexity, it has not been employed in practice.

Another method, *scale-space transform,* emphasizes extracting the spatial information from an image over a range of scales, from fine details in local areas to large features that extend across the image (Lindeberg 1994). A number of approaches are used for this type of transform; often they are grouped within the category scale-space filtering. Filters are applied repeatedly to generate multiple resolution pyramids. This kind of approach can dramatically change the range of pixel values in an image, but maintain the inherent spatial structures.

Wavelet theory provides a general mathematical framework for decomposition of an image into components at different resolution. Multiresolution analysis (MRA) can be achieved using the wavelet transform to reorganize the image information into a set of details in a sequence of closed subspaces (Mallat 1989a, 1989b). Although the mathematics of wavelet transforms are complex, an image after application of a wavelet transform can be similar to the result of simple aggregation methods and scale-space transforms, depending on the wavelet transform that is used. Within the context of wavelet decomposition, Csillag and Kabos (1996) and Csillag (1997) applied a quadtree-based image segmentation procedure to map a complex environment and argued that this procedure leads to an optimal solution for determining mapping units based on spatial variability.

Wavelet theory was developed as a pure mathematical topic in the early 1980s. However, in only a few years of existence as a theory, it has shown great potential and applicability in many fields, such as image compression (DeVore et al. 1992). Application of the wavelet analysis on remote sensing have just begun. Li and Shao (1994) showed the usefulness of wavelet in image edge detection and automated interpretation of buildings from aerial images. Ranchin and Wald (1993) illustrated the potentials of wavelet transform on the multiresolution analysis of remotely sensed images. Yocky (1996) applied wavelet decomposition approach to merge TM and SPOT data. Mohanty (1997) used wavelet transform in image enhancement. Within the framework of the wavelet decomposition, Csillag (1997) and Csillag and Kabos (1996) applied a quadtree-based image segmentation procedure to map complex environment and argued that this procedure leads to an optimal solution for determining mapping units based on spatial variability.

Figure 5 illustrates a 3-level wavelet scheme. The details of an image at the resolution R can be represented by the difference of its approximation at the resolution R and that at the lower resolution R/2. This difference is extracted by decomposing the function in a wavelet basis. At each resolution R, the image was represented by its wavelet approximation image at resolution level R/2 called context image, and the three subspaces corresponding the details structures at vertical, horizontal and diagonal orientations. The context image contains the information due to the structures which scales are greater than the current scale. The detail images illustrate all the structures having a characteristic length comprised between R and R/2.

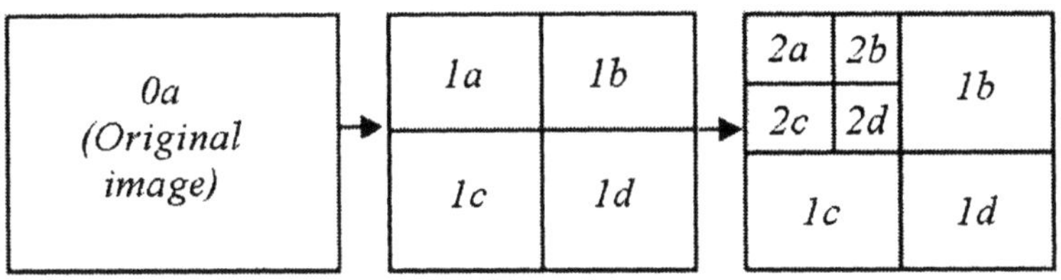

Figure 5. Scheme of a three-level hierarchical wavelet decomposed analysis.

0a: Resolution level 0 image (with the finest resolution R)

1a, 2a: Context images at resolution level 1 (2R) and 2 (4R)

1b, 2b: Horizontal sub-images at resolution level 1 and 2

1c, 2c: Vertical sub-images at resolution level 1 and 2

1d, 2d: Diagonal sub-images at resolution level 1 and 2

With the simplest Harr wavelet, the variability of the original image (*Var*) is decomposed into the three directional subimages: vertical, horizontal and diagonal. It is expressed as:

$$Var = W1b + W1c + W1d \tag{2}$$

Here *W1b, W1c, W1d* are values in *1b, 1c* and *1d* sub-images. In this way we can keep track of how much variance has lost where during aggregation. The texture images are same as those generated by the averaging method. The detail variance images at each level can be used: a) as additional texture bands and b) in evaluating the relationship between scale effect on classification accuracy and the variance.

Spatial Statistics in Image Analysis and Classification

Many spatial statistical measures have been used to establish the spatial characteristics and scene model characteristics of images and to assess the scale of spatial variation in remotely sensed data for determining:

- Image spatial structure (Woodcock and Strahler 1987, Strahler etc. 1986);
- Suitable support sizes and spatial resolutions for different applications (Woodcock et al., 1988a 1988b, Townshend and Justice 1988, Atkinson and Curran 1995 and 1997, Atkinson 1997a);
- Training data selection (Labovitz and Masuoka 1984);
- Measurement error (Atkinson 1997b, Congalton 1988); and
- Ground-sampling design (Curran and Williamson 1986, Atkinson 1991).

Several variables play a role in the success of image classification: the type and level of classification system, spatial, spectral and radiometric resolution of imagery, landscape variability, and classification algorithm. As summarized before, proper application of image classification procedures requires knowledge of the above variables of the data to determine the appropriate classification methodology and parameters to use. Values for these attributes are often selected on an *ad-hoc* basis, ignoring the spatial characteristics of the image being considered (Woodcock and Strahler 1987, Woodcock and Harward 1992). Selecting an appropriate classification routine and optimally specifying its required parameters are paramount issues to facilitate accurate use of high spatial resolution and hyper-spectral imagery and the integration of these data in GIS, as noted by Ryerson (1989), Davis et al. (1991), and Lunetta et al. (1991).

Woodcock and Strahler (1987) used the local image variance to help in selecting an appropriate image scale for forested, agricultural, and urban/suburban environments. A measure of local variance at different spatial resolutions was computed and plotted, and the location and height of the peak of the curve was used to select an optimal spatial resolution for the scene. An optimal resolution of 15 to 20 meters was determined for a scanned color infrared photograph of suburban southern California. The drawback of the local variance method is that it is being strongly related to the global variance of the image studied. Therefore, the measure is image dependent and cannot be compared between images.

Townshend and Justice (1988) applied scale variance and Fourier analyses on spatially degraded Multispectral Scanner data, from 125 m to 4 km, to investigate the required spatial resolution for global monitoring of land transformations. They advised an average resolution of 500 m as the best compromise between the detail of changes detected and the resultant data volume.

For a forested environment, Marceau et al. (1994b) tested an approach to identifying the optimal spatial resolutions for detection and discrimination of coniferous classes. The minimal intraclass variance was used as the indicator of the optimal spatial resolution. MEIS-II data with 0.5 m resolution were degraded to 29.5 m with an increment of 1 m, using an averaging window algorithm. They concluded that the optimal spatial resolution is primarily affected by the spatial and structural parameters of forest stands.

Fractal analysis is another technique used for assessing scale effects. The key concept underlying fractals is self-similarity (Lam and Quattrochi 1992). Although true fractals with self-similarity at all scales are rare, self-similarity often exists over a limited range of scales. Several studies such as Quattrochi et al. 1997, and Xia and Clarke 1997 have demonstrated that fractals are a promising tool for providing insights on spatial scale and resolution.

Spatial autocorrelation is an important factor in selecting appropriate spatial resolutions and image models (L- or H-resolution). With higher resolution systems the positive autocorrelation is even higher (Labovitz and Masuoka 1984, Congalton 1988b). The concept of spatial autocorrelation has been introduced as a basis for understanding the effect of scale. Several studies have explored spatial autocorrelation measures for the autocorrelation of pixel DNs and to determine the optimal spatial resolution of a remotely sensed application.

The semi-variogram is the most commonly used measure of autocorrelation. Since Jupp et al. (1989a, 1989b) and Curran (1988) introduced the basic methods of autocorrelation and regularization in digital images, the variogram has been widely adopted for modeling the scale variation of remote sensing applications. The semi-variogram was also suggested by Atkinson and Curran (1995 and 1997a) as a tool in choosing an appropriate spatial resolution in an agriculture application of remote sensing.

Different Strategies of Incorporating Information from Multiple Resolutions

Three strategies were developed for MLC in this research to exploit information obtained from different resolutions and thus, to improve the classification results.

Of the many approaches available, the Gaussian maximum-likelihood (GML) classifier is utilized here for developing the multi-resolution strategies, although much of the discussion will be applicable to other classifiers. This is because the GML classifier is by far the most common supervised classification method. The GML is robust and utilizes means, variances, and covariances of training site statistics, where most of the decision rules are based on simple statistics.

The ML classifier is based on an estimated probability density function for each of the classes under consideration. The class statistics are obtained from the training data. Pixels are allocated to their most likely class of membership. By means of standardizing the inverse of the variance-covariance matrix of each class, the process takes into account not only the marginal properties of the data sets, but also their internal relationships. This is one of the reasons for the great robustness of the process and for its relative insensitivity to distribution anomalies. However, the GML is a parametric classifier and training data presumed to be normally distributed.

Strategy I: Feature Layers from Different Resolutions as Layers in Classification

A simple means of using information from multiple resolutions is to incorporate them simultaneously in a classification routine. In this way the feature measures obtained from coarse resolutions are merged with those from finer resolutions. Then the ML classifier is applied. This method is easy to understand and the information from multiple resolutions is incorporated together. No other algorithms are needed to organize the data. All features from coarse resolutions need to be mapped back to the finest resolution using pixel replication. The major drawback is that computation cost may be high.

Strategy II: Comparing Posterior Probabilities from Multiple Resolutions

For this approach, the ML classifier is applied at each resolution to obtain the probability $P(k|i)$ for each pixel k as a member of class i (i=1, 2, ..., m possible classes) and converted to *a posteriori* probabilities of class membership, which are assessed as the probability density of a case for a class relative to the sum of the densities. The *a posteriori* probability of a pixel k belonging to a class i, $L(i|k)$, can be determined by the following equation (1):

$$L(i\,|\,k) = \frac{a_i P(k\,|\,i)}{\sum_{r=1}^{m} a_r P(k\,|\,r)} \qquad (3)$$

where $P(k|i)$ = the probability for a pixel k as a member of class i,

a_i = *a priori* probablity of membership of class i,
m = total number of cclclasses.

For each pixel, the *a posteriori* probabilities sum to 1.0. At each resolution, the highest *a posteriori* probability and its related class were output for each pixel.

With the above approach, the feature layers obtained from coarse resolutions do not have to map back to the finest resolution specified and the computation cost is lower than that in the first strategy. However, there is a requirement for classifying the entire image at all resolutions so this method may still not be very efficient. Pixels having the highest confidence at one resolution can be removed (e.g., masked out) at subsequent finer resolutions and this leads to the third strategy, a top-down procedure.

Strategy III: Top-Down Filtering Approach

A top-down multi-resolution procedure starts with the coarser resolution. Only when necessary the finer resolution images are used. The basic steps involved in the top-down, per-pixel classification process can be stated explicitly as follows:

(1) At each level of resolution l, select groups of training pixels for each class through the image. Use $T_l(i)$ to denote the training data set for class i at resolution level l. These training data are used for calibrating the classification routine.

(2) Classify the image at the coarsest resolution R with the maximum likelihood (ML) classifier. For each pixel k, calculate $P(k|i)$ *and* $L(i|k)$, which are the probability and *a posteriori* probability as a member of class i, respectively. Using equation *1* obtain the maximum probability $MaxP(k|i)$ and maximum *a posteriori* probability $MaxL(i|k)$.

(3) Calculate the mean and standard deviation of maximum probabilities ($Mean(P(k|i)$, $Std.(P(k|i))$ and maximum *a posteriori* probability ($Mean(L(i|k)$, $Std.(L(i|k))$ for each class i based on the training data $T_R(i)$. If the training data for each class are unbiased samples, a pixel k can be assigned to a class c with high confidence if

(i) $P(k|c) \geq P(k|i)$ *where* i= 1, 2, … m possible classes, and

$$Mean(P(k|c) - Std.(P(k|c) \leq P(k|c) \tag{4}$$

(ii) $L(c|k) \geq L(i|k)$ where i = 1, 2, 3, … m possible classes; and

$$Mean(L(c|k)) - Std.(L(c|k)) \leq L(c|k) \tag{5}$$

where $k \in T_R(c)$

So *(4)* and *(5)* are used as thresholds for deciding whether a pixel is assigned to the maximum likelihood class.

(4) All pixels with maximum probabilities not satisfying (4) and (5) are regarded as mixed pixels, or pixels which do not have identical signatures and cannot be identified at this resolution level. For these pixels the *a posteriori* probability $L_R(i|k)$ is calculated using equation (1), where $L_R(i|k)$ is the *a*

posteriori probability $L(i|k)$ at resolution level R for pixel k as a member of class i.

(5) All pixels that are already assigned to a class are excluded (masked) for subsequent processing. For other pixels, the process goes to finer resolution images. Repeat step 2. Use $L_R(i|k)$ as priori probability a_i in equation (1) to calcuate the *a posteriori* probabilty for pixel k;

(6) Repeat step 3-5.

The above sequential process stops once a) all the pixels are assigned to a class, or b) the finest resolution is reached. In case b) if there are still pixels unassigned to a class, assign a pixel to the class with the highest *a posteriori* probability.

CASE STUDY

San Diego County, California, has seen rapid urban expansion due to economic and population growth in the last twenty years. In order to assist regional planning process, the San Diego Association of Government (SANDAG) generates and updates land use/cover maps of the entire San Diego region every five years. The land use/cover data are used by environmental managers, elected officials, urban planners, businesses, academicians, and the general public to address a wide variety of issues and topics. The urban fringe area of the Del Mar quadrangle is one of the main locations of recent urban expansion in San Diego County, with large tracts of undeveloped rural land rapidly being converted into urban land uses. These land use dynamics make it difficult for planners and land managers to obtain or maintain up-to-date land cover/use information.

The multi-resolution framework was tested using simulated multi-resolution images derived from 1 m USGS CIR Digital Orthophoto Quarter Quads (DOQQ) data for a portion of the rural-urban fringe of Del Mar quadrangle. The image data have a spatial resolution of 1 m with three spectral bands (Green, Red, and NIR).

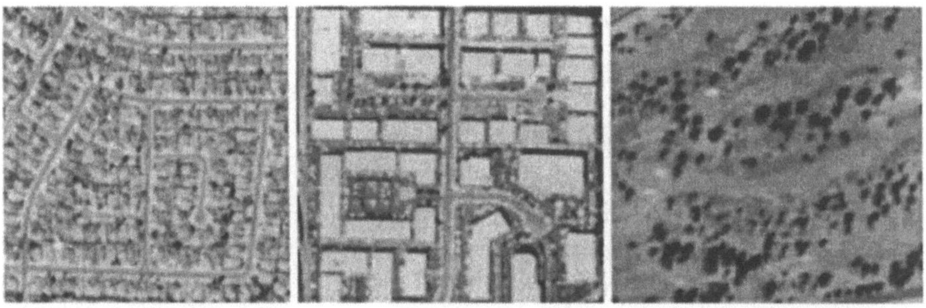

(a) Single-family residential (b) Industry (c) Irrigated grassland

Figure 6. Three USGS CIR DOQQ subsets with single land use/cover components.

DOQQ images with 1 m resolution were aggregated progressively into six nominal resolution levels (2m, 4m, 8m, 12m, 16m, and 20m) by an averaging method. Eight land use/cover classes were used, including single-family residential, multi-family residential, industrial/commercial, irrigated grassland, high-density vegetation, cleared land, undeveloped land, and agriculture land. Figure 6 shows color composites of three subsets with single land use/cover components at 4 m resolution, respectively, with Bands 1, 2, and 3 being assigned to the blue, green and red color, respectively. Initial exploratory data analysis was carried out as three trials. The first trial used histograms to determine the types of distribution exhibited by each band and resolution since MLC assumes normally distributed data. The second trial included the mean and standard deviation to assess distribution properties. The final trial established if each band offers or similar or different information, i.e. are they correlated?

Semi-variograms were calculated for systematic incremental spatial lags for representative sites of each land use/cover. The variogram is the core tool of geostatistics. (Detailed theoretical and mathematical exploration of variograms is presented in Cressie (1991)). Each value of a pixel was compared to its neighbors at varying distances (lags) and the differences were calculated. The semivariance was then calculated by averaging the summation of squared differences for pixels. In most cases the semivariance tends to increase with spatial intervals. As distance increases, the difference between the value of pixel pairs becomes larger. After reaching a maximum value, the semi-variogram flattens (called the *sill*). The lag at which the sill is reached is called the *range*. The range indicates the extent to which values sampled from a spatial process are similar (spatially related). The value of the sill indicates the variance of the entire image.

The non-directional or isotropic semi-variogram was calculated and plotted for a set of training data to assess the degree of spatial autocorrelation in respective bands. The range along each semi-variance curve was located to identify the effective range of spatial dependence. The ranges were determined by visual examination and through a comparison of piecewise slopes. The shape and range of each semi-variogram were useful for determining suitable sizes for training data, sampling interval, resolution or window sizes used for spatial feature extraction. Based on the discussion in previous section, when image resolution is close to or coarser than the range, an L-resolution scene model is generally most appropriate. Otherwise, spatial features that incorporate texture/contexture information should be generated at H-resolution.

Training data were selected by visually identifying and manually digitizing blocks of pixels. As a general rule, the length and width of small blocks for each class were close to the range obtained from the semi-variogram, so that each block was big enough to represent the spectral and spatial properties of each class. Thus, the heterogeneity or autocorrelation within each class was included in the training data. The distance between any two polygons was at least the range, so the pixels in one block were correlated, but not spatially autocorrelated with those in another block (See Chen and Stow 2001 for the detail).

Both single-resolution and multi-resolution classification were conducted. The single-resolution classification was used as a benchmark for evaluating various multi-resolution approaches.

The results were evaluated and analysed based on their classification accuracies in discriminating between eight land use/cover classes. The reference data were derived with the aid of an extant land use GIS layer, aerial photographs, and field reconnaissance. In each classification exactly the same training pixels and reference data were used. The Kappa coefficients (Bishop et al. 1975, Jensen 1996) were reported for the general study area for a series of classification maps to evaluate the agreement between the classification results and the reference data.

Table 2 presents the summarized results obtained from single-resolution and multi-resolution approaches. The Kappa values obtained from classification using three multiple strategies are greater when compared to those from single resolution image input. When compared to the accuracy results obtained from three strategies, the Strategy III resulted the highest classification accuracy (0.6483). The classification improvement from the best classification results obtained from three strategies are at 0.95 confidence level when compared with classification results obtained from a single resolution.

Table 2. Summary of classification accuracies derived from single-resolution and multi-resolution strategies. Accuracy is expressed as Kappa values. A high Kappa value indicates a better classification accuracy.

Approach	Number of classification maps	Average Kappa	Maximum Kappa	Minimum Kappa
4 m	1	0.4789	0.4789	0.4789
8 m	1	0.5181	0.5181	0.5181
12 m	1	0.5435	0.5435	0.5435
16 m	1	0.5526	0.5526	0.5526
20 m	1	0.5664	0.5664	0.5664
Strategy I	24	0.569	0.601	0.5145
Strategy II	15	0.5822	0.6342	0.5387
Strategy III	14	0.6078	0.6483	0.5557

Binary agreement/disagreement images were generated for each of the three subsets of a single land use/cover component, including single-family, industry, and irrigated grassland shown in Figure 8 by comparing the classified image pixel by pixel with the reference data. In each pixel where the classification disagreed with the reference data, a '1' was placed in the agreement/disagreement image. For every pixel where the two data sets agreed, a '0' was placed in the agreement/disagreement image. In order to examine the percentage of misclassified pixels for each subset at different resolutions, the means of the binary agreement/disagreement images were calculated. Table 3 contains information on means of these three subsets at five resolutions.

Table 3. Mean of agreement/disagreement image for the three subsets from single resolution and several examples of multi-resolution approaches.

	Single-Family	Industry	Irrigated Grassland
4 m	0.734	0.654	0.239
8 m	0.626	0.503	0.198
12 m	0.562	0.429	0.175
16 m	0.489	0.406	0.144
20 m	0.482	0.342	0.113
Strategy I (4 m and 20 m)	0.396	0.411	0.138
Strategy II (12 m and 16 m)	0.356	0.346	0.096
Strategy III (12 m and 16 m)	0.374	0.317	0.1
Strategy III (8 m + 20 m)	0.38	0.291	0.103
Strategy III (8mw3 +20mw3)	0.263	0.179	0.094

The mean represents the proportion of errors in the agreement/disagreement image. At a given resolution, the mean of the binary image for irrigated grassland was always the lowest while the mean for single-family was the highest. This means the single-family has the highest percentage of pixels that were misclassified. The percentage of errors listed in Table 4 obtained from multi-resolution approaches for three subsets was reduced compared with those from single-resolution approaches. The best classification result obtained from Strategy III has the least percentage of errors while the agreement/disagreement image obtained from Strategy I has the highest error for all three subsets.

SUMMARY

One of the fundamental considerations when using remotely sensed data for land use/cover mapping is that of selecting appropriate spatial resolution(s). With the increased availability of very high resolution multi-spectral images such as from the IKONOS and IRS satellites and ADAR and DOQQ airborne sources, spatial resolution variation will play an increasingly important role in the employment of remotely sensed imagery. The appropriate spatial resolution is a function of the type of environment, the kind of information desired, and the image analysis techniques used to extract information.

The correct application of image classification procedures for mapping land use/cover requires knowledge of certain spatial attributes of the data to determine the appropriate classification methodology and parameters to use. In general, traditional single-resolution classification procedures are inadequate for understanding the effects of the chosen spatial resolution. They have difficulty discriminating between land use/cover classes that have complex spectral/spatial features and patterns. Although a number of different approaches have been developed for classifying highly heterogeneous landscapes, current research focuses on contextual, knowledge-based, and segmentation routines using spatial and spectral information. Most of approaches developed mainly for Landsat TM and SPOT HRV images are often scene specific and untested on high resolution images (1 m to 10 m).

Following the concept of the scene model developed by Strahler et al (1986) and scale of variance of features in the image (Woodcock and Strahler 1987), the ideal situation for image analysis and classification is reached when the elements of analysis in the image correspond to the objects in the ground scene (Woodcock and Harward 1992, Marceau et al 1994a, 1994b). Since different objects (land cover/use types) are not the same size, the analysis scale corresponding to one object does not match the others. Therefore, it is difficult to achieve optimal classification performances at a single resolution.

The multi-resolution framework proposed in this chapter recognizes that image classification procedure selection should be cognizant of image spatial structure to minimize errors, and increase efficiency and information extraction from the classification process. Selection of the training scheme and classification decision rules should be guided by specification of the type of scene model (H- and L- resolution) and level of spatial variance represented by the image to be classified.

The technical basis of the multi-resolution framework and its potential advantages over commonly used single-resolution classification procedures were introduced and discussed. A variety of approaches may be used to generate multi-resolution image data sets. Different spatial analysis methods can provides the above information to allow resolution effects on individual classes examined.

Three strategies of incorporating information from multiple resolutions were described. The first strategy is to combine layers of images of varying resolution. Layers from coarse resolution are derived from and added as additional layers to the image of finer resolution. The second strategy involves comparing the posterior probabilities of each class at different resolutions. A pixel is assigned to the class with the highest posterior probability among all resolutions selected. The last strategy is based on a top-down approach starting with the coarsest resolution. Only when necessary to increase the maximum likelihood probability are finer resolution images used. The posterior probabilities obtained from coarser resolutions are used as prior probabilities at finer resolution levels.

The case study illustrated the potential of multi-resolution classification framework. Using a simulated multi-resolution data set, it was demonstrated that multi-resolution classification approaches developed in this chapter could significantly improve land use/cover classification accuracy when compared with those from single-resolution approaches. Exploratory data analysis can provide useful information to ensure that subsequent classification methods and parameters are suited to the spatial characteristics of the features (or classes). The results confirm the validity and efficiency of the proposed framework.

While the primary areas of consideration of this chapter are the spatial and radiometric resolutions of remote sensing images, it should be mentioned that incorporating images from multiple dates into a single classification, or comparing classification from multiple dates to infer changes in land use/cover over time and space are also possible.

The efficiency of multi-resolution strategies may be influenced by many factors during classification, including training strategy, classification system, feature extraction method and window size, characteristics of the landscape, classification algorithm, and accuracy assessment method. With the development of remote sensing techniques, an integrated approach using spatial analytical techniques and remote sensing is possible and necessary (Stein and etc. 1998, Treitz and Howarth 2000). Further refinement, particularly of class structures and descriptors from spatial techniques, and exploration of how different spatial techniques can detect resolution-dependent spatial characteristics of the image and can be used in the classification routine are required. More advanced classification approaches such as neural nets, fuzzy set classifiers, and spectral mixture models also be tested in the multi-resolution context to further improve classification accuracy.

REFERENCES

Atkinson, P. M., 1991. Optimal ground-based sampling mean for remote sensing investigations: estimating the regional-mean. *International Journal of Remote Sensing, 12(3):559-567.*

Atkinson, P. M., 1997a. Selecting the spatial resolution of airborne MSS imagery for small-scale agricultural mapping. *International Journal of Remote Sensing, 18(9):1903-1917.*

Atkinson, P. M., 1997b. On estimating measurement error in remotely sensed images with the variogram. *International Journal of Remote Sensing, 18(14):3075-3084.*

Atkinson, P. M., P.J. Curran, 1995. Defining an optimal size of support for remote sensing investigations, *IEEE Transactions on Geoscience and Remote Sensing,* 33(3):768-777.

Atkinson, P. M., P.J. Curran, 1997. Choosing an appropriate spatial resolution for remote sensing investigations. *Photogrammetric Engineering and Remote Sensing,* 63(12):1345-1351.

Barnsley, M.J., and S.L. Barr. 1992. Developing kernel-based spatial re-classification techniques for improved land-use monitoring using high resolution images. *Processing of the XXIX conference of the International Society for Photogrammetry and Remote Sensing (ISPRS'92), International Archieves of Photogrammetry and Remote Sensing: Commission 7,* Washington, D.C., 2-14 Augest 1992, pp, 646-654.

Barnsley, M.J., and S.L. Barr.1996. Inferring urban land use from satellite sensor images using kernel-based spatial reclassification. *Photogrammetric Engineering and Remote Sensing,* 62(8): 949-958.

Bian, L. and R. Butler, 1999. Comparing effects of aggregation methods on statistical and spatial properties of simulated spatial data. *Photogrammetric Engineering and Remote Sensing, 65(1):73-84.*

Cao, C., and N.S.N. Lam, 1997. Understanding the scale and resolution effects in remote sensing and GIS, In *Scale in Remote Sensing and GIS.* edited by Quattrochi, D.A., and Goodchild, M.F.(ed.), Lewis Publishers, pp.57-72.

Campbell, J.B., 1981. Spatial correlation effects upon accuracy of supervised classification of land cover, *Photogrammetric Engineering and Remote Sensing,* 47(3):355-363.

Chen, D., and D. Stow, 2001. The effect of training strategies on supervised classification at different spatial resolutions, *Photogrammetric Engineering and Remote Sensing. (in review)*

Campbell, J.B., 1996. *Introduction to Remote Sensing.* The Guilford Press, New York, 622pp.

Cihlar, J., 2000. Land cover mapping of large areas from satellites:status and research priorities. *International Journal of Remote Sensing,* 21(6):1097-1114.

Civco, D.L, 1993. Artificial neural networks for land-cover classification and mapping. *International Journal of Geographical Information Systems.* 7(2):173-186.

Congalton, R., 1988a. A comparison of sampling schemes used in generating error matrices for assessing the accuracy of maps generated from remotely sensed data, *Photogrammetric Engineering and Remote Sensing,* 54:593-600.

Congalton, R. G., 1988b. Using spatial autocorrelation analysis to explore the errors in maps generated from remotely sensed data, *Photogrammetric Engineering and Remote Sensing,* 54(5): 587-592.

Congalton, R. G., 1991. A review of assessing the accuracy of classifications of remotely sensed data, *Remote Sensing of Enviornment,* 37: 35-46.

Conners, R.W., M. M. Trivedi, and C. A. Harlow, 1984. Segmentation of a high-resolution urban scene using texture operators, *Computer Vision, Graphics, and Image Processing,* 25:273-310.

Collins, J.B., and C. E. Woodcock, 1999. Geostatistical estimation of resolution-dependent variance in remotely sensed image. *Photogrammetric Engineering and Remote Sensing, 65(1):41-51.*

Couloigner, I., and T. Ranchin, 2000. Mapping of Urban Areas: A Multiresolution Modeling Approach for Semi-Automatic Extraction of Streets. *Photogrammetric Engineering and Remote Sensing, 66(7):867-875.*

Cressie, N., 1991, *Statistics for spatial data.* Chichester:John Wiley.

Curran, P.J., 1988. The semi-variogram in remote sensing: an introduction. *Remote Sensing of Environment,* 24: 493-507.

Curran, P.J., and H. D. Williamson, 1986. Sample size for ground and remotely sensed data. *Remote Sensing of Environment,* 31-41.

Cushine, J.L., 1987. The interactive effect of spatial resolution and degree of interval variability within land-cover types on classification accuracies. *International Journal of Remote Sensing, 8(1):15-29.*

Csillag, F., and S. Kabos, 1996. Hierarchical decomposition of variance with applications in environmental mapping based on satellite images. *Mathematical Geology,* 28(4): 385-405.

Csillag, F., 1997. Quadtrees: hierarchical multiresolution data structures for analysis of digital images. In *Scale in Remote Sensing and GIS.* edited by Quattrochi, D.A., and Goodchild, M.F.(ed.), Lewis Publishers, pp. 247-272.

Davis, F, and etc. 1991. Environmental analysis using integrated GIS and remotely sensed data: Some research needs and priorities. *Photogrammetric Engineering and Remote Sensing,* 57: 689-697.

De Cola, L., and N. Montagne, 1993. The pyramid system for multiscale raster analysis. *Computers & Geoscience,* 19(10):1393-1404.

De Cola, L., 1994. Simulating and mapping spatial complexity using multi-scale techniques, *International Journal of Geographical Information Systems, 8(5):411-427.*

De Cola, L., 1997. Multiresolution covariation among Landsat and AVHRR vegetation indices. In *Scale in Remote Semsing and GIS.* edited by Quattrochi, D.A., and Goodchild, M.F.(ed.), Lewis Publishers, pp. 73-93.

DeVore, R.A., B. Jawerth, and B.J. Lucier, 1992. Image compression through wavelet transform coding, *IEEE Transaction of Information Theory,* 38:719-746.

Ehlers, M., D. Greenlee, T. Smith, and J. Star, 1991. Integration of remote sensing and GIS: data and data access, *Photogrammetric Engineering and Remote Sensing*, 57:669-675.

Fisher, P.F. and S. Pathirana, 1990. The evaluation of fuzzy membership of land cover classes in the suburban zone. *Remote Sensing of Environment* 34:121-132.

Flygare, A-M., 1997. A comparison of contextual classification methods using Landsat TM, *International Journal of Remote Sensing*, 18(18):3835-3842.

Foody, G.M., 1996. Approaches for the production and evaluation of fuzzy land cover classification from remotely sensed data, *International Journal of Remote Sensing*, 17:1317-1340.

Foody, G.M., 1999. The continuum of classification fuzziness in thematic mapping. *Photogrammetric Engineering and Remote Sensing*, 65(4): 443 451.

Foody, G.M. and D.P. Cox, 1994. Sub-pixel land cover composition estimation using a linear mixture model and fuzzy membership functions. *International Journal of Remote Sensing* 15(3):619-631

Foody, G. M, P.J. Curran, 1994. Scale and environmental remote sensing. In: *Environmental Remote Sensing From Regional To Global Scales.* (Eds: Foody,Giles M; Curran,Paul J) John Wiley & Sons, New York, 223-232.

Franklin, Janet; C.E. Woodcock, 1997. Multiscale Vegetation Data for the Mountains of Southern California: Spatial and Categorical Resolution. In: *Scale in Remote Sensing and GIS.* (Eds: Quattrochi,D; Goodchild, MF) Lewis Publishers, New York,

Friedl, M.A., F.W. Davis, J. Michaelsen, and M.A. Moritz, 1995. Scaling and uncertainty in the relationship between NDVI and land surface biophysical variables: An analysis using a scene simulation model and data from FIFE, *Remote Sensing of Environment*, 54:233-246.

Gong, P., and P.J. Howarth, 1990. An assessment of some factors influencing multispectral land-cover classification, *Photogrammetric Engineering and Remote Sensing,* 56(5):597-603.

Gong, P. and P.J. Howarth, 1992. Frequency-based contextual classification and gray-level vector reduction for land-use identification, *Photogrammetric Engineering and Remote Sensing,* 58(4): 423-437.

Gong, P., D.J. Marceau and P.J. Howarth, 1992. A comparision of spatial feature extraction algorithms for land-use classification with SPOT HRV data, *Remote Sensing of Environment,* 40: 137-151.

Hepner, G.F., T. Logan, N. Ritter, and N. Bryant, 1990. Artificial neural network classification using a minimal training set: comparision to conventional supervised classification. *Photogrammetric Engineering and Remote Sensing*, 56(4):469-473.

Hlavka, C.A., and G.P. Livingston, 1997. Statistical models of fragmented land cover and the effects of coarse spatial resolution on the estimation of area with satellite sensor imagery, *International Journal of Remote Sensing,* 18(10):2253-2259.

Hu, Z., Y. Chen, and S. Islam, 1998. Multiscaling properties of soil moisture images and decomposition of large- and small-scale features using wavelet transforms, *International Journal of Remote Sensing*, 19(13):2451-2467.

Li, D., and J. Shao, 1994. wavelet theory and its application in image edge detection, *Int. J. of Photogr. Remote Sensing*, 49: 4-12.

Irons, J.R., B.L. Markham, R.F. Nelson, D.L. Toll, D.L. Williams, S. Latty, and M.L. Staufer, 1985. The effects of spatial resolution on the classification of TM data, *International Journal of Remote Sensing*, 6(8):1385-1403.

Jensen, J.R., 1996. *Introductory Digital Image Processing: A Remote Sensing Perspective.* Prentice Hall Series in Geographic Information Science. Prentice Hall, Upper Saddle River, New Jersey, 316 pp.

Jensen, J.R., F. Qiu, K. Patterson, 2001. A neural network image interpretation system to extract rural and urban land use and land cover information from remote sensing. *GeoCarto International*, 16(1):19-28.

Johnson, D.D., and P.J. Howarth, 1987. The effects of spatial resolution on land cover/use theme extraction from airborne digital data. *Canadian Journal of Remote Sensing*, 13(2):68-74.

Jupp, D.L.B., A.H. Strahler, and C.E. Woodcock, 1989a. Autocorrelation and regularization in digital images I: Basic Theory, *IEEE Trans. on Geoscience and Remote Sensing,* 26(4):463-473.

Jupp, D.L.B., A.H. Strahler, and C.E. Woodcock, 1989b. Autocorrelation and regularization in digital images II: Simple image models, *IEEE Trans. on Geoscience and Remote Sensing,* 27(3):247-258.

Justice, C.O., Townshend, R.G., and Kalb, V.L. 1991. Representation of vegetation by continental datasets derived from NOAA-AVHRR data. *International Journal of Remote Sensing,* 12(5):991-1002.

Kontoes, C.C., and D. Rokos, 1996. The integration of spatial context information in an experimental knowledge-based system and the supervised relaxation algorithms – two successful approaches to improving SPOT-XS classification, *International Journal of Remote Sensing,* 17(16):3093-3106.

Labovitz, M. L. and E. J. Masuoka, 1984. The Influence of Autocorrelation in Signature Extraction- An Example from a Geobotanical Investigation of Cotter Basin, Montana, *International Journal of Remote Sensing,* 5(2):315-332.

Lam, N. S-N., and D.A. Quattrochi, 1992. On the issues of scale, resolution, and fractal analysis in the mapping sciences. *Professional Geographer*, 44(1):88-98.

Latty, R.S., R.F. Nelson, B.L. Markham, D.L. Williams, D.L. Toll, and J.R. Irons, 1985. Performance comparisons between information extraction techniques using variable spatial resolution data. *Photogrammetric Engineering and Remote Sensing*, 51(9):1159-1170.

Lillesand, T.M., and R.W. Kiefer, 1987. *Remote Sensing and Image Interpretation*, Second edition. John Wiley & Sons, New York, 721p.

Lindeberg, T., 1994. Scale-space theory: A basic tool for analysing structures at different scales. *Journal of Applied Statistics*, 21(2):225-270.

Lunetta, R.S., R.G. Congalton, L.K. Fenstermaker, J.R. Jensen, K.C. McGwire, L.R. Tinney, 1991. Remote sensing and geographic information system data integration: error sources and research issues. *Photogrammetric Engineering and Remote Sensing, 57(6):677-687.*

Mallat, S. 1989a. A theory for multiresolution signal decomposition: the wavelet representation. *IEEE Trans. on Pattern Analysis and Machine Intelligence*, 11(7): 674-693.

Mallat, S. 1989b. Multifrequency channel decompositions of images and wavelet models, *IEEE Trans. On Acoustics, Speech, and Signal Processing*, 37(12):2091-2110.

Marceau, D.J., P.J. Howarth, D.J. Gratton, 1994a. Remote sensing and the measurement of geographical entities in a forest environment 1: The scale and spatial aggregation problem, *Remote Sensing of Environment* 49:93-104.

Marceau, D.J., P.J. Howarth, D.J. Gratton, 1994b. Remote sensing and the measurement of geographical entities in a forest environment 2: The optimal spatial resolution. *Remote Sensing of Environment* 49, pp. 105-117.

Marceau, D.J, P.J. Howarth, J.M.M. Dubois, and D.J. Gratton, 1990. Evaluation of the grey-level co-occurance matrix method for land-cover classification using SPOT imagery. *IEEE Trans. On Geoscience and RS*, 28(4): 513-519.

Markham, B.L., and J.R.G. Townshend, 1981. Land cover classification accuracy as a function of sensor spatial resolution, *Proceedings of the* 15th *International Symposium on Remote Sensing of Environment*, pp. 1075-1090.

Martin, L.R.G., P.J. Howarth, and G. Holder, 1988. Multispectral classification of land use at the rural-urban fringe using SPOT data, *Can.J.Remote.Sens.* 14(2):72-79.

Mallat, S. 1989a. A theory for multiresolution signal decomposition: the wavelet representation. *IEEE Trans. on Pattern Analysis and Machine Intelligence*, 11(7): 674-693.

Mallat, S. 1989b. Multifrequency channel decompositions of images and wavelet models, *IEEE Trans. on Acoustics, Speech, and Signal Processing*, 37(12):2091-2110.

Mohanty, K.K., 1997. The wavelet transform for local image enhancement, *International Journal of Remote Sensing,* 18(1):213-219.

Moigne, J.L., and J.C. Tilton, 1995. Refining image segmentation by integration of edge and region data. *IEEE Trans. on Geoscience and Remote Sensing*, 33(3):605-615.

Qi, Y., and J. Wu, 1996. Effects of changing spatial resolution on the results of landscape pattern analysis using spatial autocorrelation indices. *Landscape Ecology*, 11(1):39-49.

Quattrochi, D.A., N.S-N. Lam, H-L. Qiu, and W. Zhao, 1997. ICAMS: A geographic information system for the characterization and modeling of multiscale remote sensing data. In *Scale in Remote Sensing and GIS.* edited by Quattrochi, D.A., and Goodchild, M.F.(ed.), Lewis Publishers, pp. 295-308.

Ranchin, T., and L. Wald, 1993. The wavelet transform for the analysis of remotely sensed data. *International Journal of Remote Sensing*, 14:615-619.

Richards, J.A., 1986. *Remote Sensing Digital Image Analysis: An Introduction.* Springer-Verlag, 281 p.

Ryerson, R., 1989. Image interpretation concerns for the 1990s and lessons from the past. *Photogramm. Eng. Remote Sens.* 55(10):1427-1430.

Schowengerdt, R. A.,1997. *Remote sensing models and methods for image processing*, Second Ed. San Diego, Academic Press, 522 p.

Stein, A., W.G.M. Bastiaanssen, S. De Bruin, A.P. Cracknell, P.J.Curran, A.G. Fabbri, B.G.H. Gorte, J.w. Van Groenigen, F.D. Van Der Meer, and A. Saldana, 1988. Integrating spatial statistics and remote sensing. *International Journal of Remote Sensing*, 19(9):1793-1814.

Strahler, A.H., C.E. Woodcock, and J.A. Smith, 1986. On the nature of models in remote sensing. *Remote Sensing of Environment*, 20: 121-139.

Toll, D.L., 1985. Effects of Landsat TM sensor parameters on land cover classification. *Remote Sensing of Environment,* 17(2):129-140.

Townshend, J.RG, C.O. Justice, 1981. Information extraction from remotely sensed data: a user view. *International Journal of Remote Sensing,* 2(4):313-329.

Townshend, J.RG, C.O. Justice, 1988. Selecting the spatial resolution of satellite sensors for global monitoring of land transformations, *International Journal of Remote Sensing, 9:187-236.*

Treitz, P., and P. Howarth, 2000. Integrating spectral, spatial, and terrain variables for forest ecosystem classification, *Photogrammetric Engineering and Remote Sensing,* 66(3):305-317

Wang, F., 1990. Improving Remote Sensing Image Analysis through Fuzzy Information Representation. *Photogrammetric Engineering and Remote Sensing,* 56(8):1163-1169.

Welch, R., 1982. Spatial resolution requirements for urban studies, *International Journal of Remote Sensing, 3(2):139-146.*

Wharton, S.W., 1982. A context-based Land-Use Classification Algorithm for high-resolution Remotely Sensed data. *Journal of Applied Photographic Engineering*, 8(1): 46-50.

Woodcock, C.E., A.H. Strahler, 1987. The factor of scale in remote sensing, *Remote Sensing of Environment*, 21:311-332.

Woodcock, C.E., A.H. Strahler, D.L.B. Jupp, 1988a. The use of variograms in remote sensing I: Scence models and simulated images. *Remote Sensing of Environment*, 25:323-348.

Woodcock, C.E., A.H. Strahler, D.L.B. Jupp, 1988b. The use of variograms in remote sensing II: Real digital images, *Remote Sensing of Environment*, 25:349-379.

Woodcock, C.E., and V.J. Harward, 1992. Nested-hierarchical scene models and image segmentation, *International Journal of Remote Sensing*, 13(16):3167-3187.

Xia, Z., and K. C. Clarke, 1997. Approaches to scalling of geo-spatial data, In Scale in Remote Sensing and GIS, edited by D.A. Quattrochi and M.F. Goodchild. Lewis Publisher, pp: 309-360.

Yocky, D.A., 1996. Multiresolution wavelet decomposition image merger of Landsat Thematic Mapper and SPOT Panchromatic data, *Photogrammetric Engineering and Remote Sensing*, 62(9):1067-1074.

Chapter 12

URBAN GROWTH IN KATHMANDU, NEPAL: MAPPING, ANALYSIS, AND PREDICTION

Barry Haack[1], David Craven[2], Susan Jampoler[3], and Elizabeth Solomon[1]

[1]*George Mason University, Department of Geography, Fairfax, Virginia 22030, USA bhaack@gmu.edu*
[2]*Development Alternatives Inc., 7250 Woodmont Avenue, Suite 200, Bethesda, Maryland* 20814, USA
[3]*GeoKnowledge Inc., 43351 Spinks Ferry Drive, Leesburg, Virginia 22075, USA*

Keywords: Kathmandu, urban, urban growth, change detection, GIS.

Abstract A significant and recent problem in the Kathmandu Valley of Nepal is the loss of very productive agricultural lands to uncontrolled urban growth. This problem is compounded by the lack of information available to the responsible organizations of the government of Nepal on where, and at what rate, this growth is occurring. A raster-based Geographic Information System (GIS) for the Kathmandu Valley was constructed with cooperating Nepali scientists to evaluate the location and extent of urban growth. The lack of current land cover information resulted in an evaluation of spaceborne data to assess urban expansion. Three dates of multispectral SPOT imagery were obtained for Kathmandu. Only one date, July, was useful (94 percent accuracy), because of the complexity of the crop calendar and confusions between cover types. Manipulation of spaceborne radar data, which has the advantage of cloud penetration, produced an urban location map with an accuracy of 87 percent. After adding the urban maps from remote sensing into the Kathmandu Valley GIS, a variety of spatial analyses were accomplished. Between 1954 and 1989, the urban areas expanded by greater than 3,500 hectares or over 300 percent. Another level of GIS analysis, and one that truly demonstrates the power of spatial data integration, is model building for the prediction of future growth, or for conducting suitability or capability analyses. A model on where future urban growth should be encouraged is presented.

INTRODUCTION

The Kathmandu Valley is one of the most productive agricultural regions of Nepal. Most of the valley floor, and much of the terraced valley walls, are intensively cultivated. The temperate climate and availability of water for irrigation allows for almost continuous crop production. However, the attractions of the national capital for business, commercial, and social interests have been inducements to migrants from other areas in Nepal and neighboring countries. This in-migration, coupled with natural population increases, has caused tremendous population growth in the valley. Much of this growth has occurred without effective planning causing serious problems including environmental pollution, rising unemployment, inadequate infrastructure facilities and conflicting land use demands. The population in 1981 was 739,000; it is currently estimated to be over 1,200,000. Governmental statistics indicate a 44 percent population increase during the decade of the 1980's (Koirala 2000).

A significant and recent problem in the Kathmandu Valley is the loss of very productive agricultural lands to the uncontrolled urban growth. The loss of prime agricultural land to residential and commercial uses has impaired the ability to feed the current valley population, a problem that previously did not exist. This problem is compounded by the lack of information available to the responsible organizations of the government of Nepal on where, and at what rate, the urban expansion is occurring.

Recent land use/cover and land use/cover change information is necessary for provision of various infrastructures and services such as transportation, utilities, medical facilities, and schools. This information is also necessary for planning for future growth or attempts to control future growth. Unfortunately, there is limited capability in the responsible Nepalese governmental agencies to collect this needed spatial information by either traditional ground surveys or the use of airborne or spaceborne remote sensing, particularly given the rapid rate of change.

This chapter describes how various Geographic Information Science (GISc) procedures collected and analyzed urban expansion data for the Kathmandu Valley. A raster-based Geographic Information System (GIS) was constructed with cooperating Nepalese scientists. The GIS provides planners with useful information on where urban growth has occurred and how different planning controls might affect future growth. As a generic GIS, it can be used for applications other than urban expansion.

The lack of current land use/cover information resulted in the use of various remote sensing systems to provide data layers for the GIS. Remote sensing has a long and successful history in mapping urban features, estimating population and delineating urban infrastructures (Jackson et al. 1980, Haack et al. 1997, Jensen and Cowen 1999). Various dates of aerial photography as well as spaceborne imagery were obtained to map urban growth as input to the

Kathmandu Valley GIS. This chapter includes a brief description of the Kathmandu Valley, the GIS, the use of remote sensing as a source of spatial data, and concluding results from the GIS analyses.

THE KATHMANDU VALLEY

The Kathmandu Valley is in the central region of Nepal and encompasses an area of approximately 30 x 35 km. The generally flat floor of the valley is at an average elevation of 1,300 m and the sides of the valley are very steeply sloping to elevations of over 2,000 m. The floor of the valley consists of two primary landforms, broad river floodplains and elevated ancient lake and river terraces, locally called *tars*. Historically, the valley has been extensively agricultural-based on excellent soils and appropriate climatic conditions coupled with a large local food demand and geographic isolation (see Figure 1). Major crops include rice, wheat, maize, potatoes, mustard, and a number of other seeds used for oil production. A large variety of vegetables are increasingly grown throughout the year providing fresh produce to the local population.

Figure 1. Typical Kathmandu Valley view of agricultural fields on the flat valley floor and terraced slopes during the dry season.

The flat floodplains are multi-cropped, often with rice grown during the wet season followed by wheat, oil seeds, or vegetables during the dry season. These areas are usually irrigated and out of production only long enough for the ground to be readied for the next crop.

There are many environmental problems within the valley. Air pollution has emerged as one of the most visible components of environmental degradation in Kathmandu. Air pollution measurements show that particulate pollution is the most significant problem. The main sources of particulate pollution are the brick industry, domestic fuel combustion, the cement industry, vehicle exhaust, and resuspention of road dust. The pollutant concentrations in Kathmandu often substantially exceed the World Health Organization's Air Quality Guidelines (Dhikari 1998). Water pollution, sewage and garbage disposal, and availability of drinking water are other serious concerns in the valley. In addition, the lack of viable control of small manufacturing activities, such as carpet factories, has created serious chemical and heavy metal contamination.

These land use/cover conflicts and environmental problems are primarily a result of population growth. The annual rate of population growth in Kathmandu has accelerated from 3 percent in the 1950s to a 5.4 percent in 2000 (United Nations 1989). In 1991, the average valley population density was 1532 persons per square kilometer compared to 126 for the country (Koiral 2000). The purpose of this research was to acquire, analyze, and distribute a variety of spatial data to better understand and mitigate these land use/cover changes and environmental concerns. This chapter presents both basic science results, particularly on the procedures and effectiveness of remote sensing, and applied science from the use of the constructed GIS to address problems within the valley.

SPATIAL DATA COLLECTION AND INTEGRATION

The intent was to construct a GIS for the valley using existing maps and other spatial data. However, it quickly became apparent that additional layers would need to be derived from remote sensing procedures. The following section describes both the basic GIS and remote sensing data extraction components.

Kathmandu Valley GIS

The Kathmandu Valley GIS was constructed using compatible hardware and software available at both George Mason University in the United States and the Nepal National Remote Sensing Centre. This compatibility allowed for easy exchange of data layers and analysis results. The GIS was provided to the International Centre for Integrated Mountain Development (ICIMOD), which is an United Nations Global Resource Information Databank subnode and the Department of Geography at Tribuvan University, both in Kathmandu, for training, demonstration, and utilization.

There was a very surprising limited amount of appropriate spatial data for the Kathmandu Valley. It was difficult to identify and locate existing maps and then, in many cases, only single copies existed, and no accurate method of reproduction was available. Additional problems encountered included varied map scales, lack of documentation (metadata) on how the maps were produced, unstated accuracies, and lack of a consistent geographic coordinate system. For example, there was a difference of about 1 km in the latitudinal grid location between the primary land use/cover and topographic maps of the valley. A serious difficulty for the existing land use/cover maps for different years was the lack of consistency, or even definitions in some cases, for the mapped classes. For many of these difficulties, no viable correction was available and the choice was to not include the data or include it with appropriate caveats. Generally, it was felt that some data, albeit questionable, were better than none.

Table 1 lists some of the various GIS layers obtained. Each data layer is raster-based, using a 70-m cell size and all layers are registered to the Universal Transverse Mercator geographic coordinate system. Seventy-meter pixels were established, based upon the size of the valley (35 km wide) and the computer screen resolution at the time of construction (500 pixels). This was also a reasonable size given the scale of most available data. A subset of the GIS for the urban core was also constructed at 20-m cells.

Table 1. Selected Kathmandu valley GIS data layers.

LAYER	***DATA SOURCE AND SCALE***
Land Use	LRMP Land Use 1:50000
Land Capability	LRMP Land Capability1:50000
Land Forms	LRMP Land Systems 1:50000
Geomorphology	LRMP Land Systems, 1:50000
Soils	LRMP Land Systems1:50000
Soils Texture	LRMP Land Systems1:50000
Depth toGround Water	LRMP Land Systems1:50000
Slope	LRMP Land Systems 1:50000
Geology	Shrestha, 1:63000
Slope Failures	Shrestha, 1:63000
Administrative Units	Nepal Census Bureau, varied scales
Population Density	Nepal Census Bureau, varied scales
Elevation	Schneider 1:50000
Road Network	Schneider 1:50000
Rivers	Schneider 1:50000
Urban Areas (1954-1989)	Varied remote sensing and others

All layers, with the exception of those derived directly from remote sensing, were manually digitized from hard copy products and then converted from a vector to a raster data structure. Significant data were obtained from the Canadian-funded Land Resources Mapping Program (LRMP) maps and a German topographic map, both at 1:50,000 scale (LRMP 1976, Schneider 1977). These data sets were both compiled from a combination of aerial photography and extensive ground surveys. Figure 2 is the LRMP land use GIS layer. It effectively indicates the size, shape, and patterns of the valley. The

GIS also included layers showing the extent of Kathmandu's built-up area for seven dates beginning in 1954.

Remote Sensing

Because of the limited quality and availability of existing spatial data for the GIS, remotely sensed data were used to provide additional information layers. Remotely sensed data are frequently an excellent source of recent, reliable spatial information. The following sections describe three remote sensing sources of data for this project: aerial photography, spaceborne optical, and spaceborne radar.

Aerial Photography

Aerial photography and ground surveys have traditionally been used to measure the extent of urban expansion needed to update existing maps and as GIS input (De Bruijn 1987, Lutchman 1987). Two dates of air photos were obtained for the Kathmandu urban area: 1:60,000 scale for 1978, and 1:15,000 scale for 1987 (Figure 2). These photos were used to create GIS layers of urban extent by manual interpretation. The differences in scale created some difficulties in producing compatible data. These difficulties included an appropriate definition of urban and use of a consistent minimum mapping unit (MMU).

Determining a remote sensing classification system is one of the least appreciated, most difficult, and most important considerations in the appropriate use of remote sensing. Frequently, there is a difference between what classifications or definitions the user or decision-maker may want, and what the data are able to accurately provide. In many cases, it is the difference between land use required by decision-makers and land cover extractable from the imagery.

In this study, the classification was simplified to urban and non-urban, but that still provided definition problems. For example, does urban include the forested areas around the major temples, soccer fields and military parade grounds, or should it be limited to areas with physical structures or in effect, built-up areas? Historical land use/cover maps used varied urban definitions resulting in some difficulty in comparison.

Even to concentrate on structures requires a determination of the number of structures to be considered urban, buildings per hectare, and a MMU. These definitions are particularly important in Kathmandu where so much of the new urban growth is a seemingly random pattern of houses interspersed with active agricultural fields, not the contiguous suburban subdivision common in many countries.

The first remote sensing mapping for this study was with the 1978 aerial photography at a scale of 1:60,000. For those photos, urban was defined as a minimum density of five structures, commercial or residential, per hectare and the MMU was six hectares. Subsequent work with larger scale 1987 photography and the satellite imagery required a redefinition of those mapping criteria. The spaceborne imagery did not allow for specific counting of structures, but delineation based on spectral tone, either visual or digital. The aerial photographs did provide excellent maps. However, in developing countries generally, air photos are of inconsistent quality, are often out of date, and in many cases, do not exist at all.

Spaceborne Optical Remote Sensing

Studies have demonstrated the potential of spaceborne sensors as a replacement of aerial photographs for monitoring urban growth (Jensen 1981, Zhao 1986, Duggin et al. 1988, Haack et al. 2000). The advantages of satellite sensors as a data source include systematic and frequent acquisition, synoptic coverage, digital data format, availability, and relatively low cost.

Data collected by the different satellite sensors currently in operation vary in terms of spatial, temporal, spectral and radiometric resolutions. Studies based on Landsat data have concluded that the 80-m spatial resolution of the multispectral scanner (MSS) and 30-m Thematic Mapper (TM) resolution can be too coarse for analysis at the urban fringe (Khorram et al. 1987). This is not always the case. Application potential is data dependent, and some authors have reported positive results using Landsat data in urban environments (Royer et al. 1988, Moller-Jensen 1990).

Research with the French satellite, Le System Pour l' Observation de le Terre (SPOT), indicates that the improved spatial resolution of 20-m in multispectral mode, and 10-m in panchromatic mode, make it potentially useful for urban applications (Martin and Howarth 1989, Ehlers 1990, Baraldi and Parmiggiani 1990, Lo and Fabe, 1998). For this study, SPOT multispectral data were acquired on the assumption that the 20-m spatial resolution was critical for the urban patterns in Kathmandu, where housing units are quite small and frequently isolated. Very recently, better spatial resolution data, 1-m panchromatic and 4-m multispectral from IKONOS, became available from space, which may be useful in urban studies.

Frequently, one of the most crucial issues in successfully utilizing remotely sensed data is the date of acquisition. Changing surface conditions as a function of precipitation, vegetation phenology and land use/cover patterns, including cropping practices, determine the ability to successfully extract information from the data. The ability to map urban growth in the Kathmandu Valley with spaceborne optical data is determined by the image date as a specific function of the agricultural practices. As has been observed elsewhere, primary confusion is between urban features and bare soil such as fallow agricultural lands. This confusion is particularly true with the limited spectral bands of SPOT. The use of indigenous building materials, such as clay tiles, brick and thatch, makes spectral confusion with bare soil a significant problem in Nepal. A primary strategy for interpreting the SPOT imagery was to acquire data with limited bare soil. Because of the intense cultivation of the valley, most bare soil will be fallow fields between crops in rotation.

Figure 2. Aerial photograph of a portion of the Kathmandu Valley from 1987. Approximate scale of 1:17,000.

There are two primary crop rotations in the valley determined by the ability to irrigate during the summer monsoon rains. Those areas of relatively flat lands that are irrigable are planted in rice from July to November, and then planted in wheat in January to May. These are primarily the broad floodplains. The upland or tar areas that are removed from sources of water for irrigation or have slopes which are too steep for irrigation, are normally planted in maize from May to September and occasionally followed by millet or mustard until

December. These areas are usually fallow from December until the planting of maize in May.

Three dates of SPOT imagery were acquired; March 22, October 19 and July 22. The utility of these images was directly related to the crop calendar and the spectral confusion between fallow agricultural fields and urban, especially new urban areas. In the March image, much of the upland tars were fallow fields that will be planted in maize. Distinguishing between the fallow, upland tar fields and new urban areas was not spectrally possible.

In the October image, the location of the agriculture components has reversed. The upland tar areas were highly vegetated fields of millet or mustard planted after the harvesting of the maize. In the valleys there were fallow fields where rice is either in advanced stages, often in senescence and occasionally harvested, and the wheat has not yet developed. These fallow fields are not separable from the building roofs.

In the July SPOT image (Figure 3), there are significant areas of very dark tones. This image was collected early in the rice cycle when there is little if any actual green rice. The dark fields are either flooded paddy or saturated soil where the rice has just been planted or will soon be planted. There are no highly reflective dry fallow fields to be confused with new urban growth. The light tones are easily identifiable urban features. The July image is within the best temporal window for mapping urban expansion. An obvious difficulty is that July is during the middle of the monsoon and there is almost continuous cloud cover limiting the opportunities for optical data acquisition.

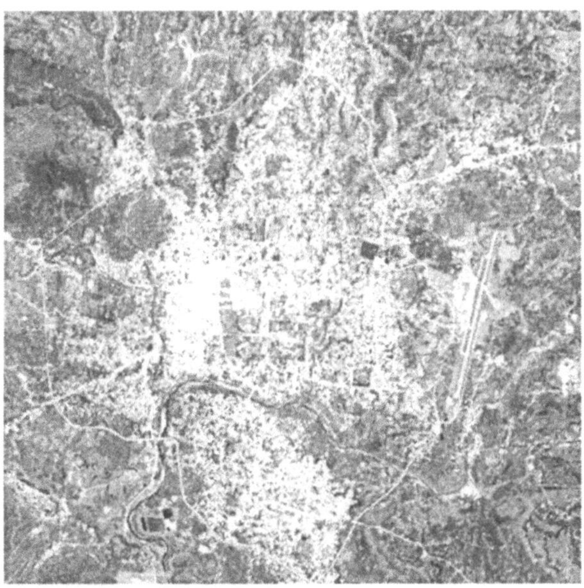

Figure 3. July 22 1989 SPOT Image in the red visible band. The light tones are urban and a few areas of river gravel, the dark tones are flooded or saturated rice fields in the early stages of growth or vegetated areas of maize of forest. Approximate scale of 1:100,000.

Even after obtaining data at the optimal temporal window to minimize confusion between the bare soil of the fallow agricultural fields and urban areas, imagery examination together with field visits identified two additional sources of spectral confusion. The riverbeds and immediately adjacent floodplains are largely comprised of highly reflective sand and gravel that are spectrally similar to the flat concrete or metal roofs of the city's newer buildings. These riverbeds are relatively large and fully exposed, because of the very active extraction of these materials for construction. On a standard image, the sand and gravel deposits and the concrete rooftops both appear as light tones.

Consequently, a number of riverbeds were misclassified as urban features in the initial examination of the data. This spectral confusion was minimized by using a GIS layer to mask out the rivers and a proximity buffer to the rivers to exclude them from future analysis. This river GIS layer was obtained from image interpretation, fieldwork, and topographic maps.

Another source of spectral confusion was related to roof type. Materials for housing construction within the valley have changed. Most older, traditional structures are brick with sloping thatched or clay roofs (Figure 4). The new construction still relies heavily on brick, but the number of concrete buildings is increasing. More importantly from an image analysis perspective, the modern roofs are flat, and made of either concrete or corrugated metal (Figure 5). These materials produce a significantly different spectral reflectance than clay or thatch.

Figure 4. Thatch is a common roofing material for older houses in Kathmandu. These roofs are generally small, sloping, and spectrally confused with wet soils.

Figure 5. Newer housing has larger, flat roofs made of concrete or metal. They are very light toned and spectrally similar to dry soils or sand and gravel.

Both types of roofing compositions can be confused with agricultural features. Modern roofs are light toned and easily mistaken for dry bare soils, especially within the tar areas; and, older natural material roofs are dark toned and can be confused with wet or flooded soils. A GIS mask was employed to eliminate or minimize the confusion between the older roofs with wet or flooded soils in the July SPOT imagery. This GIS mask removed the known areas of urban features using base maps derived from earlier aerial photography.

Two visual interpretations of 1:25,000 scale July SPOT imagery were made. The first was produced without the benefit of first-hand experience in the field, and the second during a visit to Kathmandu. These visual urban/non-urban land use classifications were traced on mylar overlays to the images and were subsequently digitized into GIS layers for manipulation with the other layers and for accuracy assessments. Computer-assisted classifications of the digital SPOT data also produced three estimates of urban extent in Kathmandu using different methods of spectral signature extraction. All digital classifications were spatially filtered for consistency with the MMU used for visual interpretations.

The five approaches to analyzing the SPOT data all identified the major areas of urban development in Kathmandu, though with varying degrees of accuracy. Accuracy assessments were obtained from a sample of 200 sites collected in the field and expressed in percent correct (Congalton and Mead 1983). The visual interpretation during field visit produced the most accurate results at 85 percent correct. The three digital techniques produced accuracies between 78 and 82 percent correct.

These accuracy assessments included classifications of the urban fringe, and thus, are indications of urban growth and not simply the ability to locate urban features. A larger sample of 500 points, not only in the urban fringe but including historic urban and non-urban regions, provided an increase in classification accuracies of 10 to 15 percent to over 90 percent correct classification of urban areas (Craven and Haack 1994). The urban map produced by the in-field visual interpretation was included in the GIS.

Spaceborne Radar

Radar sensors have special properties that make them a particularly viable alternative, or compliment, to traditional optical remote sensing techniques. For instance, the longer wavelengths of microwave energy are capable of penetrating atmospheric conditions that render traditional spaceborne optical and multispectral systems useless. Radar therefore has the ability to image through rain, fog, hail, smoke, and most importantly clouds. This all weather operational characteristic holds enormous data-collecting potential for many geographic areas around the world, especially those often obscured by adverse weather conditions such as the summer months in Kathmandu.

A difficulty with the analysis of radar as an independent sensor is that most current radar spaceborne systems only collect data at a single wavelength with a fixed polarization. Only one component of the total surface scattering is thereby being measured, while any additional information contained within the reflected radar signal is lost (Zebker and van Zyl 1991). Future systems will include an increased number of wavelengths and polarizations, but until then, the goal of improved informational content may possibly be reached through simpler methods, such as the extraction of textural measures (Henderson and Xia 1998).

Textural information may be as important as spectral information in radar, as the information content of an image resides in both the intensity (spectral) of individual pixels and the spatial arrangement of those pixels images (Anys and He 1995, Dobson et al. 1995). Standard image classification procedures, used to extract information from remotely sensed images, usually ignore this spatial information and are based on purely spectral characteristics. Such classifiers may be ineffective when applied to land use/cover classes such as residential and urban areas that are largely distinguished by their spatial, rather than their spectral, characteristics (Lee and Philpot 1991; Dobson et al. 1997).

Figure 6 is the RADARSAT image obtained for this study. This image was collected on 19 November 1998. In this scene, the vast majority of high backscatter, light tones, are the urban features on the valley floor. Clearly there are also areas of high return from the greater topography along the valley edges. There are virtually no remaining forest areas within the valley that would be expected to provide high backscatter and would likely be confused with the urban features. For this study, four land uses/covers were examined. These include the dense, older urban core; the more recent expansions of the urban areas on the margins, suburban or new urban, agriculture which is primarily fallow at this time of year, and a few areas of open grass.

The classification accuracy of radar derived land use/cover categories for the Kathmandu Valley were examined as an alternative or complement to optical spaceborne data. Methods to improve the classification accuracy from radar included speckle reduction, texture measures, post-classification smoothing, and various combinations of these techniques.

Initial examination of the RADARSAT image for Kathmandu provided poor classification results. As can be seen in Table 2, the overall classification, as well as most of the individual class accuracies, were quite low. The more unique, higher backscatter areas of the old urban class had the best producer's classification accuracy at approximately 77 percent, but provided a very low user's accuracy (42 percent). The primary confusion was between old and new urban. If these classes were combined, the results would be much better. There was, however, also considerable confusion between agriculture and new urban (1081 erroneously classified truth pixels) and thus poor results for new urban.

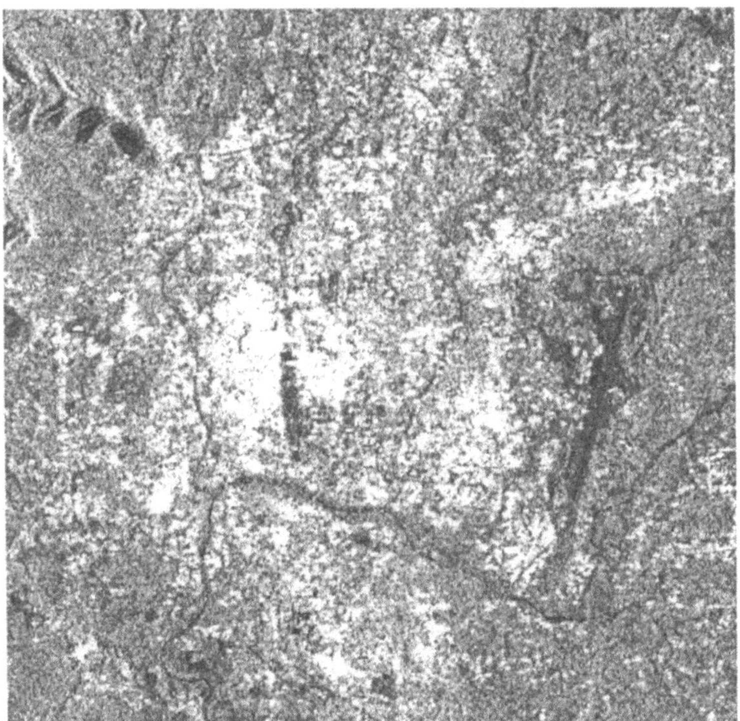

Figure 6. RADARSAT scene of Kathmandu, Nepal collected on November 19,1998. Approximate scale is 1:100,000. The light toned features, high backscatter, are primarily the urban features.

The high standard deviation associated with the RADARSAT's urban classes suggested that measures of image texture could possibly increase the discriminatory ability of radar. Measures of Variance texture were applied to the original RADARSAT data over 5x5, 13x13, and 21x21 moving window sizes. Previous analysis with similar data had determined that the Variance measure of texture was most useful (Haack and Bechdol 2000). The results indicate that texture does improve discrimination. Overall classification accuracies of 71, 75 and 69 percent by respective window size were achieved, improving over those of the original radar data by about 20 percent.

Results achieved with use of the 13x13-texture window size were best. The effectiveness of this window size is related to the variability and spacing of the ground features being observed. Table 3 contains the contingency matrix for these results. The remaining confusions are between the two non-urban and the two urban classes. Class combinations would improve results.

Table 2. RADARSAT original data classification results for the Kathmandu Valley.

	Grass	Agriculture	Urban_old	Urban_new	Total	User Accuracy
Grass	**337**	1257	0	115	1709	19.72%
Agriculture	112	**2882**	44	1081	4119	69.97%
Urban_old	0	6	**1086**	1499	2591	41.91%
Urban_new	0	185	284	**1594**	2063	77.27%
Total	449	4330	1414	4289	**10482**	
						OVERALL
Producer	75.06%	66.56%	76.80%	37.16%		**56.28%**

Several two- and three-band combinations of the radar data and the manipulations of those data were examined to assess the utility associated with an increased number of radar derived bands. Combinations examined included original radar and texture combinations, speckle filtered radar and texture combinations as well as combinations of multiple textures. Table 4 contains the results of the best of these combinations, which is the Median speckle reduction filter and a 13x13 window Variance texture extracted from the same speckle reduction scene. The accuracies are excellent, particularly when compressed to two cover types, and very similar to those obtained with the SPOT interpretation.

Table 3. RADARSAT 13x13 pixel window Variance texture accuracy for the Kathmandu Valley.

	Grass	Agriculture	Urban_old	Urban_new	Total	User Accuracy
Grass	**62**	624	88	46	820	7.56%
Agriculture	213	**3594**	2	0	3809	94.36%
Urban_old	122	112	**801**	791	1826	43.87%
Urban_new	52	0	523	**3452**	4027	85.72%
Total	449	4330	1414	4289	**10482**	
						OVERALL
Producer	13.81%	83.00%	56.65%	80.48%		**75.45%**

Table 4. RADARSAT 5 x 5 Median filter + 13 x 13 Texture on a 5 x 5 Median speckle reduction accuracy for the Kathmandu Valley.

	Grass	Agriculture	Urban_old	Urban_new	Total	User Accuracy
Grass	**241**	0	0	0	241	100.00%
Agriculture	192	**4286**	0	79	4557	94.05%
Urban_old	0	0	**1094**	674	1768	61.88%
Urban_new	16	44	320	**3536**	3916	90.30%
Total	449	4330	1414	4289	**10482**	
						OVERALL
Producer	53.67%	98.98%	77.37%	82.44%		**87.36%**

GIS ANALYSIS AND MODELING

After adding the urban maps derived from remote sensing into the Kathmandu Valley GIS, a variety of spatial analyses could be accomplished. Initially, maps showing the location and statistics of the amount of urban growth were obtained. Table 5 documents the amount of urban expansion obtained by change analysis of the various GIS layers of land use\cover beginning in 1954. The intervals between dates are not consistent and the GIS layers were obtained from varied source materials; existing maps, air photos and spaceborne imagery. The varied source materials also had different mapping parameters that somewhat limits the validity of comparisons.

However, the data generally determine quite well where and at what rate change has occurred. Often planners must document growth and these GIScience procedures provide an easily understood and visually appealing format. The urban area has expanded by over 3500 hectares, or over 300 percent, between 1954 and 1989. Koirala (2000) estimates that if these rates of urban expansion continue, the agricultural area as a percent of the valley will decline from 64 percent in 1984 to 14.5 percent in 2020.

Table 5. Kathmandu Urban Growth.

Year	Growth from Previous GIS Layer (hectares)
1954	1308*
1964	147
1971	786
1978	562
1981	1403
1987	492
1988	473

*base layer

Note: These values are from varied sources that have some inconsistencies in how they were derived.

As a second level of analysis, the urban growth was cross indexed with other GIS layers such as existing land use/cover, soil type, slope, and proximity to roads or rivers to show where the growth has occurred. For example, a comparison was made between urban expansion and land systems. The two land systems in Kathmandu, alluvial plains (floodplains) or elevated dissected and non-dissected ancient lake and river terraces (tars), are very different in their appropriateness for urban use.

The earliest map of the Kathmandu urban area (1954) indicates 18 percent of the urban area was in the valleys and 80 percent was on tars. Over time, the percent of urban expansion in the valleys has increased and on the tars decreased so that by 1989, more urban growth was on the alluvial floodplains than the uplands (Table 6). Not only does this floodplain expansion decrease agricultural productivity, structures on the floodplains are very susceptible to damaging or destroying floods and are potentially sources of pollutants for the ground water. These statistics provide planners useful information for policy development.

Table 6. Percentage of new urban growth by land unit from prior GIS layer dates.

Year	Alluvial Plain (Floodplain)	Terrace (Tars)
1954*	18	80
1964	7	93
1971	19	81
1978	18	73
1981	33	64
1987	32	65
1989	49	48

* base layer

A third level of GIS analysis, and one that truly demonstrates the power of spatial data integration, is model building for the prediction of future growth, or for conducting suitability or capability studies. A model example on where future urban growth should be encouraged is presented. That model consists of several steps involving various GIS layers. First, it was assumed that new growth should occur in proximity to existing urban areas and major roads. For the initial model, areas within approximately 1-km of existing urban areas and 0.5-km of major roads were located independently from the current urban GIS layer and the roads layer. These areas were then combined in a new GIS layer of potential growth locations.

Next, it was assumed that there were certain conditions that should exclude urban expansion. These conditions included; (1) slopes over 30 percent, (2) areas where the depth to groundwater was under 2 m, (3) protected forests and plantations, (4) existing rivers and streams and, 5) soils which were less than 50 cm to bedrock. These areas were identified from the five relevant GIS layers and combined to a new GIS layer of areas not suitable for urban development. These areas were then excluded from the urban and road proximity map of potential growth.

This layer was then overlaid with the existing land use map to identify what lands are likely to be consumed by future growth. The 7,818 hectares of current land use consumed by future growth according to this model include 5,534 hectares of existing tar agriculture, 789 hectares of valley floor agriculture, 788 hectares of terrace agriculture, and only 707 hectares of lands not currently classified as agriculture. This model suggests a relatively small amount of valley floor agriculture will be consumed by future urban growth, because areas where the depth to groundwater was under 2 m were excluded from growth.

This differs from the reality of where growth is occurring, because much of the growth is on the floodplains. Obviously, a significant challenge for planners will be to resolve this inconsistency. The ability to easily implement and alter these models is one of the great strengths of a GIS. It allows planners to easily evaluate different models, and thus assist in establishing land-planning regulations and in the anticipation of future infrastructure and service needs.

CONCLUSIONS

Reasonable estimates and locations of urban growth can be made from remotely sensed data. With certain qualifications and refinements in methodology, the techniques presented in this study produced accurate and consistent results. However, the Kathmandu case study serves to illustrate an important aspect of remote sensing studies, that there are relatively few general rules or standard procedures by which to conduct an analysis. Appropriate techniques are data and site dependent. There are good arguments to support the use of multispectral data for the delineation of urban growth, but they are only valid if the data are available for the right time.

This study clearly documents the importance of the appropriate date for data acquisition. Of three SPOT images obtained, only one was in a temporal window where useful results could be extracted. Using GIS masks for locations of spectral confusion demonstrates the synergism of multiple sources of spatial data and of combining remote sensing and GIS. Visual interpretation results were as accurate as the more complicated digital analysis. Finally, fieldwork is a necessary component to maximize the information extracted and to provide a method of accuracy assessment to validate a study. Maps produced via remote sensing should have a clearly stated accuracy assessment. Radar was very effective, but only with the appropriate radar derived procedures.

The automated GIS is an extremely useful and appropriate tool for the integration of spatial data. This study illustrated how spatial data from a variety of sources can be combined to examine the location and rate of urban growth, identify what type of lands are being consumed, and develop models of predicted urban expansion and land suitability.

Perhaps more important than the direct technical accomplishments of this project have been the interactions with Nepalese scientists and organizations. One objective of this project was to provide Nepalese scientists a better understanding of GIScience capabilities for indigenous issues. This project was very successful in meeting that objective. There is not only extensive interest, but also increased use of GIScience methods, evidenced by the improved availability of related software and hardware in Nepal and relevant case study applications (Shrestha and Pradhan 2000).

ACKNOWLEDGMENTS

This project was supported by National Science Foundation Science in Developing Countries Grant INT-86-03298, National Geographic Society Committee for Research and Exploration Grant 3972-88, and the United States National Aeronautical and Space Administration Application Development and Research Opportunity program under Grant Numbers NAGW-4970 and NAG5-4140. Cooperating scientists and organizations who greatly assisted in this effort include the staff of the Nepal Remote Sensing Centre, directed by K. B. Malla, P. B. Shah of the Integrated Survey Section of Topographic Survey, ICIMOD and the Geography Department at Tribhuvan University.

REFERENCES

Anys, H. and D. He, 1995. Evaluation of textural and multipolarization radar features for crop classification. IEEE Transactions on Geoscience and Remote Sensing, 33(5):1170-1181.

Baraldi, A. and F. Parmiggiani, 1990. Urban area classification by Multispectral SPOT images. IEEE Transactions on Geoscience and Remote Sensing, 28 (4): 674-680.

Congalton, R. and R. Mead, 1983. A quantitative method to test for consistency and correctness in photointerpretation. Photogrammetric Engineering and Remote Sensing, 49 (1): 69-74.

Craven, D. and B. Haack, 1994. Seeking the truth in Kathmandu, fieldwork as an essential component in remote sensing studies. International Journal of Remote Sensing, 15 (7): 1365-1377.

De Bruijn, C., 1987. Monitoring a large squatter area in Dar es Salaam with sequential aerial photography. ITC Journal, 87 (3): 233-238.

Dhikari, A. 1998. Urban and Environmental Planning in Nepal. IUCN-The World Conservation Union.

Dobson , M., L. Pierce, and F. Ulaby, 1997. The role of frequency and polarization in terrain classification using SAR data. IEEE Transactions on Geoscience and Remote Sensing, 4:1621-1623.

Dobson, M., F. Ulaby, and L. Pierce, 1995. Land cover classification and estimation of terrain attributes using synthetic aperture radar. Remote Sensing of Environment, 51:199-214.

Duggin, M., R. Rowntree and A. O'Dell, 1988. The application of spatial filtering methods to urban feature analysis using digital image data. International Journal of Remote Sensing, 9: 543-553.

Ehlers, M., 1990. Application of SPOT data for regional growth analysis and local planning. Photogrammetric Engineering and Remote Sensing, 56 (2): 175-180.

Haack, B., N. Herold and M. Bechdol, 2000. Radar and optical data integration for land-use/land-cover mapping. Photogrammetric Engineering and Remote Sensing, 66 (6): 709-716.

Haack, B. and M. Bechdol, 2000. Integrating multisensor data and radar texture measures for land cover mapping. Computers and Geosciences, 26:411-421.

Haack, B., S. Guptill, R. Holz, S. Jampoler, J. Jensen, and R. Welch, 1997. Urban analysis and planning, Chapter 15 in Manual of Aerial Photo Interpretation, Second Edition, American Society for Photogrammetry and Remote Sensing, 517-554.

Henderson, F. M. and Z. Xia., 1998. Radar applications in urban analysis, settlement detection and population estimation. Principles and Applications of Imaging Radar. John Wiley, New York, 733-768.

Jackson, M., P. Carter, T. Smith and W. Gardner, 1980. Urban land mapping from remotely sensed data. Photogrammetric Engineering and Remote Sensing, 46 (8): 1041-1050.

Jensen, J. R., 1981. Urban change detection mapping using Landsat digital data. The American Cartographer, 8 (2): 127-147.

Jensen, J. and D. Cowen, 1999. Remote sensing of urban/suburban infrastructure and socio-economic attributes. Photogrammetric Engineering and Remote Sensing, 65 (5), 611-622.

Khorram, S., A. Brockhaus and H. Cheshire, 1987. Comparison of Landsat MSS and TM data for urban land use classification. IEEE Transactions on Geoscience and Remote Sensing, GE 25 (2): 238-243.

Koirala, H., 2000. RS and GIS in assessing urban environment: a case study of the Kathmandu metropolitan city. http://pages.hotbot.com/edu.geoinformatince/f118.html.

LRMP, 1986. Land Resouces Mapping Project, Kenting Earth Science Limited.

Lee, J. and W. Philpot, 1991. Spectral texture pattern matching: a classifier for digital imagery. IEEE Transactions on Geoscience and Remote Sensing, 29(4):545-554.

Lo, C. and B. Faber, 1998. Automated population and dwelling unit estimation from high-resolution satellite images: A GIS approach. International Journal of Remote Sensing, 16: 17-34.

Lutchman, H., 1987. Monitoring land subdivisions of the fringe of Paramaribo using aerial photography. ITC Journal, 87 (3): 248-253.

Martin, L. and P. Howarth, 1989. Change detection accuracy assessment using SPOT multispectral imagery of the rural-urban fringe. Remote Sensing of Environment 30 (1): 55-66.

Moller-Jensen, L., 1990. Knowledge-based classification of an urban area using texture and context information in Landsat TM imagery. Photogrammetric Engineering and Remote Sensing, 56 (6): 899-904.

Royer, A., L. Charbonneau and F. Bonn, 1988. Urbanization and Landsat MSS albedo Change in the Windsor-Quebec corridor since 1972. International Journal of Remote Sensing, 9 (3): 555-566.

Shrestha, B. and S. Pradhan, 2000. Kathmandu Valley GIS Database: Bridging the Data Gap. International Centre for Integrated Mountain Development, Kathmandu.

Schneider, E. 1977. Topographic Map of the Kathmandu Valley, 1:50,000 scale. Munich, Germany.

United Nations Department of International Economic and Social Affairs, 1989. Prospects of World Urbanization-1988. Population Study 112, New York.

Zhao, R., 1986. Types of urban districts in Nanjing (China) according to Landsat. Photo Interpretation: Images Areinnes et Spatiales: 29-36.

Zebker, H. and J. van Zyl, 1991. Imaging radar polarimetry: a review., Proceedings of the IEEE, 79:1581-1606.

Chapter 13

FAO METHODOLOGIES FOR LAND COVER CLASSIFICATION AND MAPPING

John S. Latham[1], Changchui He[1], Luca Alinovi[1], Antonio DiGregorio[1], and Zdenek Kalensky[2]

[1]*Food and Agriculture Organization of the United Nations, Rome, Italy*
john.latham@fao.org
[2]*Mapsat Ltd., Ottawa, Canada*

Keywords: Classification scheme, land use, land cover.

Abstract

Information on land cover and its changes is essential for rational planning of agricultural land use, sustainable management of agricultural land, increase of crop production, and environmental protection. Yet, in most developing countries such information is non-existent or not reliable. This chapter describes how the Food and Agriculture Organization of the United Nations (FAO) assists developing countries with land cover mapping and establishment of land cover databases. FAO assistance to developing countries in the field of land cover mapping is based on a three-pronged approach:

- Development of standardized methodologies for land cover and land use classification and mapping appropriate for developing countries,
- Providing assistance with land cover and land use mapping and establishment of associated digital databases,
- Strengthening the institutional capacities for land cover and land use mapping and monitoring in developing countries.

FAO, to assist developing countries to standardize their land cover and land use classification and mapping procedures, developed the Land Cover Classification System (LCCS) and the AFRICOVER Interpretation and Mapping System (AIMS). Although both systems were developed in the framework of the AFRICOVER project, they can be used for implementation of land cover and land use mapping projects in any country.

Many developing countries have requested the FAO assistance with land cover and land use mapping and establishment of associated digital databases. Over 20 such projects have been implemented by FAO during the last 10-years. The chapter summarizes the practical experience, specific to land cover and land use mapping in developing countries, obtained by FAO. The AFRICOVER-EA project for land cover mapping at the scale 1:200,000 in the area of over 9 million km^2 in East Africa was selected as the case study for this chapter.

Strengthening the capacities for land cover and land use mapping and monitoring of the institutes cooperating in the implementation of such projects, has always been one of the FAO priorities. The successful accomplishment of this task requires the development of close linkages with relevant organizations. National officers participating in the project receive a thorough on-the-job training in all aspects of project implementation. Technical training is supplemented by appraisal workshops for government decision-makers to assure the appropriate use of project products. All project equipment and materials are left in the country after the project completion.

INTRODUCTION

Since the late of 1980s, FAO has been increasingly involved in the application of remote sensing, GIS, and GPS technologies to assessment, mapping, and monitoring of land cover and land use (LCLU) at the global, regional, and country levels (Di Gregorio 1991, FAO 1992 1993b 1994b 2000c 2001b, FAO and UNEP 1999). This reflects the need for reliable information on the current state of LCLU and their changes. Such information is essential for land use planning, sustainable management of agricultural land, forest and fresh water resources, as well as for environmental protection and rehabilitation (FAO 1993a 1995).

Effective utilization of GIS capacities for land cover and land use data integration, change assessment, and trends forecasting requires compatible geospatial databases. However, there is hardly any harmonization among specifications for geospatial data collected by different sectors in developing countries. In most developing countries, the LCLU information is collected by number of organizations using different basemaps, land cover map legends, classification and accuracy standards, and the existing LCLU databases are often outdated (Fanta 1993 1995, ECA 1996, Singh et al. 1997). There is also an increasing need to improve the compatibility among land cover and land use databases of different countries to improve the monitoring of land cover changes associated with food security, natural disasters, and environmental protection. Furthermore, the regional and global environmental initiatives, such as the Global Terrestrial Observing System (GTOS), need the land cover and land use data inputs with compatible accuracy and in harmonized formats from different countries.

The lack of reliable information on current LCLU situation is slowing the land use planning and implementation of effective measures for prevention of land degradation and deforestation, as well as mitigation of the effects of natural disasters, such as the agricultural drought, floods, and forest and rangeland fires.

Increasing population pressure, threatening the future food security prospects in many developing countries, particularly in Africa, underlines the need for reliable information on current LCLU status. While the introduction of new scientific developments and technologies is an essential requirement for a steady increase of food production and thus assurance of food security in developing countries, this has to be achieved in the framework of sustainable land management and environmental protection. Otherwise, the benefits would be short-lived and food security situation further aggravated (Kalensky 1995).

FAO, to obtain the reliable LCLU information needed for planning and implementation of its development projects, was one of the first international organizations that used the advanced geo-information technologies for LCLU assessment, mapping, and monitoring. Such technologies, including the remote sensing from Earth observation satellites (RS), geographic information systems (GIS), and global positioning systems (GPS) are particularly suitable for mapping of large, not easily accessible areas. They provide an objective, uniform, and timely record of current LCLU over large areas, not obtainable by other surveying methods. The problem is the lack of technical capacities in developing countries for the efficient application of advanced geo-information technologies (He 2000, Kalensky and Cumani 2000).

Importance of the availability of reliable LCLU information for sustainable management of the Earth's natural resources was emphasized at the UN Conference on Environment and Development (UNCED), better known as the "Earth Summit," which took place in Rio de Janeiro in 1992. At the Conference, the Governments agreed that all countries should undertake "*... a comprehensive national inventory of their land resources in order to establish a land information system in which land resources will be classified according to their most appropriate uses. Furthermore, the environmentally fragile or disasters-prone areas will be identified for special protection measures.*" The UNCED Action Plan, the Agenda 21, in its Chapter 40, recommends the use of advanced geo-information technologies, such as remote sensing, GIS, and GPS, for speeding up the completion of geospatial databases that will provide reliable information on land cover and land use. These UNCED recommendations were further strengthened at the follow-up international conferences and by the establishment of the UN Framework Environmental Conventions on Biodiversity, Climate Change, and Combating Desertification (UNCED 1993, FAO 2000a).

In the context of the above situation, the AFRICOVER project has been long overdue. While the provision of reliable land cover information in standardized digital formats is the main project objective, an equally important objective is strengthening the technical capacities of participating regional and national organizations. This will enhance the effective use of project products at the regional- and country-levels and facilitate its follow-on activities, including the maintenance of land cover databases produced by the project. The project is fulfilling this latter objective through organization of training courses for technical officers and appraisal workshops for decision-makers. Furthermore, it provides on-the-job training to national staff assigned to the project. On average, four technical officers from each country, with good knowledge of vegetation and other land cover types, are participating in the interpretation of land cover from satellite images of their respective countries. The same officers also participate in the field validation of interpretation results in the selected sampling sites. The extent of utilization of project products, and thus the impact of the AFRICOVER-EA project, will largely depend on successful achievement of this capacity-building project objective.

LAND COVER AND LAND USE

There are many definitions, and some confusion, in the scientific literature regarding the terms "land cover" and "land use." In the context of this paper, the following definitions are used:

Land cover - the type of biophysical surface layer of land. It includes vegetation, bare land, water, and human-made constructions in a specific area. It is described by the type, extent, and location of land cover classes in that area.

Land use - the purpose to which people use a specific land area. It is described by the type, extent, and location of land use classes in that area. The associated terms "land cover class" and "land use class" are defined as follows:

Land cover class - an abstract representation, based on classification criteria, of dominant natural or artificial land cover in land area corresponding to mapping unit. Some mapping units may consist of composite land cover classes.

Land use class – an abstract representation, based on classification criteria, of dominant socio-economic function of land area corresponding to mapping unit. Some mapping units may consist of composite land use classes.

Land cover conveys the type of biophysical surface layer of land, whereas the land use conveys the use of the land by people. While each area on the earth's surface can be identified by a unique land cover class, only land use classes can also identify the areas used by people. Although there is a link between land cover and land use, not all land use classes can be determined from remote sensing data alone. Supplementary information, such as socio-economic, agronomic, environmental protection measures, etc., is required for the mapping of land use (Kalensky 1996, Cihlar and Jansen 2001).

An interesting study on the relationship between land cover and land use classes has been recently completed by Cihlar and Jansen (2001). They developed a methodology for land use mapping of large areas, for which the primary data source was land cover maps produced by remote sensing. Predictably, the accuracy depended on the type of land use class. For example, the identification of land use class "abandoned fields" from land cover data alone was difficult. However, the results were improved when assumptions based on local knowledge were included in the classification procedure. Similarly, the accuracy of other land use classes was improved when their identification was based on land cover data integrated with other types of geospatial data, such as the elevation or irrigated areas data.

The urgent need for solution of increasing socio-economic and environmental problems in developing countries requires new approaches. These problems include the inadequate food security for growing population, conversion of cropland to non-agricultural uses, land degradation and desertification, deforestation, diminishing fresh water resources, lack of employment opportunities in rural areas, and migration of rural population to towns. One of the prerequisites to successful solution of these problems is the optimization of agricultural land use resulting in increased production of food and feed crops, while assuring the conservation of land and water resources and protection of the environment. Achievement of this goal is best accomplished through agro-ecological zoning (Higgins and Kassam 1981, Fischer et al. 2001, Kalensky and Van Velthuizen 2001).

The agro-ecological zoning (AEZ) enables rational planning of sustainable use of agricultural land. Its objective is development of realistic strategy for optimization of agricultural land use while preventing land degradation. To achieve this objective, the implementation of AEZ involves a comprehensive assessment of the main biophysical and other relevant factors effecting crop production. Specifically, the AEZ methodology provides a standardized framework for systematic assessment of agricultural land use options, quantification of attainable crop yields and rating of land suitability for selected crops.

Agricultural land use is largely determined by the environmental physical conditions, biodiversity conditions, and socio-economic factors such as population density, poverty, land tenure systems, markets for agricultural products, and agricultural policies. The primary inputs to AEZ are therefore the climate, soils, topography, land cover data, and socio-economic information.

Systematic use of satellite remote sensing in AEZ projects, as a primary data source on land cover, has started relatively recently. While the climate, soils, topography and socio-economic data are available in most developing countries, there is usually a need for new land cover mapping to obtain the current information on bio-diversity and land use. Reduced cost of Landsat Thematic Mapper (TM) and Enhanced Thematic Mapper (ETM) image products has served as a catalyst for the integration of remote sensing component into AEZ methodology.

FAO LAND COVER CLASSIFICATION SYSTEM (LCCS)

FAO has developed the LCCS in response to increasing need for the establishment of standardized, systematic, multipurpose, land cover classification system. The LCCS is based on systematic and comprehensive land cover characterization, regardless of land cover type, data collection method, or geographic location. It thus facilitates a comparison and integration of land cover classes over large areas. It is the only universally applicable system in operational use at present. Most of the other land cover classification systems are single-purpose systems, tailored to requirements of a specific project or based on sectoral approach. Land cover classes produced by such systems are generally not comparable. Despite large number of land cover classification systems in existence, none has been internationally recognized and selected as a standard because of the above mentioned limitations. Yet, the increasing number of regional and global programmes and projects, such as the Global Terrestrial Observing System (GTOS), Global Forest Resources Assessment project, Africa Real Time Environmental Monitoring and Information System (ARTEMIS), etc., urgently need a universally applicable land cover classification system for objective international comparisons of land cover state and its changes (Anderson et al. 1972, Howard et al. 1985, Carroll and Morain 1992, Young 1994, FAO 1996 1998, Di Gregorio and Jansen 1996 2000). At present, the LCCS represents a leading edge land cover classification methodology. As its use is increasing, it may become a worldwide land cover classification standard.

There is sometimes confusion in technical literature between land cover classification and map legend. The following definitions clarify both terms and their relationship. *Land cover classification* is the systematic grouping of land cover types into classes according to selected diagnostic criteria. The criteria should enable a clear definition of class boundaries, without overlaps. The classification system, to be universally applicable, should have the following four characteristics:

- Independent of map scale,
- Independent of data source, that is data collection methodology,
- Independent of geographic location,
- Independent of application.

Legend of land cover map is a subset of classification, including only the classes related to specific geographic area covered by the land cover map, application (the type of land cover map) and mapping scale. There are two basic approaches to land cover classification: *a priori* and *a posteriori*.

In the *a priori* classification, the definition of land cover classes is done at the outset of the project, before the data collection takes place. Its main advantages are the independence of land cover classes of the geographic area and data collection methodology. Hence, the *a priori* classification facilitates the standardization of land cover classes. Its disadvantage is its inherent rigidity in having to select all the classes in advance. Since in most projects the number of classes have to be limited, some data may be difficult to assign to pre-defined classes.

In the *a posteriori* classification, the land cover classes are defined after data collection and their clustering. Its advantages are its flexibility and direct correspondence to collected data from which it was derived. Its disadvantages are its dependence on the geographic area from which the data were collected, mapping scale, application and data collection methodology.

LCCS Design Principles

The LCCS is based on modified *a priori* classification concept (Di Gregorio and Jansen 2000). The *a priori* classification was chosen because of the increasing need to standardize land cover classes to establish linkages among land cover projects implemented in different geographic areas, for different applications, using different mapping scales and data collection methodologies. However, without its modification, an enormous amount of pre-defined classes would be needed for consistent description of land cover occurring anywhere in the world.

The modification of the *a priori* land cover classification developed by FAO is based on the following unique concept: instead of pre-defining the land cover classes, the classification criteria that uniquely identify the classes are pre-defined. This presumes that any land cover class, regardless of its type and geographic location, can be defined by a set of pre-selected independent diagnostic attributes, the classifiers. The number of classifiers used determines the level at which the land cover is classified. Thus, the larger number of classifiers is needed when more detailed classification of land cover is required.

Another unique approach to modified *a priori* land cover classification adopted by FAO was driven by pragmatic, operational considerations. Instead of attempting to use the same, large set of pre-selected classifiers compatible with land cover of large areas, such as the whole of Africa, it divided the classifiers into eight groups corresponding to eight major land cover classes representing the global land cover diversity. This has greatly reduced the number of classifiers needed for precise definition of any land cover class and thus significantly simplified the classification procedure. However, it required designing the LCCS implementation in two phases: the initial *Dichotomous Classification Phase* and the follow-up *Modular-Hierarchical Classification Phase*.

The *dichotomous phase* uses the following three classification criteria: presence of vegetation, edaphic condition and artificiality of land cover. It consists of three classification levels and results in eight major land cover classes in the third level (Table 1). These classes are then further classified during the *modular hierarchical phase*, based on eight sets of predefined classifiers. Each of the eight major land cover classes defined during the dichotomous phase has its own distinct set of classifiers, tailored to the type of land cover class. These *pure classifiers* have a pre-defined hierarchical structure, which has to be followed during classification procedure.

Each set of *pure classifiers* also includes two types of additional classification attributes, the *environmental attributes* and *specific technical attributes*, which are used when further, more detailed definition of land cover classes is required.

Table 1. Three land cover classification levels of LCCS dichotomous phase.

<table>
<tr><th>First Level
(Presence of vegetation)</th><th>Second Level
(Edaphic conditions)</th><th>Third Level
(Artificiality of land cover)</th></tr>
<tr><td rowspan="4">PRIMARILY VEGETATED</td><td rowspan="2">TERRESTRIAL</td><td>MANAGED TERRESTRIAL AREAS</td></tr>
<tr><td>NATURAL and SEMI-NATURAL TERRESTRIAL VEGETATION</td></tr>
<tr><td rowspan="2">AQUATIC or REGULARLY FLOODED</td><td>CULTIVATED AQUATIC AREAS</td></tr>
<tr><td>NATURAL and SEMI-NATURAL AQUATIC VEGETATION</td></tr>
<tr><td rowspan="4">PRIMARILY NON-VEGETATED</td><td rowspan="2">TERRESTRIAL</td><td>ARTIFICIAL SURFACES</td></tr>
<tr><td>BARE LAND</td></tr>
<tr><td rowspan="2">AQUATIC or REGULARLY FLOODED</td><td>ARTIFICIAL WATER BODIES</td></tr>
<tr><td>NATURAL WATER BODIES, SNOW and ICE</td></tr>
</table>

LCCS Dichotomous Classification Phase

This phase consists of three classification levels, defining eight major land cover classes based on the following classification criteria (Table 1):

- Presence of vegetation,
- Edaphic condition,
- Artificiality of land cover.

Modular-Hierarchical Classification Phase

In the modular-hierarchical classification phase the selection of classifiers and their hierarchical arrangement are tailored to each of the eight major land cover classes corresponding to the third level of the dichotomous classification phase. Thus, eight different classifier sets are used in this phase. The user is not obliged to use all classifiers. Depending on the required level of land cover information, the classification can be stopped at any time and the corresponding land cover class automatically determined. Each land cover class is described by three codes:

- Boolean formula, consisting of a string of classifiers used for the class definition (e.g., A3A10B2),
- Common name of land cover class (e.g., "high closed forest"),
- Numerical (GIS-friendly) code (e.g., 20006).

All defined land cover classes are hierarchically arranged in the map legend, which is automatically generated.

The composition and hierarchical structure of a set of "pure" classifiers corresponding to the (dichotomous) major land cover class "Natural and Semi-Natural Terrestrial Vegetation" is in Table 2 and the example of formation of the actual land cover classes in Table 3.

Table 2. "Pure" classifiers for the class natural and semi-natural terrestrial vegetation.

Classification levels	Classifiers			
4	Life form of main layer (e.g., woody, herbaceous)	Vegetation cover of main layer	Vegetation Heights	Spatial distribution (macro pattern)
5	Leaf type (e.g., broadleaf, needleleaf, aphyllous)		Leaf phenology (e.g., evergreen, deciduous, mixed)	
6	Vertical stratification second/third layer		Vegetation cover second/third layer	Vegetation height second/third layer

Table 3. Formation of land cover classes during the modular-hierarchical classification phase.

Classifiers	Boolean Formula	Forest Name	Num. Code
Life form (LF)	A3	Forest	20004
Vegetation cover (C)	A3A10	Closed	20005
Height (H)	A3A10B2	High closed	20006
Macro pattern	A3A10B2C1	Continuous high closed	20007
Leaf type	A3A10B2C1D1	Broadleaf cont. high closed	20095
Leaf phenology	A3A10B2C1D1E2	Decid. broadleaf cont. high closed	20097
2nd layer (LF, C, H)	A3A10B2C1D1E2 F2F5F7G2	Multilayer decid. broadleaf cont. high closed	20628
3rd layer (LF, C, H)	A3A10B2C1D1E2 F2F5F7G2 F2F5F10G2	Multi. decid. broadleaf cont. high closed with emergents	20630

Example of definitions of the classes in the above table:
A3 = Woody vegetation (trees);
A10 = Closed vegetation cover (>70%);
B2 = Vegetation height >30meters; etc.
A3A10B2 = High closed forest

The land cover classification based on the pre-selected sets of "pure" classifiers can be further extended when more detailed description of land cover is required. The LCCS includes, in addition to "pure" land cover classifiers, two other sets of optional classification attributes, which provide additional description of land cover characteristics. These attributes are not hierarchically arranged and can be combined in any user-defined order:

Environmental attributes, which influence land cover but are not essential for its definition. Therefore, they should not be confused with "pure" land cover classifiers and used only when additional, more detailed land cover description is required. The environmental attributes include climate, landform, altitude, soils, lithology, and erosion;

Specific technical attributes, which relate to specific applications. For example, these attributes may include the description of crop types in managed terrestrial areas, floristic aspect of natural and semi-natural terrestrial and aquatic vegetation, salinity of artificial, and natural water bodies, etc.

LCCS Software Modules

The LCCS software package consists of four modules: classification, map legend, field data, and translator:

Classification module performs the two-phase land cover classification based on hierarchical arrangement of classifiers as described above. Subdivision of classes is achieved by addition of new classifiers. The more classifiers used, the greater the detail of the land cover class definition. Hence, as the classification proceeds to higher levels, the detail of land cover mapping is increasing. The classes at any level are all unique and unambiguous. If a classifier at any level cannot be determined, the classification would stop and the class definition based on the last classifier used. A strict class boundary definition and internal class consistency are inherent to LCCS.

Legend module stores the land cover classes that were defined in *Classification module* and creates a legend. It usually contains only a subset of the classification, consisting of classes applicable to area to be mapped. Because the classes stored in *Legend module* represent an abstract characterization of a specific area, they can form the *mixed mapping units*. These comprise two or three land cover classes, each occupying more than 20% of mapping unit area. The order of classes in a mixed mapping unit reflects the dominance of land cover types in the area with fragmented LCLU. This module allows export of data in commonly used file formats and enables adding the user-defined names to the provided list of common names.

Field data module stores information from field validation forms providing detailed description of field validation sites. These data are automatically translated into LCCS classifiers and attributes and exported to *Classification module*. The *Field data module* contains the retrieval and edit functions that enable manual input of additional relevant information.

Translator module enables translation of existing classifications and legends into the LCCS. It also enables comparison of land cover classes from external classification systems. This module thus enhances the LCCS role as the reference classification standard.

The flowchart modeling the interrelationship among LCCS modules is in Figure 1. The advantages of the FAO Land Cover Classification System are manifold. It is a highly flexible system in which each land cover class is mutually exclusive and clearly defined, thus providing internal consistency. These characteristics are independent of the classification level. The classification can be stopped at any desired level and still result in clearly defined land cover classes corresponding to that level. Any land cover type can be readily accommodated. The system is truly hierarchical and applicable at a variety of mapping scales and in any geographic location. It can be used as a reference standard system because its diagnostic criteria allow correlation with existing classifications and legends. The LCCS thus contributes towards harmonization and standardization of land cover classification and mapping.

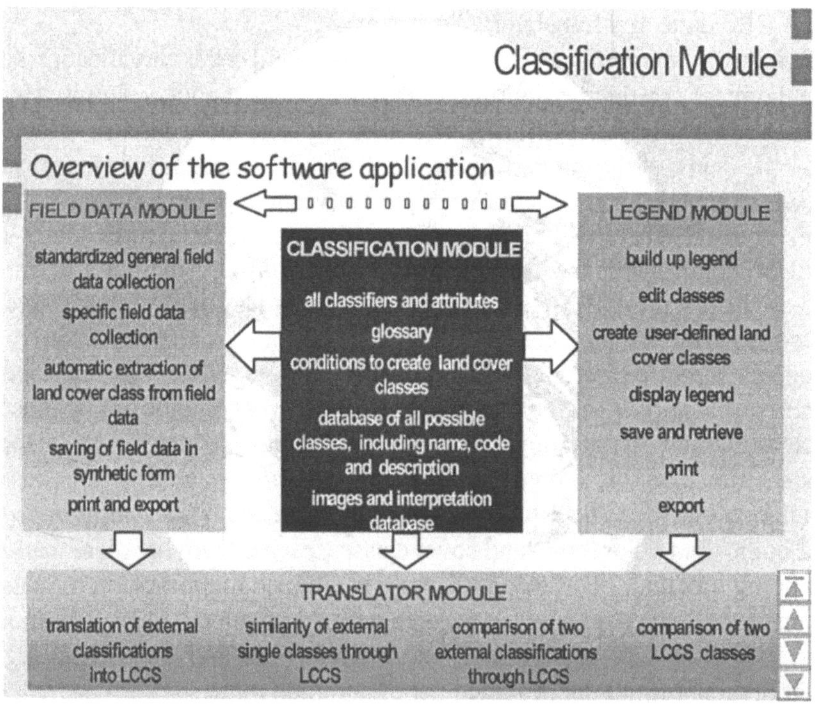

Figure 1. Overview of LCCS Modules and their Links.

LAND COVER AND LAND USE MAPPING BASED ON SATELLITE REMOTE SENSING AND GIS

Image data recorded by Earth observation (EO) satellites have become the primary input for LCLU mapping at small and medium scales, that is at the global, regional, and country levels. The land cover and land use applications of satellite remote sensing were one of the first operational applications that started soon after the first civilian EO satellite, the ERTS 1, later renamed to Landsat 1, had been launched in 1972. While the meteorological applications of satellite remote sensing had started about a decade earlier, it was the success of Landsat program that has provided the basis for the assessment, mapping, and monitoring of the Earth surface from satellites. Growing network of EO satellites, increasing range of ground detail (at present 1 m to 1 km) that can be recorded by their remote sensing systems in number of spectral bands, provide unprecedented opportunities for the operational use of remote sensing in the LCLU assessment, mapping, and monitoring projects. The efficiency of satellite remote sensing and the compatibility of satellite image data with the requirements of land cover and land use mapping at scales ranging from 1:1 million to 1:50,000 have been demonstrated by a number of projects implemented in industrialized as well as in developing countries (Di Gregorio 1991, FAO, 1992 1993b 1994, 2000c 2001b, Loveland et al. 1991, Kalensky 1996, Palko et al. 1996, EUROSTAT 1999, Thompson and Newby 2000).

The scope of mapping applications based on satellite remote sensing has been greatly enlarged by the operational availability of imagery from a new generation of EO satellites and sensors. In 1999, a new generation of EO satellites started to be launched under the framework of NASA-led international program called the Mission to Planet Earth. These satellites will eventually form the Earth Observing System (EOS) for systematic, long-term observation of the Earth's surface from space. The EOS, when fully deployed, will consist of 17 Earth observation satellites equipped with a wide range of remote sensing systems.

Of particular interest to LCLU mapping projects was the launch of EOS-1 satellite, the Landsat 7 in April 15, 1999. Its remote sensing system, the Enhanced Thematic Mapper (ETM) records the Earth's surface with 15 m resolution in one panchromatic band, 30 m resolution in six spectral bands and 60 m resolution in one thermal band. The radiometric and geometric qualities of ETM image products have been greatly improved, resulting in lower noise levels, better registration among spectral bands and geodetic accuracy (ground location of image pixel) better than 100 m, which can be further improved. The image acquisition has been scheduled on a global basis, using a cloud-avoidance algorithm and knowledge of seasonal vegetation cycles to optimize the LCLU data collection. Cost of radiometrically and geometrically corrected (systematic correction) digital image data of one Landsat-7 scene is US \$600. Considering that each Landsat-7 scene covers 185 x 170 km (31,450km^2), it is less than 2 cents per one km^2. Value-added film and digital products offered by commercial vendors could raise the price by US $1000 per scene on average that is to still very reasonable 5 cents per one km^2.

In tropical humid climate zone, for example in equatorial Africa, there are problems with obtaining good quality images of land cover by EO satellites with optical remote sensing systems, such as Landsat, because of persistent clouds. It sometimes takes years to obtain good quality images of such areas. EO satellites that use the synthetic aperture radar (SAR) remote sensing system can record images of the earth surface through clouds as well as during night, because they use the reflected microwave rather than solar radiation for image formation. Therefore, these "all-weather" EO satellites are the primary source of remote sensing imagery for LCLU mapping in countries with persistent cloud cover (ESA and FAO 1996, Davidson et al. 1997).

While several EO satellites with synthetic aperture radar (SAR) remote sensing system have been launched on experimental basis, Canada has become the first country to launch the civilian SAR remote sensing satellite, the RADARSAT-1 for operational worldwide use. It has been in operation since 1995. The unique feature of RADARSAT SAR system is the wide choice of its imaging modes. There are 25 possible choices, with the ground resolution ranging from about 10 to 100 m and corresponding size of RADARSAT scene from 50 x 50 km to 500 x 500 km. The Canadian Space Agency, to assure a continuity of RADARSAT SAR data supply, has initiated a development of RADARSAT-2, with more advanced SAR system. Its image products will be available with the ground resolution ranging from 3 to 100 m over the nominal swath width 10 to 500 kilometers. Furthermore, in order to obtain an improved discrimination of land cover and land use features, the horizontal polarization (HH) of RADARSAT-1 will be supplemented by vertical polarization (VV) and cross-polarization (HV or VH). RADARSAT-2 is scheduled for launch in 2003 (Baeggli and Brule 2001). Figure 2 illustrates a system flowchart for LCLU mapping and change monitoring based on satellite remote sensing and GIS. (Kalensky 1996).

While the main advantage of SAR remote sensing systems is their all-weather imaging capability, they also contribute to higher information content of land cover maps when the SAR and optical image data are combined and jointly analyzed. For example, the joint analysis of SAR and optical imagery enabled to differentiate rice fields from marshes in the Niger River flood plain. Since the SAR and optical RS systems are complementary, each recording different characteristics of land cover features, their combined use yields a higher level of information about the ground scene. The SAR systems record the spatial characteristics, such as the roughness of land cover surface and its spatial composition, while the Landsat-TM imagery has been the main LCLU data source for the AFRICOVER – EA project until now. The multispectral image data with 30 m pixel size have been fully compatible with project's information requirements. Supply of high quality, rectified and enhanced digital and hard copy (false-color paper prints) image products by commercial vendor has proceeded smoothly. The cost of Landsat imagery, about US \$0.05 per 1 km^2, was by far the lowest in comparison with remote sensing products from other EO satellites. Based on the above considerations, it is expected that the Landsat-TM and ETM imagery will continue to be the main data input used by the AFRICOVER project. The satellite SAR image products should complement the Landsat imagery in areas with persistent cloud cover.

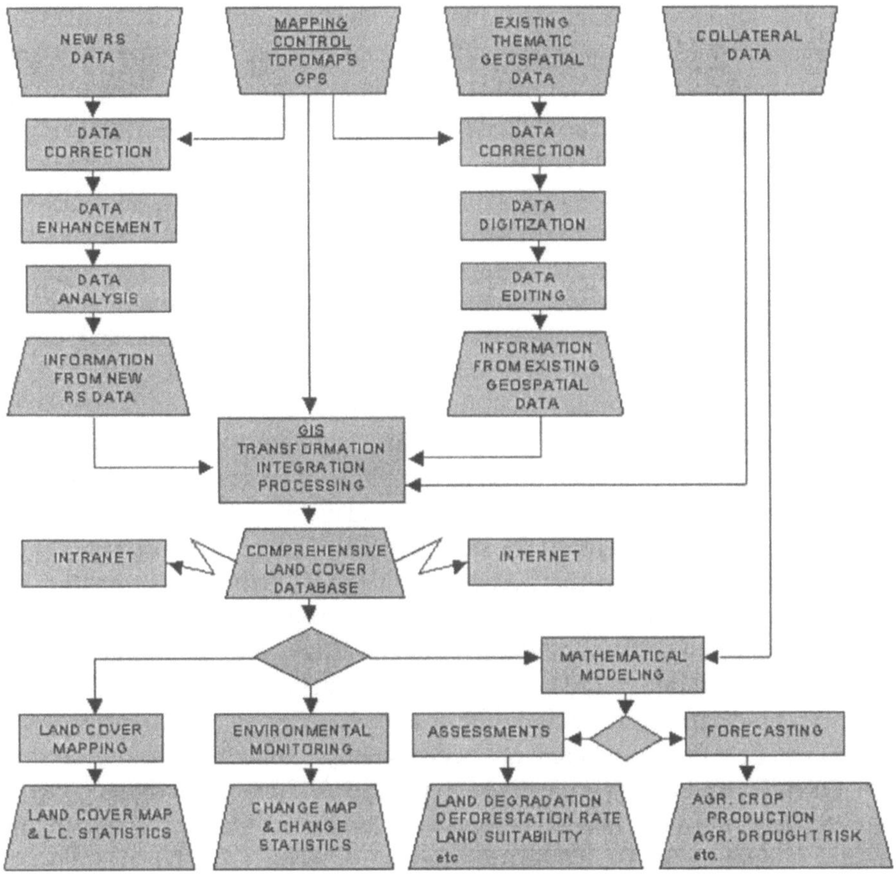

Figure 2. System flowchart for land cover mapping based on RS and GIS.

EXAMPLES OF LARGE-AREA LAND COVER AND LAND USE MAPPING PROJECTS

The implementation of LCLU mapping projects over large areas is best accomplished by satellite remote sensing and GIS. The Earth observation satellites provide a source of remote sensing data, while GIS provides the capacity for integration of geo-referenced data from different sources, their joint analysis and generation of cartographic, statistical, and modeling products (Figure 1). The integration of remote sensing data with topographic, soils, climatic, demographic, and other relevant geo-referenced data in GIS database increases the quantity and quality of derived land cover and land use information (Molenaar and Janssen 1992, Schultink 1992, Alberz and Tauch 1994, Kalensky 1994 1996).

The majority of LCLU mapping projects are sector-specific. The agriculture, forest, and water resources sectors are all producing their own LCLU maps, usually without any harmonization of their technical specifications. Hence, the LCLU maps differ in the type of their basemaps (cartographic projection and coordinate systems), scales, formats, LCLU classification methodologies, map legend, and accuracy requirements. These one-purpose inventory maps provide location-specific LCLU information that is needed by organizations with mandates for management of natural resources. Their lack of compatibility was not a major constraint when GIS capabilities and spatial data exchange were non-existent or limited.

The AFRICOVER-EA project has adopted the remote sensing and GIS – based strategy and produces a cross-disciplinary, multipurpose land cover database, and digital maps with uniform specifications for participating countries in East Africa (Allan 1994, Lund 1994, Di Gregorio and Jansen 1996 2000, Kalensky 1998). It is hoped that adequate funding will become available to extend the project implementation to the whole of Africa. Other important, complementary international land cover and land use initiatives are listed below.

One of the earliest programs for systematic land cover mapping at regional level was implemented in the framework of the CORINE (Coordination of Information on the Environment) program in Europe. It is managed by the European Environment Agency (EEA) of the European Commission. This program, which started in 1985, was originally designed for West Europe. After 1990, it was expanded to include the former Eastern block countries in the Central and East Europe where it has been recently completed. The objective of the CORINE Land Cover project was to produce a computerized inventory of Europe's land cover. The inventory was conducted by each participating country according to CORINE specifications. The CORINE land cover classification system consisted of three levels, with 44 classes in the third level. The inventory results were compiled at a mapping scale 1:100,000. The area of the smallest mapping unit was 25 hectares. The LCLU classes were interpreted from false-color images at the scale 1:100,000. The Earth observation satellites Landsat and SPOT recorded the multispectral image data. The interpreted LCLU classes were digitized, geo-referenced, transformed to a common cartographic projection, and integrated in the CORINE database. CORINE LCLU products are disseminated through the Internet-based European Information and Observation Network, EIONET (Commission of the European Communities 1987, Schneider 1989, Swedish Space Corporation 1997).

In 1996, the National Aeronautics and Space Administration (NASA) of the United States initiated the Land Cover and Land Use Change Strategy in the framework of its Earth-oriented program, the Mission To Planet Earth (MTPE). The objective of LCLU Change Strategy is to develop the capability for repeated global inventories of land cover and land use from space to increase our understanding of environmental consequences of their changes. It is expected that in the future, increasing pressure for land and food will accelerate the LCLU changes, largely driven by human societies. Better understanding of these changes is necessary for timely adoption of effective measures to safeguard the environmental protection, sustainable use of agricultural land and adequate food production. Hence, the emphases of this program are on LCLU change monitoring and trends forecasting at global level (NASA 1997).

In late 2000, the U.S. Geological Survey EROS Data Center completed the National Land Cover Database (NLCD) for the conterminous United States. The Landsat TM imagery with 30 m pixel size from early 1990s, combined with sample aerial photographs, provided the primary input data for land cover interpretation. The Landsat TM data were geometrically corrected and transformed to Albers Equal Area cartographic projection grid. Two or more Landsat scenes, recorded in different times of the growing season, were used for land cover interpretation. The output is a seamless land cover digital database consisting of 21 land cover classes. The NLCD is available on CD-ROMs or can be downloaded, free of charge, from the US Geological Survey EROS Data Center website. At present, plans are underway to update the NLCD in order to enable monitoring of land cover changes. The new NLCD will be based on Landsat-5 and -7 imagery recorded in 2000 (Vogelmann et al. 2001).

The European Commission has started a major program, the Global Monitoring of Environment and Security (GMES). It was launched in 1998 as a dedicated effort to use the Earth observation from satellites and information technologies for more effective environmental management and strengthened protection against natural disasters and effects of climate change. The European Earth observation satellites, SPOT, ERS, Envisat and Meteosat will be the main data sources. Detailed Action Plan for GMES will be prepared by the end of 2001 (European Commission 2001). The GMES potential priority areas will include:

- Assessment of land cover change in Europe between 1950 and 2000,
- Identification and mapping of land degradation in Europe, as related to environmental stress,
- Initiation of global vegetation monitoring.

The above major initiatives underline the worldwide recognition of the importance of land cover and land use mapping and change monitoring at the global and regional levels. Rational planning of new, realistic strategies for reaching the goal of adequate food supplies for growing population, while assuring the sustainable land management and protection of the environment, has to be based on reliable and timely information on current land cover and land use and their changes (Tucker et al. 1985, Townshend et al. 1991, Foody and Curran 1994, Wood et al. 1995, Kalensky 1996, Wyatt et al. 1997, Vogelmann et al. 1998, Loveland et al. 1999, Cihlar 2000, Hansen et al. 2000, Mucher et al. 2000, Zhu and Waller 2000).

CASE STUDY: THE FAO AFRICOVER PROJECT

The Food and Agriculture Organization of the United Nations designed the AFRICOVER project in the beginning of 1990s in response to requests by African countries for assistance with updating their land cover and land use databases. The required project parameters included the establishment of GIS-based land cover database compatible with the mapping scale 1:200,000 based on systematic land cover classification and uniform mapping specifications for the whole continent of Africa. The project implementation started in East Africa (AFRICOVER-EA) in 1994 (Kalensky 1992 1994, 1998, FAO 1994a 1996 1997).

The AFRICOVER-EA land cover mapping methodology is based on the application of advanced geo-information technologies, the satellite remote sensing, geographic information systems (GIS), and global positioning systems (GPS). A similar project, based on AFRICOVER classification and mapping methodologies is being formulated for land cover mapping of Asia (ASIACOVER).

Overall Project Objective

Overall objective of the AFRICOVER-EA project is to improve the availability of reliable, timely and location-specific land cover information in eleven East African countries covering an area of over 9 million km^2. They include nine riparian countries of the Nile River Basin as well as Somalia and Djibouti, to provide a complete coverage of East Africa. The reliable and timely land cover information is an essential input to food security early warning systems, sustainable management of natural resources, agro-ecological zoning and land use planning, environmental protection and rehabilitation, climate change assessment and international humanitarian and development projects.

Specific Objectives

- Development of technical specifications and methodology for land cover mapping of Africa at the basic scale 1:200,000. Derived scale is 1:1 million. This stage included two important spin-off developments: (a) development of the Land Cover Classification System (LCCS), (b) development of the AFRICOVER Interpretation and Mapping System (AIMS),

- Provision of reliable land cover information for the whole of East Africa, based on Landsat TM/ETM images and LCCS and AIMS classification and mapping methodologies. This stage includes field validation of preliminary interpretation results in sample areas,
- Development of land cover digital database, referenced to uniform cartographic projection and co-ordinate system, for East Africa.
- Production of digital land cover maps, at the scales 1:200,000 and 1:1 million on CD ROMs, for each participating country,
- Strengthening the land cover mapping and monitoring capacities, based on remote sensing, GIS and GPS geo-information technologies, of participating African regional and national organizations.

Concept of AFRICOVER Project Implementation

The main product of the AFRICOVER project is reliable information on the current status of land cover in each participating country. The information is in the form of digital land cover database and associated land cover maps at the scales 1:200,000 and 1:1 million on CD-ROMs. In addition, the project outputs include the methodologies and facilities for land cover mapping, such as the LCCS and AIMS land cover classification and mapping methodologies; GIS and image analysis software packages; trained national personnel in their operational applications; and strengthened LCLU mapping capacities of regional and national institutes that participated in project implementation. These additional project outputs are important for keeping the newly established land cover database updated and allowing the implementation of project follow-on activities, such as the land cover mapping at larger scale in selected areas. However, they are secondary to the AFRICOVER main output, that is the provision of reliable land cover data with uniform accuracy and based on standardized land cover classification and mapping methodologies.

The AFRICOVER project implementation is based on close cooperation with the relevant African regional and national organizations. Furthermore, two International Working Groups were established by the project: Africover International Working Group on Geodesy and Cartography and the Africover International Working Group on Land Cover Classification and Map Legend.

Members of both Working Groups were involved in project's preparatory activities, including the specification of land cover information requirements (land cover classes, classification accuracy and mapping scales); geodetic and cartographic specifications (geodetic reference datum, cartographic projection and co-ordinate system) and specifications for field validation of preliminary interpretation results (field validation methodology, sampling design and frequency) (Fanta 1993 1995, Iliffe 1994, Chodota 1996, FAO 1996 1997).

The project is hosted by the Regional Centre for Mapping of Resources for Development (RCMRD) in Nairobi, Kenya. Participating East African countries appointed the AFRICOVER National Coordinators to assure a close link with the project. At the meetings organized by the project, they obtain briefing on its progress and exchange experience with its implementation related to their respective countries. The project has also established a training facility at RCMRD and organized several workshops for training the national technical officers in the use of Land Cover Classification System, interpretation of satellite images and field validation of interpretation results.

Project design is modular, to allow its implementation either at the sub-regional or a country level. Its original design was based on the following five sub-regional modules:

- East Africa – 12 countries, 9,582,907 km^2 (includes Democratic Republic of Congo),
- North Africa – 4 countries, 5,017,440 km^2 (excludes Egypt, which is part of the module #1),
- Sahelian – 9 countries, 5,307,820 km^2 (includes Cape Verde),
- West and Central Africa – 15 countries, 5,066,955 km^2 (includes Sao Tome, and Principe).
- Southern Africa – 13 countries, 5,350,747 km^2 (includes the Indian Ocean countries Madagascar, Mauritius, Comoros and Seychelles).

Only the East African module has obtained funding and is being implemented at present. Although negotiations regarding the implementation of other sub-regional modules are on-going, it soon became obvious that some prospective sponsors preferred a more flexible approach, allowing them to select a country or group of countries according to their priorities and available budget. Revised concept of AFRICOVER project implementation accommodates this requirement and allows the project implementation at a country level as long as the AFRICOVER concept and technical specifications are followed. FAO prepared a template project document for implementation of the AFRICOVER project at a country level (FAO 2001a).

Five AFRICOVER scenarios were developed for project implementation at a country level. They differ according to the level of country involvement:

- Project inputs by FAO AFRICOVER staff are limited to training on AFRICOVER specifications, its Land Cover Classification System and the Interpretation and Mapping System. Implementation of all project activities is done by the country,

- FAO AFRICOVER staff selects the satellite imagery, supervises its geometric rectification and enhancement and provides training. Remaining project activities are implemented by the country,
- FAO AFRICOVER international consultant provides training and supervises the implementation of all project activities in the country,
- As above with the exception of development of GIS database, that task is contracted by FAO to an international company,
- International company implements the whole project according to AFRICOVER specifications. FAO certifies that the specifications were sustained and the AFRICOVER label can be used for the project products.

It is expected that the majority of countries will adopt FAO international consultants to provide training and supervision. Only the Republic of South Africa will likely have in-country staff perform project activities for its proposed updating of the "South African National Land Cover (NLC) Database" project. This project provides a standardized, baseline inventory of LCLU at 1:200,000 scale for the whole of South Africa, including Swaziland and Lesotho, based on Landsat TM imagery recorded in 1994-95. It was recommended that the proposed follow-up project, the "National Land Cover 2000" is based on AFRICOVER specifications, including its Land Cover Classification System (Thompson 1996, Thompson and Newby 2000).

Projects implemented at a country level, regardless under which scenario, will obtain the AFRICOVER certification only after it is asserted that they comply with the AFRICOVER specifications. This policy will assure the homogeneity of the AFRICOVER database for the whole continent of Africa and uniformity of its land cover mapping products in terms of their thematic content and accuracy.

Basic Principles of AFRICOVER Classification and Mapping Methodologies

Development of the AFRICOVER land cover classification mapping methodologies has been guided by the following principles:

- Development of appropriate methodologies and common standards for land cover classification and mapping of Africa at 1:200,000 scale. These include, in particular, development of the Land Cover Classification System (LCCS). Manual interpretation of satellite images will be supported in the project second phase by the computer-based AFRICOVER Interpretation and Mapping System (AIMS),

- Thematic content of land cover map should be applicable to such key applications as the sustainable land use planning, food security assessment and environmental protection,
- Geometric accuracy of project mapping products should be compatible with the requirements of land cover mapping at 1:200,000 scale, including the GIS data integration, processing and modeling,
- The main project products will be GIS-based land cover database and digital land cover maps at scales 1:200,000 and 1:1 million on CD ROMs. Hard copy mapping products will be generated when required. After project completion, the land cover database will be maintained at the country-level by the national organizations that participated in the AFRICOVER project implementation,
- Development of meta-database for land cover information produced by the AFRICOVER project. The meta-database will be available on the AFRICOVER Internet web-page,
- The AFRICOVER methodologies should be user-friendly to facilitate the capacity building in land cover classification and mapping based on satellite remote sensing imagery, in the regional and national organizations participating in project implementation. Provision of on-the-job training in all phases of land cover classification and mapping would be an inherent part of project implementation.

Project Outputs

The following main outputs are produced by the AFRICOVER project:

- Digital, seamless land cover database and the associated meta-database,
- Digital land cover maps at scales 1:200,000 and 1:1 million on CD-ROMs,
- Consistent land cover legend for maps at scales 1:200,000 and 1:1 million,
- Archive of satellite images used by the project and field validation results,
- Land cover classification and mapping methodologies (LCCS and AIMS),
- AFRICOVER training and reference materials,
- GIS, GPS, image analysis and cartographic equipment (hardware and software) used by the project,
- Trained national officers who participated in project implementation,

- Increased technical capacities for land cover mapping at participating African regional and national organizations,
- Established links and closer cooperation among participating African organizations.

To maximize the usefulness of AFRICOVER land cover digital and hard copy products, they have to fulfil the following four criteria in all participating African countries:

- *Accessibility*: the AFRICOVER land cover products will be accessible to all qualified users. Meta-database will be included on the AFRICOVER Internet website to facilitate the dissemination of information on the type and availability of AFRICOVER products. They will be distributed at a nominal charge covering only their actual reproduction and handling costs,
- *Consistency*: production, storage and distribution of AFRICOVER land cover products will follow the same specifications in all African countries,
- *Quality and reliability*: field validation of interpreted land cover classes will assure the thematic accuracy of AFRICOVER land cover polygons over 80%,
- *User-friendliness*: all products will be designed to facilitate their operational use.

The AFRICOVER digital land cover database consists of two main components: (1) base map layers, (2) land cover layer.

The base map layers, consisting of the cultural (man-made) features including the political and administrative boundaries, settlements, roads, dams, etc.; hydrographic features; dominant topographic features; coastlines and toponymy, are compiled from the existing topographic maps. Their updating is based on manual (visual) interpretation of Landsat TM/ETM false-color image prints at 1:200,000 scale. All information selected for inclusion in the basemap layers is edited, digitized, geometrically corrected and transformed into UTM cartographic projection.

The land cover layer is compiled from the most recent satellite multispectral digital images that meet the quality and cloud cover requirements. Ground resolution of remote sensing multispectral imagery compatible with land cover mapping at a scale of 1:200,000 is 50 m, which corresponds to 0.25 mm at image scale. Corresponding size of digital image pixel is about 35m. Spatial resolution of multispectral image products from several EO satellites (Landsat-5 and -7, SPOT-4, IRS-1C) fulfils this requirement. The AFRICOVER-EA project has selected the Landsat-TM and ETM multispectral digital image products, with 30 m pixel size and 185x170 km scene size, as the primary source of optical remote sensing data for land cover mapping. The

RADARSAT-1 standard or wide-mode SAR image products (30 m pixel size, 100 x 100 km or 150 x 150 km scene sizes) are used in areas with persistent cloud cover.

Geodetic Framework

All AFRICOVER basemap and land cover layers are referenced to a unified geodetic and cartographic framework for the whole African continent, following the recommendations by the AFRICOVER International Working Group for Geodesy and Cartography. It has been an important requirement, considering that at least seven reference ellipsoids and over twenty horizontal geodetic datums are used in Africa. As a result, there is little correspondence among coordinate systems in African countries, making a compilation of uniform basemap at 1:200,000 scale, covering the whole continent, difficult (Leatherdale 1992, Fanta 1993, Iliffe 1994, Lund 1994, Nino-Fluck 1995, Chodota 1996, Petrie 1996 1997, FAO 1997).

The AFRICOVER International Working Group on Geometry and Cartography recommended the adoption of the following geodetic and cartographic framework for the Africa-wide land cover mapping at 1:200,000 scale (FAO 1997):

- *Reference ellipsoid*: Geodetic Reference System 1980 (GRS 80),
- *Geodetic datum*: World Geodetic System 1984 (WGS 84),
- *Cartographic projection*: Universal Transverse Mercator (UTM),
- *Planimetric accuracy*: 50m (dominant base map features) to 100m (land cover polygons),
- *Map sheet layout*: based on the International Chart of the World (ICW).

Production Cycle of the AFRICOVER Land Cover Map

Compilation of the AFRICOVER land cover map has been based on the multi-phase image interpretation approach, which was successfully used by FAO in a number of projects. Examples of such projects are the land cover mapping projects of Afghanistan and Lebanon (Di Gregorio 1991, FAO 1993b). The multi-phase image interpretation approach consists of the following eight tasks:

- Preliminary delineation of distinct image patterns (interpretation units) on satellite false-color paper print images at 1:200,000 scale and/or digital imagery displayed on computer monitor,
- Field validation of 0.5% to 5% sample of interpretation units. It consists of field collection of pre-defined diagnostic attributes for

land cover classification, the classifiers, which are the key LCCS components. They uniquely define the land cover of interpretation units. Implementation of this task requires a preparation of field validation form,

- Final interpretation of mapping units (land cover polygons) based on LCCS. This task also includes a derivation of map legend, that is a selection of land cover classes that are compatible with the AFRICOVER basic mapping scale 1:200,000,
- Accuracy assessment of land cover classification based on field control data set,
- Finalization of land cover layer through editing, digitization, coding and geographic referencing of land cover polygons to facilitate their follow-up GIS processing,
- GIS processing and integration of land cover and base map layers in a comprehensive digital land cover database (Figure 1),
- Production of digital land cover maps on CD-ROMs by GIS-based digital cartographic processing,
- Development of meta-database and Internet web page for the effective dissemination of information on AFRICOVER products.

Preliminary delineation of interpretation units of similar appearance by manual (visual) interpretation of Landsat TM/ETM false-color images at 1:200,000 scale is based on their colour, contrast and pattern differences. When the scale of satellite imagery becomes larger (e.g., 1:100,000), image texture becomes an important criterion as well. Delineation is facilitated by computer display of digital imagery, which can be zoomed, enhanced and analyzed on monitor screen. Consistency during delineation of interpretation units is an important requirement. Since the task involves subjective decisions, it requires good experience and is best done by the same person. When there are more interpreters, the harmonization of their interpretation skills should be included in their training. After the preliminary delineation is completed, the interpretation units are labeled. The units of the same appearance have the same label, but no attempt for their identification is made at this stage.

The objective of field validation is to collect the pre-defined diagnostic attributes for land cover classification, the classifiers, in the areas of selected interpretation units representing different types of (unknown) land cover. The classifiers are used by LCCS for definition of land cover classes. The field validation sampling rate ranges from 0.5% to 5%, depending on the complexity of interpretation units. Obviously, the well-defined, homogeneous interpretation units require a lower rate of field validation than those with fragmented LCLU pattern. Results of field validation are recorded in the field validation form that was prepared for this purpose. One form is required for each sample spot.

Itinerary of validation missions should follow (never precede) the progress of preliminary delineation of interpretation units and be carefully planned. Practical considerations, such as the security and accessibility of the area will top the list of criterions for selecting the field validation sites. After elimination of some areas based on these practical considerations, the field validation plan is prepared. Planning of field validation missions should also include a selection of alternative validation sites. This will reduce an improvisation in changing the selected sites during field missions. The field validation forms include the topomap coordinates of planned validation sites as well as GPS coordinates of sites in which the field validation actually took place. Care should be taken to assure that both sets of coordinates are well within the same interpretation unit that was selected for field validation. Photographs, taken at four directions when it is practical, are an important part of field validation. They provide an objective record of LCLU in the validation site at the time when the mission took place. Result of field validation is a set of diagnostic attributes, the classifiers, which uniquely define the land cover class in each interpretation unit (Scepan 1999, Di Gregorio and Jansen 2000).

Final interpretation of land cover classes and definition of mapping units by LCCS is based on (a) delineated interpretation units and (b) collected land cover diagnostic attributes. This task may require a modification of interpretation units, such as their combination or splitting. Such a decision often results from the requirement that the mapping units must be identified with at least 80% accuracy. In areas with fragmented LCLU, only mixed mapping units can be delineated with adequate accuracy. Topography, soils, hydrography, density of population and the type of land use in the mapped area influence the level of homogeneity or fragmentation of land cover within the mapping unit. The predominant land cover type and the estimated percentage of other important land cover types within the mapping unit define the mixed mapping units. The LCCS guidelines provide a threshold of 80% (the second land cover class must occupy more than 20% of mapping unit area to be recognized) and the maximum of three land cover classes for the definition of mixed mapping units. Each final land cover class has a unique symbol as well as a numeric code to facilitate GIS processing.

The minimum size of mapping unit (land cover polygon) has been traditionally derived from the map scale, because it has to comply with the cartographic requirements for presentation of the map content, that is, its appearance and readability. Data from land cover database may have to be aggregated to assure their compatibility with the mapping scale. LCCS enables automated derivation of land cover classes as long as the required diagnostic attributes, the land cover classifiers, are included in the database.

The LCCS guidelines recommend the adoption of variable minimum size of mapping unit, depending on the type of land cover classes. The recommended minimum size of mapping units at the map scale varies from about 3 mm^2 for artificial terrestrial surfaces and artificial as well as natural water bodies; 6 mm^2

for cultivated areas (terrestrial and aquatic); 8 mm^2 for (semi)natural aquatic vegetation; 12 mm^2 for (semi)natural terrestrial vegetation and 24 mm^2 for bare land. Considering that the AFRICOVER basic mapping scale is 1:200,000 the corresponding sizes of ground areas are about 10, 25, 30, 50, and 100 hectares respectively.

The magnitude of the AFRICOVER project is documented by the following examples. Land cover classification in Kenya resulted in 30,000 mapping units (final land cover polygons), representing 100 land cover classes. The land cover pattern of Somalia consists also of 30,000 mapping units, representing 73 land cover classes. The tasks of such magnitude would require much longer time to complete if the LCCS had not been developed. Similarly, it is expected that the AIMS will further increase the AFRICOVER mapping productivity, after its completion later in 2001.

The final interpretation of land cover classes is followed by an accuracy assessment in sample areas to assure that it complies with the AFRICOVER specifications. The accuracy assessment is based on field checks in randomly selected control sites. Use of GPS receivers has been essential to correct identification of field validation sites. All land cover classes have to be identified with accuracy of at least 80% to be included in the map legend. The finalized land cover layer is referenced to a base map, which serves as the geographic reference. The referencing is done through control points provided by dominant features that can be accurately identified in both, the satellite image and the base map.

Completion of the AFRICOVER-EA project is planned in December 2002. Inventory of potential users of its products has been completed and a potential client network established. International projects that will benefit from its products include the Nile Basin Water Resources Project, Lake Victoria Water Resources Project, Lake Chad Basin Project, African Rangelands Assessment and Monitoring Project, Tropical Forest Resources Assessment Project, Africa Real Time Environmental Monitoring Information System (ARTEMIS) and the Global Land Cover Mapping Project 2000 of the E.U. The AFRICOVER-EA project, when successfully completed, will contribute to more effective and efficient management of renewable natural resources and environmental protection in East Africa.

AFRICOVER INTERPRETATION AND MAPPING SYSTEM (AIMS)

The AFRICOVER Interpretation and Mapping System (AIMS) is based on desktop computer, Windows NT operating system and MultiScope NT image analysis and GIS software. It combines the advantages of digital image classification and traditional manual (visual) interpretation in a synergic way. Its key component is software for "on-screen" image interpretation directly on computer monitor. It eliminates the need for digitization of interpreted map units (land cover polygons), facilitates their geometric referencing to base map and the interactive use of ancillary data stored in database. Drawing of polygons and lines, cursor positioning, panning and zooming are done in a seamless way.

The AIMS software will become part of the Multipurpose AFRICOVER Database for Environmental Resources (MADE), which will significantly enhance the GIS processing of environmental data, when its development is completed in the latter part of 2001 (Alinovi et al. 1999, FAO, 2000d). In addition to AIMS, the MADE will include the following software packages:

- *Land Cover Classification System (LCCS),* previously described,
- *AFRICOVER Database Gateway (ADG)*, an advanced browsing tool that will provide an Internet-based multi-user data network. It will allow the qualified users fast access to land cover databases and thus maximize their potential benefits,
- *AFRICOVER Interactive Database (AID)*, an interactive database that will be used to guide and homogenize the on-screen interpretation of land cover from displayed satellite imagery.

The AIMS is directly linked with LCCS and is set up to follow all the phases of the interpretation process according to the AFRICOVER procedures. AIMS will improve the efficiency of the interpretation and interactive use of ancillary data. The software will be extensively utilized to update the land cover and topographic databases, and to revitalize existing ancillary data to produce specific applications.

AIMS Software Modules

AIMS is based on digital image analysis system, Multiscope. The AIMS software package consists of the following modules:

- Photo-interpretation module,
- Georeferencing module,
- Land cover classification module,
- Map edit module,
- Work flow module,
- Stereo module,
- Hydrology and geology module.

AIMS Main Features

AIMS will fully integrate the Multiscope image analysis and LCCS land cover classification modules. It will facilitate the computer-aided updating of land cover changes, road infrastructure and hydrographic network. In particular, the AIMS software will facilitate the image analysis of a large number of optical as well as radar (SAR) satellite images. The system will be able to produce map sheets at selected scales and accommodate different cartographic systems.

The LCCS data files, stored in a common exchange file, can be easily imported into the AIMS image processing core module. Furthermore, the user will be able, while delineating land cover polygons, to select a particular land cover class by its common name, numerical (GIS) code or by its classifiers. The interface between LCCS modules and AIMS image processing core module is implemented by the AFRICOVER Legend Interface, which permits the visualization of the legend and selection of the appropriate land cover class. Table 4 summarizes the major features of AIMS.

Table 4. AIMS summary table.

Major Charact-eristics	Basic image processing and interactive (on-screen) image interpreta-tion system with an extension for vector capabilities
Standard Use of the Software	All image processing functions, plus vector capabilities
Major Use for AFRICOVER	Interactive image interpretation with on-line use of ancillary information when required, such as aerial photographs, new satellite products (e.g. Ikonos), maps, etc.
Major Advantages	Effective linkage of LCCS with AIMS. Combined processing of vector and raster data.
Limits	Limited editing vector functions including the imposs-ibility of creating the topology.

CONCLUSIONS

This chapter described the FAO effort to improve the quality and harmonization of land cover and land use databases in developing countries. Reliable land cover and land use databases are needed for timely monitoring of LCLU changes and trends forecasting. Such information is essential for rational land use planning, sustainable management of agricultural land, increased crops production and environmental protection. Yet, most developing countries in Africa do not have the capacities nor the resources needed for the production of reliable land cover maps and establishment of digital databases.

In response to this situation, the geographic information systems have been introduced to a large number of African countries under the various national and international development assistance programs in recent years. Hopes have been raised that GIS will help to modernize the management of natural resources and enable more effective protection against natural disasters and famine. However, many international development projects for LCLU mapping in developing countries have prioritized the tools rather than the product. They have established modern GIS laboratories and provided training but produced only limited new data sets. The new facilities are often underutilized after the project termination. Building the LCLU mapping and monitoring capacities in developing countries is necessary, but without assuring that they are utilized, it is a waste of effort and funding. Furthermore, the effectiveness of GIS depends on the availability of reliable input data. The integrated use of satellite remote sensing and GIS, as outlined in this chapter, provides the most effective and efficient methodology for land cover and land use mapping and change monitoring.

Sustainable management of natural resources and environmental protection are global tasks. The future of all mankind depends on their successful fulfilment. The availability of reliable, timely and location-specific land cover and land use information is essential to planning of realistic sustainable management and environmental protection strategies. It will empower the governments of developing countries to develop rational policies for sustainable management of their agricultural, forestry and water resources. Such policies are essential to protect the livelihood of people depending on renewable natural resources and assure that they will support the future generations as they have supported the past ones. Yet, for most developing countries, timely availability of such information remains an elusive goal. There is a need for a more effective international coordination and cooperation in the planning and implementation of land cover mapping and environmental monitoring projects.

Mapmakers have traditionally been at the forefront of economic development, providing charts and maps for exploration of new territories. Let's keep this tradition alive and join forces with our colleagues from developing countries to enable them to complete land cover maps and establish reliable digital geographic databases. Such baseline geographic information is urgently needed for planning the sustainable economic development of their countries, improving the food security through rational land use planning and environmental protection. The international cooperation in the field of land cover and land use mapping in developing countries should continue to be a part of our agenda for sustainable economic development in the beginning of the new millennium.

REFERENCES

Albertz, J. and R. Tauch. 1994. Mapping from Space - Cartographic Applications of Satellite Image Data. GeoJournal, Vol.32, No. l, pp .29-37.

Allan, J.A. 1994. Unpublished FAO background documents for the Technical and Donor Consultations on the Africover Project, held in Addis Ababa, Ethiopia. 43p.

Alinovi, L., A. Di Gregorio and J.S. Latham. 1999. Africover – Eastern Africa Module. Slide presentation. FAO. Rome, Italy. 21 slides.

Anderson, J.R., E.E. Hardy and J.T. Roach. 1972. A Land-Use Classification System for Use with Remote-Sensor Data. U.S. Geological Survey Circular 671. Washington, U.S.A. 16p.

Baeggli, H. and L. Brule. 2001. RADARSAT-2 Mission Update. Proceedings, 23rd Canadian Symposium on Remote Sensing, Sainte-Foy, Quebec, Canada. Pp. 1-4.

Carroll, C.S. and S.A. Morain. 1992. Defining Biophysical Land Units for Resource Land Management. Photogrammetric Engineering and Remote Sensing, Vol. 58, No. 8, pp. 1239-1244.

Chodota, M.W.L. 1996. The FAO Africover Project and a Possibility of a Unified Geodetic Datum for Africa (UGDA). Presented at the Ninth United Nations Cartographic Conference for Africa. United Nations Economic Commission for Africa (ECA), Addis Ababa, Ethiopia. ECA/NRD/CART.9/ORG.3O Add.1. 8p.

Cihlar, J. 2000. Land Cover Mapping of Large Areas from Satellites: Status and Research Priorities. International Journal of Remote Sensing, Vol.21, Nos.6&7, pp.1093-1114.

Cihlar, J. and L.J.M. Jansen. 2001. From Land Cover to Land Use: A Methodology for Efficient Land Use Mapping over Large Areas. Professional Geographer, Vol. 53, No. 2, pp. 275-289.

Commission of the European Communities. 1987. CORINE Land Cover Manual. CEC Directorate-General for the Environment. Brussels, Belgium. 163p.

Davidson, D., R. van Wyngaarden, D. Lantieri, J. Latham, A. Di Gregorio and R. Brown. 1997. An Evaluation and Demonstration of Mapping Africa Using RADARSAT for the UN-FAO Africover Project. Presented at the International Symposium "Geomatics in the Era of RADARSAT (GER `97)". Ottawa. Canada. 27p.

Di Gregorio, A. 1991. Technical Report on Land Cover Mapping of Lebanon. FAO project document NECP/LEB/001/SAU. FAO, Rome, Italy. 38p.

Di Gregorio, A. and L.J.M. Jansen. 1996. Africover Land Cover Classification. FAO Working Paper for the Dakar Meeting of Africover Working Group on Legend and Classification. FAO, Rome, Italy. 25p and 27p of appendices.

Di Gregorio, A. and L.J.M. Jansen. 2000. Land Cover Classification System (LCCS). Classification Concepts and User Manual. Food and Agriculture Organization of the United Nations (FAO). Rome, Italy. 80p and 99p of appendices.

ECA. 1996. Status of Mapping Coverage and Programmes in Africa. Presented at the Ninth United Nations Regional Cartographic Conference for Africa. United Nations Economic Commission for Africa (ECA), Addis Ababa.18p.

ESA and FAO. 1996. ERS Radar Data for Operational Land Cover Mapping. An ESA/ESRIN - FAO joint pilot study for Africover. European Space Agency (ESA) publication. 31p and 4 maps.

European Commission. 2001. A European Approach to Global Monitoring for Environment and Security (GMES). Joint document from Commission Services and the European Space Agency. Commission Staff Working Paper. SEC(2001)993. 31p.

EUROSTAT. 1999. Land Cover and Land Use Information Systems for European Union Policy Needs. Seminar proceedings. Luxembourg. EUROSTAT, Studies and Research, Theme 5. 193p.

Fanta, A. 1993. Mapping in Africa: Constraints and Prospects. Earth Resources Mapping in Africa. Newsletter of the Regional Centre for Services in Surveying, Mapping and Remote Sensing, Nairobi, Kenya. Issue No.1/1993, pp.l-7.

Fanta, A. 1995. Issues Relating to Data Access in Africa. Presented at the International Symposium MARISY 95, Ifrane, Morocco. Special issue of Geo-Observateur "MARISY 95", Royal Centre for Remote Sensing, Rabat, Morocco. Pp.225-228.

FAO. 1992. FAO Project Terminal Statement: Land Cover Map of Barbados based on Orthophotomap. Rome, Italy. 7p.

FAO. 1993-a. Guidelines for Land-Use Planning. FAO Development Series 1. Rome, Italy. 96p.

FAO. 1993-b. FAO Project Terminal Statement: Land Cover Map and Digital Database of Afghanistan. Rome, Italy. 12p.

FAO. 1994-a. Land Cover Map and Digital Database of Africa “AFRICOVER”. FAO project document. FAO Remote Sensing Centre. 142p.

FAO. 1994-b. FAO Project Terminal Statement: Land Cover Map of Belize. Rome, Italy. 6p.

FAO. 1995. Planning for Sustainable Use of Land Resources. Towards a new approach. FAO Land and Water Bulletin 2. Rome, Italy. 60p.
FAO. 1996. AFRICOVER Land Cover Classification Scheme. Working Paper for the Dakar Meeting of the AFRICOVER International Working Group on Land Cover Classification and Map Legend. Rome, Italy. 26p.
FAO. 1997. Summary Report of the AFRICOVER International Working Group on Geometry and Cartography. Rome, Italy. Part I, 26p; Part II, 18p; Part III, 23p.
FAO. 1998. Land Cover Classification System (LCCS). SD Dimensions electronic publication. FAO Sustainable Development Department. Rome, Italy. 3p.
FAO. 2000-a. The State of Food and Agriculture 2000. FAO Agriculture Series, No. 32. Rome, Italy. 329p.
FAO. 2000-b. Lebanon Ministry of Agriculture Information Management System. Project Inception Report. FAO project TCP/LEB/8924. 51p.
FAO. 2000-c. FAO Project Terminal Statement: Development of RS & GIS – based Land Cover Mapping Capacity in Bulgaria. Rome, Italy. 8p.
FAO. 2000-d. AIMS: Africover Interpretation and Mapping System. Rome, Italy. 6p.
FAO. 2001-a. Template Project Document for Implementation of the FAO AFRICOVER Project at a Country Level. Rome, Italy. 25p.
FAO. 2001-b. FAO Project Terminal Statement: Land Cover Map of Azerbaijan. Rome, Italy. 7p.
FAO and UNEP. 1999. The Future of Our Land. Facing the Challenge. FAO and UNEP guidelines for integrated planning for sustainable management of land resources. FAO Lands and Water Development Division. Rome, Italy. 71p.
Fischer, G., M. Shah, H. van Velthuizen and F.O. Nachtergaele. 2001. Global Agro-Ecological Assessment for Agriculture in the 21st Century. Joint publication of the International Institute for Applied Systems Analysis (IIASA), Laxenburg, Austria and Food and Agriculture Organization of the United Nations (FAO), Rome, Italy. 33p.
Foody, G.M. and P.J. Curran, eds. 1994. Environmental Remote Sensing from Regional to Global Scales. John Wiley & Sons, England. 232 p.
Hansen, M.C., R.S. Defries, J.R.G. Townshend and R. Sohlberg. 2000. Global Land Cover Classification at 1km Spatial Resolution Using a Classification Tree Approach. International Journal of Remote Sensing, Vol.21, Nos.6&7, pp.1331-1364.
He, C. 2000. Filling Digital Gap-Use of Geo-Information and Decision Support Tools in Sustainable Development. Invited paper. Nineteenth Congress of the International Society for Photogrammetry and Remote Sensing. Amsterdam, the Netherlands. 12p.
Higgins, G.M. and A.H. Kassam. 1981. The FAO Agro-Ecological Zone Approach to Determination of Land Potential. Pedologie, Ghent, Belgium, Vol.31, No.2, pp.147-168.
Howard, J.A., Z.D. Kalensky and F. Blasco. 1985. Concepts for Global Monitoring of Woody Vegetation using Remote Sensing Data. Invited Paper. Ninth World Forestry Congress. Mexico City, Mexico. FAO Remote Sensing Centre publication No.32. 14p.
Iliffe, J. 1994. Africover - Geodesy and Map Projections. Unpublished FAO background paper for the Technical and Donor Consultations on the Africover project. Addis Ababa, Ethiopia, 4-11 July 1994.
Kalensky, Z.D. 1992. Overview of FAO Remote Sensing Activities in Africa. Presented at the Fourth Meeting of the World Bank Advisory Committee on Environment Information Systems in Sub-Saharan Africa. Kampala, Uganda. 4p.
Kalensky, Z.D. 1994. Land Cover Mapping and Monitoring by Satellites. Presented at the International Workshop on Global Mapping, organized by the Geographical Survey Institute of Japan Izumo, Japan. GSI Technical Report A. 1, No.173, pp.112-125.
Kalensky, Z.D. 1995. Use of Space Technology to Enhance Food Security and Economic Stability in Developing Countries. Proceedings of the United Nations & European Space Agency Symposium on Space Technology for Improving Life on Earth. Graz, Austria. Pp.61-71.

Kalensky, Z.D. 1996. Regional and Global Land Cover Mapping and Environmental Monitoring by Remote Sensing. Invited paper. Eighteenth Congress of the International Society for Photogrammetry and Remote Sensing. Vienna, Austria. Proceedings, Vol.XXXI, Part B4, Commission IV, pp.413-426.

Kalensky, Z.D. 1998. AFRICOVER Land Cover Database and Map of Africa. Canadian Journal of Remote Sensing, Vol.24, No.3, pp.292-297.

Kalensky, Z.D. and R. Cumani. 2000. Use of Advanced Geo-Information Technologies in Land Use Planning and Management. Presented at the Geomatics 2000 Conference. Montreal, Canada. 12p.

Kalensky, Z.D. and J.S. Latham. 1998. The Establishment of Environmental Information Systems (EIS) in Developing Countries. Geomatica, Vol. 52, No. 4, pp.474-480.

Kalensky, Z.D. and H. van Velthuizen. 2001. Remote Sensing Inputs to Agro-Ecological Zoning. Presented at the 23rd Canadian Symposium on Remote Sensing. Laval University, Sainte-Foy, Quebec, Canada. 16p.

Leatherdale, J. 1992. Prospects for Mapping and Spatial Information Management in Developing Countries. ITC Journal, 1992, No.4, pp.343-347.

Loveland, T.R., J.W. Merchant, D.O. Ohlen, and J.F. Brown, 1991. Development of Land Cover Characteristics Database for the Conterminous U.S. Photogrammetric Engineering and Remote Sensing, Vol.57, No.11, pp.1453-1463.

Loveland, T.R., Z. Zhu, D.O. Ohlen, J.F. Brown, B.C. Reed and L. Yang. 1999. An Analysis of the Global Land Cover Characterization Process. Photogrammetric Engineering and Remote Sensing, Vol.65, No.9, pp.1021-1032.

Lund, G.H. 1994. Considerations for the Africover Project. Unpublished FAO background document for the Technical and Donor Consultations on the Africover Project, held in Addis Ababa, Ethiopia. 12p.

Molenaar, M. and L.L.F. Janssen. 1992. Integrated Processing of Remotely Sensed and Geographic Data for Land Inventory Purposes. EARSeL Advances in Remote Sensing, Vol.1, No.3, pp.113-121.

Mucher, C.A., K. Steinnocher, F. Kressler and C. Heunks. 2000. Land Cover Characterization and Change Detection for Environmental Monitoring of Pan-Europe. International Journal of Remote Sensing, Vol.21. Nos.6&7, pp.1159-1181.

NASA. 1997. Land-Cover and Land-Use Change Strategy. NASA Mission-To-Planet-Earth web page (www.hq.nasa.gov/office/mtpe). 19p.

Nino-Fluck, O. 1995. Cartography and Land Development. Presented at the International Symposium MARISY 95, Ifrane, Morocco. Special issue of Geo-Observateur "MARISY 95", Royal Centre for Remote Sensing, Rabat, Morocco. Pp.179-200.

Palko, S., L. St-Laurent, T. Huffman and E. Unrau. 1996. The Canada Vegetation and Land Cover: A Raster and Vector Data Set for GIS Applications – Uses in Agriculture. Published in GIS Applications in Natural Resources 2. Edited by M. Heit, H.D. Parker and A. Shortreid. GIS World Inc. Fort Collins, Colorado, U.S.A. Pp.185-191.

Petrie, G. 1996. Topographic Mapping from Satellite Imagery in Africa. Presented at the Ninth United Nations Regional Cartographic Conference for Africa. United Nations Economic Commission for Africa (ECA), Addis Ababa, Ethiopia. ECA/NRD/CART.9/ORG.13. 23p.

Petrie, G. 1997. The Current Situation in Africa Regarding Topographic Mapping and Map Revision from Satellite Images. ITC Journal, No. 1, pp.49-63.

Scepan, J. 1999. Thematic Validation of High Resolution Global Land Cover Data Set. Photogrammetric Engineering and Remote Sensing, Vol.65, No.9, pp.1051-1060.

Schneider, G. 1989. CORINE. Directorate-General for the Environment of the Commission of the European Communities. Brussels, Belgium. 32p.

Schultink, G. 1992. Integrated Remote Sensing, Spatial Information Systems, and Applied Models in Resource Assessment, Economic Development, and Policy Analysis. Photogrammetric Engineering and Remote Sensing, Vol.58, No.8, pp.1229-1237.

Singh, A.; E.A. Fosnight and R. Rykhus. 1997. Spatial Data Sets for Environmental Assessment: Towards Bridging the Data Gap. United Nations Environment Program (UNEP) - Environment Information and Assessment Technical Report. UNEP/DEIA/TR.97-4. 67p.

Swedish Space Corporation. 1997. Swedish CORINE Land Cover. Annual Report of the Remote Sensing Technology Division. Solna, Sweden. Pp. 6-10.

Thompson, M. 1996. A Standard Land-Cover Classification Scheme for Remote Sensing Applications in South Africa. South African Journal of Science, Vol.92, No.1, pp.34-42.

Thompson, M. and T. Newby. 2000. National Land Cover 2000. A Strategy for Updating the South African National Land-Cover Database. Conceptual business plan. Environmentek, CSIR. Pretoria, South African Republic. 9p.

Townshend, J.; C. Justice; W. Le; C. Gurney and J. McManus,1991. Global Land Cover Classification by Remote Sensing: Present Capabilities and Future Possibilities. Remote Sensing of Environment, Vol.35, Nos.2&3, pp.243-255.

Tucker, C.J., J.R.G. Townshend and T.E. Goff. 1985. African Land Cover Classification Using Satellite Data. Science, Vol.227, No.4685, pp.369-375.

UNCED, 1993. Agenda 21: Programme of Action for Sustainable Development. United Nations, New York, U.S.A. 294 p.

UNEP. 1993. Vegetation Classification. Report of the UNEP-HEM/WCMC/GCTE Preparatory Meeting. Charlottesville, Virginia, U.S.A.. GEMS Report Series publication. No.19. 21p.

Vogelmann, J.E., T.L. Sohl, P.E. Campbell and D.M. Shaw. 1998. Regional Land Cover Characterization Using Landsat Thematic Mapper Data and Ancillary Data Sources. Environmental Monitoring and Assessment, No.51, pp.415-428.

Vogelmann, J.E., S.M. Howard, L. Yang, C.R. Larson, B.K. Wylie and N. Van Driel. 2001. Completion of the 1990s National Land Cover Data Set for the Conterminous United States from Landsat Thematic Mapper Data and Ancillary Data Sources. Photogrammetric Engineering and Remote Sensing, Vol. 67, No.6, pp. 650-662.

Wood, W.B., P.H. Freeman and J.A. Miller, eds. 1995. EARTMAP Design Study and Implementation Plan. The Global Environment and Technology Foundation. U.S.A. 57p.

Wyatt, B.K. et al. 1997. Guidelines for Land Use and Land Cover Description and Classification. Draft Final Report. ITE Huntingdon/ITC Enschede/WCMC Cambridge/UNDP Nairobi. 134p.

Young, A. 1994. Towards an International Classification of Land Use. UNEP/FAO Consultancy Report. FAO Remote Sensing Centre. Rome, Italy. 43p.

Zhu, Z. and E. Waller. 2000. Global Forest Cover Mapping for the United Nations Food and Agriculture Organization Forest Resources Assessment 2000 Program. Consultancy Report. FAO Forest Resources Assessment Programme. Rome, Italy. 18p.

Chapter 14

SPATIAL EXPLICIT LAND USE CHANGE SCENARIOS FOR POLICY PURPOSES: SOME APPLICATIONS OF THE CLUE FRAMEWORK

Tom Veldkamp[1], Peter H. Verburg[1], Kasper Kok[2], Free De Koning[3], and Welmoed Soepboer[1]

[1]*Laboratory for Soil Science and Geology, Wageningen University, P.O. Box 37, 6700 AA Wageningen, The Netherlands. Tom.Veldkamp@geomin.beng.wau.nl*
[2]*International Centre for Integrative Studies (ICIS), Maastricht University, Kapoenstraat 2, 6211 KV Maastricht, The Netherlands*
[3]*GTZ-PPF, Casilla 17-21-1925, Quito, Ecuador*

Keywords: land use change, modeling, multi-scale, policy, scenarios, scales, drivers.

Abstract Worldwide Land Use and Land Cover Change (LULCC) plays an important role in global climate change. To be able to describe and analyze possible LULCC in the near future, spatially explicit modeling approaches are necessary. In this paper, applications of a LULCC-model, CLUE (Conversion of Land Use and its Effects), are demonstrated. The model uses quantitative information on the main drivers of land use change at different spatial scales, derived from historical and actual land use patterns. For a base-resolution (usually in the order of several square kilometers), georeferenced data for the study areas were collected and stored, and aggregated to different resolutions to create a series of artificial spatial scales. The collected data contains information on biophysical and socio-economic variables that were considered theoretically important land use drivers; the relative contribution at various spatial scales remained to be determined. By means of stratified, multi-scale, spatial statistical analyses, the relative importance of the independent drivers in explaining land use patterns were quantitatively determined for different spatial scales. The multi-scale information on land use drivers was implemented in the dynamic CLUE model that simulates LULCC, using time steps of one year. It consists of a demand and allocation module. The demand module estimates the demanded physical output from agricultural land use systems at the highest aggregation level considered. In a multi-scale iteration-algorithm, the allocation module calculates spatially explicit LULCC at the level of individual cells (for different resolutions), considering (changes in) national demands as well as site-specific values of biophysical and socio-economic drivers.

We will demonstrate applications for a large area such as Central America and for smaller regions like the Zona Atlantica (Costa Rica) and Sibuyan Island (Philippines). For all these applications the model was calibrated with historical data. Validation took place by comparing model output with data from an independent data set. The validation results indicate good agreement of modeled land use dynamics with observed dynamics. Through scenario studies, the impacts of changing conditions in the near future at different levels were studied, such as the effects of changing national demands for food products, national policies concerning market liberalization or nature protection, but also regional and local developments like migration and land degradation. Time horizons of 15 to 20 years were used in the scenario studies. The evaluation of different scenarios, based on different policies and development trajectories can help to assess these spatial patterns and their effects on nutrient balances and erosion, food production, and other ecosystem functions. In this way, spatial modeling can improve land use planning and inform policymaking.

Introduction

Worldwide Land Use and Land Cover Change (LULCC) plays an important role in global climate change. Human induced conversions of land cover (e.g. deforestation) as well as human land use practices (e.g., fertilizer use, irrigation, cropping patterns) have a clear influence on the functioning of the functioning of the earth system (Houghto 1994, Meyer and Turner 1994, Turner et al. 1995, Vitousek et al. 1997). Land use is relevant for climate change, because of its influence on CH_4 and N_2O emissions, on sources and sinks of CO_2, and on water and energy balances. The Kyoto protocol has clearly underlined the importance of land use for policy decisions concerning global change. Furthermore, land use changes affect biodiversity.

Land use change is driven by the interaction in space and time between biophysical and human dimensions (Turner et al. 1995). This has stimulated research in the understanding of land use change and its main causes and effects. As a result numerous land use change models are or have been developed (e.g., reviews by Lambin 1994, Riebsame et al. 1994, Kaimowitz and Angelsen 1998, Veldkamp and Lambin 2001). Models are tools that help us to conceptualize the complexity of the causes of land use and cover change. Furthermore, models help to understand the relative importance of individual drivers and processes for larger scale land use changes. Finally, through scenario studies models, can be used to explore near future changes in land use.

The integration of biogeophysical and socio-economic variables at the appropriate scale is still a major challenge in land use studies. Some propose the identification of separate disciplinary hierarchies and finding the appropriate, coherent, levels at which these hierarchies can be linked (O'Neill 1988, Turner et al. 1995). Others propose the complete integration of all information, and then look for the most important processes at different nested scales (Veldkamp and Fresco 1997a).

Social scientists have a long tradition in studying individual behavior in human-environment interactions at the micro-scale, mostly by narrative approaches (Bilsborrow and Okoth Ogondo 1992, Bingsheng 1996). At higher levels of aggregation, geographers and ecologists have studied land use change either by direct observation using remote sensing and GIS. Other studies have applied the systems/structures perspective to find understanding in the organization and institutions of society and landscapes that establish opportunities and constraints on decision making (Wiens 1989, Lambin 1994).

A specific methodological niche for land use change research can be found at the landscape level. Landscapes are characterized by their structure (the spatial relationships between landscape elements), function (the interaction between spatial elements), and change (the alteration in structure and function through time) (Forman and Godron 1986, Turner and Gardner 1991). In an agro-ecological landscape structure, function, and change are the result of an interaction between humans and their natural environment. Such agro-ecological landscapes are shaped by a set of interrelated and non-linear processes, causing complex linkages and interactions between different locations in the landscape (Milne 1991, Rigon et al. 1994). This means that simple aggregation of locally observed developments to a whole region is not realistic (Easterling 1997). The approach for land use change research presented in this chapter specifically uses the observed structure of the land use pattern to study the drivers of change. The dynamic land use model that is presented is also based on understandings of the structure of land use and a systems perspective on the functioning of land use. The objective of this chapter is to present this methodology and illustrate the implementation of the different concepts through a number of applications at three different regional scales in different parts of the world.

THE CLUE MODEL

The CLUE model was first applied to Costa Rica (Veldkamp and Fresco 1996) and new developments based on the same concepts were made for later applications (Verburg et al. 1999a). See also our webpage: (http://gissrv.iend.wau.nl/~clue/clue.html). The CLUE methodology is made up of two parts. The first part aims at establishing relations between land use and its driving factors, explicitly taking scale dependencies into account. The second part aims at dynamic modeling.

Multi-scale Analysis of the Driving Factors of Land Use Change

An empirical analysis of land use patterns is used to identify the most important biophysical and socio-economic drivers of land-use, as well as the quantitative relationships between these drivers and the surface area of the different land use types.

The georeferenced data used are (proxies of) factors that are considered potentially important for determining land use patterns in the study area (Turner et al. 1995). The resolution of these data (that are integrated in a geographical information system) should be in correspondence with the resolution of the base grid. In general, biophysical data are derived from existing maps and/or satellite images. Data used are for example, soil fertility indicators, soil physical properties, soil erosion susceptibility, precipitation, temperature, extreme weather events, altitude, relief, geology etc. Examples of data on infrastructure are distance to cities, markets, rivers, and roads. Typical socio-economic data are (rural and urban) population density, (agricultural) labor force, income, illiteracy etc.

The land use and cover data are derived from agricultural censuses, sometimes complemented and/or updated with satellite images. These data sources yield surface areas of land use types and separate crops, crop yields, number of animals and their production, and management data such as irrigation practices, fertilizer use, and mechanization.

Grid cells are not treated as homogenous units. For example with respect to land use types, a cell can consist of 10% grassland, 35% arable crops and 55% forest.

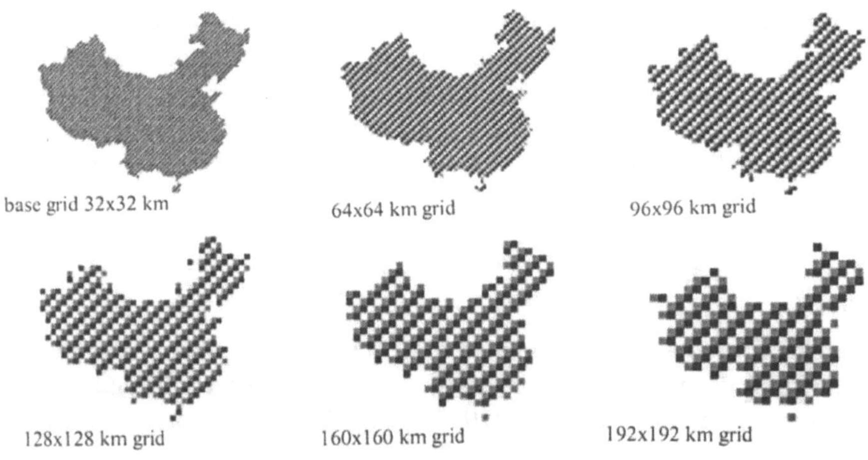

Figure 1. Six aggregation levels for China.

To investigate scale effects the data are organized in a series of artificial nested scales. The spatial organization of biophysical and socio-economic units only rarely coincides, and therefore processes and drivers do not overlap in space. To avoid this discrepancy, artificial scales are constructed on the basis of a series of aggregation steps, starting with a uniform geographical grid. The grid approach and the chosen resolution are related to the resolution of biophysical as well as socio-economic input data. For example, in the study for Ecuador (de

Koning et al. 1998 1999a) the highest resolution grid consisted of 5 x 5 minute cells (9.25 x 9.25 km). This base grid contained almost 3000 cells. Higher aggregation levels were made by aggregating the grid data into larger grid cells, composed of respectively 2 x 2 base cells, 3 x 3 base cells, 4 x 4 base cells, 5 x 5 base cells, and 6 x 6 base cells. This way 6 nested scale levels were created. Figure 1 demonstrates the same principle for China starting at a base grid of 32 x 32 km (Verburg et al. 1999b). Once data are converted into these basic grid cells, similar and equally sized units can be compared without any spatial aggregation problem. An additional advantage is that grid data can be aggregated into many different scales, while, for example, data grouped in administrative boundaries can only be aggregated into a few predetermined scales.

The resulting database is used to investigate statistically the potential biophysical and socio-economic drivers or their proxies of past and actual land use with multivariate statistics such as multiple-regression methods. To take scale dependencies into account, this analysis is performed independently at the different artificial aggregation levels.

The most important land use drivers (independent variables) for different land use types (dependent variables) are, for each aggregation level, selected from the basic set of theoretically important variables by means of stepwise regression. The selected significant variables are then used to construct multiple regression models. An example of such a model for a specific land use type at a specific scale is:

$$\% \text{ permanent crops} = \beta_0 + \beta_1 * (\text{rural population}) + \beta_2 * (\text{precipitation}) + \beta_3 * (\text{steep\_slopes}) + \beta_4 * (\text{rural poverty}) + e$$

In this example, the percentage of permanent crops in a certain cell is determined by a constant (β_0), the density of the rural population, the total annual precipitation, the percentage of the cell that has steep slopes, the percentage of the rural population living in poverty, and an error term (e). For all land use types such regression models are made at different aggregation levels. The validity of the relations is checked with expert knowledge and corrections are made if it is supposed that causality is violated.

Modeling Methodology for Land Use Change

The CLUE-model uses time steps of one year, and consists of a demand module and an allocation module (Figure 2). In the demand module, the national demands for agricultural products are calculated on the basis of population size, consumption patterns, import/export developments, and agricultural productivity. Annual changes in national demands drive sub-national land use changes, assuming that total land use meets the demand. Sub-national changes are modeled in a multi-scale allocation module (Verburg

1999a). The allocation is based on dynamic modeling of the competition between land use types on the basis of:

- the suitability of a cell for a certain land use type compared to the suitability of all other cells within the country.
- the suitability of a cell for a certain land use type compared to the suitability of that same cell for all other land use types.

These suitabilities are, for each year, determined by the actual biophysical and socio-economic conditions in each cell (Figure 2) and the relation between these conditions and land use are determined in the statistical analysis for the reference year (de Koning et al. 1999b). Changes in demand, determined at the national level, can also influence the allocation of the different land use types through its influence on the competition between land use types: cells are not converted to the most suitable land use type if there is not sufficient demand for that land use type.

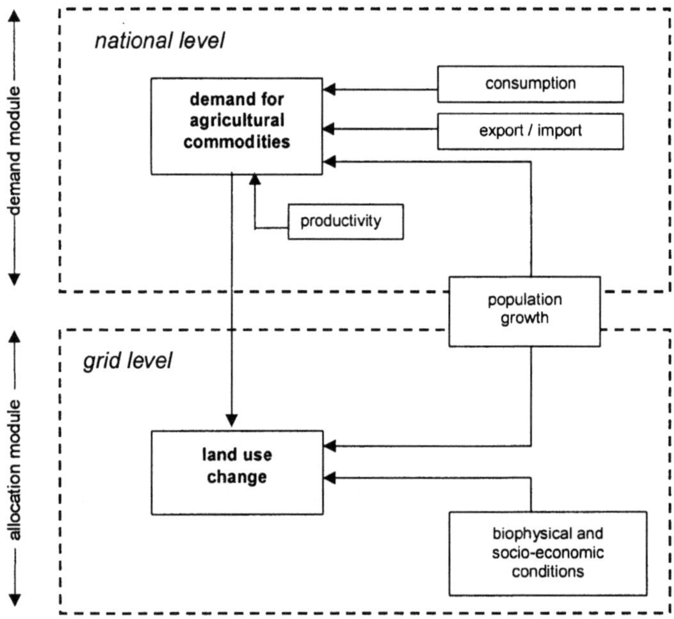

Figure 2. General structure of the CLUE-modeling framework.

The allocation procedure calculates changes in the different land use/covers for multiple scale levels simultaneously. Through an iterative procedure, it is ensured that the suitability of a location (grid cell) is not only determined by the local driving factors, but also by the driving factors at higher aggregation levels. Thus, the method allows for both top-down and bottom-up dynamics and can thus solve "conflicts" between driving factors at different scales. At a highly aggregated level, an area may seem very suitable for a certain land use type, while the same area viewed in more detail, may not be suitable at all or contain only a few grid cells with high suitability. This apparent conflict is solved by a feedback mechanism between the different scales. A more detailed description of the allocation procedure is given by Verburg et al. (1999a).

The multi-scale competition of land uses often leads to spatial pattern dynamics referred to as spatial connectivity. Connectivity implies that changes at one location might induce changes at other locations that are spatially distant. High competition pressure at one location can affect a completely different region within a country, which seems not directly related to this location. The establishment of a nature reserve at one location might affect land use at other locations as a tradeoff due to migration. The multi-scale structure of the model allows the analysis of this type of spatial interactions.

The CLUE model simulates multi-scale land use change dynamics under certain assumptions of possible future developments, defined as scenarios. Therefore "what if" situations can be described. Results of CLUE are not meant to be predictions, but rather possible spatial outcomes of feasible land-use developments. Scenarios can be focused on the changes in demand (quantity of change) or on the spatial explicit allocation (location of change). Examples of the latter application are the effects of national parks, regional migration policies, impacts of over-use or erosion.

MODEL APPLICATIONS

CLUE Application for Central America

The main objective of this application was to develop various plausible near future pathways (1996 until 2010) of land use change at the national level and to analyze the consequences on spatial patterns in Central America (Guatemala, El Salvador, Honduras, Nicaragua, Costa Rica, and Panama). It offers the possibility to study various countries simultaneously. Although largely similar in climatic and biophysical land use potential and recent political history (Schumann and Partridge 1989), the six countries display large economic (World Bank, 1998) and political (Brockett 1988, Pelupessy and Weeks 1993) differences. Besides, biophysical, climatic, and socio-economic gradients (mountain ranges, rainfall, and population density) are steep over small distances within the countries, which induces strong variation in land use over relatively small areas. Details can be found in Kok and Veldkamp (2001).

Scenario Development

Data were largely extracted from the FAOSTAT databases (FAO, 1998). Economic variables were obtained from World Bank (World Bank 1998).

Scenarios were calculated for the period 1997 until 2010. Analysis of past changes encompassed 1975 until 1996. Scenarios were developed for ten commodities that translate into five land uses. Together with natural vegetation, for which no specific demand was calculated, those six land use types were used as input in the allocation module of CLUE.

A number of scenarios were defined and developed in collaboration with members of the World Bank, UNEP, CIAT, and ministries of the various countries in Central America during a series of workshops (Farrow and Winograd 2001). Results of the Base scenario and an Optimistic scenario will be discussed here. Main assumptions are: 1) Base scenario: Historical trends continue until 2010. This means that cultivated land continues to increase on the expense of natural vegetation. Deforestation continues at about 1% per year, mainly as a result of expansion of grasslands. Export and import also increase slightly. Three sub-scenarios are discerned, differing in GDP growth at national level. Base1% assumes a 3% annual GDP growth which is line with the recent development. Base1% and Base5% assume 1% and 5% annual GDP growth respectively. Because GDP is used as a proxy for income, mainly the domestic consumption per capita differs between the three sub-scenarios; and

Table 1. Areas (km^2) of six land use types in Central America in different scenarios.

Area	*Annuals*	*Bananas*	*Coffee*	*Sugar Cane*	*Pasture*	*Natural Vegetation*
1996	41,600	1,600	7,700	3,300	104,700	284,800
2010 Base 1%	51,950	1,800	8,500	4,050	103,100	274,300
2010 Base 3%	49,300	1,800	8,600	4,100	107,800	272,100
2010 Base 5%	45,900	1,850	8,850	4,200	120,500	262,400
2010 Optimisitc	35,950	1,950	8,800	3,900	92,500	300,600

2) Optimistic scenario: It is assumed that various measurements will be taken that decrease the demand for agricultural area and thus decrease the pressure on the remaining natural vegetation. Yields are projected to increase to the maximum attained in 1996 anywhere in the region; export and import increase to ensure cultivation of more “appropriate” crops; waste losses are reduced to the minimum reported in 1996. Furthermore, GDP growth will be 5% per year, under the reasoning that the assumptions will only be realized if income growth is high. Summarizing data are given in Table 1.

Allocation Under Given Scenarios: Base Scenario

Maps of hot spots of change (1996 to 2010) under the Base scenario with 3% GDP growth are given in Figure 3. The spatial pattern of land use is projected to change considerably, even though changes at the national level are relatively small and largely follow historic developments. Besides, changes within a country are rarely unidirectional. Strong increases and decreases often occur within a short range.

There are considerable differences between countries. Dynamics in Honduras are very low, and to a lesser extent the same holds for Panama and El Salvador. Particularly dynamic changes occur in Guatemala. Differences in total demand changes between countries account for part of the observed variation. Demand for bananas in Costa Rica for example, is projected to rise 25%, which explains the patterns to be concentrated in that country. On the other hand, annuals in Honduras increase at the same rate and comparable quantity as annuals in Nicaragua, but dynamics in the latter country are far larger. Of all land use types, annuals show most vigorous changes. Often, (very) strong increases are coupled with strong decreases.

This highly dynamic modeled behavior of annual crops is in line with the temporal character of the main crops classified as annuals, namely maize, beans, and rice. Unlike permanent crops, their location can potentially be changed within a year. The spatial pattern of bananas shows remarkable changes in Costa Rica. Plantations are shifted from the Atlantic to the pacific side. Those projected new locations coincide with prior locations of banana plantations in the 1960s and 1970s (Goluboay and Vega 1988), when cultivation in the Atlantic zone was temporarily ceased due to a fungus infection. However, the projected shift to the south is not very realistic. An omission from the set of land use drivers that was used (Kok and Veldkam 2001) is the transportation from port to final destinations in Europe and the US. In this case, particularly the costs to pass through the Panama Canal make banana cultivation in the south unprofitable, despite the proximity of a port. The spatial patterns of coffee and sugar cane change largely. Especially in Guatemala and El Salvador, present plantations of both land uses are projected to shift to various other spots spread out over the countries. Whether such shifts of land uses with a permanent character are to be expected, depends on variables like the existence of local corporations, and should be analyzed using more detailed models. Model results indicate a strong potential for sugar cane and coffee cultivation outside present locations. There are clear hot spots of increase and of decrease of pasture. Areas of high increase are present in the south of Costa Rica and in the central highlands of Guatemala. Both are areas with a relatively high forest cover in the proximity of roads and people. Areas of pasture decrease are found in most countries, and are mostly located on poor soils, in relatively remote areas. Hot spots of deforestation are scattered over all countries, except Panama. Remarkably, the three large continuous forested areas in the south of Panama, the north of Guatemala and the east of Honduras remain largely intact. Some deforestation takes place in the north of Guatemala, but deforestation elsewhere in the country is more severe. This tendency is even stronger in Honduras. The general trend is one of deforestation of remaining small forest stands.

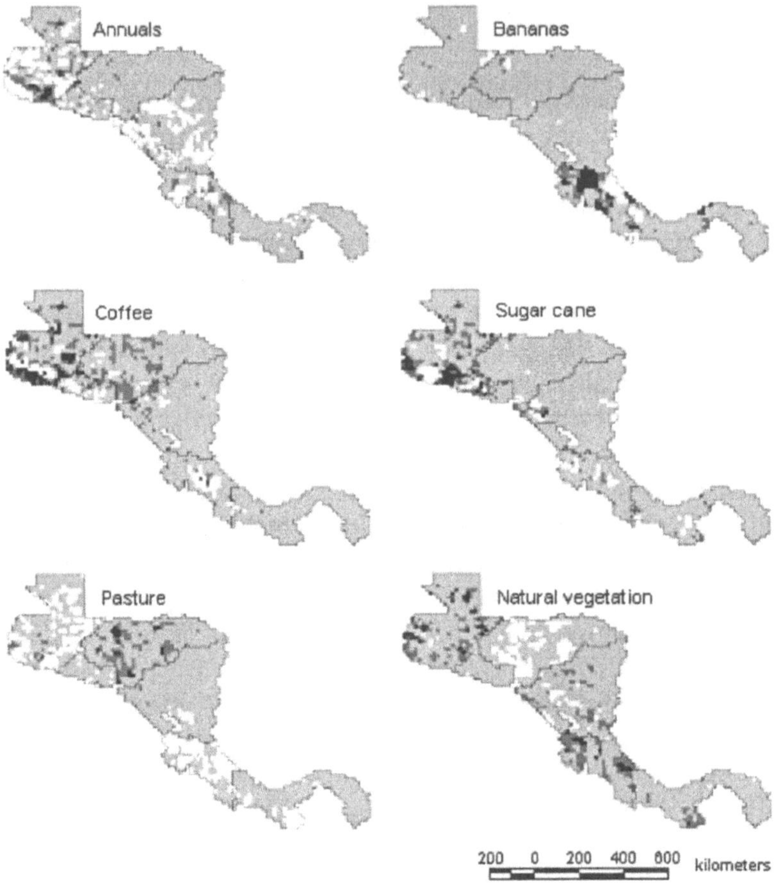

Figure 3. Classified differences between land use in 1996 and 2010 under the Base3% scenario. White areas indicate decreasing cover percentage; (very) dark shades indicate a (strong) increase. Medium gray shades indicate no substantial change. Classification is crop-dependent, but hot spots usually indicate > 5% change per cell.

Hot spots of land use changes in the Base1% and Base5% scenario were not different from the Base3% results to justify separate evaluation. Distributions of the three permanent land uses, that are generally influenced stronger by export and import than by changes in domestic intake, were virtually identical in all variants of the Base scenario. Patterns of annual crops and pasture are influenced stronger by changes in income and domestic consumption, but most important hot spots of change remained. Deforestation hot spots likewise did not change, except for a slight acceleration in the eastern part of Honduras.

Optimistic Scenario

Hot spots of land use change between 1996 and 2010 under the "optimistic" scenario are given in Figure 4.

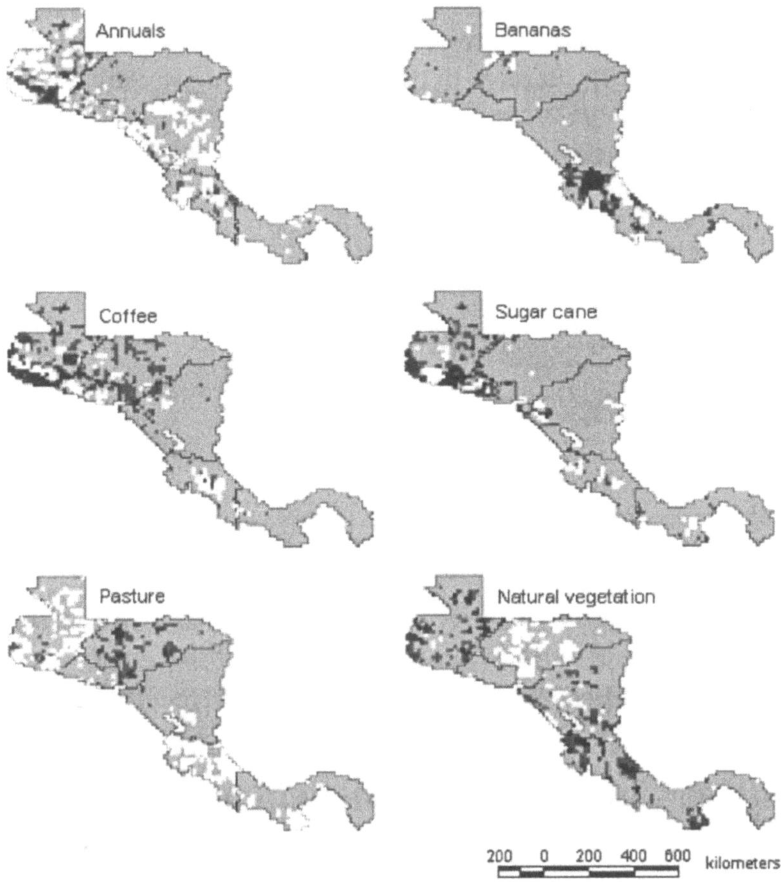

Figure 4. Classified differences between land use in 1996 and 2010 under the Optimistic scenario. White areas indicate decreasing cover percentage; (very) dark shades indicate a (strong) increase. Medium gray shades indicate no substantial change. Classification is crop-dependent, but hot spots usually indicate > 5% change per cell.

Again, land use changes are highly dynamic, but there are differences with the Base3% scenario. Annuals decrease especially in Nicaragua and Guatemala. The overall lower cover is caused by assumed yield increases of annuals. Because Nicaraguan yields are the lowest in the region, the effects of sustainable management are largest in this country. In Costa Rica hot spots of annuals become more dynamic, possibly caused by a competition for land with bananas and pasture. Hot spots of change in banana plantations are predominantly found in Costa Rica, similarly to Base3%. The location of increasing cover, however, is different. New plantations are projected in the northern part instead of the south. Projected changes are unlikely to happen in such short term, due to the lack of proper infrastructure and a distribution system on this side of the country, but the potential for banana expansion in Costa Rica is indicated. Sugar cane and coffee both display hot spots of change that are very comparable to the Base3% scenario. Yields were already very high and other measures that are assumed in the Optimistic scenario did not strongly affect the total area of both land uses (Table 2). Because of projected increasing stocking rates and decreasing demand, large areas of grassland are afforested in Costa Rica, Guatemala, and to a lesser extent in Nicaragua. The opposite process of increased deforestation to satisfy the demand for pasture (mainly beef) can be observed in Honduras. In all countries, changes are almost unidirectional, which indicated that new pastures are developed where no strong competition with other land uses takes place, and that abandoned pastures are almost always converted to secondary forest.

CONCLUSIONS

Differences between countries were large in both scenarios. Given these differences and the different processes that take place, we conclude that the country is the largest organizational unit that can be modeled. Treating the whole of Central America as one unit can produce erroneous results. A radically different and hardly plausible pathway of land use change has to be assumed (in the Optimistic scenario) to generate differences in land use patterns. This persistence of local circumstances stresses the importance of spatially explicit modeling at different scales.

CLUE Application for the Northern Atlantic Zone of Costa Rica

To study the Central American LUCLC in more detail, a study was carried out for the Northern Atlantic Zone (NAZ) in Costa Rica (Figure 5). Previous CLUE simulations and land use change analyses for the whole country demonstrated a highly dynamic LULCC in Costa Rica both in quantity and location (Veldkamp and Fresco 1997a and 1997b). These results were confirmed by other modeling approaches such as the GEOMOD2 model (Pontius et al. 2001). The Atlantic Zone is characterized by a typical deforestation pattern commonly observed in Latin America. The landscape is predominantly flat, its climate is perhumid without a pronounced dry period (Gomez 1986).

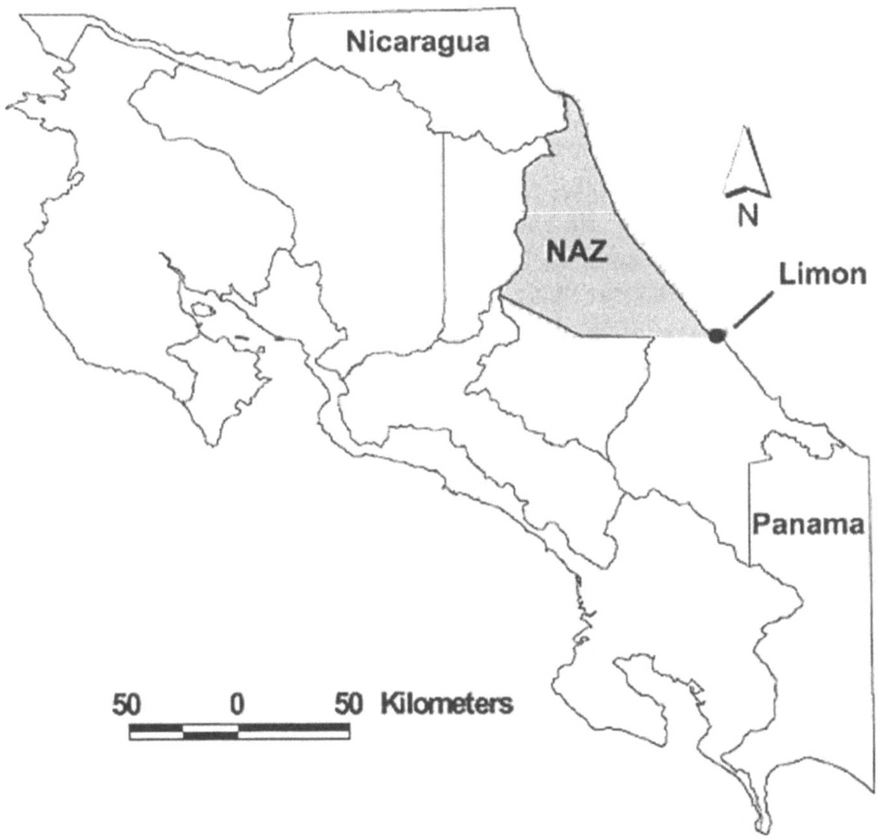

Figure 5. Location of Northern Atlantic Zone (NAZ) in Costa Rica.

The soils in the region are predominantly classified as Andosols and Inceptisols (Wielemaker and Vogel 1993). Major influence of humans on land cover started 100-years ago with the construction of a railroad from the Central Valley to the harbor of Limón, but accelerated during the last two decades (Pelupessy and Weeks 1993). By the mid 1980s more than 50 percent of the original tropical rainforest had disappeared and though the absolute deforestation rate of Costa Rica is decreasing, this does not apply to NAZ. In recent years the expansion of the export crops bananas and palm heart is noteworthy, though the main agricultural land use is pasture for cattle both for meat and milk production. Like in other countries in Central America, development of agricultural land use in the recent past depended directly on the market expansion in developed countries (Goluboay and Vega 1988). Like all other CLUE studies a set spatially explicit biophysical and socio-economic data was collected and statistically analyzed to determine the main drivers of land use change. This analysis demonstrated that forest was largely restricted to the poor biophysical conditions while banana plantations have claimed the best soils in flat, non-flooded areas. Annuals were thus restricted to the somewhat poorer circumstances, while pasture remained in the steep, stony, wet areas. Extended results can be found in (Kok and Veldkamp 2000).

Scenario Development

Results for two scenarios that were formulate for the period 1997 until 2005 are discussed: a Base scenario and a scenario in which agricultural practices inside national parks are completely prohibited (Park protection scenario). The assumptions of the Base scenario are largely the same as in the Central American application: data were largely extracted from the FAOSTAT databases. Economic variables were extracted from long-year investigations in the area (see Bouman et al 2000). Analysis of past changes encompassed 1984 until 1996. Scenarios were developed for ten commodities that translate into five land uses.

Assumptions Base Scenario

Based on studies dealing with past changes in Costa Rican land use (e.g., Easterly et al. 1997, May and Bonilla 1997), a Base scenario was developed, where historical trends continue until 2005. Annual GDP growth was set at 3% and similar import/export growth as in the recent past was assumed. The main driving force is the increasing demand of a growing population.

Table 2. Quantitative effects of Base scenario on the area with major kinds of land use in the Northern Atlantic Zone (in ha).

	Annuals	Banana	Pasture	Palm Heart	Forest	Rest[1]
1996 Actual	484	51 330	243 298	2000	123 369	65 496
2005 Base	891	54 962	262 701	4700	92 727	69 996

[1]The rest group constitutes of other plantations, urban areas, lakes, and a large area of secondary vegetation.

Assumptions Park Protection Scenario

A large part of the NAZ has a protected status and although only 5% of the NAZ are considered to be within a national park, a further 25% has some kind of protected status. Maintaining 30% of the area under complete protection in the future will have large consequences for land use. Full protection would be difficult to realize, considering the extent of the area and means available. This is suggested by the fact that despite their status and despite all efforts undertaken, every year a part of the area within the national parks is deforested (Ramirez and Madonado 1988). Nevertheless, international aid and/or pressure might change this situation. Thus, the effect of total protection of national parks and partial protection of other protected areas were examined in an adapted version of the Base scenario.

Results

Results of the Base scenario for major kinds of land use are given in Figure 6, including a comparison with the initial 1984 situation.

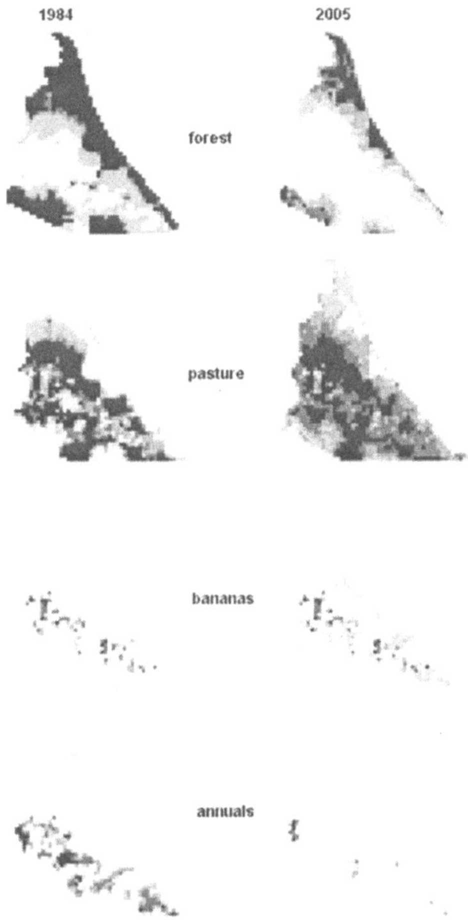

Figure 6 (above, on preceding page). Input data and results of Base scenario for four land use types. Left side: situation in 1984; right side: situation in 2005. Dark colors: high percentage coverage of land use type within grid cell; light colors: low percentage coverage of land use type within grid cell. Palm heart did not exist in 1984.

Base Scenario

In the Base scenario, forest will become restricted entirely to remote and areas unsuitable for agriculture. Demand for agricultural land in potentially suitable areas for forest reduced natural cover in large areas to almost zero. Noteworthy is the persistence of forest in areas within the boundaries of national parks, even though these areas do not have a protected status in the Base scenario. Pasture increasingly becomes the dominant land use in the entire zone. Besides some isolated pixels with banana plantations and the already mentioned areas with forest, pasture becomes the dominant cover everywhere. Expansion of banana plantations will predominantly occur near areas where banana was already present. In particular close to Limón (in the southeastern corner of the study area) new plantations are modeled. The location of annual crops will be restricted to a few well-defined areas. Those areas do not necessarily correspond to the original areas in 1984. Palm heart was not present in the zone in 1984, but its projected presence is in the area just south of the main banana zone in the slightly higher locations.

Park Protection Scenario

Restricting land use allocation through protecting parks considerably influences land use patterns. Figure 7 shows the distribution of forest with and without protection of national parks. The pressure on forest *outside* parks is enhanced almost up to the point of complete disappearance. Only in the most remote area close to the border with Nicaragua, forest is still present in reasonable quantities; the remainder of the NAZ is almost exclusively devoted to agricultural uses.

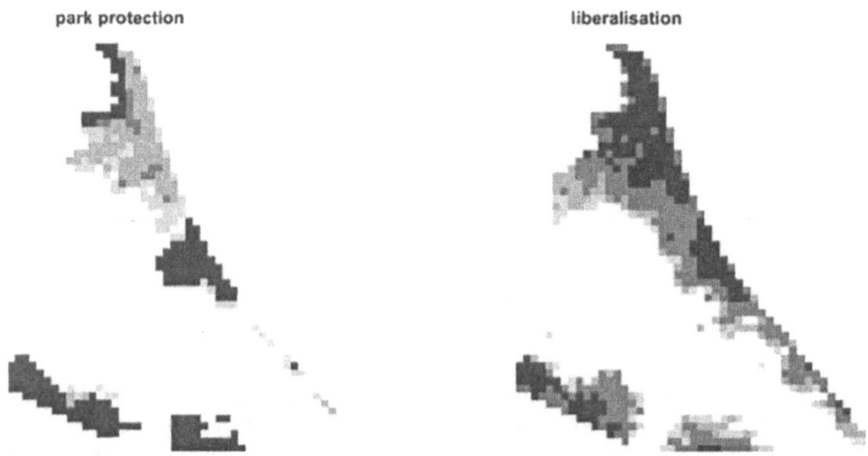

Figure 7. Distribution of forest with (left) and without (right) protection of national parks. Percentages indicate forest cover after 21-year of simulation.

Sub-national Conclusions

CLUE was successfully applied at sub-national level. In comparison with various other methodologies that were applied for the same region (see Bouman et al. 2000), it has the advantage that it is cheap, quick, and has relatively low data requirements. Scenarios that alter allocation possibilities changed the speed of land use changes in the remaining area, but hardly influenced the pattern of those changes.

CLUE Application for Sibuyan Island, Philippines

The CLUE model is applied to the island Sibuyan in the province of Romblon, Philippines. The island measures 28-km east to west at its widest point and 24-km north to south, with a land area of approximately 456 km^2 surrounded by deep water. In the center of the island lies a large protected area marked by the NIPAS border in Figure 9 (Mount Guiting-Guiting Natural Park). It is characterized by its steep mountain slopes, covered with forest canopy. The land surrounding the high mountain slopes gently to the sea and is used for natural and plantation forest and agricultural, mining and urban activities.

The island is believed to be completely covered by forest until the 1940's. From then on the forest has been cleared from the footslopes. Highest on the footslopes are the grassland derived from deforestation, used for pastures. They are regularly burnt to stimulate new grass growth. Rice paddies are common at low-lying land. Most cleared areas are however used for coconut plantation. The island is surrounded by some mangrove forests, sandy beaches and coral reefs. See also Figure 8A.

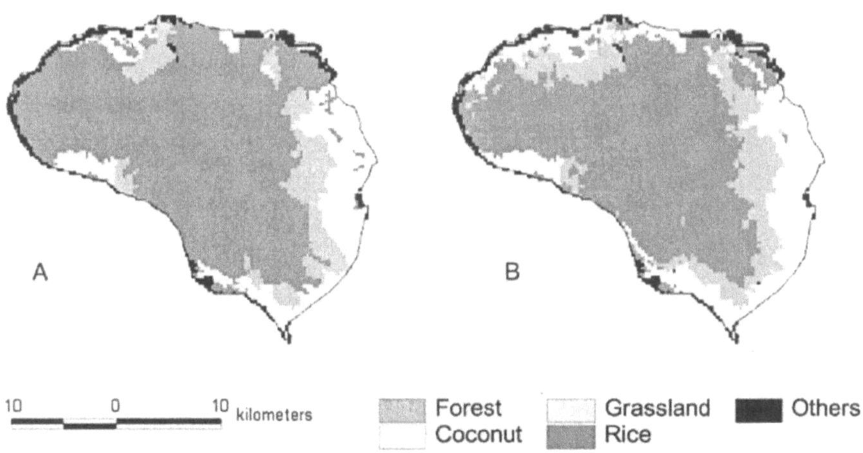

Figure 8. Land use patterns in Sibuyan 1997 (A) and 2012 (B).

The CLUE high resolution model version (CLUE-S) evaluates a base line scenario for Sibuyan for 15 years (1997-2012). The higher resolutions demand a different data representation and modeling approach. Where the low-resolution data made use of non-homogeneous grid cells, this model version is based on homogeneous grid cells. The scale is 250x250 m.

A statistical analysis based on logistic regression is used to find the most important determinants of the land use patterns. Five to ten driving factors are necessary to explain the four dynamic land use types, coconut, forest, grass and rice. The population density, which acts over large distances, is taken into account by using an upscaling of grid information by a focal function.

The demands that are used are the linear extensions of the development between 1940-1997. The assumption is that the island is first opened up from 1940 on and that a linear development has occurred until 1997.

The model predicts the further development along the footslopes of the mountains especially in the west and the north (Figure 8B). Especially the coconut plantations expand towards the northern part of the island. The northwestern part consists of very steep slopes, too steep for coconut, but some patches of grassland are developing. Some of the rice paddies will move to the south-west, and also more new paddies will develop here.

The black line in Figure 9B indicates the NIPAS-area. This is the border of the Mount Guiting-Guiting National Park. Without taking the border into account, the park will slowly be invaded for agricultural purposes, especially on the East Side. However, it also shows that new agricultural developments mainly occur at the West Side of the island.

The model is sensitive for the pathways of change. In Figure 9 the differences in Land use pattern is given for two different demand scenarios. The initial and final demands are the same only the changes in time are different, linear (base scenario) or logarithmic. This nonlinear behavior of land use change is a typical characteristic of complex systems and implemented in the model structure through a dependence of future land use change on the present condition of the land.

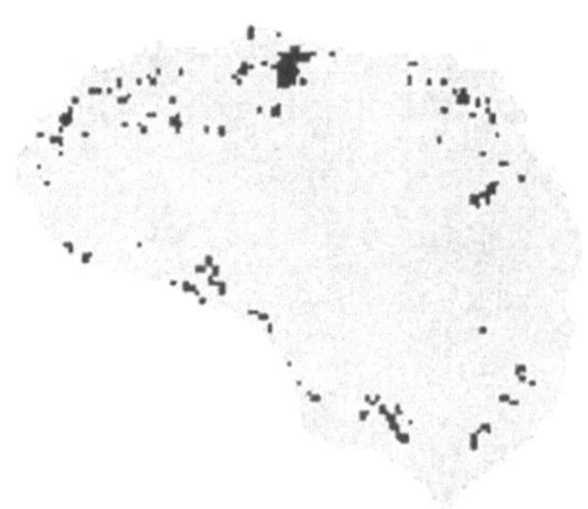

Figure 9. Difference in the allocation pattern between a linear and logistic demand development. Initial and end demand is similar, only the pathway differs.

The high-resolution version of CLUE resembles at first view LULCC models based on cellular automata (Wolfram 1984 1999). On closer inspection, the main difference is that CLUE is still based on an empirical and reproducible analysis instead of arbitrary expert rules. This first application for Sibuyan Island demonstrates that remote sensed data as collected in many LULCC studies (e.g., Walsh et al. 1999, 2001) is now directly suitable input for CLUE simulations.

High ResolutionConclusions

The CLUE methodology can also be applied to high-resolution dominant cell data. The high-resolution version is more sensitive for the pathways of change and shows typical non-linear dynamics.

CLUE Validation

Uncertainties in predicted land use change can arise from both non-realistic scenario conditions and allocation errors. Scenario conditions can generate the wrong demand quantities, which will yield unrealistic patterns. Still, when the simulated quantities are correct the allocation has to be validated independently (see also Pontius and Schneider 2001). The allocation module of the CLUE model has been validated successfully in a number of cases: Ecuador (de Koning *et al.* 1999a; Verburg *et al.* 1999c; Kok *et al.* 2000). The validations were made through the simulation of historic, documented land use changes. However, any land use change scenario and model has a high, inherent, uncertainty due to the complexity of the system addressed. Therefore, results of land use change models should never be treated as predictions for future land use but rather as exploration of the potential dynamics of the land use system. To demonstrate a validation of the allocation algorithm, we will demonstrate a validation for the regional application of the Northern Atlantic Zone. Two different land use maps we available for the region, 1984 and 1992. The demand was derived by subtracting the two maps. The model was parameterized for 1984 and run from 1984 to 1992. The model outcome was consequently compared with the 1992 map. Correlation between the actual and predicted map was analyzed by aggregating major kinds of land use to administrative unit averages. The performance of these aggregated results was examined, as we were interested in capturing patterns of change and not detailed grid to grid changes. Figure 10 shows the aggregated results of the validation run of CLUE.

CLUE proved to be capable of modeling land use patterns in 1992 from the 1984 situation. Especially the location of the dominant land use categories, pasture and forest were modeled very well, with a correlation coefficient between actual and predicted land use of 0.87 and 0.95, respectively. For bananas, the corresponding correlation coefficient was 0.74. The modeling of location of annual crops was less successful, as reflected in a correlation coefficient of 0.36. For forest, pasture and annuals, the regression line between actual and predicted land use did not differ significantly from the 1:1 relationship. For banana, cover was overestimated in districts with little banana, and underestimated in districts with large banana plantations.

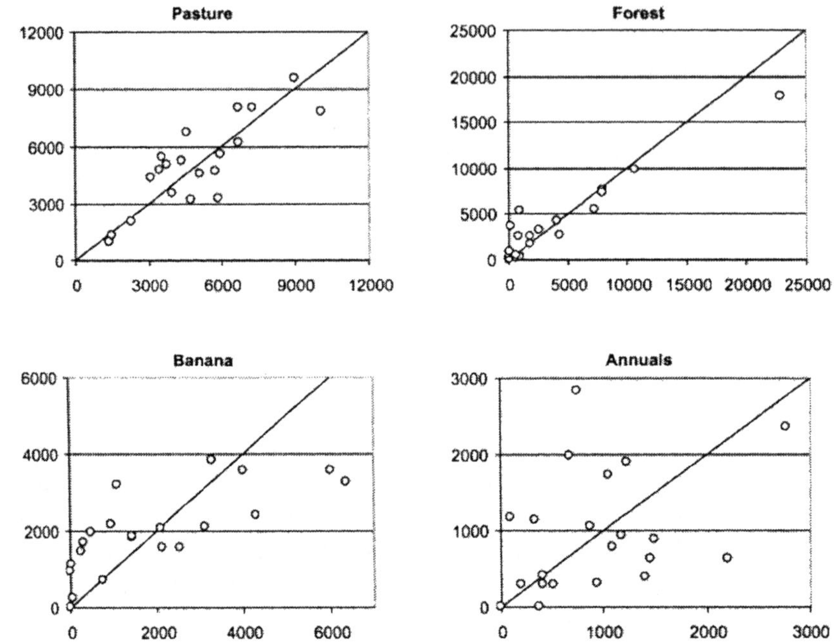

Figure 10. Modeled (y-axis) versus actual (x-axis) cover in hectares for 1992 of four major kinds of land use, aggregated per administrative unit. Lines are the 1:1 relationship (y = x).

Validation Conclusions

Changes in land use types that are omnipresent, or that do not change abruptly, are reproduced successfully. Less satisfying results are obtained for especially permanent crops that are cultivated on highly clustered large plantations. The manner in which competition is incorporated in CLUE might overrate the competitive power of large land use types. The various studies on the validity of CLUE in Latin America and Asia demonstrate the ability of the model to handle very distinct land use change processes.

GENERAL DISCUSSION AND CONCLUSIONS

In the context of global change policies information on LULCC is of crucial importance. However, quantification of near future land use change and of the factors driving this change still remains a major scientific challenge. Insights in the main drivers of land use change, analysis of possible land use change scenarios, and the detection of hot-spots of land use change and their effects offer opportunities to assess the possible impacts of land use on climate change. These impacts take place through emission of CH4 and N2O, through changes in sources and sinks of CO2 and through influences on energy and water balances. Applications of CLUE results are the estimation of CH4 emissions based on spatial explicit land use cover maps of China and Java (Indonesia) (Denier van der Gon 2000, Denier van der Gon et al. 2001). The multi-scale approach of CLUE is not only a more realistic description of complex land use systems, but also offers the possibility to address the influence of land use drivers and policy decisions at different organizational levels, from local to global, and therefore relates to decisions makers at these levels.

For impact assessment, links can be made with other models. This was demonstrated with a link to a nutrient flux assessment model, with which land use change was related to estimated changes in nutrient balances (de Koning et al. 1999b). Other applications that directly use spatially explicit information on land use change are erosion/sedimentation models, water resource assessments (Wear et al. 1998), and biodiversity (Lebel et al. 1998; van der Meer et al. 1998).

Maintaining biological diversity depends on the spatial arrangement of land uses on the landscape, because of the importance of fragmentation and habitat size on the risks of extinction. The spatial arrangement of land uses also has implications for ecosystem function, and hence the goods and services. For example, the way in which residual forest is distributed relative to the stream margin can have an impact on soil erosion and hydrology of a catchment. Finally, how natural disturbances like fires, pests and diseases propagate through a heterogeneous landscape clearly depends on its structure. The evaluation of different scenarios based on different policies and development trajectory can help to assess these spatial patterns and their effects on biodiversity, food production and other ecosystem functions (Figure 11). This way spatial modeling can improve land use planning and inform policymaking.

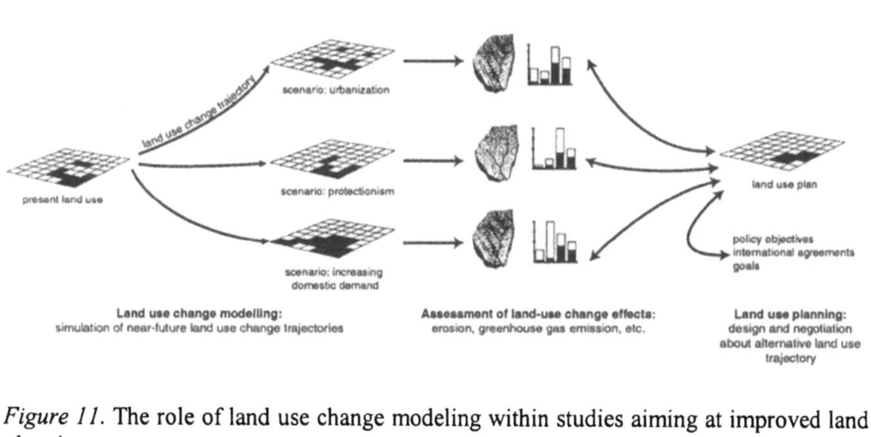

Figure 11. The role of land use change modeling within studies aiming at improved land use planning.

Apart from these biophysical assessments, land use change studies can also help to identify vulnerable people and places in the face of global change. Differential impacts and abilities to respond create winners and losers in this change. Whether involving land fragmentation, degradation of agricultural productivity, declines in economic well being, or involuntary human migration, land use/land cover change plays an important role. The results of spatially explicit land-use change models allow a straightforward identification of areas that are likely to face high rates of land use change in the near future, so-called 'hot-spots' of land use change. Based upon results of the CLUE modeling framework, (Verburg *et al.* 2000b) identified regions in China that are lagging behind in economic development, while, at the same time, these regions are faced with deteriorating land resources. The identification and understanding of the dynamics underlying these regional differences will help to focus in-depth research and policy intervention at the most threatened regions.

Another example of the role a land use change model such as CLUE can play is demonstrated by the study for Central America that was executed in collaboration the International Center for tropical Agriculture (CIAT). CIAT is developing a set of rural sustainability indicators that can be used for monitoring the state of the rural environment. These indicators are meant to aid policy makers and planners in their understanding of the development process. Although economic and social indicators are widely adopted, indicators to assess, monitor and evaluate changes and impact in the state and quality of land resources and the environment are lacking and needed to be developed. CIAT has used results from CLUE as a source of information for the construction of indicators (Farrow and Winograd 2001). Different scenarios were run for Central America, and these scenarios were defined on the basis of panel discussions with policy makers in the region. The spatially explicit outputs of the scenarios were re-classified to reduce the number of output maps. By comparing these maps with potential land use maps potential near future land use conflicts can be estimated. The results of such cross-comparisons can be presented in the form of maps or tables for policy makers. The presentation of maps to decision-makers is appropriate to communicate results and provoke discussions between policy makers and scientist on the seriousness of the foreseen changes (Goodchild, 2000).

REFERENCES

Bilsborrow, R.E., Okoth Ogondo, H.W.O., 1992. Population-driven changes in land use in developing countries. Ambio 21: 37-45.

Bingsheng, K., 1996. Regional inequality in rural development. In: Garnaut, R., Shutian, G., Guonan, M. The third revolution in the Chinese countryside. Cambridge University Press, Cambridge. pp: 245-258.

Bouman, B.A.M., Jansen, H.G.P., Schipper, R.A., Nieuwenhuyse, A., Hengsdijk, H. (Eds.), 2000 Tools for land use analysis at different scale levels, with case studies for the Atlantic Zone of Costa Rica. Kluwer Academic Publishers, Dordrecht, The Netherlands. 274 pp.

Brockett, C.D., 1988. Land, power, and poverty: agrarian transformation and political conflict in Central America. Allen & Unwin Inc., Winchester.

Denier van der Gon, H.A.C., 2000. Changes in CH4 Emissions from Rice Fields from 1960 to the 1990s. I. Impacts of modern rice technology,Global Biogeochemical Cycles 14: 61-72.

Denier van der Gon, H.A.C., P.H. Verburg, S. Houweling, P.M. van Bodegom and N. van Breemen, 2001. Combining Upscaling and Downscaling of Methane Emissions From Rice Fields: Methodologies and Preliminary Results, Nutrient Cycling in Agro-ecosystems (in press).

Easterling, W.E., 1997. Why regional studies are needed in the development of full-scale integrated assessment modelling of global change processes. Global Environmental Change 7: 337-356.

Easterly, W., Loayza, N., Montiel, P., 1997. Has Latin America's post-reform growth been disappointing? Journal of International Economics 43: 287-311.

FAO, 1998. The FAOSTAT databases. http://apps.fao.org/lim500/agri_db.pl or via the FAO homepage: http://www.fao.org/

Farrow, A., Winograd, M., 2001. Land use modelling at the regional scale: an input to rural sustainability indicators for Central America. Special Issue "predicting land use change" Agriculture Ecosystems and Environment 85: 249-268.

Forman, R.T.T., Godron, M., 1986. Landscape Ecology. Wiley, New York, USA, 619 pp. Fotheringham AS. 2000. Context-dependent spatial analysis: A role for GIS? Journal of Geographical Systems 2: 71-76.

Goluboay M., H. Vega R. 1988. La actividad bananera en Costa Rica. p. 131-161 In: FLACSO/CEDAL/FES (eds.), 'Cambio y continuidad en la economía bananera', FLACSO/CEDAL/FES, San José, Costa Rica. 258 p.

Gomez, L.D., 1986. Vegetación de Costa Rica. First edition, EUNED, San José, Costa Rica. 323 pp.

Goodchild MF. 2000. The current status of GIS and spatial analysis. Journal of Geographical Systems 2: 5-10.

Houghton, R.A., 1994. The worldwide extent of land use change. BioScience 44: 305-313.

Kaimowitz, D., Angelsen, A., 1998. Economic Models of Tropical Deforestation: A Review. Center for International Forestry Research, Bogor.

Kok, K., Veldkamp, A., 2000. Multi-scale land use change modelling using the CLUE framework. In: Bouman, B.A.M., Jansen, H.G.P., Schipper, R.A., Hengsdijk, H., Nieuwenhuyse, A. (Eds.), Tools for land use analysis at different scales. With case studies from Costa Rica. Kluwer Academic Publishers, Dordrecht, Chapter 3. P. 35-63.

Kok, K., Farrow, A., Veldkamp, A., Verburg, P.H., 2001. A method and application of multi-scale validation in spatial land use models. Special Issue "predicting land use change" Agriculture Ecosystems and Environment 85: 223-238.

Kok, K., Veldkamp, A., 2001. Evaluating impact of spatial scales on land use pattern analysis in Central America. Special Issue "predicting land use change" Agriculture Ecosystems and Environment 85: 205-222.

Koning, G.H.J. de, Veldkamp, A., Fresco L.O., 1998. Land use in Ecuador: a statistical analysis at different aggregation levels. Agriculture Ecosystems and Environment 70: 231-247.

Koning, G.H.J. de, Verburg, P.H., Veldkamp, A., Fresco, L.O., 1999a. Multi-scale modelling of land use change dynamics in Ecuador. Agricultural Systems 61: 77-93.

Koning, G.H.J. de, Veldkamp, A., Fresco L.O., 1999b. Exploring changes in Ecuadorian land use for food production and their effects on natural resources. Journal of Environmental Management 57: 221- 237.

Lambin, E.F., 1994. Modelling Deforestation Processes, A Review. Trees, Tropical Ecosystem Environment Observation by Satellites, Research Report No. 1. European Commission Joint Research Centre/European Space Agency.

Lebel, L., Murdiyarso, D. Modelling Global Change Impacts on Tropical Landscapes and Biodiversity. 5. 1998. Bogor, BIOTROP-GCTE/Impacts Centre for Southeast Asia (IC-SEA). IC-SEA Report.

May, P.H., Bonilla, O.S., 1997. The environmental effects of agricultural trade liberalization in Latin America: an interpretation. Ecological Economics 22: 5-18
Meyer, W.B., Turner II, B.L., 1994. Changes in land use and land cover: a global perspective. Press Syndicate of the University of Cambridge, UK, 507 pp.
Milne, B.T., 1991. Heterogeneity as a multi-scale characteristic of landscapes. In: Kolasa J,Pickett STA. (Editors) Ecological Heterogeneity. Ecological studies 86. Springer-Verlag. P 69-84.
O'Neill, R.V., 1988. Hierarchy theory and global change. In: Rosswall, T., Woodmansee, R.G., Risser, P.G. (Eds.), Scales and global change. Spatial and temporal variability in biospheric and geospheric processes. SCOPE 35, John Wiley & Sons Ltd., Chichester. pp. 29-45.
Pelupessy, W., Weeks, J., 1993. Economic maladjustment in Central America. Houndmills/MacMillan, Washington, USA.
Pontius, R.G. Jr, Schneider, L.C., 2001. Land-cover model validation by an ROC method for the Ipwisch watershed, Massachusetts, USA. Special Issue "predicting land use change" Agriculture Ecosystems and Environment 85: 239-248.
Pontius, R. G. Jr., Cornell, J.D., Hall, C.A.S., 2001. Modeling the spatial pattern of land-use change with GEOMOD2: application and validation for Costa Rica. Special Issue "predicting land use change" Agriculture Ecosystems and Environment 85: 191-203.
Ramirez S.A., Madonado U., 1988. Desarollo socioeconomico y el ambiente natural de Costa Rica: situacion actual y perspectivas. Editorial Heliconia, San Jose, Costa Rica. 160 p.
Riebsame, W.E., Meyer, W.B., Turner II, B.L., 1994. Modelling land use and cover as part of global environmental change. Climate Change 28: 45-64.
Rigon, R., Rinaldo, A., Rodriguez-Iturbu, I., 1994. On landscape self-organisation. Journal of Geophysical Research vol 99: no b6 11,971-11,993.
Schoorl, J.M., Veldkamp, A. 2001. Linking land use and landscape process modelling: a case study for the Alora region (SE Spain). Special Issue "predicting land use change" Agriculture Ecosystems and Environment 85: 281-292.
Schumann, D.A., Partridge, W.L. (Eds.), 1989. The human ecology of tropical land settlement in Latin America. Westview special studies on Latin America and the Caribbean, Westview Press, Boulder Arizona
Turner II, B.L., Skole, D., Sanderson, S., Fischer, G., Fresco, L., Leemans, R., 1995. Land-use and land-cover change, Science/Research plan, IGBP report no. 35, HDP Report no. 7., 132 pp.
Turner, M.G., Gardner, R.H., 1991. Quantitative methods in landscape ecology: an introduction. In: Turner, M.G., Gardner, R.H. (Eds.), Quantitative methods in landscape ecology. Ecological Studies 82, Springer Verlag, New York, USA, pp. 3-14.
Van der Meer J, van Noordwijk M, Anderson J, Ong C, Perfecto I. 1998. Global change and multi-species agroecosystems: concepts and issues. Agriculture, Ecosystems and Environment 67: 1-22.
Veldkamp, A., Fresco, L.O., 1996. CLUE-CR: an integrated multi-scale model to simulate land use change scenarios in Costa Rica. Ecological Modelling 91: 231-248.
Veldkamp, A., Fresco, L.O., 1997a. Exploring land use scenarios, an alternative approach based on actual land use. Agricultural Systems 55: 1-17.
Veldkamp, A., Fresco, L.O., 1997b. Reconstructing land use drivers and their spatial scale dependence for Costa Rica (1973 and 1984). Agricultural Systems 55: 19-43.
Veldkamp A., Lambin, E.F., (Eds) 2001. Predicting land use change. Special Issue "predicting land use change" Agriculture Ecosystems and Environment 85: 1-6.
Verburg, P.H., De Koning, G.H.J., Kok, K., Veldkamp, A., Bouma, J., 1999a. A spatial explicit allocation procedure for modelling the pattern of land use change based upon actual land use. Ecological Modelling 116: 45-61.
Verburg, P.H., Veldkamp, A., Fresco, L.O., 1999b. Simulation of changes in the spatial pattern of land use in China. Applied Geography 19: 211-233.
Verburg, P.H., Veldkamp, A., Bouma, J., 1999c. Land use change under conditions of high population pressure: the case of Java. Global Environmental Change 9: 47-56.

Verburg PH, Chen YQ, Veldkamp A. 2000. Spatial explorations of land-use change and grain production in China. Agriculture, Ecosystems and Environment. 82: 333-354.
Vitousek, P.M., Mooney, H.A., Lubchenco, J., Melillo, J.M., 1997. Human domination of earth's ecosystems. Science 227: 494-503.
Walsh, S.J., Evans, T.P., Welsh, W.F., Entwisle, B, Rindfuss R.R., 1999. Scale-dependent relationships between population and environment in northeastern Thailand. Photogrammetric engineering and remote sensing 65: 97-105.
Walsh, S.J., Crawford, T.W., Welsh, W.F., Crews-Meyer, K.A., 2001. A multiscael analysis of LULC and NDVI variation in Nang Rong district, northeast Thailand. Special Issue "predicting land use change" Agriculture Ecosystems and Environment 85: 47-64.
Wear DN, Turner MG, Naiman RJ. 1998. land cover along an urban-rural gradient: implications for water quality. Ecological applications 8: 619-630.
Wielemaker, W.G., Vogel, A.W., 1993. Un sistema de información de suelos y tierras para la Zona Atlántica de Costa Rica. AZP, phase 2, report no. 22. CATIE/MAG/AUW, Guápiles. 82 p.
Wiens, J.A., 1989. Spatial scaling in ecology. Functional Ecology 3: 385-397.
Wolfram, S., 1984. Cellular Automata as Models of Complexity. Nature 311: 419-424.
World Bank, 1998. World Development Indicators, 1997. World Bank CD-ROM. World Bank, Washington D.C.
Wu, F., 1999. GIS-base simulation as an exploratory analysis for space-time processes. J. Geograph. Syst. 1: 199-218.

Index

MIX
Papier aus verantwortungsvollen Quellen
Paper from responsible sources
FSC® C105338

If you have any concerns about our products,
you can contact us on
ProductSafety@springernature.com

In case Publisher is established outside the EU,
the EU authorized representative is:
Springer Nature Customer Service Center GmbH
Europaplatz 3, 69115 Heidelberg, Germany

Printed by Libri Plureos GmbH
in Hamburg, Germany